PETROLOGY
Igneous, Sedimentary, and Metamorphic

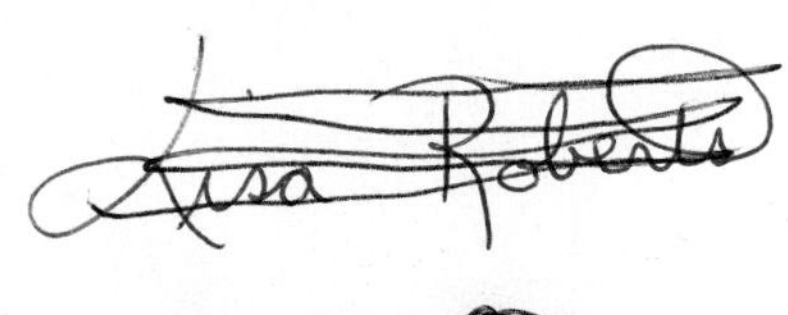

PETROLOGY
Igneous, Sedimentary, and Metamorphic

Ernest G. Ehlers
Ohio State University

Harvey Blatt
University of Oklahoma

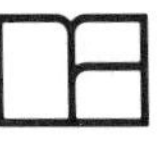

W. H. Freeman and Company
San Francisco

Project Editor: Pearl C. Vapnek
Copyeditor: Sean Cotter
Designer: Ron Newcomer
Production Coordinator: Bill Murdock
Illustration Coordinator: Cheryl Nufer
Compositor: Allservice Phototypesetting Company
Printer and Binder: The Maple-Vail Book Manufacturing Group

Library of Congress Cataloging in Publication Data

Ehlers, Ernest G.
Petrology : igneous, sedimentary, and metamorphic.

Includes bibliographies and index.
1. Petrology. I. Blatt, Harvey. II. Title.
QE431.2.E38 552 81–12517
ISBN 0–7167–1279–2 AACR2

Printed in the United States of America

4 5 6 7 8 9 MP 1 0 8 9 8 7 6 5 4 3

Skilled work, of no matter what kind, is only done well
by those who take a certain pleasure in it,
quite apart from its utility, either to themselves
in earning a living, or to the world through its outcome.

Bertrand Russell

Contents

Preface

The subject of petrology has undergone a revolution within the last decade as a result of the recently developed awareness by geologists of the role of plate tectonics in petrogenesis. Field examinations have included studies—by means of deep sea cores, dredge samples, and submersibles—of igneous rocks formed at mid-ocean rifts. The distribution and origin of igneous and metamorphic rocks at convergent plate junctions have been more closely defined in terms of plate tectonics. The occurrence of sedimentary materials can now be related to plate tectonic regimes. Laboratory studies are now able to duplicate most of the conditions upon and within the earth. Mineral syntheses have defined the general pressure, temperature, and compositional limits within which most of the common rock-forming minerals can exist. Isotopic studies can now furnish a date of formation of igneous and metamorphic rocks, as well as the temperatures at which many organisms lived in sedimentary environments. In addition, improved classification of igneous and sedimentary rocks now yields more meaningful subdivisions and is more closely related to the method of origin than were earlier systems.

All of this information (as well as much more) exists within the literature of petrology, but relatively little of it has found its way into undergraduate petrology textbooks. Our aim has been to present a summary of the more significant portions of both the older and the more recent literature at a level appropriate to the college sophomore or junior who is exposed to a first course in petrology. Literature sources are given at the end of each chapter for those who wish to pursue some of the subjects presented. It is assumed that the student has an understanding of elementary chemistry, crystallography, and mineralogy, as well as some facility with the petrographic microscope (although this is not critical to an understanding of the general concepts).

We are indebted to all of the investigators referred to in the text, as well as those who have granted us permission to use their materials. In addition, we would like to thank those individuals who have inspired and/or critically reviewed various portions of the text; these include D. S. Barker, A. V. Cox, B. Derek, W. R. Dickinson, R. V. Dietrich, R. N. Ginsburg, J. F. Hubert, R. V. Ingersoll, G. V. Middleton, G. E. Moore, R. C. Murray, C. H. Schulz, L. Snee, and J. F. Sutter. Their comments have materially improved our approach to the subject. Many thanks also go to our patient typists, Kathleen Wuichner, Marjorie Starr, and Sallie Orr. Errors of commission and omission are our own.

July 1981

Ernest G. Ehlers
Harvey Blatt

PETROLOGY
Igneous, Sedimentary, and Metamorphic

Introduction: Why Petrology?

Rocks are the stuff of which the earth is made. Because of this, the study of rocks, petrology (from the Greek *petra,* "rock," and *logos,* "discourse or explanation"), occupies a central position in geology. It is the source of nearly all of our ideas about the history of our planet. Knowledge about rocks, their origins, and their distribution is potentially capable of solving or contributing to the solution of a wide variety of problems that run the gamut of geologic interests.

1. Most evolutionary biologists believe that living matter evolved from nonliving matter in a reducing atmosphere because primitive cells are believed to have been defenseless against oxidation (that is, chemical dissolution). This implies an absence of oxygen in the atmosphere of the early earth, a supposition supported by the absence of free oxygen in the atmospheres of the other planets in our solar system. However, hematitic iron ores are abundant in the earliest Precambrian rocks, a fact often interpreted to mean that the early atmosphere contained oxygen. How much oxygen was present in the atmosphere of the primitive earth?

2. The relative abundance of sedimentary materials forming at present is drastically different from rock abundances found in the geological record.

   a. Dolomite in Precambrian rocks is three times as abundant as limestone; at present dolomite formation is rare, restricted to unusual environments such as the Persian Gulf and the Netherlands Antilles.

   b. The Middle Precambrian stratigraphic column (about 2.5 billion years old) contains about 15% fine-grained silica in the form of chert; at present chert formation is trivial outside of the deep ocean basins.

   c. Evaporites are extremely rare in the Precambrian as compared to more recent times. Why? Has the composition of sea water changed during the last

2.5 billion years? Must the doctrine of uniformitarianism be abandoned in the face of these data?

3. Igneous rocks have a wide variety of textures and compositions. Yet the great bulk of these rocks consists of two chemical and textural types. One of these is coarse-grained, silica-rich granitic rocks (about 75% silica), and the other is fine-grained, silica-poor basaltic rocks (about 55% silica). What does this tell us about the processes that form the earth's crust?

4. Metamorphic rocks of identical composition are often composed of greatly differing mineral assemblages. Can this observation be used to determine the pressures and temperatures that existed in the earth's crust during their formation?

5. Some metamorphic rocks consist of randomly oriented minerals and others are strongly foliated. How can this be used to interpret the conditions of stress that existed during the time of their formation?

6. On a regional scale, the distribution pattern of igneous and metamorphic rocks is beltlike and often parallel to the borders of the present continents. For example, the granitic rocks that form the core of the Appalachian Mountains parallel the east coast of the United States, and those of the Sierra Nevada parallel the west coast. What is the origin of this distribution?

7. Some sandstones contain only quartz; others, 30% feldspar; still others, 90% volcanic rock fragments. Can these data be used to infer the types of rocks exposed on the earth's surface at different times and in different geographic locations? Is the mineral composition of sandstones related to tectonic processes and crustal evolution?

THE MAJOR ROCK TYPES

Rocks are naturally occurring aggregates of minerals or mineraloids (such as coal, glass, opal, and so on), and most rocks consist of several mineral species. Rocks are traditionally divided into three groups: igneous, sedimentary, and metamorphic. In most outcrops or hand specimens it is not difficult to use these pigeonholes, and they serve the useful purpose of dividing rocks on the basis of conditions in which they were formed. The American Geological Institute's *Glossary of Geology* defines each group of rocks as follows:

Igneous rock: A rock that solidified from molten or partly molten material; that is, from a magma.

Sedimentary rock: A rock resulting from the consolidation of loose sediment or chemical precipitation from solution at or near the earth's surface; or an organic rock consisting of the secretions or remains of plants and animals.

Metamorphic rock: Any rock derived from pre-existing rocks by mineralogical, chemical, or structural changes, especially in the solid state, in response to marked changes in temperature, pressure, and chemical environment at depth in the earth's crust; that is, below the zones of weathering and cementation.

This fundamental classification scheme, as is true of all human classification schemes, contains a flaw. Nature is a continuum; it is not segregated into discrete parts for our convenience. Hence, borderline or transitional rocks exist and are thrown into one or another of the three groups because of historical precedence or the whim of the classifier. For example, volcanic tuffs are rocks that originate in volcanoes. After explosive ejection as fragments into the atmosphere, they settle either on land surfaces or in water. If they settle into layers, should they then be classified as sedimentary? In most classification schemes, they are classified as igneous. A second example is the rock serpentinite, a rock that is mainly composed of minerals of the serpentine group. This material begins its history in many instances as a molten silicate rock that cools to produce a crystalline aggregate of olivine and pyroxene. Such material would of course be classified as igneous. However, during cooling, water vapor reacts with the previously formed crystalline aggregate, converting much of it into serpentine, which implies that the rock should now be classified as metamorphic. Although there are some questionable types of rocks, the general threefold subdivision is satisfactory, well accepted by most geologists, and shall be used here.

After accepting such a primary classification scheme, the next problem is to learn how to put rocks into each of the three slots. Following that, additional subdivisions can be described within each of the three categories. Subdivision into the three types can be accomplished to some extent by watching some of the rocks being formed and then extrapolating to others of similar nature whose formation cannot be seen; in other cases pigeonholing can be done by examination of the various structures and textures within the rock, and in still other cases the rocks can be subjected to various experimental tests that are designed to indicate the limits of stability and hence condition of formation of the various constituent minerals. We shall examine the classification, characteristics, and origins of each of the major rock types, starting with igneous.

In order to facilitate the subdivision of rocks into one of the three major subdivisions, general characteristics of each type are listed in Table I–1. This table lists some of the outcrop characteristics and structures, followed by textures and characteristic minerals. Note that any single characteristic may not be sufficient to categorize the rock, but rather a number of such features may have to be used. Unfamiliar terms are defined in the text.

RELATIVE SURFACE ABUNDANCES OF ROCK TYPES

The earliest geologic maps date from about 1800, and most parts of the earth have now been mapped at a useful scale. Using these maps, we can determine the relative abundances of igneous, sedimentary, and metamorphic rocks on the continents. The

Table I–1
General Characteristics of Igneous, Sedimentary, and Metamorphic Rocks

Outcrop characteristics and structures		
Igneous	Sedimentary	Metamorphic
1. Volcanoes and related lava flows	1. Stratification and sorting	1. Distorted pebbles, fossils, or crystals
2. Cross-cutting relations to surrounding rocks, as in dikes, veins, stocks, and batholiths	2. Structures such as ripple marks, cross bedding, or mud cracks	2. Parallelism of planar or elongate grains common over large areas
3. Thermal effects on adjacent rocks, such as recrystallization, color changes, reaction zones	3. Often widespread and interbedded with known sediments	3. Located adjacent to known igneous rocks, occasionally as a zoned aureole
4. Chilled (finer-grained) borders against adjacent rocks	4. The shape of the body may be characteristic of a sedimentary form, such as a delta, bar, river drainage system, and so on	4. Typically located in Precambrian or orogenic terranes
5. Lack of fossils and stratification (except for pyroclastic deposits)	5. The rocks may be unconsolidated or not	5. Rock cleavage related to regional structures
6. Generally structureless rocks composed of interlocking grains		6. Progressive change in mineralogy over a wide area
7. Typically located in Precambrian or orogenic terranes		7. Some are massive hard rocks composed of interlocking grains
8. Characteristic shapes and sizes, as in laccoliths, lopoliths, sills, stocks, batholiths, and lava flows		
Textures		
Porphyritic, glassy, vesicular, amygdaloidal, graphic, pyroclastic, or interlocking aggregate	Fragmental, fossiliferous, oolitic, pisolitic, stratified, interlocking aggregate	Brecciated, granulated, crystalloblastic, or hornfelsic

Table I–1 (*continued*)

Characteristic minerals		
Amphibole	Abundant quartz, carbonates (especially calcite and dolomite), or clays	Amphibole
Feldspar abundant	Anhydrite	Andalusite
Leucite	Chert (microcrystalline quartz)	Cordierite
Micas	Glauconite	Epidote
Nepheline	Gypsum	Feldspar
Olivine	Halite	Garnet
Pyroxene		Glaucophane
Quartz		Graphite
		Kyanite
		Sillimanite
		Staurolite
Glass		Tremolite–actinolite
		Wollastonite
		Micas
		Quartz

method used to obtain the data is to generate random latitudes and longitudes by computer, and then to examine existing maps to determine the frequency of igneous, sedimentary, and metamorphic outcrops. The result is 66% sedimentary rocks and 34% combined igneous and metamorphic rocks. The bulk of the 34% is probably igneous, but large areas of continents are mapped as "undifferentiated igneous and metamorphic rocks," so exact percentages are uncertain.

The data for sedimentary rocks reveal the extent to which geologic information is lost with time (see Figure I–1). The data indicate that half of all outcropping sedimentary rocks are younger than 130×10^6 years (that is, Cretaceous or younger). Interpretations based on older rocks must be less comprehensive than those based on more recent rocks.

Crustal Abundances of Rocks

The earth's *crust* is usually defined as the outer shell of the earth above the Mohorovičić discontinuity. The terms crust and lithosphere are not synonyms. The *lithosphere* contains both the crust and the upper portion of the mantle to a depth of about 100 km. The base of the lithosphere is marked by a zone, generally called the *asthenosphere,* in which seismic velocities are lower than in the regions directly above and below. Geophysical data reveal that the mean thickness of the crust is 17 km, dominated by granitoid rocks on the continents but by basaltic rocks in the ocean

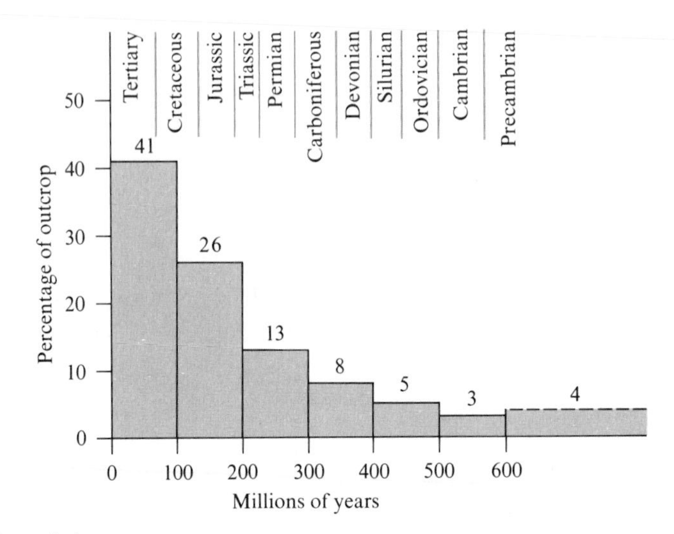

Figure I–1
The relationship between the age of sedimentary rocks and the amount of outcrop area. [From H. Blatt and R. L. Jones, 1975, *Geol. Soc. Amer. Bull., 86,* Fig. 1.]

basins. These data also indicate that the mean sediment thickness on the continental blocks averages about 1.8 km and that the blocks occupy 35% of the earth's surface area. In the ocean basins the sediment cover is very thin, approximately 0.3 km. Combining these data, the average thickness of sediment on the earth's surface approximates 0.8 km and forms about 4.8% of the total crust, and only 0.013% of the whole earth.

As the earth has a radius of 6,371 km, only the outer 0.27% is crust (0.74% of the volume). Nevertheless, the crust is the only part of the earth directly exposed for examination, so the subject matter of petrology is concerned largely with the analysis of its composition, structure, and origin. As we shall see, problems of crustal origin are intimately related to the character of the material beneath the crust—the mantle. And some properties of the mantle depend on heat trapped in the core. It is not too farfetched an analogy to state that petrologists are often in the position of examining the flea to infer some properties of the dog. It is sobering to keep this in mind.

FURTHER READING

Amer. Assoc. Petroleum Geol. and U.S. Geol. Surv. 1967. Basement map of North America between latitudes 24° and 60° N. U.S. Geol. Surv. Map.

Bates, R. L., and J. A. Jackson, eds. 1980. *Glossary of Geology,* 2d Ed. Washington, D.C.: Amer. Geol. Institute, 749 pp.

Blatt, H. 1970. Determination of mean sediment thickness in the crust: a sedimentologic method. *Geol. Soc. Amer. Bull., 81,* 255–262.

Blatt, H., and R. L. Jones. 1975. Proportions of exposed igneous, metamorphic, and sedimentary rocks: *Geol. Soc. Amer. Bull., 86,* 1085–1088.

Dietrich, R. V., and B. J. Skinner. 1979. *Rocks and Rock Minerals.* New York: John Wiley and Sons, 319 pp.

Parker, R. L. 1967. Composition of the Earth's crust, *Data of Geochemistry,* Chap. D. U.S. Geol. Surv. Prof. Paper 440–D, 19 pp.

Periodicals containing significant petrologic literature include *American Journal of Science, Contributions to Mineralogy and Petrology, Journal of Geology, Journal of Petrology, Journal of Sedimentary Petrology,* and *Lithos.*

I

IGNEOUS ROCKS

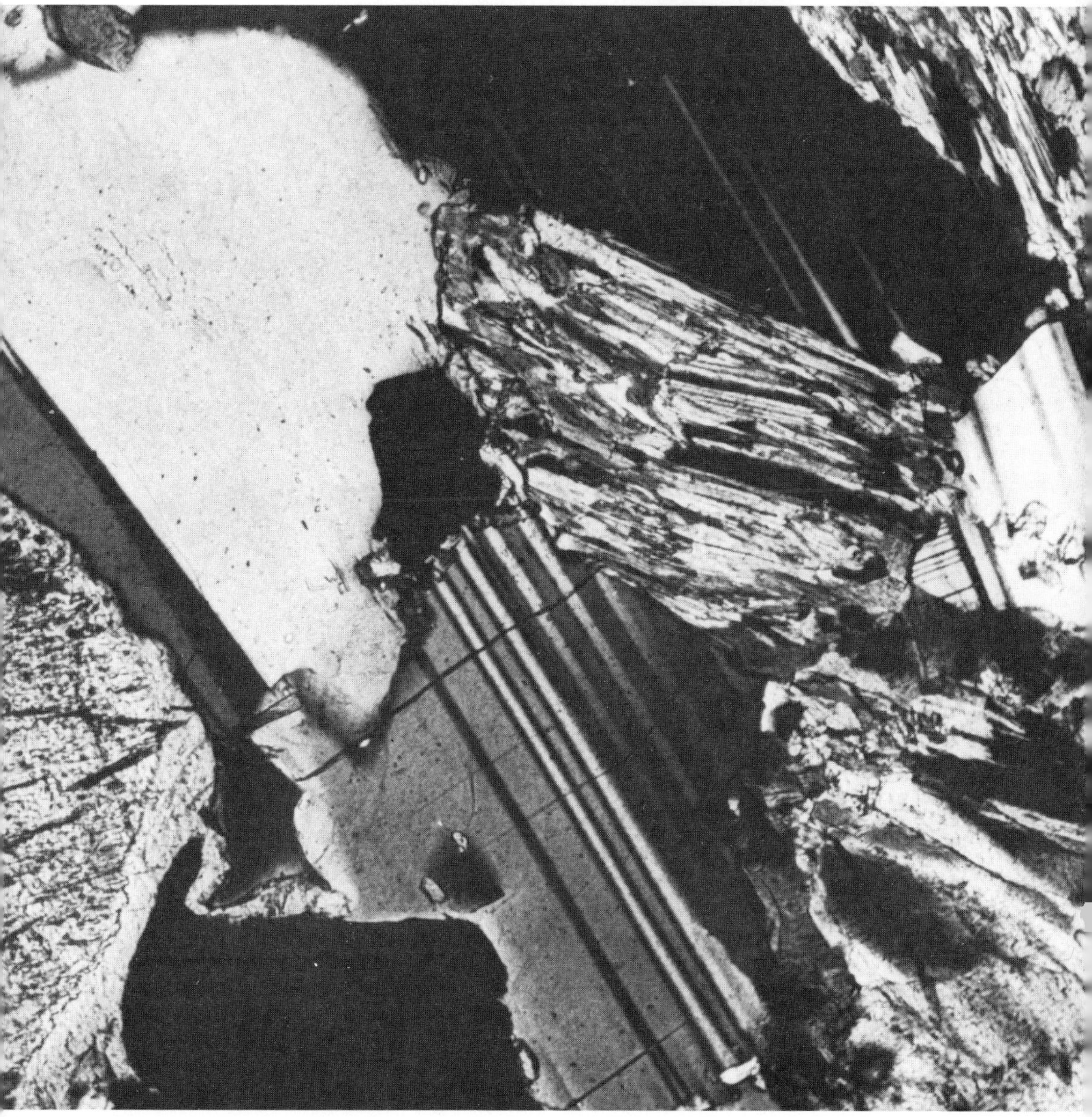

1

Field Observations of Igneous Rocks

When faced with an outcrop of igneous rock, we require a number of observations, ranging in scale from kilometers down to individual grains. All of this information contributes to an understanding of the extent and shape of the igneous mass, the method of emplacement, the relationship to adjacent country rock, the chemistry and mineralogy of the igneous body, its cooling history, and perhaps ultimately its origin. We begin with a consideration of the sizes and shapes of igneous masses, their relationship to the country rock, and progress to methods of emplacement and smaller-scale features.

LARGE-SCALE FEATURES: INTRUSIVE VERSUS EXTRUSIVE

One of the first things to determine about an igneous rock is whether it is intrusive or extrusive; that is, whether it has formed below or at ground level. The origin of extrusive rocks has been clear for hundreds of years because the formation of these rocks can be readily observed at active volcanic sites; but the origin of intrusive igneous rocks was a subject of debate as late as the 1870s. Many geologists considered intrusive rocks to be buried lava flows, and proof was needed to establish their intrusive character.

The decisive proof of the magmatic origin of intrusive rocks was provided by Gilbert in the 1870s in his classic studies of the Henry Mountains of Utah. This area contains relatively flat-lying sediments, within which are located mushroom-shaped igneous bodies whose compositions are midway between granites and basalts. These intrusions, called *laccoliths* in their idealized shape (see Figure 1–1), are floored by more or less horizontal sediments; the roof consists of domed sedimentary rocks. Gilbert concluded that magma forced its way between the sedimentary strata and

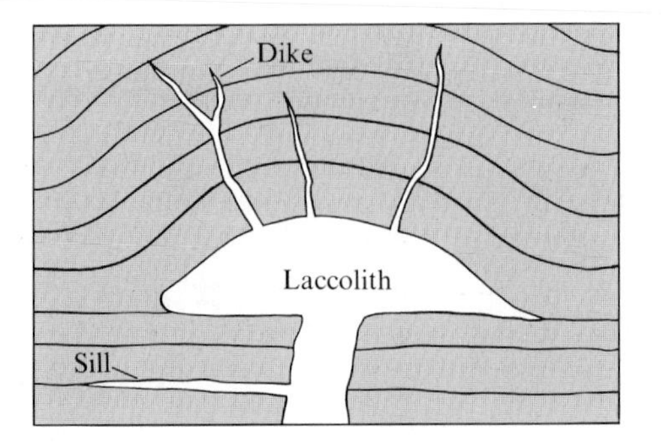

Figure 1–1
Cross section of an idealized laccolith. Horizontal extent is generally 1–8 km, and thickness is about 1,000 m.

bulged the upper layers. His proof that the igneous rocks were intrusive rather than extrusive included several lines of evidence.

1. Some of the igneous rocks cut across sedimentary bedding, indicating a later igneous origin.
2. The sedimentary rock on both the floor and the roof of the intrusion was baked as a result of the heat of the intrusion.
3. No gas cavities or fragmentation of the upper surface of the igneous rock were found.
4. No fragments of igneous rocks were found in the overlying sediments, as would be expected in a buried volcanic terrane.
5. The arching of the overlying sedimentary rock often exceeds the angle of repose (the maximum angle at which sediments can be deposited on a sloping surface).

Although most scientists accepted this overwhelming array of evidence, in 1877 a geologist named Neumayer felt that the conclusion was so surprising that further evidence was needed. Ten years later another geologist, Ryer, continued to insist that the laccoliths were buried surface eruptions. Old ideas die hard.

Assuming the magmatic origin of both intrusive and extrusive igneous rocks, let us consider the type of information available to distinguish between the two types.

Extrusive igneous rocks are formed by the flow of lava or the deposition of ash over preexisting surfaces. Usually these surfaces are irregular as a result of erosion. The contact surface between the two rock types is therefore also irregular. The contact may in some places be parallel to the bedding or foliation of the country rock, in which case it is called *concordant;* more typically the contact is at an angle to the bedding or foliation of the country rock and is called *discordant*. In the case where an extrusive igneous rock is located over an irregular surface of flat-lying sediments, the contact may be concordant in some places and discordant in others.

In addition to observing the geometric relationships, you should examine both rock types carefully at the contact. The country rock below an extrusion may show the effects of earlier weathering, such as soil formation, oxidation, or hydration; such findings establish that the igneous rock was extrusive in origin. If loose weathered material was present at the contact surface, some of it may be found within the lower part of the lava or ash. These foreign materials within an igneous rock, either extrusive or intrusive, are called *inclusions*. Inclusions consisting of rock fragments are called *xenoliths* (see Figure 1–2), and inclusions of crystals are called *xenocrysts*. Both inclusions and country rock may show some effects from the heat treatment they have received; with extrusive rocks these effects are relatively minor, and usually consist of baking (which produces some color change and perhaps a slight increase in hardness due to recrystallization).

The upper surface of a buried lava or ash is typically irregular as a result of erosion; some lavas may have irregular surfaces due to the breakup of a solidified crust during flowage. The contact in the typical case is discordant with the overlying sedimentary rock. The upper portion of a buried flow might be indicated by the

Figure 1–2
Xenoliths of hornblende diorite (dark fragments) in light-colored granodiorite, near Copper Lake, North Cascades National Park, Washington. [From M. H. Staatz, 1972, *U.S. Geol. Surv. Bull.*, *1359*, Fig. 15–B.]

presence of *vesicles*—small- or medium-sized gas bubbles frozen into the lava. Vesicular structure is common in lavas, particularly in the upper portions, as the entrapped gas tends to move upward through the denser melt. Erosion of the lava or ash may have taken place before the deposition of a covering sediment. A soil layer may be present as a result of chemical weathering. Detailed examination of the overlying sediment might reveal igneous fragments from the igneous rock below, in which case the extrusive origin of the igneous rock is established.

Intrusive igneous rocks have solidified in preexisting rocks beneath the earth's surface. The contact of an intrusive igneous rock may be generally concordant or discordant with the country rock (see Figure 1–3). If the behavior of the country rock is brittle during igneous intrusion, emplacement of the igneous body is accompanied by fracturing, faulting, and forcing apart of the country rock. This usually leads to discordant contacts, as the bedding or foliation is truncated at the edge of the igneous body. This type of contact generally occurs at shallow levels in the crust where brittle behavior of rocks is the rule. At greater depths (several kilometers or more) the pressures and temperatures result in plastic or ductile behavior of country rock. A magma (or even a rock mass of different density from the surrounding rocks) may move into place by squeezing aside and deforming the surrounding rock in a similar manner to the shaping of wet clay by a potter's fingers. Bedding or foliation present in the country rock is commonly squeezed so as to be parallel or essentially parallel to the intruding pluton. Intrusion of this type is called diapiric, and the vertically moving melt or rock mass is called a *diapir.* Contact with the country rock in this case is concordant. Concordant contacts may also be found at shallow levels of intrusion, where the intrusive force of the magma is able to dome or force apart country rock along bedding planes without appreciable fracturing.

As with any classification of this type, many cases are not clear-cut. Many intrusives, seemingly concordant at isolated outcrops, are broadly discordant when the contacts and country rock structures are mapped on a large scale. Concordancy is often a function of the scale of observation. An example of this occurs if the igneous body has never left the region in which it formed. During orogenesis the pressures and temperatures of metamorphic rocks (near the base of the crust) may be brought to the point where partial or complete melting occurs. This process, called *anatexis,* produces magma that may leave its site of origin or remain in place. If the magma cools in the region in which it originated and is later exposed through erosion, the igneous and metamorphic portions will be seen to have concordant contacts in some areas and discordant contacts in others, depending upon the amount of movement of the melted material.

Examination of the contact will reveal it to be either sharp or diffuse. A sharp, precisely defined contact indicates lack of reaction between the magma and country rock. This may be due to either a relatively nonreactive country rock (such as quartzite) or a rapid chilling of the intrusion against relatively cool country rock. A large temperature difference might be indicated by a decrease in grain size within the igneous rock as the contact is approached. This would be expected if the magma were a shallow intrusive body or if it were an extrusive flow.

(A) (B)

Figure 1-3
(A) A vertical (dark) dike cutting a nearly horizontal segment of an earlier dike. The later dike is discordant to the earlier. Norfolk County, Massachusetts. [From N. E. Chute, 1966, *U.S. Geol. Surv. Bull., 1163–B,* Fig. 4.] (**B**) Concordant contact of the Beech Granite (massive rock at the upper left) with the Cranberry Gneiss (foliated rock on lower right). Avery County, North Carolina. [From B. H. Bryant, 1970, U.S. Geol. Surv. Prof. Paper 615, Fig. 20.]

Xenoliths and xenocrysts should also be looked for and examined carefully. They are often located only near the edges of the igneous body, but occasionally are present throughout. These foreign bodies are representative of the rocks through which the magma has moved or of the rocks surrounding the final crystallization site of the melt. Certain igneous intrusions have originated at depths of over 100 km and contain fragments (such as diamonds) brought from the source of magma formation as well as materials collected during the vertical ascent. In other cases xenolithic fragments were never present or have been dissolved by the melt. These fragments, where present, provide a clue to magmatic origin and should always be sought during field examination.

Mode of Occurrence

One of the most important things to determine in the field is the extent of the igneous body. This can be accomplished by "walking the contact"—a pleasant process on a beautiful spring day, but perhaps not too much fun when trudging around in subzero weather in Antarctica. Fortunately, the labor can be considerably eased by the use of topographic maps, standard aerial photographs, remote sensing techniques, or perhaps Landsat imagery.

Topographic maps and aerial photographs often prove useful because many igneous rocks and their surrounding country rocks show marked differences in weathering characteristics, resulting in differences in elevation and drainage patterns.

The first Earth Resources Technology Satellite, ERTS-1 (later renamed Landsat-1), was launched in 1972. Landsat-2 was launched in 1975. These satellites have acquired hundreds of thousands of images that cover the entire earth, except for the poles. The images, taken at several different wavelengths, permit the production of either black and white prints or various color composite images, and they are extremely valuable for mapping purposes. The images not only reveal large-scale structural patterns, but also differences in rock types not normally resolved with standard aerial photography. The usefulness of the technique can be seen by examination of Figure 1–4.

Before considering the smaller-scale features of igneous rocks we will first examine the various types of intrusive and extrusive bodies, bearing in mind that the forms taken by igneous materials are often transitional between the various listed types.

Types of Intrusions

Sills

Sills are concordant, tabular bodies that are emplaced essentially parallel to the foliation or bedding of the country rock (see Figure 1–5). They are typically thin and shallow and are located in relatively unfolded country rock. A high degree of fluidity is necessary to produce this sheetlike form, and because basaltic magmas are more fluid than granitic ones, most sills are basaltic in composition. Sills may be simple, multiple (more than one injection), or differentiated. A differentiated sill is one in which the (usually) heavier, early-formed crystals have settled to the chilled base, with the result that the composition of the sill varies from upper to lower surface. Sill thicknesses may vary from a few to several hundred meters. The Triassic Palisades Sill, located along the Hudson River near New York City, is about 300 m thick, and crops out over an area 80 km long and 2 km wide. The Peneplain Sill of the Jurassic Ferrar diabase swarm in Antarctica is up to 400 m thick and crops out over at least 20,000 km^2.

Laccoliths

Laccoliths are concordant, commonly mushroom-shaped intrusions that range in diameter from about 1 to 8 km, with a maximum thickness on the order of 1,000 m. They occur in relatively undisturbed sediments at shallow depths. Laccoliths are created when magma rising upward through essentially horizontal layers in the earth's crust encounters a more resistant layer, under which the magma spreads out laterally, forming a dome in the overlying strata. If it encounters low horizontal resistance to spreading, the laccolith may grade into a sill. Most laccoliths are silicic or intermediate in composition. Silicic magmas have a higher viscosity than basic magmas, and this may account for the general lack of lateral spreading of laccoliths as compared to sills. Increase of viscosity due to chilling at the thin edges results in a greater upward thrust.

Many of the laccoliths of the Henry Mountains, Utah, deviate from the ideal mushroom shape and are tongue-shaped forms. The common situation in each of the Henry Mountains involved injection of a large central mass of magma several kilometers in diameter and about a thousand meters high. Forceful intrusion resulted in the creation of a wide zone of intensely shattered country rock, and bulging of the roof and walls. Fracturing and weakening of adjacent rocks outside the shattered zone resulted in the injection of peripheral laccoliths and a wide variety of other intrusive bodies around the central cross-cutting mass. The laccoliths probably originated as irregular bulges at the side of the central intrusion, and were then injected radially from the central mass, arching the overlying sediment to form tongue-shaped intrusions. In a few cases, intense bulging raised the overlying rocks by faulting. Typical features associated with laccoliths are shown in Figure 1–6.

Lopoliths

A *lopolith,* as described originally by Grout, consists of a large, lenticular, centrally sunken, generally concordant, basin or funnel-shaped intrusive mass. Most lopoliths are found in unfolded or gently folded regions. The thickness is generally one-tenth to one-twentieth of the width. The diameter may range from tens to hundreds of kilometers, with thickness up to thousands of meters. Lopoliths are usually mafic to ultramafic and a few have upper layers of fairly silicic differentiates. Commonly cited examples are the intrusive complexes at Duluth (Minnesota), Sudbury (Ontario), and the Bushveld (South Africa). Two of these, the Sudbury and Bushveld intrusions, may have been intruded as a consequence of large-scale meteorite impacts. If so, the intrusion would have been formed from melting as a result of both a decrease in pressure caused by rapid loss of surface materials and the heat produced by shock waves. The Sudbury "irruptive" (so called due to the controversy over its origin) is seen in plan and cross section in Figure 1–7. The cross section indicates that the body may be basin-shaped. Since the nature of the lower surfaces of lopoliths is not well known, it is possible that many are funnel-shaped.

Phacoliths

Phacoliths are intrusive concordant bodies associated with folded rocks. When occurring within an anticline they are doubly convex upward; when in the trough of a syncline they are doubly convex downward. It is generally assumed that the intrusion

Figure 1–4
Landsat image 8114801282500, showing petrological and structural features in the eastern Philabara region of western Australia. Color composites made from several spectral bands bring out even more features than those shown here in black and white. Scale: 1 cm = 1 km. [From R. P. Viljoen, M. J. Viljoen, J. Grootenboer, and T. G. Longshaw, 1972, *Minerals Sci. Engng., 7,* Fig. 9. The original photographic data were obtained by NASA and reproduced by the USGS EROS Data Center, Sioux Falls, South Dakota.]

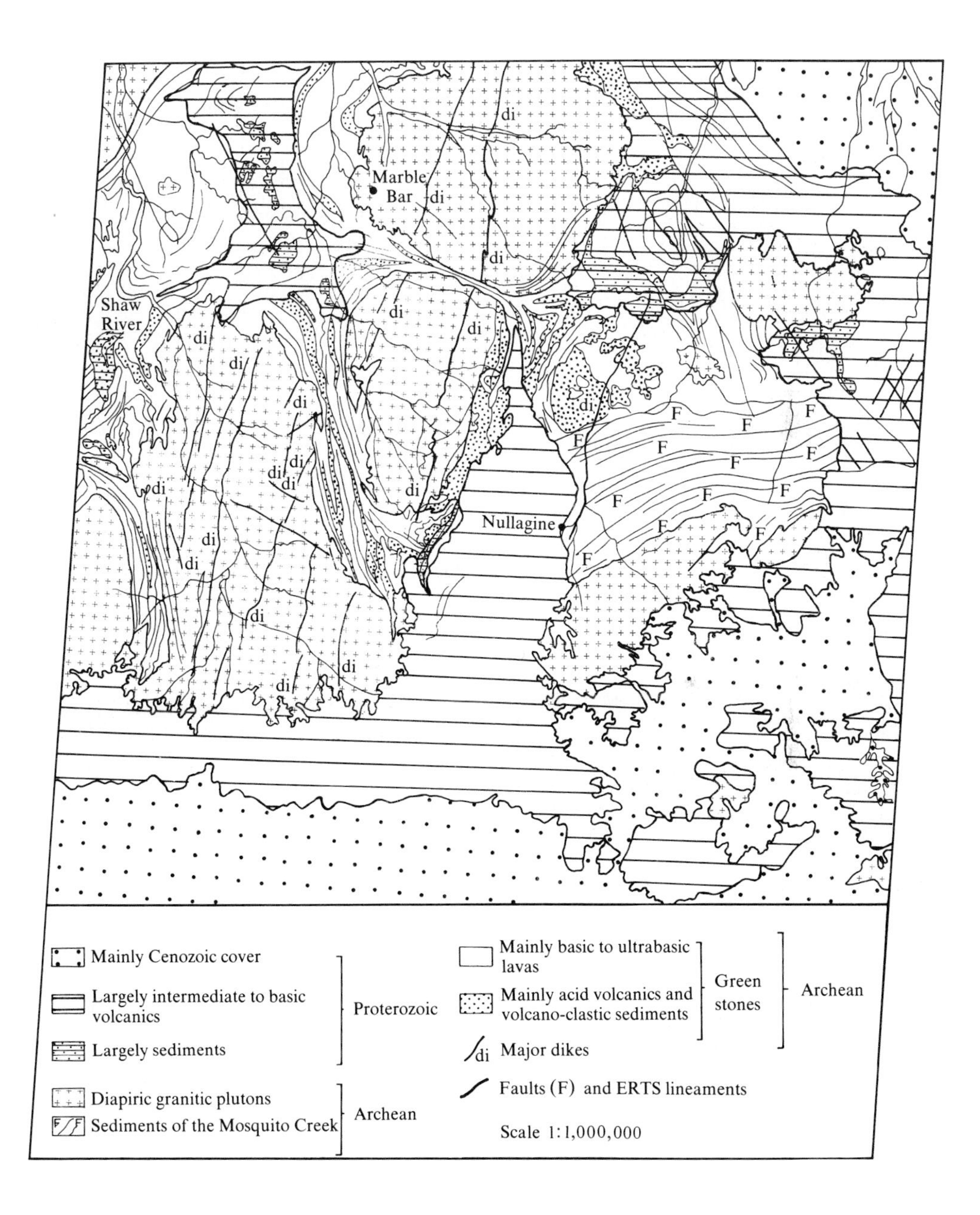

Marble Bar
Shaw River
Nullagine
di
F
Mainly Cenozoic cover
Largely intermediate to basic volcanics
Largely sediments
Proterozoic
Diapiric granitic plutons
Sediments of the Mosquito Creek
Archean
Mainly basic to ultrabasic lavas
Mainly acid volcanics and volcano-clastic sediments
Green stones
Archean
Major dikes
Faults (F) and ERTS lineaments
Scale 1:1,000,000

Figure 1–5
Columnar-jointed (light-colored) Cape Aiak gabbro sill intruded between flat-lying lavas and sediments of the Unalaska formation. The sea cliff is about 400 m high. Unalaska Island, Aleutian Islands, Alaska. [U.S. Geological Survey photo by G. L. Snyder, 1954.]

of a phacolith is passive—that is, magma fills (and perhaps enlarges) the open or potentially open areas that develop at crests and troughs during folding. Openings or low-pressure volumes are developed due to the sliding of incompetent beds between more competent strata, or the sliding of one competent bed against another. Phacoliths are generally emplaced in relatively deep zones, and have sharp to gradational contacts. Any foliation within the phacolith is generally subparallel with the contact or parallel with the axial plane of the fold. Rock composition is variable, and areal extent may be tens of kilometers.

Dikes and Veins

Dikes are tabular discordant plutons that cut across the foliation or bedding of the country rock. They are typically emplaced into preexisting joint systems and may occur singly or in swarms (see Figures 1–3A and 1–8). In some areas dikes may be associated closely with volcanic necks or shallow (hypabyssal) intrusions, in which case they may occur as radiating swarms. Many dikes are more resistant to erosion than the surrounding country rock, and may form residual ridges (see Figure 1–9). Occasionally vertical- or outward-dipping *ring dikes* or inward-dipping *cone sheets* may be found distributed in oval or circular patterns about an intrusion. This appears to be related to fracturing associated with doming of an igneous body and release of

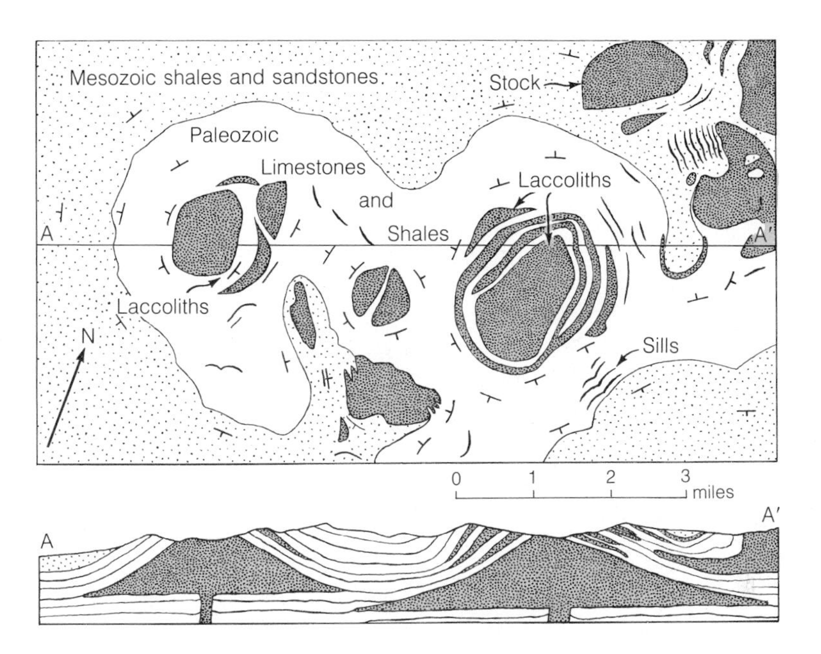

Figure 1–6
Map and cross section of laccoliths and associated features in the Judith Mountains, Montana. [From *Earth* by F. Press and R. Siever. W. H. Freeman and Company. Copyright © 1978. After W. H. Weed and L. U. Pirsson, U.S. Geol. Surv.]

pressure upon extrusion, followed by subsidence of the upper country rock due to partial emptying of the magma chamber. *Veins* are small tabular or sheetlike mineral fillings of a fracture in a host rock, often associated with some replacement of the host rock.

Batholiths

Batholiths (see Figure 1–10) are large intrusive plutons with steeply dipping walls, typically lacking any known floor. Batholiths, which are commonly composed of silicic rocks, range (as outcrops) in size from about one hundred to several thousand square kilometers in geosynclinal belts. Similar to batholiths in shape and composition are *stocks* (or *bosses*), which are defined as having a maximum surface outcrop of 100 km^2. Stocks and batholiths range from being completely concordant to completely discordant. Many batholiths, although broadly concordant to the regional structure, are highly discordant when mapped in detail. Large silicic plutons are often described as granites during field mapping, although compositions are often granodiorite or quartz monzonite.

The problem of attempting to categorize plutons of batholithic dimensions stems from the fact that these rock masses are located in different environments and may

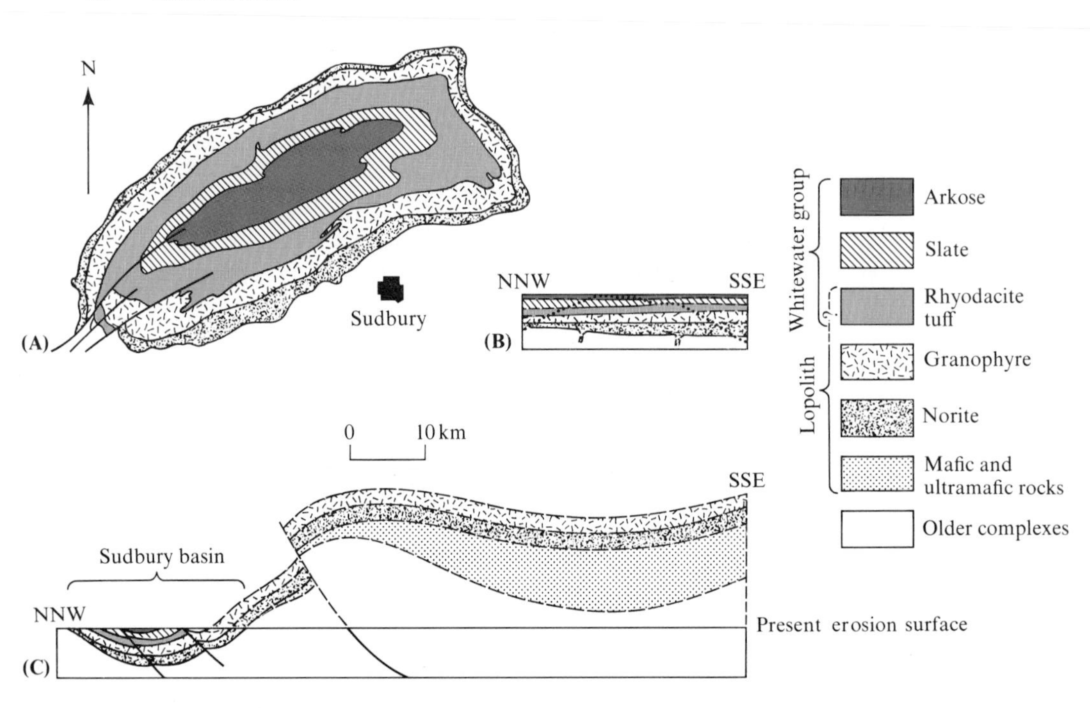

Figure 1-7
Map and cross sections of the Sudbury lopolith, Ontario. (**A**) Map of the Sudbury lopolith. (**B**) Cross section prior to deformation. (**C**) Present cross section showing preerosion extent of the lopolith. [From W. B. Hamilton, 1960, *Canad. Miner., 6,* Fig. 1.]

have formed in different ways. Many granitic rocks within Precambrian shield areas may have been created very early in the earth's history and may have been precipitated from residual liquids derived from the differentiation of basalts, which originated by partial melting within the mantle. Most of the more recent granites are not associated with large amounts of basalts and ultrabasic rocks and are therefore not considered to be related to crystallization of basaltic melt. Many petrologists feel that the more recent granites originated by partial melting of preexisting rocks. If the anatectic source zone is exposed by uplift and erosion, many of the granitic plutons are seen in a generally concordant association with surrounding highly metamorphosed rocks. Other granite magmas may have risen considerable distances from their source areas and may have quite different relationships with the country rocks.

Several early geologists classified granite batholiths according to depth zones, based upon the associated metamorphic minerals and textures in the surrounding country rocks. There was a firm feeling that granites in close association with high-grade metamorphic rocks must have come from great depths. More recent work, however, has shown that, in at least some places, there appears to be no consistent relationship between the level of metamorphism of the country rock and the depth of

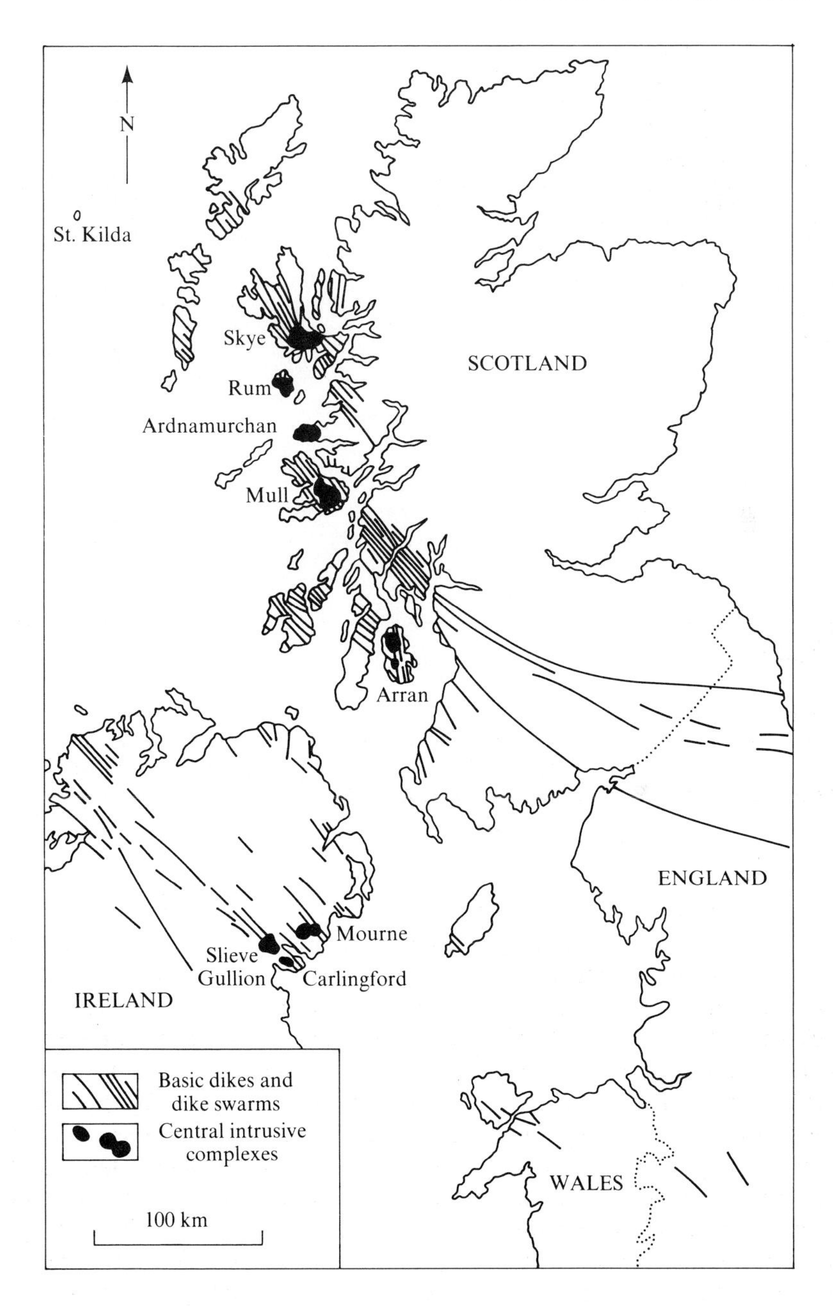

Figure 1-8
Regional dike swarm in western Scotland. [After J. E. Richie, 1961, *Scotland: The Tertiary Volcanic Districts,* 3d Ed., revised by A. G. MacGregor and F. W. Anderson (Edinburgh: Dept. Sci. Ind. Res., Geol. Survey and Museum), 119 pp.]

Figure 1-9
Dike and volcanic neck exposed by erosion at Ship Rock, New Mexico. The exposed dike is about 8 km long. [From *Geology Illustrated* by J. S. Shelton. W. H. Freeman and Company. Copyright © 1966. Fig. 19.]

granite formation. Tentative pressure–temperature gradients have been determined for a number of metamorphic areas. An estimated temperature of 600°C has been calculated to occur at a depth of 20 km in some cases, and at depths as shallow as 5 km in other cases.

Although any classification of granite batholiths is an oversimplification, Buddington (1959) found that most plutons could be classified according to general structural characteristics into three broad categories: catazonal, mesozonal, and epizonal. Catazonal plutons are very large bodies associated with metamorphic rocks whose minerals and textures indicate very high levels of pressure and temperature. These are plutons that are found at or near their former anatectic location in the root zones of the great mountain belts. Epizonal plutons result from the cooling of magma that has left its place of origin and migrated to shallow and cooler portions of the crust, whereas mesozonal plutons have intermediate characteristics.

Catazonal plutons are surrounded by extensive metamorphic rocks indicative of high pressures and temperatures. Contacts are diffuse, and broad zones of migmatites are present (see Figure 1-11). *Migmatites* (from the Greek *migma,* "mixture") are "mixed rocks," parts of which look igneous and parts of which look metamorphic. Foliation within these plutons and their country rock tends to be parallel, indicating

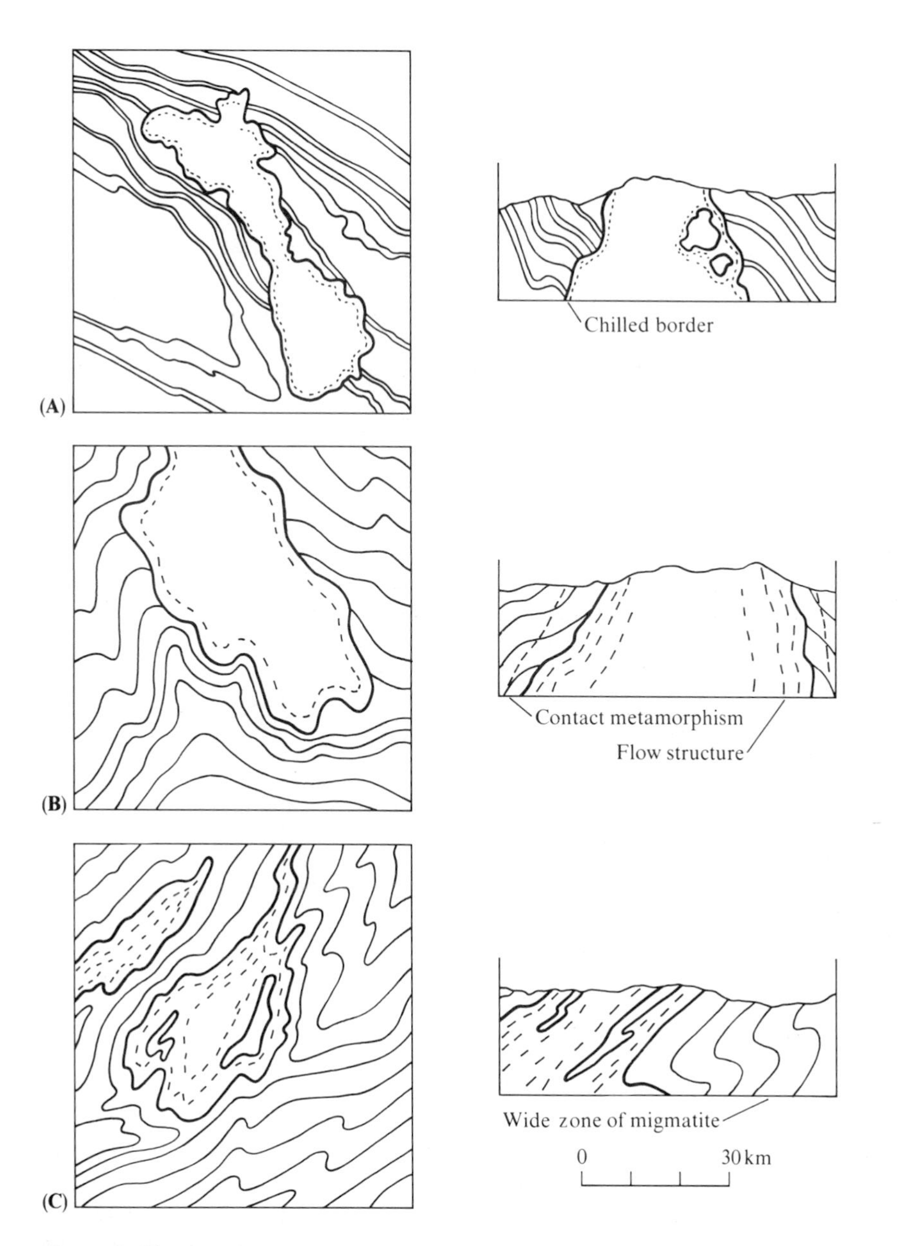

Figure 1–10
Structural patterns of batholiths and stocks (viewed from above and in cross section). (**A**) Epizonal, with strongly discordant contacts and chilled border. Average size about 10–100 km^2. (**B**) Mesozonal. Contacts are partially discordant and partially concordant. A zone of contact metamorphism is present, as well as flow structure within the intrusion. Average size 100–500 km^2. (**C**) Catazonal. Contacts are generally concordant, but may be discordant if the magma has moved from its site of anatexis. Flow structure or gneissic banding parallels the contact. A wide zone of migmatite is present. Average size 50–1,000 km^2.

Figure 1–11
Migmatite, showing typical wavy foliation. Dark layers are biotite and plagioclase. Light layers are mainly quartz, microcline, and plagioclase. Summit County, Colorado. [From M. H. Bergendahl, 1971, U.S. Geol. Surv. Prof. Paper 652, Fig. 6.]

that tectonism and pluton formation occurred simultaneously (syntectonic). *Mesozonal plutons* are surrounded by metamorphic rocks whose minerals and textures are indicative of a fairly low grade of metamorphism. Contacts are fairly sharp and may be concordant or discordant. Migmatites are minor or absent. Moderate deformation may be present in the country rock, and flow structure may be present within the granite. *Epizonal plutons* are largely discordant with their country rock and with regional structure, and most do not possess internal flow structure. Contact with the country rock is sharp, and metamorphism of the country rock tends to be minimal. The outer borders of the pluton are often chilled, as indicated by their relatively small grain size. Volcanic rocks and collapse structures are commonly associated with epizonal plutons.

Batholiths are the igneous intrusions that form the cores of the major mountain systems of the world. These intrusions may extend for hundreds of kilometers along the major structural direction, and some are tens or even hundreds of kilometers wide. Examination of the major mountain chains has indicated that these igneous cores do not represent a single large intrusion, but rather consist of many hundreds of related intrusions, which differ in size, composition, and time of emplacement. Together they represent one of the major products of the orogenic cycle.

TYPES OF EXTRUSIONS

Extrusive rocks take on a wide variety of forms, depending upon the nature and amount of the erupted material as well as their relationships to the country rock. Magmas that have a low gas content and a relatively low viscosity usually produce lava flows (the term lava is used for both the molten extrusion and the rock that has solidified from it). Many lavas include minor amounts of *pyroclastics* (clastic rock material ejected from a volcano).

Most *lava flows* are of basaltic and, to a lesser extent, intermediate composition. Subaerial lavas may be classified as pahoehoe (pronounced pa-'ho-e-'ho-e), aa (pronounced ah-ah), or block lava (pronounced block lava). *Pahoehoe lavas* have glassy, smooth, billowy, and occasionally ropy surfaces (see Figure 1–12). The shape of the upper surface often resembles irregular folds in cloth; this results from distortion of an upper chilled but still plastic surface zone by movement of an underlying, more fluid material that is within the flow. Many pahoehoe flows contain large areas that are relatively smooth.

Aa lavas have rough, fragmented surfaces (see Figure 1–13). The upper portions of aa flows are vesicular and resemble furnace clinkers, whereas the interior portions are fairly dense. The clinker fragments are extremely rough and very irregular in shape; projecting spines may cover the surface. Clinker fragments are usually less than 15 cm in diameter, but may be as large as a few meters. Space between the larger fragments is partially filled with finer material produced by abrasion of the larger fragments during their movement. Below the surface zone the fragments are usually welded together.

Block lavas (see Figure 1–14) are composed of fragments that are relatively smooth as compared to the irregular aa clinkers. Surfaces of block lava flows are quite irregular, with relief of up to several meters. Often ridges are present perpendicular to the direction of flow (which may reflect surges of material from the vent). Although a central massive layer is generally present, this is much less extensive than in aa lavas. Flow movement is slow and irregular with a definite tendency for some layers to shear over others, producing considerable amounts of crushed fragmental material. Very high viscosity combined with occasional rapid flow causes breakage into angular fragments that become rounded or fragmented. The rigidity of the flow causes some sliding of the lava mass over the surface below.

Figure 1–12
Cascade of pahoehoe lava, Halemaumau, Kilauea, Hawaii. Foreground width is about 20 m. [From J. Green and N. M. Short, 1971, *Volcanic Landforms and Surface Features* (New York: Springer-Verlag), Plate 110A (Photo courtesy Geophysical Laboratory, Carnegie Inst., Washington, D.C.).]

A complete gradation exists between pahoehoe, aa, and block lavas. Commonly flows issue from the vent as pahoehoe and change to aa or block lavas during movement downslope. This change is related to increase in viscosity and the amount of agitation of the flowing lava. Downslope movement results in cooling, loss of gas, and partial solidification, all of which serve to increase the viscosity and convert the pahoehoe to either aa or block lavas. Aa or block lavas may be produced directly from the vent as well. Block lavas are usually more siliceous and are associated with greater amounts of pyroclastics.

Lavas produced during a single eruption may consist of several outpourings of material that are essentially contemporaneous. These may vary locally in type, but all are considered as a single flow unit. *Flow units* (see Figure 1–15) may vary from a few centimeters in thickness to about 200 m, although most are considerably less than 100 m; the thickness of individual flow units in the Columbia River basalts in Wash-

Figure 1-13
Aa lava surface at Mount Vesuvius, Italy. [From J. Green and N. M. Short, 1971, *Volcanic Landforms and Surface Features* (New York: Springer-Verlag), Plate 140 (Photo courtesy Geophysical Laboratory, Carnegie Inst., Washington, D.C.).]

ington and Idaho varies from 14 to 35 m. The total thickness of the multiple flow units of the Columbia River basalts occasionally reaches almost 2 km, with a total volume of close to 160,000 km^3. The areal extent of the flows of the Deccan basalt in India (see Figure 1–16) covers 160,000 km^2.

Pillow lavas (see Figure 1–17) consist of masses of ellipsoidal or pillow-shaped bodies. The base of each pillow takes the shape of the pillows below; the structure of each pillow consists of a glass crust with radial jointing. This contrasts with the concentric structures (caused by a radial arrangement of vesicles) occasionally found in pahoehoe lavas. Pillows generally range in size from 10 cm to 6 m. Their composition is basaltic, andesitic, or altered sodic basalt (spilite). Observations of underwater lava flows by divers indicate that the pillow structure results from contact of lava with sea water. As pillow lavas are the most common type of lava on the sea floor and the oceans cover about 70% of the earth's surface, this type of lava is the most abundant of the various types of flows.

Lava may emerge from either a central eruption or a long narrow fissure. The fissure eruptions often produce huge amounts of relatively fluid basaltic magma that spread out to form lava plateaus; occasionally less fluid lavas along a fissure may form chains of volcanoes.

Figure 1–14
Block lava of the Yatsugatake Volcano, Japan. The average fragment diameter is about ½ to 1 m. [From J. Green and N. M. Short, 1971, *Volcanic Landforms and Surface Features* (New York: Springer-Verlag), Plate 139b (Photo courtesy H. Takeshita).]

Figure 1–15
Columnar basalt exposed in the upper 200 m of the east wall of Grand Coulee, Washington. Flow units came different distances from various vents. Some erosion and soil layers are present. Average thickness of each flow unit is about 15 m. [From *Geology Illustrated* by J. S. Shelton. W. H. Freeman and Company. Copyright © 1966. Fig. 316.]

Eruptions from a central vent give rise to a variety of volcanic forms, depending upon the degree of fluidity of the melt and the amount of pyroclastic material thrown from the vent. At one extreme are the *shield volcanoes;* these are broad flat cones (resembling in cross section ancient Germanic rounded shields) composed mainly of lava flows. The sides usually have slopes from 2° to 10°, and the base makes a gradual transition to the surroundings. Shield volcanoes may be small (with elevations of about 1,000 m) or as large as some of the Hawaiian shield volcanoes, whose summits are several kilometers above the sea floor. Many of the shield volcanoes are not circular in plan view but are somewhat elongate; elongations are controlled in direction by the rift zone beneath. The crest of the volcano may be capped by a cinder or spatter cone. The summits of the Hawaiian shields often contain a large depressed central area—a caldera—formed as a result of collapse after depletion of the magma chamber. Eruptions may fill the caldera with molten material, forming a lava lake; other eruptions occur on the flanks of the shield.

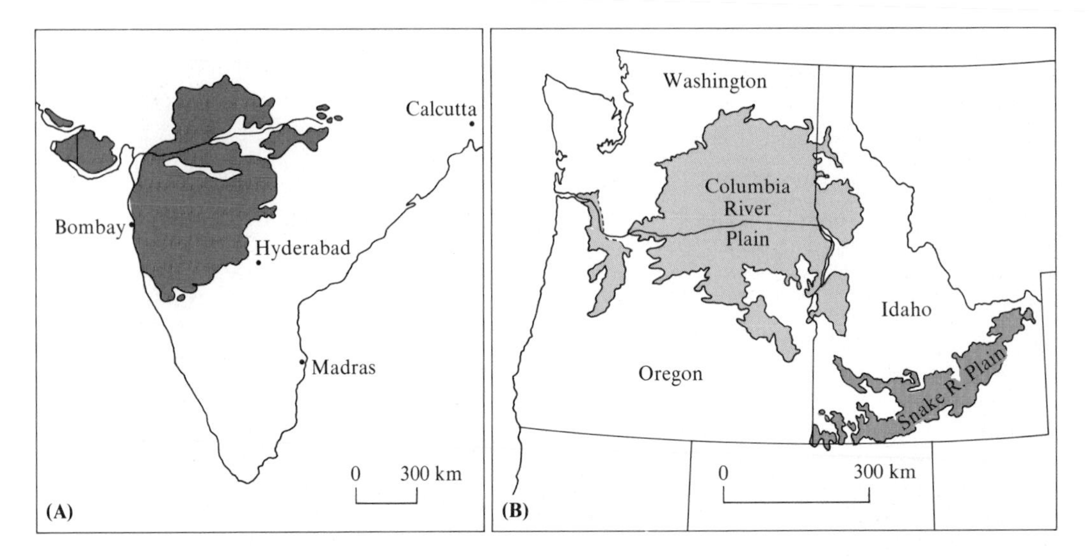

Figure 1–16
Maps showing the extent of (**A**) the Deccan basalts of India and (**B**) the Columbia and Snake River basalts of the northwestern United States. [From G. A. Macdonald, 1972, *Volcanoes* (Englewood Cliffs, N.J.: Prentice-Hall), Fig. 11–2, after H. S. Washington, 1922, *Geol. Soc. Amer. Bull., 33,* 765–803, and A. C. Waters, 1955, Geol. Soc. Amer. Special Paper 62, 703–722.]

Volcanic eruptions may be quite violent, with the result that rock fragments and molten magma are thrown from the top of the vent. Forceful ejection of liquids that fall and solidify immediately adjacent to the vent produces *spatter cones* (see Figure 1–18); these are commonly found associated with the eruption of very fluid basalts. Solids thrown from the vent are called ejecta; accumulations of this material are known as pyroclastic rocks or *tephra*. Many volcanic structures consist of mixtures of both lava and tephra. Volcanoes produced by both flows and tephra are described as *composite* or *stratiform*. The vent is usually smaller than that of a shield volcano and the sides are considerably steeper. Well-known examples are Mount Rainier in Washington and Mount Shasta in California. Tephra or *cinder cones* are more or less circular cones resulting from the accumulations of cinders about the central vent; the larger sizes and amounts of fragments accumulate adjacent to the vent. The sides slope outward at about 30° (close to the angle of repose for loose irregular materials) and the crater is generally bowl-shaped, having been formed by moderately explosive eruptions.

Pyroclastic materials are classified as to fragment size: bombs and blocks (greater than 32 mm), lapilli (4–32 mm), and ash (less than 4 mm). *Blocks* are solid when ejected and form blocky fragments, whereas *bombs* are fluid when ejected, and their shapes are modified during aerial flight. Although it is tempting to speculate on special forms, such as ribbon and cow-dung bombs, the great majority of bombs are simply irregular vesicular lumps, which are known as cinder, scoria, or pumice. *Scoria,* essentially synonymous with cinder, is dark, relatively heavy mafic rock with

Figure 1–17
Pillow lava at Nemour, Hokkaido, Japan. Foreground width is about 8 m. [From J. Green and N. M. Short, 1971, *Volcanic Landforms and Surface Features* (New York: Springer-Verlag), Plate 155B (Photo courtesy K. Yagi).]

abundant vesicles, and is most common as crusts on cooled flows. *Pumice* is a light-colored vesicular glass of silica-rich composition that is often light enough to float on water. *Lapilli* are formed either from rock fragments or by consolidation of very fluid drops of lava that solidified in the air; they are usually drop-shaped, but may be spherical or rod-shaped. *Ash* is usually composed of glass (vitric), but may also consist of rock or mineral fragments, in which case it is called a lithic or crystal ash. Consolidated layers of ash are called tuff. *Vitric tuff* commonly consists of triangular fragments with concave sides, formed by rapid production and breaking up of gas bubbles in the melt (see Figure 1–19). As ash is deposited mainly by vertical fall, it forms a continuous blanket over the surrounding countryside, as contrasted to flows, which follow the topographic lows of the area. Ash and tuff commonly show some size sorting, both vertically and laterally; their distribution is dependent upon the size of fragments as well as prevailing wind velocities and directions. A particularly violent eruption, that of Krakatoa (Java) in 1883, threw ash into the upper atmosphere, from which it was carried around the earth, producing brilliant sunsets throughout the world. Distribution of an ash fall in Alaska is shown in Figure 1–20. Ash falls

Figure 1–18
Spatter cone forming on road during the 1955 eruption of Kilauea volcano, Hawaii. Cone height is about 7 m. [U.S. Geological Survey photo by G. A. Macdonald.]

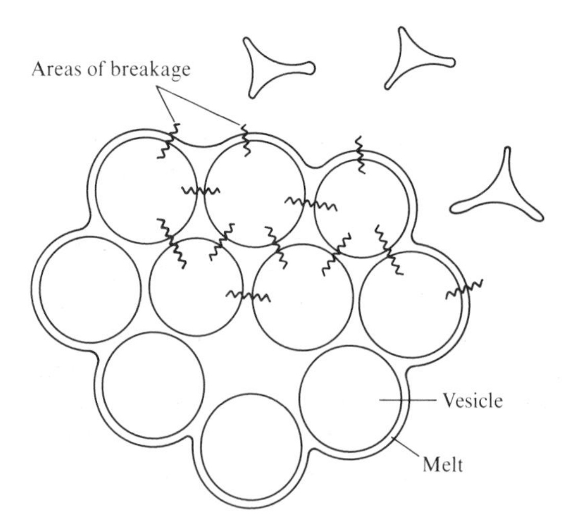

Figure 1–19
The formation of glass shards. Rapid vesiculation of a melt produces triangular viscous masses with concave sides. If these are deposited after solidification, their triangular shapes are retained.

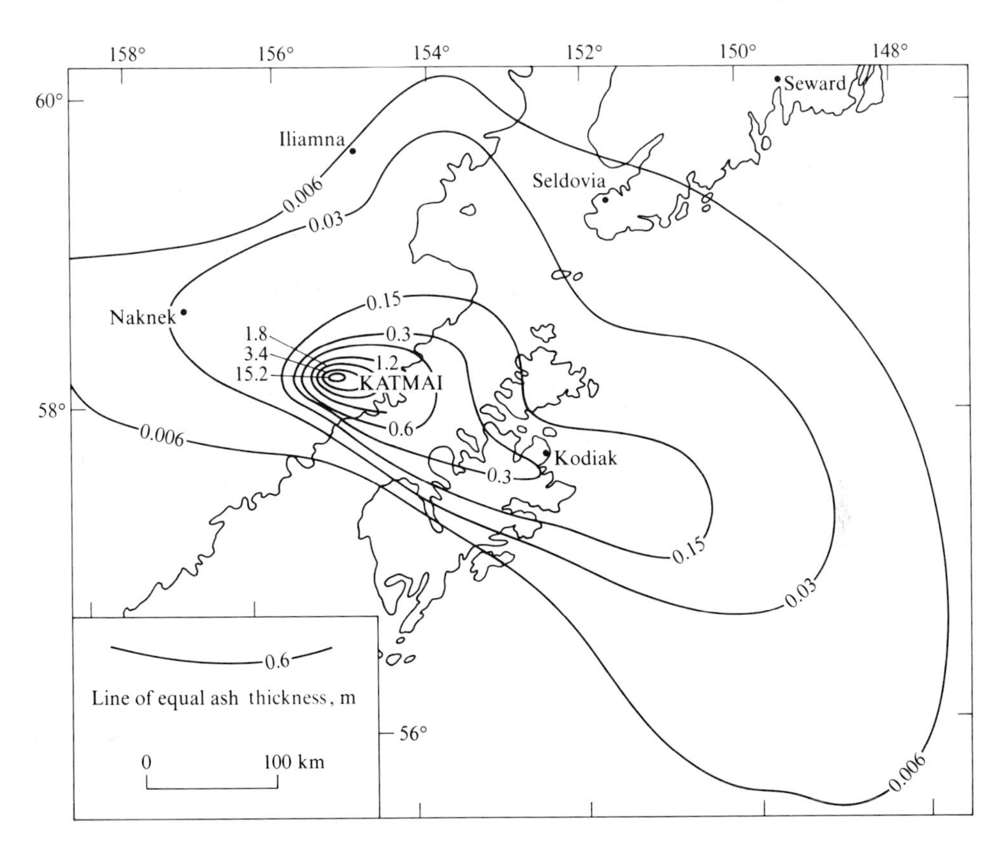

Figure 1–20
Map showing the thickness and distribution of ash deposited by the Katmai eruption of 1912. [From G. A. Macdonald, 1972, *Volcanoes* (Englewood Cliffs, N.J.: Prentice-Hall), Fig. 7–5, modified from R. E. Wilcox, 1959, *U.S. Geol. Surv. Bull., 1028–N,* 409–476.]

may be spread over thousands of square kilometers; they are usually deposited within a few days, and thus provide an excellent "time plane" for correlation of ancient rock sequences.

Intermediate in behavior between a lava flow and an ash fall is the glowing avalanche or *nuée ardente.* A nuée ardente caused the total destruction of the city of St. Pierre on the island of Martinique in 1902 (see Figure 1–21), killing about 30,000 inhabitants. A glowing cloud of incandescent rock fragments and gases at temperatures between 700°C and 1000°C swept down from Mt. Pelée at about 150 km/hr, destroying everything in its path. A small nuée ardente photographed on Martinique in 1931 is shown in Figure 1–22. Several small nuée ardentes were observed in the May 1980 eruption of Mount St. Helens, Washington; these descended the south flank and flowed into the forest below in less than three minutes. Release of pressure

Figure 1–21
St. Pierre, Martinique, shortly before and after the 1902 nuée ardente eruption. [From T. A. Jagger, 1945. *Volcanoes Declare War* (Honolulu: Paradise of the Pacific, Ltd.), Plates 26a, 26c.]

Figure 1–22
Very small nuée ardente photographed at close range in 1931 on the flank of a volcanic dome, Martinique. [From F. A. Perrett, 1937, *The Eruption of Mt. Pelée 1929–1932* (Washington, D.C.: Carnegie Inst.), Fig. 33.]

in a magma chamber that contains a volatile-rich magma causes instant vesiculation of the melt. If an extreme amount of vesiculation occurs, the molten material will be blasted from the cone to form an ash fall. In less extreme cases a cloud of ash is produced, with the simultaneous formation of a glowing (basal) avalanche. The glowing avalanche consists of gas, frothy particles, liquids, and solids. The particles are presumed to be continually releasing gas, which maintains the turbulence and prevents particle-to-particle abrasion. The emission of gas allows the mass to behave as a fluid; it proceeds downslope by gravity for long distances with little loss of heat from the interior. The nuée ardente may arise either by explosions shooting the material out at low angle or, as in the case of Mt. Soufrière on the island of St. Vincent, by material falling from the edges of a thick ash cloud. When the avalanche or ash flow comes to rest, the particles may be largely liquid or solid. The deposit formed from mainly solid particles is largely unconsolidated and is called an *ash flow tuff*. If enough heat has been retained to weld the still-plastic fragments together, the deposit is called a welded ash flow tuff. The term *ignimbrite* is used for any rock formed by an ash flow, regardless of its degree of welding.

The fragments within ash falls and ash flow tuffs consist mainly of glass shards, which when erupted may be of irregular shape or may look somewhat triangular with concave sides (as shown in Figure 1–19). If the ash fall or ash flow tuff is unconsolidated, these shapes are retained. However, in thick deposits where welding occurs,

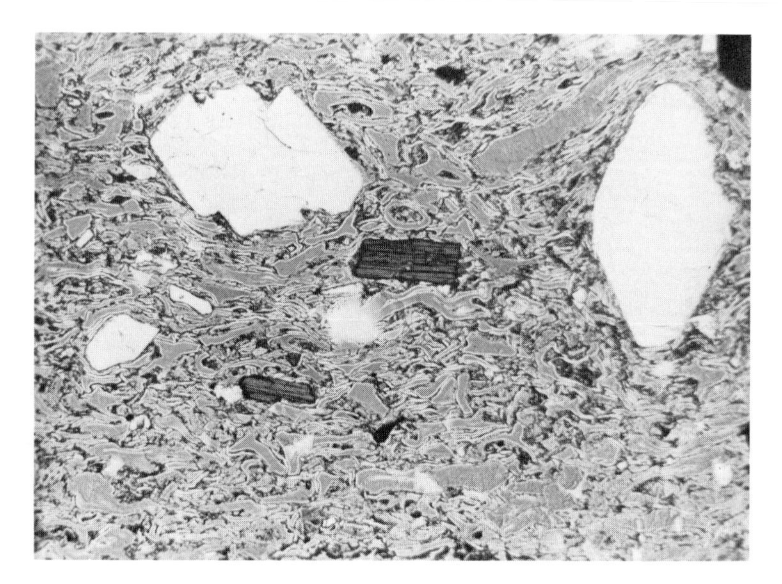

Figure 1–23
Compaction of plastic shard fragments deforms and welds the mass together. Photomicrograph (plane-polarized light) shows the deformed shards and fragmented crystalline material. [From J. C. Olson, 1968, *U.S. Geol. Surv. Bull., 1251–C,* Fig. 9.]

these shapes are distorted by compaction of the partially plastic fragments (see Figure 1–23). Irregular masses of pumice are often compacted in the same way, producing elongate fragments containing elongate vesicles. Multiple layers of ash flow tuffs can often be distinguished and mapped on the basis of degree of welding. The base of each unit is often chilled and is typically nonwelded. The degree of welding increases above the base, as do secondary reactions from the presence of vapors. The amount of welding decreases toward the top of the simple cooling unit. The upper layer, being unconsolidated, is often quickly removed by erosion, and is often absent from multiple cooling units.

METHODS OF EMPLACEMENT OF IGNEOUS ROCKS

The methods of emplacement of intrusive igneous rocks have been touched upon earlier in this chapter. Here we will review some of the evidence that can be observed in the field. The two methods for which most evidence is available are forceful injection and magmatic stoping. Formation of igneous melts by anatexis and retention of the melts within the metamorphic complex in which they were created is not really emplacement, as the melt has remained at its place of origin.

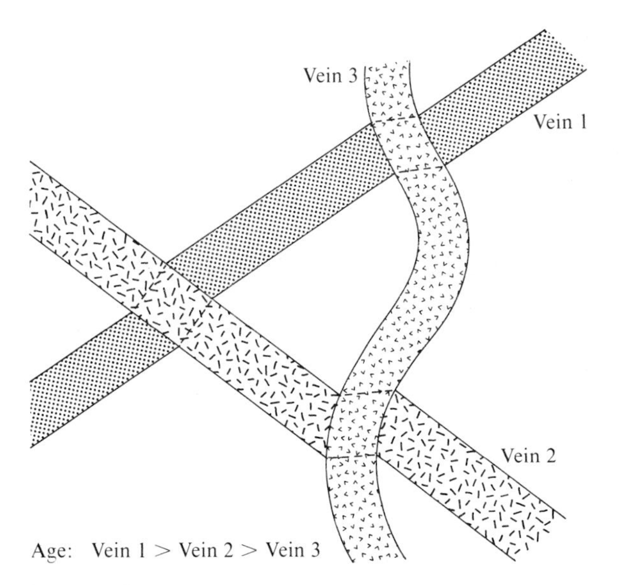

Figure 1–24
Schematic sketch of dilation veins. The displacement of earlier veins by later veins indicates the sequence of intrusion.

Evidence of forceful injection is abundant around many intrusive bodies. This is often easily seen at small dikes, where the intruding magmas have clearly forced apart the adjacent walls. The thing to look for here is the presence of one or more structures that cross the intrusion at other than 90°. Dilation or forcing open by intrusion will displace these features in the wall (see Figure 1–24). Even better is to find two country rock structures crossing the intrusion at different angles; when this is found it is obvious that the two wall segments have not been subjected to faulting along with dilation. Matching of the particular contours of the opposite sides of the intrusion aids the argument as well. Forceful injection can also be established by fracturing, doming, and faulting of overlying rock (as in the Henry Mountains, Utah).

Occasionally an igneous intrusion may be almost completely crystallized when emplaced. In this case a shear zone may be developed at the country rock contact, with apparent displacement and intermixing of country rock and solidified igneous material near the edges of the intrusive body. In other cases magmatic pressures may be sufficient to raise the roof of the chamber, forming an inward-dipping conical fracture. Magma injected along the fracture will crystallize to form a cone sheet.

Another method of intrusion is *magmatic stoping*. In mining terminology "overhand stoping" describes a process whereby miners tunnel under an ore body and mine it from below, permitting the roof rocks to be easily removed. Magmatic stoping is initiated by fracturing of surrounding country rock during intrusion; fragments sink into the magma chamber, allowing the magma to move upward into the vacated space. The foreign fragments (xenoliths) may show some reaction with the melt, indicated by mineralogical changes within the xenolith as well as differences in mineral composition between the area around the xenolith and the rest of the igneous mass. Total assimilation of xenoliths may result in diffuse bands or clots of material that have a somewhat different composition than the bulk of the igneous rock.

Stoping that takes place by incorporation of many small blocks of country rock is often called piecemeal stoping. In some cases a large mass of the chamber roof may collapse and sink into the magma below. Magma moving upward along the break will crystallize to form a more or less vertically dipping ring dike. The most extreme case of magmatic stoping is that of roof foundering. Here the magma is large, close to the surface, and has a thin roof. Collapse of the roof can result in inclusion of huge blocks of country rock, which have been tilted at different angles and perhaps have partially settled into the magma below. This mechanism has been postulated to explain the unroofing of the Idaho batholith.

SMALL-SCALE FEATURES

Small-scale features of an igneous rock can be determined in a hand specimen with either the unaided eye or a standard hand lens.

The first distinction to be made is whether or not individual mineral grains can be seen. If the mineral grains can be seen the rock is classified as *phaneritic;* if not, it is *aphanitic.* Care should be taken not to consider a rock as phaneritic if it contains some crystal-filled cavities in a fine-grained matrix, as is common with some extrusive rocks. The grains in igneous rocks form an interlocking texture (in contrast to the granular texture found in rocks such as sandstone).

Examination of grain sizes quickly reveals that most intrusive rocks are phaneritic and most extrusive rocks are aphanitic. Furthermore, it has often been noted that both intrusive and extrusive rocks commonly show a finer grain size as the country rock contact is approached. Some contacts (of extrusive rocks) may even be noncrystalline, and consist of glass. It is obvious from this that (at least in a general way) grain size can be correlated with cooling rate, with the coarser grain sizes related to a slower rate of cooling. In some igneous rocks some crystals are conspicuously larger than others. These larger crystals, called *phenocrysts,* are generally considered to represent crystallization during a period of slower cooling than the remainder of the rock. Further examination of the phenocrysts shows that they usually consist of only one or two mineralogical types; the finer *groundmass* or matrix materials are often composed of a larger variety of minerals. This observation leads to the idea that all of the minerals in an igneous rock have not formed simultaneously; certain minerals have begun crystallization before others. This relationship of sequential and simultaneous crystallization will be discussed in more detail later.

If crystals are visible in the hand specimen, the next thing to notice is whether or not their orientation is random. A random orientation is usually interpreted as indicating that crystallization occurred while the magma was at rest. Preferred orientations are developed if the magma has flowed either during or after partial crystallization. Flow structures can sometimes be detected by the presence of elongated vesicles or parallelism of inclusions or phenocrysts.

Care must be taken in the proper interpretation of flow textures. Rocks formed from volcanic ash deposits often show a parallelism of grains as a result of their method of deposition and later compaction. During crystallization, some large intrusions permit the sinking or floating of grains that form horizontal layers in which the crystals show a preferred orientation; here it is necessary to determine the location of zones of preferred orientation in relation to the whole intrusion. Preferred orientation as a result of flowage is most commonly found near the walls of intrusive bodies (where chilling has encouraged early crystallization) or near the top or bottom of flows (for the same reason). Parallelism of early-formed phenocrysts may be found anywhere within an igneous body.

The color of an igneous rock is a result of the colors of the various minerals present. If a fresh specimen is obtained the color is directly related to the igneous minerals present. The color of a weathered fragment is a consequence of both the original igneous minerals and later weathering processes (usually oxidation and hydration). Weathering processes commonly tend to destroy igneous minerals partially or completely and to form other minerals that are stable under atmospheric conditions (see Chapter 11). A result of weathering is the creation of a dull-looking rock, which has a dull sound when struck with a hammer. When unaltered, most igneous rocks present a somewhat vitreous appearance and produce a ringing sound when struck.

Assuming that you can obtain a fresh specimen, it then becomes possible to make a crude subdivision on the basis of color. Rocks that are rich in silica usually contain considerable quartz and light-colored feldspar. Rocks rich in Mg and Fe tend to contain the mafic minerals (olivine, pyroxene, amphibole, and mica). These minerals are usually strongly colored. On the basis of the amount of color versus white (or clear), a color index can be set up that relates in a rough way to the composition. A rock that contains less than 30% colored ferromagnesian minerals is considered light colored and may be called *leucocratic*. With 30–60% ferromagnesian minerals it is *mesocratic*, and with more than 60%, *melanocratic*. More commonly light-colored rocks are called *felsic*, and dark ones *mafic*.

The identification of the minerals present in an igneous rock can be carried out very successfully in a phaneritic rock using a hand lens to observe color, cleavage, and grain shape. The composition of aphanitic rocks can only be guessed at on the basis of the color index and the presence of any phenocrysts that might be present.

SUMMARY

A rock outcrop can be identified as igneous on the basis of both large-scale and small-scale features. Its extent, shape, and contact relationships with country rock can be used to determine whether the igneous rock is intrusive or extrusive. Small-scale features such as grain shapes, sizes, and manner of aggregation can be used to verify the method of origin. A more detailed description of small-scale features and their origin is given in the following two chapters.

FURTHER READING

Buddington, A. F. 1959. Granite emplacement with special reference to North America. *Geol. Soc. Amer. Bull., 70,* 671–747.

Carmichael, I. S. E., F. J. Turner, and J. Verhoogen. 1974. *Igneous Petrology.* New York: McGraw-Hill, 739 pp.

Coates, R. R., R. L. Hay, and C. A. Anderson. 1958. *Studies in Volcanology.* Geol. Soc. Amer. Mem. 116, 678 pp.

Green, J., and N. M. Short, eds. 1971. *Volcanic Landforms and Surface Features.* New York: Springer-Verlag, 519 pp.

Grout, F. F. 1918. The lopolith: an igneous form exemplified by the Duluth gabbro. *Amer. Jour. Sci., 46,* 516–522.

Hunt, C., P. Averitt, and R. L. Miller. 1953. *Geology and Geography of the Henry Mountains Region, Utah.* U.S. Geol. Surv. Prof. Paper 228, 234 pp.

Newell, G., and N. Rast, eds. 1970. *Mechanism of Igneous Intrusion.* Liverpool: Gallery Press, 380 pp.

Parker, D. C., and M. F. Wolff. 1965. Remote sensing. *International Science Tech., 37,* 20–31.

Viljoen, R. P., M. J. Viljoen, J. Grootenboer, and T. G. Longshaw. 1975. ERTS-1 imagery: an appraisal of applications in geology and mineral exploration. *Minerals Sci. Engng., 7,* 132–168.

Walker, K. R. 1969. *The Palisades Sill, N.J.: A Reinvestigation.* Geol. Soc. Amer. Spec. Paper 111, 178 pp.

2

Experiments with Molten Silicates: Unary and Binary Systems

The magmatic origin of the rocks we call basalt and rhyolite was first recognized in the late 1700s in Italy and France as a result of observations of existing volcanoes and correlation of the rocks formed with similar-looking ancient counterparts. As obvious as this origin seems to us now, one earlier belief held that these (volcanic) rocks were chemically precipitated from a primeval ocean. This general failure to observe is a common defect in field studies and is probably no less common today than in the late 1700s. The vision we get with hindsight is always 20/20.

The observations in Italy and France greatly impressed James Hutton (a Scotsman) with the importance of underground heat. He had observed a coal seam that had been baked adjacent to basaltic dikes and had also noted an example of forceful injection of basalt into overlying sediments. In order to obtain additional information on the nature and origin of basalt, James Hall (an Englishman) heated basalt in his laboratory and found that it melted at between 800°C and 1200°C. Rapid cooling of the melt yielded a glass. With slower cooling a fine-grained crystalline aggregate (similar to the basalt seen by Hutton) was produced.

Subsequent studies of igneous rocks revealed that most are composed of silicate minerals whose individual melting points are well over 1000°C. Investigations into the melting of igneous rocks have indicated that complete melting of basalt is usually achieved at about 1100°C, whereas that of granite is usually 150–200°C lower.

The initial information about igneous rocks comes from observations made in the field. These observations present the questions that we must answer to unravel the history of the rock. The types of things we notice rapidly are the color of the rock, whether it extends the length of the outcrop or pinches out, its thickness, whether it is layered, the nature of the contact with adjacent rocks, and whether the grains in the rock are visible without use of a hand lens. Before leaving the outcrop we try to determine the mineral composition of the rock. The purpose of making as many observations as possible at the outcrop rather than waiting to do everything in the laboratory is that areal changes in rock type and mineral composition are important clues to rock origin. An accurate sense of geologic setting is more clearly felt standing at the outcrop than sitting at a bench in a laboratory.

From the mineralogic point of view, perhaps the most striking thing about igneous rocks is the dominance of feldspars. Fully two-thirds of the minerals in igneous rocks consist of feldspars, mostly potassic and sodic in light-colored rocks, and mostly calcic in dark-colored ones. For example, an average *granite* contains chiefly alkali-rich feldspars (orthoclase or microcline), sodium-rich plagioclase (usually oligoclase), and quartz, with minor ferromagnesian minerals. An average *basalt* is composed chiefly of plagioclase (chiefly labradorite) and clinopyroxene, with lesser quantities of olivine, orthopyroxene, nepheline, or quartz. It is obvious that if we wish to understand igneous rocks we must study feldspars in some detail. There are two ways to do this: (1) visually, using hand specimens and thin sections; and (2) chemically, by conducting laboratory experiments.

HAND SPECIMENS AND THIN SECTIONS

In hand specimens of granites, potassium-rich feldspars and plagioclase are readily seen with a hand lens and can be distinguished most easily by color difference and twinning. Potassic feldspar is usually pink or salmon colored, and plagioclase is usually white to gray; often polysynthetic twinning is visible as parallel striations on fresh cleavage surfaces of plagioclase. Sometimes the twinning is not seen, either because the grain is altered, because the grain has been rubbed by too many greasy hands, or because the exposed surface is parallel to the twin's composition plane. Some feldspar grains may show perthitic texture—elongate patches of one color within a larger grain of a different color.

Observations of thin sections of granitic rocks will reveal additional features. Perthitic feldspar grains that seem irregularly colored in hand specimens are now clearly seen to be a mixture of two feldspars in the same grain. Usually the grain consists mainly of microcline or orthoclase that contains irregularly shaped patches or blebs of plagioclase (usually oligoclase); this compound crystal is called a *perthite.* If the situation is reversed, and plagioclase is seen to contain patches of a potassium feldspar, the grain is called *antiperthite* (see Figure 2–1).

Plagioclase crystals commonly show a difference in extinction angle between the center and edge of the grain. This difference in extinction angle is a response to differences in optical orientation within the crystal, which in turn is related to differences in composition in different portions of a single crystal. These differences in extinction position may be gradational or abrupt, or even oscillatory, indicating gradual, discontinuous, or constantly variable changes in composition during growth. Precise determination of composition by extinction angle methods or refractive index measurements usually indicate that the core of the crystal is more calcic than the rim.

Figure 2–1
Perthitic feldspar in granite. Perthite usually consists of sodium-rich plagioclase exsolved from microcline. These minerals are determined usually on the basis of their relative indices of refraction. Crossed nicols. Width of photo is about 0.8 mm.

This compositional arrangement is so common that it is called *normal zoning* (as contrasted to the much less common *reverse zoning*) (see Figure 2–2).

In some granites a peculiar intergrowth of potassium feldspar and quartz occurs in a pattern resembling ancient Near Eastern cuneiform inscriptions; this has been called *graphic texture* (see Figure 2–3). How are these and other variations in textures and compositions produced in igneous rocks? Observations of natural rocks cannot provide the answer. Further information must be sought in laboratory experiments.

LABORATORY EXPERIMENTS

The experiments performed in 1792 by James Hall are one of the earliest examples of experimental igneous petrology. Hall's studies were, of necessity, restricted to the melting of existing rocks because he was unable to synthesize rock compositions from chemical reagents. Present chemical knowledge permits such syntheses to be carried out and the results analyzed with great precision. In modern geochemical laboratories it is possible to produce the entire range of compositions, temperatures, and pressures that might exist during the formation of magmas.

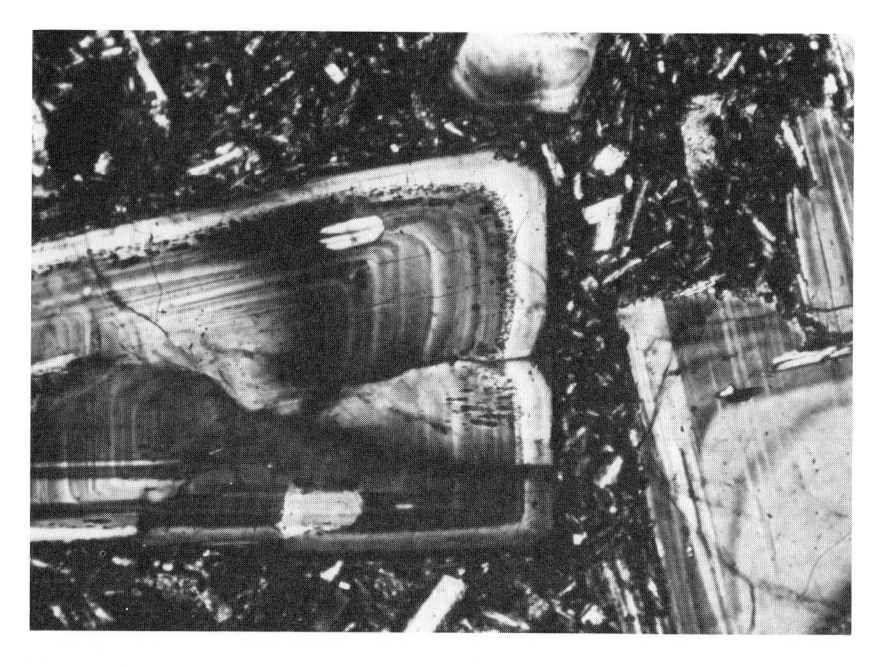

Figure 2–2
Oscillatory zoning of plagioclase phenocrysts in andesite. Compositional zoning as revealed by differences in extinction angle shows changes in grain shape during growth. Polysynthetic twin bands parallel the longer grain edges. Washoe, Nevada. Crossed nicols. Width of photo is about 0.3 mm.

In principle the experimental method is simple and appears analogous to cooking up a stew for supper—developing the initial mixture of ingredients, heating, perhaps testing at various times, and finally enjoying the result. In fact, though, the procedures are much more complicated and carefully thought out. The initial starting compositions are made from either chemically pure reagents or analyzed minerals, carefully weighed and homogenized. If the experiments are to be carried out at high temperatures the samples are commonly fused, cooled to form a glass, and ground to a fine grain size in order to insure homogeneity of composition. Experiments at low temperatures often use chemical batches made of gels in order to increase reaction rates. Occasionally, crystalline seeds are added in order to encourage crystallization of slow-forming materials. The heating equipment is varied, and may range from platinum-wound furnaces for high-temperature work to autoclaves (hydrothermal reactors) capable of maintaining a gas phase in contact with the sample during the experiment. Pressures and gas compositions are controlled as well as temperature. In other cases, high confining pressures are needed; devices capable of maintaining pressures and temperatures equivalent to those found within the earth's mantle are utilized on a routine basis in many research laboratories.

Figure 2–3
Graphic texture in granite shown by intergrowth of quartz (dark) and feldspar (light).

After subjecting the sample to the desired pressure–temperature conditions for an adequate period of time (usually days or weeks), pressures and temperatures are usually rapidly decreased to room conditions in order to "freeze in" the high-temperature, high-pressure mineral and liquid assemblage. Supercooled liquids will often solidify without crystallization to form a glass. The sample is then examined with a petrographic microscope and/or x-ray diffractometer to determine the nature and amounts of crystals and glass. More detailed examination may require single crystal x-ray analyses of crystalline phases or detailed chemical analyses by the electron microprobe or other methods. The resultant data may then be plotted graphically; a plot showing the stability regions of phases as a function of composition, temperature, and/or pressure is called a phase diagram. Thousands of such diagrams have been created for geological and industrial purposes. The American Ceramic Society has published three volumes containing 5,000 phase diagrams of interest to ceramic engineers and geologists. The U.S. Geological Survey has published others in the sixth edition of *Data of Geochemistry,* and the Geological Society of America has published a partial compilation in Memoir 97, *Handbook of Physical Constants;* in spite of this, many of the pertinent diagrams remain scattered throughout the literature. We will examine a few phase diagrams in order to gain some understanding of melting and freezing processes.

PHASE RELATIONS AND TEXTURES

One-Component Systems and the Phase Rule

We can describe the equilibrium relationships of minerals most easily in terms of phase diagrams. A *phase* is defined as a portion of a system (here a magma) that is physically distinct and mechanically separable from the other parts of the system. A magma may contain a gas phase, one or more liquid phases, and a number of solid phases (such as included crystals of olivine, pyroxene, and plagioclase). A phase diagram (see Figure 2–4) is a graphic means of showing the stability conditions of the various phases, usually in terms of the variables pressure, temperature, and composition. In the one-component (unary) SiO_2 system, seven phases (six solids and one liquid) are indicated within the pressure–temperature limits of the diagram. All of these phases have the composition SiO_2. The six solids are different structural ar-

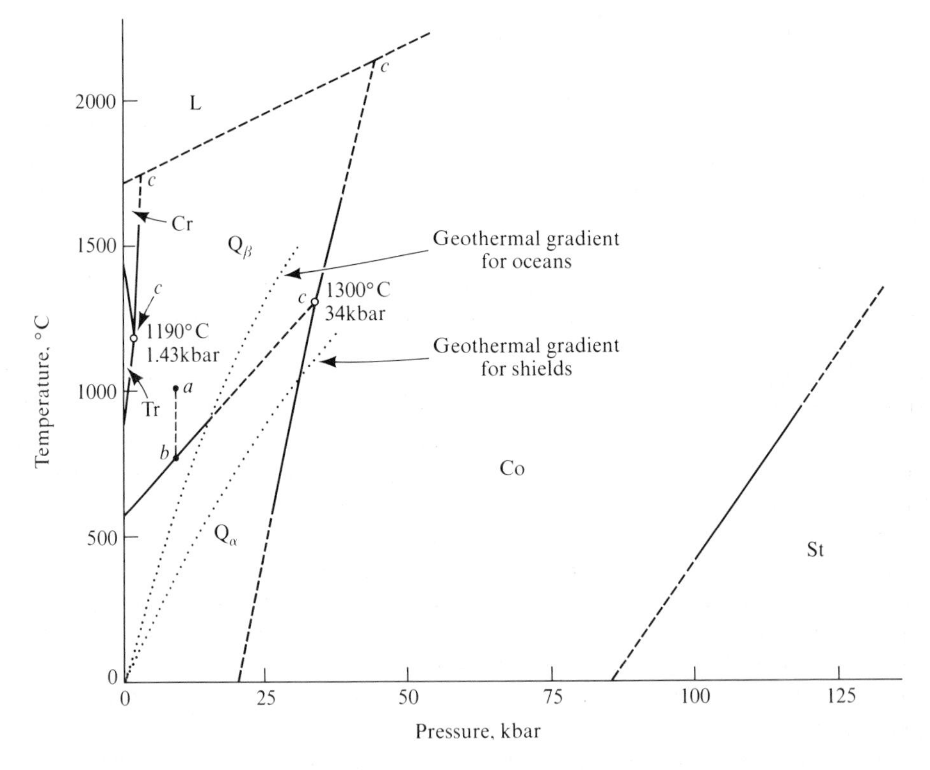

Figure 2–4
The one-component system SiO_2. At point *a* only β-quartz is present. At *b*, α- and β-quartz exist together. At points such as *c*, three phases exist together. L = liquid, Cr = cristobalite, Tr = tridymite, Q_α = α-quartz, Q_β = β-quartz, Co = coesite, St = stishovite. [After I. A. Ostrovsky, 1967, *Geol. Jour., 5,* Fig. 2.]

rangements of SiO_2, and are called *polymorphs*. Each labeled area within the diagram is separated by lines that define the limits of stability of the particular silica phase in terms of the pressure–temperature coordinate system. These lines are called phase boundaries. Dashed lines indicate experimental uncertainty.

Figure 2–4 shows the result of experiments conducted using only one component, silica (SiO_2). *Components* are defined as the minimum number of chemical constituents necessary to describe the composition of all of the solids, liquids, and gases possible in the system. Most igneous rocks have at least 10 to 12 components. Note that a component is not a mineral. The component SiO_2 is not necessarily the phase quartz. It may be molten SiO_2, or any of the six major crystalline varieties. Similarly a component such as Al_2O_3 is not synonymous with corundum, as Al_2O_3 can exist in a variety of forms. Thus the material being examined (the system) is always described in chemical terms. In this and the following chapter we will use as many as three components to illustrate how the more complex rock systems behave. But as a start, let's look at what happens at various temperatures and pressures when only a single component, SiO_2, is present.

The point *a* on the diagram, at a pressure of 10 kilobars (kbar; 1 kbar = 1,000 bar $\simeq$ 1,000 atm) and a temperature of 1000°C, falls within the area labeled β-quartz, which means that if equilibrium conditions prevail (that is, the system is at its lowest energy state consistent with the applied pressure and temperature), the component SiO_2 should exist as a solid in the particular structural arrangement that has been named β-quartz. Normally this is the case, but it may be that in certain circumstances, such as might occur with a rapid change of pressure (P) or temperature (T), the SiO_2 might consist of a different phase, such as the high-pressure polymorph stishovite, or a glass. If this occurs, the system is in a state of nonequilibrium and is not described by the phase diagram, which applies only to equilibrium conditions. A phase is unstable if it is in the process of changing, or is just on the verge of doing so. It is called metastable if it is not in its lowest energy state, but tends to persist in the state it is in. Many dense minerals, such as diamond, kyanite, and coesite, are stable only at pressures higher than one atm, but tend to persist metastably at low pressures and temperatures because of the extreme slowness of reaction rates.

Returning to Figure 2–4, consider what happens when the stable β-quartz at a point *a* is lowered in temperature while the pressure is maintained at 10 kbar. The β-quartz remains stable until the phase boundary at *b* is encountered. Here β-quartz begins a polymorphic inversion to form α-quartz. If the temperature were held constant at *b* both phases would exist, but with a decrease in temperature, complete conversion to α-quartz occurs. Alpha-quartz persists as the stable phase to room temperatures.

The points labeled *c* define particular *PT* conditions where three stability areas intersect—liquid, cristobalite, and β-quartz; tridymite, cristobalite, and β-quartz; liquid, β-quartz, and coesite; or α-quartz, β-quartz, and coesite. These phases will be present if the system is in an equilibrium state. Phase boundary intersections of this type are called *invariant points,* because neither pressure nor temperature can be changed without the elimination of one or two of these phases. The phase boundaries that define the limits of each mineral stability field are called *univariant curves;* along these curves two phases will exist together; the two-phase assemblage can be maintained if one of the variables (P or T) is changed independently, while simultaneously changing the second dependently (in such a way as to keep the PT values on the curve). The areas where single-phase assemblages exist are called *divariant regions,* as both variables can be independently changed, while still maintaining the stability of a single phase. Point a represents a state of divariant equilibrium, point b is univariant, and points c are invariant.

The general relationships among phases, components, pressure, and temperature are described by the Phase Rule, propo:ed by J. Willard Gibbs in the 1870s. The phase rule provides a means of classifying systems. In addition it describes the maximum number of phases that can exist in a system of any complexity, thus providing us with a method of testing whether the minerals in a rock are coexisting in stable equilibrium.

Derivation of the Phase Rule can be made in terms of variance or degrees of freedom (F). The term F = unknown relationships − known (dependent) relationships. The unknown relationships within a system include the number of phases and their compositions. Assume that there are p phases, and each phase contains some of each of the components c that constitute the system. As the composition of each phase can be given in atomic or mole fractions or weight percentages of each of the components, the composition of each phase can be taken as $c - 1$ (as all of the components add up to 100% of the total, and the last component can be found by difference); as p phases are present the number of chemical variables is therefore $p(c - 1)$. In addition to chemical variables the system is subjected to a particular (but in general unspecified) pressure and temperature that is the same for all of the coexistent phases. Thus the number of unknown variables in the system is $p(c - 1) + 2$.

As for known or dependent variables or restrictions, it has been established by thermodynamics that the tendency of each component to react (known as the chemical potential) is the same in all coexistent phases. If it were not, the system would not be in equilibrium, and chemical reactions should occur. If the chemical potential of a component were determined in one phase, it would also be known for all other phases in the system. Hence for each component there are $p - 1$ dependent relationships. As c components are present in the systems, the total number of dependent relationships is $c(p - 1)$.

Subtracting the known or dependent relationships from the unknown gives the Phase Rule—a statement of the variance of a system:

$$F = p(c - 1) + 2 - c(p - 1)$$
$$F = c - p + 2$$

If a restriction is placed on the system, such as fixed pressure (the isobaric condition), the number of unknown relationships is reduced by 1, and the Phase Rule is taken as $F = c - p + 1$.

The above discussion assumes that the system is closed—with no material entering or leaving. If the system is open to the movement of mobile components, the number of dependent variables is increased by the number of mobile components (m). Thus, $F = p(c - 1) + 2 - [c(p - 1) + m] = c - m - p + 2$.

Consider the application of the Phase Rule to the one-component system SiO_2 (see Figure 2–4). If it is stated that three phases coexist together within the system (as at points c), the Phase Rule indicates that the assemblage is invariant:

$$F = c - p + 2$$
$$F = 1 - 3 + 2$$
$$F = 0$$

Without recourse to the diagram it is obvious that a three-phase assemblage has no degrees of freedom, and can therefore exist only at a specific pressure and temperature; this is true, of course, for all three-phase assemblages. As we cannot have negative degrees of freedom, stable coexistence of more than three phases is impossible.

A two-phase assemblage (as at point b) is shown to be univariant:

$$F = c - p + 2$$
$$F = 1 - 2 + 2$$
$$F = 1$$

The assemblage is maintained with change of one variable (either P or T) if a dependent change is made of T or P, respectively.

Divariant equilibrium is present in the one-phase regions of the diagram (point a) as

$$F = c - p + 2$$
$$F = 1 - 1 + 2$$
$$F = 2$$

Here both P and T can be changed independently and the same one-phase assemblage is maintained.

Two-Component Systems—Congruent Melting

Another type of phase diagram of interest to petrologists is a two-component (binary) system that melts congruently. *Congruent melting* means that the pure solid phase when brought to its melting point, melts completely to produce a liquid of identical composition to the solid. Figure 2–5 shows the binary system $CaAl_2Si_2O_8$–$CaMg(SiO_3)_2$, which could be considered as an idealized basalt at atmospheric pressure. Temperature increases upward on the vertical coordinate and composition is

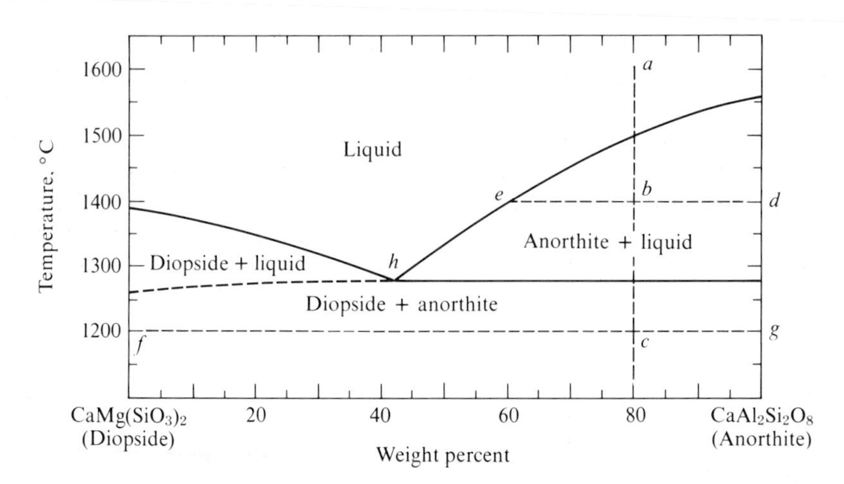

Figure 2–5
The two-component system $CaAl_2Si_2O_8$–$CaMg(SiO_3)_2$ at 1 atm pressure. The melting point of diopside, $CaMg(SiO_3)_2$, is 1392°C; that of anorthite, $CaAl_2Si_2O_8$, is 1553°C; and the eutectic point is 1274°C. The labeled points are described in the text. The slight variation in temperature of the solidus line (see dashed portion) occurs because the system is not truly binary for diopside-rich compositions; a small amount of Al enters diopside crystals. [After E. F. Osborn, 1942, *Amer. Jour. Sci., 240,* Fig. 4.]

plotted horizontally. Any point within the diagram indicates a particular temperature (T) and composition (X). The percentages of components are given in weight percent, as they are on most diagrams, unless otherwise stated.

In this diagram a pressure coordinate is not given, as the system was examined at atmospheric pressure—1 atm; that is, pressure is constant. A binary system shown with the third variable (P) would require an extra coordinate. The variable P is often set up at right angles to the other two and shown as a three-dimensional perspective representation.

The phase assemblages in different areas of the diagram are labeled in Figure 2–5 for different combinations of T and X. Composition *a* of 80% $CaAl_2Si_2O_8$ and 20% $CaMg(SiO_3)_2$ at 1600°C consists of a homogeneous liquid. At 1400°C this same composition *b* is composed of anorthite and liquid; and at 1200°C (*c*) it is a mixture of crystals of diopside and anorthite. The determination of the phase assemblage on a binary phase diagram is simply a question of finding the proper point on the diagram in terms of T and X, and reading off the description of the phase assemblage for the area in which it is located.

Carrying the process one step further we can proceed to determine the percentages and compositions of each of the phases present at points *a*, *b*, and *c*. The first point *a* at 1600°C is located in the one-phase field called liquid. As the entire batch consists of liquid it is obvious that the composition of this liquid must be that of the given composition—namely 80% $CaAl_2Si_2O_8$ and 20% $CaMg(SiO_3)_2$. Point *b* at 1400°C consists of a two-phase assemblage—liquid and anorthite. The percentage of each of these phases can be determined. It can be seen that if the point *b* representing the bulk composition were further to the right it would be closer to the composition of anorthite (*d*), $CaAl_2Si_2O_8$, and would therefore contain more anorthite; and if it were moved to the left, the composition would be closer to that of the liquid *e* and the batch would contain a high percentage of liquid. From this nonrigorous approach, the method of determining percentages can be deduced. Draw a horizontal (constant temperature) line through the point of interest (here point *b*) in the two-phase field to the two extremities of the two-phase field. This line is called a *tie line*. The intersections of the tie line at the limits of the two-phase field are points that indicate the compositions of the two coexisting phases. In this case, the bulk composition is given by *b* [80% $CaAl_2Si_2O_8$, 20% $CaMg(SiO_3)_2$], the liquid composition is *e* [62% $CaAl_2Si_2O_8$, 38% $CaMg(SiO_3)_2$], and the anorthite composition is *d* (100% $CaAl_2Si_2O_8$). The percentages of liquid *e* and anorthite *d* are given by the location of the bulk composition *b* on the line *de*.

$$\frac{\text{length } be}{\text{length } de}(100) = \%\ \text{anorthite } (d)$$

$$\frac{\text{length } bd}{\text{length } de}(100) = \%\ \text{liquid } (e)$$

Using measurements from Figure 2–5, this comes to

$$\frac{1.7\ \text{cm}}{3.4\ \text{cm}}(100) = 50\%\ \text{anorthite}$$

$$\frac{1.7\ \text{cm}}{3.4\ \text{cm}}(100) = 50\%\ \text{liquid}$$

Similarly, if we wanted to know the phase percentages represented by the bulk composition at *c*, the tie line would be drawn to the limits of the two-phase field at *f* (diopside) and *g* (anorthite).

$$\frac{\text{length } cf}{\text{length } fg}(100) = \%\ \text{anorthite}$$

$$\frac{\text{length } cg}{\text{length } fg}(100) = \%\ \text{diopside}$$

The lower limit of the area described as liquid is indicated by the two curved, sloping lines. These are called *liquidus lines*. The upper limit of the region that consists of only solid phases is a single horizontal line called the *solidus*. Between the liquidus and solidus lines are the areas in which both liquid and solid exist in equilibrium.

The temperature at which crystallization begins is given for various compositions by the liquidus lines. The freezing temperatures of the pure components $CaAl_2Si_2O_8$ and $CaMg(SiO_3)_2$ are maxima; the addition of the second component has the effect of decreasing the temperature at which the first crystals begin precipitating from the melt. This decrease continues to a minimum liquid temperature at point *h*, which is called the *eutectic*. This point is defined by both composition and temperature.

The eutectic is an invariant point, as indicated by the Phase Rule. Recall that the system is isobaric (fixed pressure), so that the Phase Rule is

$$F = c - p + 1$$
$$F = 2 - 3 + 1$$
$$F = 0$$

The decrease in freezing temperatures by admixture of a second component can be understood in terms of the tiny ionic structures that exist within the liquid. The size and number of these structural units increase as the temperature is lowered toward the freezing point. Due to the thermal energy of the system, atomic and molecular vibrations are present, and these units are in a constant state of change, with atoms constantly being lost to the disorganized liquid and other atoms forming or adding on to the structural units. A dynamic state of transition exists, and the ratio of the number of atoms combining to the number dissociating determines the number and size of the structural units at a particular temperature and composition. At high temperatures (see Figure 2–6) the dissociation rate is high, so the number and size of

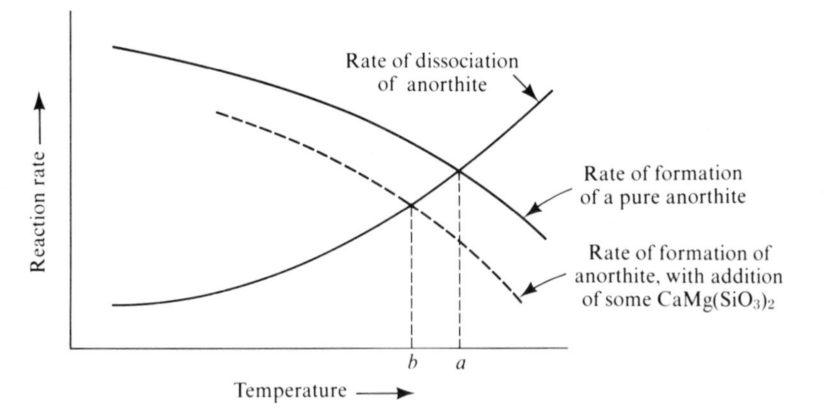

Figure 2–6
Schematic diagram showing the rate of formation and dissociation of anorthite (solid lines) as a function of temperature. Intersection of these lines at *a* defines the freezing point of anorthite. The dashed line shows the lowered rate of formation of anorthite when the second component is present. Point *b* is the decreased initial freezing point of the mixture.

such units is low. As the temperature is lowered, the combining rate is increased. Below the freezing temperature the combining rate exceeds the dissociation rate; the intersection at temperature *a* defines the freezing point. These small units can be thought of as crystal nuclei—tiny seeds upon which crystal growth proceeds.

If the melt consisted of a single component such as $CaAl_2Si_2O_8$, every atom in the liquid could add to the nuclei to form crystals of anorthite. If, on the other hand, 20% of the liquid consists of a second component such as $CaMg(SiO_3)_2$, the rate of formation of anorthite nuclei is decreased because a large number of atoms or molecular groups (characteristic of diopside) would not be able to attach themselves to the anorthite nuclei. The anorthite nuclei would grow at a decreased rate; but the rate of dissociation of these nuclei would remain essentially the same as in a pure $CaAl_2Si_2O_8$ melt. Thus, the addition of the second component to the melt has the effect of decreasing the number and size of anorthite nuclei. In order to crystallize anorthite in the presence of a second component, it now becomes necessary to lower the temperature below the freezing temperature of pure anorthite in order to allow the formation rate to exceed that of dissociation (temperature *b*). Increasing the amount of the $CaMg(SiO_3)_2$ component leads to continued decrease in the freezing point of anorthite. This decrease proceeds to the eutectic, where both anorthite and diopside crystallize simultaneously. This decrease in freezing temperatures by addition of a second component is well known in cold climates by the systems "H_2O–antifreeze" and "salt–ice."

The crystallization of a liquid of composition *a* can now be considered (see Figure 2–7). The easiest approach is to draw a vertical line downward through the composition of interest; this line of constant composition and variable temperature is called an *isopleth*. The isopleth goes through the various phase fields of the diagram, and the sequence of phase assemblages can be read in each area. At any particular temperature in a two-phase field the percentages of phases present can be determined by use of tie lines. The assemblages encountered along the isopleth are liquid, anorthite + liquid, diopside + anorthite + liquid (at the solidus), and finally diopside + anorthite. These will be the assemblages produced during crystallization.

The liquid *a* is cooled to the liquidus at *b*. Here it enters the field of liquid + anorthite, which means that anorthite begins to crystallize. The anorthite composition is *c*. As the anorthite has the composition $CaAl_2Si_2O_8$, depletion of this component from the melt means that the melt composition becomes relatively enriched in the remaining component, $CaMg(SiO_3)_2$. This change of liquid composition requires a change in freezing temperature as well. With cooling and precipitation of anorthite the liquid composition changes along the liquidus curve in the direction of the eutectic. When the sample has cooled to temperature *d*, the tie lines indicate that the initially homogeneous liquid has changed to a mixture of anorthite *e* and liquid of composition *f*. Analysis of the tie lines through this two-phase region shows an increasing amount of anorthite with decreasing temperature.

As cooling continues to *g* the anorthite continues precipitating and the liquid composition changes to that of the eutectic *h*. The eutectic is the minimum temperature

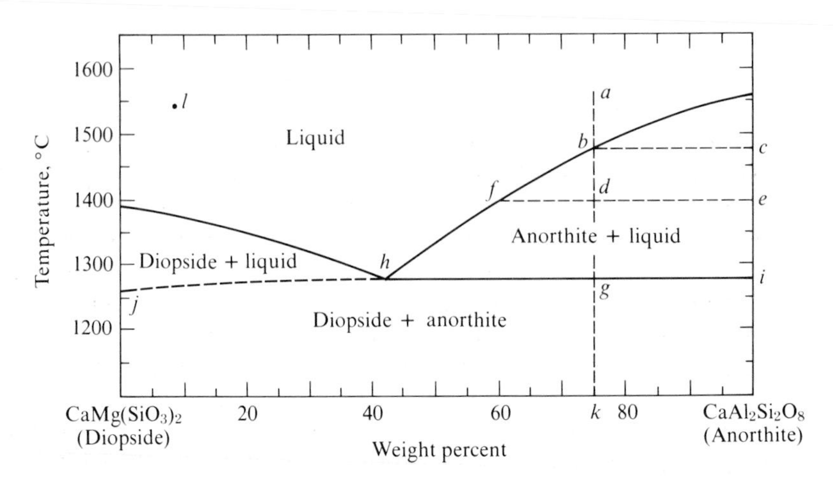

Figure 2–7
Crystallization within the system $CaAl_2Si_2O_8$–$CaMg(SiO_3)_2$. The labeled points are described in the text. [After E. F. Osborn, 1942, *Amer. Jour. Sci., 240,* Fig. 4.]

at which liquid can exist. The eutectic liquid is composed of a mixture of both $CaAl_2Si_2O_8$ and $CaMg(SiO_3)_2$ components. Both components crystallize (as anorthite and diopside) at a fixed temperature until the liquid is consumed. The ratio of the precipitating phases can be found by consideration of the segments of the tie line *jh* and *hi*. The process occurring at the eutectic can be deduced by observing that at a temperature just above *g* the phases indicated on the isopleth are liquid + anorthite; and just below *g* the phases indicated are anorthite and diopside. The anorthite and diopside cool stably to room temperature.

Melting of an anorthite–diopside mixture proceeds in an opposite manner. The mixture (composition *k*) is brought to the eutectic temperature *h*. At the temperature *h*, addition of heat causes both solid phases to melt simultaneously to form a liquid *h* of eutectic composition. All of the diopside and some of the anorthite are melted at this temperature. At a temperature infinitesimally above the eutectic (just above *h*) the batch will consist of liquid *h* and anorthite *i*. Additional heating causes the remaining anorthite to melt; this results in a change in liquid composition along the liquidus from *h* through *f* to *b*. At *b* the last crystal of anorthite is eliminated and the liquid composition (on the isopleth) remains constant with additional heating. Note that for any composition composed of the two components, such as *k*, melting will begin initially at the eutectic temperature, rather than at the melting point of either of the pure phases. Anorthite or diopside will only melt or freeze at its "listed" melting temperature when in the pure state. This occurs at a fixed temperature rather than over a temperature range, as is the case within a binary or more complex system.

Textures

The cooling and heating sequences described occur under the equilibrium conditions indicated by the phase diagram. Slow cooling of a melt allows sufficient time for atomic migration, with the result that the cooled material is completely crystalline. If, however, the cooling rate is extremely fast the melt can be brought to room temperature without crystallization; this results in the formation of metastable glass. Volcanic glasses such as obsidian are formed by supercooling as a result of extrusion of a melt at the earth's surface. The metastable character of glass is indicated by the fact that volcanic glasses are almost unknown on earth in pre-Tertiary rocks. Given sufficient time glass will tend to devitrify (crystallize). Detailed examination of glassy rocks often shows the presence of minute curved and often concentric cracks that result from unequal expansion due to subsequent hydration. This is called *perlitic* texture (see Figure 2–8). Devitrification often begins adjacent to these cracks, which suggests that migration of water may be one of the controlling factors in the crystallization of these glasses. The presence of very ancient glass in anhydrous

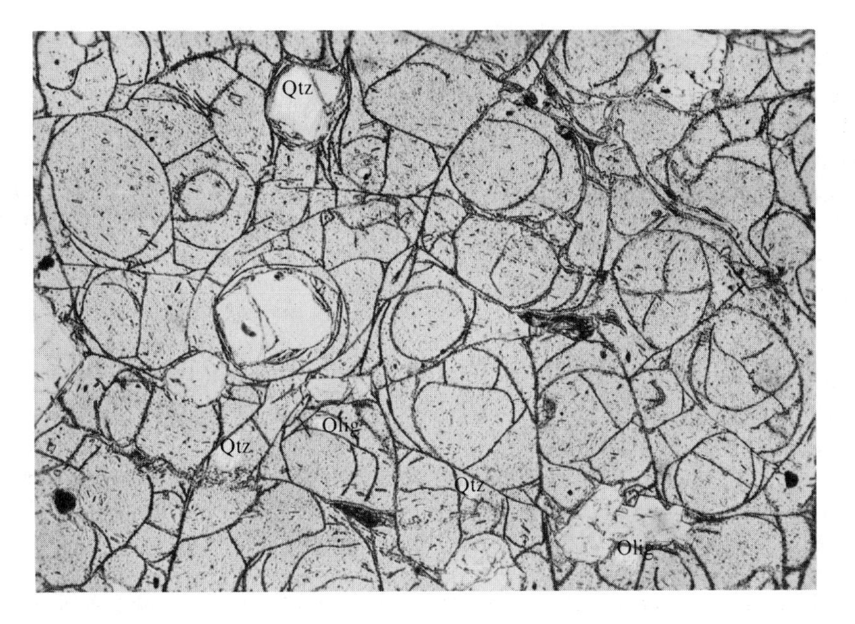

Figure 2–8
Perlitic texture in dacitic glass. The photomicrograph (plane-polarized light) shows somewhat concentric perlitic cracks in (gray) volcanic glass. A few crystals of quartz (Qtz) and oligoclase (Olig) are indicated. Width of photo is about 2.5 mm. The specimen is from Saipan, Marianna Islands. [From T. G. Schmidt, 1957, U.S. Geol. Surv. Prof. Paper 280–B, Plate 26–B.]

lunar rocks supports this contention. Another feature commonly considered to result from fairly rapid cooling of a melt is the formation of *spherulites*. Spherulites form in somewhat undercooled viscous melts when nucleation rate is low and growth rate is high. Here, a particular point (perhaps a minute crystal) within the melt acts as a seed, from which needlelike crystals grow in a radial manner. These small spherical bodies can be seen macroscopically or microscopically. When viewed in the microscope spherulites produce an extinction cross that remains in fixed position during rotation of the stage under crossed polars (see Figure 2–9).

Rapid cooling might produce a rock that, although mainly glass, contains tiny incipient crystals. The smallest of these are called *crystallites* (see Figure 2–10); they may be branched, circular, or rod-shaped, but all have the characteristic of being so small that they can be seen only with a microscope, and are of insufficient thickness to produce interference colors. Rapid cooling may also produce *microlites* within a glass matrix; microlites are tiny needlelike crystals, which can be seen only with a microscope and do produce interference colors. Some can be identified as to general mineral composition (see Figure 2–11).

Certain textures (see Figure 2–12), although common, are of somewhat questionable origin. These include *myrmekite,* which is a wormy-looking intergrowth of quartz and oligoclase. This texture is probably related to the presence of hot, corrosive, water-rich solutions during the later stages of cooling. *Graphic intergrowth* is a regular intergrowth of quartz and alkali feldspar. The quartz is commonly subtriangular in shape and located within large feldspar grains, producing an appearance resembling cuneiform writing. Suggested origins are replacement or crystallization at a eutectic (by analogy with metallurgical systems where similar textures are produced). *Alpite* is a fairly fine and equigranular rock that is more or less free of dark minerals; the term is used as a rock name or texture (aplitic). Aplitic materials appear to be related to crystallization of a water-rich silicate melt during a period in which excess volatiles are boiled off in the last stages of crystallization.

Crystal size is influenced by many variables, including the rate of nucleation, crystal growth, and cooling. It has been established experimentally that rapid cooling generally produces many crystal nuclei, with the effect that a large number of small crystals are produced—an aphanitic texture. With a slower cooling rate, a smaller number of crystal nuclei are produced, which are able to grow by diffusion processes within the melt to produce larger crystals—a phaneritic texture. Grain sizes can be further subdivided as very coarse ($>$ 3 cm), coarse (5 mm–3 cm), medium (1–5 mm), and fine ($<$ 1 mm).

Differences in the cooling rate of various portions of a magma produce grain size differences. A common example of this results from the chilling of the borders of an igneous intrusive body against colder country rock. The edges of many igneous bodies are aphanitic, whereas the more slowly cooled central areas are phaneritic. Rapid cooling of viscous siliceous lavas often produces aphanitic equigranular texture, whereas equally rapid cooling of the less viscous mafic magmas usually permits formation of macroscopically visible plagioclase laths. When the space between these

(A)

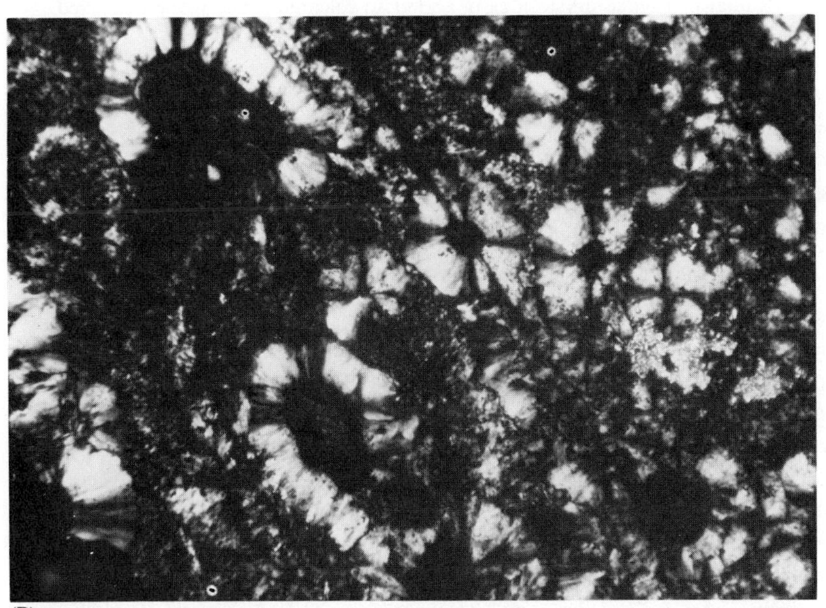

(B)

Figure 2-9
Spherulites (gray) are formed in undercooled viscous melts, or in devitrified glass. (**A**) Spherulites in obsidian may be formed by radial crystal growth as a result of devitrification of glass. (**B**) When viewed microscopically with crossed nicols, spherulites show a black extinction cross (upper right) that remains in a fixed position during rotation of the stage. Crossed nicols. Width of photo is about 0.8 mm.

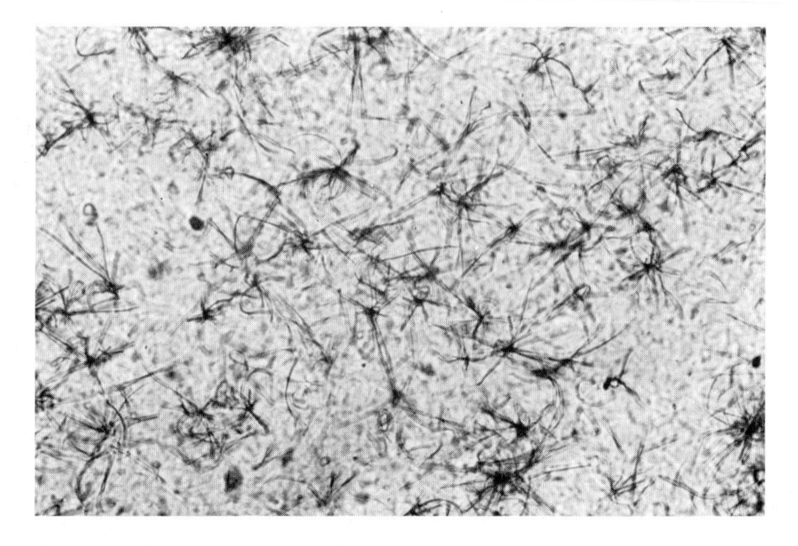

Figure 2–10
Radially oriented crystallites in an acidic volcanic glass. Plane-polarized light. Width of photo is about 0.3 mm.

Figure 2–11
Plagioclase microlites in basalt. Sample collected on the Mid-Atlantic Ridge. Crossed nicols. Width of photo is about 0.8 mm.

Figure 2–12
Myrmekite (in center of photo), consisting of an intergrowth of plagioclase and vermicular (wormlike in form) quartz. The larger twinned crystals are potassium feldspar. Hornblende is present in the lower left (light gray) and upper right (almost black). Crossed nicols. Width of photo is about 0.8 mm.

laths is filled with finer crystals the texture is called *intergranular;* when filled with glass, the texture is *intersertal.*

Changes in cooling rate during crystallization produce inequigranular texture (crystals of different sizes) rather than an even-sized granular texture. A melt may cool slowly at depth to produce a number of large crystals. Following this the magma may be extruded as lava or intruded at shallow levels, significantly increasing the rate of cooling to produce an aphanitic or glassy matrix between the earlier-formed larger crystals. When such a hiatus in grain size (a *hiatal texture*) is present, the larger crystals are called *phenocrysts,* and the rock is said to be *porphyritic.** When the phenocrysts comprise less than 50% of the rock, this is indicated in the rock name as a prefix; for example, porphyritic andesite. If the phenocrysts exceed 50% (or are conspicuous in appearance), this is described with a suffix; for example, andesite porphyry. Occasionally phenocrysts may be found in clusters, in which case they are

*In some instances phenocrysts may grow after emplacement of the magma. During magmatic cooling, before a significant increase in viscosity has occurred, there is a period in which the number of nuclei is small, but the growth rate is rapid. If the magma is cooled slowly through this interval, phenocrysts may develop.

described as *glomerocrysts*. Magma may also crystallize as it moves slowly upward, so that an unbroken series of crystal sizes is produced in the rock (*seriate texture*).

The first-formed crystals in a magma grown in an environment with few or no adjacent crystals are often able to develop well-formed crystal faces. Such crystals are described as *euhedral* (see Figure 2–13). As crystallization proceeds, mutual interference of growing crystals occurs, and the crystals formed may have only a partial development of faces (*subhedral*), or no faces (*anhedral*). To some extent the order of crystallization may be recognized by the degree of euhedralism, but this technique is of questionable validity, as early-formed crystals may not show good face development in spite of uncrowded conditions. In addition, late-stage processes may result in recrystallization that could obliterate crystal faces on early-formed crystals, and form faces on later ones.

In the case of crystallization of a rock that produces mainly a mixture of plagioclase and pyroxene, particular textures may develop depending upon the ratio of the two components that yield these minerals. In Figure 2–7, an anorthite-rich melt *a* was considered. During the cooling period from *b* to *g*, only crystals of anorthite are formed. At temperature *i* the liquid has reached the eutectic *h*, where simultaneous

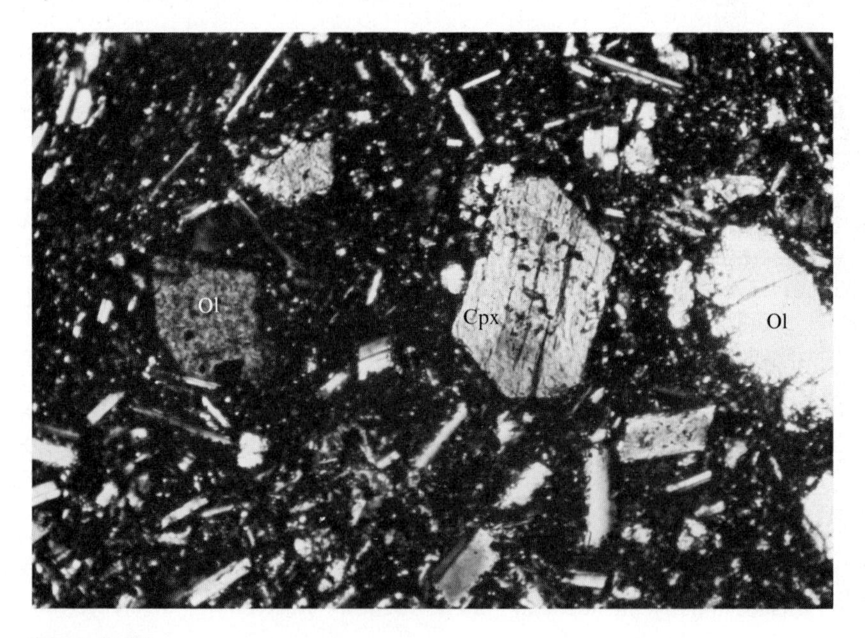

Figure 2–13
Euhedral clinopyroxene phenocryst (Cpx) and subhedral olivine crystals (Ol) in a cryptocrystalline matrix containing plagioclase microlites. Crossed nicols. Width of photo is about 0.8 mm.

crystallization of diopside and anorthite occur at fixed temperature. As the pyroxene is present in minor amounts, it will comonly be located in the interstices between plagioclase crystals. Most commonly in rocks the pyroxene is anhedral augite, which is located between laths of labradorite; this is called *diabasic texture.* The opposite texture, called *ophitic,* results when the melt is enriched in pyroxene components (as with a composition such as *l* in Figure 2–7); following initial crystallization of pyroxene, later simultaneous growth of both pyroxene and plagioclase results in enclosure of lathlike plagioclase crystals within pyroxene (see Figure 2–14). The more common *subophitic* texture is developed when subhedral plagioclase laths are partly surrounded by pyroxene crystals.

Movement of the magma may occur during crystallization, causing the formation of *flow texture*—a wavy pattern of prismatic or platy minerals that are oriented along planes of lamellar flowage. A variety of this, called *trachytoid texture,* develops when a more or less uniform parallel alignment of microlites is formed within a subparallel or randomly aligned matrix.

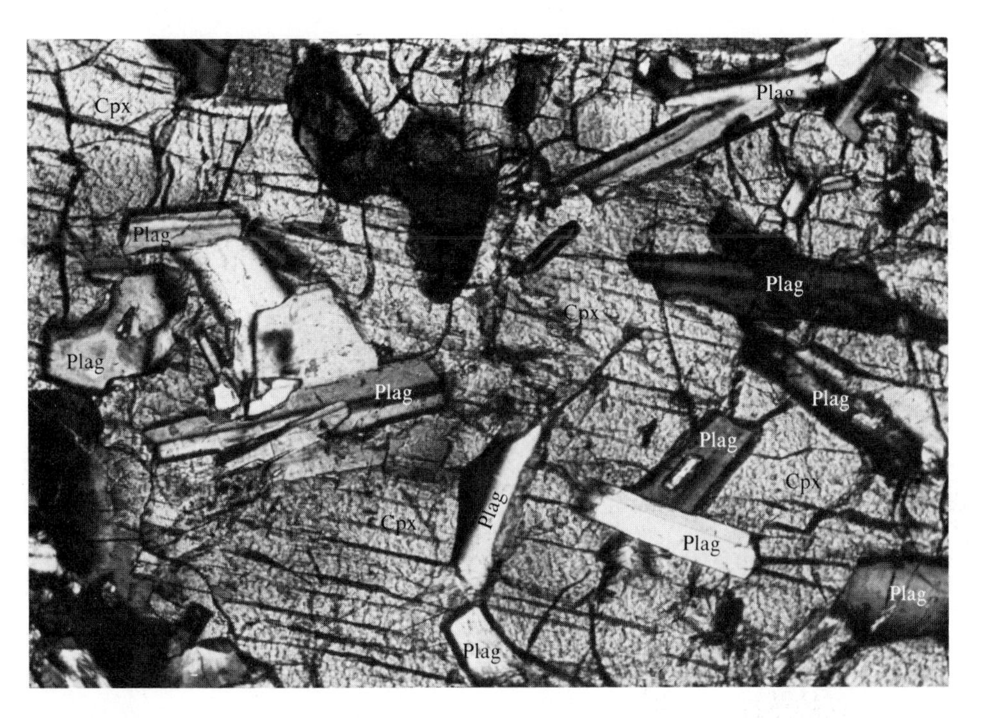

Figure 2–14
Ophitic texture, showing plagioclase laths (Plag) enclosed by clinopyroxene (Cpx). Diabasic texture is the opposite of this, with pyroxene crystals enclosed by plagioclase. Crossed nicols. Width of photo is about 0.8 mm.

Two-Component Systems—Incongruent Melting

In the case of *incongruent melting,* a solid phase breaks down to form both a solid and liquid phase. The composition of neither of these two phases is identical to the incongruently melting solid; but together, of course, their composition must equal that of the original solid.

Figure 2–15, the system $KAlSi_2O_6$–SiO_2, illustrates both congruent and incongruent melting behavior. First note that an intermediate compound, potassium feldspar

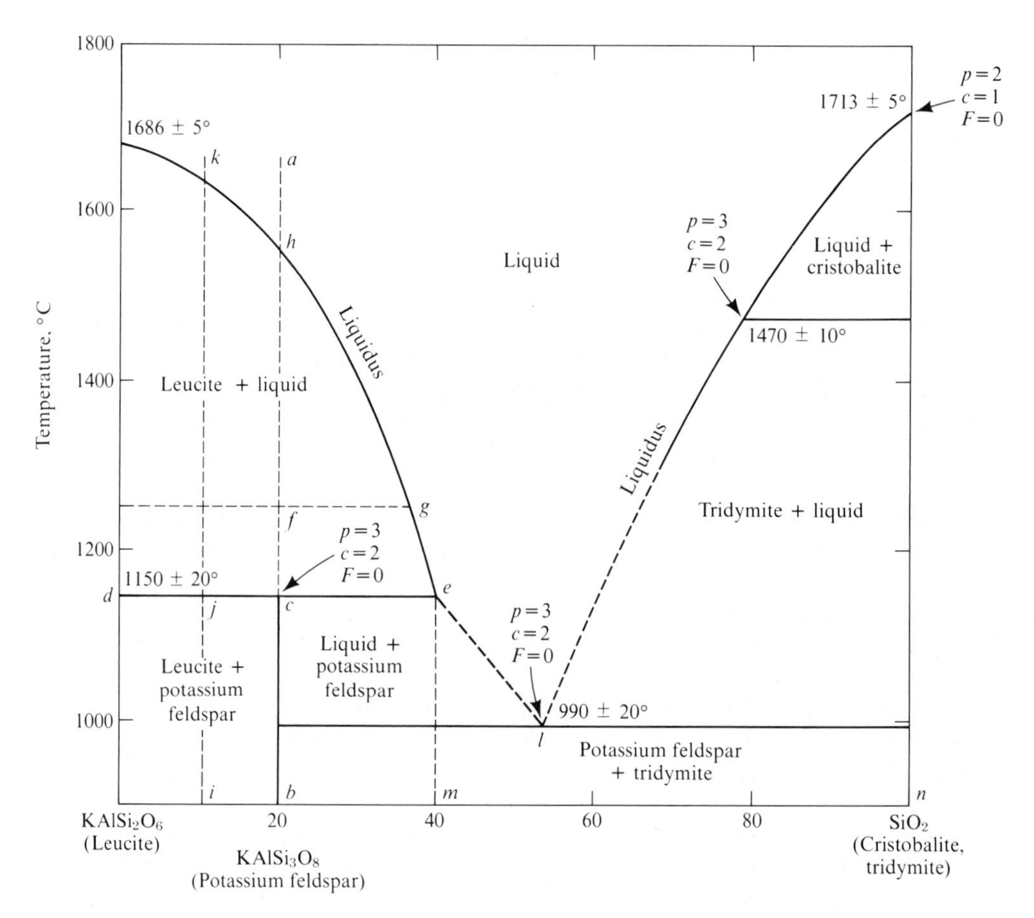

Figure 2–15
The binary system $KAlSi_2O_6$–SiO_2 at 1 bar pressure. Dashed liquidus lines indicate experimental uncertainty. Potassium feldspar ($KAlSi_3O_8$) melts incongruently at 1150 ± 20°C. Various points are described in terms of the Phase Rule, where p = number of phases, c = components, and F = degrees of freedom. Other designations are described in the text. [From J. F. Schairer, and N. L. Bowen, 1948, *Comm. Geol. Finlande Bull., 140,* Fig. 3.]

($KAlSi_3O_8$), is present; the composition and maximum stability limit of this phase are given by the intermediate vertical solid line. The possible stable solid phase assemblages for various compositions are indicated as leucite, leucite + potassium feldspar, potassium feldspar, potassium feldspar + tridymite, or tridymite. Notice that the diagram indicates that under equilibrium conditions the feldspathoid leucite and the silica polymorph tridymite have no field where they can coexist. A mixture of these two phases will react to produce either a mixture of potassium feldspar and tridymite, leucite and potassium feldspar, or just potassium feldspar (depending upon the proportions of the mixture).

Both leucite and the SiO_2 polymorph melt congruently to produce a liquid of identical composition to the solid. Potassium feldspar, however, melts incongruently. The melting behavior can be considered by drawing an isopleth through the potassium feldspar composition; it extends through *a* in the diagram. A batch of potassium feldspar (at *b*), when heated, remains stable to *c* (at a temperature of about 1150°C). Here the isopleth enters the field of leucite + liquid. This indicates the potassium feldspar melts at this temperature to produce a mixture of leucite and liquid. The temperature remains constant until the process is complete. The compositions of the two phases formed are found by means of a tie line through *c*. The leucite composition is at *d* and the liquid composition at *e*. The percentages of each of these are found in the usual way. Continued heating causes partial melting of the leucite. As a result of this, at temperature *f* the liquid composition has shifted to *g*. The leucite continues to melt with increasing temperature, and at temperature *h* the entire batch is molten, and the liquid composition *h* is the same as the original composition *b*.

The presence of incongruently melting compounds can be very significant in the formation of magmas whose compositions are greatly different from that of the original source rock. Assume that a rock is of composition *i*—a mixture of leucite and potassium feldspar. Heating of this rock to *j* causes the potassium feldspar (*c*) to melt incongruently. Upon completion of this melting and elimination of potassium feldspar (*c*) the two phases present are liquid *e* and leucite *d*. The amount of leucite is large (length *ej*) as compared to liquid (length *dj*). If heating of the rock continued, the leucite would melt over a temperature range from 1150°C to about 1650°C to produce a liquid of composition *k*—the same composition as the starting batch. Consider what might happen if the melting process is disturbed at temperature *j* such that the liquid *e* was physically removed from the leucite crystals (perhaps by compression of the liquid–crystal batch, with escape of the liquid through a fissure). The liquid *e* could be considered as a separate system whose freezing behavior is independent of the residual leucite crystals. This separation is a nonequilibrium process with respect to the stable relations indicated by the diagram, but the diagram permits one to deduce further equilibrium behavior.

Consider that the liquid *e* (now separated from the residual leucite) is permitted to cool and crystallize. An isopleth downward from *e* shows the phase assemblages that will form. The immediately adjacent field is liquid + potassium feldspar. Cooling of the melt *e* produces potassium feldspar while the liquid changes in composition along the liquidus to the eutectic at *l*. At the eutectic the liquid is consumed by simultaneous crystallization of potassium feldspar and tridymite, the final proportions indicated by lengths *mn* and *bm*.

What was accomplished in this total process? The original rock (*i*) contained leucite and potassium feldspar. The presence of leucite indicates that this rock is incompatible with the existence of crystalline SiO_2. Separation during the melting process has resulted in the formation of one rock consisting of pure leucite and a separate smaller batch consisting of potassium feldspar and tridymite. A tridymite-containing rock has been produced by partial melting of a silica-deficient reactant. Fractionation processes of this type are very common in petrologic systems during both melting and freezing, and result in a large diversity of magma types.

An example of the same type of fractionation occurring during cooling is shown in Figure 2–15. The now-familiar liquid *a*, having the composition of potassium feldspar, when cooled under equilibrium conditions should yield only potassium feldspar. Interruption of the sequence at the appropriate time can change this. Cooling of *a* to the liquidus causes precipitation of leucite. With continued cooling to *c* the liquid shifts in composition to *e* while the coexisting leucite is of composition *d*. The point *e* on the diagram is called a *peritectic;* it represents the temperature and melt composition that should exist for the freezing of an incongruently melting compound. Under equilibrium conditions the liquid *e* should react at constant temperature with leucite *d* to produce a completely crystalline batch of potassium feldspar. If, however, the leucite and liquid were physically separated (as by a crystal–liquid filter-pressing situation described earlier, or perhaps by rising of leucite crystals to the top of the magma chamber so that the liquids in the lower portions could not react with them), then the liquid *e* would have to crystallize without reaction with leucite. As described earlier, a physically separated liquid *e* would cool to crystallize a mixture of potassium feldspar and tridymite. Note that this type of process can occur at a peritectic, but not at a eutectic. In a system containing only a eutectic, removal or separation of early-formed crystals has no effect on the crystallization sequence, although it may change the texture.

Textural relations in thin section often reveal the presence of a compound that melts incongruently. The system Mg_2SiO_4–SiO_2 (see Figure 2–16) contains one intermediate compound, $MgSiO_3$ (enstatite), which melts incongruently to yield forsterite *c* and liquid *d*. Crystallization and melting in this system can be deduced by use of isopleths and tie lines in the manner discussed earlier. The only exceptional aspect of the diagram is a region of two immiscible liquids near the high silica portion, which can also be treated by consideration of isopleths and tie lines.

A liquid *a* cools to produce, initially, forsterite. Continued cooling to *b* permits the peritectic reaction, which yields a mixture of forsterite and enstatite. This crystallization sequence is sometimes revealed by the textures. The liquid *a* under conditions of slow cooling will first produce forsterite crystals that are often euhedral to subhedral (similar to some phenocrysts in basalts, as shown in Figure 2–13). When the forsterite–liquid batch cools to the peritectic temperature *b*, it consists of liquid *d* and

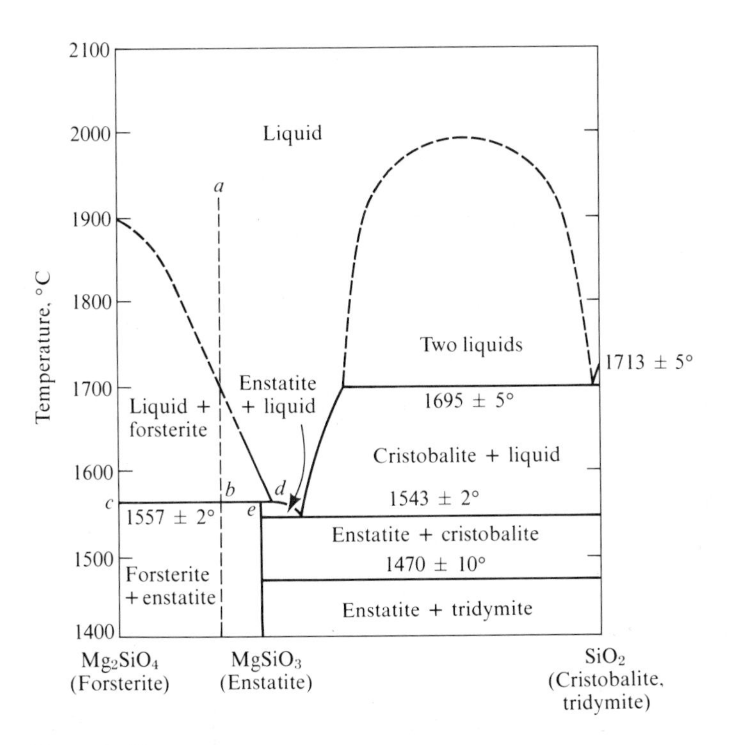

Figure 2–16
The binary system Mg_2SiO_4–SiO_2 at 1 bar pressure. The pyroxene enstatite ($MgSiO_3$) melts incongruently to yield a mixture of forsterite and liquid. A two-liquid (immiscibility) region is present in the silica-rich portion of the system. The crystallization of liquid *a* is described in the text. [From N. L. Bowen and O. Andersen, 1914, *Amer. Jour. Sci.*, [*4*], *37*, Fig. 1, modified by J. W. Grieg, 1927, [*5*], *13*, Fig. 3.]

forsterite *c* (in proportion to length *cb* and length *bd*). At this temperature all remaining liquid reacts with some of the forsterite to produce a final mixture of enstatite *e* and forsterite *c*. As seen from the tie line ratios *bd*/*cd* and *be*/*ce*, the relative amount of forsterite in the batch has been decreased by reaction with the liquid. The reaction forsterite + liquid → enstatite may destroy much of the original euhedral shape of the olivine. In a normal rock system this results in a *reaction rim* (or corona) about partially corroded or embayed olivine crystals, the rest of the rock being occupied by other minerals (see Figure 2–17). In the case of the simpler two-component system this results in corroded olivine crystals in a matrix of pyroxene. The reaction rim is produced by equilibrium processes during cooling, and is not indicative of abrupt changes in magmatic conditions.

The reaction relation described above should not be confused with a similar effect due to the increase in oxygen content of the melt (which might result from the

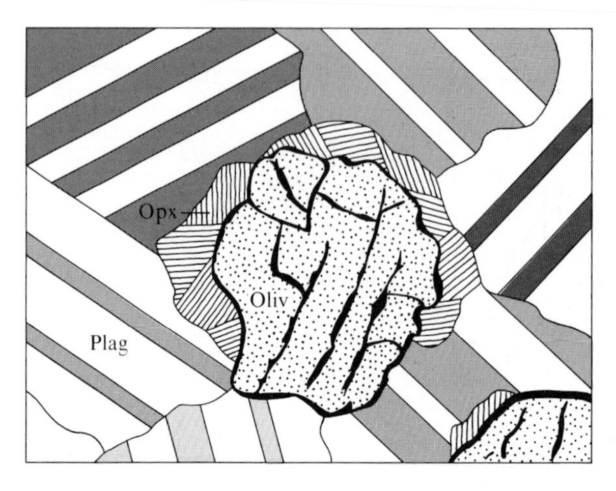

Figure 2–17
Reaction rim of orthopyroxene (Opx) about olivine (Oliv) in basalt. Plag = plagioclase. [Redrawn from E. Wahlstrom, 1947, *Igneous Minerals and Rocks,* (New York: John Wiley and Sons), Fig. 1.]

addition and dissociation of water). Here, oxygen causes oxidation of the ferrous iron in the olivine. As ferric iron cannot be accommodated in the olivine structure due to its smaller size and higher charge, it is exsolved from the structure. This in turn causes a relative increase in the Si/(Mg, Fe) ratio in the olivine, and results in conversion of some of the olivine to pyroxene. This reaction can be identified, as the pyroxene rim will contain large amounts of magnetite ($FeO \cdot Fe_2O_3$) (see Figure 2–18).

Two-Component Systems—Solid Solution

The systems discussed so far contained no solid solution between the various mineral phases. But most minerals show at least partial or even complete solid solution. The system $NaAlSi_3O_8$–$CaAl_2Si_2O_8$ (the plagioclase system) shows complete substitution between albite ($NaAlSi_3O_8$) and anorthite ($CaAl_2Si_2O_8$). Any composition intermediate between the two end-members will produce (under equilibrium conditions) a single solid phase of that composition. (A small exsolution region that is present near the sodium end-member at low temperature will be ignored here.) Such intermediate compositions are described in terms of the end-member phases, as $Ab_{10}An_{90}$, or more simply An_{90}. The system at 1 bar as shown in Figure 2–19 contains one-phase regions at high and low temperatures, and an intermediate crystal–liquid "solid-solution loop." A melt of composition *a* is a single phase of composition An_{65}; the point *b* is a single plagioclase of composition An_{65}; the point *c* lies in the two-phase region, and by

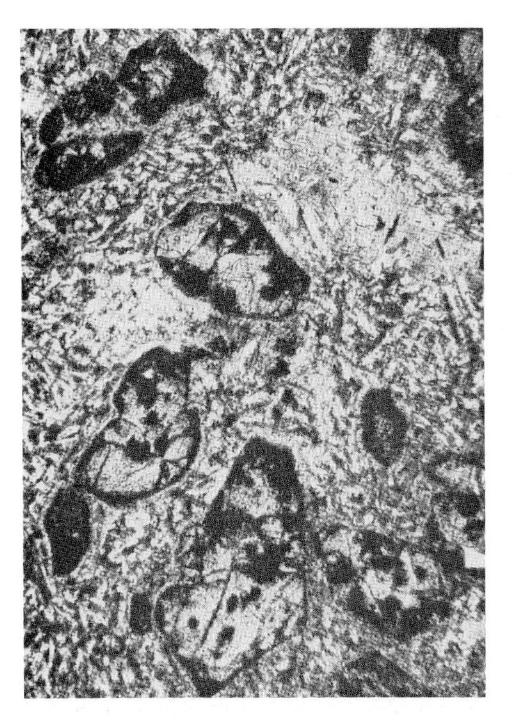

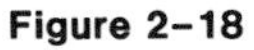

Figure 2–18
Oxidized olivine phenocrysts clouded mainly with magnetite surrounded by a thin rim of hypersthene in a groundmass of plagioclase, clinopyroxene, glass, and magnetite. [From W. Schreyer, in H. S. Yoder, Jr., and C. E. Tilley, 1962, *Jour. Petrology, 3,* Plate 5A. Plane-polarized light. Width of photo is 2.8 mm. © Oxford University Press.]

use of the horizontal tie line is seen to consist of liquid *d* and plagioclase *e* in almost equal amounts.

Consideration of the tie lines drawn from the isopleth through the two-phase region (from *f* to *i*) shows that with decreasing temperature the liquid changes composition from *f* to *h*, while simultaneously the coexisting plagioclase changes composition from *g* to *i*. The liquid cooled to *f* begins crystallization with the precipitation of plagioclase of composition *g*. As the crystals contain a larger amount of $CaAl_2Si_2O_8$ than the original liquid, the remaining liquid is relatively enriched in $NaAlSi_3O_8$. This causes a depression of the freezing point of the liquid, which follows the path shown by the liquidus line. The liquid, now of a more sodic composition, can only coexist with plagioclase that is also more sodic (that is, liquid *d* with plagioclase *e*). The previously formed crystals of plagioclase continue to grow while reacting with the liquid by diffusion and becoming more sodic. Simultaneously the liquid crystallizes new and equally sodic plagioclase. Both liquid and coexisting crystals follow compositional paths along the liquidus and solidus curves until the plagioclase is of the same composition (*i*) as the original melt (*a*). Here the last bit of liquid is eliminated, as seen by consideration of the tie lines. During melting the same sequence occurs in reverse.

This process should result in a batch of plagioclase crystals that are uniform in composition from centers to edges if the diffusion process has gone to completion.

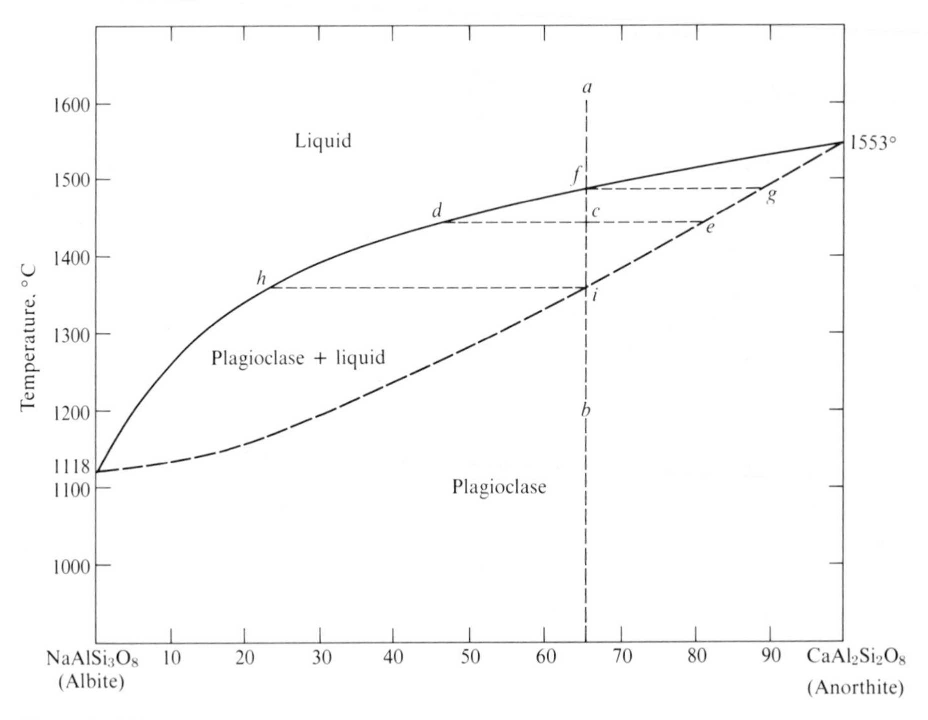

Figure 2–19
The binary system $CaAl_2Si_2O_8$–$NaAlSi_3O_8$ at 1 bar pressure. Designations are explained in the text. A small region of exsolution (not shown) is present at lower temperatures. The system Mg_2SiO_4–Fe_2SiO_4 (forsterite–fayalite) shows the same behavior of complete solid solution, with end-members melting at 1890°C and 1205°C respectively. [Based on N. L. Bowen, 1913, *Amer. Jour. Sci.*, [*4*], *35*, Fig. 1.]

Under the slow cooling conditions of plutonic rocks this situation is often realized. However, with rapid cooling of hypabyssal or extrusive rocks, the continuous diffusion process necessary to keep the crystal composition in equilibrium with the melt is often not sufficiently fast; only the outer portions of the crystals may be in equilibrium with the melt and a more calcic core remains. This results in normal zoning, and is easily observed in thin section, as differences in compositions are seen as differences in extinction angle. In some circumstances the zoning may be reversed (calcic rims and sodic centers). The zoning may be continuous or discontinuous, depending upon whether the compositional changes are gradual or abrupt. An extreme case is oscillatory zoning, where the crystals have large numbers of bands with alternating Na/Ca ratios. Oscillatory and reverse zoning have been considered to be due to either (1) movement of the crystals during growth by convection currents to different parts of the magma chamber or (2) growth in a single location within the magma during changes of P_{H_2O}. Such changes result in rise and fall of liquidus and solidus surfaces, with consequent supersaturation and undersaturation. It now appears that oscillatory zoning is diffusion controlled. Examination of basaltic glass in the vicinity of a zoned crystal by means of electron-microprobe analysis reveals a change in glass composition as the crystal interface is approached (see Figure 2–20). The glass shows a decrease in Al content, a rise in Si, Mg, and Fe, and absence of a Na and Ca gradient. Diffusion in the melt (with the exception of the more mobile Na and Ca) is unable to keep pace with the rate of crystal growth. The crystal, being richer in Ca and Al than the melt, utilizes the immediately adjacent Al, causing a decrease in amount as the interface is approached. The growing crystal contains less Si and Na than the melt, which leads to an increase of Si near the interface. As Mg and Fe are not incorporated in plagioclase, they are rejected by the growing crystal. The result of these compositional gradients is a growth zone of steadily decreasing anorthite content. Growth rate decreases or ceases until the immediately adjacent melt becomes supersaturated once again. Compositional zoning is illustrated in Figure 2–2.

A common textural feature in igneous rocks is *perthite,* elongate blebs of plagioclase within potassium feldspar (or the reverse). The meaning of this feature may be explained by experiments in the binary system $KAlSi_3O_8$–$NaAlSi_3O_8$. The system is shown in Figure 2–21. Above a temperature of about 650°C there is complete solid solution between the two end-member minerals. Potassium and sodium are completely interchangeable in the feldspar crystal structure and the feldspar (called an-

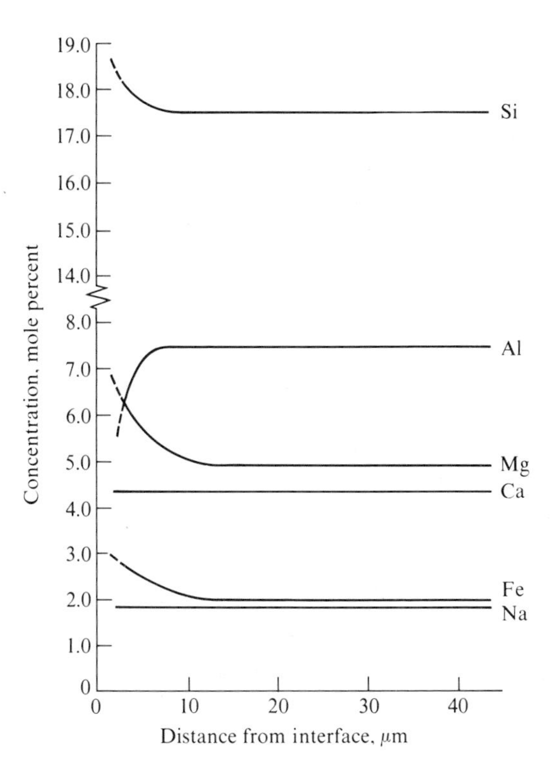

Figure 2–20
Concentration gradients in a basaltic glass near bytownite phenocryst. With the exception of Na and Ca, diffusion within the melt was unable to keep pace with the growing crystal. Microprobe analyses were recalculated to mole percent assuming that the sum of the oxides is equal to 100%. [From Y. Bottinga, A. Kudo, and D. Weill, 1966, *Amer. Mineral., 51,* Fig. 3.]

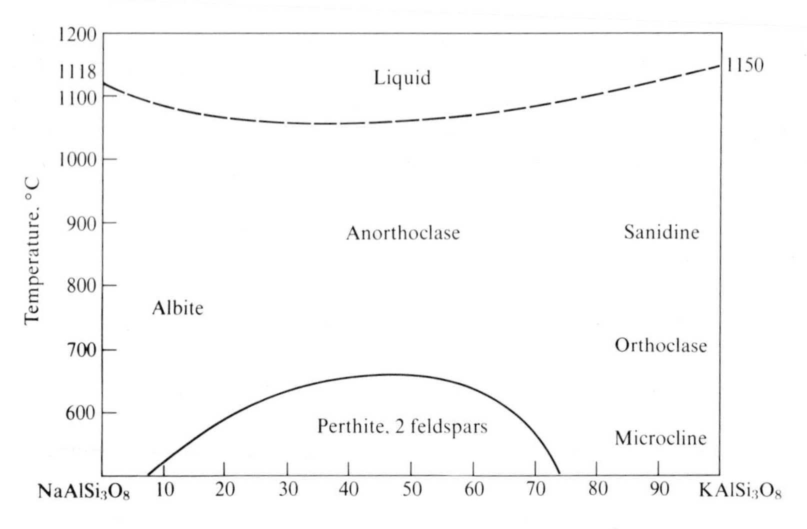

Figure 2–21
Subsolidus relations between potassium feldspar and albite, showing the equilibrium compositions of the two unmixed phases as a function of temperature. Dashed liquidus line is indicated to simplify the incongruent melting of potassium feldspar. [After O. F. Tuttle and N. L. Bowen, 1958, Geol. Soc. Amer. Memoir 74, Fig. 17.]

orthoclase) appears homogeneous in thin section. This is possible because at high temperatures crystal structures expand, permitting ions of greatly different size to coexist in the structural site (potassium has a radius 33% larger than sodium). But as the temperature drops and the crystal structure shrinks, the ions must separate into separate phases, to form the texture we term perthite. In most siliceous magmas, the common perthite is dominantly microcline or orthoclase with blebs of oligoclase. The texture may be visible in hand specimen (macroperthite), be present on a scale visible only if a polarizing microscope is used (microperthite), or be detectable only using X-ray techniques (cryptoperthite). The scale is determined by the rate of cooling (the time available for unmixing), as well as the presence of a vapor phase (which increases the rate of chemical reaction). As an alternative to the unmixing process, it has been demonstrated that some perthites have been created by partial metasomatic replacement of one feldspar by another.

SUMMARY

The behavior of phases within a chemical system can be described in terms of the Phase Rule. The phase assemblage in one- and two-component systems is determined by the particular combination of pressure and temperature to which the system is subjected. Within systems of more than one component, melting may be congruent or incongruent. Melting and freezing relationships are influenced by the amount of solid solution between solid phases and the formation of compounds of intermediate composition. Textures in igneous rocks are a result of the chemical composition of the melt (and its resultant phase relationships), the pressures and temperatures to which the rock has been subjected, and the rate at which the system has been cooled.

FURTHER READING

Bottinga, Y., A. Kudo, and D. Weill. 1966. Some observations on oscillatory zoning and crystallization of magmatic plagioclase. *Amer. Miner., 51,* 792–806.

Ehlers, E. G. 1972. *The Interpretation of Geological Phase Diagrams.* San Francisco: W. H. Freeman and Company, 280 pp.

Ernst, W. G. 1976. *Petrologic Phase Equilibria.* San Francisco: W. H. Freeman and Company, 333 pp.

Levin, E. M., C. R. Robbins, and H. F. McMurdie. 1964. *Phase Diagrams For Ceramists.* Columbus, Ohio: The Amer. Cer. Soc., 601 pp.; 1969 Supplement, 625 pp.; 1975 Supplement by E. M. Levin and H. F. McMurdie, 513 pp.

Morey, G. W. 1975. Phase equilibrium relations of the common rock-forming oxides except water. In M. Fleischer (ed.), *Data of Geochemistry,* 6th Ed., Chap. 2. U.S. Geol. Surv. Prof. Paper 440-L. Washington, D.C.: U.S. Gov't. Printing Office.

Wahlstrom, E. E. 1950. *Introduction to Theoretical Igneous Petrology.* New York: John Wiley and Sons, 365 pp.

3

Experiments with Molten Silicates: Ternary Systems

The previous chapter covered textural development and phase relationships in unary and binary systems. In this section we will cover some aspects of ternary systems, the effects of different types of pressure on rock systems, and the crystallization and emplacement of magmas. More detailed discussions of ternary systems are found in the book by Ehlers (1972).

SIMPLE TERNARY SYSTEMS

Ternary systems present a special problem in representation. The compositions of the three components are usually represented on an equilateral triangle (see Figure 3–1). Each corner represents 100% of one component; points along the edge consist of a mixture of the two components indicated at the ends of the line; points within the triangle represent mixtures of the three components.

The temperature coordinate is perpendicular to the compositional triangle; that is, perpendicular to the paper. Liquidus surfaces are contoured with lines of equal temperature (isotherms) in the same manner that a topographic map is contoured. The observer is thus looking down on the upper surfaces of a three-dimensional object. This arrangement seen in perspective as a solid object is shown in Figure 3–2. The points *X*, *Y*, and *Z* represent the three components; lines joining these points produce the compositional triangular base. Temperature is indicated on the vertical coordinate. Each side is seen to be a now-familiar binary system. The upper, shaded portion represents the liquidus surfaces. When these liquidus surfaces are contoured, one obtains the ternary representation that is normally given in the literature. Such a system is shown in Figure 3–3, the system $CaMg(SiO_3)_2$–Mg_2SiO_4–$CaAl_2Si_2O_8$, at

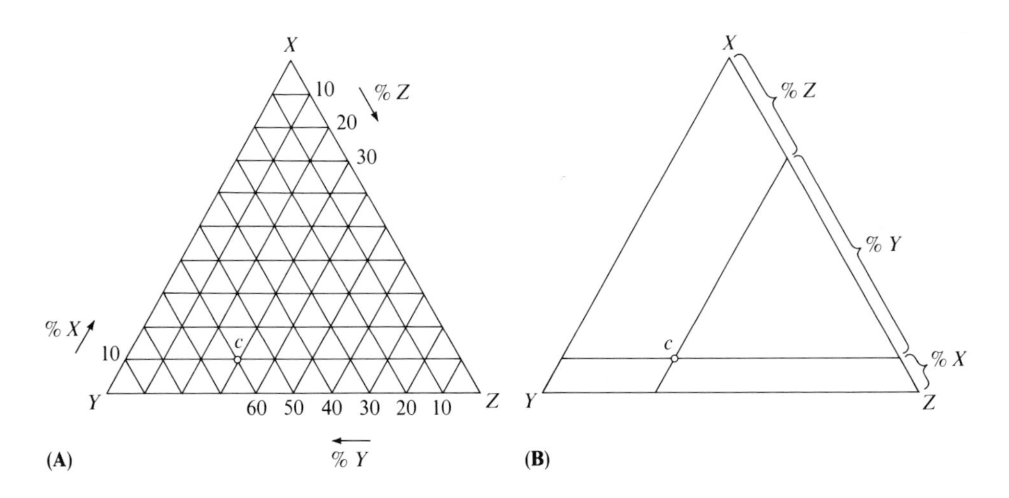

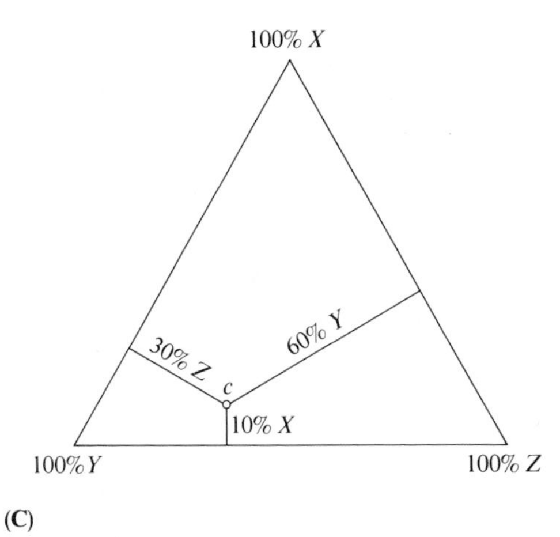

(C)

Figure 3–1
(**A**) The point c on triangular graph paper. The percentages of components X, Y, and Z are read directly from the numbered coordinate axes. (**B**) Determination of point c by the two-line method. Two lines are drawn through c parallel to any two of the sides of the triangle (here XY and YZ). The intersection of these two lines with the third side (XZ) divides that side into three parts whose lengths are proportional to the relative amounts of components X, Y, and Z at point c. (**C**) Determination of point c by construction of perpendicular lines through c to each of the triangle's sides. The relative lengths of these three perpendicular lines yield the percentages of components X, Y, and Z.

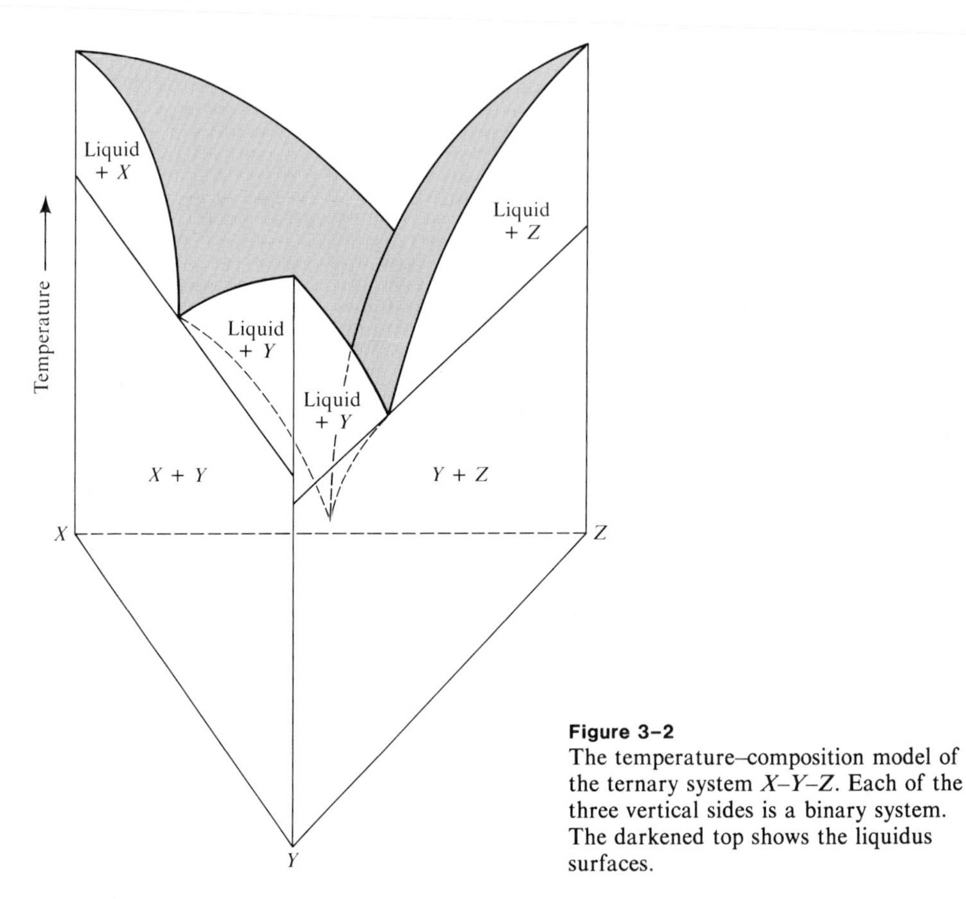

Figure 3–2
The temperature–composition model of the ternary system *X–Y–Z*. Each of the three vertical sides is a binary system. The darkened top shows the liquidus surfaces.

1 bar, which we can consider similar to an idealized olivine basalt.* The simplified contoured triangle in part B shows the three liquidus surfaces of diopside, forsterite, and anorthite separated by heavy lines where the liquidus surfaces intersect. These lines are called *boundary curves.* Each of the liquidus surfaces is labeled according to the name of the solid phase that represents the first crystalline product of any liquid located within this area. A liquid of composition *a* when cooled to intersect the liquidus surface at 1700°C will begin crystallization with the precipitation of forsterite. A liquid *b* cooled to 1400°C will precipitate anorthite. These liquidus surfaces are called primary-phase fields, as they show the first crystals to precipitate during cooling. A liquid of composition *c,* lying on the boundary curve between the primary phase fields of diopside and forsterite, will begin crystallization with the simultaneous

*This system contains a liquidus surface for the mineral spinel ($MgAl_2O_4$). As the composition of this mineral cannot be represented by any mixture of the three components making up the corners of the triangle, the system is therefore not truly ternary, but merely pseudo-ternary, being a slice of the more complex quaternary system $CaO–MgO–Al_2O_3–SiO_2$. As spinel is eliminated in the final products of crystallization, its field of crystallization has been eliminated in the simplified system, and the system is discussed as if it were truly ternary.

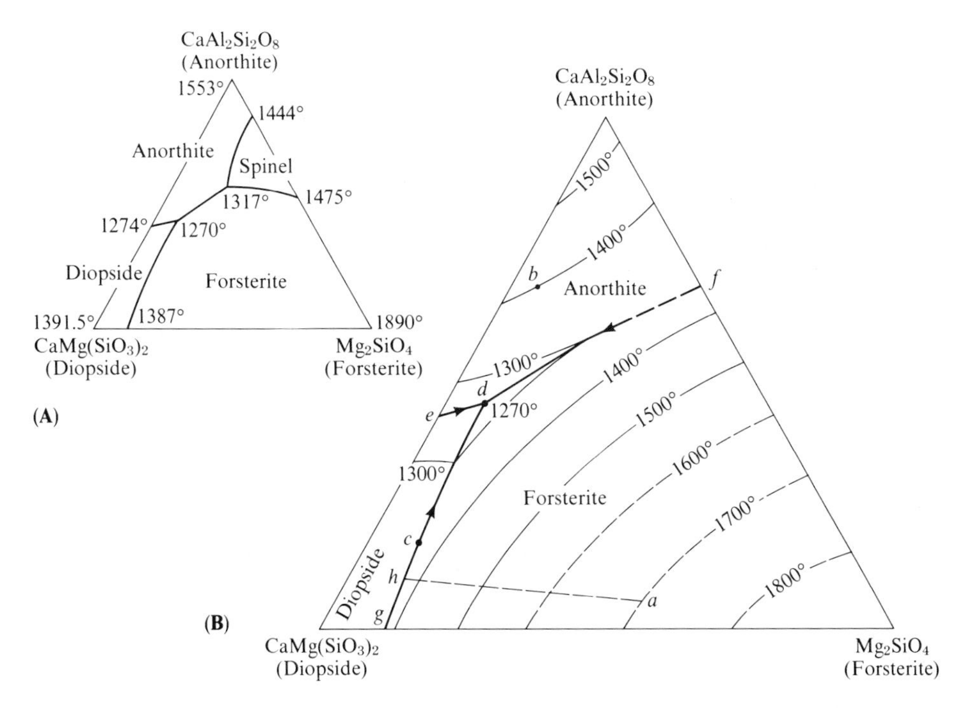

Figure 3–3
(**A**) The ternary system $CaMg(SiO_3)_2$–$CaAl_2Si_2O_8$–Mg_2SiO_4 at 1 bar pressure. [From E. F. Osborn and D. B. Tait, 1952, *Amer. Jour. Sci., Bowen Volume,* Fig. 4.] (**B**) The same system modified by elimination of the field of spinel. Labeled points are described in the text.

precipitation of both of these solid phases. A liquid lying at the point *d,* the intersection point of the three primary-phase fields, will simultaneously precipitate forsterite, anorthite, and diopside during cooling. This minimum temperature point is a ternary eutectic. From this point the three boundary curves rise thermally to the binary eutectics at *e, f,* and *g.* The boundary curves are thus extensions of the binary eutectics into the ternary system. Note that the isotherms point upward (thermally) when crossing a boundary curve in the same way that topographic contours point upstream on a map. Arrows are usually added to the boundary curves to indicate the direction of decreasing temperature.

Consider the cooling of a ternary melt such as *a.* The melt remains completely liquid until it encounters the liquidus surface at 1700°C. As it lies in the primary field of forsterite, forsterite is the first solid phase to precipitate. The liquid is depleted in those chemical constitutents that constitute this mineral, and therefore must change in composition directly away from the point representing the composition of forsterite—namely the lower right corner of the triangle. The path of liquid compositional migration is found by extending a line from the forsterite composition point through the composition of the liquid *a* that is crystallizing. As the melt cools the liquid continues to precipitate forsterite and changes in composition along this line until the boundary curve at *h* is encountered. At *h* the primary field of diopside is encountered, and both forsterite and diopside precipitate together with decreasing temperature along the boundary curve. The liquid changes in composition away from the Mg_2SiO_4–$CaMg(SiO_3)_2$ side of the triangle, as both of these components are being removed. The particular direction taken by the boundary curve depends upon the ratio of these two solids being removed; curvature of the boundary curve indicates a changing ratio. Cooling and crystallization of forsterite and diopside cause the liquid to migrate in composition down the boundary curve to the ternary eutectic at *d*. Here forsterite, diopside, and anorthite precipitate together at the eutectic temperature until the remaining liquid is consumed. Changes occurring in the solid state with decreasing temperature are not indicated in this type of diagram.

Melting follows the opposite sequence of cooling. Consider composition *a*. When solid, it consists of a mixture of forsterite, diopside, and anorthite (the proportions found by the techniques described in Figure 3–1). When heated, simultaneous melting of the three solids begins at 1270°C, and the first liquid is of eutectic composition. As this eutectic liquid is more than 40% (by weight) $CaAl_2Si_2O_8$, and less than 10% anorthite was present in the initial mixture (as seen by the location of composition point *a*), continued heating results in elimination of anorthite. Additional heating causes simultaneous melting of diopside and forsterite and the liquid changes composition along the boundary curve to *h*. At *h*, all of the diopside is eliminated. Continued heating melts the remaining forsterite as liquid composition migrates toward the forsterite composition point. At *a*, the original composition, the last crystal of forsterite is eliminated and liquid composition ceases to change. In order to determine a melting sequence in a system of this sort, first determine the freezing sequence and then reverse the procedure. Observe that all ternary liquids migrate in composition during freezing to the ternary eutectic, and all mixtures of three solids begin melting at the ternary eutectic and follow a compositional path that brings them back to the bulk composition point. Removal of previously formed crystals during a crystallization sequence in this simple ternary system does not change the path of liquid compositional migration, as the cooling liquid does not in any way react to change the composition of earlier-formed crystals. In other systems, where the melt reacts with previously precipitated crystals, the path of liquid compositional migration is influenced by the presence or absence of earlier-formed solid phases.

Studies of melting and freezing of basaltic rock (containing plagioclase, pyroxene, and olivine, as well as other minerals) have confirmed the general correctness of the above freezing pattern. Yoder and Tilley (1962) observed that the first silicate mineral to crystallize from a basaltic melt was either an olivine, a pyroxene, or a plagioclase. With additional cooling, the primary phase was joined by additional silicate phases at lower temperatures. One significant difference in the more complex basalt system is that the ternary eutectic of the simple system is converted to a univariant line, resulting in the crystallization of the major silicate phases over a range of temperatures, rather than at a fixed temperature.

COMPLEX TERNARY SYSTEMS

Ternary systems can be considerably more complicated than the system described above. Some of these complications result from the presence of compounds of intermediate composition, incongruent melting, and the effect of solid solution. An introduction to these problems can be gained by consideration of the three-component system $CaMg(SiO_3)_2$–$CaAl_2Si_2O_8$–$NaAlSi_3O_8$. In the temperature–pressure range characteristic of the earth's crust the minerals diopside and plagioclase crystallize from such a melt.

As shown by the three two-component boundary systems (whose combination generates the three-component phase triangle), there is complete solid solution between the albite and anorthite end members but none between plagioclase of any composition and diopside (see Figure 3–4). As a result of the plagioclase solid solution, any liquid composition within the system should cool to produce a two-phase solid assemblage, diopside, and a single plagioclase of appropriate composition. Thus a liquid *a* (see Figure 3–5) will produce a mixture of diopside and plagioclase of composition An_{75}; liquid *b* will produce a mixture of diopside and a plagioclase of composition An_{50}. The final proportions of diopside and plagioclase are calculated (as with binary systems) by considering the relative lengths of the tie lines between the original liquid composition and the composition points of the two phases crystallized.

Figure 3–6 shows the liquidus surface at 1 bar pressure with isotherms, and the single boundary curve between the primary fields of diopside and plagioclase. As the system is isobaric, any specific liquid composition on the boundary curve can only exist with a fixed composition of plagioclase ($F = c - p + 1$, $F = 3 - 3 + 1$, $F = 1$); thus liquid *a* can exist in equilibrium with diopside and plagioclase *b*; liquid *c* coexists with diopside and plagioclase *d*, and so on. (Compositions of plagioclase coexisting with liquids on the boundary curve must be determined by experiment.) From this relationship it can be understood that a liquid such as *e* or *f* (which cools to produce diopside and plagioclase of composition *b*) must have its final liquid composition at *a*.

Consider the crystallization of liquid *f*. Cooling to the liquidus surface results in the initial precipitation of diopside; liquid composition is changed directly away from the diopside composition point to the boundary curve at *g*. The experimentally determined tie lines indicate that plagioclase of composition *h* now coprecipitates with diopside. As the liquid changes in composition along the boundary curve, diopside continues crystallizing, along with plagioclase of increasingly sodic composition; earlier formed plagioclase is continuously changed in composition to conform to the more sodic liquid compositions. The last drop of liquid is consumed at *a*, and the final crystalline mixture consists of diopside, and plagioclase of composition *b*.

Crystallization of an original liquid *e* also produced the same two phases, but in different proportions. Cooling produces initial crystallization of a very calcic plagioclase such as *i* (as determined by experimentation). With additional cooling this initial plagioclase reacts with the liquid to become more sodic, while simultaneously

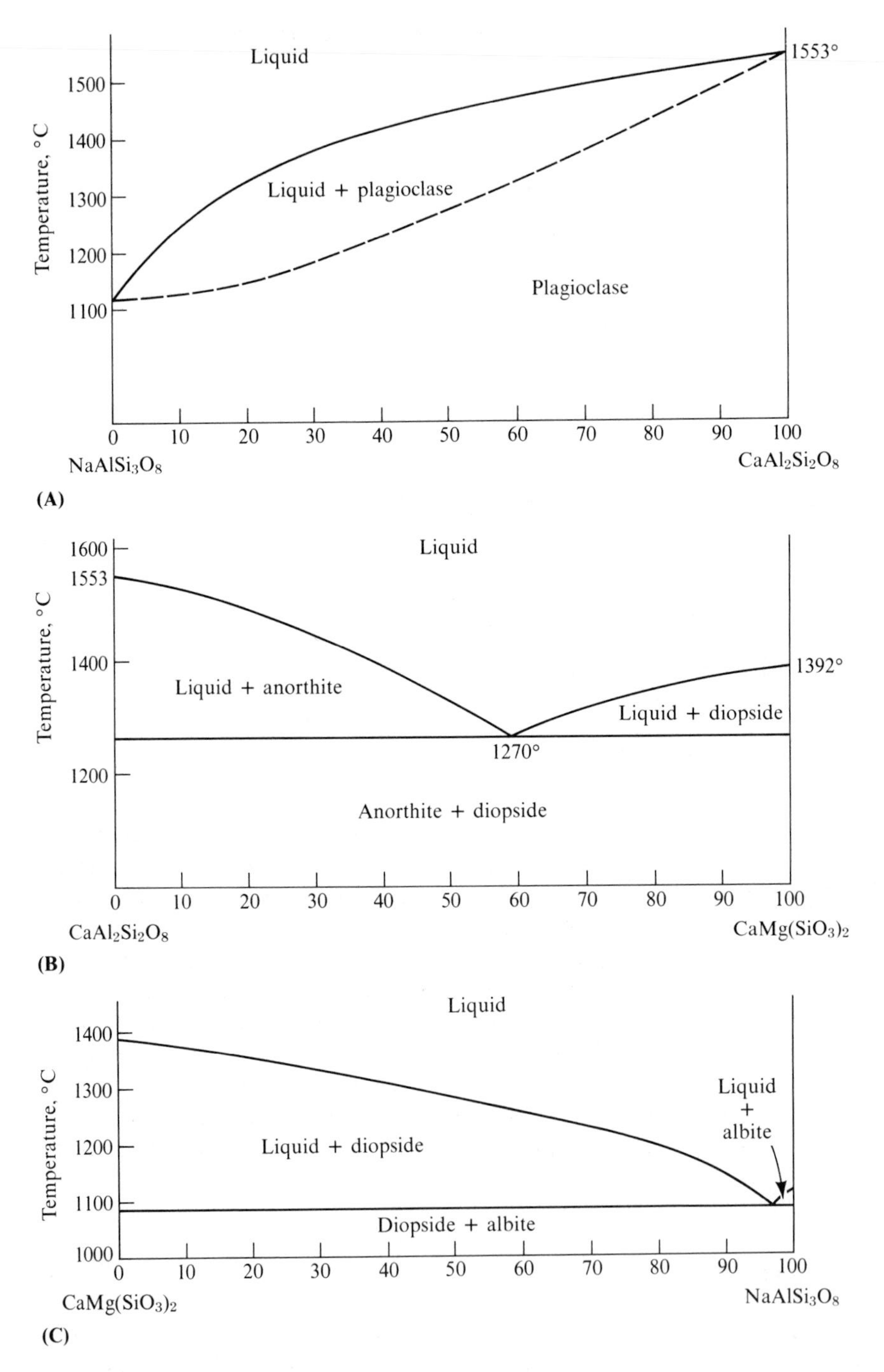

Figure 3–4
The binary systems $NaAlSi_3O_8$–$CaAl_2Si_2O_8$,$CaAl_2Si_2O_8$–$CaMg(SiO_3)_2$, and $CaMg(SiO_3)_2$–$NaAlSi_3O_8$ at 1 bar pressure. [From N. L. Bowen, 1913, *Amer. Jour. Sci.*, [*4*], *35*, Fig. 1; E. F. Osborn, 1942, *Amer. Jour. Sci.*, *240*, Fig. 4; and N. L. Bowen, 1915, *Amer. Jour. Sci.*, [*4*], *40*, Fig. 3.]

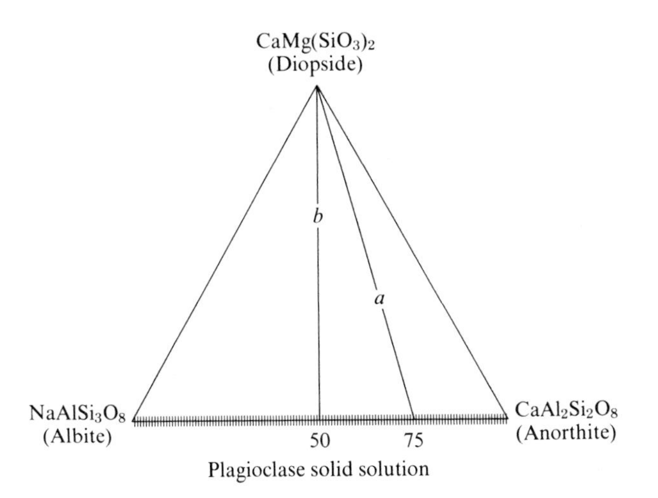

Figure 3-5
The compositional triangle $CaMg(SiO_3)_2$–$NaAlSi_3O_8$–$CaAl_2Si_2O_8$. Complete solid solution is present between the plagioclase end members $NaAlSi_3O_8$ and $CaAl_2Si_2O_8$, as indicated by hatching. A liquid composition *a* crystallizes to yield diopside and plagioclase (An_{75}); liquid *b* crystallizes to yield diopside and plagioclase (An_{50}).

the liquid precipitates equally sodic plagioclase. This double process of reaction and precipitation causes the liquid composition to follow a curved path to the boundary curve at *j*. When the liquid arrives at the boundary curve at *j* the coexistent plagioclase is of composition *k*.* Continued crystallization of diopside and plagioclase results in a change in plagioclase composition to *b*, and elimination of the liquid at *a*.

In both of the above examples continuous reaction between liquid and plagioclase is assumed. With incomplete reaction (due to rapid cooling, removal of early-formed crystals, and so on) a variety of plagioclase compositions will result (some more calcic and some more sodic than the predicted final composition). This is seen as compositional zoning. The liquid will change in composition during nonequilibrium crystallization further along the boundary curve than its predicted final composition under equilibrium crystallization.

PRESSURE EFFECTS

Rock systems are subject to significant changes in phase relationships as a result of pressure. As a result of burial, rocks are subjected to a pressure that is dependent

*The original liquid *e* is located on the experimentally determined two-phase tie line *jk*. When the liquid corresponds in composition to a point on the boundary curve such as *j*, the coexistent plagioclase must be at *k* in order for the composition of these two phases to add up to that of the original composition *e*.

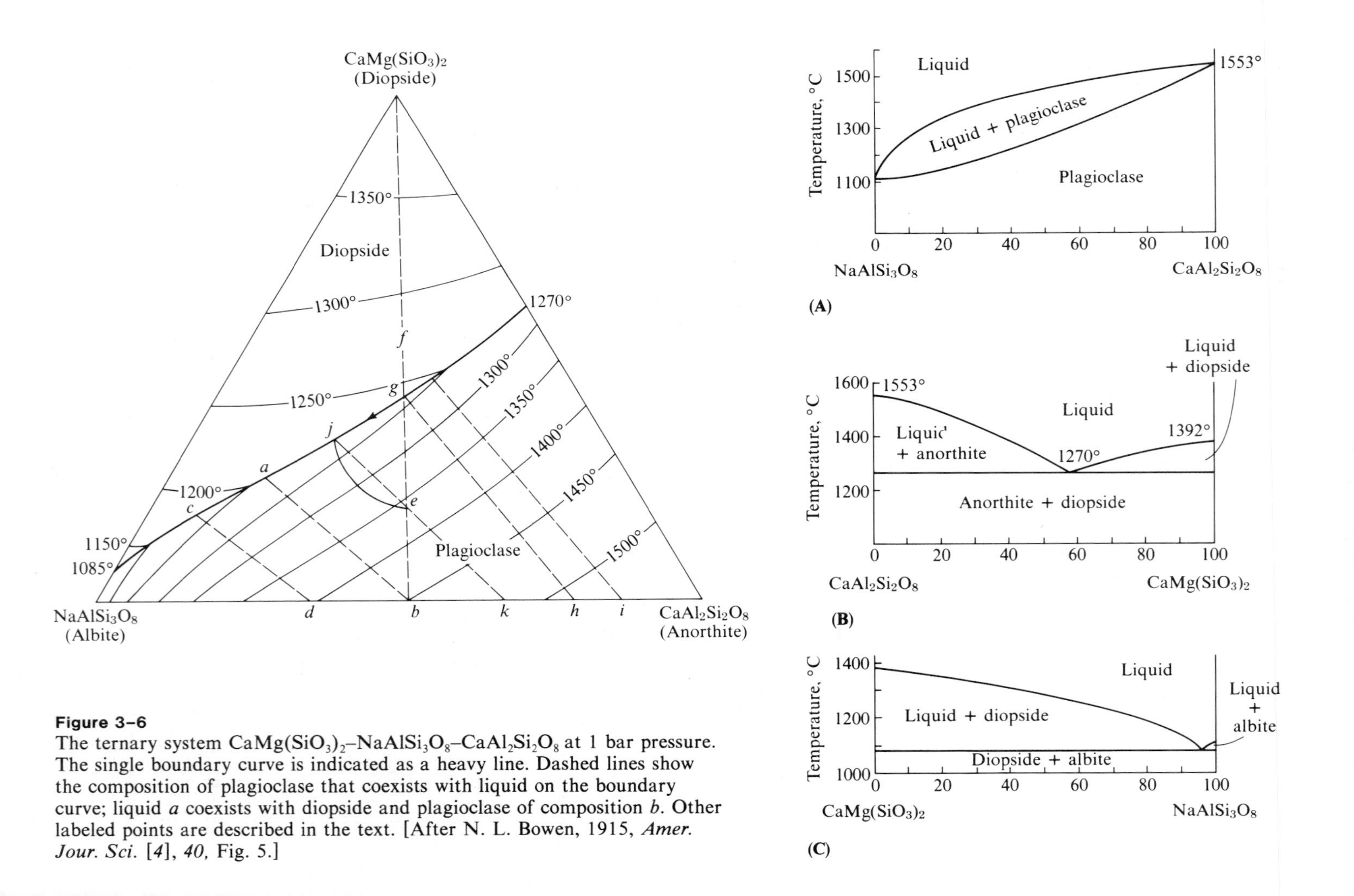

Figure 3–6
The ternary system $CaMg(SiO_3)_2$–$NaAlSi_3O_8$–$CaAl_2Si_2O_8$ at 1 bar pressure. The single boundary curve is indicated as a heavy line. Dashed lines show the composition of plagioclase that coexists with liquid on the boundary curve; liquid *a* coexists with diopside and plagioclase of composition *b*. Other labeled points are described in the text. [After N. L. Bowen, 1915, *Amer. Jour. Sci.* [4], *40*, Fig. 5.]

upon the overlying mass of materials. This lithostatic (load, or burial) pressure increases with depth at a rate of about 250–300 bar/km as a function of the density of the rocks above. At a depth of a few kilometers it is usually assumed that this vertical stress exceeds the strength of the rock so that the rock is capable of flow. Below this depth the load pressure becomes hydrostatic (that is, applied equally in all directions, similar to an object at depth in the sea). The hydrostatic or load pressure P below this level increases as a function of $P = g\rho h$, where g is the acceleration of gravity, ρ is the density of the overlying materials, and h is the depth of burial.

Consider the situation of a rock buried at a shallow depth, where it will retain some strength. Cavities within the rock might contain a hydrous or magmatic fluid. As the rock is not able to flow, the pressure of the fluid within the cavities is independent of the lithostatic pressure; that is, it might be greater, less, or equal to this pressure. The fluid pressure could be a function of the adjacent physical and chemical conditions, such as the presence of magma. Alternatively, if the cavities were interconnected to the earth's surface, the fluid pressure would be dependent upon the fluid load above. However at greater depths where rock strength is exceeded, fluid pressures are not usually known, and it is generally assumed that fluid pressure is equal to lithostatic pressure ($P_{litho} = P_{fluid}$).

The pressures within a magma chamber have both lithostatic and fluid aspects also, but the relationships are considerably more complicated than the pressures existing within rocks. Different portions of a magma are subjected to various load pressures as a function of the mass of the overlying rocks and melt (if any). As the confining pressure of solid materials directly above the magma chamber is less than that on its sides, and the density of molten materials is generally less than that of the surrounding country rock, a gravitational instability is present and the lighter magma has a tendency to rise through the overlying rocks. This tendency is increased if the magma is subjected to lateral tectonic compression. Most magmas contain some volatile components, and fluid pressure can be expected as well as lithostatic pressure. Volatiles may be dissolved in the silicate melt, or (if saturation limits are exceeded) may exist as a separate gas phase in the upper portions of the chamber.* As all gases are able to mix with each other, no more than one vapor phase may exist; it will commonly be composed of a mixture of components, such as H_2O, O_2, CO_2, SO_2, Cl_2, and so on, although H_2O is generally predominant, as verified by analyses of volcanic gases and of fluids that are often found within minerals as microscopic inclusions. Each of the components exerts its own partial pressure. The sum of the partial

*A distinction is often made between the terms vapor and fluid. Vapor may be taken to mean a gas phase that exists below critical conditions; that is, there is a distinct break in properties between liquid and gas phases. Fluid refers to supercritical conditions, where a substance has gradational changes between a liquid or vapor as a result of changes in P and T. In other definitions the term fluid applies to any liquid or gaseous substance. We will use the terms *vapor pressure* or *fluid pressure* (P_{fluid}) as any gas pressure produced by volatile components. Use of P_{fluid} is valid in most cases, as water exists in a supercritical state at pressures above 221 bar; this pressure is usually attained at the pressures expected at depths in the earth greater than 1 km.

pressures (P) of these components is the total fluid pressure of the system. If H_2O is the only fluid component, then $P_{fluid} = P_{H_2O}$.

Fluid pressure in a magma may be less, equal to, or greater than the lithostatic pressure. A magma formed by fusion or partial fusion of relatively anhydrous rocks (as in the mantle) will have a low H_2O content, whereas a magma formed by fusion of hydrous crustal rocks could be expected to have relatively large percentages of water. In addition, during the crystallization of a silicate melt, formation of anhydrous phases concentrates the volatile fraction in a decreasing volume, leading to a rise of P_{fluid} relative to P_{lith}. These varied conditions lead to situations where the melt is water-undersaturated, saturated, or oversaturated. Examples of each situation will be discussed in the text.

The Effects of Pressure on the Melting and Crystallization of Magma

Lithostatic Pressure

Melting temperatures of anhydrous compounds are increased by an increase in rock confining pressure. This occurs because increase in lithostatic pressure increases the size of stability regions of denser phases in relation to those of less dense phases. Such systems adjust to the increased pressure by occupying less volume. With the exception of a few materials (such as Bi and Ga), liquids occupy a greater volume than the same amount of anhydrous solid of identical composition. Therefore, the temperatures of the melting points, liquidus lines, and invariant points are raised (see Figure 3–7). In contrast, an increase of lithostatic pressure causes a slight decrease in the melting temperature of many hydrous compounds; the same is true with H_2O itself. This is because most hydrous compounds have slightly lower densities than their corresponding melts.

Fluid Pressure

Increased water pressure causes a decrease in stability of anhydrous phases in favor of hydrous phases. This is an example of Le Châtelier's Principle. The system attempts to nullify the effect of a change in conditions by eliminating anhydrous phases and forming phases that contain water. An example would be a favoring of the formation of hornblende over augite with increased water pressure because amphiboles contain OH-groups and pyroxenes do not.

Increased water pressure lowers the melting point of anhydrous solids. This occurs because of the presence of charged structural units within the melt (such as SiO_4^{-4}, AlO_4^{-5}, Na^{+1}, Mg^{+2}, and so on). When no H_2O is present these units tend to combine (polymerize), which promotes the crystallization of the various anhydrous silicate minerals. In the presence of H_2O, which is partially dissociated into ions, some of

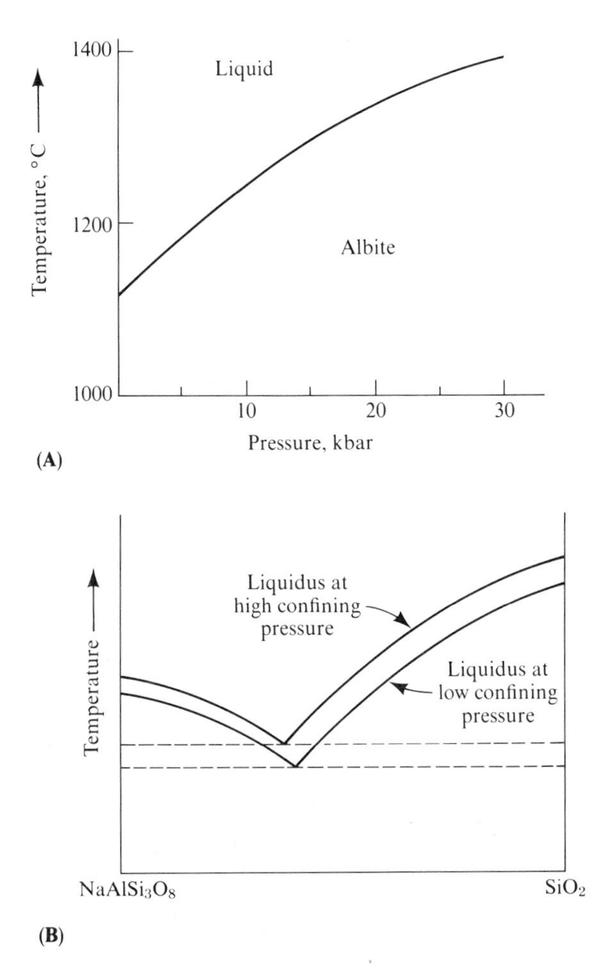

Figure 3–7
(**A**) The rise in the melting point of albite ($NaAlSi_3O_8$) as a function of increasing confining pressure. [From P. M. Bell, 1964, *Carnegie Inst. Washington Yearbook, 63,* Fig. 65.] (**B**) Schematic sketch of the binary system $NaAlSi_3O_8$–SiO_2 showing rise of liquidus surfaces with increase in confining pressure. Rise of the liquidus surface of a SiO_2 polymorph at a greater rate than that of albite results in a shift of the eutectic to the left. Alternatively, when subjected to increase in fluid pressure liquidus surfaces of anhydrous compounds are decreased in temperature.

the charges on these structural units are satisfied. This decreases the extent of polymerization (and consequently melting points) until much lower liquid temperatures are reached during cooling. Melting points, liquidus lines, and invariant points are decreased (an effect opposite to the changes shown in Figure 3–7).

As might be expected, the effect of water pressures has an opposite effect on hydrous compounds—that is, as the water pressure is increased, the stability ranges of the hydrous compounds are increased. Thus in the crystallization of a magma with a high water content, hydrous minerals such as the micas and hornblende will form at higher temperatures than anhydrous phases such as feldspar and quartz. On the other hand, in a relatively dry magma feldspar and quartz will crystallize early in the sequence and the hydrous minerals will crystallize later at lower temperatures.

For the reasons given above, it is obvious that the temperatures of crystallization given in the phase diagrams earlier in this chapter are only a starting point when investigating the crystallization of magmas. The diagrams are all for "dry" melts—magmas lacking water—whereas normal magmas contain varied amounts of water, which could significantly change crystallization sequences and temperatures.

In many, if not most, magmas the pressure is due to a combination of lithostatic pressure and fluid pressure. As indicated above, an increase in lithostatic pressure can have an effect directly opposite to that of an increase in fluid pressure. In this book we are only able to scratch the surface of the complexities of magmatic crystallization.

CRYSTALLIZATION OF MAGMA

The foregoing sections discuss the crystallization of rocks mainly from the standpoint of laboratory studies. We have made the tacit assumption that duplication of a rock texture and mineralogy by a laboratory procedure proves that the method of origin has been determined. This is, of course, not necessarily the case, as a variety of processes might produce the same result. It is therefore necessary to look at the rocks themselves to check the applicability of laboratory procedures.

Crystallization at depth cannot be examined directly by present techniques (send a submersible vehicle into a magma chamber?), but it is possible to observe crystallization at surface and near-surface conditions. This has been accomplished in a number of studies by the U.S. Geological Survey on the island of Hawaii and elsewhere.

Following a period of inflation of the summit of Kilauea (as measured by a network of tiltmeters), an eruption occurred in March of 1965 that yielded 15 million m^3 of basalt. A lava lake 83 m deep was formed in a pit within Makaopuhi crater. The eruption was accompanied by collapse of the Kilauea summit. During the eruption, lava samples were collected by inserting ceramic tubes into small pahoehoe flows at the edge of the rising lake. Flow temperatures of 1100–1160°C were determined by means of thermocouples and optical pyrometers. Samples of pumice (air-quenched glass) were determined to be 30–40° hotter, as indicated by the presence of only 3% crystals of olivine and pyroxene, in contrast to the flows, which contained 10% crystals. Cooling of the flow first produced a solid crust. A small, lightweight drilling rig set on the crust was able to obtain samples that extended to the still-molten lava below at about 9 m. Cores of the relatively solid tholeiitic basalt were obtained at rock temperatures up to 1079°C. Above that temperature the lava is essentially liquid; samples of the higher-temperature melt were obtained up to 1135°C in ceramic and steel sampling tubes pushed into the melt through the drill hole. Samples collected in this manner cooled to a black glassy rock when brought to the surface.

Petrographic examination of the samples collected at the various temperatures permitted the crystallization sequence to be determined. Figure 3–8 shows a plot of sample temperature versus the amount of melt. From this it is clear that crystallization begins at about 1200°C, and is completed just below 1000°C. Chrome spinel and olivine begin the crystallization sequence at about 1200°C, with the spinel slightly

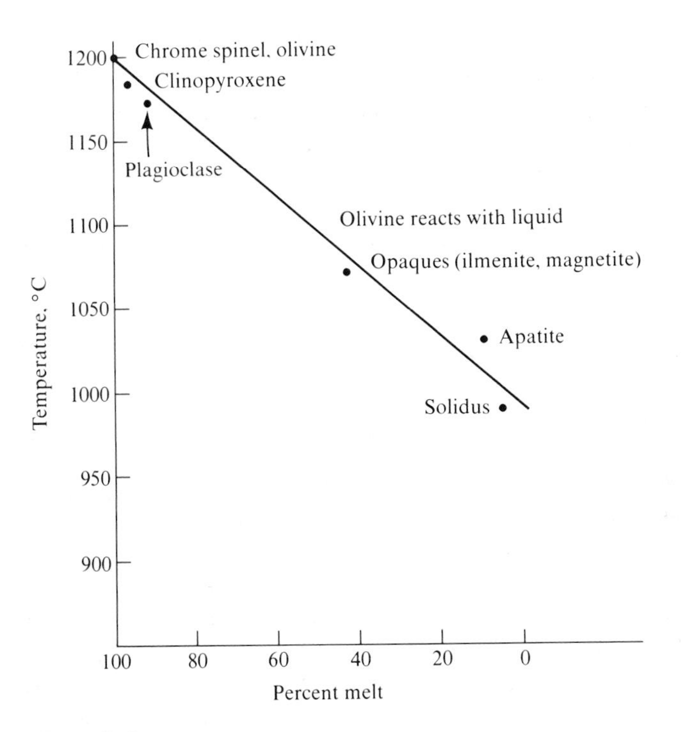

Figure 3–8
The crystallization of a tholeiitic basalt lava at Makaopuhi crater, Kilauea, Hawaii as determined from the surface and core samples. Minerals are listed at their initial temperatures of crystallization. Percent melt is shown as a function of temperature. [Data from Thomas L. Wright, 1967, U.S. Geological Survey news release.]

earlier than olivine, as indicated by spinel inclusions within olivine. At 1185°C crystallization of clinopyroxene begins. This is joined by plagioclase at 1175°C, followed by opaque minerals (ilmenite and magnetite) at 1070°C, apatite at 1030°C, with complete solidification below 1000°C. The general sequence is seen in the photomicrographs of Figure 3–9.

Notice first that the three silicate phases olivine, pyroxene, and plagioclase all begin crystallization within about 25°C, between about 1200°C and 1175°C. According to the ternary system diagram in Figure 3–3, it would seem that the melt composition must be fairly close to the eutectic. But the eutectic of the pure system $CaMg(SiO_3)_2$–$CaAl_2Si_2O_8$–Mg_2SiO_4 is at 1270°C; furthermore, this diagram indicates that the simultaneous crystallization of the three phases should begin and end at 1270°C. By contrast, in the tholeiitic basalt the three phases begin simultaneous crystallization at about 1175°C, with completion of crystallization at about 990°C—a freezing interval of almost 200°C. Why the discrepancy?

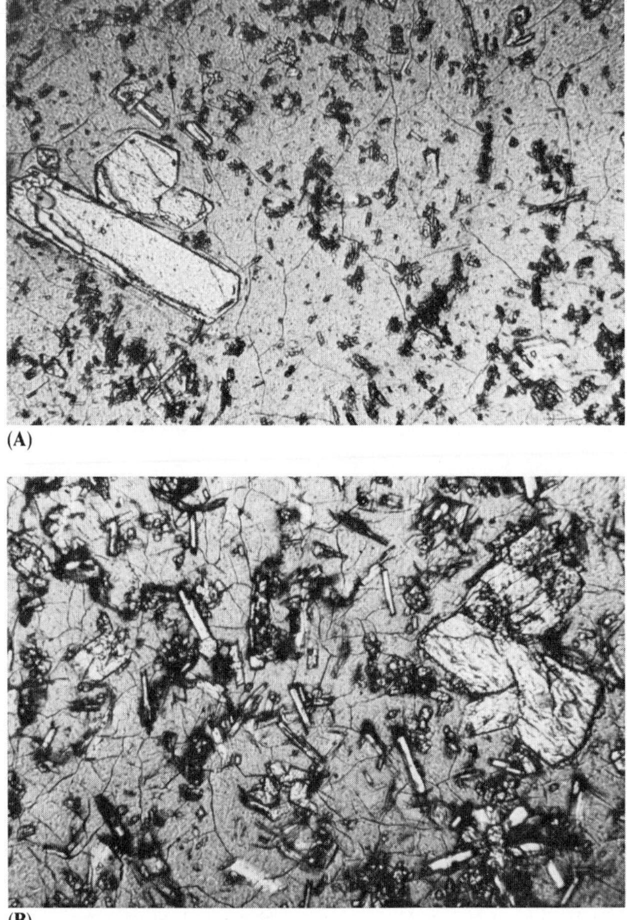

Figure 3-9
The photomicrographs of this group each cover an area approximately 0.9 × 1.2 mm. Olivine crystals have a maximum size of 2 mm and average 0.5 mm. Pyroxenes are small (0.1 mm) and equant. Plagioclase and ilmenite laths are usually 0.1 by 0.01 mm. The temperature at which the sample was collected and cooled is given in each case. [U.S. Geological Survey photos by T. L. Wright, 1967.] (**A**) 1170°C. A cluster of euhedral olivine phenocrysts is present. Small equant crystals are pyroxene. Plagioclase, which has just begun crystallization, is present as laths (near the olivine) and a flat plate (in upper left). (**B**) 1130°C. Olivine phenocrysts are at the right. The number of small equant pyroxene crystals has increased. Plagioclase laths have increased in both number and size. (**C**) 1075°C. An assemblage of equant pyroxene and plagioclase laths. The glass matrix is darker (red brown) as a result of increase in iron and titanium content. (**D**) 1065°C. Plagioclase and pyroxene are the same as the previous sample. A few opaque blades of ilmenite have crystallized. (**E**) 1020°C. An interlocking network of pyroxene, plagioclase (clear), and opaque ilmenite. The interstices are filled with slightly colored (brown) glass that contains needles of apatite.

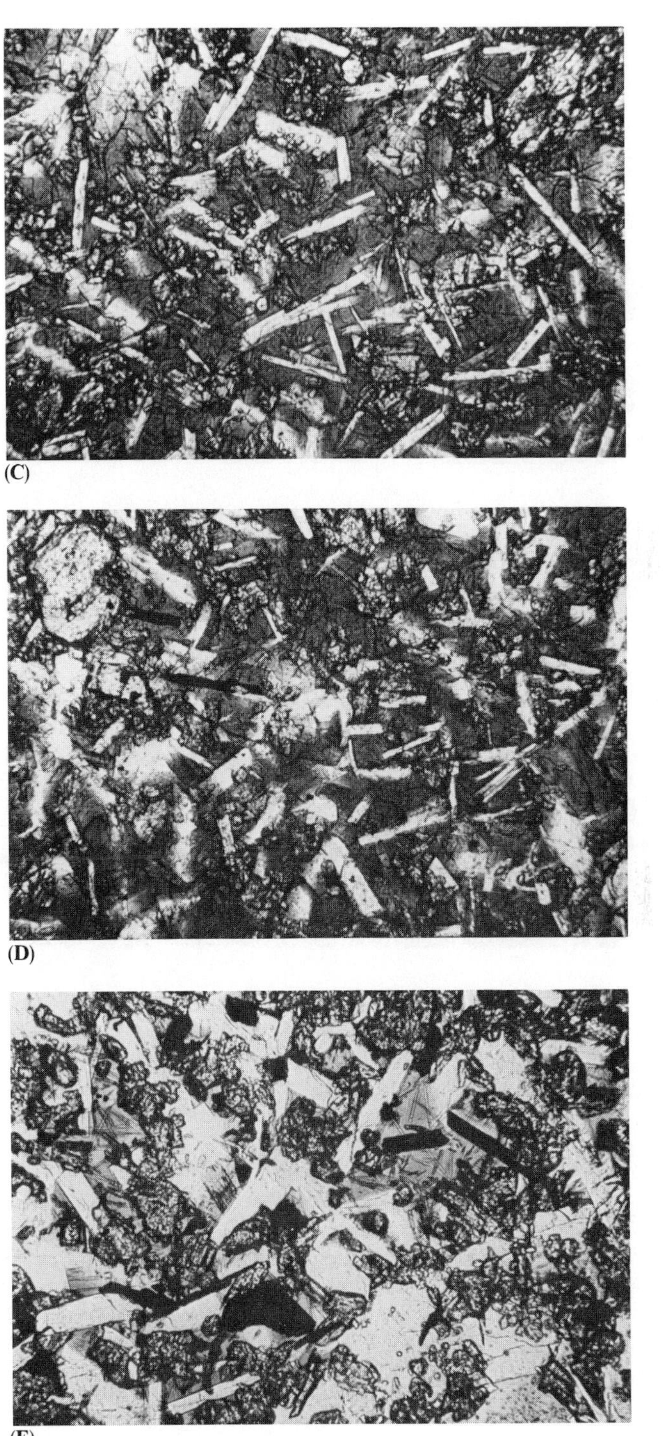

(C)

(D)

(E)

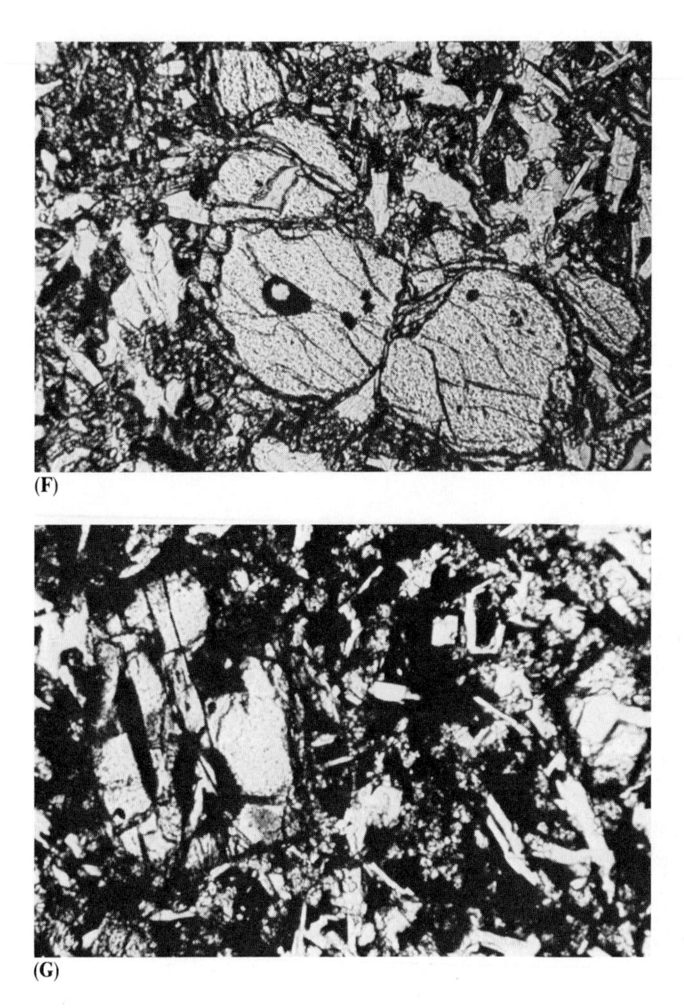

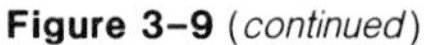

Figure 3–9 (*continued*)
(F) ~890°C, below the solidus. The large olivine phenocryst that dominates this photo is fractured and partially resorbed (see part A), but not altered. **(G)** ~750°C. The large olivine crystal at the left is partially altered (dark rim) to what is probably hematite. Other silicate phases are metasomatically replaced by unidentified opaque phases.

Here we must adopt the prime rule of the experimenter, "If it happens, it's possible," and then attempt to explain *how* it is possible. The answer in this case is simple and lies in the chemical differences between the phase diagram (which only permits compositions composed of mixtures of CaO, MgO, Al_2O_3, and SiO_2) and the rock, whose much more complex analysis is given in Table 3–1. Each of the additional chemical components in the lava has its effect on freezing temperatures. For example, the plagioclase that crystallized in the rock is not anorthite but rather an intermediate composition (An_{65}). Also both the olivine ($Fo_{80}Fa_{20}$) and pyroxene contain some Fe as well as minor amounts of other elements such as Mn and Ti. In addition

Table 3-1
Chemical and Mineralogical Composition of Makaopuhi Lava

Chemical composition (in weight percent)[a]	
SiO_2	50.24
Al_2O_3	13.32
Fe_2O_3	1.41
FeO	9.85
MgO	8.39
CaO	10.84
Na_2O	2.32
K_2O	0.54
TiO_2	2.65
P_2O_5	0.27
MnO	0.17
	100.00
Mode (in volume percent)[b]	
Olivine	5
Clinopyroxene	51
Plagioclase	30
Opaque minerals	9
Glass	5
	100

[a]Average chemical composition (recalculated to dry weight) from analyses of pumice collected during the March 1965 eruption.

[b]Average mode of the crystalline rock from Makaopuhi lava lake.

Source: T. L. Wright, 1967, U.S. Geological Survey News Release.

to not having end-member silicate phases, freezing points are also depressed by the presence of other chemical components (such as Fe^{+3} and P_2O_5) that could not enter these silicate structures.

The natural rock system is one in which many chemical components were present and apparently responsible for the significant differences from the results indicated by simplified laboratory studies. No competent petrologist, of course, would pretend that the two are directly comparable. The ternary eutectic of the pure system is no longer an invariant point in the real world but rather is extended into a line or a region within which several phases crystallize over a large temperature interval. This does not mean that the experimental approach is invalid. Many of our ideas about magmatic crystallization have been derived from this approach. Instead it indicates that equilibrium diagrams of this type furnish only approximate results for application to real rock systems. With the passage of time experimental studies have begun to approach more closely the complexities of natural rock systems; but the process is a slow one, and it is always necessary to examine the rocks themselves in order to verify any conclusions about their origins that have been derived by indirect approaches.

Time and Crystallization

One of the obvious problems in laboratory studies of igneous rock-forming processes is the time factor. Laboratory studies are conducted for periods of days, weeks, or months, whereas cooling of large masses of magma may take place over many thousands of years. Fortunately this seems to have little effect on the determination of melting and freezing temperatures, as reaction rates at liquidus temperatures are quite fast; in fact it is often difficult to prevent a melt from crystallizing during cooling, and special temperature-quenching techniques are required to preserve the liquid phase as a quenched glass. Even slow processes such as crystal settling and floating can be duplicated to some extent. Difficulties are encountered at subsolidus temperatures, because atomic and molecular diffusion rates are slowed due to the lower thermal energy of the system. Processes such as unmixing of solids, polymorphic inversions, and interchange of materials among phases may be indicated in short-term experiments, but some do not proceed to completion. A benefit of these slow reaction rates is that the chemical and mineralogical characteristics of the high-temperature assemblages are often relatively well preserved for study.

In order to get some idea of the time involved in the cooling of igneous bodies, consider first the cooling of a lava. Initially a large temperature difference exists between the lava and the relatively cool ground below and the atmosphere above. This naturally leads to a chilled base and upper surface. The chilled edges are solidified first, and relatively slower cooling occurs in the interior of the flow. A study was made of the cooling of the 600,000-m^3 Alae basaltic lava lake that formed from the 1963 eruption of Kilauea, in Hawaii. The lake had a maximum depth of 15 m. It had an initial temperature of 1140°C and became completely solidified at about 980°C. Complete crystallization occurred in a little more than a year—between August 1963 and September 1964. Temperatures were measured within the flow by means of core samples taken through the solidified crust (by techniques described earlier in this chapter). The relationship between isotherms and depth within the lava lake is shown as a function of time in Figure 3–10. The base of the upper, more or less solid crust on the lake is at a temperature of about 1065°C. Examination of this isotherm in Figure 3–10 shows the downward migration of the base of the crust as a function of time. The isotherms within the upper crust increased in depth as a function of the square root of time. From these types of data it has been calculated that thermal diffusivity (a measure of heat flow through time) is consistent with a value of 0.0071 cm^2/s. The maximum temperature two years after eruption was 690°C; this was found at a depth of 11.5 m, about three-fourths of the distance to the base of the lake. Measurement of the physical parameters involved in cooling permits calculation of cooling rates of other flows of similar composition but different thickness.

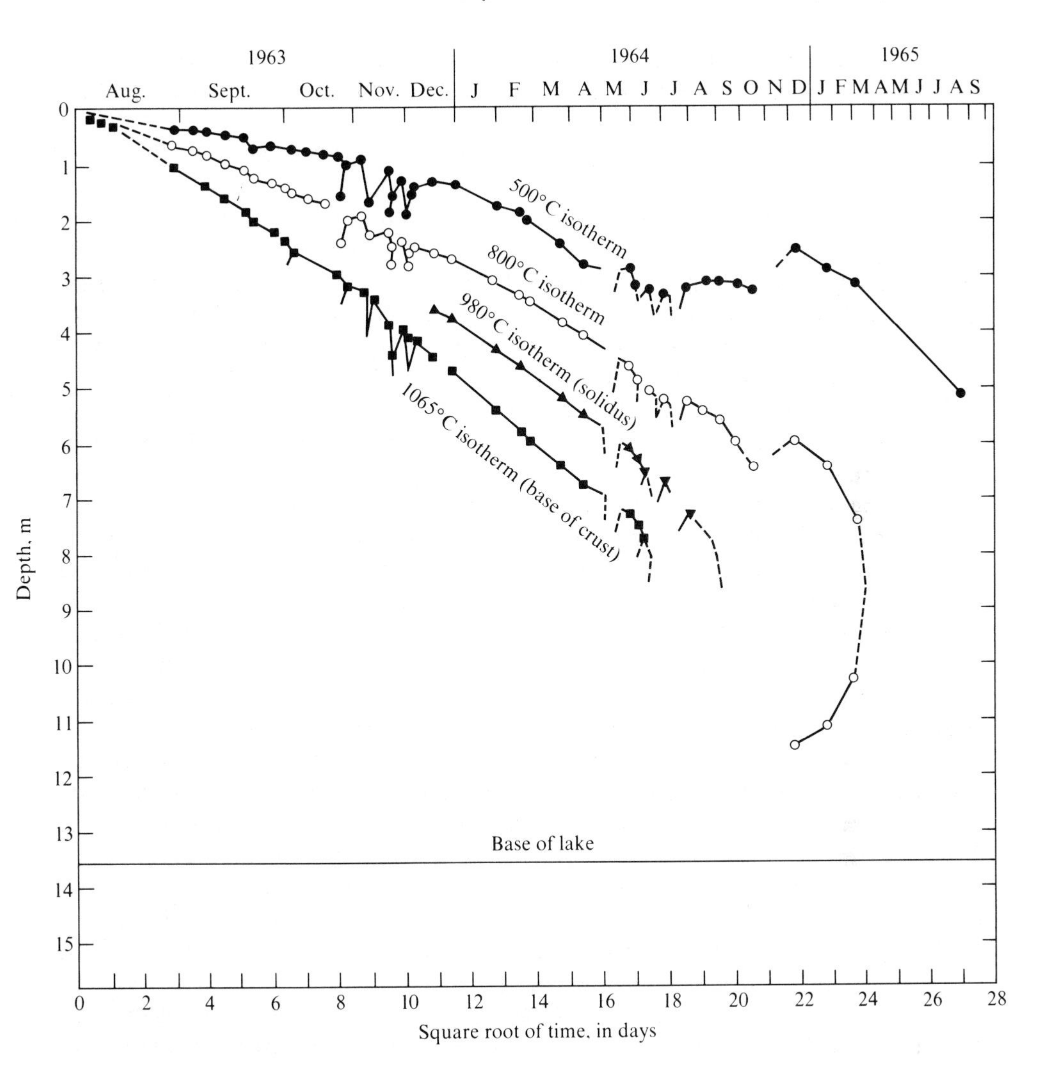

Figure 3–10
The cooling of 600,000-m^3 Alae Lava Lake, Hawaii, based on drill core samples. The various isotherms decrease in a generally regular manner as a function of time. Complete crystallization was achieved in 13 months. [From D. L. Peck, T. L. Wright, and J. G. Moore, 1966, *Bull. Volc., 29,* Fig. 3.]

Analysis of the cooling of an intrusive body is considerably more complicated because direct measurements cannot be made. Rates of cooling are dependent upon the size, shape, and thermal behavior of the intrusive body, as well as the thermal behavior and initial temperature of the surrounding country rock. Nevertheless, calculations have been made and we have a pretty good general idea of cooling rates. One of the most important factors is the size of the pluton; a larger intrusive has a greater heat content than a smaller one. Consequently small intrusives have a narrow zone of baked rock surrounding them, whereas larger intrusives, taking longer to dissipate their heat, are often surrounded by a large thermally metamorphosed zone. The composition of the intrusive is also of importance. One would assume that basaltic melts should show greater thermal effects on the country rocks than granitic melts, as basaltic melts are several hundred degrees hotter. In fact this is not the case. The effects of granitic intrusions are generally more widespread because granitic melts have a higher volatile content than basaltic melts. The volatiles are dissipated into the country rock, causing significant temperature increases and mineralogical changes.

In an example of one theoretical cooling model, Jaeger (1968) has calculated the time necessary for crystallization of a cooling vertical intrusive sheet of magma in cold country rock. Several simplifying assumptions were made. Consider a magma that intrudes at 1100°C and undergoes crystallization to 800°C by means of conductive heat loss through the walls. Assuming reasonable values for thermal diffusivity and other factors, solidification at 800°C is completed in 0.014 D^2 years, where D is the thickness of the sheet in meters. For various thickness we obtain the following times for completion of crystallization:

Thickness	Time, in years
1 m	0.014 (5 days!)
10 m	1.4
100 m	140
1 km	14,000
10 km	1,400,000
100 km	140,000,000

The numbers, of course, give only a general order of magnitude for any specific case. It is obvious, however, why thin dikes and sills inevitably have fine grain sizes, whereas stocks and batholiths are coarse-grained.

Rock Types and Mode of Occurrence

Studies of the frequency of occurrence of igneous rocks indicate that the most abundant volcanic rock type is basalt and the most abundant intrusive (plutonic) type is granite. Why does this relationship exist?

Three factors appear to control the environment in which a magma will crystallize. One of these is the availability of an escape route to the earth's surface. Perhaps, in the environment in which certain types of magmas are created, easy upward movement is prevented and in other cases enhanced. Another factor could be termed the eruptive force of the magma. As molten silicate melts are generally lighter than the rocks from which they have been derived, there exists a gravitational instability that tends to force the melt upward. This effect can be enhanced by the presence of volatiles in the melt (such as H_2O, CO_2, O_2, and so on). A third factor is the viscosity of the melt. Clearly a very fluid, low-viscosity melt has a better chance of reaching the surface than a more viscous material.

Consider what is known about basaltic and granitic melts. The density of granitic magma is less than that of basaltic magma (2.4 g/cm^3 at 800°C versus 2.65 g/cm^3 at 1200°C), which should favor a rise in the granitic melt. (Note that when heated to 1200°C granitic magma will have a density even less than 2.4 g/cm^3.) Furthermore, granitic liquids have a higher volatile content than basaltic melts, which should enlarge the difference. Most granitic magmas are thought to be created in the crust, whereas most basaltic melts originate at greater depths in the mantle. These factors, which should favor the rise of granitic magmas to the surface, are apparently not sufficient to do the job; granitic magmas normally crystallize at depth. Other factors must be considered.

Two properties favor the general rise of basaltic melts to the surface. One of these is the low viscosity of basaltic melts [because of low $(SiO_4)^{-4}$ content] as compared to granitic ones. Under conditions of equivalent water saturation and crustal pressure (1–2 kbar) the viscosity of basaltic melts is only 10^3 g/cm·s, whereas granitic melts are 10^6 g/cm·s, a very significant difference, resulting in a fluid basaltic melt and a viscous granitic melt when the two are subjected to similar conditions. A second and perhaps even more important factor results from the volatile content of basaltic and granitic magmas, which affects the slope of the melting curves.

As pointed out earlier, the melting curves of anhydrous materials will rise as a function of increasing lithostatic pressure and decrease as a function of increasing fluid pressure. Thus there are two types of melting curves, depending on the type of pressure to which the rock is subjected (see Figure 3–11).

Examination of volcanic gases and study of the metamorphic effects around intrusives have indicated that volatiles are present in both granitic and basaltic melts. Although estimates vary, it is thought that granitic melts contain about 6% H_2O by weight, and basaltic melts about 1%. The effect of the volatile component is to create a depression of the freezing point of the magma as the water pressure is built up to the 6% (granitic melt) and 1% (basaltic melt) levels. If the water content is kept at these values, additional pressure on the system can only be lithostatic. At these higher lithostatic pressures the melting point rises, giving the melting curves shown in Figure 3–12. It is obvious from the curves that the granitic melting curve is much more depressed than that of the basaltic melt, as a result of the higher H_2O content of the granitic melt. The difference between the two curves has a drastic effect on the cooling history.

Consider a granite melt at *a* (in Figure 3–12A) containing 6% H_2O (by weight) at depth and at a temperature just above the beginning of crystallization. If the melt begins to rise toward the surface the lithostatic pressure decreases. The temperature, however, remains almost constant because the surrounding rocks at great depth are

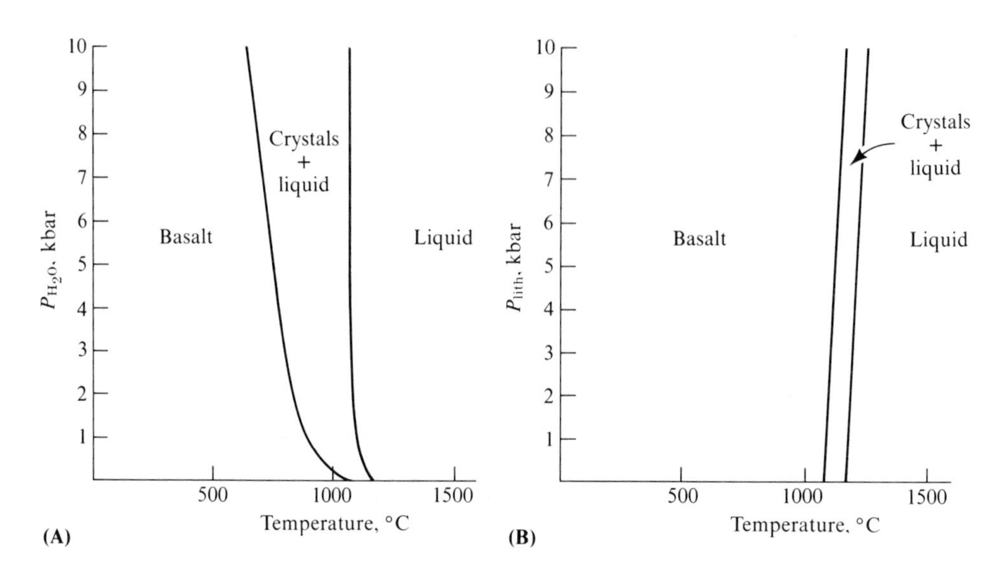

Figure 3–11
The change in the melting temperature of a basalt as a function of P_{H_2O} (part A) versus P_{lith} (part B). [Diagrams are constructed from the data of H. S. Yoder, Jr., and C. E. Tilley, 1962, *Jour. Petrology, 3,* 342–532; L. H. Cohen, K. Ito, and G. C. Kennedy, 1967, *Amer. Jour. Sci., 265,* 475–518; and D. L. Hamilton, C. W. Burnham, and E. F. Osborn, 1964, *Jour. Petrology, 5,* 21–39.]

only slightly cooler than the melt. The melt encounters the crystallization curve at *b*. Crystallization begins, and by the time the melt has reached point *c* crystallization is completed and the magma has turned to rock without reaching the surface. On the other hand, a basaltic melt beginning at a similar point *a* (in Figure 3–12B) rises to the surface (*a* to *b* to *c*) without encountering the crystallization curve, and is extruded with no crystallization. Obviously there are exceptions to this general arrangement, as basaltic melts occasionally may cool at depth to yield gabbro; this may happen for a variety of reasons, such as partial crystallization due to slow ascent, resistance by overlying rocks, and so on. Granitic melts may reach the surface to form rhyolitic lavas and ash flows if the thin overlying layer of rocks is fractured or faulted. But the relations shown by the graphs are generally valid. From the graphs we can see that the lower the water content, the more likely it is that the magma will get to the surface without crystallizing.

Andesite is an aphanitic rock of composition and water content intermediate between granite and basalt. The depression of the freezing curve of an andesitic melt is usually not sufficient to permit the melt to crystallize at depth (that is, andesite is much more common than its coarse-grained equivalent, diorite). The final freezing point of andesitic melts is thought to be reached just below the earth's surface. However, the volatile content often is sufficient to cause the crystallizing melt to break

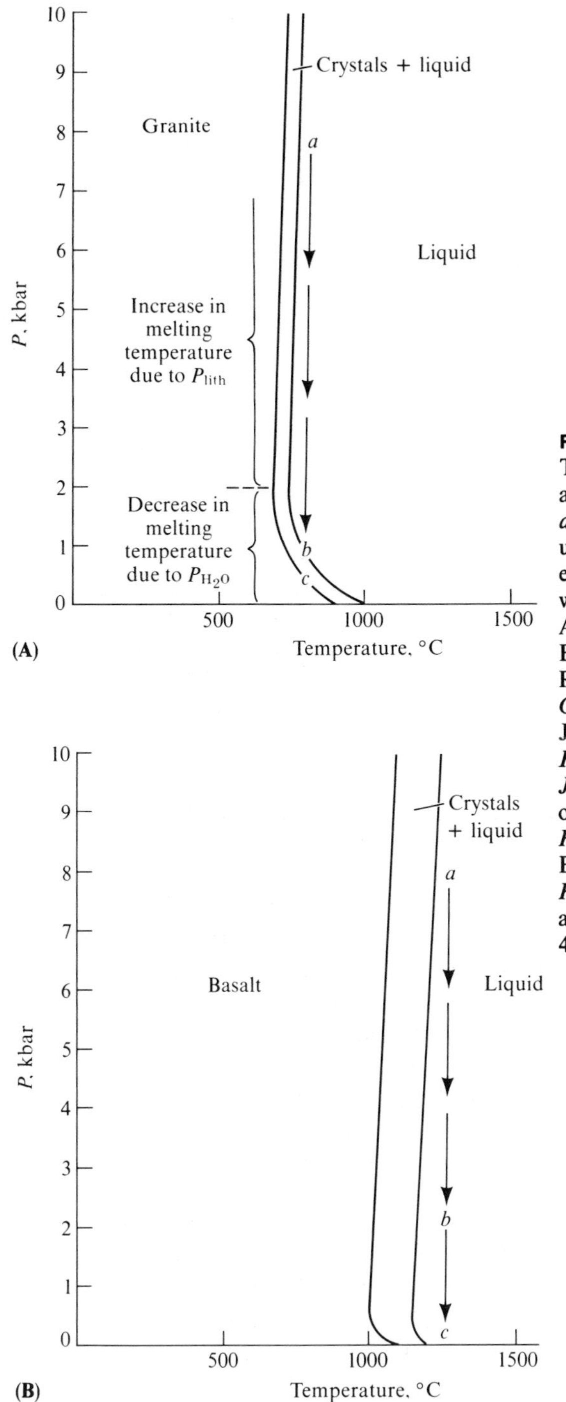

Figure 3–12
The freezing curves of granite (with 6% H_2O) and basalt (with 1% H_2O). A granite melt (point *a*) when rising (and decreasing in pressure) usually will crystallize before reaching the earth's surface, whereas a basalt melt usually will reach the surface before crystallizing. [Part A from the data of O. F. Tuttle and N. L. Bowen, 1958, Geol. Soc. Amer. Memoir 74; F. R. Boyd and J. L. England, 1963, *Jour. Geophys. Res., 68,* 311–323; W. C. Luth, R. H. Jahns, and O. F. Tuttle, 1964, *Jour. Geophys. Res., 69,* 759–773; and A. J. Piwinskii, 1968, *Jour. Geol., 76,* 548–570. Part B from the data of H. S. Yoder, Jr., and C. E. Tilley, 1962, *Jour. Petrology, 3,* 342–532; D. L. Hamilton, C. W. Burnham, and E. F. Osborn, 1964, *Jour. Petrology, 5,* 21–39; and L. H. Cohen, K. Ito, and G. C. Kennedy, 1967, *Amer. Jour. Sci., 265,* 475–518.]

through the brittle crust and be ejected as a solid–liquid–gas mixture at the surface.

Finally, let us consider why a magma leaves its place of origin and rises toward the surface. The ascent of magmas is often ascribed to two reasons (although a wide variety of other mechanisms have been proposed). The first of these is the buoyancy hypothesis, which maintains that the igneous melt is lighter than the surrounding rocks and moves upward as a result of gravitational instability. This argument works well with siliceous melts such as granitic magmas, as the melt density is about 2.4 g/cm^3, less than that of the various metamorphic and sedimentary rocks through which it passes.

Basaltic melts, on the other hand, have a density of 2.65 g/cm^3. This is less than the density of the basalts and peridotites of the oceanic crust; consequently the buoyancy hypothesis is applicable in this environment. In continental crustal environments, the density of sedimentary and metamorphic rocks is usually either equal to or less than that of a basaltic melt, and there would be no tendency to rise based on density differences alone. A second mechanism suggested for the rise of mafic and ultramafic melts is based on volume expansion during melting. Melting in the mantle to produce a basaltic melt results in an increase in volume on the order of 5%, which will result in an increase of pressure on the surrounding rocks. If the pressure is sufficient this will result in ductile yield of the surrounding rocks with a consequent rise of the magma. The level the magma reaches is mainly determined by the amount of volumetric expansion on melting and the space available above the area of melting. If the magma solidified while moving upward (consequently decreasing in volume), available space would be filled by creep of the surrounding rocks. If later melting occurred, the process could be repeated, and the magma could reach the surface in a series of steps.

Termination of vertical diapiric movement followed by solidification at depth yields a concordant intrusive. In the shallower, more brittle areas of the crust, different mechanisms for emplacement are required. The magma may be passively injected into fractures or joints that were opened by tectonic activity or magmatic stoping. Alternatively the magma may have a considerable eruptive force and may forcefully inject itself into the adjacent or overlying rocks by faulting, doming, or folding them.

SUMMARY

Igneous melts, arising from some type of melting process within the earth, are driven upward by a gravitational potential or volumetric expansion on melting. The magma (with or without a dissolved vapor phase) has a lower density than the rocks from which it was developed. Its upward migration is controlled by many variables—the amount of magma, the density difference between the magma and its surrounding rocks, the degree of fluidity, the depth of origin, the strength of the surrounding rocks, the degree of crystallinity of the melt, the pressure distribution around the magma chamber, the temperature difference between the melt and its surroundings, and other factors.

During magmatic rise each of these variables changes, making it difficult to generalize why a magma will crystallize in its particular circumstances. A few relationships are, however, clear. Of the extrusive rocks, the great majority are basaltic in nature. This common occurrence of basaltic flows appears to be mainly due to (1) the greater fluidity of a mafic melt than that of a silicic one, and (2) the slope of the freezing curve as a function of pressure. Silicic melts have a considerably higher viscosity than mafic melts and a strong depression of the freezing curve due to a high volatile content. They are thus retarded in upward migration, so most silicic magmas have formed intrusive rocks.

FURTHER READING

Carmichael, I. S. E., F. J. Turner, and J. Verhoogen. 1974. *Igneous Petrology.* New York: McGraw-Hill, 739 pp.

Ehlers, E. G. 1972. *The Interpretation of Geological Phase Diagrams.* San Francisco: W. H. Freeman and Company, 280 pp.

Jaeger, J. C. 1968. Cooling and solidification of igneous rocks. In H. H. Hess (ed.), *Basalts,* vol. 2, pp. 503–536. New York: Wiley Interscience.

Newell, G. and N. Rast, eds. 1971. *Mechanism of Igneous Intrusion.* Liverpool: Gallery Press, 380 pp.

Peck, D. L., T. L. Wright, and J. G. Moore. 1966. Crystallization of tholeiitic basalt of Alae lava lake, Hawaii. *Bull. Volcan., 29,* 629–656.

Wright, T. L., W. T. Kinoshita, and D. L. Peck. 1968. March 1965 Eruption of Kilauea volcano and the formation of Makaopuhi lake. *Jour. Geophys. Res., 73,* 3181–3205.

Wyllie, P. J. 1971. *The Dynamic Earth.* New York: John Wiley and Sons, 416 pp.

Yoder, H. S., Jr. 1976. *Generation of Basaltic Magma.* Washington, D.C.: National Acad. Sciences, 264 pp.

Yoder, H. S., Jr., and C. E. Tilley. 1962. Origin of basalt magmas: an experimental study of natural and synthetic rock systems. *Jour. Petrology, 3,* 342–532.

4

The Classification and Description of Igneous Rocks

The classification of igneous rocks, although seemingly simple, is beset with problems. Petrologists have been attempting to identify, characterize, and classify igneous rocks for hundreds of years. Igneous rocks have been named on the basis of mineral content, chemistry, locality, or texture. Some of the classification schemes have been arranged in a rectangular layout, others as triangles, a few as tetrahedra, and even one in a circular pattern. Because of this proliferation (almost one per famous igneous petrologist), a huge nomenclature exists for igneous rocks—in the form of rock names, textural varieties, and so on. Much of this earlier approach has been summarized by Johannsen (1931, 1937, 1938) in his four-volume set of books entitled *A Descriptive Petrography of the Igneous Rocks*. Johannsen ably summarized the problem as follows:

> Many and peculiar are the classifications that have been proposed for igneous rocks. Their variability depends in part upon the purpose for which each was intended, and in part from the characters of the rocks themselves. The trouble is not with the classifications but with nature which did not make things right. (1931, p. 51)

The obvious approach to the classification of igneous rocks is one built around mineralogy and texture. These features supply a great deal of information about the origin and history of the rock, are easily described in the field, and will only be refined rather than overturned by later thin section study. The mineralogic data can be elaborated, if deemed useful, by chemical analyses. In the case of microcrystalline rocks such as basalts and rhyolites, chemical analyses are usually required for classification.

The accuracy of mineralogic identification from field and hand specimens is quite variable. It depends on the size of the crystals, their degree of alteration, and the quality of observation by the geologist. This last factor is often as important as the others because differences in ability among geologists are as great as the difference in crystal size between rhyolite and granite. In any event, it is certain that descriptions such as "myrmekitic albite–riebeckite granite" are not based only on hand specimen study. Thin sections are required for accurate classification. In the field we often use descriptions such as "basalt(?)" for a dark-colored microcrystalline rock or "lithic rhyolite tuff(?)" for a light-colored, lightweight, layered rock containing fragments of uncertain character. Your characterization of rocks will improve in direct proportion to your knowledge of mineral characteristics, mineral associations, and experience.

Most igneous rocks contain a few minerals in large amounts, as well as a variety of minor ones. Because of ease in identification as well as their significance in petrogenesis, the more common minerals are usually chosen as the basis of classification. The most abundant crustal igneous rocks contain significant amounts of feldspars, along with a silica mineral such as quartz, or (to a considerably lesser extent) a feldspathoid mineral (indicative of a silica deficiency). Igneous rocks rich in these light-colored minerals are often referred to as *felsic*. On the other hand, many igneous rocks contain abundant dark-colored minerals such as pyroxene, amphibole, olivine, or biotite—the *mafic* minerals, which are rich in Mg and Fe. Most igneous classifications are based upon the relative amounts of these minerals, as well as grain size, which reflects cooling rate and manner of emplacement.

THE IUGS CLASSIFICATION SYSTEM

In order to answer the need for a single rational system of classification of igneous rocks for world use, Albert Streckeisen, after extensive correspondence with geologists in many countries, published a generally accepted classification scheme in 1967. At a later date the International Union of Geological Sciences (IUGS) elaborated Streckeisen's proposal, so that there now exists an internationally accepted, comprehensive system for igneous rock classification that permits nomenclatural pigeonholing to any desired degree of precision. We will cover that portion of the IUGS system that applies to the more commonly occurring rocks.

To classify a rock correctly on the basis of its mineral composition, the percentages of five minerals (or mineral groups) must be determined: quartz, plagioclase (anorthite content $> An_5$), alkali feldspar, ferromagnesian minerals, and feldspathoids. Note that alkali feldspar includes some albite (0–5% anorthite). In hand specimen work, quartz is identified by its translucency, vitreous luster, and lack of obvious cleavage; plagioclase by its cleavages and polysynthetic twinning; potassic feldspar (orthoclase, microcline, sanidine) by its cleavages, lack of twin-produced striations, and common pink to tan color; ferromagnesian minerals by their dark color; and feldspathoids by one or more of their individual physical characteristics. The only difficult one to recognize is nepheline, which is easily mistaken for quartz in hand specimen work. In thin section studies this difficulty disappears.

The mineralogy of a rock reflects its chemical composition (see Table 4–1). Rocks that contain free quartz are relatively rich in silica (granite); those with plagioclase as the dominant feldspar are usually high in calcium (diorite and gabbro); those dominated by mafic minerals contain considerable magnesium and ferrous iron (peridotite).

The IUGS classification distinguishes the common rocks first on the basis of grain size. Phaneritic rocks (whose grains are sufficiently coarse to be individually distinguishable) are classified as plutonic, and aphanitic rocks (whose grains are too small to be individually distinguishable) are classified as volcanic. Within each of these broad categories the rocks are named on the basis of mineral percentages. The classification categories assigned are based upon common usage and abundance of natural mineralogic groupings; that is, most rocks encountered will fit well within their pigeonholes.

The common plutonic rocks are best shown in a triangular arrangement (see Figure 4–1A) that permits a rock to be classified in terms of three constituents—quartz (Q), alkali feldspar (A), and plagioclase (P). The triangle permits classification of plutonic rocks that contain as little as 10% Q, A, and P; the rest of the rock consists of mafic minerals. The technique used is to determine the percentages of the Q, A, and P minerals, along with the mafic constituents. Assume that a rock has 50% mafic minerals, 15% Q, 20% A, and 15% P. As the mafic minerals are not included in the classification triangle, the Q, A, and P minerals are recalculated so as to equal 100%—giving us 30% Q, 40% A, and 30% P. The point is plotted within the field of granite in the triangle. The procedure for finding a point within a triangular graph is given in Figure 4–2. If a plutonic rock contains a small amount of mafic minerals this can be indicated by the prefix leuco- (as in leuco-granite); with large amounts of mafic minerals it is prefixed by mela- (as in mela-granite). Alternatively, if a mafic mineral or minerals are abundant and can be determined they can be used in the name, as in hornblende biotite granite; the least abundant of the two minerals is placed first in the name.

One additional point must be made relative to the QAP triangle. The plagioclase-rich areas of the diagram (lower right corner) contain several rock types that must be distinguished. An anorthosite is a plutonic igneous rock that contains more than 90% plagioclase. Between gabbro and diorite (and their volcanic equivalents basalt and andesite) the distinction is based on several criteria. The plagioclase in gabbro has a composition more calcic than An_{50}, whereas the plagioclase in diorite is more sodic than An_{50}. The same rule applies in distinguishing monzogabbros and monzodiorites. If the rock is to be classified by means of hand specimens, distinction between gabbro and diorite cannot be based on plagioclase composition. Instead the amount of mafic minerals can be used. Gabbro usually contains more than 35% mafic minerals by volume; these are usually augite, hypersthene, or olivine. Diorites contain less than 35% mafic minerals by volume, which are usually hornblende or hypersthene (± augite).

A small number of plutonic igneous rocks are low in silica and contain a feldspathoid (foid) rather than quartz. A second classification triangle is used for these rocks

Table 4–1
The Composition of Some Common Igneous Rocks[a]

	Granites (calc alkalic) (72)	Alkali granites (48)	Quartz diorites (58)	Diorites (50)	Pyroxene gabbros (38)	Alkali gabbros (38)	Peridotites (23)	Nepheline syenites (80)
SiO_2	72.08	73.86	66.15	51.86	50.78	43.94	43.54	55.38
TiO_2	0.37	0.20	0.62	1.50	1.13	2.86	0.81	0.66
Al_2O_3	13.86	13.75	15.56	16.40	15.68	14.87	3.99	21.30
Fe_2O_3	0.86	0.78	1.36	2.73	2.26	4.35	2.51	2.42
FeO	1.67	1.13	3.42	6.97	7.41	7.80	9.84	2.00
MnO	0.06	0.05	0.08	0.18	0.18	0.16	0.21	0.19
MgO	0.52	0.26	1.94	6.12	8.35	9.31	34.02	0.57
CaO	1.33	0.72	4.65	8.40	10.85	12.37	3.46	1.98
Na_2O	3.08	3.51	3.90	3.36	2.14	2.32	0.56	8.84
K_2O	5.46	5.13	1.42	1.33	0.56	0.92	0.25	5.34
H_2O+[b]	0.53	0.47	0.69	0.80	0.48	0.66	0.76	0.96
P_2O_5	0.18	0.14	0.21	0.35	0.18	0.44	0.05	0.19
CO_2								0.17

[a]Totals for each rock type equal 100%. The numbers beneath each rock name refer to the number of analyses of each rock type.
[b]Refers to water that is chemically combined.
Source: After S. R. Nockolds, and R. Allen, 1954, *Geol. Soc. Amer. Bull., 65,* 1007–1032.

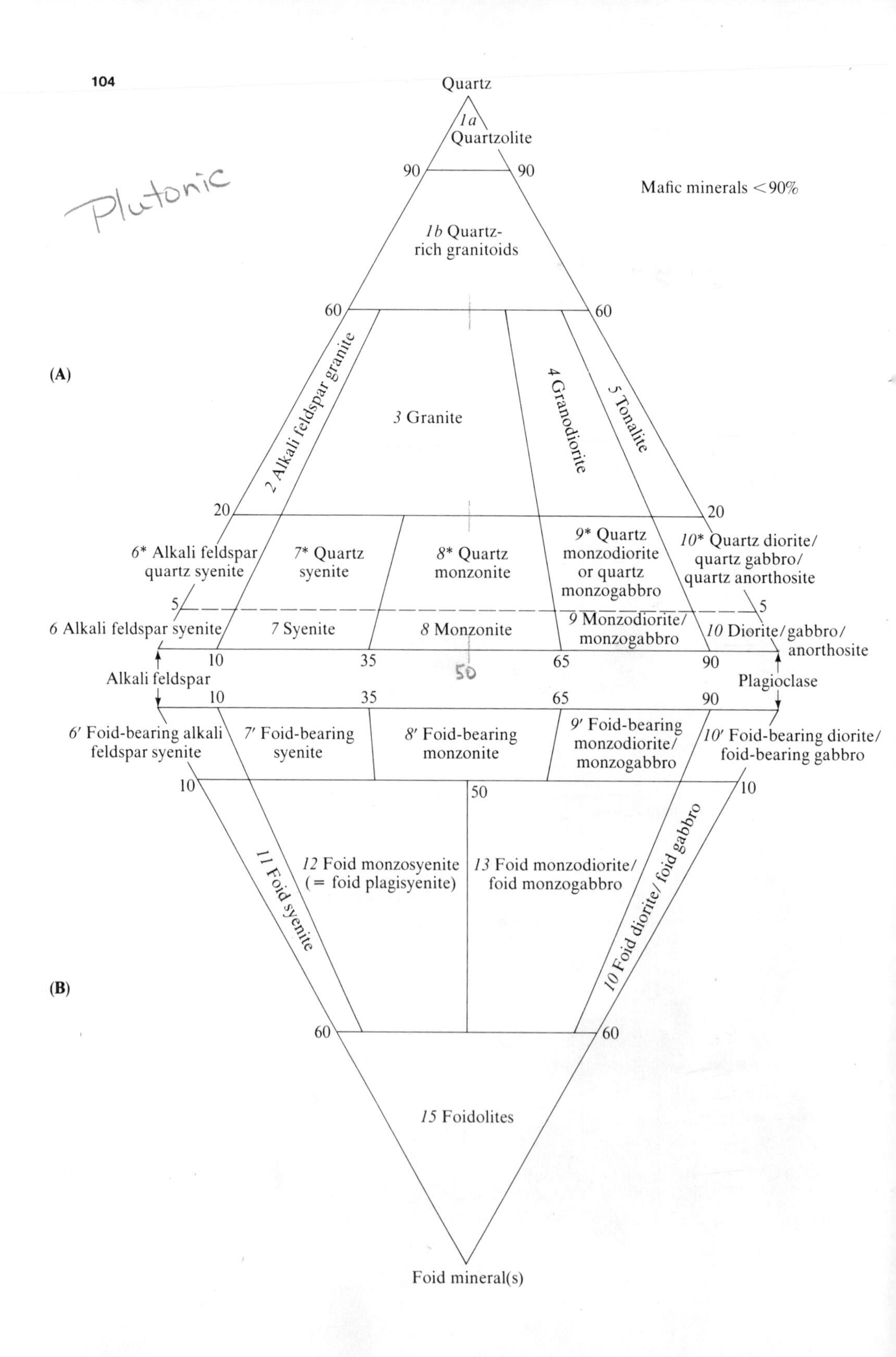
Quartz
1a Quartzolite
90
90
Mafic minerals <90%
Plutonic
1b Quartz-rich granitoids
60
60
(A)
2 Alkali feldspar granite
3 Granite
4 Granodiorite
5 Tonalite
20
20
6* Alkali feldspar quartz syenite
7* Quartz syenite
8* Quartz monzonite
9* Quartz monzodiorite or quartz monzogabbro
10* Quartz diorite/ quartz gabbro/ quartz anorthosite
5
5
6 Alkali feldspar syenite
7 Syenite
8 Monzonite
9 Monzodiorite/ monzogabbro
10 Diorite/gabbro/ anorthosite
10
35
65
90
50
Alkali feldspar
Plagioclase
10
35
65
90
6′ Foid-bearing alkali feldspar syenite
7′ Foid-bearing syenite
8′ Foid-bearing monzonite
9′ Foid-bearing monzodiorite/ monzogabbro
10′ Foid-bearing diorite/ foid-bearing gabbro
10
50
10
11 Foid syenite
12 Foid monzosyenite (= foid plagisyenite)
13 Foid monzodiorite/ foid monzogabbro
10 Foid diorite/ foid gabbro
(B)
60
60
15 Foidolites
Foid mineral(s)

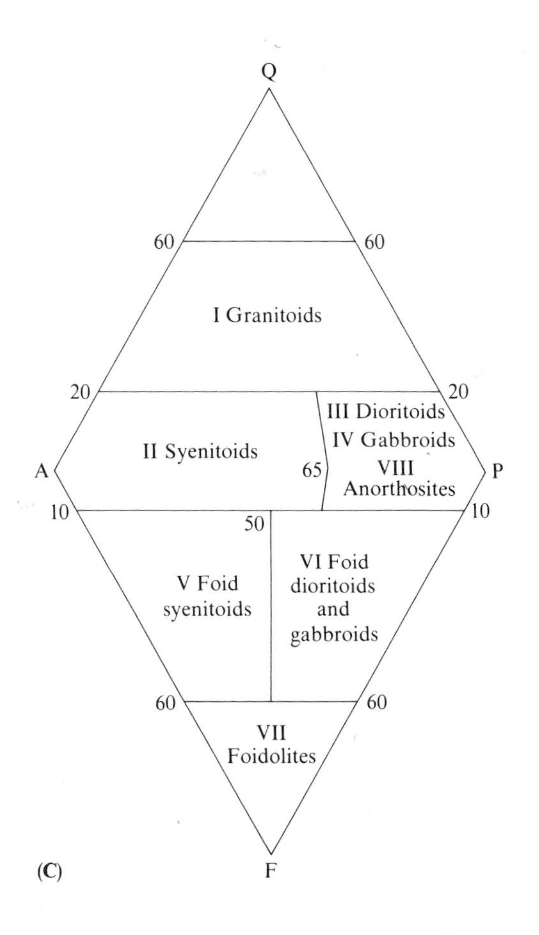

Figure 4–1
(**A, B**) The classification of plutonic rocks. In order to be included in these triangles, the igneous rock must be phaneritic. The rock must contain at least 10% plagioclase, alkali feldspar, and either quartz (triangle A) or feldspathoid (triangle B). The relative amounts of these minerals are recalculated to 100% and plotted within the appropriate triangle by the technique shown in Figure 4–2. Appropriate modifying terms are based upon mafic mineral composition or distinctive texture. (**C**) Generalized group names (for field use) when mineral percentages cannot be determined with precision. In fields II, III, and IV the qualifier "foid-bearing" should be used when feldspathoids are present. [From A. L. Streckeisen, 1976, *Earth Sci. Rev., 12,* Fig. 1a.]

(see Figure 4–1B). A rock cannot appear on both triangles, because quartz and feldspathoid are chemically incompatible; when mixed they will react to form a compound (feldspar) of intermediate silica content.

The volcanic igneous rocks are named on the basis of a similar triangular arrangement (see Figure 4–3). Distinction between basalt and andesite is made mainly on the basis of silica content (a rock with more than 52% SiO_2 is andesite, and a rock with less than 52% SiO_2 is basalt, as shown in part D), or less accurately on plagioclase composition (a rock with a plagioclase composition more sodic than An_{50} is andesite).

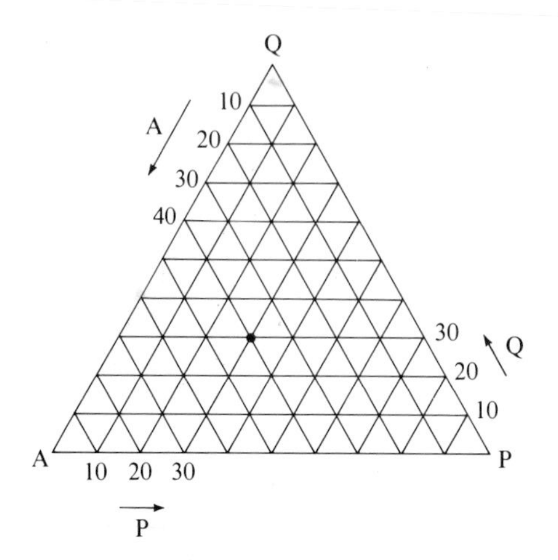

Figure 4–2
The technique of finding a point with a triangular coordinate system. Lines parallel to the base indicate percentages of quartz (Q). Lines parallel to the left side yield percentages of plagioclase (P), and those parallel to the right side, percentages of alkali feldspar (A). Q, A, and P for every point plotted within the triangle must total 100%. Points plotted on the edges indicate the relative percentages of only two of the constituents. The point plotted indicates 30% Q, 40% A, and 30% P. This falls within the granite field shown in Figure 4–1.

Plagioclase composition is difficult to use because of the very common presence of compositional zoning. Basalt, which falls near the P corner of the triangle, is commonly subdivided into tholeiitic types, which fall mainly in areas 10 and 10*, high-Al basalts, area 10, and alkali basalt, area 10′; the characteristics of the different varieties are shown in Table 4–2. Note that many basalts cannot be placed in a particular category by microscopic analyses, but may require chemical analyses or calculation of normative phases (see below). The common feldspathoid-bearing rocks may be rich in alkali feldspar (phonolite), plagioclase (tephrite), or nepheline or leucite (nephelinite or leucitite).

Ultramafic rocks (containing more than 90% mafic minerals) are classified by alternate methods. Fine-grained varieties are rare. The more common types are defined as follows:

Peridotite: A general term for a rock containing 40–100% olivine, with the remainder mainly of pyroxene and/or hornblende.

Dunite: A peridotite containing 90–100% olivine, with most of the remainder composed mainly of pyroxene, or less commonly hornblende.

Table 4–2
Types of Basalt

Varietal names	Pyroxenes	Other minerals[a]	Remarks
Tholeiite	Augite and hypersthene or pigeonite	Interstitial siliceous glass or quartz may be present	Oceanic varieties have > 2.5% TiO_2; continental varieties ≃ 1% TiO_2
Olivine tholeiite	Augite and hypersthene or pigeonite	Considerable olivine may be present, which may have reaction rims of pyroxene	Oceanic varieties have > 2.5% TiO_2; continental varieties ≃ 1% TiO_2
High-Al_2O_3 basalt	Augite, rarely pigeonite or hypersthene	Typically nonporphyritic; olivine common, with or without reaction rims of pyroxene	Al_2O_3 > 17%, TiO_2 ≃ 1%; intermediate characteristics between tholeiites and alkali types
Alkali basalt	Augite or titanaugite (slight purple pleochroism)	Feldspathoids, alkali feldspar, phlogopite, or kaersutite may be present; no reaction rims of pyroxene on olivine	Total alkali content higher and silica content lower than types above

[a]Plagioclase is always present in addition to the minerals listed in this column.

Pyroxenite: A rock whose major constituent is pyroxene, with the remainder composed mainly of olivine and/or hornblende.

Hornblendite: A rock whose major constituent is hornblende, with the remainder consisting mainly of pyroxene and/or olivine.

Other Aspects of Classification

Unfortunately, the IUGS classification does not yet cover all igneous rocks. Aside from making a distinction between phaneritic and aphanitic rocks, there is little said about texture. Yet a few igneous rocks are named on the basis of texture, with the mineral content being a secondary consideration (see Figure 4–4). These include the following:

Pegmatite: A very coarse-grained rock (most grains are > 1 cm) with interlocking grains. The composition is usually granitic, and the rock may occasionally contain abundant rare minerals with normally uncommon elements.

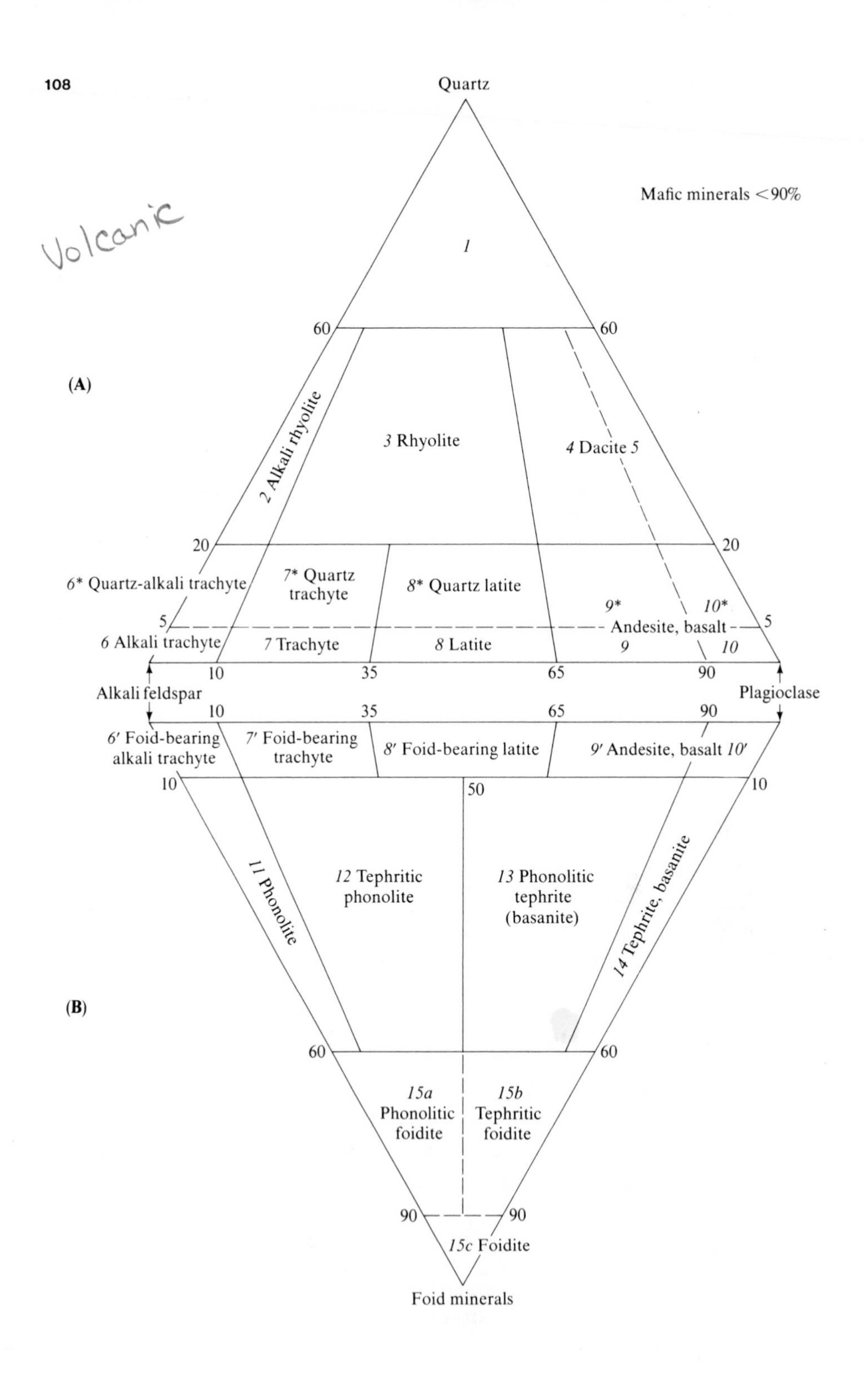
Quartz
Mafic minerals <90%
Volcanic
1
60
60
(A)
2 Alkali rhyolite
3 Rhyolite
4 Dacite 5
20
20
6* Quartz-alkali trachyte
7* Quartz trachyte
8* Quartz latite
9*
10*
5
Andesite, basalt
5
6 Alkali trachyte
7 Trachyte
8 Latite
9
10
10
35
65
90
Alkali feldspar
Plagioclase
10
35
65
90
6′ Foid-bearing alkali trachyte
7′ Foid-bearing trachyte
8′ Foid-bearing latite
9′ Andesite, basalt 10′
10
50
10
11 Phonolite
12 Tephritic phonolite
13 Phonolitic tephrite (basanite)
14 Tephrite, basanite
(B)
60
60
15a Phonolitic foidite
15b Tephritic foidite
90
90
15c Foidite
Foid minerals

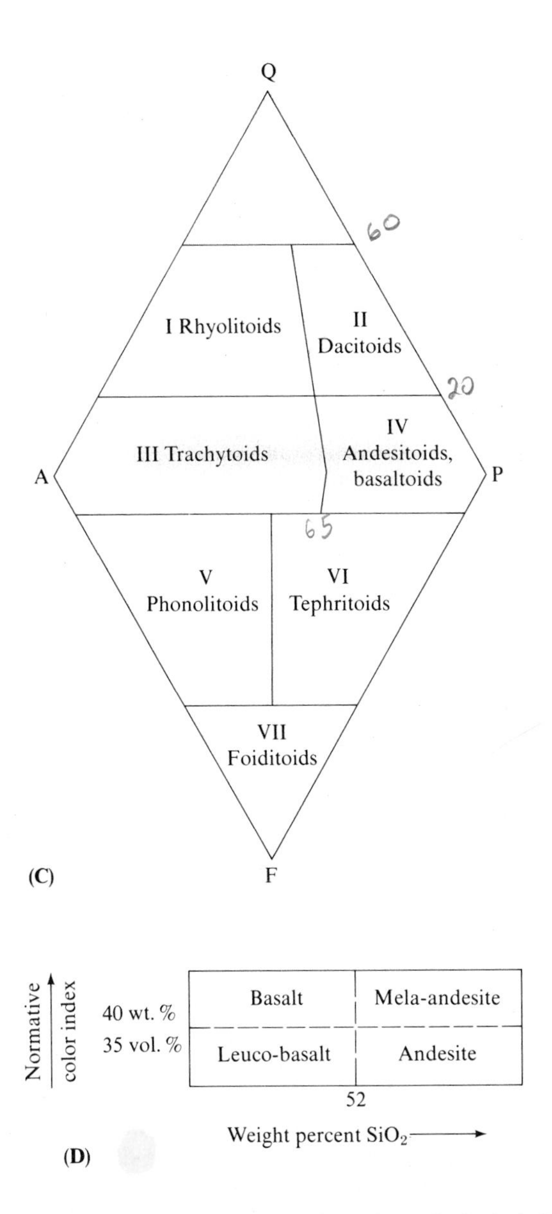

Figure 4–3
(**A, B**) The classification of volcanic igneous rocks. In order to be included within the triangles an igneous rock must be aphanitic. The rock must contain at least 10% plagioclase, alkali feldspar, and either quartz or feldspathoid (foid). The relative amounts of these three minerals are recalculated to 100% and plotted within the appropriate triangle, as shown in Figure 4–2. Appropriate modifying terms are used based upon mafic mineral composition or distinctive texture. In the case of those rocks whose matrix is too fine for determination, a tentative classification can be based upon minerals present in phenocrysts. (**C**) Generalized group names (for field use). (**D**) Distinction between basalt and andesite is based on color index (volume percentage of mafic minerals) and silica content. [From A. L. Streckeisen, 1979, *Geology, 7,* Figs. 1, 2; and A. Streckeisen, 1979, personal communication.]

Obsidian: A black or dark-colored, essentially nonvesicular volcanic glass, usually of rhyolitic composition. The fracture is conchoidal.

Tuff: A compacted deposit of volcanic ash and dust that may contain up to 50% sedimentary material.

Breccia: Similar to tuff, but containing large angular fragments (> 2 mm) in a finer matrix.

Some well-recognized igneous rocks are commonly found in a highly altered state. The alteration is related to their method of origin, rather than to later weathering processes. These include:

Spilite: An altered basalt, commonly vesicular, and exhibiting pillow structure. The feldspar has been converted to albite, and is usually accompanied by chlorite, calcite, epidote, chalcedony, or prehnite.

Serpentinite: A rock consisting almost entirely of serpentine, derived from alteration of preexisting olivine and pyroxene.

Kimberlite: An altered porphyritic mica peridotite containing olivine (commonly converted to serpentine or a carbonate mineral) and phlogopite (commonly converted to chlorite); some contain diamonds.

CHEMICAL CLASSIFICATION

A great variety of chemical classifications of igneous rocks has been devised. Some of them are based on the complete chemical analysis of the rock, others on only a portion of the rock chemistry.

As igneous rocks crystallize from melts, it is convenient to think of the magma in terms of being oversaturated with a particular mineral. If the cooling magma is oversaturated with respect to a particular mineral, that mineral will precipitate and be found in the resultant rock. This concept is mainly used for silica. Thus a rock may be described as *silica-oversaturated* if it contains quartz (or an equivalent SiO_2 mineral). It is *silica-undersaturated* if it contains silica-deficient minerals that cannot exist in the presence of quartz (feldspathoids, olivine, and corundum). In the presence of silica, olivine (Mg_2SiO_4) will combine with SiO_2 to form two molecules of the pyroxene enstatite ($MgSiO_3$):

$$Mg_2SiO_4 + SiO_2 \rightarrow 2MgSiO_3$$

Feldspathoids will form feldspars: for example, $KAlSi_2O_6 + SiO_2 \rightarrow KAlSi_3O_8$. A *silica-saturated* rock contains neither free silica nor silica-deficient minerals.

Another approach stems from the now-discarded idea that igneous rocks are the crystallization products of silicic acids. Thus a rock with a high silica content is called an acid rock, and a rock low in SiO_2 a basic or ultrabasic rock. This approach is summarized in Table 4–3. The term ultrabasic is often used interchangeably with ultramafic, but the meaning of the two terms is quite different, as ultramafic refers to a rock with a very high content of mafic minerals.

The most popular classification that uses the great majority of the chemical constituents of a rock is the system described by Cross, Iddings, Pirsson, and Washington in 1902. In the *CIPW system* the chemical analysis is recalculated to a set of standardized anhydrous normative minerals. The rock *norm* is the resultant calculated mineralogical analysis of the rock, as contrasted to the actual mineralogy, which is the *mode*. The details of a norm calculation are given by Johannsen (1931).

There are many uses for a CIPW calculation. Chemical analyses of fine-grained or glassy igneous rocks can be recalculated in order to determine the minerals that might have formed under conditions of slower cooling. Determination of the norm of highly altered igneous rocks, metamorphic rocks, or even sediments can indicate the general nature of the anhydrous igneous equivalent. The method has also been used by experimentalists on the melts produced in phase equilibria studies of crustal and mantle petrology.

CHARACTERISTICS OF THE COMMON IGNEOUS ROCKS

Brief descriptions of the more common igneous rocks are given below. Alteration reactions are written to show typical reaction products, and are not intended to be read as standard chemical formulas. Terms used in these descriptions and not used elsewhere in the book will be defined here. *Iron-ores* or *ores* refers to (generally unidentified) opaque minerals found in thin section, generally including magnetite and ilmenite. *Accessory minerals* refers to minerals generally present in such small quantities that a prefix notation to the rock name is not justified. *Sodium-rich amphiboles* refers to riebeckite and arfvedsonite. *Sodium-rich pyroxenes* are aegerine or aegerine–augite. *Orbicular texture* refers to areas of concentrically banded light and dark minerals, thought to result from magmatic reaction with an inclusion. *Interstitial* refers to a mineral that has grown between preexisting minerals, whereas *intersertal* refers to a porphyritic rock whose groundmass consists mainly of randomly oriented feldspar laths with a minor amount of glassy or partially crystalline

Table 4–3
Igneous Rock Classification Based on Silica Percentage[a]

Percent SiO_2	Rock type	Example
> 66	Acidic	Granite, syenite, quartz diorite
52–66	Intermediate	Granodiorite
45–52	Basic	Diorite, gabbro, basalt, andesite
< 45	Ultrabasic	Peridotite, dunite

[a]This classification, although commonly used, has fallen into disfavor as it is based upon solution chemistry and pH.

material filling the interstices. *Trachytoid* (or *trachytic*) texture refers to a subparallel orientation of microlites in the groundmass of a phaneritic volcanic rock. *Poikilitic* refers to a mineral or crystal that contains abundant randomly oriented crystals of another mineral. *Holocrystalline* means that the rock is entirely crystalline; *hypocrystalline* refers to a rock containing both crystals and glass; *holohyaline* means that the rock is completely composed of glass.

Plutonic Types

Granites and alkali granite. (See Figures 4–5, 4–6, and 4–14A.) Potassium feldspar is usually orthoclase or microcline (both commonly cryptoperthitic or microperthitic). Plagioclase is usually oligoclase. Both types of feldspar are subhedral to anhedral. In porphyritic varieties, the phenocrysts are usually potassium feldspar. Quartz is usually anhedral and assumes shapes of surrounding grains; fluid inclusions are common. Muscovite and green or dark brown subhedral to euhedral plates of biotite are common; zircon inclusions often show pleochroic halos. Green to brown subhedral hornblende prisms are common. Sodium-rich amphiboles are common in alkali granite. Pyroxenes are rare, except for aegirine or aegirine–augite in alkali granites. Accessory minerals include apatite (tiny prisms and needles), iron ores, zircon, sphene, tourmaline, and occasional metamorphic minerals. Common alterations are biotite → chlorite, sphene; potassium feldspar → sericite, kaolinite; plagioclase → epidote, zoisite, sericite, kaolinite. The general texture is subhedral granular; other textures include quartz–feldspar graphic intergrowths, phenocrysts, myrmekite, and orbicular texture.

Syenite and alkali syenite. (See Figures 4–8 and 4–15A.) The principal alkali feldspar may be subhedral to tabular sanidine, orthoclase, microcline, or anorthoclase; microperthite or cryptoperthite is common. Plagioclase is generally subhedral with composition An_{20-40}. Other minerals include muscovite and brown or green biotite, subhedral prismatic green hornblende, augite or diopsidic augite, sodium-rich amphiboles and pyroxenes, and minor quartz or olivine. Quartz may be interstitial, micrographic, or myrmekitic. Accessory minerals include apatite, zircon, iron ores, sulfides, euhedral sphene, calcite, and zeolites. Common alterations are feldspar → kaolinite, sericite; biotite → chlorite, sphene; pyroxene and amphibole → chlorite, limonite, calcite. The typical texture is subhedral granular; syenites may be porphyritic and trachytic.

Monzonite. (See Figures 4–9A and 4–15B.) Plagioclase is usually subhedral to euhedral. Composition ranges from oligoclase to andesine. Lath shapes and zoning are common. The potassium feldspar is usually orthoclase (rarely microcline); zoning, poikilitic, and microperthitic textures are common. Other minerals include light brown biotite, green hornblende, and pale green augite. Minor quartz, hypersthene, or olivine may be present. Accessory minerals often include zircon, sphene, apatite, and iron ores. Alteration products include epidote, chlorite, sericite, kaolinite, calcite, and serpentine. The general texture is subhedral granular. Monzonite porphyries generally contain zoned phenocrysts of plagioclase, or less commonly orthoclase or perthite.

Diorite. (See Figures 4–11 and 4–16.) Plagioclase is commonly subhedral and zoned, with an average composition of An_{40-45}. Biotite is often present in brown, euhedral, somewhat poikilitic plates. Green or brown hornblende is subhedral or prismatic. Orthopyroxene or augite, when present, is often subhedral, prismatic, and poikilitic. Minor amounts of orthoclase, quartz, muscovite, or olivine may be present. Accessory minerals often include apatite, zircon, sphene, iron ores, sulfides, and occasionally metamorphic minerals. Common alteration reactions are pyroxene → biotite, amphibole; hornblende → biotite; biotite → chlorite, sphene; plagioclase → epidote, zoisite, kaolinite, sericite. The general texture is subhedral granular. Minerals may be porphyritic, poikilitic, and/or myrmekitic.

Gabbro. (See Figures 4–12 and 4–17.) Plagioclase varies in composition from An_{50} to An_{100} with an average of An_{65}; it is usually unzoned, anhedral, and equant. Occasionally it may be subhedral and tabular. Hypersthene is a common orthopyroxene; it is usually prismatic and shows exsolution of clinopyroxene on {100} or {001}. Augite is the most common pyroxene; it is anhedral to subhedral, and commonly shows twinning and exsolution. Subhedral magnesian olivine may be present. Hornblende, when present, is green or brown and anhedral. It is often poikilitic and rims pyroxene. Less abundant minerals are the micas, quartz, potassium feldspar, nepheline, or analcite. Accessory minerals include the iron ores, sulfides, apatite, green or brown spinel, sphene, rutile, and garnet. Common alterations are plagioclase → sericite, zoisite, epidote, calcite, albite, chlorite, actinolite, orthoclase; olivine → serpentine, talc, amphibole; orthopyroxene → amphibole, chlorite, talc, serpentine; biotite → chlorite, sphene, epidote, prehnite. The various textures are subhedral–granular, subophitic, poikilitic, perthitic, gneissoid, and banded.

Ultramafic rocks. (See Figures 4–13 and 4–18.) Olivine is usually magnesium-rich; it may be euhedral to anhedral; irregular fractures are common; occasionally cleavage or deformation twinning occurs. Pyroxenes may occasionally be relatively large and poikilitic; the common pyroxenes are enstatite and colorless Al-rich diopside; chrome diopside is pale green. The amphibole is usually green or brown hornblende, but sodium-rich types may be present. Pale brown phlogopite and pyropic garnet are common. The accessory minerals include several varieties of spinel, apatite, sulfides, calcic plagioclase, and iron ores, including chromite. Minerals produced by alteration include serpentine, brucite, magnesite, talc, magnetite, various amphiboles and pyroxenes, and chlorite. The common texture is anhedral–granular. The rocks may be cataclastic or schistose and show large-scale layering.

Foid syenite. (See Figure 4–8D.) Alkali feldspar may be orthoclase, sanidine, microcline, or anorthoclase (probably as microperthite or cryptoperthite), with albite. Alternatively, albite may occur with nonperthitic grains of orthoclase or microcline. Both are usually subhedral, but may have a tabular habit. Nepheline varies from anhedral to stubby euhedral prisms, and may be poikilitic. Sodalite is usually present as euhedral dodecahedra, but may be anhedral and interstitial. Sodium-rich, subhedral, prismatic or interstitial amphibole and pyroxene are common, along with lesser analcite, green to brown biotite, iron-rich olivine, and calcite. Accessory minerals include zircon, apatite, sphene, iron ores, sulfides, muscovite, and fluorite. Common alteration reactions include biotite → muscovite; nepheline → cancrinite, muscovite, sodalite, kaolinite, analcite, natrolite, and calcite; sodalite → analcite, cancrinite, calcite; potassium feldspar → sericite, kaolinite, sodalite, analcite, cancrinite, nepheline, calcite. The general texture is subhedral granular, and may show cataclastic, fluidal, or trachytoidal textures. Phenocrysts may be common.

Volcanic Types

Siliceous volcanics. (See Figures 4–7 and 4–14C.) Glass is abundant. Rhyolitic types of glasses have refractive indices of about 1.49 and are reddish to dark green. Alkali feldspars are generally phenocrystic and lath-shaped and are sanidine, orthoclase, or rarely anorthoclase. The common plagioclase is oligoclase or andesine. Silica minerals may be quartz, cristobalite, or tridymite. These may be intergrown with feldspar, in phenocrysts, or in cavity fillings. Quartz is usually bipyramidal and embayed. Biotite phenocrysts are common as euhedral tablets. Other minerals include augite, hypersthene, iron-rich olivine, or hornblende. Sodium-rich amphiboles and pyroxenes are common in alkali varieties. Accessory minerals include apatite, zircon, iron ores, topaz, and fluorite. Alteration reactions are devitrification, silicification, potassium feldspar → albite; feldspars → kaolinite, sericite, montmorillonite; ferromagnesian minerals → chlorite, epidote, sericite, calcite. The general texture varies from holohyaline to holocrystalline. Rocks may be vesicular, cryptocrystalline, microcrystalline, micropoikilitic, microspherulitic, or porphyritic. Perlitic cracks, crystallites, microlites, and flow structure may also be present.

Trachyte. (See Figure 4–14B.) The potassium feldspar is usually sanidine, and less commonly orthoclase or anorthoclase. It may be perthitic and zoned, or occur as phenocrysts; they vary from large tabular laths to microlites. Plagioclase is usually oligoclase, but may be albite in alkali trachytes; plagioclase varies in size from phenocrysts to microlites. Rarely, some glass may be present. Biotite is often present as euhedral tablets. Fairly common minerals are hornblende, quartz, tridymite, or iron-rich olivine, or sodium-rich amphiboles and pyroxenes in alkali trachytes. Common accessory minerals are apatite, zircon, iron ores, and sphene. Alteration reactions noted are potassium feldspar → albite; biotite → iron ores; amphiboles → chlorite, carbonates, and iron ores. Textures vary from holocrystalline to hypocrystalline. Phenocrysts are common, as well as trachytic texture.

Latite. (See Figures 4–9B and 4–15D.) Plagioclase phenocrysts are commonly euhedral and zoned and have an average composition of andesine. Groundmass plagioclase laths and microlites are oligoclase to andesine. Potassium feldspars, sanidine, orthoclase, or anorthoclase, are usually present as groundmass laths and microlites. Glass and biotite phenocrysts are common. Minor amphibole may be present as hornblende or oxyhornblende. Pyroxene may be present as diopsidic augite, or less commonly as aegirine–augite, or hypersthene. Olivine is uncommon. Accessory minerals are quartz or tridymite (when less than 5% is present), iron ores, apatite, and zircon. Alteration minerals include chlorite, sericite, calcite, quartz, and limonite. The texture is usually porphyritic, with a holocrystalline to hypocrystalline matrix.

Andesite. (See Figures 4–10, 4–15C, and 4–16.) Plagioclase varies in composition from anorthite to oligoclase, but averages about An_{40}. Highly zoned phenocrysts are common. Plagioclase in the groundmass as laths or microlites is generally more sodium-rich than phenocrysts. Pyroxenes consist of diopsidic augite, augite, pigeonite, or hypersthene; these are present as phenocrysts and in groundmass, and may be zoned. Hornblende or oxyhornblende may be present as prismatic phenocrysts. Less common phases are glass, magnesium-rich olivine, and biotite. Accessory minerals include apatite, zircon, iron ores, hornblende, biotite, and garnet. Alteration reactions are plagioclase → albite, epidote, sericite; hypersthene → serpentine, chlorite, calcite, epidote; clinopyroxene → chlorite, carbonate, epidote, serpentine, and iron ores; biotite, hornblende → chlorite, sphene, calcite, iron ore; olivine → serpentine, chlorite, epidote, calcite. The texture is generally porphyritic, and the groundmass is commonly holocrystalline. Andesites are often vesicular, amygdaloidal, trachytic, and ophitic or subophitic.

Basalt. (See Figures 4–12 and 4–17.) The principal pyroxene is generally augite or diopsidic augite, in phenocrysts and groundmass. Euhedral to subhedral pigeonite and/or hypersthene are more common in the groundmass. Olivine (euhedral to anhedral) is often present and occasionally zoned; it may be present in phenocrysts and/or groundmass. Plagioclase composition varies from bytownite to labradorite, with An_{55} as an average. Plagioclase in phenocrysts and groundmass is euhedral to subhedral, often as laths, and is often zoned. Minor phases may include glass, alkali feldspar, cristobalite or tridymite, or nepheline, and perhaps analcite. Accessory minerals are usually iron ores and apatite. Alteration reactions are olivine → iddingsite, serpentine, chlorite; pyroxene → serpentine, chlorite, actinolite, carbonate; calcium-rich plagioclase → albite, chlorite, kaolinite; iron ores → sphene; glass → chlorite, palagonite, montmorillonite. The texture may be holocrystalline to hypocrystalline, porphyritic, intersertal, vesicular, and ophitic to subophitic.

Phonolite. The alkali feldspar is sanidine, anorthoclase, or orthoclase (often microperthitic), as phenocrysts, laths, or microlites, and is occasionally zoned. Nepheline is usually present as euhedral grains in the groundmass, and rarely as phenocrysts. Plagioclase (albite) is rare. Other minerals present may include sodalite, haüyne, analcite, iron-rich olivine, sodium-rich amphibole or pyroxene, biotite, leucite, and phlogopite. Accessory minerals are sphene (common), apatite, corundum, zircon, and iron ores. The principal alteration is nepheline → analcite, zeolites, leucite, kaolinite, nosean, haüyne, or sodalite. Most phonolites are porphyritic–holocrystalline. The groundmass may be trachytic or granular, rarely glassy or vesicular.

Examples of the Common Igneous Rock Types

Figures 4–4 through 4–13 show hand specimens, and Figures 4–14 through 4–18 are photomicrographs. Figure 4–4 shows rocks named mainly on the basis of texture; the other figures are arranged in a sequence of increasing mafic content. It should be understood that there is no substitute for examining a suite of similar rocks by means of thin sections.

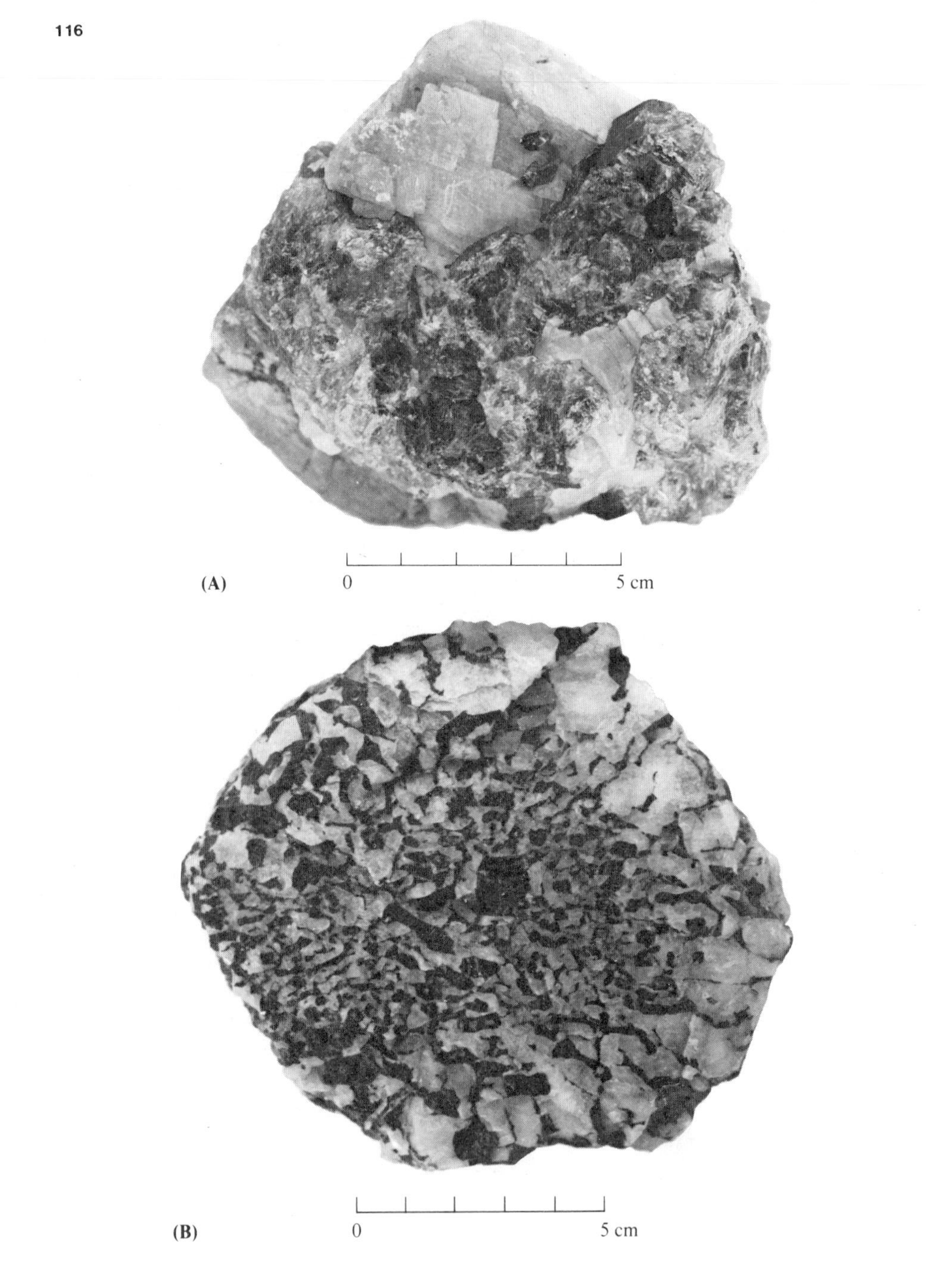

Figure 4–4
Rocks that are named mainly on the basis of texture. (**A**) Pegmatite. Pegmatites are named on the basis of their coarse texture. The typical composition is similar to granite. This specimen consists of crystals of microcline (light crystals showing cleavage) with darker quartz and mica. (**B**) A pegmatite, consisting mainly of quartz (light) intergrown with amphibole (dark). The amphibole can be identified in hand specimen by the characteristic cleavage intersections at 56° and 124°. Graphic texture is dominant in this specimen.

(**C**) Obsidian, a dark-colored volcanic glass. The glassy nature of the specimen can be determined by the vitreous luster and conchoidal fracture. This specimen has a weathered front surface (with a somewhat dull appearance); the weathering has emphasized the well-developed flow structure. See also Figure 2–9A. (**D**) Vitrophyre. The name *vitrophyre* is given to a rock having quartz and potassium feldspar (lighter areas) in a distinctly glassy matrix. The color of the glass is usually quite dark. The composition of the rock is granitic.

Figure 4–4 *(continued)*
(**E**) Scoria. This is an extrusive highly vesicular rock having the appearance of a clinker. Scoriaceous texture is usually found in ejected volcanic fragments, or within crusts of lava flows. (**F**) Perlite. This is a glassy volcanic rock that is silica-rich. It is named on the basis of perlitic texture (irregular and spheroidal cracks formed during cooling). See also Figure 2–8.

Figure 4–5
A variety of granites. (**A**) Polished surface of a granite from West Chelmsford, Massachusetts. The darker areas are mainly quartz. Quartz, being relatively clear and colorless, appears darker than the light-colored feldspars. The light-colored areas consist of both pink potassium feldspar and light gray plagioclase. The relative amounts of these can be easily distinguished in hand specimen on the basis of color. (**B**) Biotite granite. The black areas are biotite, easily distinguished by reflections from cleavage surfaces. The whiter areas are mainly potassium feldspars, and the gray areas are mainly quartz. The feldspar can be determined by the presence of two perpendicular cleavage surfaces, in contrast to the quartz, which shows conchoidal fracture.

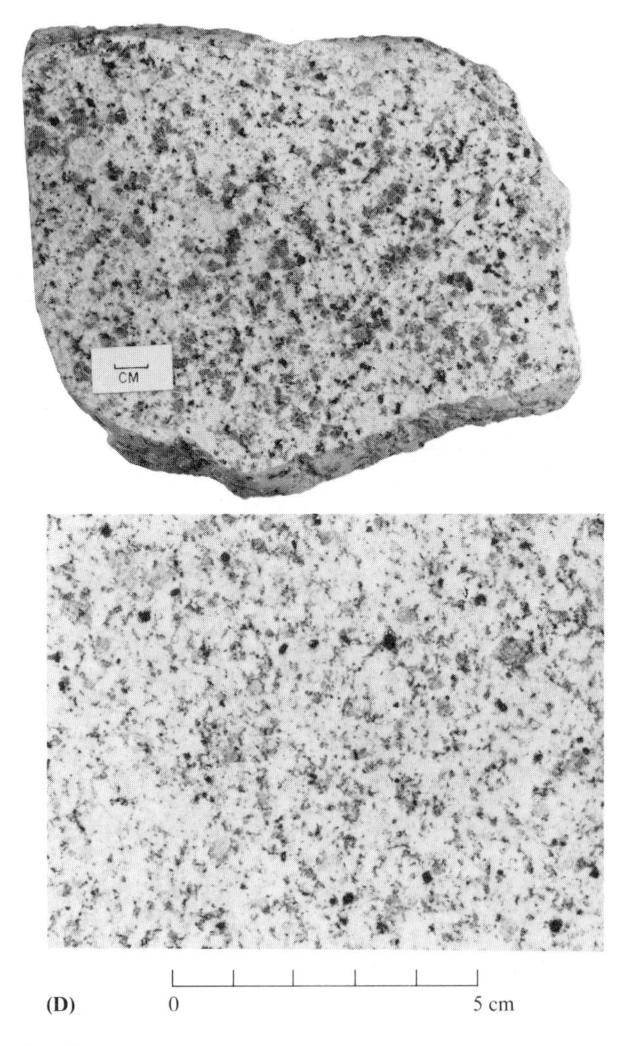

Figure 4–5 *(continued)*
(**C**) A somewhat finer-grained granite than the earlier examples. The quartz, because it is more transparent than feldspar, is seen as gray areas. The black specks are biotite, and the lighter areas are mainly potassium feldspar. [From A. Hietanen, 1963, U.S. Geol. Surv. Prof. Paper 344–D, Fig. 14–A. (**D**) Polished surface of a granite. White areas are mainly potassium feldspar, gray is quartz, and dark areas are minor ferromagnesian minerals. Note that in spite of the particular color shown by each mineral, the degree of color intensity is greater for ferromagnesian minerals than for quartz and feldspar.

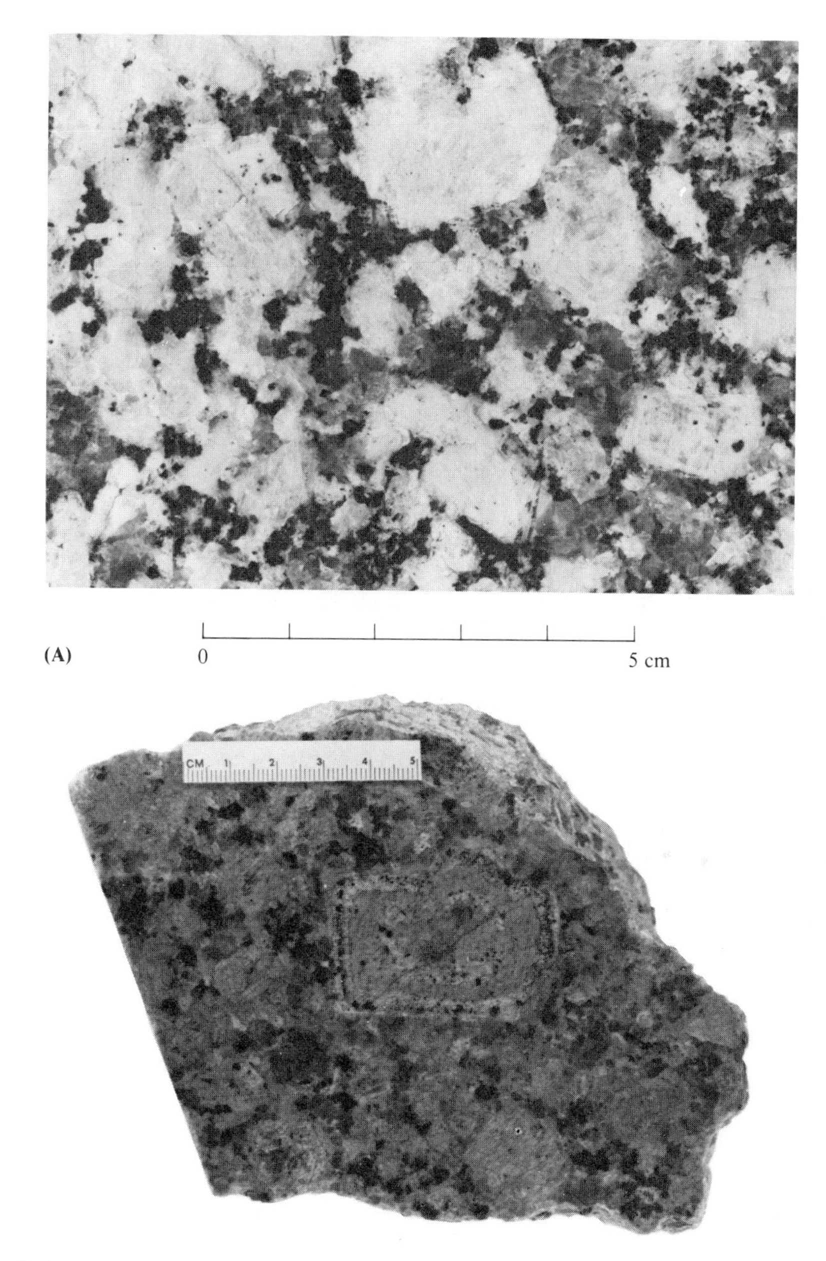

Figure 4–6
Some textures shown by granitic rocks. (**A**) Polished slab of granite porphyry. The majority of the rock consists of large phenocrysts of potassium feldspar (white) in a finer matrix of quartz (gray) and ferromagnesian minerals (black). Some compositional zoning is evident in the phenocrysts as slight differences in color from center to edge of grains. (**B**) A coarse-grained granite with large (gray) grains of potassium feldspar and smaller grains of quartz (dark gray) and ferromagnesian minerals (black). A large potassium feldspar phenocryst shows zoning and incorporation of ferromagnesian minerals.

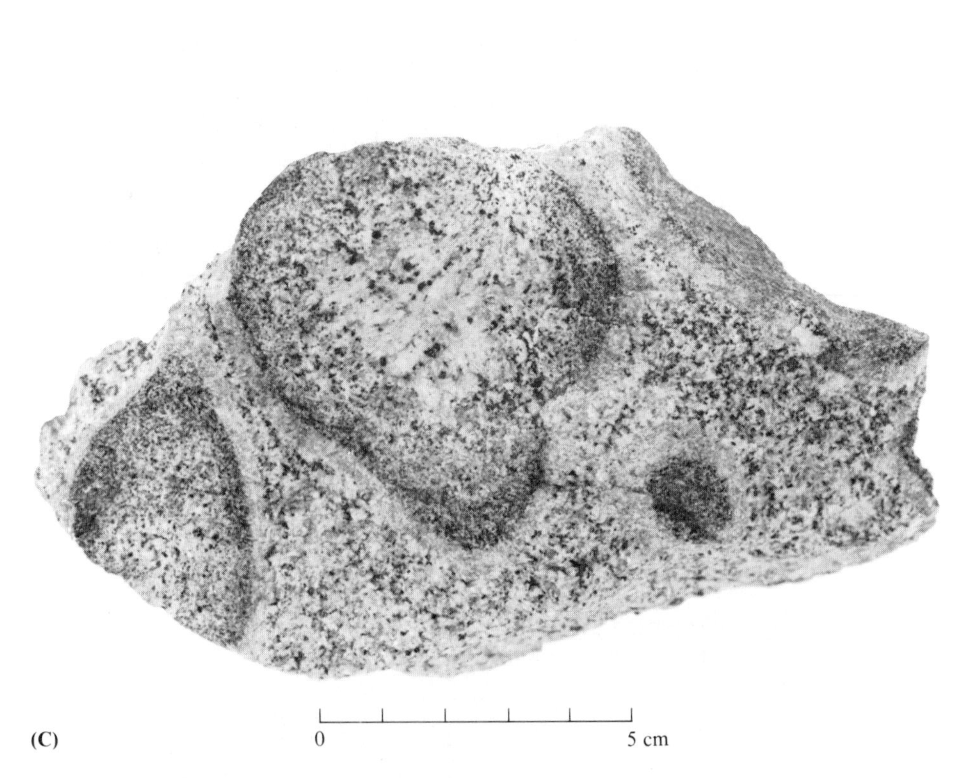

Figure 4–6 (*continued*)
(C) Orbicular texture in granite. The rounded masses consist mainly of biotite and plagioclase, whereas the granite is composed of quartz, potassium feldspar, and biotite. The origin of orbicular texture is somewhat questionable, but it is generally considered to result from incomplete assimilation of xenoliths.

Figure 4–7
(**A**) Rhyolite. This is a light-colored aphanitic rock of granitic composition. The composition can only be guessed at in hand specimen (and often with thin section examination as well). The volcanic character of this rock is verified by a small number of vesicles that are present, as well as by the presence of flow structure. (**B**) Rhyolite prophyry. This light-colored aphanitic rock can be identified as a rhyolite on the basis of the phenocrysts present. The phenocrysts consist of quartz and potassium feldspar (identified as such by the absence of striations on cleavage surfaces). The somewhat darkened area on the right side of the specimen results from slight weathering.

Figure 4–8
A variety of syenites. (**A**) Hornblende syenite. The darker grains are hornblende, and the lighter grains are mainly potassium feldspar with minor plagioclase. As is usual the relative amounts of the feldspars are determined on the basis of slight color differences and the presence or absence of striations. The absence of quartz eliminates the possibility that this rock could be a granite. (**B**) A syenite that consists mainly of potassium feldspar (light) and minor biotite and hornblende (dark). The large dark xenoliths are black and aphanitic, and are identified as basalt fragments. (**C**) The polished surface of a porphyritic syenite. Large

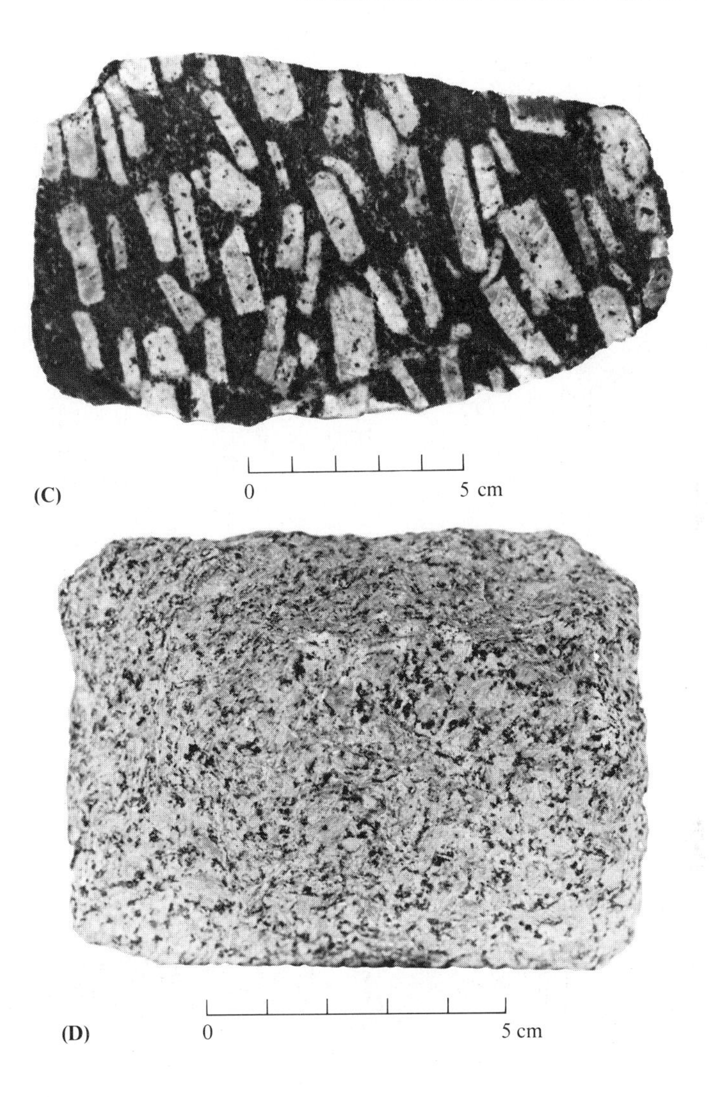

phenocrysts of orthoclase show a definite flow structure. The fine-grained matrix is composed mainly of orthoclase and biotite. **(D)** Foid (nepheline) syenite. The dark grains are biotite and pyroxene, and most of the lighter areas are mainly potassium feldspar. Nepheline, present among the feldspar grains, is very difficult to detect macroscopically. The presence of nepheline can often be surmised by the presence of trace amounts of sodalite (blue) or cancrinite (yellow). Microscopic identification of nepheline is preferred.

Figure 4–9
(**A**) Quartz-bearing monzonite prophyry. The large light-colored phenocrysts are composed of both potassium feldspar and plagioclase. The plagioclase is distinguished by a slightly lighter color and striations on some cleavage surfaces. The matrix is a mixture of feldspars with minor amphibole. (**B**) Latite, obtained in a dike in Salt Lake County, Utah. The euhedral to subhedral phenocrysts are composed of an equal mixture of potassium feldspar and plagioclase. The mineralogy of the aphanitic groundmass cannot be determined in hand specimen. [From W. J. Moore, 1973, U.S. Geol. Surv. Prof. Paper 629–B, Fig. 27–A.]

Figure 4–10
(**A**) Andesite porphyry. The groundmass cannot be determined in hand specimen. The anhedral phenocrysts are composed of plagioclase and minor mafic phases. The matrix, which is dark green rather than black, would favor calling this rock an andesite rather than a basalt. (**B**) Hornblende andesite. The darker phenocrysts can be determined as hornblende rather than pyroxene, based on the cleavage angle intersections. This, combined with a gray-green color of the matrix, indicates that this rock should be called an andesite. The dark area on the right is a weathered surface.

Figure 4–11
(**A**) Quartz diorite. The dark minerals, hornblende and biotite, show a parallel orientation. The light-colored constituents are quartz and plagioclase. Clearwater County, Idaho. [From A. Hietanen, 1963, U.S. Geol. Surv. Prof. Paper 433–D, Fig. 5–A. (**B**) Hornblende biotite quartz diorite from the Bucks Lake pluton, Plumas County, California. Dark minerals are hornblende and biotite; light-colored minerals are plagioclase and quartz. [From A. Hietanen, 1973, U.S. Geol. Surv. Prof. Paper 731, Fig. 27.]

Figure 4–12
Rocks of basaltic composition. (**A**) Hypersthene gabbro. Light areas are plagioclase. Dark areas are augite and hypersthene. Pyroxenes are generally determined on the basis of their usual dark color and intersecting cleavage angles of essentially 90°. (**B**) Vesicular basalt. The groundmass is black and aphanitic. Cavities are partially or completely filled with epidote.

Figure 4–12 (*continued*)
C) Diabase porphyry. Diabase is a medium-grained rock having the composition of basalt, with an ophitic texture. The large light-colored phenocrysts in this black rock are composed of calcium-rich plagioclase. (**D**) Fragments of basalt showing the typical type of columnar semiparallel jointing that develops during cooling. This jointing, which usually occurs on a large scale, is seen also in Figure 1–5.

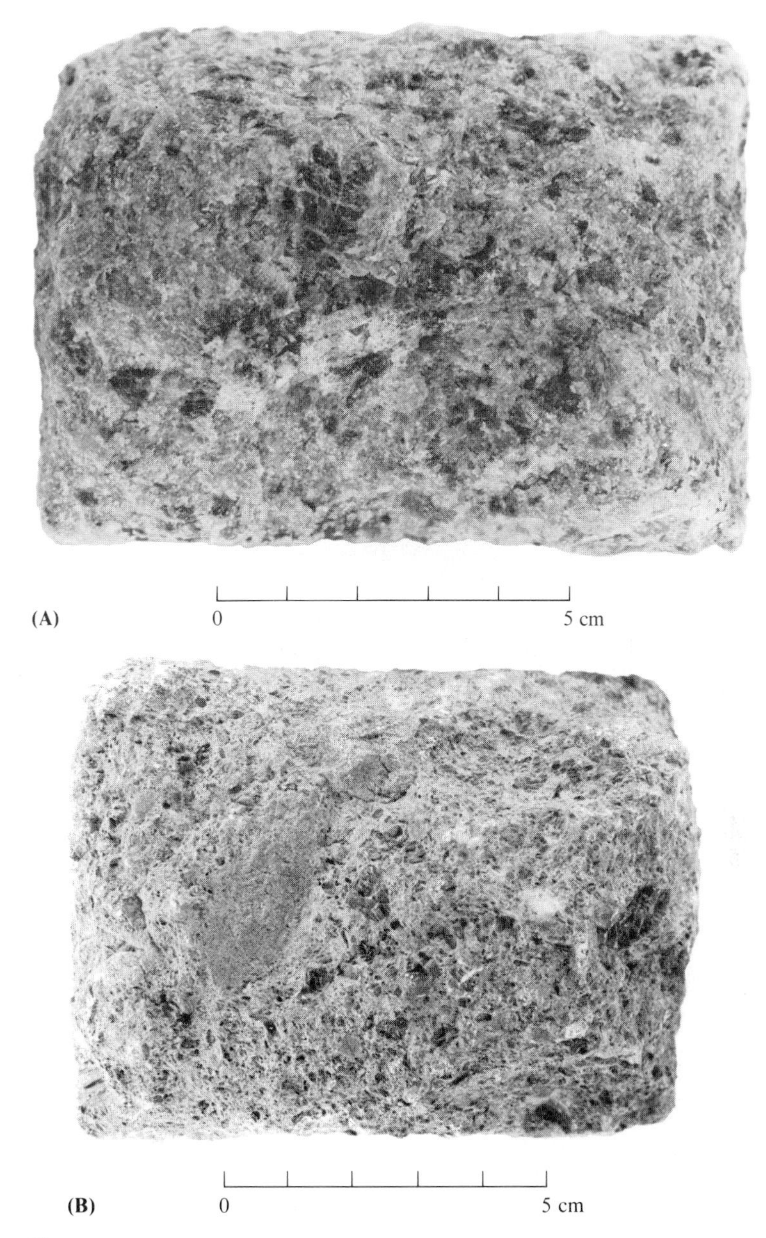

Figure 4–13
Ultrabasic and ultramafic rocks. (**A**) Anorthosite. Both light and dark-colored areas consist of pale green crystals of calcium-rich plagioclase in various orientations. Essentially no mafic minerals are present. (**B**) Kimberlite. This is an altered prophyritic peridotite containing olivine and phlogopite. Most of the darker areas are inclusions of country rock acquired during the ascent; these can be of any composition and usually show only minor thermal effects.

Figure 4–13 (*continued*)
(**C**) Hornblende-bearing pyroxenite. The dark masses are hornblende, and the matrix consists of phaneritic pyroxene of a dark green color. Fine-grained varieties of ultramafic rocks are extremely rare. [From A. Hietanen, 1963, U.S. Geol. Surv. Prof. Paper 344–D, Fig. 4–A. (**D**) Pyroxenite. The rock is almost entirely augite.

(A)

Figure 4–14
Silica-rich rocks. (**A**) Hornblende granite. The larger masses consist of perthite. Examination with plane-polarized light would reveal (by Becke tests) that this consists of one type of feldspar within another. Clear quartz (Qtz) grains are intergrown with and included within the micro-perthite. A few (almost opaque) grains of hornblende (Hbd) are indicated in the lower right of the photograph. Crossed nicols. Width of photo is about 5.3 mm. Dover, New Jersey. [From P. K. Sims, 1958, U.S. Geol. Surv. Prof. Paper 287, Plate 16–B.]

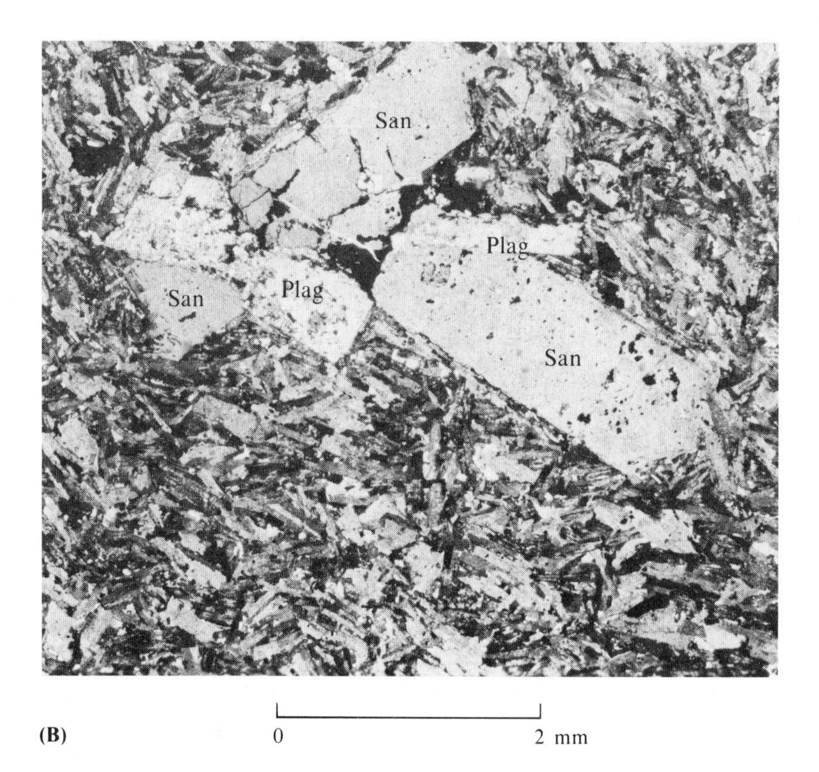

Figure 4–14 (*continued*)
(**B**) A sanidinite trachyte porphyry. The larger crystals are mainly sanidine (San). The common Carlsbad twinning is not present in this photograph. The two lighter phenocrysts are sericite pseudomorphs, probably after plagioclase (Plag). The groundmass consists primarily of lath-shaped sanidine, with minor quartz. Crossed nicols. Gilpin County, Colorado. [From W. A. Braddock, 1969, U.S. Geol. Surv. Prof. Paper 616, Fig. 15–F.]

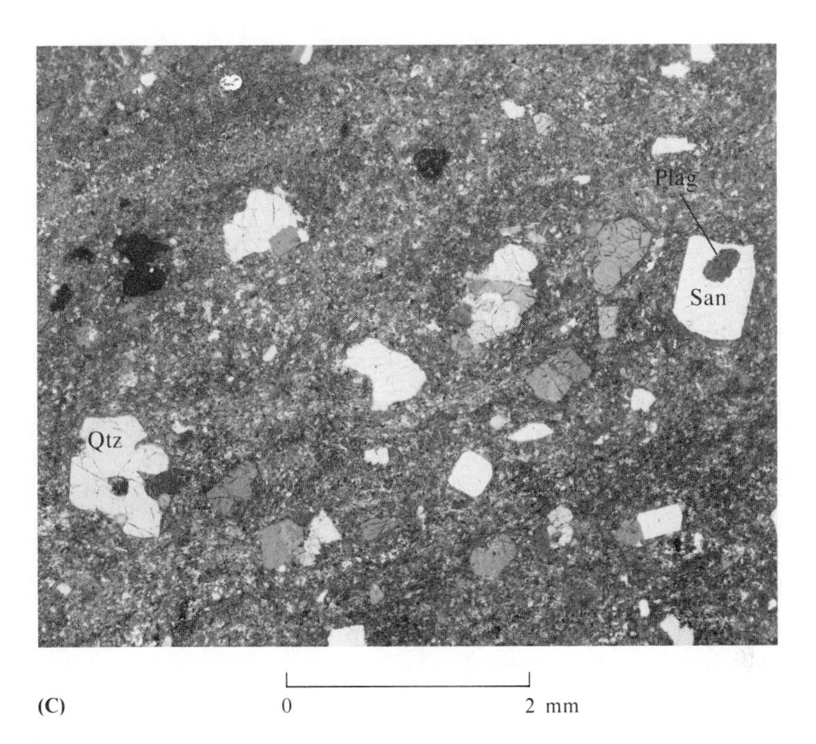

(C)

(C) Rhyolite. Phenocrysts consist of embayed quartz, sodic sanidine (Or_{40}), and sparse plagioclase, biotite, and Fe–Ti oxides. In thin section the quartz can be distinguished by lack of cleavage and uniaxial positive interference figure. Feldspars will usually show at least one good direction of cleavage, and are commonly twinned. The groundmass is microcrystalline and devitrified. Weak flow layering is present. Rio Grande County, Colorado. Crossed nicols. [From P. W. Lipman, 1975, U.S. Geol. Surv. Prof. Paper 852, Fig. 54–B.]

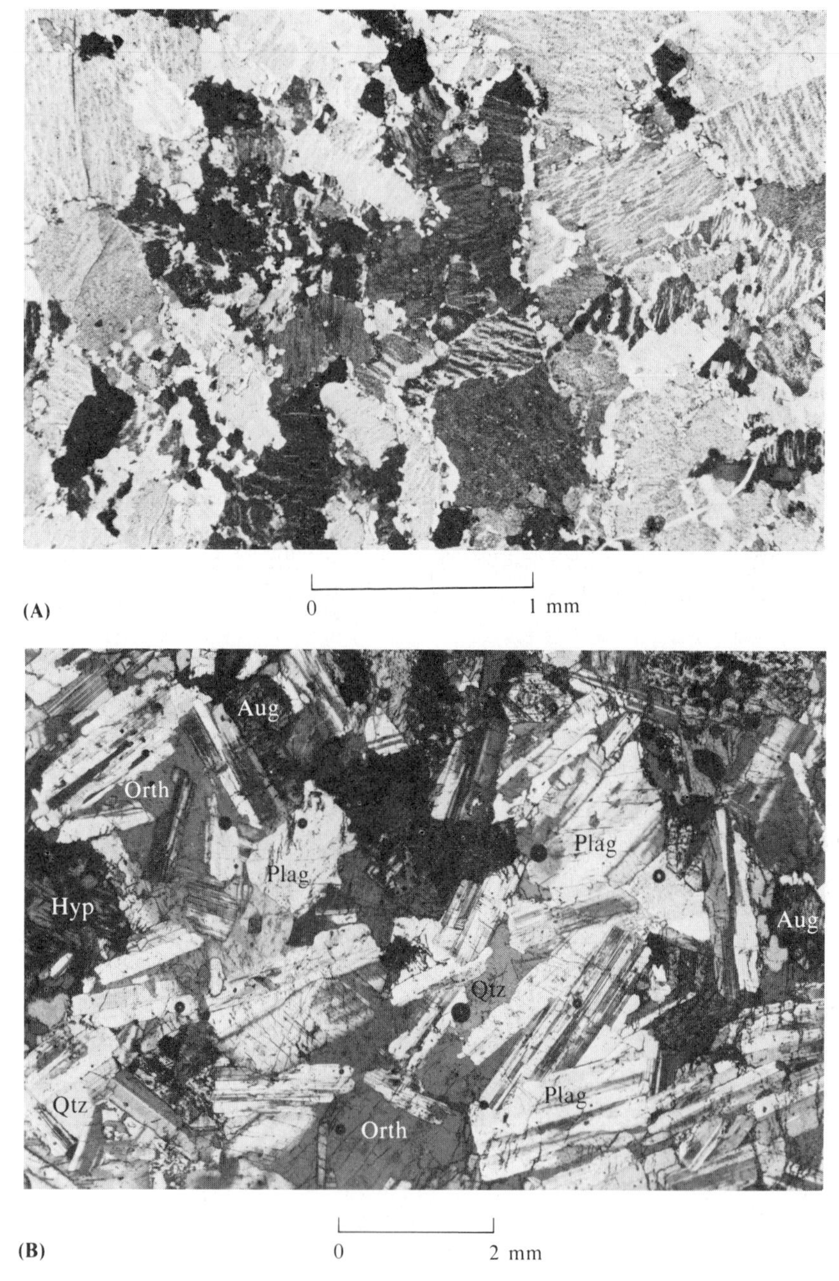

Figure 4–15
Rocks of intermediate silica content. (**A**) Syenite. The rock consists almost entirely of large grains of perthite, which are separated by rims of granular albite (white). Crossed nicols. Gilpin County, Colorado. [From W. A. Braddock, 1969, U.S. Geol. Surv. Prof. Paper 616, Fig. 15–D.] (**B**) A very coarse equigranular monzonite. Euhedral laths of plagioclase (Plag; about An_{40}) and darker augite (Aug) and hypersthene (Hyp) (both fringed by opaque alteration products) are surrounded by large orthoclase (Orth), quartz (Qtz), and biotite grains as large as 5 mm in diameter. Archuleta County, Colorado. [From P. W. Lipman, 1975, U.S. Geol. Surv. Prof. Paper 852, Fig. 49–A.]

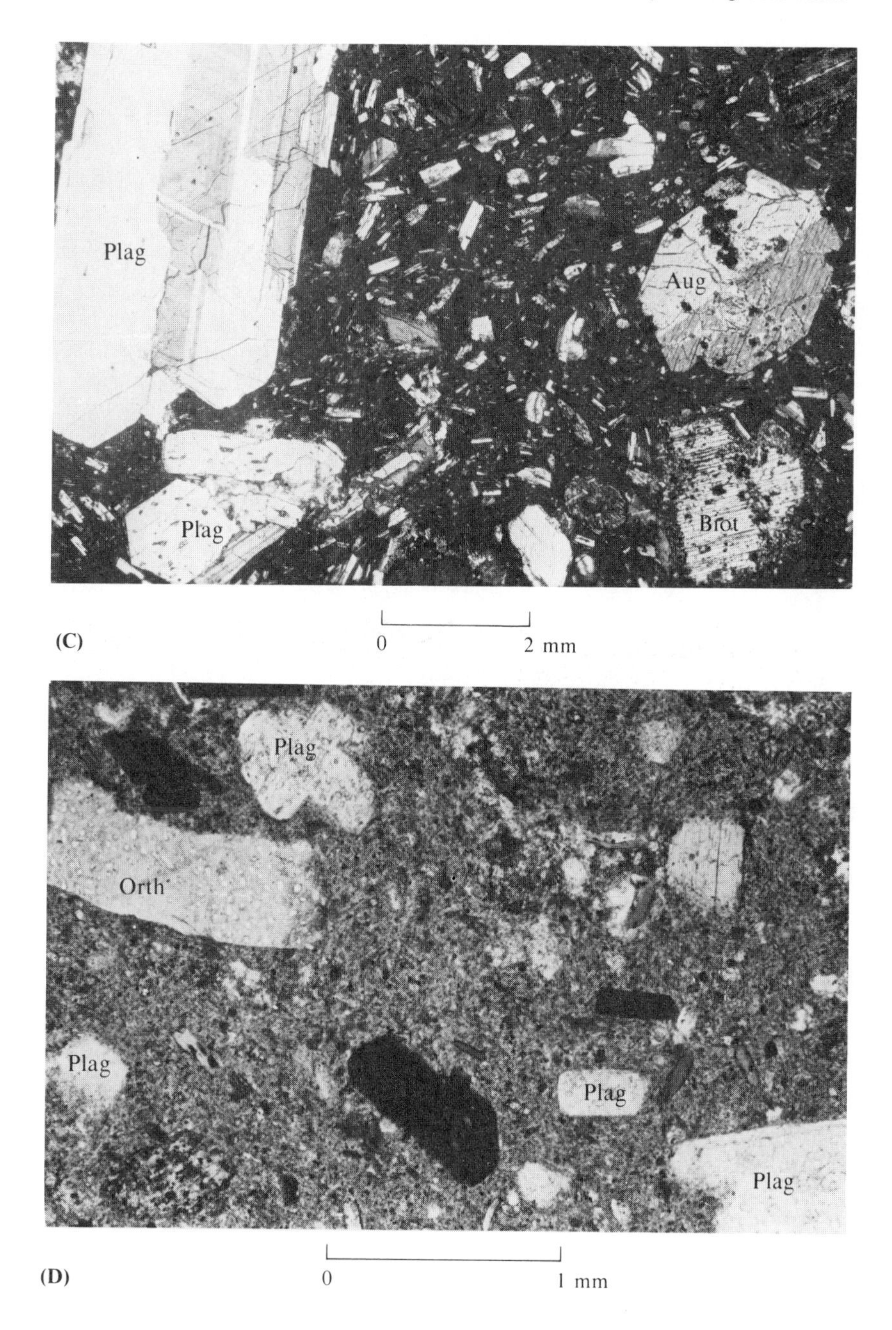

(C) Dacite. The large phenocryst at the left is plagioclase (An_{35-40}). At the right are phenocrysts of augite (Aug) and biotite (Biot) that have oxidized margins. The matrix contains microphenocrysts of the same minerals, within a cryptocrystalline groundmass. Rio Grande County, Colorado. Crossed nicols. [From P. W. Lipman, 1975, U.S. Geol. Surv. Prof. Paper 852, Fig. 43–A.] **(D)** Latite dike rock. Subhedral plagioclase, orthoclase, and biotite (black) phenocrysts in an aphanitic groundmass. Plane-polarized light. Salt Lake County, Utah. [From W. J. Moore, 1973, U.S. Geol. Surv. Prof. Paper 629–B. Fig. 28–A.]

Figure 4–16
Rocks of intermediate silica content (andesites and diorite group). (**A**) Porphyritic quartz diorite. The groundmass consists of quartz, plagioclase, biotite, and hornblende. The phenocrysts consist of plagioclase. The phenocryst at the lower left has been broken, and biotite has formed along the fracture. Crossed nicols. Clear Creek County, Colorado. [From W. A. Braddock, 1969, U.S. Geol. Surv. Prof. Paper 616, Fig. 9–C. (**B**) Diorite. The large irregularly shaped grains are clinopyroxene. These grains have been partially altered to hornblende (as indicated by dark rims), a process known as uralitization. The lighter areas are plagioclase. Mt. Hope, Maryland. Partially crossed nicols. Width of photo is about 2.3 mm. (**C**) Andesite containing abundant phenocrysts of strongly zoned plagioclase, obscure augite and hypersthene, and serpentine pseudomorphs after olivine. The groundmass, averaging about 0.1 mm in diameter, consists mainly of plagioclase, pyroxene, and Fe–Ti oxides. Conejos County, Colorado. Crossed nicols. [From P. W. Lipman, 1975, U.S. Geol. Surv. Prof. Paper 852, Fig. 38–B.]

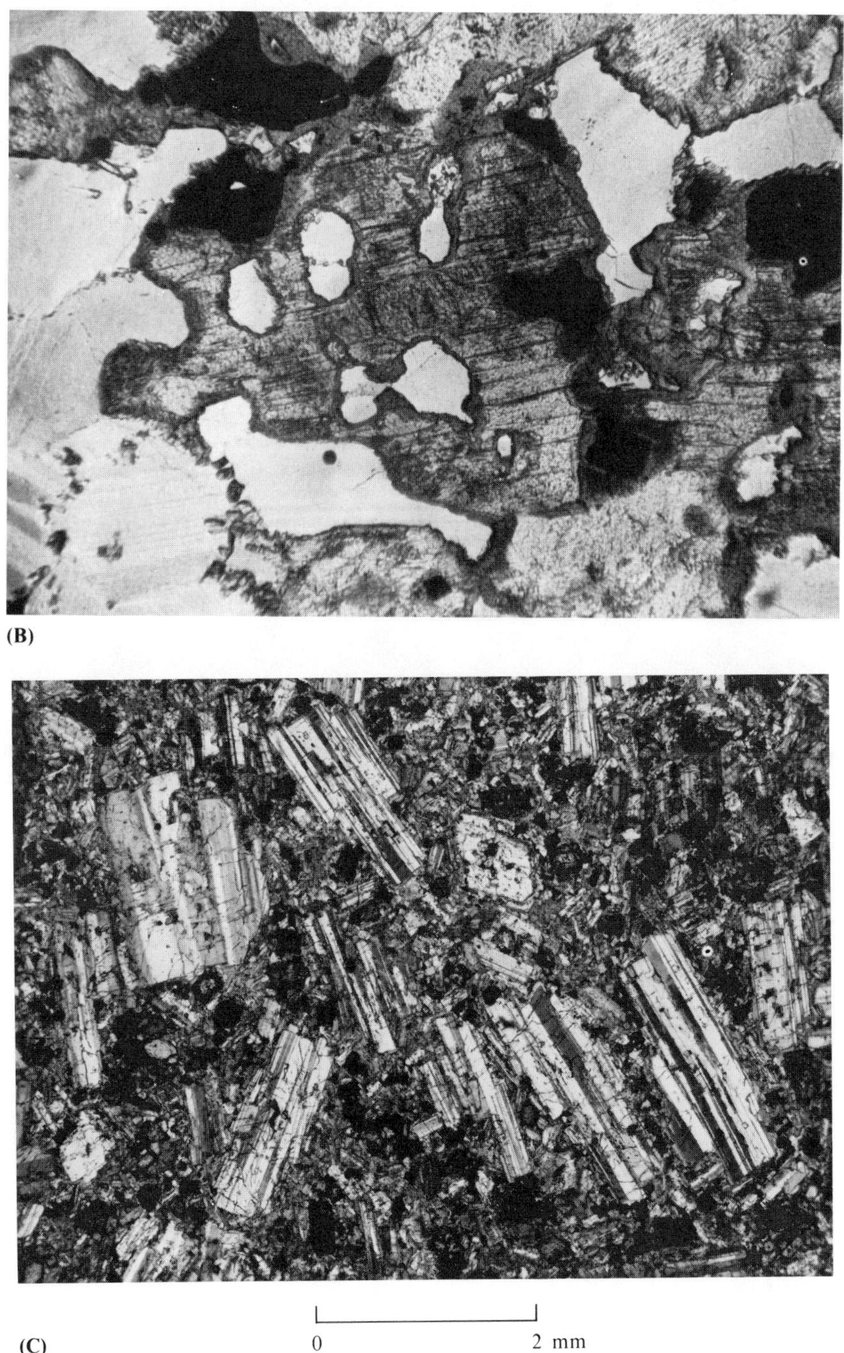

(B)

(C)

0 2 mm

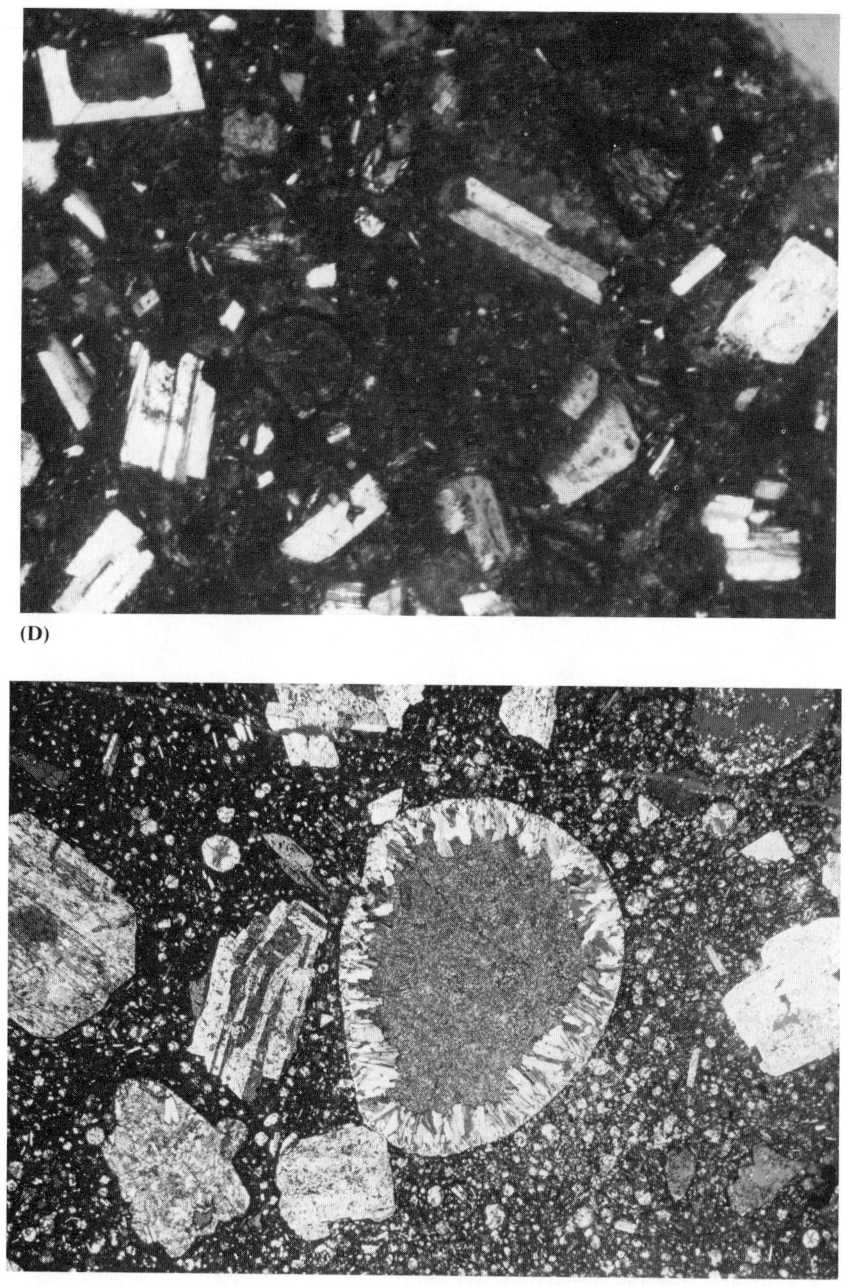

(D)

(E)

Figure 4–16 *(continued)*
(D) A typical andesite. The phenocrysts consist of (light) zoned and unzoned plagioclase feldspars, and (dark) hornblende, which is rimmed by essentially opaque iron oxides. This rim, known as opacite, is caused by dehydration and oxidation of hornblende or biotite upon extrusion. Crossed nicols. Width of photo is about 2.3 mm. **(E)** Andesite. The central feature is an amygdule lined with crystals of secondary albite, and filled with barite and chlorite. The phenocrysts consist of somewhat altered plagioclase. The groundmass is cryptocrystalline. Crossed nicols. Width of photo is about 6.7 mm. Shasta County, California. [U.S. Geological Survey photo by J. P. Albers.]

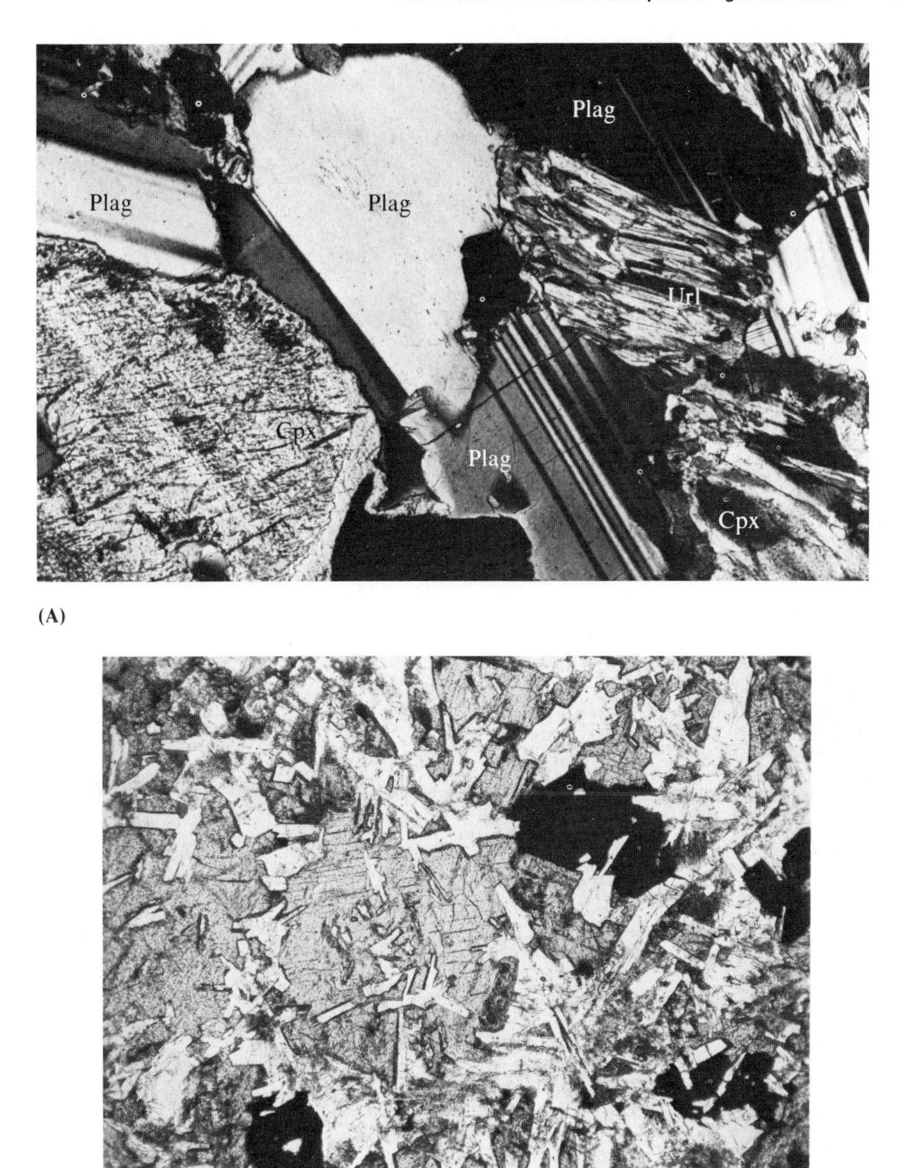

Figure 4–17
Rocks of basaltic composition. (**A**) Gabbro. The large twinned and clear grains are calcium-rich plagioclase (Plag). The grain at the lower left is clinopyroxene (Cpx) with a rim of hornblende. The other fibrous-looking grains are composed of hornblende (Url) that has formed by alteration of pyroxene (uralitization). Crossed nicols. Width of photo is about 1.2 mm. (**B**) Gabbro. The white grains are laths of labradorite. The larger gray crystals are pale brown augite. Black grains are magnetite, and cloudy areas are fine-grained chlorite and sericite. Picture area is 3 cm long. Plane-polarized light. [From W. B. Hamilton, 1956, *Geol. Soc. Amer. Bull., 67,* Plate 2.]

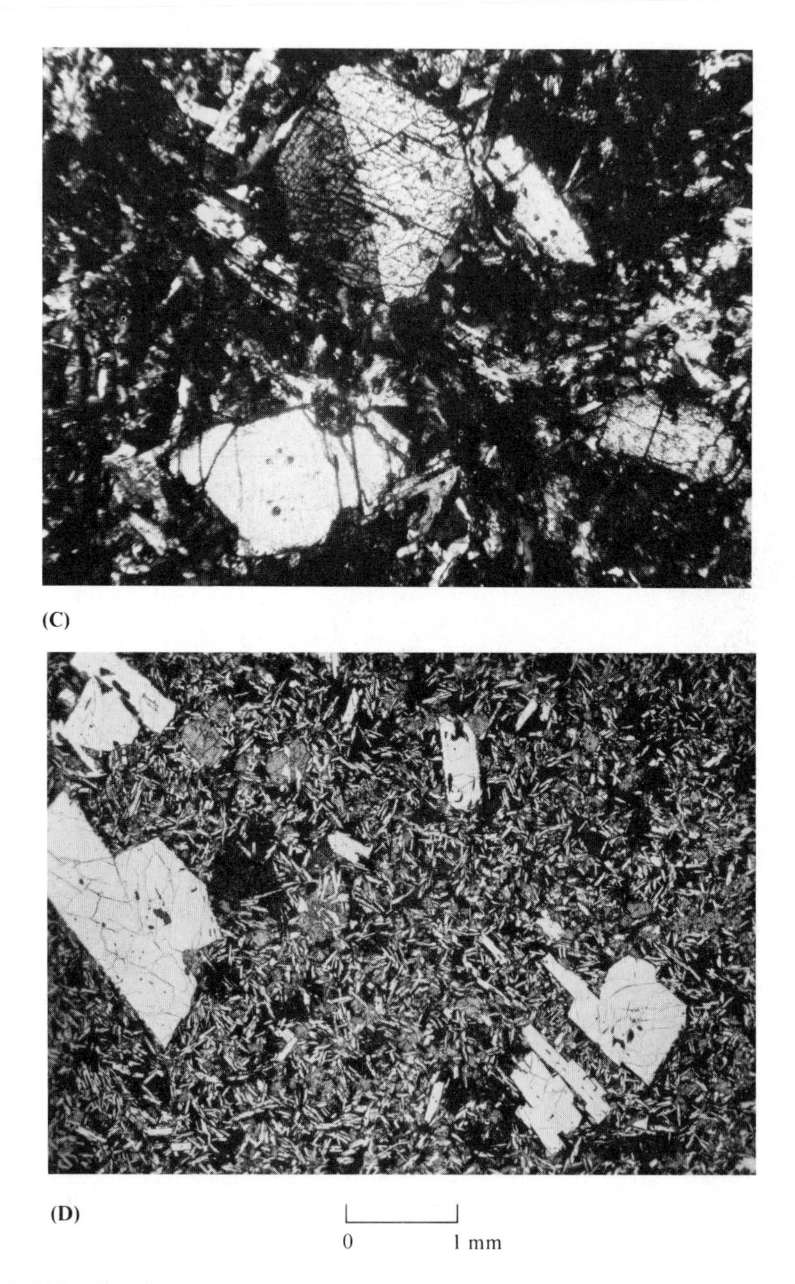

Figure 4–17 *(continued)*
(C) Basalt. Phenocrysts are euhedral pyroxene (upper center) and subhedral olivine (lower center). The matrix consists mainly of plagioclase laths (lighter), pyroxene (darker), and disseminated opaque oxides. Crossed nicols. Width of photo is about 2.3 mm. **(D)** Porphyritic basalt. The phenocrysts consist of plagioclase. The groundmass contains microlites of plagioclase enclosed in (gray) pyroxene, with minor opaque oxides. Power County, Idaho. Plane-polarized light. [From W. B. Hamilton, 1964, *U.S. Geol. Surv. Bull.*, 1141–L, Fig. 8–A.]

(A)

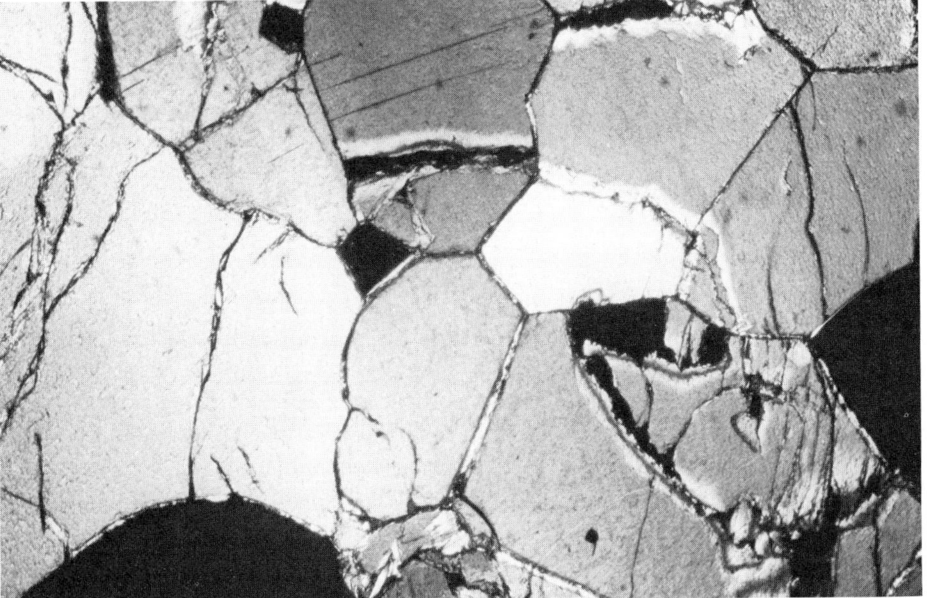

(B)

Figure 4–18
Ultrabasic rocks. (**A**) Peridotite. The rock consists mostly of olivine (rounded grains in upper left and right showing no cleavage) and clinopyroxene (showing cleavage). Crossed nicols. Width of photo is about 2.3 mm. (**B**) Dunite. All of the grains are magnesium-rich olivine. Note the common intersection of three grains forming triple point junctions. The black areas are the result of plucking out of mineral grains during thin section preparation. Crossed nicols. Width of photo is about 3.0 mm.

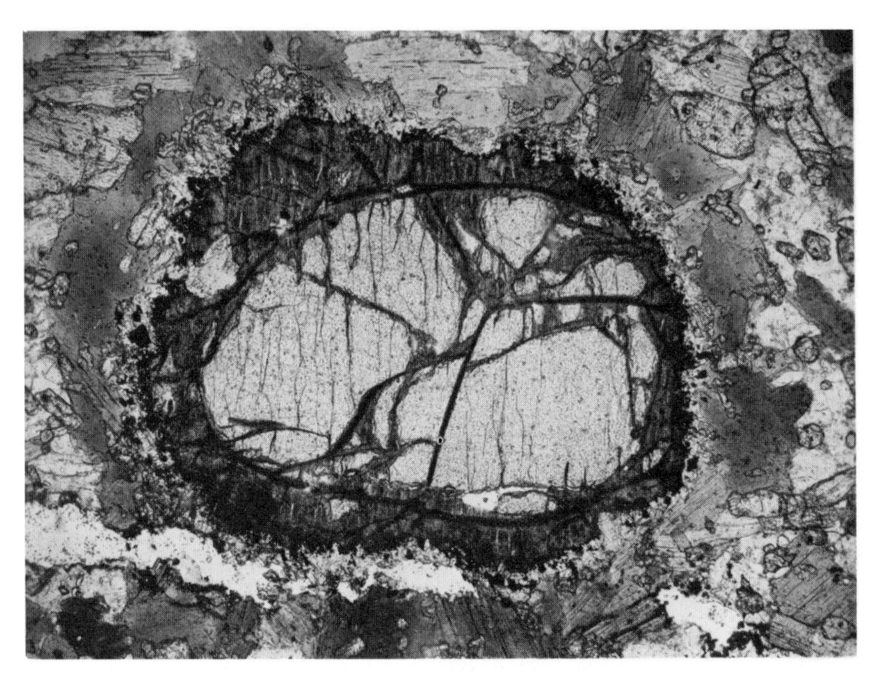

(C)

Figure 4–18 *(continued)*
(C) Peridotite. A large central olivine grain is within a matrix composed mainly of pyroxene The olivine grain is partly altered, and surrounded by a fine-grained aggregate that consists of magnetite, antigorite, chrysotile, iron oxides, and chlorite. Little Belt Mountains, Montana. [U.S. Geological Survey Photo by I. J. Witkind, 1964.]

SUMMARY

Igneous rocks are classified mainly on the basis of texture and mineralogy. The IUGS classification first subdivides igneous rocks on the basis of grain size. Phaneritic rocks are classified as plutonic; aphanitic rocks are volcanic. Following this, finer subdivision is made based on the relative amounts of the principal minerals. Siliceous to mafic rocks are mainly subdivided on the relative amounts of quartz (or other SiO_2 polymorphs), plagioclase, alkali feldspar, or feldspathoid (foid) minerals. Rock names are based on both common usage and natural groupings that have a high frequency of occurrence. Rocks that have very small amounts of felsic minerals are subdivided on the basis of mafic minerals such as olivine and pyroxene. Some rocks (such as pegmatites and tuffs) do not fall into the IUGS classification, and are classified on a textural rather than primarily mineralogical basis.

A variety of chemical classifications has been devised; the most commonly used is the CIPW system. This classification permits calculation of a group of anhydrous minerals based upon the chemical analysis. These normative minerals commonly do not correspond to the actual (modal) minerals, but are useful in making comparisons of analyses and in roughly predicting the crystallized equivalents of glassy rocks.

A wide variety of terms is used to describe the textural relations of igneous rocks. Although tedious to learn, most of these terms are useful inasmuch as textural relations not only reveal considerable information about the conditions of origin of the rock, but also permit detailed comparisons among the various igneous rocks.

FURTHER READING

Johannsen, A. 1931, 1937, 1938. *A Descriptive Petrography of the Igneous Rocks.* Chicago: University of Chicago Press, 4 vols.

Streckeisen, A. L. 1976. To each plutonic rock its proper name: *Earth Sci. Rev., 12,* 1–34.

Streckeisen, A. L. 1978. Classification and nomenclature of volcanic rocks, lamprophyres, carbonatites and melilitic rocks. *Neues. Jahrb. Mineral. Abh., 134,* 1–14.

Williams, H., F. J. Turner, and C. M. Gilbert. 1955. *Petrography.* San Francisco: W. H. Freeman and Company, 406 pp.

5

The Evolution of Magmas

Two features have been quickly noted in most places where igneous rocks have been studied. One of these is that the typical igneous rock is not uniform in composition or texture. In some cases this lack of uniformity can be observed within a single outcrop, and in other cases variations are observed on the scale of hundreds of meters or greater. The second feature noted is that most igneous rock terranes usually consist of a variety of different igneous rock types. The obvious question that geologists have asked about such occurrences is whether or not the various igneous rocks in a single region have a related origin. Is there a common parent magma from which they were somehow derived? Could there have been two parent magmas that were partially mixed together to form intermediate compositions? Does the variation in rock types result from different amounts of interaction of a magma with the surrounding wall rocks? Or are the various igneous rock types perhaps unrelated in origin, being derived from melting of different source rocks in the crust and mantle?

One method of suggesting a common origin for a group of igneous rocks is use of similarities in mineralogy (such as the presence of alkali-rich amphiboles or pyroxenes); but considering the broad range in composition of associated igneous rock types, it is often more effective to use uniqueness of trends in rock chemistry. The associated rocks might all be particularly high or low in one or more elements (such as Zn, Mo, or Zr). Or it may be that the ratios of a pair of elements or even stable isotopes might indicate a common origin. In areas where there are no special aspects of rock chemistry, a common origin (consanguinity) may be indicated by chemical trends of associated rock types. In this case each of the associated rocks is analyzed and its composition is plotted in terms of several components; if the rock compositions form relatively regular distributions, consanguinity is indicated.

As an example of chemical trends we can consider the work of McBirney and Aoki (1968), who studied the igneous rocks on the island of Tahiti in the South Pacific. In addition to its many other fine attributes, Tahiti consists of two highly eroded volcanic areas about 35 km apart. Although most of the island consists of lavas, the volcanic core contains a variety of more or less chemically equivalent plutonic rocks. Compositions run from alkali basalts through more siliceous alkali-rich lavas. Rock analyses of both plutonic and volcanic rocks, plotted in terms of three components, are given in Figure 5–1. The analyses form distinct chemical trends on these diagrams (as indicated by the solid lines). In such cases, although a common origin is indicated, the particular mechanism by which this occurred is not obvious from the analyses alone. Additional evidence is needed from field and textural relationships; such evidence must be verified in turn by theoretical and experimental work.

As a result of many investigations it has become obvious that associated igneous rocks are related to each other by three principal mechanisms. These are differentiation, assimilation, and magmatic mixing.

DIFFERENTIATION

Differentiation is any process whereby a single homogeneous magma is able to produce a variety of chemically different igneous rocks. The most important method of differentiation results from fractional crystallization (fractionation). Less important processes are liquid immiscibility and gaseous transfer.

Fractional Crystallization

We have seen earlier (in the discussion of phase relationships) that the various minerals precipitating from a melt do not usually all crystallize simultaneously. Generally a single mineral precipitates first; with cooling this is joined by a second, third, and so forth. As the melt and precipitating crystals are not usually of the same composition, the melt changes composition during crystallization and will follow a "liquid line of descent" towards the invariant point (degrees of freedom = 0) of the system. In most silicate-rich melts, earlier-formed minerals react with the ever-changing melt composition. When these reactions are permitted to go to completion (as indicated by the relationships shown in phase diagrams), the process is called *equilibrium crystallization.* No opportunity for differentiation is present; a melt of basaltic composition will yield a basalt or gabbro, and a melt of granitic composition will yield a rhyolite or granite. If, however, the crystals somehow are completely or partially prevented from reacting with the melt, petrologists call the process *fractional crystallization;* the final melt composition will be different from that predicted by equilibrium crystallization, and magmatic differentiation will occur. Note that some confusion arises from the term fractional crystallization, as it is used by chemists to describe the sequential precipitation of salts from a solution.

The process of magmatic differentiation by crystal fractionation was profoundly emphasized by N. L. Bowen in his classic book, *The Evolution of the Igneous Rocks* (1928). Bowen, as a result of both laboratory experimentation and field studies, felt

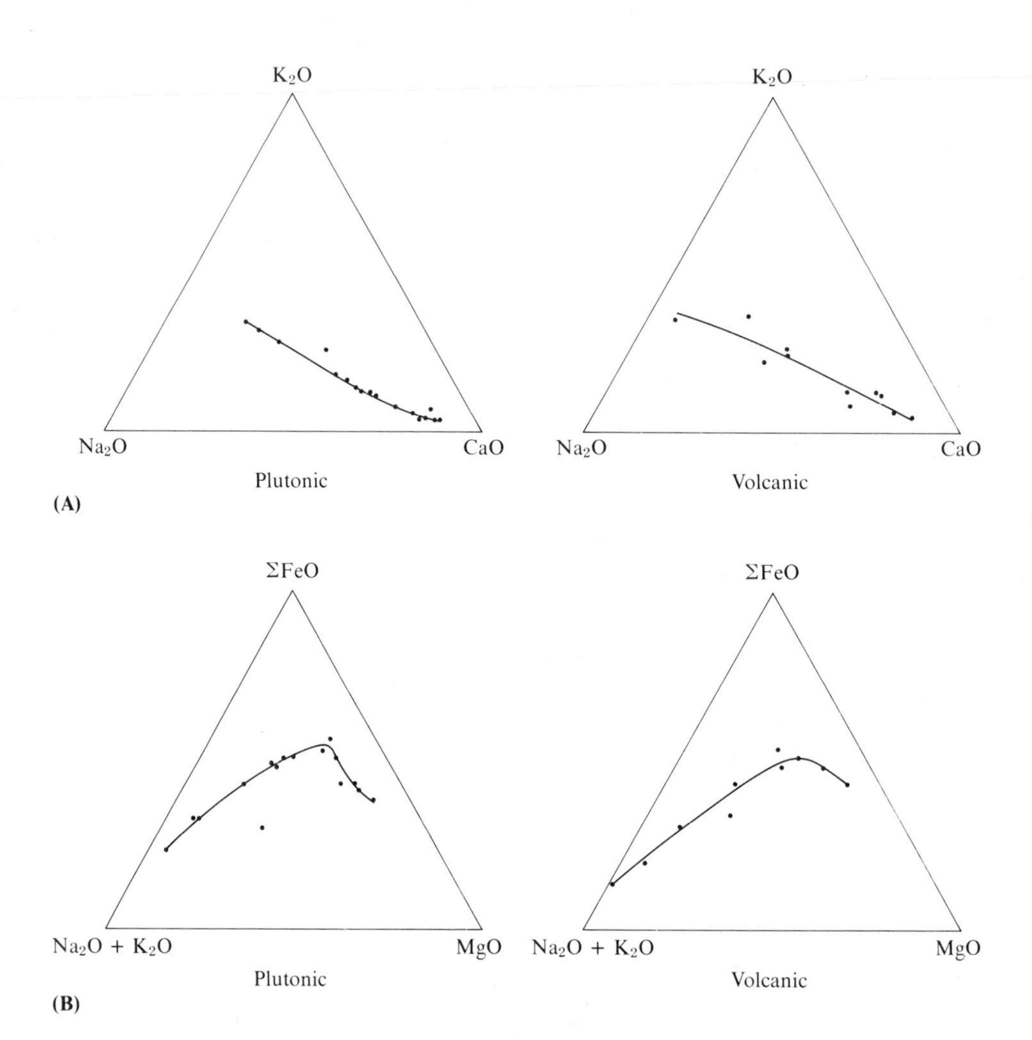

Figure 5–1
The composition of intrusive and extrusive rocks on the island of Tahiti. (**A**) Rock analyses in terms of CaO, K_2O and Na_2O. Rock analyses near the CaO corner represent a CaO-rich gabbro. A distinct chemical relationship is indicated by the linear arrangement of the analyses. (**B**) The same rocks plotted in terms of MgO, total iron as FeO, and $Na_2O + K_2O$. Again, a distinct compositional trend is indicated by the analyses. [From A. R. McBirney and K. Aoki, 1968, in R. R. Coats, R. L. Hay, and C. A. Anderson (eds.), Geol. Soc. Amer. Mem. 116, Fig. 6.]

that there was a single parent magma of basaltic composition from which all other magma types were evolved. The mechanism by which this occurs was summarized in Bowen's reaction series (see Figure 5–2). It is now known that all igneous rocks are not derived by differentiation from a basaltic melt (and a variety of reaction series are considered likely), but the processes discussed by Bowen are of great significance in relating the wide variety of igneous rock types.

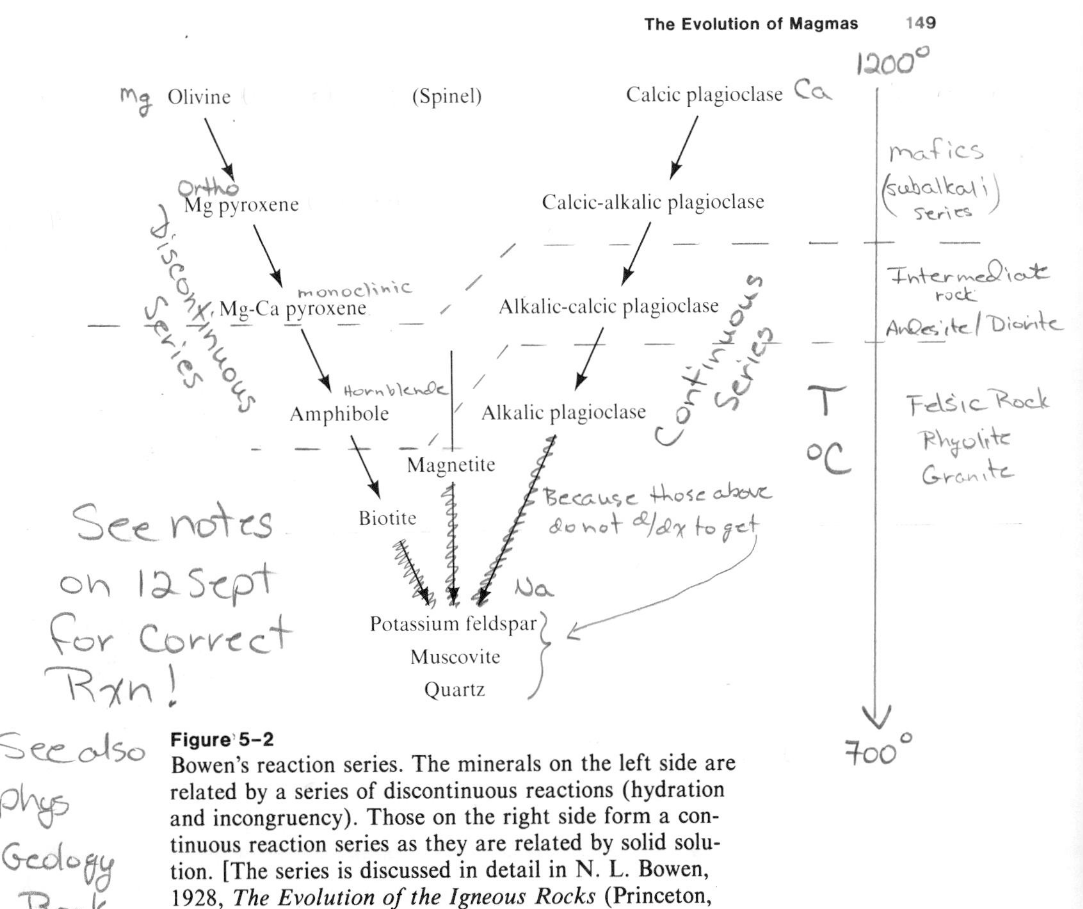

Figure 5–2
Bowen's reaction series. The minerals on the left side are related by a series of discontinuous reactions (hydration and incongruency). Those on the right side form a continuous reaction series as they are related by solid solution. [The series is discussed in detail in N. L. Bowen, 1928, *The Evolution of the Igneous Rocks* (Princeton, N.J.: Princeton Univ. Press).]

Bowen's reaction series consists of two branches. The plagioclases on the right form a continuous reaction series, as plagioclases grade into each other in both composition and crystallization temperatures (see Figure 2–19); during crystallization the crystals react continuously with the melt, changing their composition toward the $NaAlSi_3O_8$ end member. The left branch consists of a variety of mineral types that are compositionally and structurally distinct, and form a discontinuous reaction series. For example, there cannot be solid solution between an amphibole and biotite, in contrast to the case of plagioclase feldspar. Reactions between crystals and melt occur only during certain portions of the cooling sequence. Notice that the minerals at the upper portion of the series are characteristic of basalts, and crystallize at high temperatures. Minerals lower in the series crystallize at lower temperatures. The minerals characteristic of granite occur at the bottom of the diagram and have the lowest freezing temperatures.

Let's consider how the reaction series works. First we start off with a basaltic magma that should (with equilibrium crystallization) yield mainly pyroxene and a plagioclase of somewhat calcic composition. Cooling of the magma may result in initial crystallization of a calcium-rich plagioclase, and/or olivine, and perhaps a mineral of the spinel group. Under equilibrium crystallization the calcium-rich plagioclase should react with the melt and become more sodic; in addition, at some point in the cooling sequence the olivine crystals should react with the melt and be eliminated in favor of an Mg-rich pyroxene such as enstatite or hypersthene. If these reactions were permitted to occur (with equilibrium crystallization) the melt would probably complete its crystallization with the formation of additional plagioclase and pyroxene (with augite probably joining in), and produce a basalt or gabbro (as predicted). The shortage of silica and H_2O in the original melt would not permit the formation of the more silica-rich minerals lower in the series, such as hornblende or biotite. If, on the other hand, the first-formed crystals of Ca-rich plagioclase and olivine were somehow not permitted to react with the melt, the composition of the melt would change to a considerably greater extent than if reaction had been permitted. Nonreaction of silica-poor minerals permits the remaining melt to become more silica-rich. Crystallization and nonreaction of Ca-rich plagioclase results in a relative increase in Na in the melt. In fact, any element that is not present in the crystallizing phases (or is present in lesser amounts than the liquid from which it is crystallizing) is necessarily enriched in the remaining portion of magma.

The remaining differentiated melt continues crystallization with the formation of minerals lower in the reaction series (a more sodic plagioclase, for example). If nonreaction or incomplete reaction continues, with removal of the crystallizing anhydrous phases, the H_2O content of the melt could be increased to the point where hydrous phases are precipitated from the melt (or formed as reaction rims around earlier anhydrous phases). Thus amphiboles and later biotite and muscovite are precipitated. Nonreaction with the earlier-formed phases can lead to sufficient enrichment of alkalies and silica for the final liquids to crystallize potassium feldspar and quartz ($\pm$ muscovite) and thereby form granitic rocks. In summary, the trend of these magmatic liquids is to become enriched in alkalies and silica and to decrease in iron and magnesium. Minerals crystallizing in a solid solution series tend to enrich the residual melt in the lower temperature end-member of that series. Note also that a variety of differentiation trends is possible, depending upon the initial magma composition, the composition and amounts of precipitating phases, the pressure, and the oxygen content of the magma. Other reaction series, such as that of the Skaergaard intrusion in Greenland, yield final liquid residues highly enriched in iron, with only modest increases in silica. Bowen's reaction series is only one of several postulated differentiation trends (see Chapter 8). Recent data also indicate that rapid cooling may permit metastable crystallization, with the result that the sequence of crystallization deviates from that predicted by phase diagrams.

Evidence for the validity of the reaction principle is abundant in both the study of phase equilibria and textural evidence in igneous rocks. Texturally we notice the common compositional zoning of plagioclase with calcic cores and less calcic rims. Olivines are noted with rims of pyroxene. Pyroxenes in turn often possess a rim of amphibole. It should be understood, however, that although the mechanism of

Bowen's reaction series has been validated in many areas, it does not follow that all igneous rocks have originated by this mechanism. Simple proof of this is seen in the fact that the amounts of granitic and basaltic rocks exposed at the earth's surface are about equal. If all granites were produced by Bowen's reaction series, there should only be one part of granitic melt produced for every 20 parts of basaltic melt; in addition, it would follow that granitic rocks should be associated with large amounts of ultramafic rocks, which is not the case. Clearly a different mechanism of origin exists for most granites. Nevertheless, the reaction principle is of prime importance as a magmatic evolutionary procedure.

Fractional crystallization depends upon nonreaction or incomplete reaction of the magma with its crystalline products. A number of mechanisms have been suggested to prevent reaction, the most significant of which appears to be *gravitative differentiation.* Most minerals are denser than the melt from which they are derived, and a few (such as leucite) are lighter. In the early stages of crystallization of a fluid melt it is common for the early-formed crystals such as olivine and pyroxene to be prevented from reacting with the bulk of the liquid by settling to the bottom of the magma chamber. The concept is not new. Charles Darwin, when describing some of his observations made during the voyage of the *Beagle* (1828–1833), reviewed contemporary ideas on igneous rocks (especially those of Von Buch) and suggested that some of the lavas on the Galápagos Islands had been somewhat differentiated by the sinking of feldspar phenocrysts to the lower layers.

Crystal settling was demonstrated experimentally by Bowen in 1915. In one experiment Bowen made up a sample of 56% diopside and 44% forsterite and heated it in a platinum crucible until it was completely molten. The temperature was then brought down to 1405°C and held there for periods ranging from 15 to 80 minutes. At this temperature the sample (which forms a simple binary system) partially crystallized to form 4% forsterite crystals. After a suitable time interval the sample was rapidly quenched to room temperature. Liquid was converted to glass, and forsterite crystals were frozen in their original positions within the crucible. Thin sections were made at different levels within the crucible (see Figure 5–3). Examination of these revealed that after 15 minutes most of the olivine had sunk 1–2 cm to form an accumulation at the bottom of the crucible. In experiments with other systems, sinking of pyroxenes and floating of tridymite were demonstrated. Confirmation of gravity settling has been abundantly demonstrated in field studies of the crystallization of mafic rocks, particularly in sills. The effects are not as obvious in silicic and intermediate rocks, perhaps because of the higher viscosity of the original melts.

One of the best-known examples of gravity settling is the Palisades Sill, a hypabyssal intrusive about 330 m thick that outcrops along the Hudson River west of New York City. The sill has an average composition of tholeiitic basalt and is intruded into feldspathic sandstones of Triassic age. The average composition is approximately 50% augite, 47% labradoritic plagioclase, 1% olivine, and 2% other minerals.

Intrusion of the sill first resulted in chilling and crystallization of the upper and

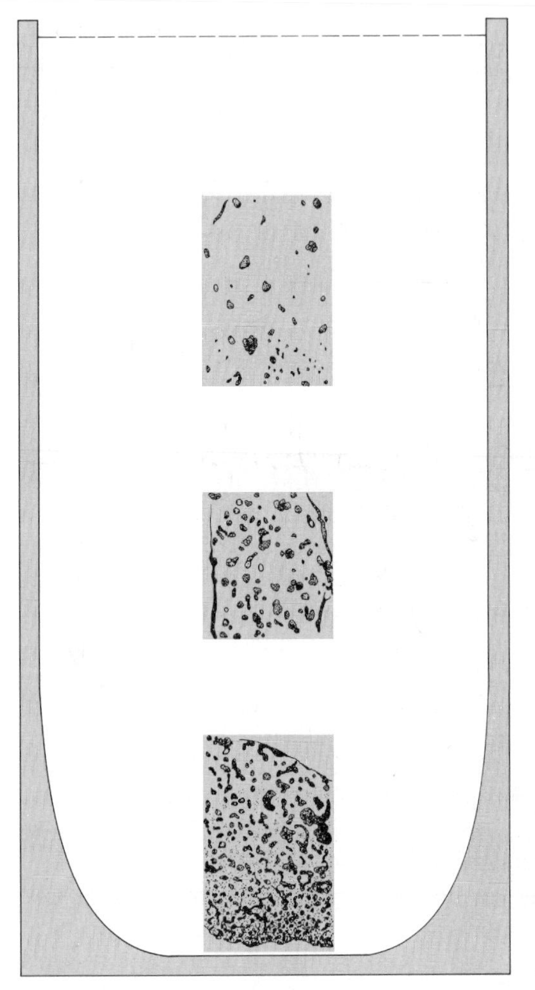

Figure 5–3
Bowen's experiment illustrating the sinking of olivine crystals within a melt. A melt containing crystals was quenched after 15 minutes. Thin sections made at different levels within the crucible revealed that most of the olivine crystals had sunk 1–2 cm to form an accumulation at the base. [From N. L. Bowen, 1915, *Amer. Jour. Sci.* [*4*], *29,* Fig. 1.]

lower margins against the colder country rock. This fine-grained margin has a maximum thickness of about 15 m, and consists of minerals in essentially the same abundance as the average sill composition. Directly above the lower chilled portion is an olivine-rich layer (see Figure 5–4A) about 5 m thick. This layer contains about 20–25% olivine. Above this layer olivine is absent until the upper chilled border is reached. In addition to segregation of olivine, plagioclase becomes more sodic and abundant toward the upper chilled zone, and pyroxenes become more iron-rich. A relatively coarsely grained "pegmatitic" zone is present about two-thirds of the way up the sill; here the silica, iron, and alkali contents are higher than in other parts of the sill.

Intrusion of the Palisades Sill was followed by chilling of the contact zone and slow cooling of the remaining melt. Olivine, present either as phenocrysts or nucleated

(A)

Figure 5–4
(A) The lower part of the Palisades Sill, New Jersey. The chilled base of the sill is a few feet below the area shown. The olivine-rich layer (emphasized by weathering) is visible just above the center of the photograph. [From *Layered Igneous Rocks* by L. R. Wager and G. M. Brown. W. H. Freeman and Company. Copyright © 1967. Fig. 2.]

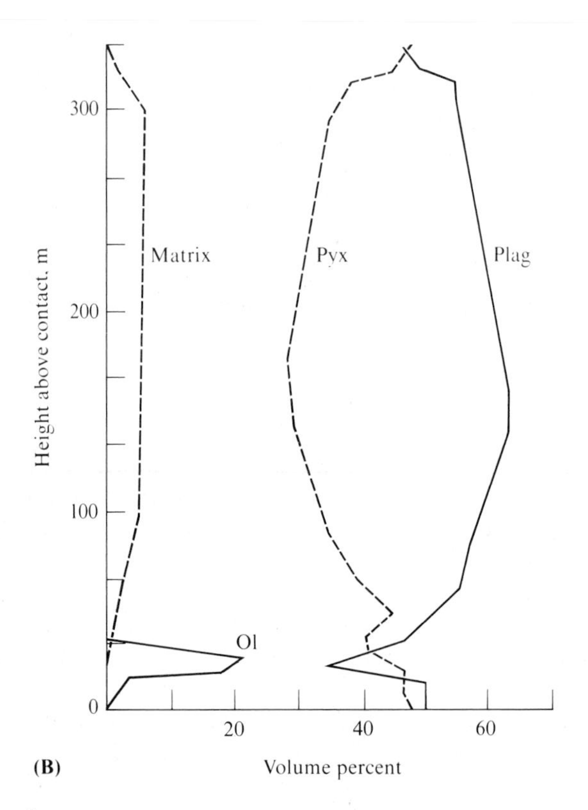

Figure 5–4 *(continued)*
(B) Variation in mineral content in the Palisades Sill (Englewood section). Plag = plagioclase, Pyx = pyroxene, Ol = olivine. [From J. C. Jaeger, 1968, in H. H. Hess and A. Poldervaart (eds.), *Basalts,* vol. 2 (New York: Wiley Interscience), Fig. 6c.]

within the melt, was the densest silicate phase present (specific gravity > 3.22) and settled to the chilled base along with calcic (labradorite) plagioclase. Depletion of calcium and increase in silica content in the remaining melt resulted in the crystallization of plagioclase richer in sodium, and pyroxene. Continued depletion of these phases resulted in a small amount of quartz-rich late differentiate (see Figure 5–4B).

In contrast to mafic rocks, silicic rocks show little direct evidence of a gravitative differentiation as a result of fractionation processes. We do not find mineralogic layering at the base of silicic intrusions. Part of the reason for this is the relative infrequency of large silicic sills, and the fact that the floors of most stocks and batholiths are not usually exposed. In addition, the amount of gravitative differentiation is probably decreased by the higher viscosity of silica-rich melts as compared to silica-poor ones. Evidence for fractionation is present, however, in the presence of compositionally zoned feldspars and contrasts in composition between chilled borders and the centers of large intrusions. The best evidence for fractionation comes from the comparison of the experimental study of the "simplified granite system" with the frequency of occurrence of silicic rocks.

The simplified granite system describes the crystallization of the alkali feldspars and silica in the presence of water vapor. This is the quaternary system $NaAlSi_3O_8$–$KAlSi_3O_8$–SiO_2–H_2O. We have already examined the anhydrous system $NaAlSi_3O_8$–$KAlSi_3O_8$ (see Figure 2–21). The other two sides of the anhydrous base, $NaAlSi_3O_8$–SiO_2 and $KAlSi_3O_8$–SiO_2 can be considered as simple binary systems (ignoring here the incongruent melting of K-feldspar to form leucite and melt). The simplified granite system has been studied at a variety of water pressures, as granites typically contain hydrated minerals and are often associated with hydrothermal activity.

Consider first the alkali feldspar join ($NaAlSi_3O_8$–$KAlSi_3O_8$) in the presence of H_2O (see Figure 5–5). As discussed earlier, under anhydrous conditions the alkali feldspars show complete solid solution at elevated temperatures between $NaAlSi_3O_8$ and $KAlSi_3O_8$. At low temperatures unmixing occurs. An anhydrous melt of intermediate composition, *a*, will cool to produce a single feldspar phase, which unmixes (at *b*) to create the alkali feldspar intergrowth known as perthite. At high water pressures the liquidus and solidus surfaces are depressed (with crystallization occurring at lower temperatures). At water pressures near 5 kbar the solidus and the unmixing curve (the solvus) intersect, eliminating the single alkali feldspar solid solution field (anorthoclase). Note that during crystallization of a melt at $P_{H_2O} = 5$ kbar the amount of perthitic exsolution is quite limited as compared to an anhydrous melt, due to the fact that two alkali feldspars are produced directly from the melt rather than one (see Figure 5–5). The presence or absence of perthitic texture in feldspars thus gives us a clue as to the water pressure that existed during the crystallization of a melt containing alkali feldspars.

The simplified granite system ($NaAlSi_3O_8$–$KAlSi_3O_8$–SiO_2–H_2O) is shown in Figure 5–6. The three components $NaAlSi_3O_8$, $KAlSi_3O_8$, and SiO_2 are shown in these compositional triangles. No H_2O is present in part A, and an excess of H_2O at 5 kbar is assumed in part B. In part A, the anhydrous system, liquidus surfaces exist for silica polymorphs and a single alkali feldspar. These surfaces slope thermally downward to a central boundary curve *ab*, which has a thermal minimum at *c*. Melts within this system, by crystallization of quartz, alkali feldspar, or both, will migrate in composition toward the thermal minimum at *c*.

At high water pressures shown in part B, freezing temperatures are depressed. Liquidus surfaces are present for the two alkali feldspars as well as for quartz. The diagram superficially resembles the simple ternary shown earlier in Figure 3–3. Melts within the simplified granite system will tend to migrate in composition toward the thermal minimum at *a*.

In both the anhydrous and hydrous simplified granite systems, crystallizing melts tend to change in composition toward the thermal minimum. Because of solid solution in the alkali feldspars, combined with possible limitations of original melt com-

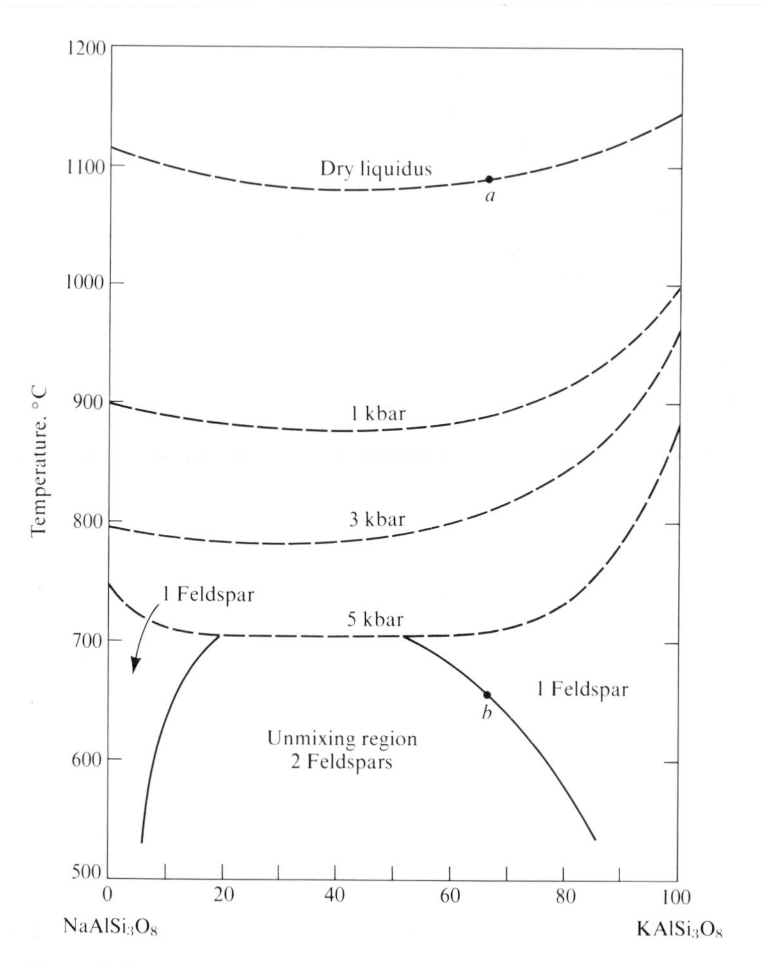

Figure 5–5
The system $NaAlSi_3O_8$–$KAlSi_3O_8$–H_2O. Liquidus surfaces have been simplified, and the field of leucite eliminated. Liquidus surfaces decrease in temperature as a function of increasing P_{H_2O}. [After O. F. Tuttle, and N. L. Bowen, 1958, Geol. Soc. Amer. Mem. 74, Fig. 17; and S. A. Morse, 1970, *Jour. Petrology, 11,* Fig. 2.]

positions, the crystallizing melt often becomes solidified before the thermal minimum is reached. However, if differentiation of the cooling melt occurs, the liquid will achieve a closer approach, and may well reach the minimum point. Thus the degree of approach of the melt to the thermal minimum has been intepreted as a measure of the effectiveness of fractionation processes.

A proof of the effectiveness of fractionation in the simplified granite system was first given by Tuttle and Bowen (1958) and was enlarged upon by Winkler and von

Platen (1961). Mineral analyses of 1,190 granitic rocks were plotted and contoured within the compositional triangle albite–orthoclase–quartz (see Figure 5–7). It has been argued that the intense clustering of analyses near the thermal minimum point of the diagram (compare Figure 5–6) is a strong indication that fractionation processes are very effective in this system; zoned crystals in rocks of granitic composition tend also to verify the direction of differentiation. This argument, although valid, is weakened by the fact that many melts of granitic composition form by a partial melting process that results in a melt similar to granite.

Concentration of H_2O

Many magmas, particularly those of silicic composition, contain several percent or more H_2O in addition to other minor volatile components. As seen above, the presence of dissolved H_2O in the siliceous melt causes a depression of the freezing point of anhydrous phases, as well as permitting the formation of hydrous phases (such as hornblende and micas). The crystallization of hydrous minerals may remove some of the dissolved H_2O from the melt, but in many cases this is insufficient to prevent a buildup of H_2O in the ever-decreasing amount of crystallizing melt. As a result, the water-saturation point of the melt may be reached, causing the evolution of a vapor phase. This has been referred to as retrograde, resurgent, or second boiling; this is a boiling that occurs because of cooling, and results from the increase in vapor pressure of confined gases in an increasingly confined volume. The amount of this hydrous fluid will vary as a function of initial water content, melt composition, and overall pressure.

Crystallization within the water-saturated silicate melt and the water-rich phase are completely different. The water-saturated silicate melt produces a fine-grained aggregate of quartz and feldspar; this material is similar to the late-stage veins (known as aplite) that are often found within and adjacent to acidic intrusives. Crystallization of quartz and feldspar from the water-rich phase produces relatively large, well-formed crystals. These larger crystals are perhaps due to the high fluidity and diffusion rates of chemical constituents within the water-rich medium. As seen in the field, crystallization from the water-rich phase has produced the coarse-grained late-stage dikes and segregations known as pegmatites. Pegmatites occasionally produce enormous crystals; examples include a spodumene crystal at the Etta Mine in South Dakota that had dimensions of 12.7 m by 1.7 m and weighed more than 8,000 kg, and a microcline crystal from Maine over 6 m across.

Evolution of a fluid phase from the cooling silicate melt permits the partitioning of other volatile and nonvolatile components in addition to H_2O. Consequently pegmatites, although chiefly quartz and alkali feldspar ($\pm$muscovite and/or biotite), are often high in the volatile components, in addition to a variety of elements that have been excluded from crystallization within the normal silicate structures; these elements include lithium, beryllium, niobium, tantalum, tin, uranium, thorium, tungsten, zirconium, and rare earths.

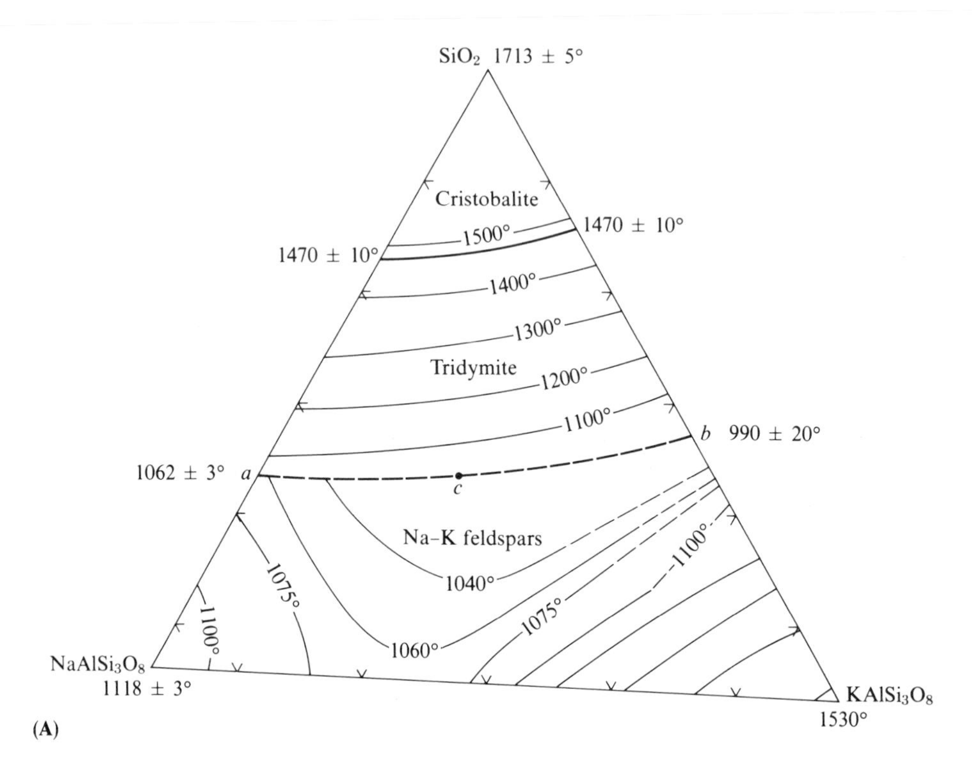

Figure 5–6
The simplified granite system ($NaAlSi_3O_8$–$KAlSi_3O_8$–SiO_2–H_2O). (**A**) The anhydrous system. A single boundary curve exists from *a* to *b* with a thermal minimum at *c*. The field of leucite has been eliminated for simplification. [From J. F. Schairer, 1957, *Jour. Amer. Cer. Soc., 40,* Fig. 29.]

Aplite and pegmatite may be found within the same dike, separated by a sharp or gradational contact. These may represent either simultaneous or sequential crystallization of both the melt and water-rich phases.

Fractionation of Trace Elements

Fractionation processes are not limited to the major elements oxygen, silicon, aluminum, iron, magnesium, calcium, sodium, and potassium, which make up 98% (by weight) of the crust. In addition, the next most abundant elements titanium, phosphorous, and manganese (called the minor elements) as well as the less abundant trace elements are involved. The distribution of trace elements in a crystallizing magma can be predicted on the basis of crystal–chemical principles. An example of

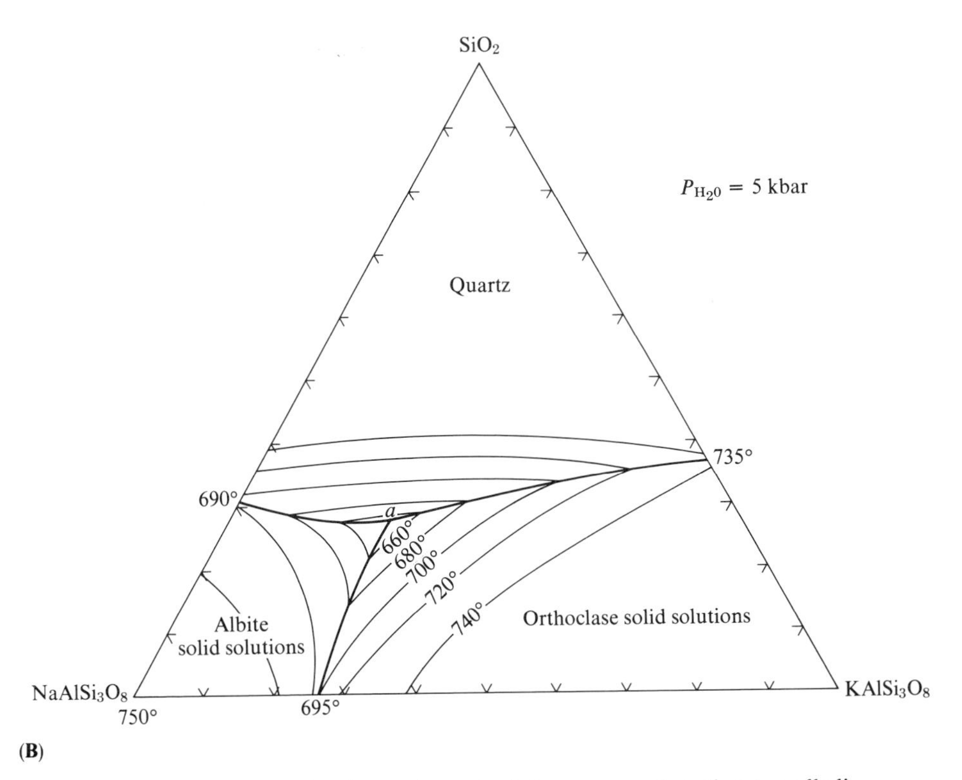

(B)

(B) The system at $P_{H_2O} = 5$ kbar has a thermal minimum at *a*. Note that two alkali feldspar primary fields are present at high P_{H_2O}, whereas only one is present in the anhydrous system. [From W. C. Luth, R. H. Jahns, and O. F. Tuttle, 1964, *Jour. Geophys. Res., 69,* Fig. 2.]

this was given by Curtis (1964), who showed that the sequence of depletion (uptake) of transition elements in the Skaergaard intrusion, Greenland, matched that predicted by theory. Divalent ions and elements concentrated in later differentiates are in turn Ni, Cu, Co, Fe, Mn, and Ca and Zn; elements that can exist in a trivalent state concentrate in the sequence Cr, Co, V, Ti, Fe, and Sc and Ga. Elements (such as Ga, Ge, Sn, Zr, Hf, P, As, Mo, Nb, and Ta) that normally would coordinate tetrahedrally with oxygen are prevented from doing so by the presence of large amounts of silica (which preferentially joins with oxygen); the elements not coordinated with oxygen are concentrated in the fractionated residual liquids or form separate phases. Knowledge of minor and trace element fractionation trends provides information about differentiation processes and permits correlation of comagmatic rocks of widely diverse types.

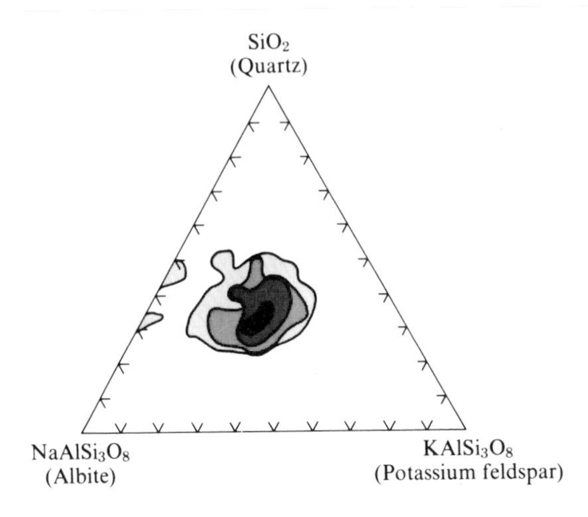

Figure 5–7
The frequency distribution of normative quartz, albite, and potassium feldspar in 1,190 granitic rocks. The outermost field encloses 86% of all granitic rocks; inner fields enclose 75%, 53%, and 14%, respectively. The frequency maximum lies within the black field. [From H. G. F. Winkler and H. von Platen, 1961, *Geochim. Cosmochim. Acta, 24,* Fig. 2.]

Other Differentiation Mechanisms

As mentioned above, the ability of a magma to differentiate is dependent upon fractional crystallization, which results in differences in composition between crystals, melt, and/or vapor. The various phases have different specific gravities, which permit crystal settling (or less commonly, floating), a process called *gravitative differentiation.* This is thought to be the major mechanism of differentiation.

Another mechanism of differentiation, which can occur when the magma is almost completely crystalline, is *filter pressing.* Deformation of the magma chamber can squeeze the crystalline mush together, forcing out much of the remaining melt, which is injected into the adjacent country rock. Alternatively, but probably less commonly, the magma chamber may be subjected to tension, causing the formation of liquid-filled voids within the densely crystallized (but unconsolidated) portion of the chamber.

A method of differentiation that has been suggested with various degrees of enthusiasm over the years is *liquid immiscibility.* This is based on the idea that not all liquids are completely soluble with each other (because of differences in behavior of different types of molecules, such as polar versus nonpolar). When the limits of solubility of one liquid within another are exceeded, the molecules of each type of liquid tend to be more attracted to each other than to the other liquid. This leads to the formation of regions of unique composition; if specific gravity differences exist between the two types of liquids, one will rise and the other will sink, to produce a two-layer liquid possessing a distinct interface. At one point geologists thought that cooling a magma of intermediate silica content could produce liquid unmixing, resulting in separate silicic and mafic layers, which could cool separately to produce distinct but related rock types. Although some evidence of immiscibility in synthetic silicate melts exists, there is little reason to believe that this is a major rock-forming process. Evidence for liquid immiscibility has been found in the differentiation of the Skaergaard intrusive in Greenland (see Chapter 9), of sulfide-rich layers within silicate magmas, and of carbonate liquids within mafic alkaline magmas. Evidence of immiscibility is fairly common in lunar rocks. In general, relatively uncommon magma compositions are required before immiscibility becomes an important process.

The upward streaming of gas bubbles within a magma, called *gaseous transfer* or *volatile streaming,* may be of importance under shallow conditions during a period of gas formation (vesiculation). Vesiculation may occur as a result of crystallization of anhydrous phases or a decrease in confining pressure. After considerable crystallization has occurred, the magma chamber is partially filled with solid phases. As most of the solids are anhydrous, the dissolved volatile components are concentrated in an ever-decreasing volume of melt. A gradual buildup of vapor pressure could result in the formation of a fluid phase through vesiculation if the melt is supersaturated with a particular volatile component. The fluid, being lighter than the melt, moves upward within the chamber. Vesiculation can also be produced by a decrease in confining pressure through fracturing of country rock above the chamber. If the process is gradual the fluid is able to migrate upward through the melt. Alternatively, if the pressure release is large and abrupt, vesiculation begins almost simultaneously throughout the whole chamber. Rapid upward movement may carry considerable amounts of silicate melt as well as fragments of earlier-crystallized material, resulting in the formation of a volcanic pipe and a hydrothermal explosion breccia. Whether the vapor-streaming process is fast or slow, escape of volatiles (which are often rich in alkalies, CO_2, and halogens, as well as water) serves to change the original magma composition.

MAGMATIC MIXING AND ASSIMILATION

The mixing of two or more magmas has been verified from several field occurrences. This has been established by the discovery of nonhomogeneous rocks, as for example clots of one rock type within another, or basalt masses with glassy rims within a siliceous rock.

Effective mixing of magmas should occur best at depth before extensive crystallization of either melt has occurred. This mixing might be proven by the presence of anomalous phase relationships—incompatible phases relatively close to each other, and deviations from the normal order of crystallization.

Assimilation of country rocks by magma has been suggested as a significant factor in differentiation. Although chemical differences between magma and wall rock do exist, the effects of assimilation are usually considered to be minor in causing large-scale changes in magma composition. The reason for this is that most magmas do not possess superheat—heat above that required to make the magma completely liquid. This is indicated by the common presence of solid phases within magmas (crystals derived either from the source area or from crystallization during ascent). Heat-absorbing processes such as melting of or reaction with country rock necessarily result in precipitation of other crystalline material from the melt, and an increase in the total amount of solidification of the melt.

The general processes can be described in terms of Bowen's reaction series (see Figure 5–2). Recall that minerals in the upper part of the series have been experimentally demonstrated to crystallize at higher temperatures than those lower in the series. Basaltic minerals are at the top, and granitic minerals at the bottom. The general rule is that melts whose compositions place them higher in the series are capable of dissolving rocks lower in the series. Thus a basaltic melt might be able to dissolve some granite, but not vice versa. Bear in mind, however, that only a limited amount of granite can be dissolved because such dissolution results in simultaneous crystallization of the melt. Alternatively, if the wall rock contains minerals whose melting temperatures are higher than that of the magmatic liquidus, a process of chemical exchange will occur such that the wall rock mineral compositions are brought into closer equilibrium with the melt. For example, suppose a granitic magma intrudes a gabbro. As the melt crystallizes, calcium ions will diffuse outward from the plagioclase crystals in the gabbro and into the melt, and be replaced by sodium ions. In this way, the composition of the wall rock is made closer to that of the magma and the composition of the magma is simultaneously changed. In an analogous fashion, olivine and pyroxene in the wall-rock gabbro can be converted into hornblende and biotite. Replacement reactions can frequently be seen caught in the act in xenoliths, fragments of wall rock that have fallen into the melt (see Figure 5–8).

Assimilation of limestone or dolomite by granitic magma has been suggested as a method of desilicating the magma. The idea is that silica from the magma will react with calcite and dolomite to form minerals such as wollastonite, grossularite, tremolite, or diopside. Some Ca and Mg will enter the magma simultaneously, making its general composition more basic. Although such reactions do occur, the composition of the melt is not changed to a drastic extent, as extensive reactions are limited by crystallization of the melt.

SUMMARY

Most igneous terranes contain a variety of igneous rocks that are related mineralogically and chemically. Such consanguinity results primarily from the sequential pattern of mineral precipitation in multicomponent systems, an example of which is Bowen's reaction series. During fractional crystallization when reaction between crystals and melt is prevented (as by crystal settling or floating, or filter-pressing), the melt changes in composition (differentiates) in a manner not predicted by equilibrium crystallization. A variety of different but related rock types can be crystallized from these differentiated melts. Differentiation permits concentration within the final

Figure 5–8
Angular limestone fragments near the contact of the Last Chance (siliceous) stock. Narrow (dark) rims of clinopyroxene have formed reaction rims about these xenoliths. Salt Lake County, Utah. [From W. J. Moore, 1973, U.S. Geol. Surv. Prof. Paper 629-B, Fig. 4.]

residual liquids of elements that have not been depleted during the earlier crystallization sequence. Magma compositions may also be modified by liquid immiscibility, assimilation of country rock, and magmatic mixing, but these processes are not considered to be of major significance in the formation of most igneous rocks.

FURTHER READING

Bowen, N. L. 1928. *The Evolution of the Igneous Rocks.* Princeton, N.J.: Princeton Univ. Press, 333 pp. Reissued by Dover Publications.

Curtis, C. D. 1964. Applications of crystal field theory to the inclusion of trace transition elements in minerals during magmatic crystallization. *Geochim. Cosmochim. Acta, 28,* 389–402.

Hunter, D. 1978. The Bushveld Complex and its remarkable rocks. *Amer. Scientist, 66,* 551–559.

McBirney, A. R., and K.-I. Aoki. 1968. Petrology of the island of Tahiti. In R. M. Coats, R. L. Hay, and C. A. Anderson (eds.), *Studies in Volcanology,* pp. 523–556. Geol. Soc. Amer. Mem. 116.

Poldervaart, A. 1944. The petrology of the Elephant's Head dike and the New Amalfi sheet (Matatiele). *Trans. Royal Soc. S. Afr., 30,* 85–119.

Tuttle, O. F., and N. L. Bowen. 1958. Origin of granite in the light of experimental studies in the system $NaAlSi_3O_8$–$KAlSi_3O_8$–SiO_2–H_2O. Geol. Soc. Amer. Mem. 74, 154 pp.

Wager, L. R. 1968. Rhythmic and cryptic layering in mafic and ultramafic plutons. In H. H. Hess and A. Poldervaart (eds.), *Basalts,* Vol. 2, pp. 573–622. New York: Wiley Interscience.

Wager, L. R., and G. M. Brown. 1967. *Layered Igneous Rocks.* San Francisco: W. H. Freeman and Company, 588 pp. (see pp. 1–32).

Walker, F. 1940. Differentiation of the Palisades Diabase, New Jersey. *Geol. Soc. Amer. Bull., 51,* 1059–1106.

Walker, K. R. 1968. A mineralogical, petrological and geochemical investigation of the Palisades Sill, New Jersey. In L. H. Larsen, M. Prinz, and V. Manson (eds.), *Igneous and Metamorphic Geology,* pp. 175–188. Geol. Soc. Amer. Mem. 115.

Winkler, H. G. F., and H. von Platen. 1961. Experimentelle Gesteinsmetamorphose; IV, Bildung anateklischer Schmelzen aus metamorphisierten Grauwacken. *Geochim. Cosmochim. Acta, 24,* 48–69.

Yoder, H. S., Jr., ed. 1979. *The Evolution of the Igneous Rocks, Fiftieth Anniversary Perspectives.* Princeton, N.J.: Princeton Univ. Press, 588 pp.

6

The Interior of the Earth

A great deal is known about the earth's interior on the basis of indirect evidence, in spite of its inaccessibility. Precise knowledge of the earth's mass gives an overall density of 5.5 g/cm^3, which is considerably higher than that of the observable rocks at the surface. The velocities and directions of seismic waves reveal also that the earth is concentrically layered. The composition and nature of these layers is indicated by seismic data, analyses of meteorites, inclusions in melts from great depth, and high-pressure experimentation.

EVIDENCE FOR THE EARTH'S COMPOSITION AND MINERALOGY

Seismic Data

A seismic event, such as an earthquake or nuclear explosion, generates various types of waves. Some of these travel parallel to the earth's surface, and others penetrate the interior. If the earth's interior were homogeneous and of fixed density these penetrative waves would follow straight-line paths at constant velocity. Observation reveals that straight-line paths are not followed. Reflection and refraction occur at various depths (much in the manner of light waves), indicating inhomogeneities within the earth (see Figure 6–1). From this type of data the density distribution within the earth can be calculated. Constraints on the method are that the calculated density distribution must match the correct values for the earth's average density and moment of inertia.

Waves that penetrate the interior are of two types. P waves are compressive waves (similar to the waves produced by a vibrating tuning fork) and represent the propaga-

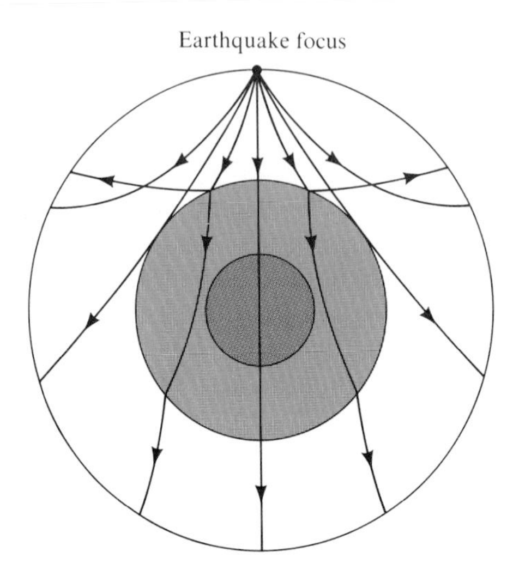

Figure 6–1
Schematic view of the earth showing typical reflection and refraction patterns of seismic waves. [After P. J. Wyllie, "The Earth's Mantle." Copyright © 1975 by Scientific American, Inc. All rights reserved. Fig. 1.]

tion of a volume change. S waves are transverse or shear waves (of the type produced by shaking the end of a rope). The P waves are capable of going through both solids and liquids. As liquids are not rigid and hence have no shear strength, S waves are transmitted only through solids. Although considerable variation exists in wave transmissions in the near-surface portion of the earth, those waves that penetrate the deep interior show a consistent worldwide pattern that definitely establishes a concentrically layered structure within the earth. A plot of P and S wave velocities (see Figure 6–2) shows the reason for subdividing the earth into the three major units of crust, mantle, and core. A thin layer present at the earth's surface is called the *crust*. This layer is about 6–8 km thick beneath the oceans, and changes to 30 km or more beneath the continents. The density of this layer is about 2.8 g/cm^3. The seismic velocities in the continental areas are commonly variable and complicated, but the general consensus is that the upper portion of the crust consists of rocks with densities similar to those of granitic materials (sial), and the lower portion of rocks with densities similar to those of basaltic materials (sima). Under the oceans the seismic velocities and densities of the crust are consistent with those of basaltic rocks.

The base of the crust is defined by a very distinct change in seismic velocity called the Mohorovičić discontinuity. From this discontinuity to a depth of 2,900 km is a region called the *mantle*. Within the mantle, at a depth of about 70 km from the surface of the earth, there is commonly found a small but distinct drop in seismic velocities. This region, about 50–75 km thick, is often called the *low-velocity zone* or *asthenosphere*, and the crust and mantle above, the *lithosphere*. The term asthenosphere is occasionally used for all rocks below the outer relatively strong brittle rocks of the lithosphere. A sharp decrease in P wave velocity and termination of S waves at the base of the mantle defines the *core*, which extends from a depth of 2,900 km to the earth's center at 6,370 km. Termination of the S waves at the core–mantle

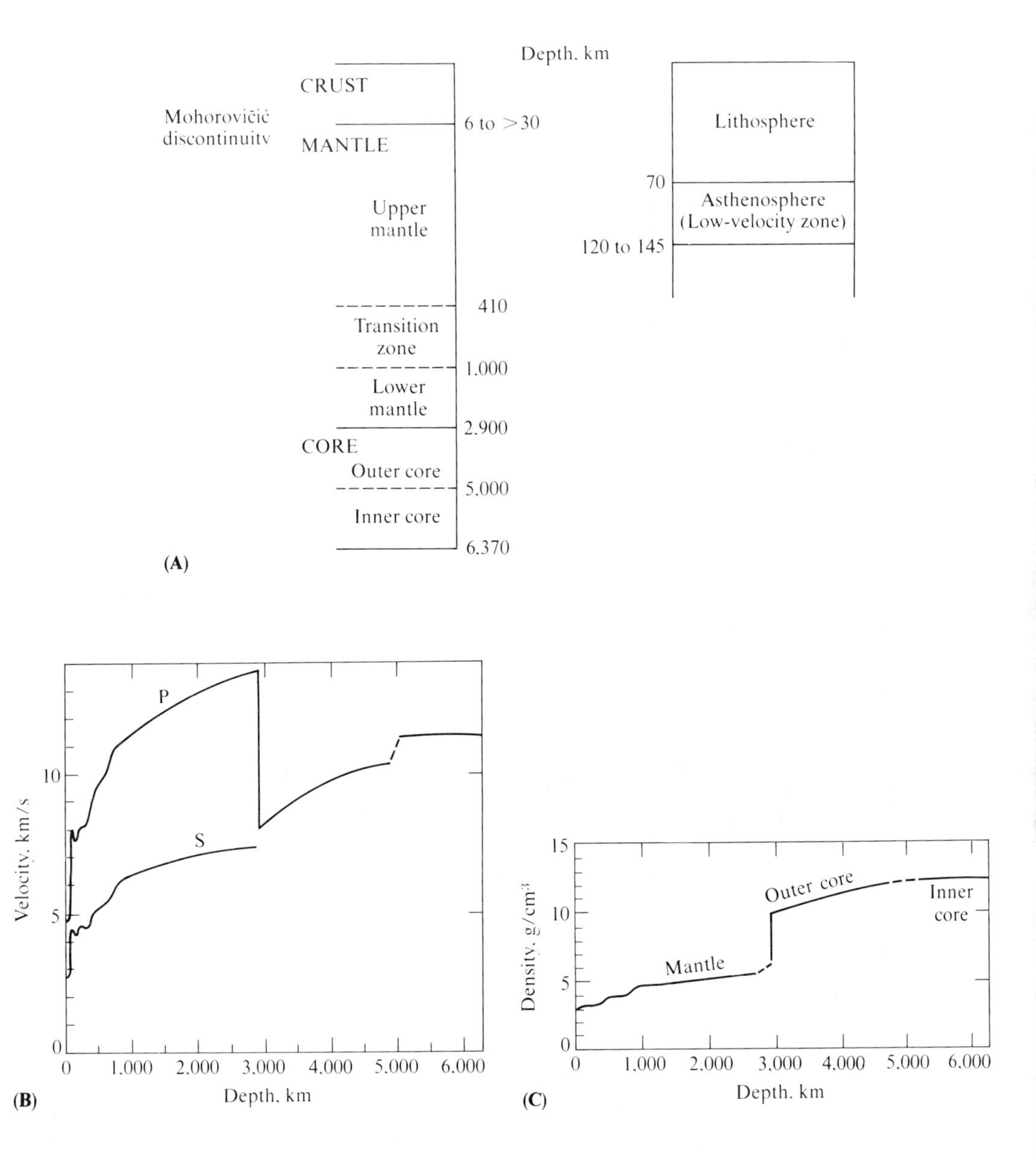

Figure 6–2
(**A**) The major zones within the earth. The low-velocity zone is of variable depth and may be absent beneath some continental areas. Not to scale. (**B**) Seismic velocities within the earth. (**C**) Density change as a function of depth.

boundary indicates an outer liquid portion of the core. Refraction of the P waves at 5,000 km favors the idea of a solid inner core.

Calculated values of densities within the mantle, based on seismic velocities, indicate that (with the exception of the low-velocity zone) mantle densities increase as a function of depth, from 3.5 g/cm^3 to 5.5 g/cm^3. Examination of the wave velocities shows that the increase is not regular. It is known that every material is somewhat compressible with pressure, and that greater depths of burial should result in a regular increase in density as a function of depth. This seems to be the case for the upper mantle (30–410 km) and the lower mantle (1,000–2,900 km); but the intermediate portion (410–1,000 km) is characterized by greater than normal velocity gradients; this region, known as the transition zone, reflects either a change in composition or a structural rearrangement of the atoms that comprise the minerals.

The Geothermal Gradient

Knowing the density variation with depth from seismic velocities, it is then possible to calculate the mass of materials within the various layers of the earth, which in turn yields the gravity and pressure distribution as a function of depth (see Figure 6–3). Pressure increases as a function of depth, whereas gravitational acceleration

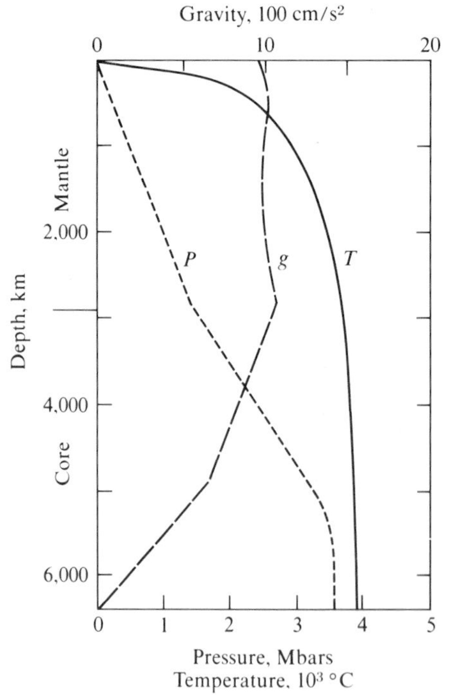

Figure 6–3
Variation in pressure (P), gravitational acceleration (g), and temperature, (T) with depth. A megabar is 1×10^3 kbar. [From P. J. Wyllie, 1971, *The Dynamic Earth* (New York: John Wiley and Sons), Fig. 3-11b, from various sources as summarized in J. A. Jacobs, R. D. Russell, and J. T. Wilson, 1959, *Physics and Geology* (New York: McGraw-Hill).]

reaches a maximum at the core–mantle boundary and then decreases to zero at the earth's center.

The next problem is to determine the geothermal gradient—the change of temperature as a function of depth. Here we pass into a labyrinth of speculation, conjecture, and dreams (see Figure 6–4). The reason for the variety of different answers to the problem is that the geothermal gradient is determined by the earth's origin and history, and the chemical and physical properties of materials comprising the earth.

Temperatures at and near the earth's surface can be determined in deep wells and by means of heat flow data. These temperatures can be extrapolated to depths of a few tens of kilometers. Below about 100 km it seems likely that significant cooling of the earth has not taken place because of the low conductivities of rocks. Deeper temperatures are therefore dependent upon gravitational energy converted to heat during the earth's accretion, and later events such as radioactive decay, earth tides, and possible convective overturn that resulted in the formation of internal layering. Modifying processes include convection within the mantle and direct transport of heat to the surface by magmas (which is considered to be minor).

With all of these variables it is easy to see why a variety of interpretations exists. Some proposed geothermal gradients are based upon interpretations of the earth's early thermal history, others on physical properties (seismic wave velocities and electrical conductivity), convective processes, geochemical and petrological inferences,

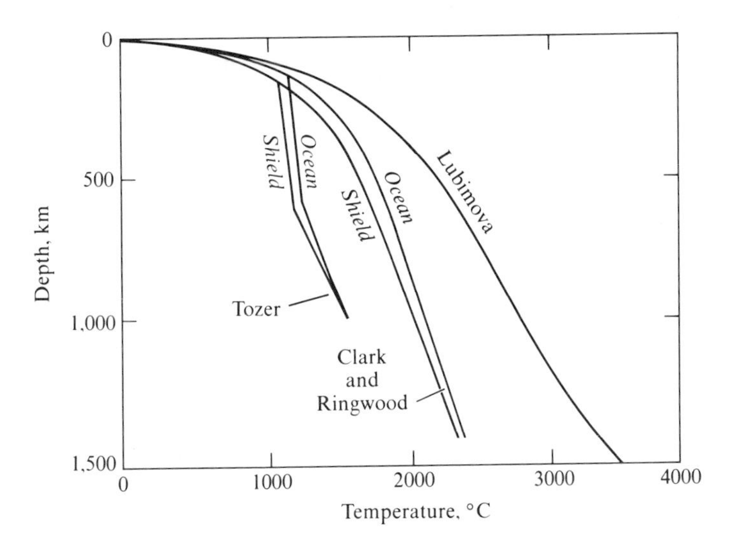

Figure 6–4
Estimates of the geothermal gradient (temperature as a function of depth). [From P. J. Wyllie, 1971, *The Dynamic Earth* (New York: John Wiley and Sons), Fig. 3–12, from various sources.]

and combinations of these. Some constraints to theory do exist, however, so all is not lost. It is known, for example, that melting occasionally takes place in the upper mantle; mantle compositions, melting points, and structural changes are known to some extent, so upper limits can be set on that basis. Other limits are set by geophysical and chemical limitations.

The commonly accepted geothermal gradients proposed by Clark and Ringwood are used in this text. Note that two gradients are given, one for oceanic areas and the other for continental shield areas. These two gradients were suggested because the heat flow data are similar in both regions; as the oceanic crust is lacking a sialic layer (which is relatively high in heat-producing radioactive elements) and is subjected to convective processes, it is presumed that the gradient is higher within the oceanic mantle. At greater depths both gradients converge.

Meteorites

In order to deduce the composition of the inaccessible mantle and core, a number of different types of evidence have been used. One of these is meteorites. It is now generally accepted that meteoritic bodies were created at about the same time as the earth from gravitational collapse of the same materials that comprise the earth. The paths of meteorites indicate that they come from the asteroid belt and follow elliptical orbits that can result in collision with the earth. Although meteorites exist in a wide variety of chemical and mineralogical types, they can be subdivided into two major varieties—the stony meteorites and the iron meteorites. The *iron meteorites* consist mostly of iron with minor amounts of nickel and iron sulfide. Estimates of the density of this meteoritic iron–nickel alloy at pressures and temperatures thought to exist in the earth's core are consistent with the hypothesis that the core is probably composed of this type of material. Stony meteorites, on the basis of corresponding data, are considered to be representative of rocks comprising the earth's mantle.

Stony meteorites, which often contain some iron–nickel alloy, are composed mainly of silicates. The relative abundances of nonvolatile elements in the stones (Mg, Si, Ca, Al, and Fe) are similar to those of the sun and other stars. It probably is not simply coincidence that the relative volumes of mantle and core in the earth (83.5% and 16.2%) are similar to the percentages of stony to iron–nickel material in observed meteorite falls. Using meteorites as a basis, mantle compositions are estimated to be about 90% SiO_2, MgO, and FeO; that is, the mantle is composed of ultramafic material. The remaining 10% is mainly Al_2O_3, CaO, and Na_2O.

Xenoliths and Xenocrysts

An additional source of information about the types of rocks within the earth comes from xenoliths and xenocrysts. Vertically ascending magma may carry up fragments from the magmatic source area (cognate xenoliths), or fragments acquired from the wall rock at higher levels. If the fragments are brought from great depths at a slow rate of ascent, it is likely that they will undergo change during the rise. Such changes may include reaction with the surrounding melt, phase changes in which dense phases convert to less dense structures, or conversion of high-temperature minerals to low-temperature equivalents. Ideally then, in order to have samples truly representative of mineral assemblages characteristic of the mantle and lower crust, the xenoliths should be brought rapidly to the surface. Fortunately, the variety of peridotite known as kimberlite furnishes such a mechanism. Kimberlites contain diamonds, known from laboratory experiments to be stable only at very high pressures. It is, therefore, assumed that kimberlitic magma originates at very great depth (150–300 km). Kimberlite magmas reach the surface very rapidly by an explosive process, and commonly contain xenoliths brought from all levels; those of deep origin are generally able to maintain the high pressure assemblages and phase compositions characteristic of the upper mantle. The most common types of deep-seated xenoliths are peridotites and eclogites. Recall that peridotite is a rock consisting mainly of olivine and pyroxene, whereas eclogite (containing pyroxene and garnet) is the high-pressure equivalent of basalt. The mineralogy of the peridotites is similar to the mantle compositions deduced by the meteoritic approach. Peridotite and eclogite nodules are also brought to the surface by some basaltic lavas, which are known from geophysical data to have originated in the shallow levels of the upper mantle. Although the earth's mantle is considered by some investigators to consist mainly of peridotite composition, it has been estimated that the upper portions of the suboceanic mantle (between 80 and 150 km) may contain up to 50% eclogite. Lateral and vertical zoning in this region with respect to both composition and mineralogy suggest the presence of a variety of peridotites, eclogite, and dunite.

High-Pressure Experimentation

Another approach to mantle mineralogy has been through high-pressure experimentation. A variety of materials have been synthesized that fulfill two basic criteria. The first is that they have appropriate densities for the various pressure–temperature conditions of the mantle. Secondly, their compositions must be such that they will furnish basaltic liquids when partially melted (20–40% of the total) under mantle conditions. This second condition is necessary because basaltic melts are known to be produced by partial melting processes in the upper mantle at rift zones. This approach also yields a material that is a variety of peridotite.

MANTLE PETROLOGY

One of the synthetic varieties of peridotite has been termed pyrolite (pyroxene–olivine rock). A typical composition (in weight percent) suggested by Ringwood (1966) is the following:

SiO_2	45.20	MnO	0.14
TiO_2	0.71	NiO	0.20
Al_2O_3	3.54	MgO	37.48
Cr_2O_3	0.43	CaO	3.08
Fe_2O_3	0.48	Na_2O	0.57
FeO	8.04	K_2O	0.13

Figure 6–5 shows the experimentally determined changes in pyrolite petrology as determined by Green and Ringwood (1970) to pressures of about 50 kbar. The phase boundaries between the various solid assemblages are controlled mainly by differences in pressure rather than temperature. Note that two types of changes occur: (1) the loss or gain of a mineral phase, and (2) the change of composition of various mineral phases within their stability field (as shown here by the percentage of Al_2O_3 in orthopyroxene within the garnet pyrolite field). These mineralogical changes, as well as additional ones at greater depth, occur mainly through a variety of adjustments of the constituent minerals to higher pressure. With increasing pressures minerals are compressed; the most compressible minerals are favored over the least compressible ones, and compositional rearrangements may take place to increase the amount of the most compressible phase. Secondly, minerals may undergo adjustments into more compact arrangements by a shifting or rotation of structural units (as in the case of quartz converting to the denser phase coesite due to pressure). Finally, materials may change their coordination state as a result of increased pressure; as the larger atoms (typically oxygen) are more compressible than the smaller, the relative sizes of the different types of atoms are changed and new coordinations are developed. Thus at high pressures it is possible to coordinate six oxygen atoms around silicon, rather than the usual four observed under crustal conditions (an example being the high-pressure conversion of coesite, with a coordination number of 4, to stishovite, with a coordination number of 6).

The phase changes shown in Figure 6–5 are the following. At low pressures the batch consists of olivine, Al-poor pyroxene, and plagioclase (plagioclase pyrolite). With higher pressures this converts to olivine, Al-rich pyroxene, and spinel (spinel pyrolite); at still higher pressures it consists of olivine, Al-poor pyroxene, and garnet (garnet pyrolite). Wyllie (1975) has summarized the work of a number of investigators on this subject; Figure 6–6 shows the experimentally determined changes to a depth of 700 km. At low pressures the Al-rich phase is plagioclase. With increasing pressure plagioclase and olivine react to form aluminous pyroxenes and spinel. At higher pressures spinel reacts with orthopyroxene to yield pyropic garnet and olivine.

All of the above changes have been demonstrated in laboratory experiments under carefully controlled conditions of known composition, temperature, and pressure. But do we have any indication that these changes actually occur in the mantle? The high-pressure laboratories at the earth's surface are far removed from the depths of the mantle. Can a few pinches of material in a laboratory reveal the mineralogy of rocks several hundred kilometers below in the mantle? The answer at the moment is run-

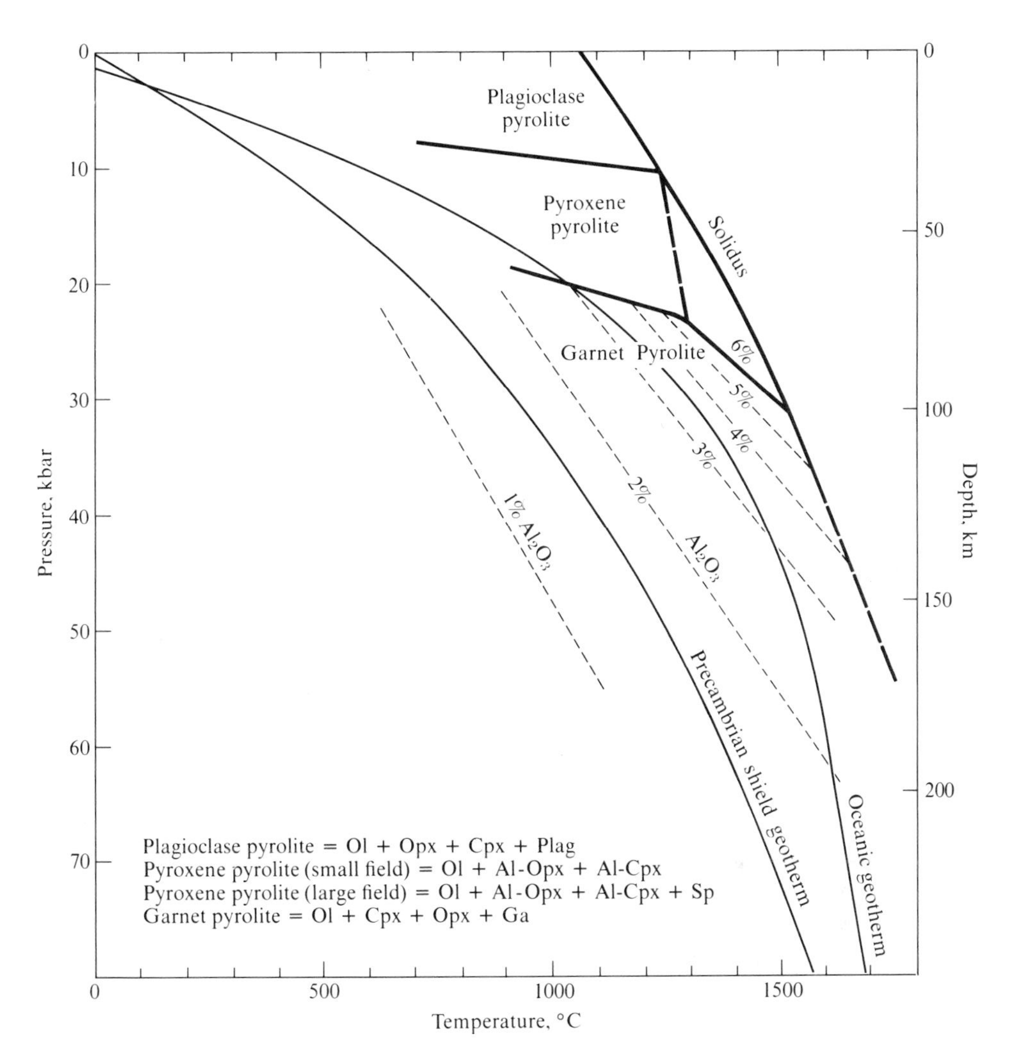

Figure 6–5
The stability fields of different mineral assemblages produced from a pyrolite composition. Dashed lines refer to the weight percentage of Al_2O_3 in orthopyroxene that is in equilibrium with garnet. Ol = olivine, Opx = orthopyroxene, Cpx = clinopyroxene, Plag = plagioclase, Ga = garnet, Sp = spinel, prefix Al = Al-rich. [From D. H. Green and A. E. Ringwood, 1970, *Phys. Earth Planet. Interiors, 3,* Fig. 2. Geotherms from D. H. Green and A. E. Ringwood, 1967, *Earth Planet. Sci. Letters, 3,* 151.]

ning strongly in favor of the experimenters. Studies of xenoliths reveal many examples of high-pressure assemblages similar to those produced in the laboratory. A second line of evidence comes from the interrelationship of experimentally determined phase changes, geothermal gradients, and seismic discontinuities.

Consider Figure 6–6. The phase changes undergone by peridotite are shown in terms of depth and temperature. These transitions occur over broad zones whose

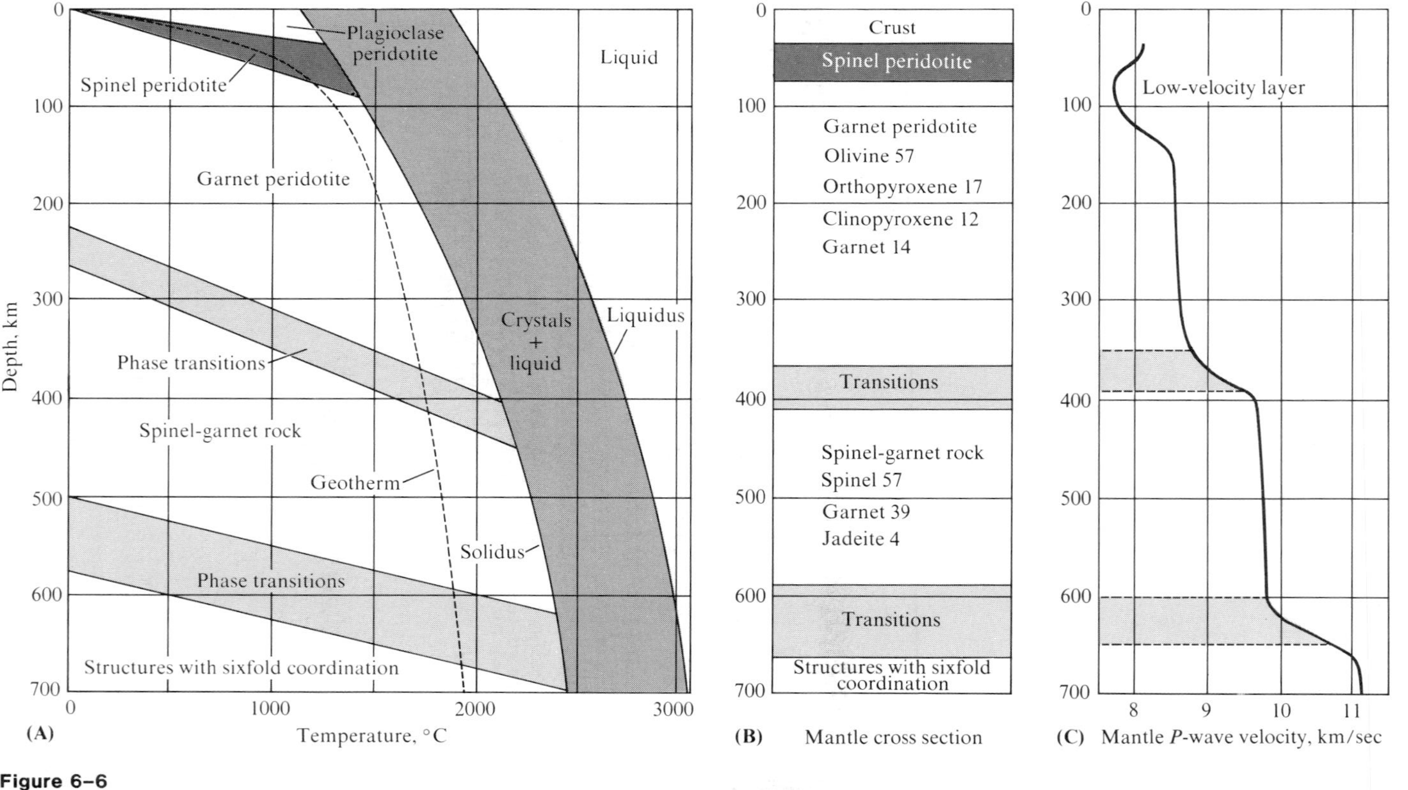

Figure 6–6
(**A**) The phase diagram of peridotite in terms of lithostatic pressure (depth) and temperature (extended to 700 km). The dashed line is an average geotherm. (**B**) Rock types and mineral percentages that are encountered along the geotherm in part A. (**C**) Velocity of P-waves as a function of depth. The depth at which significant changes in velocity of P-waves occur coincides with the depth of expected phase changes in the mantle. [From P. J. Wyllie, "The Earth's Mantle." Copyright © 1975 by Scientific American, Inc. All rights reserved.]

positions are a function of pressure and temperature. An average geothermal gradient (dashed line) is superimposed on the diagram in part A. The gradient crosses phase transition zones at about 400 km and 650 km. The transition at 400 km has been related to two phase changes. One of these is the pyroxene–garnet transformation; pyroxenes, which comprise more than 25% of the peridotite at this level, convert to the garnet structure with an increase in density of about 10%. The garnet structure has a considerable portion of the silicon atoms in sixfold coordination. The second reaction at about this same level is a conversion of olivine through spinel to β-Mg_2SiO_4 (a spinel-like structure), which involves a density increase of about 8%. The other major transition zone occurs at depths of about 650 km (230 kbar), which is just beyond the capability of most conventional static high-pressure devices. Structural changes in minerals have been studied at this level by using shock wave techniques, a recently developed ultra-high-pressure diamond cell capable of 1.5 Mbar pressure (similar to the core–mantle boundary), or by squeezing germanate analogs of the silicates. (As germania is slightly larger than silica, germania structural analogues of silicates undergo conversions at lower pressures than silicates.)

The phase changes at both 400 and 650 km produce significant changes in volume and mineralogy, which in turn create differences in density and elastic properties. These differences should be indicated as discontinuities in seismic velocities. A glance at Figure 6–6C shows that this is indeed the case. The agreement provides a strong indication of the validity of the seismic velocity technique, high-pressure experimentation, and the estimated geothermal gradient; if any of these three approaches were incorrect, the results would not agree. The transition zone of the mantle (410–1,000 km) can thus be regarded as a region of more or less fixed chemical composition in which phase transitions are present. Small changes in the Fe/Mg ratio and silica content do not invalidate this conclusion.

The Low-Velocity Zone

Studies of anhydrous materials at various confining pressures and temperatures indicate reasons for discontinuous increases in seismic velocities in the mantle as a function of increasing depth. But such studies do not explain the decrease in seismic velocity encountered in the asthenosphere, near the top of the upper mantle. This decrease of velocities usually begins at about 70 km in oceanic regions and reaches minimum velocities at about 120 to 150 km. In shield areas the low-velocity zone, when present, is deeper, and minimum values are reached at about 200–250 km depth. In addition to velocity decrease, seismic waves are more rapidly damped when traveling through this layer than through the lithosphere above or the mantle below. The low-velocity zone should not be present with an anhydrous mineral assemblage and a typical geothermal gradient. Therefore an alternate mechanism is called for—either a local increase in what is considered an average geothermal gradient, or decrease in temperature of the melting curves. A decrease in melting temperatures sufficient to intersect a standard gradient would cause the rocks at this level to be partially melted, which would in turn cause both damping of seismic waves and decrease in S velocities. Most investigators feel that the geothermal gradient is essentially correct, and that the melting curves are depressed by the presence of a small amount of fluid in the mantle.

The mantle is known to contain some water, as seen by the occasional presence of the hydrous minerals phlogopite (Mg-rich biotite), amphibole, or titanoclinohumite, $Mg_9(SiO_4)_4(F,OH)_2$, in some mantle-derived peridotite and eclogite nodules (Clark and Ringwood, 1964). Water is known to be carried into the mantle in subduction zones, and must have been present during an early stage of earth history. The presence of even a slight amount of water is sufficient to cause a significant decrease in the initial melting temperature of peridotite. In addition, some CO_2 is present in the mantle as seen by the presence of CO_2-filled fluid inclusions in peridotites. Recent experimental studies have indicated that if CO_2 were present in the upper mantle it would cause partial melting and thus account for the low-velocity zone.

Wyllie (1971a) has reviewed the effect of small amounts of water on the melting relations of both gabbro and peridotite. Both would behave in a similar manner when subjected to mantle conditions. Figure 6–7A (Wyllie, 1971b) shows a somewhat schematic diagram of the melting relations of peridotite to 600 km under dry conditions. Figure 6–7B shows the same system in the presence of 0.1% water. When a limited amount of water is present in the system, as in part B, a region of slight melting is developed at temperatures below the (anhydrous) solidus (in the lightly shaded area). The amount of liquid produced is on the order of a few percent, and it does not vary greatly within this zone as a function of temperature or pressure. This region can be thought of as a zone of incipient melting, rather than one in which magma is present. The average geotherm intersects the region of incipient melting between 100 and 300 km. This roughly coincides with the low-velocity zone and is considered to be the reason for its existence.

SUMMARY

Seismic data have established that the earth is a layered body. Studies of meteorites indicate that the earth's mantle is composed of peridotite, and the core is iron-rich. Xenoliths brought from the mantle agree with this. High-pressure experimentation with peridotites of appropriate composition reveal that mantle mineralogy changes as a function of pressure and temperature. Using average geothermal gradients, these phase changes can be related to seismic discontinuities. The presence of a minor amount of H_2O (and/or CO_2) in the upper mantle results in a very small amount of melting. Using an average geothermal gradient, this zone of incipient melting is found to occur at the approximate depth of the low-velocity zone.

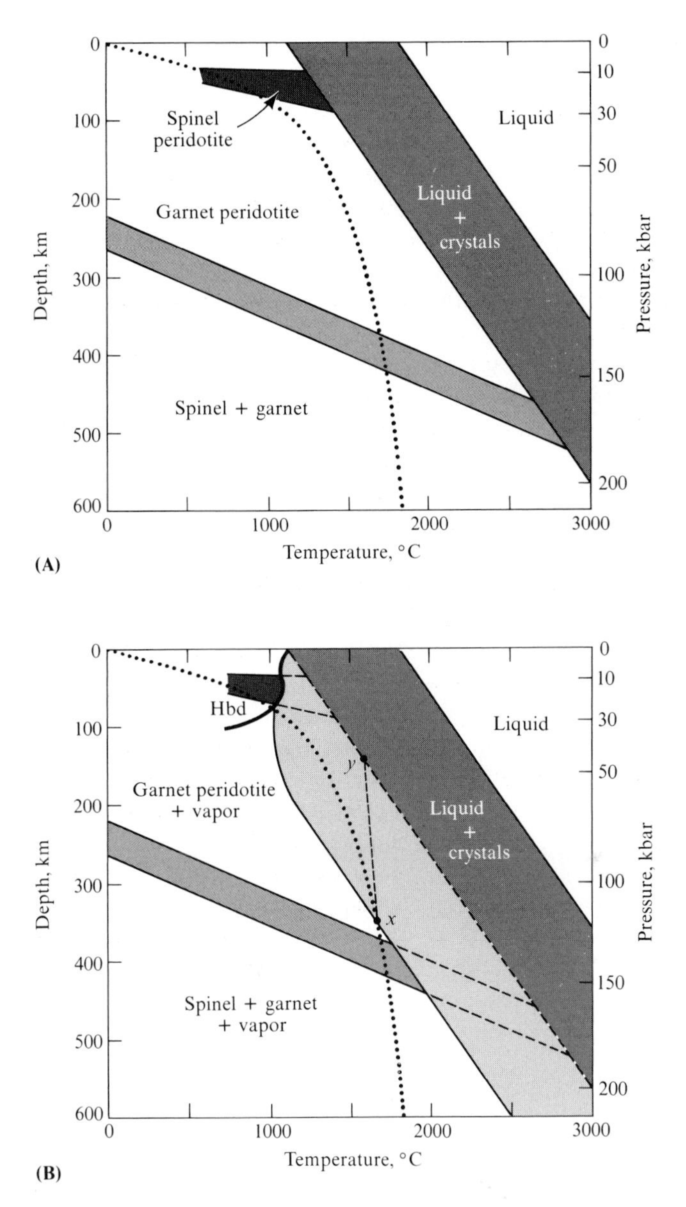

Figure 6–7
Schematic phase diagrams for phase changes of peridotite composition occurring in the mantle, extrapolated to a depth of 600 km. The dotted line represents a hypothetical isotherm. (**A**) Dry peridotite. (**B**) Peridotite with 0.1% H_2O. The lightly shaded band indicates a *PT* region of incipient melting; the darker parallel band is almost equivalent to melting under anhydrous conditions. The thick line indicates the maximum stability of hornblende (Hbd). [From P. J. Wyllie, 1971, *Jour. Geophys. Res., 76,* Figs. 2, 7.]

FURTHER READING

Anderson, D. L. 1970. The plastic layer of the earth's mantle. In *Continents Adrift.* San Francisco: W. H. Freeman and Company, pp. 28–35.

Bullen, K. E. 1970. The interior of the earth. In *Continents Adrift.* San Francisco: W. H. Freeman and Company, pp. 22–27.

Clark, S. P., and A. E. Ringwood. 1964. Density, distribution and constitution of the mantle. *Rev. Geophysics, 2,* 35–88.

Green, D. H., and A. E. Ringwood. 1970. Mineralogy of peridotitic compositions under upper mantle conditions. *Phys. Earth Planet. Interiors, 8,* 359–371.

Jordan, T. H. 1979. The deep structure of the continents. *Scientific Amer., 240,* 92–100, 103–107.

Liu, L. 1977. Mineralogy and chemistry of the earth's mantle above 1000 km. *Geophys. Jour. Royal Astr. Soc., 48,* 53–62.

Ringwood, A. E. 1966. The mineralogy of the upper mantle. In P. M. Hurley (ed.), *Advances in Earth Science,* pp. 357–399. Boston: MIT Press.

Ringwood, A. E. 1977. Composition of the core and implications for origin of the earth. *Geochem. Jour., 11,* 111–135.

Ringwood, A. E. 1970. Phase transformations and the constitution of the mantle. *Phys. Earth Planet. Interiors, 3,* 109–155.

Ringwood, A. E. 1975. *Composition and Petrology of the Earth's Mantle.* New York: McGraw-Hill, 618 pp.

Wyllie, P. J. 1971a. Role of water in magma generation and initiation of diapiric uprise in the mantle. *Jour. Geophys. Res., 76,* 1328–1338.

Wyllie, P. J. 1971b. *The Dynamic Earth.* New York: John Wiley and Sons, 416 pp.

Wyllie, P. J. 1973. Experimental petrology and global tectonics—a preview. *Tectonophysics, 17,* 189–209.

Wyllie, P. J. 1975. The earth's mantle. *Scientific Amer., 225,* 50–63.

7

The Formation of Magmas: Rift Zones

A *magma* can be defined as a large, partially molten mass of rock material whose major constituent is usually a silicate liquid; the melt may contain rock fragments that represent unmelted refractory residues from the original solid source rock, rock fragments dislodged from the walls of the magma chamber, crystals formed by partial crystallization of the melt, and in some cases a separate gas phase.

Seismic data indicate that the earth, with the exception of the outer core, is essentially solid. Magmas, originating in the earth's crust or mantle, are therefore formed by the melting or partial melting of rocks, rather than from the mobilization of a preexisting primitive melt. The question of magma formation must therefore deal with the circumstances under which rocks are melted.

MELTING PROCESSES

A number of melting mechanisms have been suggested, some of which are listed below:

1. Release of lithostatic pressure.
2. Migration of rocks to areas of lower lithostatic pressure.
3. Transport of rocks into hotter zones.
4. Addition of fluids.
5. Heat produced from radioactive decay.
6. Heat produced by friction.
7. Escape of heat from Hell through vertical fissures.

As several of these processes may be involved in the melting process, we shall consider magma formation in terms of the principal tectonic environments, starting in this chapter with divergent plate junctions.

In order to build a reasonable picture of what is happening at divergent plate junctions such as the Mid-Atlantic Ridge it is necessary to take the facts into account; this limits the number of possibilities that one can devise, but also manages to cut down adverse criticism. The facts are as follows:

1. The rate of heat flow is high at mid-ocean ridges as compared to the normal ocean bottom and continental edges.
2. Volcanism is common at mid-ocean ridges.
3. The type of lava produced is basalt. Although basaltic compositions are variable, basalts contain few radioactive elements as compared to silicic rocks.
4. Based upon seismic data and high-pressure experimentation, the type of rock present in the upper mantle is peridotite. Xenoliths of upper mantle peridotite are poor in radioactive elements.
5. Partial melting experiments with peridotites at high pressures have demonstrated that the first-formed liquids are of basaltic composition.

It is reasonable to conclude that the high heat flow at the mid-ocean ridges is a result of basaltic melts at and near the surface. The basaltic magma is probably formed by partial melting of peridotite in the upper mantle. The next question is why this happens. As both the basalts and peridotites at the mid-ocean ridges are low in heat-producing radioactive elements such as U, K, and Th, melting does not occur because of unusually high concentrations of these elements.

Melting is thought to result from convective processes. Convection not only can cause melting of rocks, but also accounts for divergence of lithospheric plates.

The idea behind convection ultimately relates to the cold accretion theory of earth origin. This states that the earth was formed by the accretion of solid particles of diverse composition at temperatures of a few hundred degrees Centigrade. The accretionary process resulted in heating of the earth, with formation of large amounts of melt. Under the physical, chemical, and gravitational constraints, a layered body was created. It is now thought that the differentiation of the earth was incomplete, with the result that compositional differences exist within the various layers of the earth's mantle. These differences include heat-producing radioactive elements; the result is that some portions of a particular level within the earth are hotter than other portions. It has been noted, for example, that seismic velocities and electrical conductivity values of the mantle are different beneath the western and eastern portions of North America as a result of temperature differences. Calculated values of geotherms under shield and oceanic areas indicate differences in temperature of a few hundred degrees at depths of about 100 km. Considering the possible viscosity of the mantle and the unequal distribution of heat sources, lateral temperature differences should be common within the mantle. Even small temperature differences should be sufficient to cause density differences between similar rocks at the same level. The lighter rocks will migrate upward by plastic flow while heavier ones simultaneously move downward, causing the formation of a convection cell. The depth and lateral extent of convective cells are not well known, and present theories range from convective cells that extend to the base of the mantle to others that are limited to depths slightly below the low-velocity zone. The current idea is that convective cells rise at the divergent plate junctions, move out horizontally away from these junctions, and sink at subduction zones near continental edges.

Figure 6–7, discussed earlier, shows the stability regions of the various mantle peridotites. The dotted line is a possible geothermal gradient that passes through the fields of spinel peridotite, garnet peridotite, and spinel–garnet rock. These changes correspond in depth to seismic discontinuities. As the minerals are mainly anhydrous (with the possible exception of some hornblende or phlogopite), the solidus and liquidus lines rise in temperature with pressure. In the presence of a minor amount of water (0.1%) the system is modified to include a zone in which a very small amount of melting can occur (the zone of incipient melting). This zone roughly corresponds in depth to the low-velocity zone. We will consider how these conditions can lead to the formation of magma.

Beneath the mid-ocean ridges we can assume the presence of vertically rising hot solid material. This could be due to convective movement as discussed above, or if that idea is not pleasing the argument can be reversed; it could be argued that the separation of the plates results in lower pressures beneath the divergent junction, which in turn leads to uprising of solid mantle material. After vertical rise of mantle rocks is initiated, the temperature difference between these hotter, rising, solid diapirs and the surrounding cooler rocks is increased as the diapirs move into higher levels. The rising diapirs will encounter the zone of incipient melting (at a point such as x in Figure 6–7B). Here a slight amount of melt is created, which tends to facilitate the rise of the essentially solid diapir. Continued rising brings the diapir to the (anhydrous) solidus at y, where large amounts of melting can occur. When the melting has increased to a significant extent (perhaps 20–40%), the melt, having a significantly lower density than the surrounding rocks, can rise to the surface as a magma. Continuation of this process results in a change in the isotherms at the rift zones. Possible temperature distributions are shown in Figure 7–1. The left edge of each diagram represents the rift zone; horizontal distances away from the rift increase to the right and depth increases downward. Based upon similar calculated heat distributions, Wyllie (1973) has presented a schematic section (see Figure 7–2) showing the distribution of rock types and magma at an active mid-ocean ridge. Rise of the isotherms in the rift zones has resulted in increasing the depth of phase changes within the mantle. This diagram also indicates that there is a considerable thickening of the zone of melting as the ridges are approached. In fact the dry solidus is exceeded within a vertical and horizontal zone greater than 200 km.

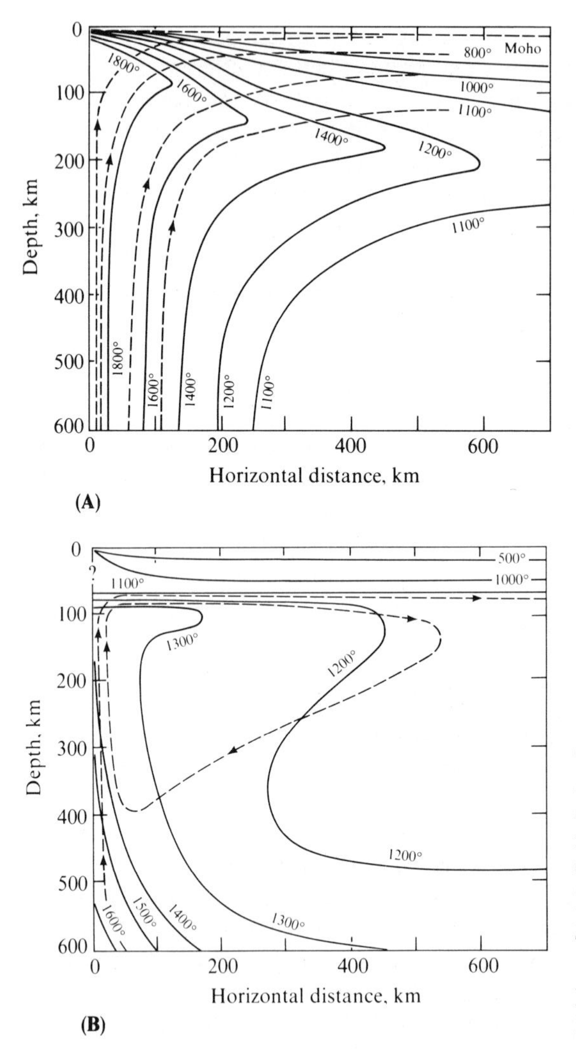

Figure 7–1
Thermal models for the mantle beneath an ocean ridge. The dashed lines with arrows show the directions of particle motion. [Part A after E. R. Oxburgh and D. L. Turcotte, 1968, *Jour. Geophys. Res., 73,* Figs. 11, 12; part B after K. E. Torrance and D. L. Turcotte, 1971, *Jour. Geophys. Res., 76,* Fig. 2, and D. W. Forsyth and F. Press, 1971, *Jour. Geophys. Res., 76,* Fig. 12.]

Partial Melting

It has been demonstrated by a wide variety of studies of both natural and synthetic systems that the initial liquids produced from a single composition (such as pyrolite) differ in composition as a result of the *PT* conditions under which melting occurs, as well as the amount of melting. This can be illustrated by consideration of simple binary systems. Consider a hypothetical anhydrous binary system (see Figure 7–3A) under both atmospheric pressure as well as high confining pressures. At atmospheric pressure a mixture of crystals (composition *a*) begins melting to produce a eutectic

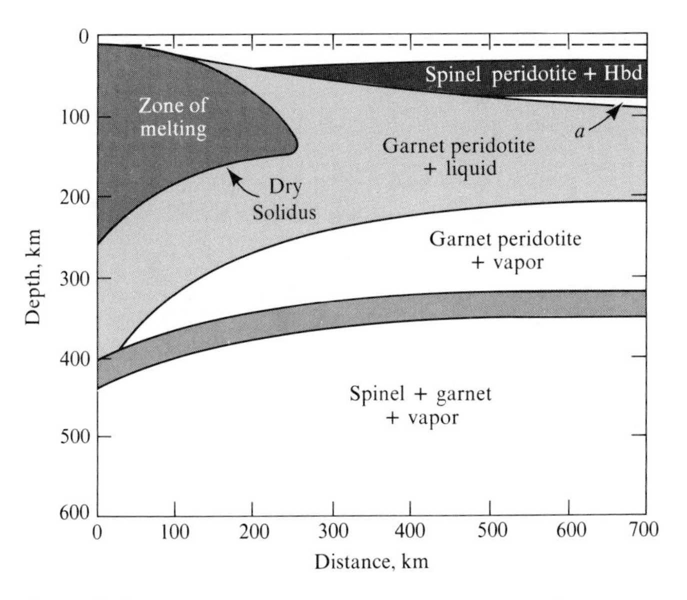

Figure 7–2
Schematic diagram showing the effect of an ocean ridge thermal regime (of Figure 7–1A) on peridotite phase changes and melting. A large area of melt is present and phase changes are vertically depressed. Hdb = hornblende; area *a* consists of garnet peridotite + vapor. [From P. J. Wyllie, 1971, *Jour. Geophys. Res.*, *76*, Fig. 12.]

liquid of composition *b*. In fact any crystalline assemblage made up of *X* and *Y* will begin melting at the eutectic to produce a first liquid of composition *b*. Continued addition of heat to the assemblage will result in more melting of the *X* and *Y* mixture in eutectic proportions at the eutectic temperature until all of the *X* crystals are consumed, leaving only *Y* crystals and a liquid of eutectic composition. Additional heating causes the remaining *Y* crystals to melt. This melting causes change in the melt composition and therefore the melting point, and the liquid changes temperature and composition along the liquidus line until the composition *c* is reached. Here the last crystal is dissolved and the liquid is the same composition as the original batch. Note that the melting process produces a liquid of fixed composition until one phase is eliminated; additional heating causes a change in composition of the melt.

The same mixture of minerals *X* and *Y* would melt differently under high-pressure conditions. Assuming anhydrous solids, the melting temperatures would be increased with higher lithostatic pressure. The rise in liquidus temperatures need not be uniform for the whole system; hence the melting temperature of mineral *X* might rise at a greater rate than that for mineral *Y*, which could result in a shift in the position of the eutectic (see Figure 7–3B). Melting at high pressure of mixture *a* would produce

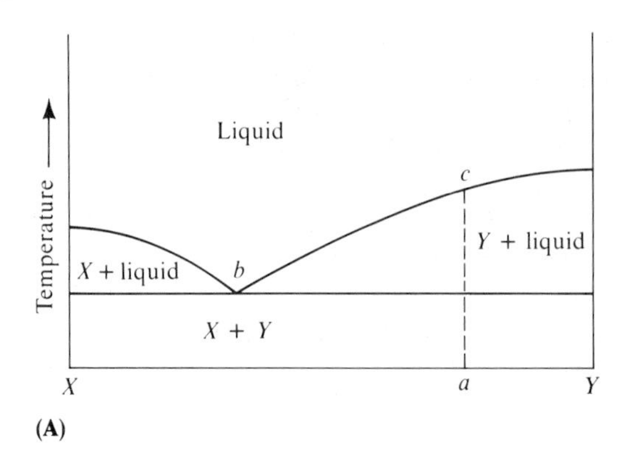

(A)

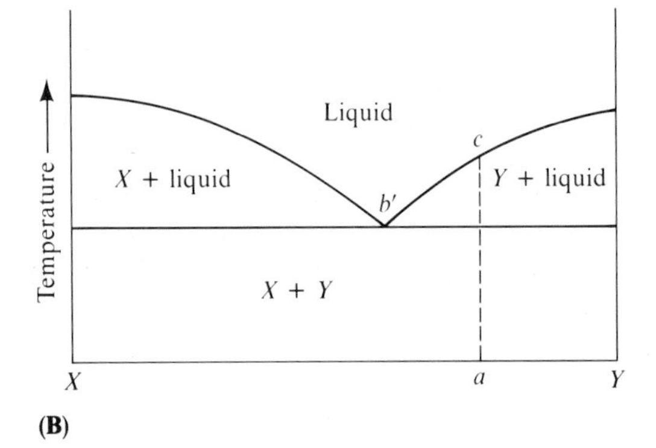

(B)

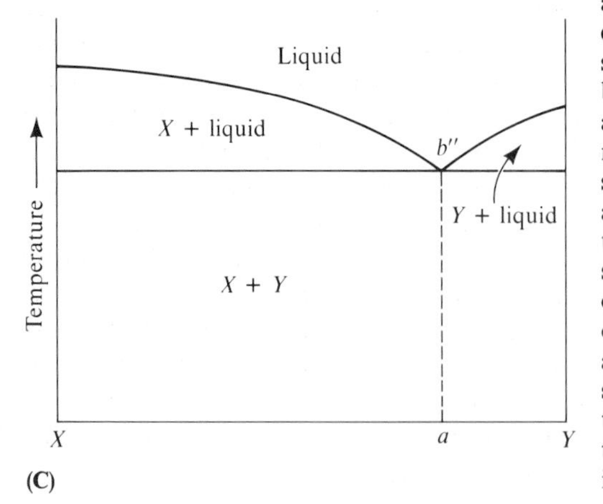

(C)

Figure 7–3
Melting relationships in a simple anhydrous binary system at three different confining pressures. (**A**) The system at atmospheric pressure. A batch of composition *a* melts to produce a liquid whose initial composition and melting temperature is *b*. (**B**) The same system at higher pressure. Both liquidus and solidus curves are raised to higher temperatures, and the eutectic has shifted to *b′*. Melting of a batch of composition *a* produces an initial liquid composition at *b′*. (**C**) The system at still higher pressure. Liquidus and solidus curves are raised, and the eutectic composition *b″* now corresponds to the batch composition *a*. Melting of *a* is initiated and concluded at *b″*.

initial liquids of composition b' (the eutectic composition), which is different in composition than liquids produced at the eutectic b under low-pressure conditions. Continued heating would cause complete melting of solid X, with the remaining liquid following the liquidus line to c as the remaining crystals of Y melted.

The system at still higher pressures might resemble Figure 7–3C, where the composition a coincides with the eutectic. Here initial melting of a produces a liquid (b'') of eutectic composition. Additional heating produces more and more liquid until the crystalline batch is consumed. The composition of the liquid remains fixed and all melting occurs at a single temperature. Note that when partial melting occurs for the same original mixture in these three examples, both the composition and the temperature at the beginning of melting are different.

A more interesting situation is shown in Figure 7–4A. The system X–Y now has an intermediate compound XY that melts incongruently at high pressures and congruently at low pressures. Melting of composition a (a mixture of X and XY) at high pressures produces a first liquid at the peritectic b (due to $XY \xrightarrow{T} X$ + liquid b). Continued heating causes melting of crystals of X, and the liquid migrates in composition to c with complete melting. But consider what would happen if, instead of complete melting, the liquid produced at b is separated from the remaining crystals of X and brought into a region of atmospheric pressure (see Figure 7–4B). If cooled at atmospheric pressure, the liquid b would intersect the liquidus at d. Crystals of XY would be produced, and the liquid would migrate to the eutectic at e, where the remainder would crystallize to yield a crystalline assemblage of XY and Y. If we liken this to melting relations in the upper mantle, it is possible to consider the original rock composed of X and XY as a peridotite. The liquid b might be a basaltic melt that crystallizes under crustal conditions to produce a basalt ($XY + Y$), leaving behind in the mantle residual refractory crystals of X (depleted peridotite or eclogite). The final assemblages of X and $XY + Y$, although separated geographically, must add up to equal the original batch composition a.

It should be noted that fusion processes can be subdivided into equilibrium fusion and fractional fusion. During *equilibrium fusion* the melt produced on heating continually reacts and reequilibrates with the residual crystals. During *fractional fusion* the melt is immediately isolated when formed, preventing further reaction with the residual crystals. Fractional fusion differs from fractional crystallization in that the liquids produced may be discontinuous in composition, in contrast to the compositionally continuous series of melts produced during fractional crystallization.

Melting within the mantle, although much more complicated than the examples given above (as a result of complex chemical compositions and amount and depth of melting), operates in similar manner. Invariant points and boundary curves shift with differences in lithostatic and fluid pressures; new minerals are formed and others are eliminated; the limits of solid solution are changed. In spite of this vast array of problems, some general relations of phase chemistry have been established, and a great deal is known about the melting relations of peridotites.

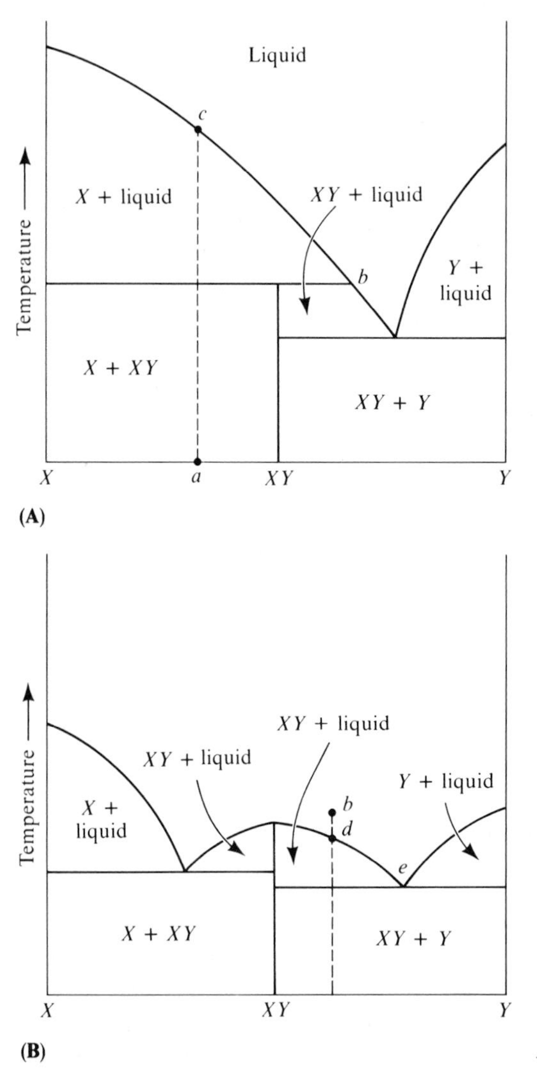

Figure 7–4
A binary system X–Y with an intermediate compound XY. At high pressures (part A) XY melts incongruently, and at low pressures (part B) XY melts congruently. The batch a heated at high pressure produces an initial melt of composition b. If this liquid is removed from the residual crystals and brought to an environment of low pressure (part B), liquid b will cool, migrate in composition to eutectic e, and produce a crystalline batch of XY and Y.

A melting model for pyrolite composition by Green and Ringwood is shown in Figure 7–5. The mineralogy of pyrolite under subsolidus conditions is shown at various depths in the lowermost of each set of blocks. The adjacent higher-level blocks indicate the type and amount of liquids produced by partial melting, as well as the residual crystalline materials. From this diagram it can be seen that melting of pyrolite at different depths produces mafic melts of different compositions. Diapirs rising in rift zones will encounter the solidus at different depths and similarly produce melts of different composition. Another variable brought out by this diagram is that the melt composition changes (see 60-km level) as a function of the amount of melt produced. Thus melt composition in the rift zone is partially controlled by the amount of melt present when the melt leaves the diapir above the solidus.

The separated melt may rapidly rise to the surface and retain its original composition or it may rise slowly or intermittently; the slower rise allows cooling and partial crystallization during ascent; crystallization at lower pressures during the rise permits considerable fractionation of the melt as well as interchange of materials with the country rock. Figure 7–6 provides an illustration of this. An olivine tholeiite melt separated from the pyrolite host at 35–70 km (box at lower right) could rise directly to the surface with no fractionation and crystallize completely to yield olivine tholeiite. It could partially crystallize at the 35–70 km level to produce alkali–olivine basalt melt, or with still more fractionation, an olivine basanite melt. [An olivine basanite is a basaltic rock characterized by the presence of calcic plagioclase, augite, olivine, and a feldspathoid (nepheline or leucite).] Another possibility is that the original olivine tholeiite melt (which is low in Al_2O_3) could fractionate during ascent to the 15–35 km level and produce an Al-rich olivine tholeiite melt. Still another possibility (not shown) is that the original olivine tholeiite could fractionate at levels above 15 km, where crystal segregation would yield quartz tholeiite. No doubt this model will undergo considerable revision as additional information is gained, but it is clear that high-pressure experimentation has made a great leap forward in accounting for some of the varieties and possible mechanisms of formation of the various basic rocks.

OBSERVATIONS AT THE MID-OCEAN RIDGES

Recent evidence on the nature of mid-ocean rift zones has come from the French-American Mid-Ocean Underseas Study (known as project FAMOUS). Many of the reports on various aspects of this investigation are published in the April and May 1977 issues of the Geological Society of America Bulletin, as well as in Langmuir et al. (1977). The project area covered a fairly well-known portion of the Mid-Atlantic Ridge, near the Azores Islands. The studies were both regional in extent (the greater FAMOUS area) as well as quite detailed in a smaller area (the FAMOUS area) as shown in Figure 7–7. Studies included bathymetry, magnetics, gravity, seismic refraction, bottom currents, heat flow, and water temperature. In addition, many dives were made with American and French submersibles in order to acquire an extensive collection of samples from very specific locations (see Figure 7–8).

The Mid-Atlantic Ridge in the FAMOUS area includes a central rift valley (see Figure 7–9) 1.5–3 km wide and 100–400 m deep. The valley is essentially symmetrical in cross section and contains a central discontinuous ridge (100–240 m high and 800–1,300 m wide), or occasionally a central trough (200–600 m wide).

It has been established by various studies that the central ridge is located along the major rifting line and contains the youngest volcanic rocks; these are fresh glassy tholeiitic basalts with pillows, lava tubes, and phenocrysts of olivine, clinopyroxene,

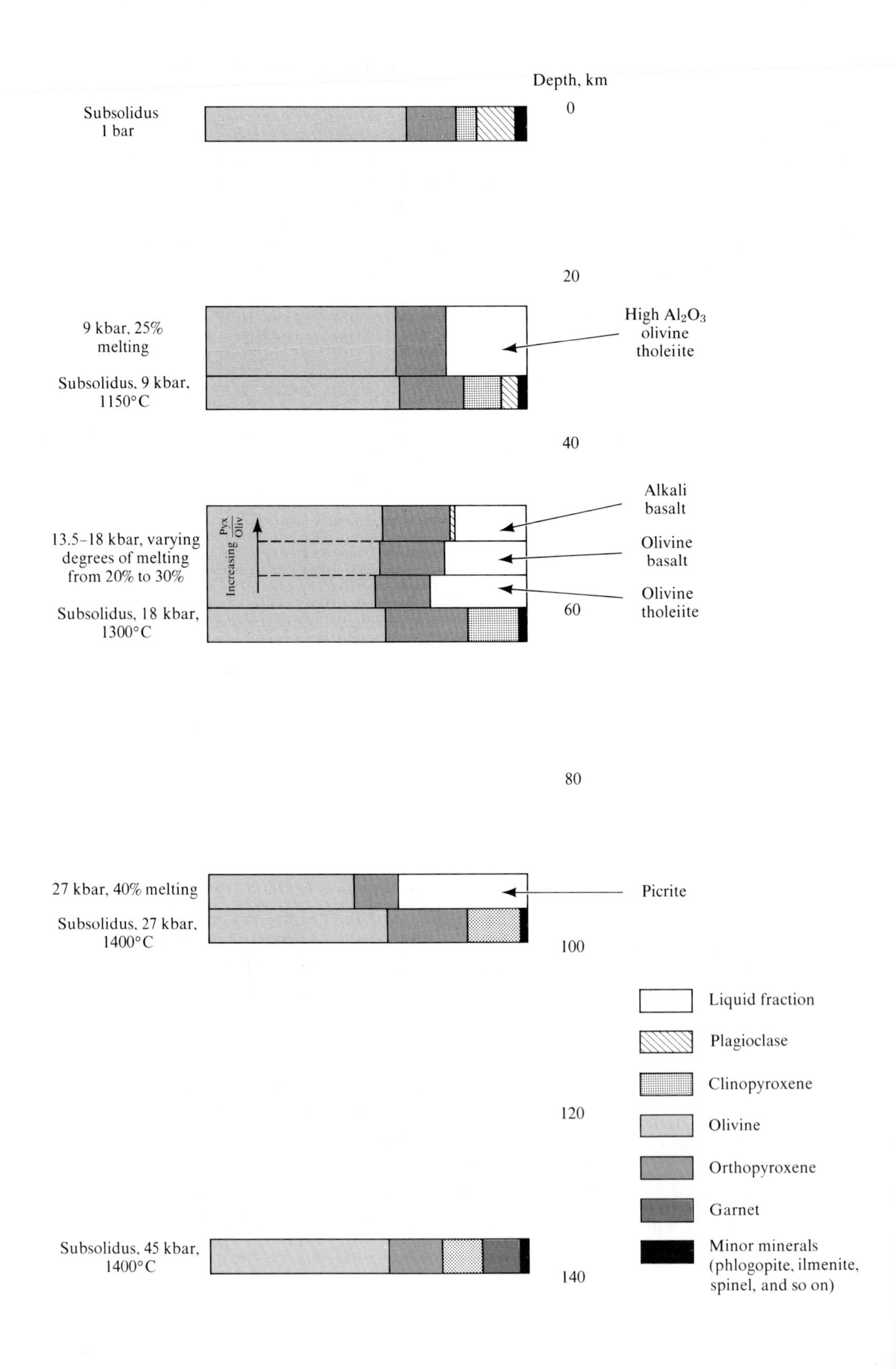

Depth, km
0
20
40
60
80
100
120
140
Subsolidus
1 bar
9 kbar, 25%
melting
Subsolidus, 9 kbar,
1150°C
High Al_2O_3
olivine
tholeiite
13.5–18 kbar, varying
degrees of melting
from 20% to 30%
Subsolidus, 18 kbar,
1300°C
Increasing $\frac{Pyx}{Oliv}$
Alkali
basalt
Olivine
basalt
Olivine
tholeiite
27 kbar, 40% melting
Subsolidus, 27 kbar,
1400°C
Picrite
Subsolidus, 45 kbar,
1400°C
Liquid fraction
Plagioclase
Clinopyroxene
Olivine
Orthopyroxene
Garnet
Minor minerals
(phlogopite, ilmenite,
spinel, and so on)

Figure 7–5 (*facing page*)
The melting relations of pyrolite. The lowermost of each group of boxes shows pyrolite mineralogy at temperatures just below the solidus. Boxes above show mineralogy and melt compositions with various amounts of partial melting. Note that the melt composition changes as the amount of melting changes. Magma of picrite composition crystallizes at the earth's surface to form a rock with abundant olivine, along with pyroxene, biotite, possibly amphibole, and less than 10% plagioclase. [From D. H. Green and A. E. Ringwood, 1969, in Geophysical Monograph 13 (Washington, D. C.: Amer. Geophys. Union), Fig. 1.]

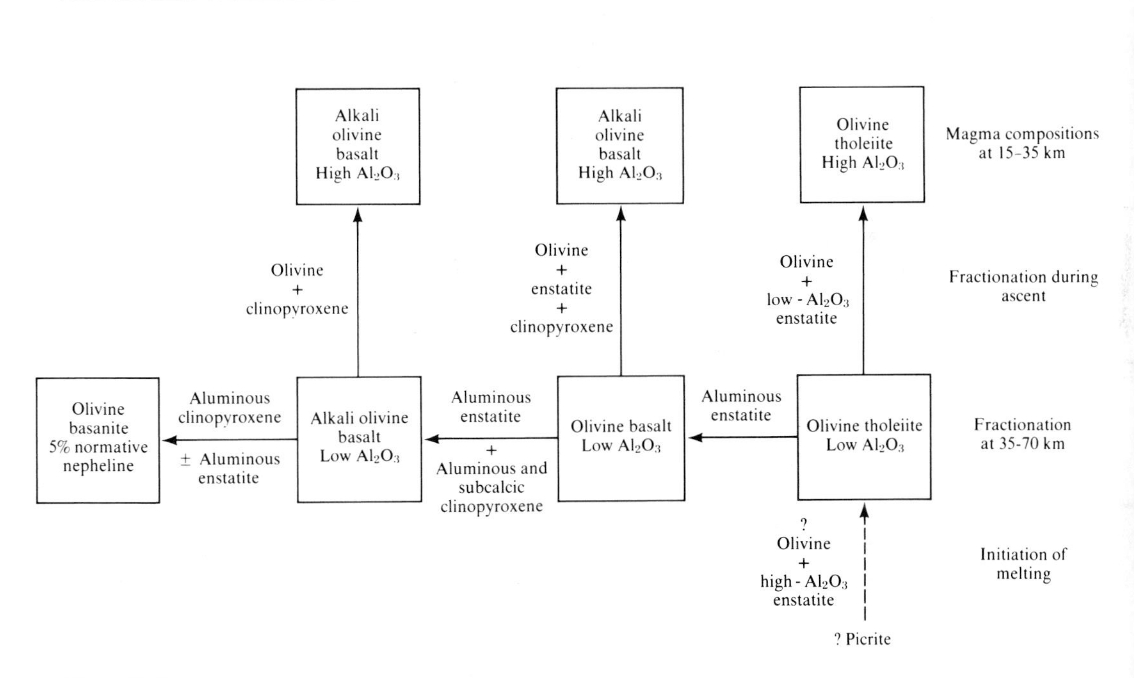

Figure 7–6
Fractionation of basaltic liquids at moderate to high pressures. An original olivine tholeiite melt (lower right) could fractionate at the 35–70 km level to produce the melts shown to the left. Rise with fractionation of these melts to higher levels (15–35 km) produces another group of derivative melts. [From D. H. Green and A. E. Ringwood, 1969, in Geophysical Monograph 13 (Washington, D. C.: Amer. Geophys. Union), Fig. 2.]

and plagioclase. After extrusion of these lavas the central ridge undergoes subsidence, perhaps due to both rifting and depletion of the magma chamber below. The roof of the magma chamber is inferred to be "floating" on the magma beneath, which is thought to be less than 2 km below the surface.

In the outer portion of the rift valley the reverse occurs, with faulting and elevations of the valley walls. Extrusion of the central lava flows is episodic, and the location of the volcanic axis varies through time (see Figure 7–10). Recurrent volcanism often occurs adjacent to previous extrusions.

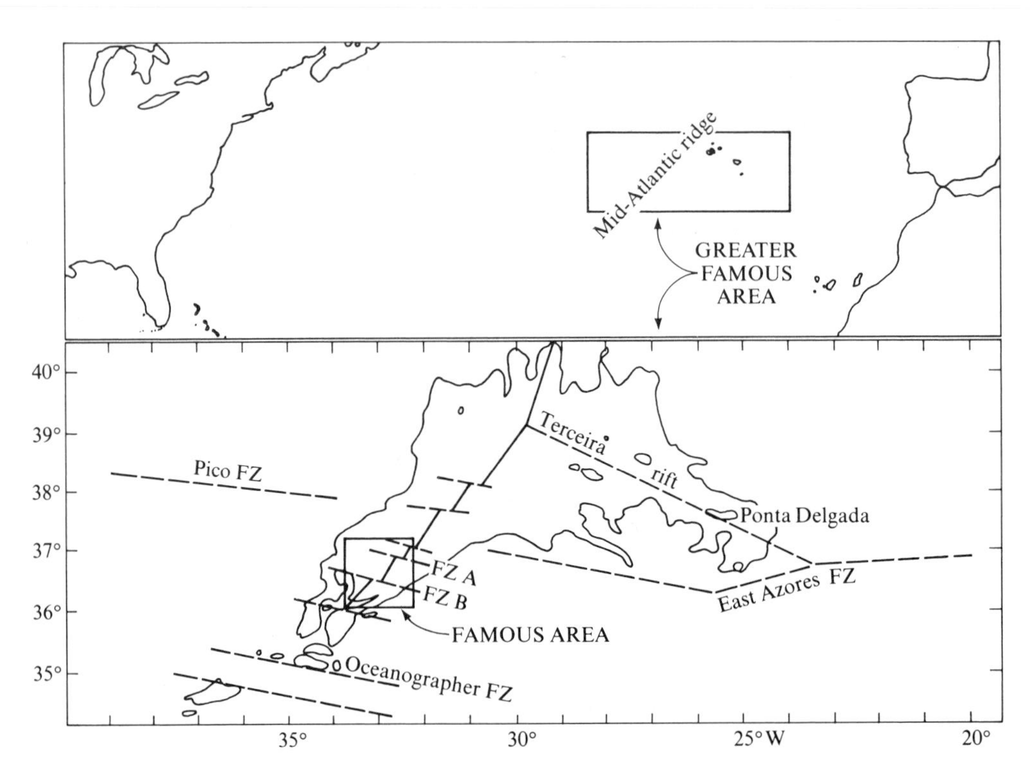

Figure 7–7
The FAMOUS and greater FAMOUS areas of the Mid-Atlantic Ridge. Fault zones (FZs) are indicated as dashed lines. The 2,000-m isobath is shown by a solid line. [From J. R. Heirtzler and T. H. van Andel, 1977, *Geol. Soc. Amer. Bull., 88,* Fig. 2.]

Figure 7–11 shows diagrammatic cross sections of the inner rift valley by two sets of authors: Ballard and van Andel (1977) and Bryan and Moore (1977). One thing to notice immediately on both diagrams is that igneous activity is not limited to the central volcanic axis; numerous vertical dikes and flows are present within the rift valley floor and adjacent to the rising valley walls. The relative ages of the various flows were determined based upon the thickness of the surface weathering rind of basalt glass (palagonite), as well as the thickness of encrustations of Mn–Fe oxides. The ages determined indicate that in general the central lavas are younger than the flank lavas; the flank lavas in turn are considerably younger than the crust they intrude (based upon the assumed spreading age of the crust on which they occur).

A second factor that emerged in this investigation is that the composition of the various lavas is not identical. Regular compositional variation exists from the center of the rift valley out to the valley walls. The central lavas have a higher ratio of phenocrysts of olivine relative to those of plagioclase and clinopyroxene. In the flank

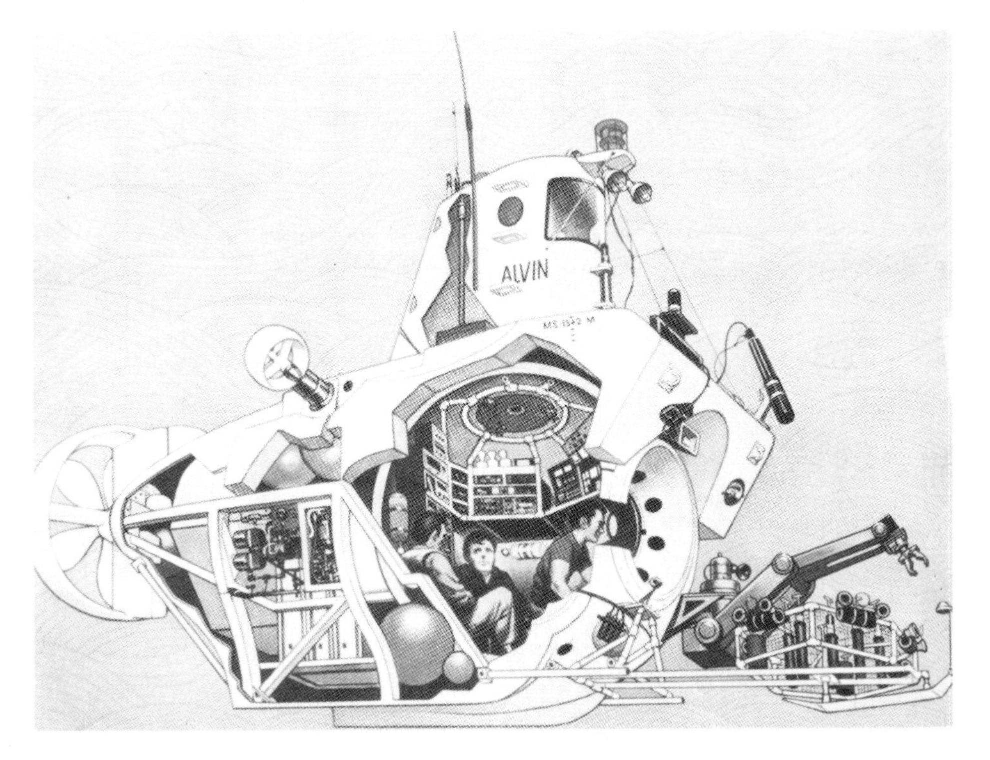

Figure 7–8
The submersible Alvin, a modern geological observational and collecting tool. [After R. D. Ballard and T. H. van Andel, 1977, *Geol. Soc. Amer. Bull., 77,* Fig. 1.]

lavas, volcanic glasses contain higher amounts of SiO_2, K_2O, H_2O, and a higher FeO/MgO ratio. This is ascribed to removal of plagioclase, olivine, and clinopyroxene from the flank lavas. The process involved in the eruption of these diverse lavas is the tapping of a zoned magma chamber. The cooler portions at the edges (see Figure 7–11) constantly change in composition as a result of phenocryst removal. Differentiation is accomplished by the settling of olivine, clinopyroxene, and spinel to form a cumulate layer on the magma floor. As the magma is particularly low in H_2O (and therefore has a relatively high specific gravity), the plagioclase phenocrysts will float and remain at the walls and roof of the magma chamber rather than join the cumulate at the base. This plagioclase will become part of the laterally solidifying differentiated magma, which will produce a fairly structureless gabbro. The process continues during rifting with episodic extrusion of undifferentiated basalt in the center, dikes and extrusions of differentiated material in the flanks, and differentiation and solidification on the outer edges of the magma chamber. As cooling continues at

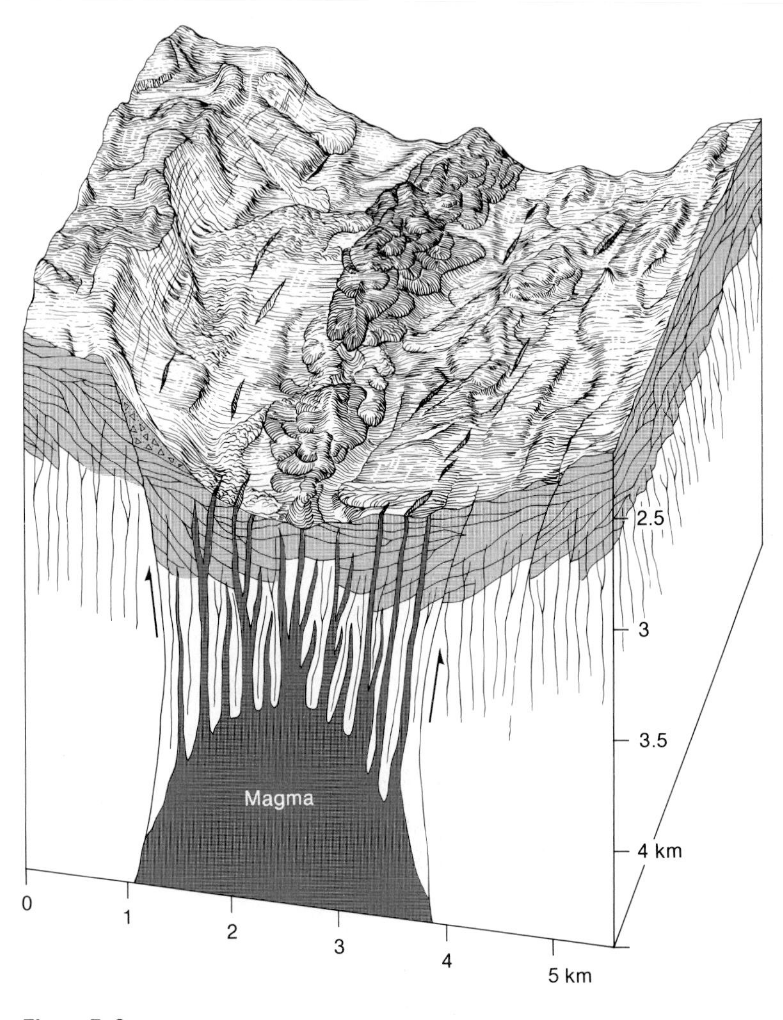

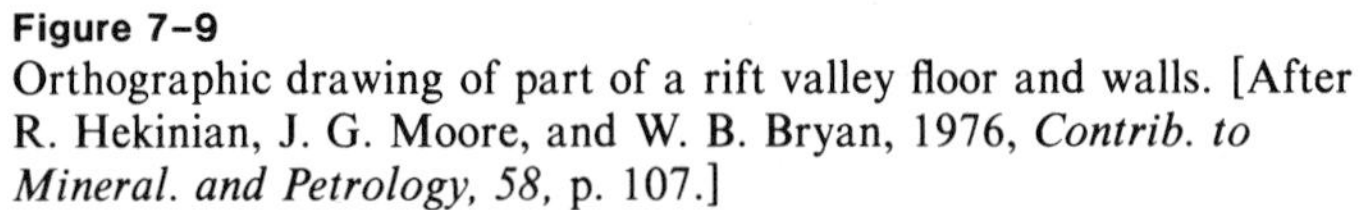

Figure 7–9
Orthographic drawing of part of a rift valley floor and walls. [After R. Hekinian, J. G. Moore, and W. B. Bryan, 1976, *Contrib. to Mineral. and Petrology, 58,* p. 107.]

the magma edges, the solidified materials become a part of the outward-moving plate edges. New magma enters the central area, new fractures are developed, and the multilayered plates continue to move outward, gradually acquiring a relatively thin layer of sedimentary material. Reaction of some of the upper basaltic material with sea water can cause the formation of spilite. The general characteristics of oceanic plates are given in Table 7–1. This sequence is very similar to ophiolite sequences in various parts of the world.

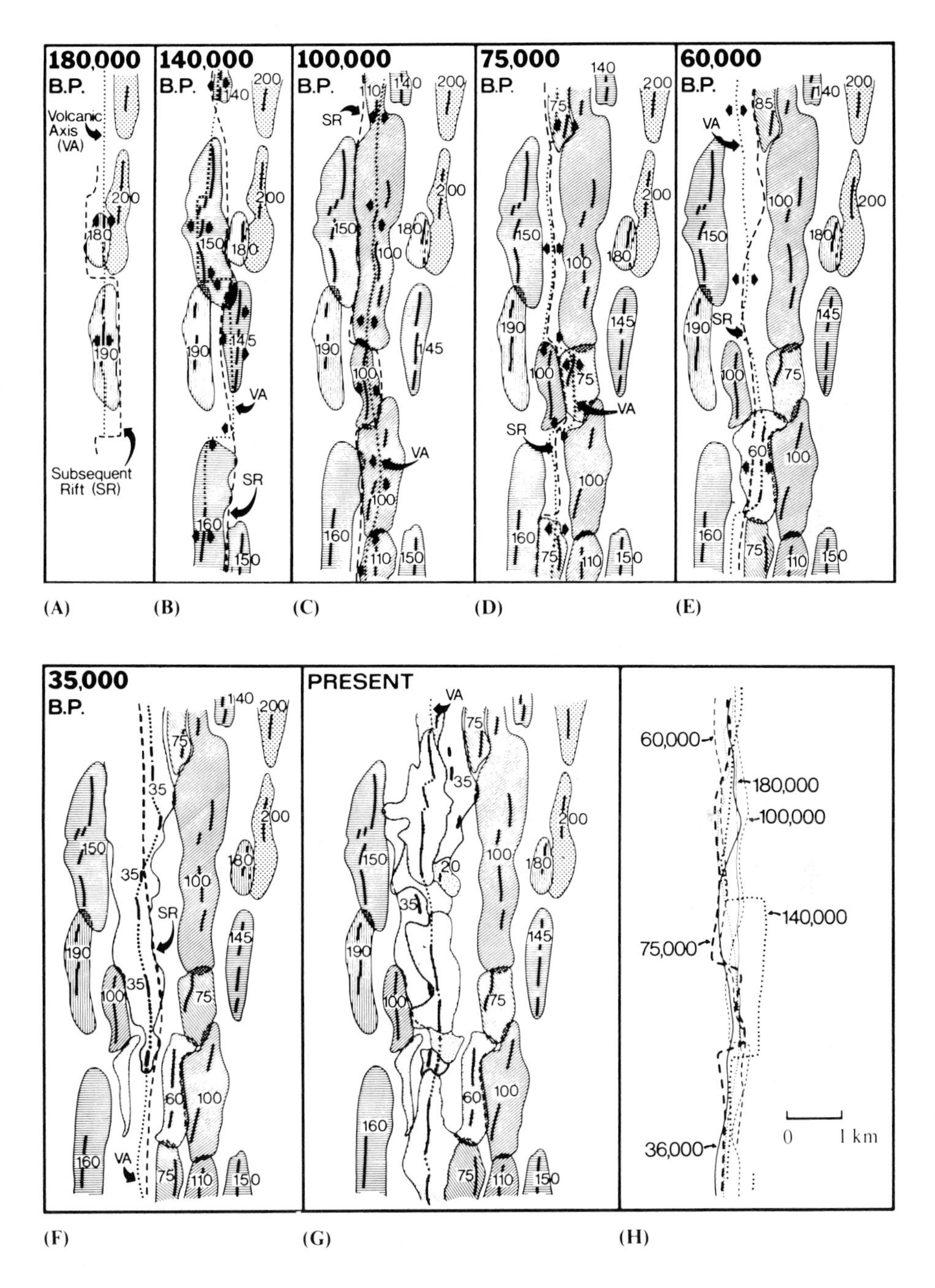

Figure 7–10
Plan view of an inner rift valley in the FAMOUS area, showing changes in the active volcanic axis (VA) and subsequent rift (SR) over a 180,000-year period. Part H shows all rifts superimposed. Numbers on all diagrams are the dates of the flows (years before present). [From R. D. Ballard and T. H. van Andel, 1977, *Geol. Soc. Amer. Bull., 88,* Fig. 17.]

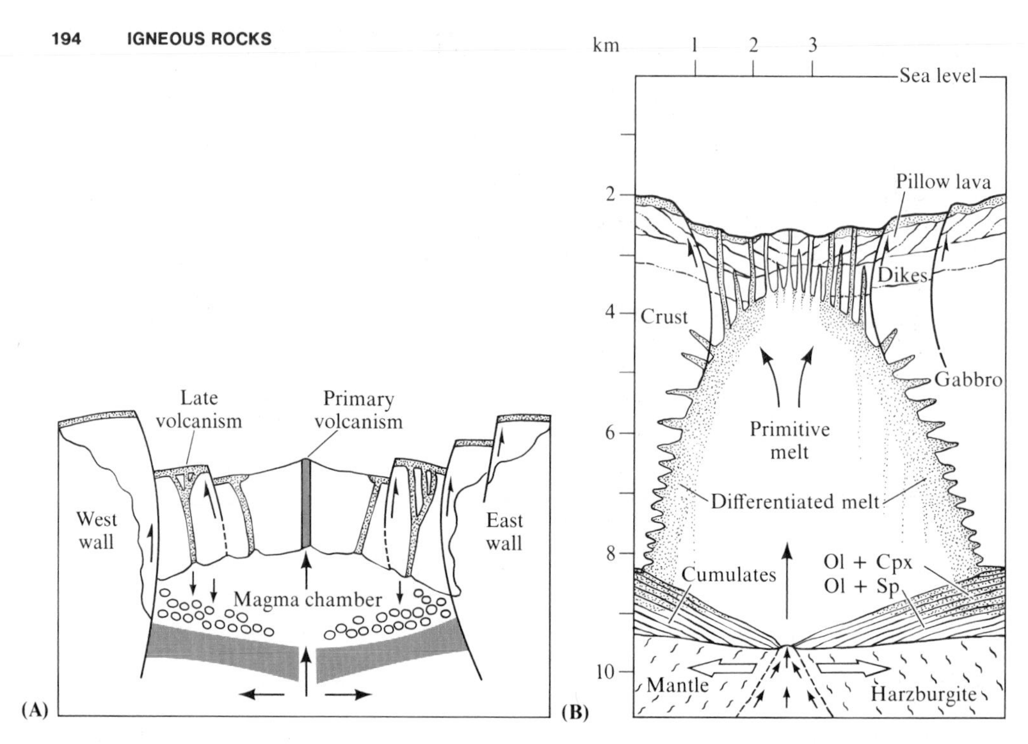

Figure 7–11
Two models of the inferred magma chamber beneath the median valley. Lateral magmatic differentiation is indicated by variable thickness of the cumulate layer at the base and by differentiated melt along the relatively cool edges. Ol = olivine, Sp = spinel, Cpx = clinopyroxene. [Part A from R. D. Ballard and T. H. van Andel, 1977, *Geol. Soc. Amer. Bull., 88,* Fig. 19; part B from W. B. Bryan and J. G. Moore, 1977, *Geol. Soc. Amer. Bull., 88,* Fig. 16.]

SUMMARY

Magmas of generally basaltic composition are produced by partial melting within the upper mantle at rift zones. The composition of the melt produced depends upon many variables, including the following: (1) the depth of melting, (2) the amount of melting, (3) the composition and presence of a fluid, (4) the rate of ascent, (5) the level and degree of fractionation during ascent, (6) possible reactions with wall rock, and (7) original differences in composition of source material.

Recent examination of the Mid-Atlantic Ridge by project FAMOUS has indicated the presence of magma less than 2 km below the surface of the rift-valley floor. Samples taken of lavas erupted near the center of the valley have essentially the composition of primitive magma from the mantle, whereas volcanic rocks obtained near the valley walls are somewhat differentiated. From this (and other information) it is concluded that the ocean crust is layered. The sequence from the surface downwards is pillow lava, sheeted diabase dikes, structureless gabbro, cumulates, and depleted peridotite of harzburgite (olivine and orthopyroxene) composition.

Table 7-1
Ocean Crust and Mantle

	Crust
Layer 1 (1 km thick)	Thin chert, lutite, pelagic limestone
Layer 2 (2 km thick)	Pillow tholeiite, alkali basalt, and spilites; these are dominantly extrusive basic volcanic rocks; manganese nodules and encrustations are common
Layer 3	Gabbroic rocks with minor diorite; thermal and hydrothermal alterations have occasionally produced low-grade metamorphism; numerous vertical basic dikes in upper portion; ultrabasic cumulates are present at the base in varying thicknesses
	Mantle
	Periodotite mantle, perhaps of harzburgite composition

FURTHER READING

Ballard, R. D., and T. H. van Andel. 1977. Morphology and tectonics of the inner rift valley at lat. 36°50′ N on the Mid-Atlantic Ridge. *Geol. Soc. Amer. Bull., 88,* 507–530.

Bryan, W. B., and J. G. Moore. 1977. Chemical variations of young basalts in the Mid-Atlantic Ridge rift valley near lat. 36°49′ N. *Geol. Soc. Amer. Bull., 88,* 556–570.

Forsyth, D. W. 1977. The evolution of the upper mantle beneath mid-ocean ridges. *Tectonophysics, 38,* 89–118.

Geological Society of America. April–May 1977 issue of the *Bulletin.* Various articles on project FAMOUS.

Green, D. H., and A. E. Ringwood. 1969. The origin of basaltic magma. In P. J. Hart (ed)., *The Earth's Crust and Upper Mantle,* pp. 489–495. Geophysical Monograph 13. Washington, D.C.: American Geophysical Union, 735 pp.

Langmuir, C. H., J. F. Bender, A. E. Bence, G. N. Hanson, and S. R. Taylor. 1977. Petrogenesis of basalts from the FAMOUS area: Mid-Atlantic Ridge. *Earth Planet. Sci. Letters, 36,* 133–156.

Wyllie, P. J. 1971. *The Dynamic Earth.* New York: John Wiley and Sons, pp. 105–137.

Wyllie, P. J. 1973. Experimental petrology and global tectonics—a preview. *Tectonophysics, 17,* 189–209.

8

Igneous Rocks at Continental Margins

Continental margins have been the major sites of sedimentary deposition as well as profound igneous and metamorphic activity. Many investigators (but certainly not all) regard the present continental edges as modern geosynclines—similar to those that have formed in the geological past. In the classical view, geosynclines were regarded as subsiding troughs, stretching for thousands of kilometers, and containing sedimentary material, often thousands of meters in thickness. Contemporary regions that can be regarded as being similar to older geosynclines are shown in Figure 8–1 in relation to submarine trenches and major rift zones. The modern geosynclinal regions have been subdivided into Atlantic, Andean, island arc, and Japan Sea types.

The Atlantic-type geosyncline (which includes much of the eastern coast of North America and South America, and both African coasts) is shown in Figure 8–2. The diagram shows the shallow continental shelf and the more steeply inclined slope, followed by a gently sloping continental rise. No igneous activity is present. Shallow water sediments deposited on the shelf overlie continental sialic crust. Deep water sediments are deposited seaward on the continental rise over oceanic crust. Examination of ancient geosynclines reveals a common lateral division of different types of sedimentary materials in the oceanward and landward sides of the geosyncline. The oceanward portion (often highly deformed) has been called the eugeosyncline, and mainly consists of deep water sediments (turbidites) and cherts. The landward portions of ancient geosynclinal sediments (called the miogeosyncline) consist of shallow water sediments (limestone, abundant organic material, fine sands, silts, and so on) typical of alluvial plains and shoreline areas. The present Atlantic-type coastline is thought to be the primary stage in the formation of geosynclines similar to those found in the geological record.

In contrast to the Atlantic-type shoreline, the Andean, island arc, and Japan Sea types (see Figure 8–3) are actively volcanic. The Andean type, as characterized by

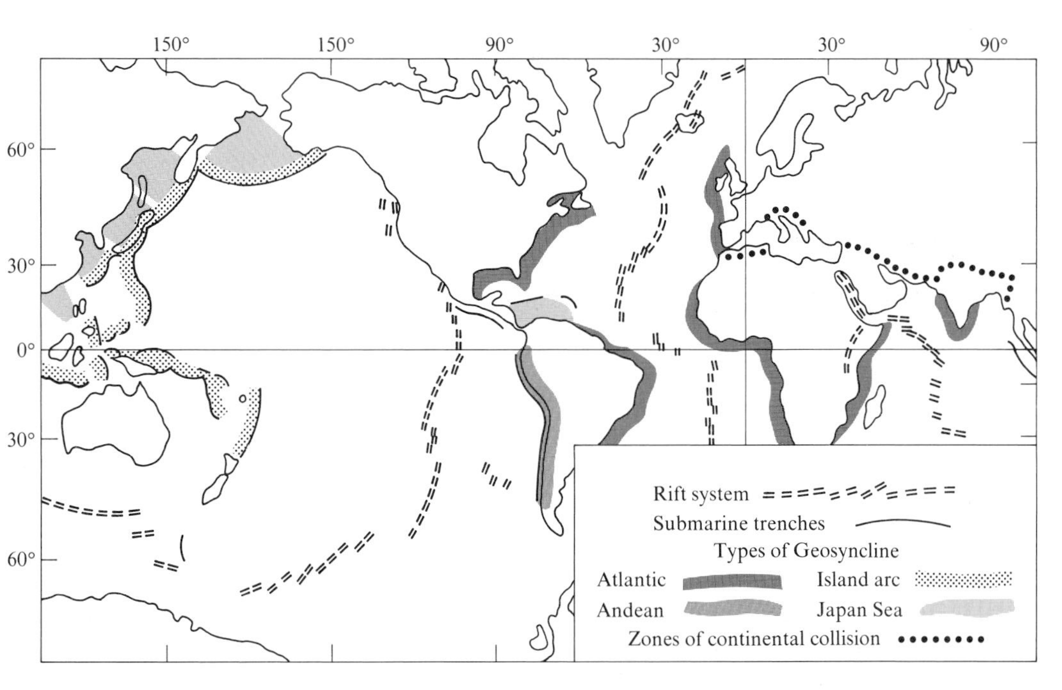

Figure 8–1
Positions of modern geosynclines in relation to rifts and submarine trenches. [From A. H. Mitchell and H. G. Reading, 1969, *Jour. Geol., 77,* Fig. 1.]

the western border of South America, contains a mountain belt bordered by a submarine trench. The trench contains mainly turbidites derived from erosion of the Andean chain. Deformation in the trench and continental margin deposits, volcanism, emplacement of batholiths, and block faulting are related to the subduction of an oceanic plate beneath the continental margin.

The island arc and Japan Sea types are similar, inasmuch as they consist of an island chain with an adjacent submarine trench produced by subduction of oceanic crust. The island arc type may be remote from a continental border, but the Japan Sea type is separated from the continent by a small ocean basin. Belts of sediments are found in both the small ocean basin and in the trench adjacent to the island arc. In the small ocean basin the sediments may be largely of continental (Asian) origin, but in the trench adjacent to the volcanic islands, the detritus will be mainly of volcanic origin. Within this material are found pyroclastics, submarine lavas, and thin pelagic sediments. These deposits may form a continuous belt, which could be considered as an island arc–type eugeosyncline.

A wide variety of igneous rocks is found adjacent to both modern and ancient continental margins—including the ophiolite suite, calc-alkalic volcanics, tholeiitic basalts, and massive granitic batholiths. These will be described in relation to current theories of plate tectonics and experimental petrology.

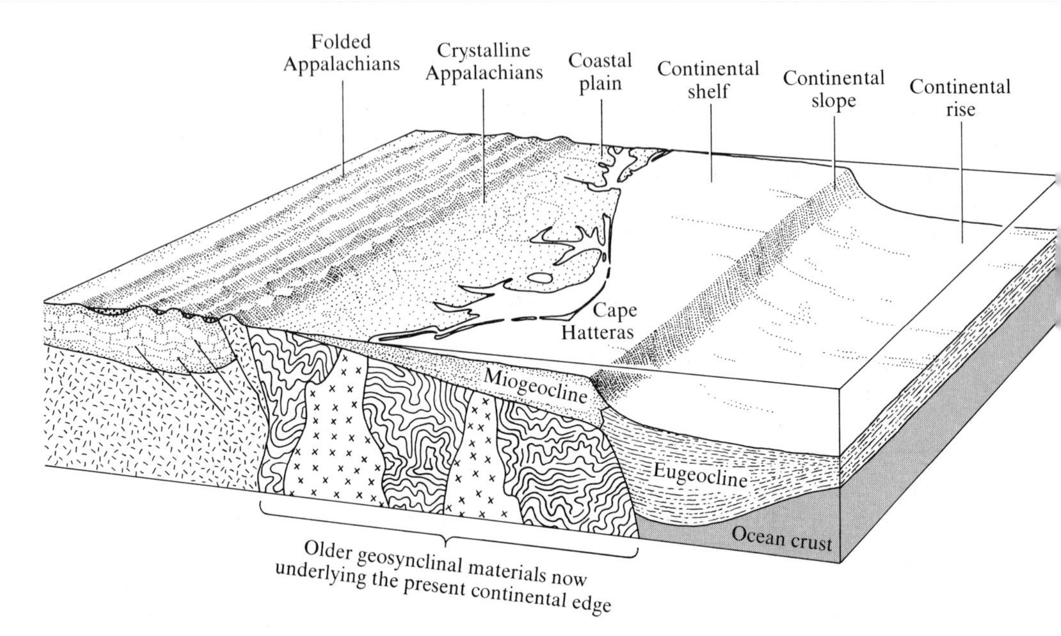

Figure 8–2
The Atlantic type geosyncline. The continental shelf is composed of miogeosynclinal sediments, and the continental rise prism consists of eugeosynclinal material. Older geosynclinal material (now converted to igneous and metamorphic continental rocks) underlies the present continental shelf and coastal areas. [From "Geosynclines, Mountains and Continent-building" by R. S. Dietz. Copyright © 1972 by Scientific American, Inc. All rights reserved.]

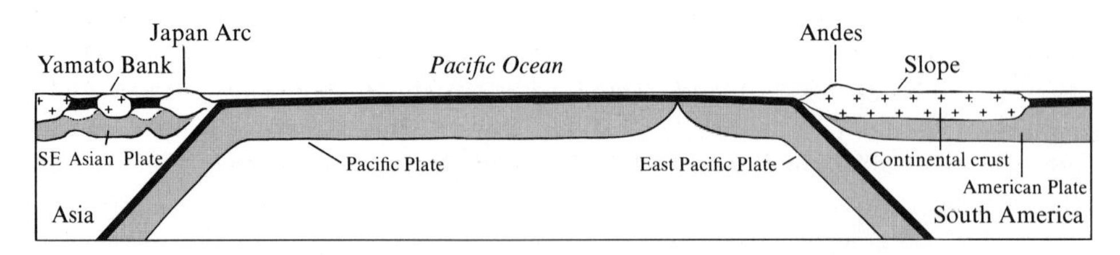

Figure 8–3
Schematic sections showing Andean and Japan Sea types of continental margins. [From J. F. Dewey and J. M. Bird, 1970, *Jour. Geophys. Res., 75,* Fig. 2.]

THE OPHIOLITE SUITE

The term *ophiolite,* as used at present, refers to a distinctive rock assemblage containing ultramafic, gabbroic, and basaltic rocks, often capped by layers of deep sea sediments. This distinctive association of rocks resembles oceanic crust and upper mantle in both thickness and sequence. For this reason many geologists now feel that ophiolites are fragments of oceanic plates that have been somehow thrust into a subaerial environment. Well-described ophiolite suites are located at Bay of Islands in Newfoundland, Troodos in Cyprus, Semail in Oman, and eastern Papua in New Guinea.

Assuming an oceanic origin for the ophiolites, consider how these rocks might become emplaced. Emplacement may occur as a result of collision between continental and oceanic plates, or between two continental plates.

In an Andean-type margin, compression of adjacent oceanic and continental areas leads to the formation of a trench and the beginning of plate descent. The usual underthrusting mechanism causes the denser oceanic plate to slide beneath the lighter continental plate. However, in the initial stages of collision the reverse might occur, with the oceanic plate overriding the continental plate. This process, which results in a seaward-dipping subduction zone, is called *obduction.* As obduction occurs the oceanic plate is thrust over the continental plate (see Figure 8–4). Obduction is basically an unstable mechanism, and the process is followed by a reversal, with the oceanic plate descending under the continental plate. In the absence of obduction, the normal subduction process (an oceanic plate underthrusting a continental plate) results in considerable fracturing of the oceanic plate. Blocks of oceanic material may be torn from the descending plate, to rise tectonically into the mélange above.* Tectonic movement of this type is indicated in the Franciscan Formation in California, where ophiolites and serpentinized peridotites ("Alpine peridotites") are closely associated with highly deformed areas.

One of the best-studied ophiolites is that of Bay of Islands, Newfoundland (see Figure 8–5). The complex trends northeast for about 96 km and is 24 km wide. The whole allochthonous (transported) sequence overlies (1) Grenville crystalline Precambrian basement, 800–1000 m.y.b.p. (million years before present), (2) a Cambrian–Ordovician clastic to limestone shallow-water sequence that rests unconformably on the basement, and above that (3) allochthonous slices of clastics of equivalent age, thought to have been deposited originally to the east of the ancient continental margin. Thin, discontinuous metamorphosed mafic or volcanic rocks are present at the base of the Bay of Islands ophiolite. The complex was emplaced by westward transport during the Ordovician Period (~460–500 m.y.b.p.).

---

*Mélange, as defined in the AGI Glossary of Geology, is "a mappable body of deformed rocks consisting of a pervasively sheared, fine grained, commonly pelitic matrix, thoroughly mixed with angular and poorly sorted inclusions of native and exotic tectonic fragments, blocks, or slabs (of diverse origins and geologic ages) that may be as much as several kilometers in length."

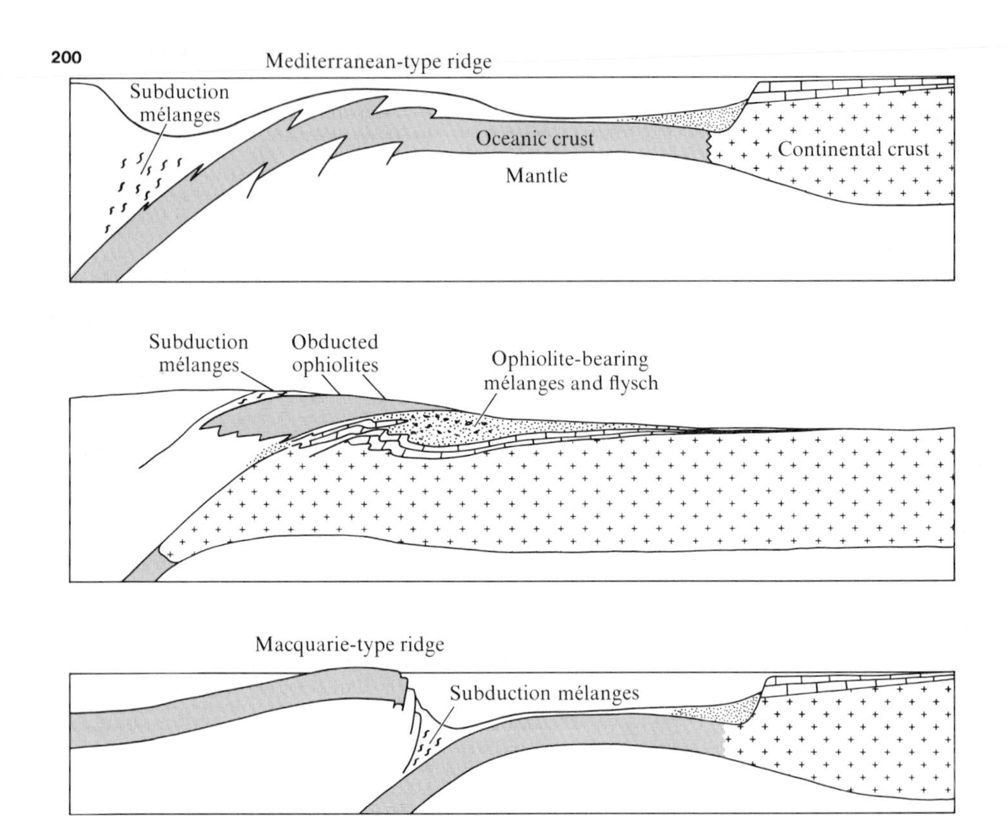

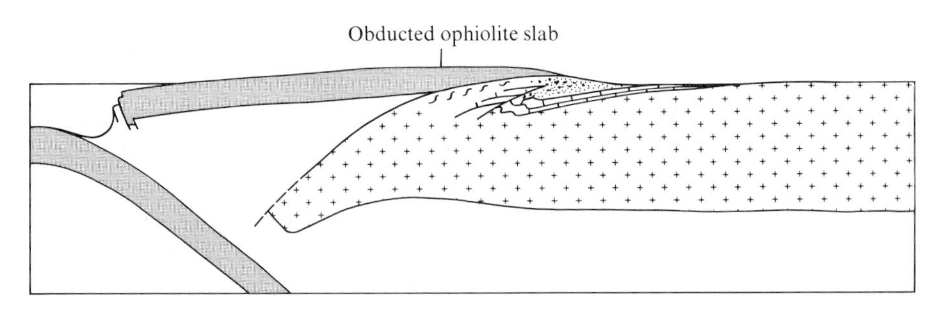

Figure 8–4
Possible mechanisms for obduction of ophiolite sheets onto continental margins. [From J. F. Dewey and J. M. Bird, 1971, *Jour. Geophys. Res., 76,* Fig. 6.]

The sequence of rocks in the ophiolite complex (see Figure 8–6 and Table 8–1) starting at the base is as follows:

A. Metamorphic aureole. These are moderately metamorphosed rocks containing pyroxene, garnet, amphibole, and plagioclase that show considerable deformation. This is thought to represent the layer within which the obduction of hot oceanic crust and mantle occurred.

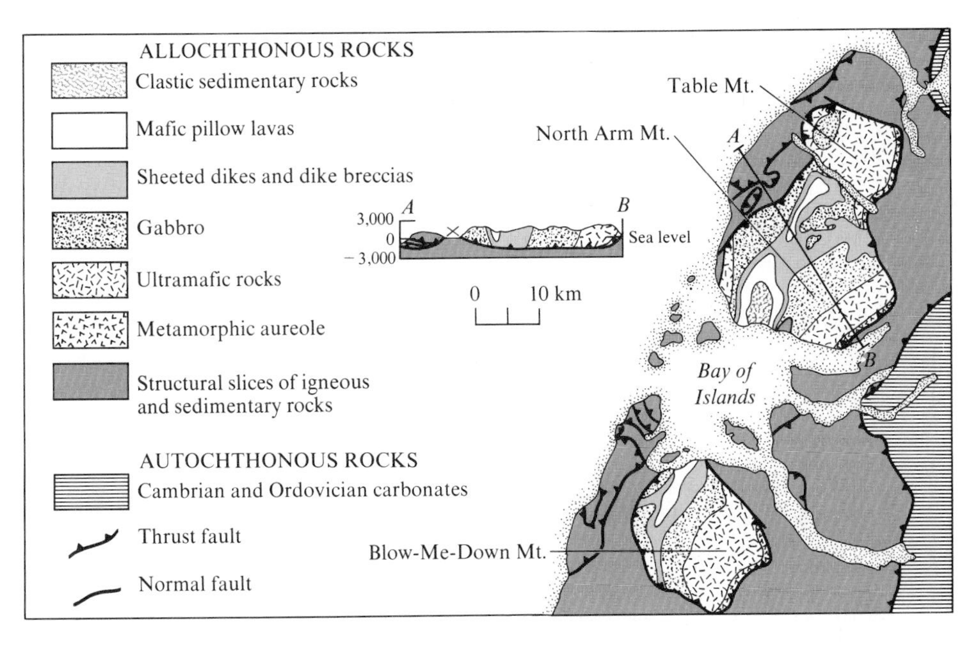

Figure 8–5
Geological map of the Bay of Islands ophiolite, Newfoundland. [From R. G. Coleman, 1977, *Ophiolites* (New York: Springer-Verlag), Fig. 64, from earlier sources.]

B. Immediately above the metamorphic rocks are found ultrabasic rocks thought to consist of both primary and "depleted" mantle—the refractory residue from partial melting processes that produced basaltic melt. These rocks consist of harzburgite (75% olivine and 25% orthopyroxene) and dunite (mostly olivine), with lherzolite at the base; lherzolite is composed of a mixture of olivine, clinopyroxene, and orthopyroxene. A distinct mineralogical banding is present in these rocks. Considerable deformation and serpentinization is evident. A mantle origin is indicated by spinel and the high aluminum content in the clinopyroxene (see Figure 6–5).

C. Above the ultrabasic rocks is a transition zone characterized by the first appearance of plagioclase and a distinct cumulate texture. Crystal settling resulted in cumulate layers rich in anorthosite, dunite, troctolite, or clinopyroxene. The general rock type is composed of calcic plagioclase, clinopyroxene, and olivine, with minor chromite, magnetite, and sulfides. *Troctolite* is chiefly composed of calcic plagioclase (usually labradorite) and olivine with little or no pyroxene. Overlying the cumulate transition zone are fairly structureless gabbros containing plagioclase (An_{70-80}), clinopyroxene, and olivine (Fo_{80}). The Mg/Mg+Fe ratio decreases upward within the gabbro, indicating some fractionation.

D. Sheeted diabase dikes are found above the gabbro, some of which are strongly brecciated. Dikes are found within the gabbro below, indicating that the

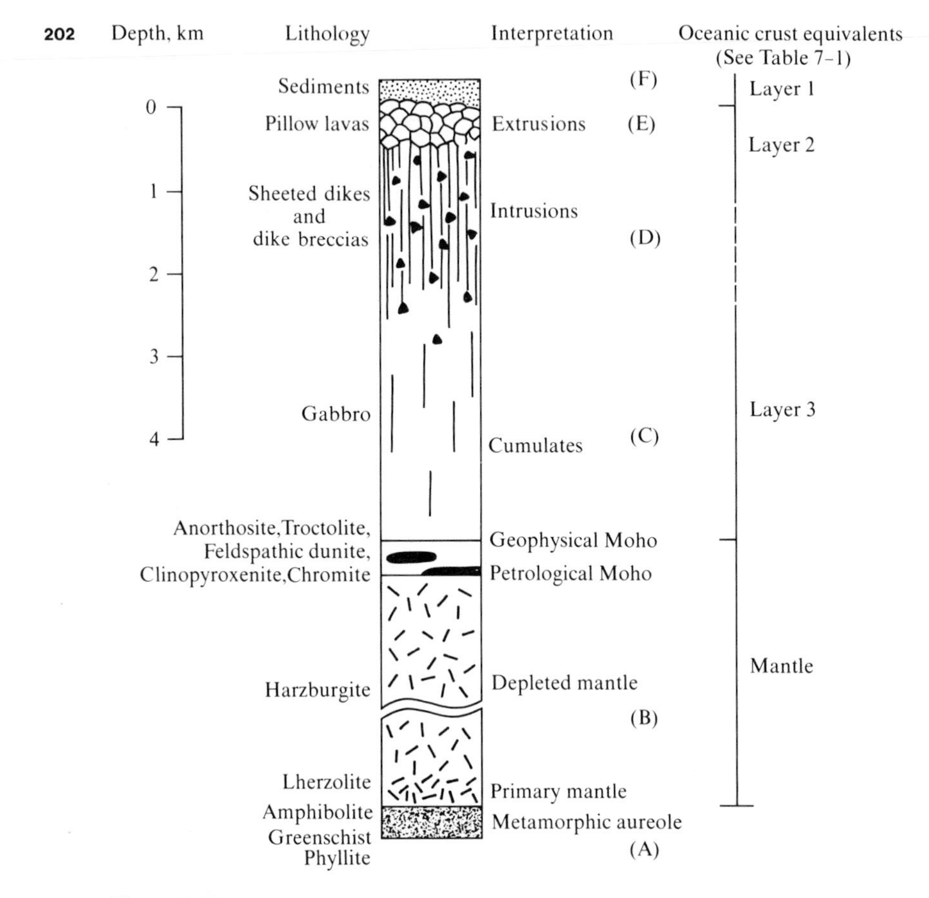

Figure 8–6
The sequence of rocks found in the Bay of Islands ophiolite, Newfoundland, as compared to oceanic crust and mantle. The letter symbols are described in the text. [From R. G. Coleman, 1977, *Ophiolites* (New York: Springer-Verlag), Fig. 65 and Table 15, from earlier sources.]

dikes are younger than the gabbro; as the contact is approached the number of dikes increases upward until the point is reached where the total mass consists of dikes. Minerals present in the sheeted dikes are plagioclase, clinopyroxene, and magnetite. The texture is fine-grained, diabasic, and porphyritic. Slight but pervasive metamorphism is present (greenschist or prehnite–pumpellyite–metagraywacke facies).

E. The next layer consists of pillow basalts, which are somewhat altered. A few dikes are present within the basalt.

F. The final layer is not abyssal sediments interlayered with volcanics (as is found in some other ophiolites), but a Lower Ordovician clastic sequence of conglomerate, sandstone, shale, and siltstone.

Table 8–1
Average Chemical Composition and CIPW Norms of the Various Units Within the Bay of Islands Ophiolite, Newfoundland[a]

	Peridotites		Feldspar dunite cumulates	Olivine gabbro cumulates		Gabbro	Sheeted dikes	Pillow lavas
	B	B	C	C	C	C	D	E
SiO_2	43.05	41.78	41.85	46.62	47.81	49.43	51.80	50.22
Al_2O_3	2.52	2.58	6.82	21.74	21.32	18.78	15.79	16.14
FeO	7.82	7.91	8.95	4.47	5.47	6.77	9.31	9.56
MgO	45.00	45.60	36.73	9.63	8.18	8.05	8.00	6.09
CaO	0.80	1.24	4.21	15.79	13.87	13.11	9.37	10.58
Na_2O	—	0.11	0.16	1.41	2.69	2.99	3.79	4.59
K_2O	—	—	—	—	—	0.21	0.53	0.64
TiO_2	—	—	0.28	0.15	0.52	0.62	1.05	1.82
P_2O_5	—	—	0.01	—	—	—	0.11	0.11
MnO	0.11	0.13	0.14	0.05	0.08	0.01	0.21	0.21
Cr_2O_3	0.39	0.33	0.60	0.11	0.04	0.02	0.03	0.02
NiO	0.31	0.33	0.25	0.03	0.02	0.01	0.01	0.01
				Normative minerals				
Quartz	—	—	—	—	—	—	—	—
Corundum	1.1	0.2	—	—	—	—	—	—
Orthoclase	—	—	—	—	—	1.2	3.1	3.8
Albite	—	1.0	1.3	8.2	15.5	20.0	32.1	25.8
Anorthite	4.0	6.1	18.0	53.0	46.1	37.2	24.5	21.5
Nepheline	—	—	—	2.0	3.9	2.9	—	7.1
Diopside	—	—	2.3	20.3	18.3	22.6	17.5	25.0
Hypersthene	15.2	3.7	4.1	—	—	—	2.6	—
Olivine	79.1	88.6	72.8	16.0	15.1	14.9	17.9	13.1
Chromite	0.6	0.5	0.8	0.2	0.1	0.03	0.05	0.03
Ilmenite	—	—	0.5	0.3	1.0	1.2	2.0	3.4
Apatite	—	—	0.03	—	—	—	0.3	0.3

[a]All analyses normalized to 100% after removal of H_2O, CO_2, and all iron recalculated as FeO. The letters at the top of each column correspond to the approximate locations indicated in Figure 8–6.

Source: From R. G. Coleman, 1977, *Ophiolites* (New York: Springer-Verlag), Table 15, based on earlier sources.

The pillow basalts indicating submarine extrusion, sheeted dikes establishing extensive tensional fracturing, gabbro and cumulates from magmatic crystallization, and peridotite compositions indicative of a depleted mantle fit in almost perfectly with the present view of oceanic crust and mantle as described in the previous chapter.

THE CALCALKALINE AND THOLEIITE GROUPS

One of the abundant groups of igneous rocks produced at continental margins is the calcalkaline suite. This assemblage of associated rocks has been referred to by various authors as the basalt–andesite–rhyolite association, the andesitic association, orogenic suite, the hypersthenic rock series, or the orogenic volcanic series. The term calcalkaline derives from Peacock's (1931) classification of igneous rock groups (see Figure 8–7 and Table 8–2). The calcalkaline suite consists of a variety of rocks—basalt, andesite, dacite, and rhyolite. The basalts are of the high-alumina type. The orthorhombic pyroxene hypersthene is usually found in all members except rhyolite. Island arcs and active continental margins are often mainly composed of this group and its plutonic equivalents. Early workers, such as Bowen (1928), regarded the calcalkaline suite as a progressive crystallization series, beginning with a basaltic

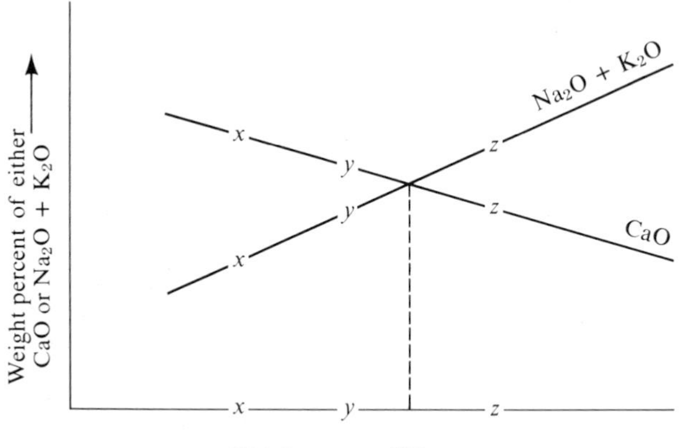

Figure 8–7
Peacock's method of igneous rock classification. A rock suite consists of rocks *x*, *y*, and *z*. Analyses of CaO, SiO_2, and Na_2 + K_2O weight percent when plotted as shown will form two intersecting lines. The weight percentage of silica that corresponds to the intersection point defines the alkalinity of the suite (see Table 8–2). [From M. A. Peacock, 1931, *Jour. Geol., 39,* pp. 54–67.]

Table 8–2
Peacock's Igneous Rock Classification

Rock series	$\%SiO_2$[a]	Typical rock association
Calcic	> 61	Basalt, andesite, rhyolite
Calc-alkalic (Calcalkaline)	56–61	Basalt, andesite, rhyolite
Alkalic-calcic	51–56	Basalt, trachyte, phonolite
Alkalic	< 51	Basalt, trachyte, phonolite

[a]The percentage of SiO_2 listed is that at which the percent of CaO equals the combined percent of $Na_2O + K_2O$. This is shown in Figure 8–7.
Source: After M. A. Peacock, 1931, *Jour. Geol., 39,* 54–67.

melt as the primary source. The series is now known to be related by melting and crystallization processes, but not in the manner envisioned by Bowen, as the major rock types are usually intermediate to silicic, with andesite (containing 54–63% SiO_2) dominant.

A second group, found to a lesser extent at continental margins and island arcs, is the tholeiitic series. This series is a group of rocks considered to be related to each other by crystal fractionation processes. In this series most of the rocks are mafic (basalt containing < 54% SiO_2 and basaltic andesite), with some intermediate compositions, and few silicic varieties. The Peacock classification is somewhat inadequate in distinguishing the calcalkaline from the tholeiitic group, as they both may have alkali–lime indices as high as 56 to 67. Mafic members of the calcalkaline group contain orthopyroxene in the groundmass, whereas tholeiites contain augite and pigeonite. The most obvious method of distinguishing between the two groups is by means of their differentiation trends.

The differences between differentiation trends in the calcalkaline and tholeiitic series are shown well by FMA diagrams (see Figure 8–8); chemical analyses of the various rocks in the series are plotted in terms of F (total Fe as FeO), M (MgO), and A (alkalies, $Na_2O + K_2O$). Rocks in the calcalkaline series plot along dashed lines within the shaded area, and rocks within the tholeiitic series plot along lines such as *a, b, c,* or *a', b', c'*. The rocks of the calcalkaline series (beginning near *aa'*) show changes in composition that indicate an increase in magma composition of alkalies and silica, with simultaneous decrease in the total Fe content. This is caused by separation of phases having a higher Fe/Mg ratio than the melt; these melts also have a higher silica content (at the same Fe/Mg ratio) than parent tholeiitic magmas. The magmas of the tholeiitic series (starting again at *aa'*) show an initial rise in total iron content (due to separation of olivine and pyroxene having Fe/Mg ratios lower than the melt) followed by a decrease in the amount of iron in the more alkalic members; silica simultaneously shows little to no increase. Many island arc systems contain suites of rocks of both tholeiitic and calcalkaline differentiation trends as well as continuous gradations between these two types.

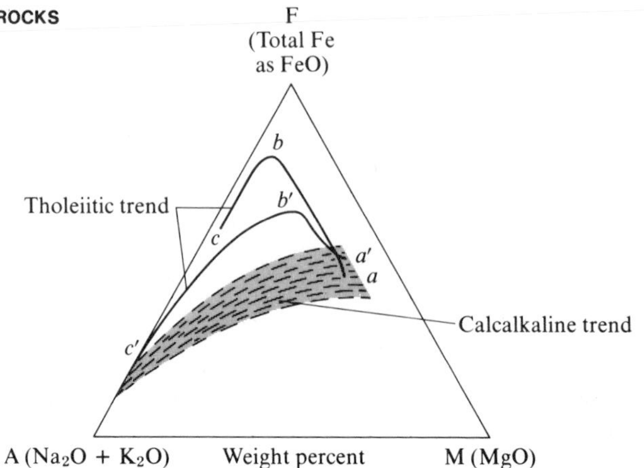

Figure 8–8
FMA diagram showing tholeiitic trends (as solid lines *abc* and *a′b′c′*) and calcalkaline trends (within the gray area). The tholeiitic trends are derived from differentiation of the Skaergaard intrusion, Greenland (upper curve), and Thingmuli volcano, Iceland (lower curve). Calcalkaline trends are from the Cascades, Aleutians, and New Zealand. [From A. E. Ringwood, 1974, *Jour. Geol. Soc., 130,* Fig. 2, from earlier sources.]

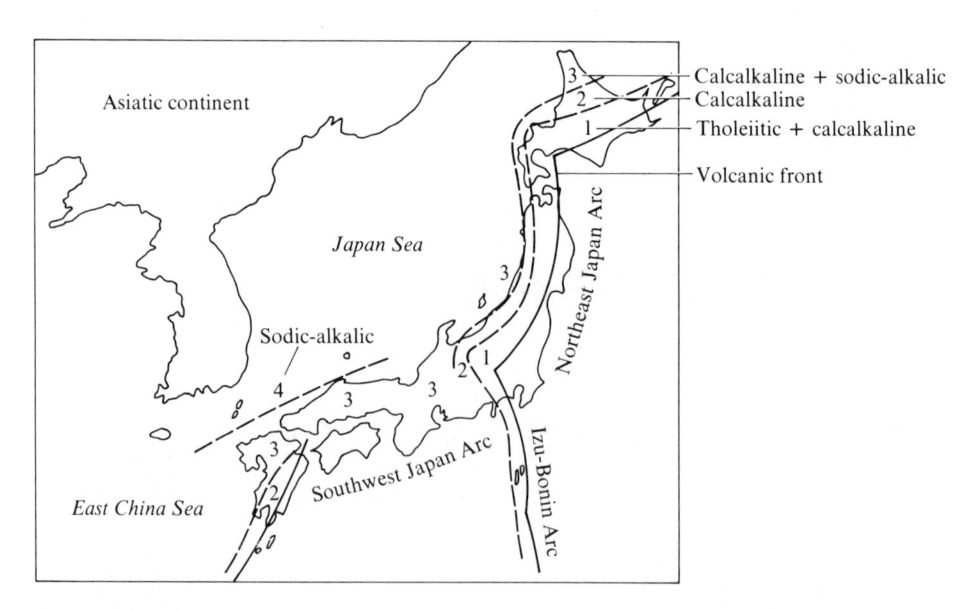

Figure 8–9
Petrographic provinces for the Quaternary volcanic rocks of Japan. The solid lines show the ocean-side limits of volcanic rocks. 1 = tholeiitic and calcalkaline rocks, 2 = calcalkaline rocks, 3 = calcalkaline and sodic-alkalic rocks, 4 = sodic alkalic rocks. [From A. Miyashiro, 1972, *Amer. Jour. Sci., 272,* Fig. 4.]

Examination of island arcs indicates in a general way that the first stage of volcanism is often of the tholeiite trend (with basalts and basaltic andesites predominating). Continued tholeiitic volcanism is joined at a later stage by lavas having the calcalkaline trend (with andesite dominant). In still later stages, alkalic or shoshonitic lavas are produced. Rocks of the *shoshonite* series are hypersthene-normative, contain 6.5–7.0% total alkalies by weight, and have a K_2O/Na_2O ratio greater than 1. Silica is relatively high (50–54%). Granites, by contrast, contain about 8–9% alkalies and more than 70% SiO_2.

The spatial distribution of volcanic rock series in island arcs and continental margins also follows a fairly general pattern (although this has recently been questioned by Arculus and Johnson, 1978). This is shown well in the Quaternary volcanic rocks of Japan (see Figure 8–9). An active subduction zone is located beneath the Japanese Islands with a trench to the east. Immediately to the west of the trench is a region of nonvolcanic rocks—the arc trench gap. Westward, the sequence of volcanic rocks encountered is (1) tholeiitic and calcalkaline rocks, (2) calcalkaline rocks, (3) calcalkaline and sodic-alkalic rocks, and (4) sodic-alkalic rocks. Any theory of origin must account for both the spatial and temporal distribution of rocks in island arcs and continental margins.

Ringwood (1974, 1977) has suggested that the formation of these volcanic rocks is controlled by fractionation processes involving mantle pyrolite, as well as amphibole-rich and high-pressure metamorphic rocks. The proposed mechanism calls for the presence of water in the subduction zone. This could be furnished by hydrated metamorphic minerals in the subducting oceanic plate, serpentinite, spilitic basalts, and possibly hydrated sedimentary material derived from the ocean floor.

Possible distributions of isotherms in a subduction zone are shown in Figure 8–10. The descending cold ocean plate depresses the isotherms, but gradually warms at depth. The stability limits of several hydrated minerals are shown in relation to the proposed isotherms. At various depths these minerals are decomposed, releasing H_2O, which in turn promotes melting in the hanging wall mantle and within the subducting slab.

Ringwood's scheme (see Figure 8–11) suggests that the oceanic basaltic crust is largely converted to amphibolite (a hornblende–plagioclase metamorphic rock) before reaching a depth of about 80–100 km. Here the amphibole has reached its stability limit (mainly due to rise in temperature). The amphibole dehydrates, leaving behind an eclogite (a dense anhydrous rock having the composition of basalt). The released H_2O rises into the hanging wall, causing partial melting of mantle peridotite (pyrolite) and diapiric rise. Fractionation during ascent results in the production of the island arc tholeiitic series.

But the process is not yet finished. Although some melting has been initiated by the dehydration of amphibole, other hydrous minerals (serpentine and dense magnesian silicates) remain within the descending slab. A complex sequence of events occurs (see Figure 8–12):

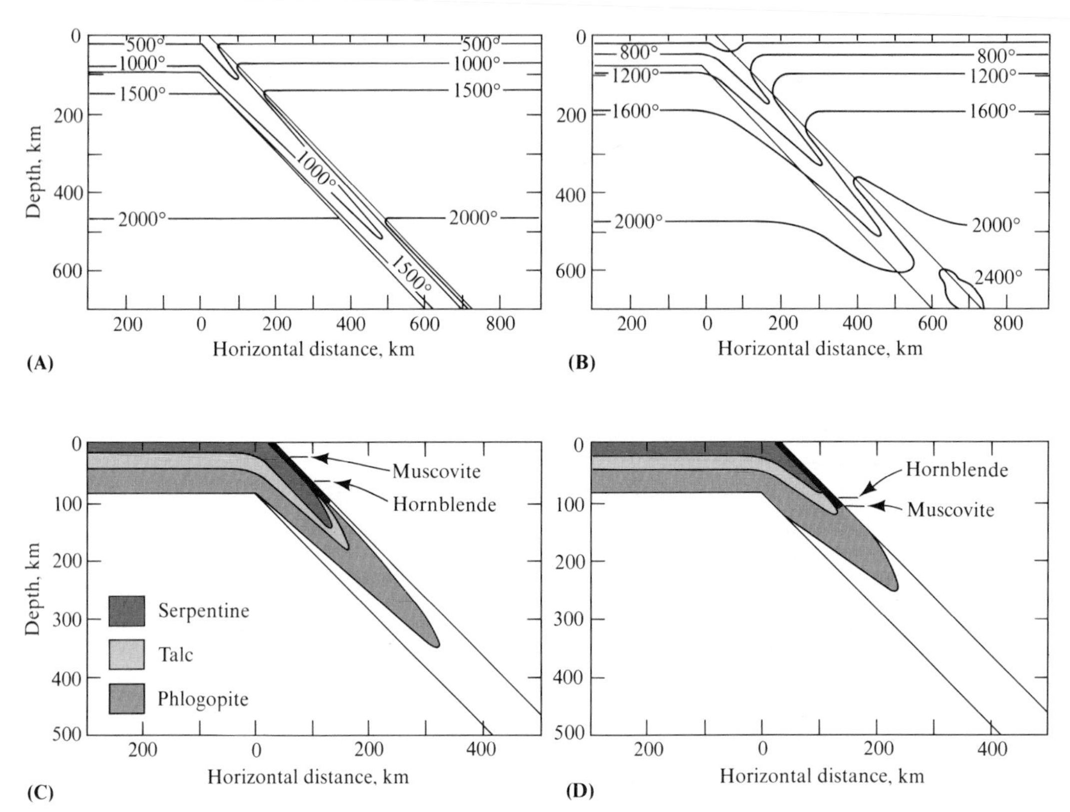

Figure 8–10
(**A, B**) Thermal models for the crust and mantle in a subduction region. [From J. W. Minear, and M. N. Toksöz, 1970, *Jour. Geophys. Res., 75,* Figs. 6, 10.] (**C, D**) The stability regions of serpentine, talc, and phlogopite according to models A and B respectively. [From P. J. Wyllie, 1973, *Tectonophysics, 17,* Fig. 7.]

1. Dehydration occurs at 100–300 km.
2. The presence of high water pressure causes partial melting of the eclogite in the descending slab (forming rhyodacite–dacite magma).
3. The magma formed reacts with the overlying mantle material to form a garnet–pyroxene rock that is less dense than its surroundings.
4. The garnet–pyroxene rock rises diapirically.
5. Diapiric rise initiates partial melting.
6. Fractionation of the melt during ascent produces calcalkaline melts.

Considering the number of variables in this model, it is reasonable to expect a wide variation in the composition of calcalkaline melts; this is certainly the case. This

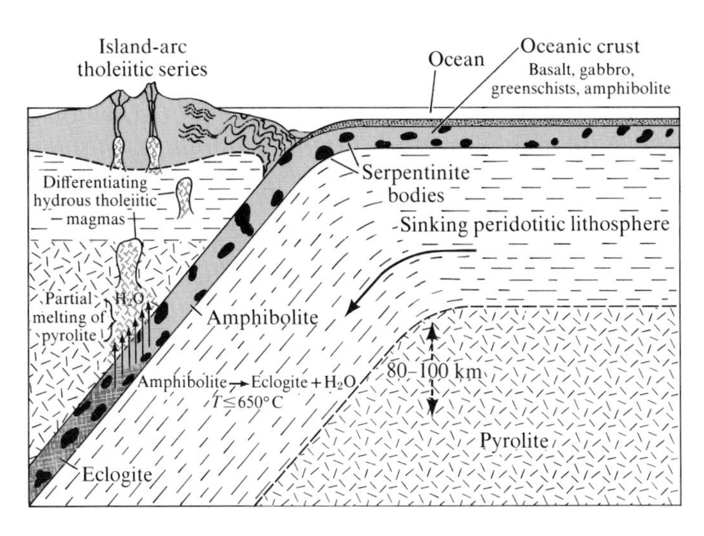

Figure 8-11
The early phase of island-arc development, involving dehydration of amphibolite, introduction of water into overlying rocks, and formation of the tholeiitic magma series. [From A. E. Ringwood, 1974, *Jour. Geol. Soc., 130,* Fig. 9.]

model, as well as other proposed mechanisms, is complicated. Although experimental evidence has been obtained for each step, modification will certainly occur as additional information becomes available. Ringwood (1975) has suggested that if this process has occurred as described, lithospheric plates that have sunk below 200 km have become irreversibly differentiated. Perhaps 30–60% of the mantle may have passed through this process. It should be noted that Ringwood's model is open to some controversy. Nepheline-bearing rocks may be found associated with the tholeiitic rocks of the island arcs. Data based on the primitive character indicated by strontium isotope ratios of some extrusions indicate that these melts were derived from the hanging wall mantle rather than the subducting slab.

The mineralogy of calcalkaline rocks shows abundant evidence of nonequilibrium relationships. Phenocrysts are the general rule. In the more basic members (andesite and basalt) plagioclase (comprising 50–70% of the rock) is the most common phenocryst; quartz and plagioclase phenocrysts are usually found in the silicic members (dacite to rhyolite). Phenocrysts of augite, hypersthene, and hornblende, with occasional biotite, olivine, sanidine, or garnet, are common. Many phenocrysts show evidence of resorption in the form of irregular corroded margins or interior zones. Normal, reversed, or oscillatory zoning is common. Olivine phenocrysts may show reaction rims of hypersthene. In the mafic members the groundmass is usually crystalline, whereas in silicic types it is commonly glassy.

The tholeiitic series has as its dominant magma types basalt or basaltic andesite, in contrast to the calcalkaline rocks, in which andesite is dominant. During crystalliza-

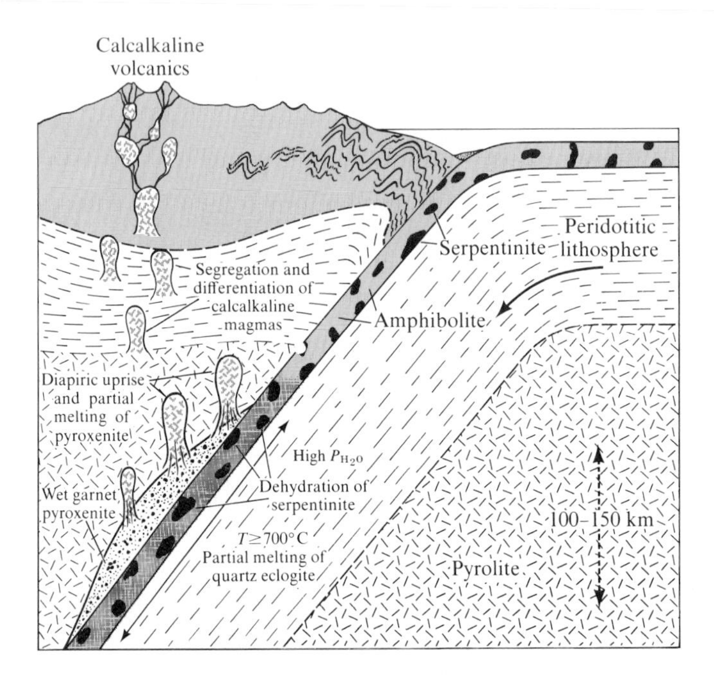

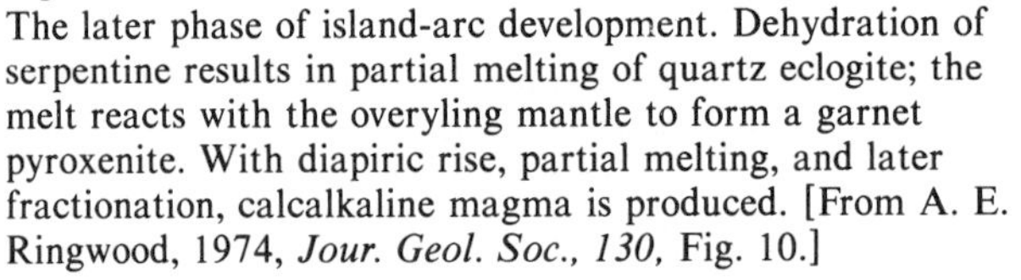

Figure 8–12
The later phase of island-arc development. Dehydration of serpentine results in partial melting of quartz eclogite; the melt reacts with the overyling mantle to form a garnet pyroxenite. With diapiric rise, partial melting, and later fractionation, calcalkaline magma is produced. [From A. E. Ringwood, 1974, *Jour. Geol. Soc., 130,* Fig. 10.]

tion there is little to no increase in SiO_2, but rather a large increase in Fe_2O_3 and FeO in the early and major part of its trend of crystallization. Olivine and pyroxene are the principal mafic minerals, with little or no amphibole or biotite.

The calcalkaline and tholeiitic suites are but a portion of the total problem of the origin of igneous rocks at subduction zones. A relationship has been quantitatively established between the potash content of igneous rocks at the surface and the depth of the Benioff zone below. With increasing depth to the Benioff zone the potash content of igneous rocks rises (although it should be noted that there are some exceptions to this generalization). As shown in Figure 8–13 (and earlier in Figure 8–9), calcic rocks produced at relatively shallow depths give way in turn to rocks that are calcalkaline, high-potassic calcalkaline [a notation used by Keith (1978)], alkalic-calcic, and alkalic. These various types of igneous rocks not only form rough belts parallel to the trench as a function of the depth of the Benioff zone, but their geographic position will therefore also relate to the dip of the subduction zone. A steep subduction zone compresses the distribution, whereas a flat dip spreads them to greater distances from the trench.

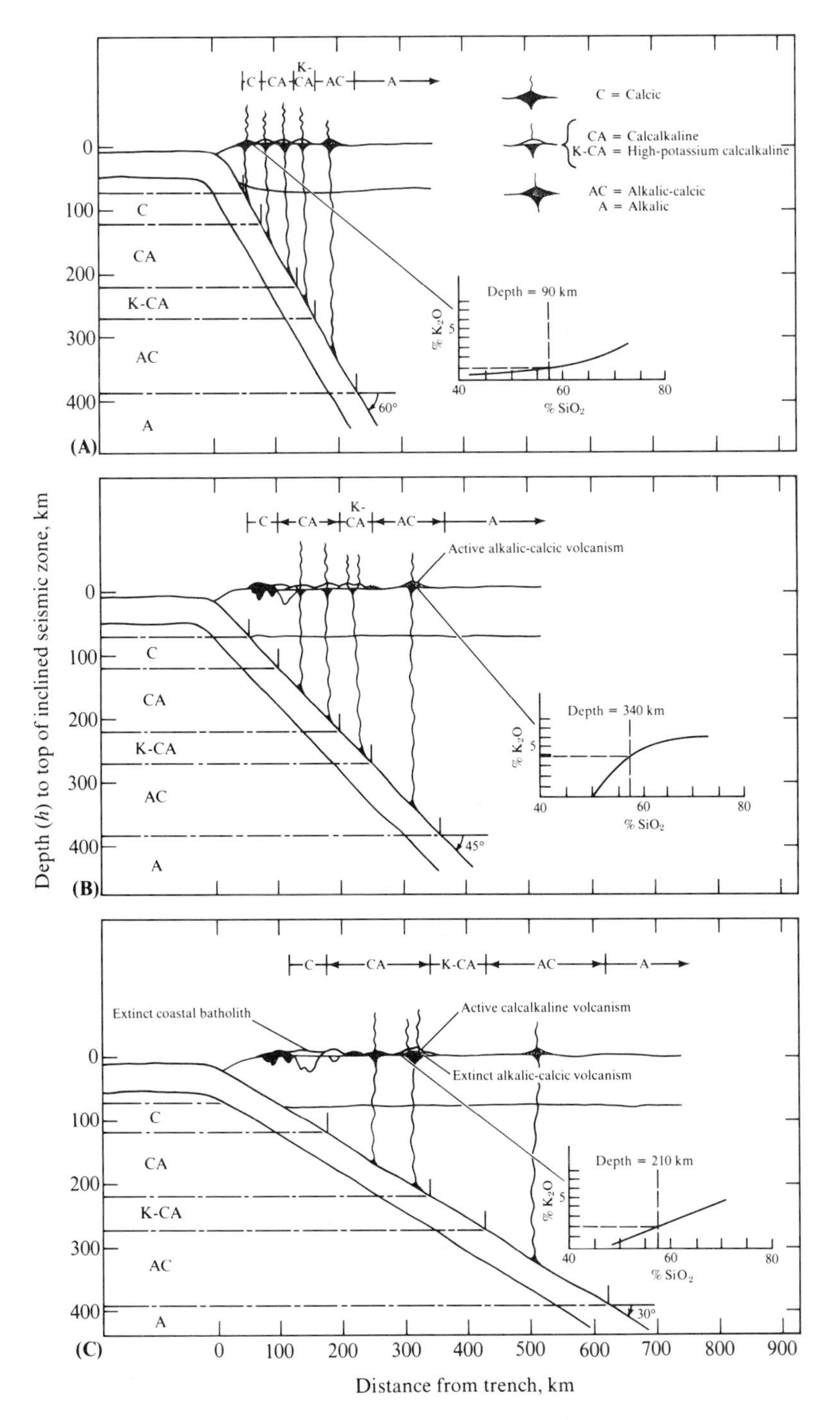

Figure 8–13
Diagrams showing the relationship between subduction dip angle and the location of various igneous rock types. Inserts show K_2O/SiO_2 variation plots that show trend lines typical of calcic, alkalic-calcic, and calcalkaline associations in parts A, B, and C respectively. [From S. R. Keith, 1978, *Geology, 6,* Fig. 1.]

It follows from this relationship that if the dip of the subduction zone remains constant through time, the same rock types will be found consistently at the same distances from the trench. With a change in subduction zone dip, the various belts will migrate through time either toward or away from the trench. A flattening of the dip will tend to bring calcalkaline rocks into a region where alkalic rocks had been found earlier, and vice versa. Using this approach in southwestern North America, Keith (1978) and others have suggested significant changes in the subduction dip angle between 140 and 15 m.y.b.p.

PLUTONIC ROCKS

The calcalkaline suite not only includes volcanic members, but also the gabbro–diorite–granodiorite–granite plutonic series that occupies a significant portion of the orogenic belts at continental margins. The silicic members are significantly more abundant, and may form batholithic chains stretching for hundreds of kilometers. The origin of these rocks is considerably more complex than the volcanic rocks, inasmuch as most of them have probably arisen from ultrametamorphism and melting in a deep crustal environment. Although the subject is discussed in some detail in Chapter 19, the generalities can be briefly considered here.

It has been generally agreed (see above) that most calcalkaline magmas arise from the Benioff zone at depths of about 150 km. Diapirs and melts rising and cooling adiabatically through the mantle and crust provide a significant source of heat to surrounding rocks. This results in various amounts of metamorphism of the crust above, as well as providing a general rise in isotherms. Inferred pressures and temperatures are shown in Figure 8–14. The probable *PT* curve for the volcanic area is shown in part A.

The continental crust is made up of a complex mixture of igneous, metamorphic, and sedimentary rocks. A variety of studies has indicated that partial melting (anatexis) of some of these rocks can occur under conditions of elevated temperature as a result of subduction. The combination of some water from the mantle, rising solid diapirs and melts, gradual accumulation of radioactive heat, and subsidence of crustal materials to the levels of the upper mantle as a result of compression often suffices to produce some melting in lower crustal materials (often gneisses containing quartz, potassium feldspar, and Na-rich plagioclase). The melt compositions are a function of source rock composition, depth, temperature, and type and amount of fluids. The first melts forming at lowest temperatures from a variety of source rocks tend to be granitic (granite to quartz diorite) in composition; at higher temperatures and with greater amounts of melting, the melts become more mafic. Data derived from trace element and isotopic studies indicate that both crustal and mantle constituents are present in the melt. Rise of these typically silicic melts may be accomplished by forceful emplacement, stoping, diapiric rise, or zone melting (a process whereby magmatic heat causes melting of the roof rocks while lower portions of

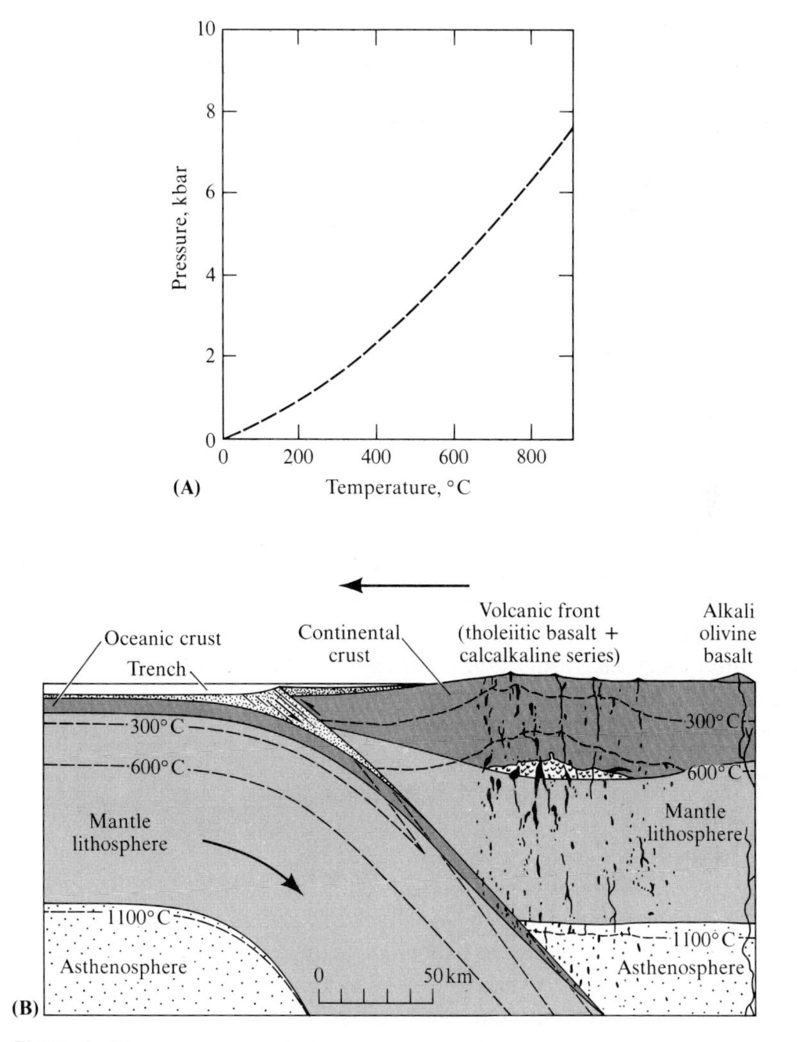

Figure 8–14
(**A**) A possible geothermal gradient for the region labeled "Volcanic front" in part B. (**B**) Cross section of a subduction zone, showing rise of isotherms as a result of magmatic ascent. Partial melting is indicated near the base of the crust. [After *Petrologic Phase Equilibria* by W. G. Ernst. W. H. Freeman and Company. Copyright © 1976. Figs. 6.46, 5.17. Thermal structure based on K. Hasebe, N. Fujii, and S. Uyeda, 1970, *Tectonophysics, 10,* 335–355; E. R. Oxburgh and D. L. Turcotte, 1968, *Jour. Geophys. Res., 73,* 2643–2661; D. L. Turcotte and E. R. Oxburgh, 1972, *Ann. Rev. Fluid Mech., 4,* 33–68; and D. T. Griggs, 1972, in E. C. Robertson, J. F. Hays, and L. Knopoff (eds.), *The Nature of the Solid Earth* (New York: McGraw-Hill), pp. 361–384.]

the magma are simultaneously crystallizing). The last processes could result in significant change in the composition of the original melt during ascent. The general process is indicated in Figure 8–14B. Alternative granite-forming processes include diffusion in the solid state (granitization) and the production of a granitic residue by differentiation of a basaltic primary magma. These processes are discussed in other chapters.

Batholiths Related to Subduction Zones

In this section we will briefly examine the general character of batholithic intrusions of Southern and Lower California and those of the Lachlan Mobile Zone in southeastern Australia.

Batholith of Southern and Lower California

The Mesozoic plutonic rocks of Southern and Lower California, as well as those of the Sierra Nevada to the north, were generated as a result of subduction during the Nevadan orogeny. The exposed portion of the batholith of Southern and Lower California extends southward from the vicinity of Riverside, California, more than 550 km into Lower California (see Figure 8–15). If completely exposed the batholith would be over 1,500 km long. The width averages about 100 km, and the surface area is about 32,000 km^2. Although at least 20 rock types are present in the many separate plutons that comprise the batholith, four types are predominant (see Table 8–3). Most of the rocks in this batholithic complex are to the west of the andesite or quartz–diorite line shown in Figure 8–15, which separates more potassic and silicic rocks to the east from more mafic rocks to the west. The general sequence of formation involves initial deposition of Triassic sediments that were folded, slightly metamorphosed, and eroded. Following this a great thickness of volcanic and sedimentary materials was deposited. After intense folding and metamorphism, many bodies of magma were injected. This occurred over a long time span (perhaps 100 million years), with the consequence that earlier magmas were often completely crystalline before injection of later magmas.

The magmas of the eastern portion are largely tonalites with lesser amounts of granodiorites. To the west the major type is granodiorite, with lesser amounts of gabbro (see Figure 8–15). All of the rocks of the calcalkaline trend are represented. Of these the general sequence of intrusions was first gabbro, followed by tonalite, granodiorite, and then granite. The fact that the earlier intrusions were largely crystallized before later intrusions occurred is indicated by sharp contacts, xenoliths in later intrusions, and small dikes within older rocks. Occasionally, a wide zone of mixed rock provides evidence that the earlier intrusion was not completely crystalline. Chilled borders are uncommon. Commonly the granodioritic and granitic bodies

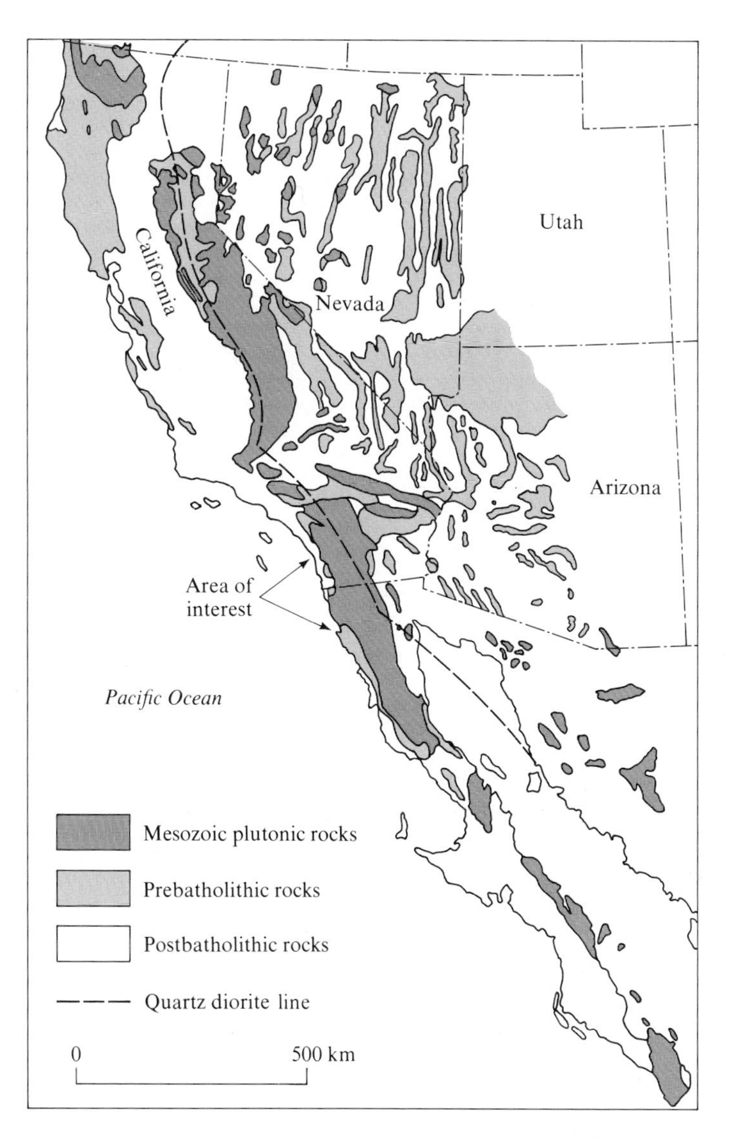

Figure 8–15
The general distribution of Mesozoic plutonic rocks in California and Lower California. The Peninsula Ranges are located in Southern and Lower California. The Sierra Nevada are in east-central California. [From E. S. Larsen, 1948, Geol. Soc. Amer. Mem. 29, Fig. 3. The dashed quartz–diorite line is based on J. G. Moore, 1959, *Jour. Geol., 67,* 198–210.]

Table 8–3
Predominant Rock Types and Average Chemical Compositions of the Batholith of Southern and Lower California

Percentage of rock types		
	1[a]	2[b]
Gabbro	14	7
Tonalite	50	63
Granodiorite	34	28
Granite	2.5	2

Average chemical composition			
	1[a]	2[b]	3[c]
SiO_2	65.0	64.1	62.0
TiO_2	.5	.6	.7
Al_2O_3	16.3	16.5	16.7
Fe_2O_3	1.2	1.3	1.4
FeO	3.8	3.8	4.5
MnO	.05	.04	.06
MgO	2.3	2.4	2.7
CaO	5.1	5.2	5.8
Na_2O	3.3	3.3	3.4
K_2O	2.1	2.0	1.6
H_2O	.5	.5	.6
P_2O_5	.06	.08	.09
S	.06	.03	.09
BaO	.06	.06	.05
SrO	tr.	tr.	tr.
ZrO_2	tr.	tr.	tr.

[a]Average composition in the Corona, Elsinore, and San Luis Rey quadrangles.
[b]Average composition in the Corona, Elsinore, San Jacinto, San Luis Rey, and Ramona quadrangles: a strip across the batholith about 70 miles long and 55 miles wide.
[c]Average composition of the Bonsall and Lakeview Mountain tonalites of the San Luis Rey, Elsinore, and Corona quadrangles.
Source: E. S. Larsen, Jr., 1948, Geol. Soc. Amer. Mem. 29.

show a definite banded zone near the contact, presumably indicating some flowage after initiation of crystallization. Some plutons display a border zone that is more mafic than the central portion of the intrusive. As this is present even when the country rock is siliceous, it does not result from assimilation, but rather from a changing magmatic composition during injection, or perhaps differentiation after emplacement.

The general texture within most of the intrusives is relatively uniform. Some cataclastic, gneissoid texture may be present near the contacts. The texture is generally characteristic of the particular intrusive and independent of the size of the body or proximity to the contact.

Chemical compositions of some of the rocks of this batholith show that these are a related group of intrusions, with rise of SiO_2, K_2O, and Na_2O correlated with a corresponding decrease in Al_2O_3, CaO, FeO, and MgO. Essentially the same trend is shown by rocks of the Sierra Nevada, indicating a similar type of origin for both batholiths. The mineralogy of the Southern and Lower California rocks (see Figure 8–16) shows a related trend; with a rise in SiO_2, K_2O, and Na_2O, the mafic minerals decrease in quantity and the salic minerals increase. Inclusions are present in most of the plutonic rocks, with the exception of gabbro. They constitute less than 1% of any of the plutons. The inclusions vary as to rock type and degree of assimilation, but are never present in large amounts.

The method of emplacement varies with individual plutons. Some evidence is present to indicate magmatic stoping, forceful injection, and occasionally replacement. Little evidence is present to favor large-scale assimilation of wall rock.

Lachlan Mobile Zone of Southeastern Australia

The Lachlan Mobile Zone is part of a larger disturbed area, the Tasman Mobile Zone, that stretches between latitudes 10° S and 42° S in eastern Australia. The

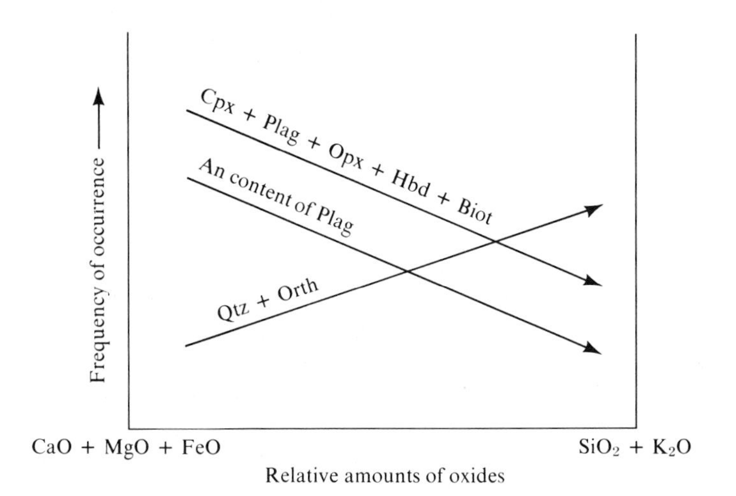

Figure 8–16
Schematic diagram showing changes in mineralogical composition of rocks of the Peninsula Ranges as a function of oxide composition. Only general trends are indicated, as the diagram is not to scale. [From the data of E. S. Larsen, 1948, Geol. Soc. Amer. Mem. 29, pp. 150, 151.]

maximum width of the granitic belt within the larger zone is about 300 km, with a general outcrop area of about 160,000 km^2. South of 32° S, within the Lachlan Mobile Zone, the granites generally vary in age from Silurian to Devonian and have an outcrop area of about 44,000 km^2 (see Figure 8–17). The sediments in the area were deposited from the Cambrian through the Devonian Period; most of these are Ordovician slates and sandstones.

The granitic rocks (granitoids) in the region have been classified into three types (White et al., 1974): (1) regional-aureole granites, surrounded by schists and gneisses of regional extent; (2) contact-aureole granites, surrounded by a small aureole measured in hundreds of meters; and (3) subvolcanic granites, surrounded by a narrow contact aureole measured in tens of meters, and intimately associated with volcanic rocks and tuffs. These three categories generally correspond to Buddington's categories of catazonal, mesozonal, and epizonal plutonic rocks (1959). As granites of different types are occasionally intruded simultaneously, this classification probably represents variations in the thermal gradient perpendicular to the contact rather than differences in the depth at which crystallization occurred. This type of relationship was suggested earlier for other areas by Buddington.

Typical granites of the regional-aureole type are composed mainly of quartz and potassium feldspar, with lesser plagioclase, biotite, and muscovite. Metamorphic minerals present in both these granites and surrounding country rock are sillimanite, andalusite, and cordierite, minerals that are stable under conditions of high temperature and low pressure. The common presence of migmatites and abundant xenoliths of high-grade metamorphic rocks suggests that these granites were derived from the surrounding metamorphic rocks by anatexis and still remain in contact with them. This is confirmed by correspondence of the Sr^{87}/Sr^{86} ratios of both granite and country rock.

The granites of the contact aureole type comprise most of the batholiths of the Lachlan Zone. The batholiths are composite, and consist of large numbers of plutons; these plutons are circular to elliptical in shape (see Figure 8–18), in contrast to the more irregular regional aureole type. Contact aureoles are about 500 m wide. Adjacent country rock consists of hornfels—a fine-grained, relatively structureless metamorphic rock, typical of many igneous contact situations. Xenoliths present are either similar to those within the regional-aureole granite or else contain hornblende. Both types are considered to be residual material brought by the magma from its source area. The granitic rocks of eastern Australia have been subdivided into I-types and S-types, depending upon whether their anatectic source is igneous or sedimentary, respectively. These are distinguished on the basis of chemical, mineralogical, or field criteria (Chappell and White, 1974).

The intrusion mechanism of the contact aureole granites was not rapid forceful injection, as little evidence of deformation of the country rock is present. Assimilation of country rock is not indicated. Magmatic stoping is eliminated from consideration, because xenoliths of typical adjacent country rock are not present. The intrusion was

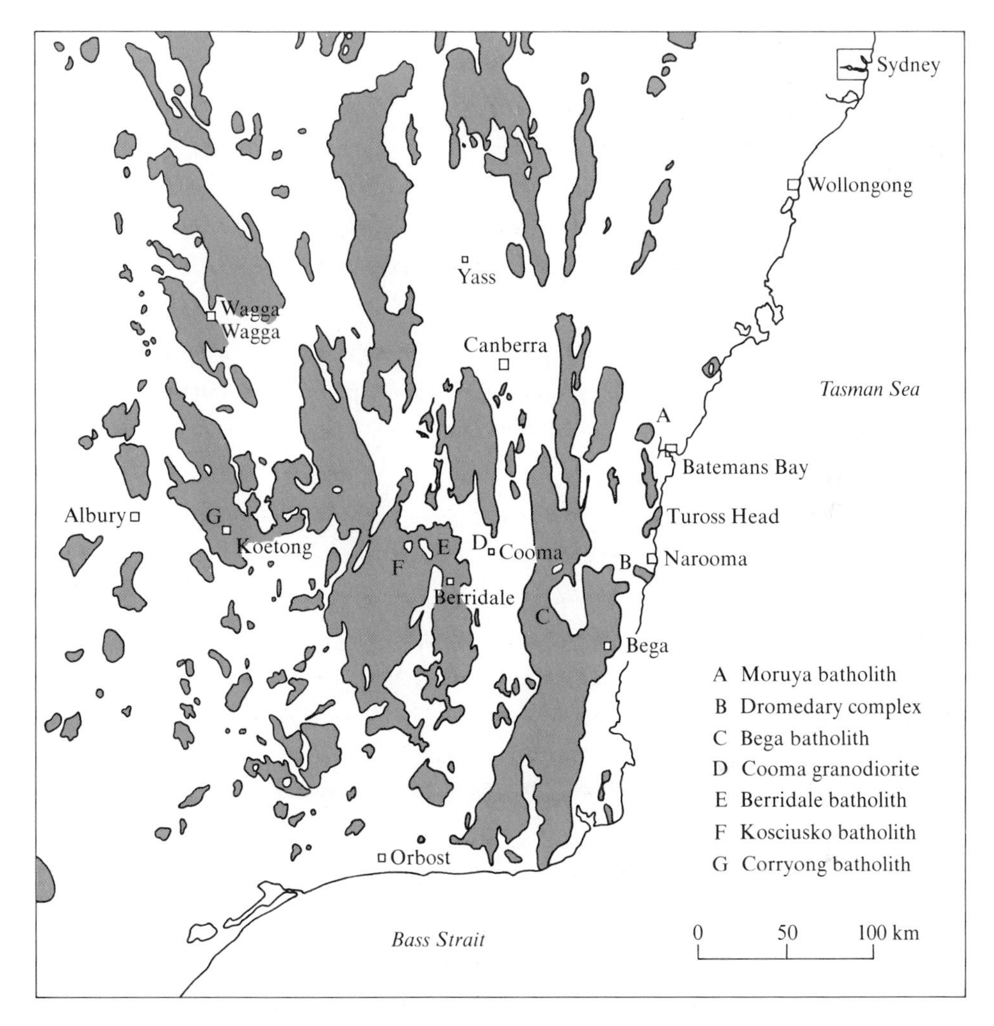

Figure 8–17
The distribution of granitic rocks (gray) in the Lachlan Mobile Zone of southeastern Australia. [Modified from B. W. Chappell and A. J. R. White, 1976, *Plutonic Rocks of the Lachlan Mobile Zone,* 25th International Geological Congress, Excursion Guide No. 13, Fig. 1.]

probably accomplished by diapirism that uplifted the (presently absent) roof.

It should be stressed once again that batholithic origins are complex. A batholith is usually composed of a significant number of intrusions that were emplaced over a long period of time in a variety of circumstances. To describe a batholith normally requires detailed examination and correlation of all of its many constituent parts.

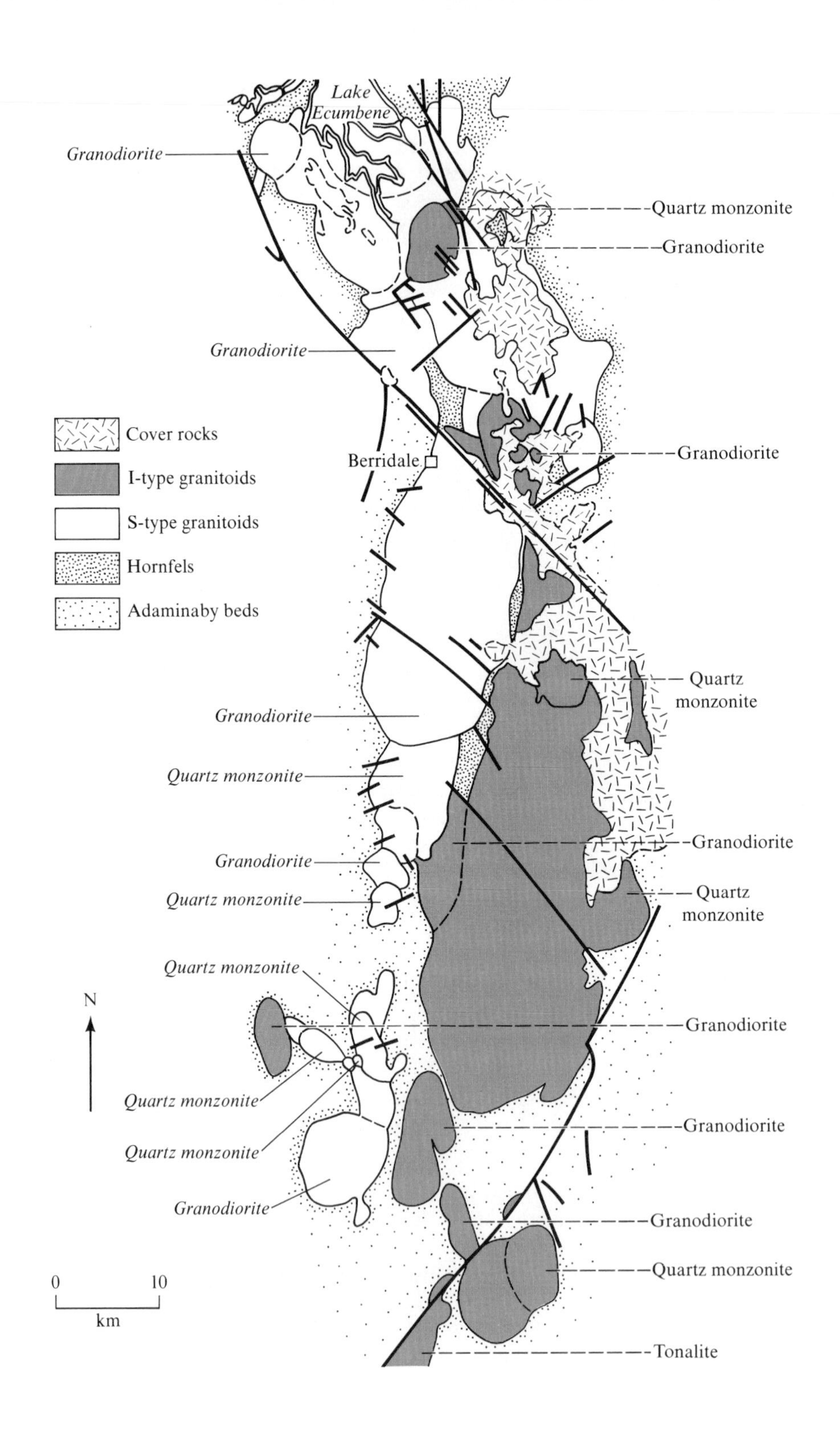
Lake Ecumbene
Granodiorite
Quartz monzonite
Granodiorite
Granodiorite
Cover rocks
I-type granitoids
S-type granitoids
Hornfels
Adaminaby beds
Berridale
Granodiorite
Quartz monzonite
Granodiorite
Quartz monzonite
Granodiorite
Granodiorite
Quartz monzonite
Quartz monzonite
Quartz monzonite
N
Granodiorite
Quartz monzonite
Granodiorite
Quartz monzonite
Granodiorite
Granodiorite
0
10
km
Quartz monzonite
Tonalite

Figure 8–18 (*facing page*)
Simplified geologic map of the Berridale batholith within the Lachlan Mobile Zone (southeastern Australia). The batholith consists of many small plutons, many of which are petrographically and chemically dissimilar. I- and S-type granitoids have originated from partial melting of igneous or sedimentary source rocks respectively. [Modified from B. W. Chappell and A. J. R. White, 1976, *Plutonic Rocks of the Lachlan Mobile Zone,* 25th International Geologic Congress, Excursion Guide No. 13, Fig. 8.]

SUMMARY

Those continental shorelines characterized by the presence of a subduction zone are actively volcanic. In the initial stages of convergence of continental and oceanic plates, obduction may occur. This process thrusts oceanic crust and mantle onto the continental edge; the distinctive association of ultramafic, gabbroic, and diabasic rocks, often capped by deep sea sediments, is known as the ophiolite suite. Examples found in Newfoundland, Cyprus, and other localities indicate that metamorphism of the sea floor is quite common.

Both tholeiitic and calcalkaline rocks are related in origin to subduction zones, with the tholeiitic rocks close to the trench area, and the calcalkaline and sodic intrusives further away, above deeper portions of the subduction zone. The fractionation trend of calcalkaline magmas is toward enrichment in alkalies and silica with simultaneous decrease in total Fe content, whereas tholeiites show considerable enrichment in Fe with little increase in silica.

Lavas produced in and near the subduction zone are considered to form by a complex group of processes related to increase in temperature with depth, followed by progressive dehydration of the subducting plate. Melting occurs at several depths both within and above the subducting plate, leading to a variety of both tholeiitic and calcalkaline lavas.

Ascent of lava through continental mantle and crust results in a rise in regional geothermal gradients and anatexis of deeply buried crustal materials. These anatectic melts may rise to upper crustal levels, where they form large and complex granitic plutons. These plutons, often composed of many smaller intrusions, are often differentiated, and may show the effects of wall rock assimilation. Metamorphism of country rock varies in extent with the circumstances of intrusion.

FURTHER READING

Arculus, R. J., and R. W. Johnson. 1978. Criticism of generalized models for the magmatic evolution of arc-trench systems. *Earth Planet. Sci. Letters, 39,* 118–126.

Baker, P. E. 1968. Comparative volcanology and petrology of the Atlantic island arcs. *Bull. Volcan., 32,* 189–206.

Bowen, N. L. 1928. *The Evolution of the Igneous Rocks.* Princeton, N.J.: Princeton Univ. Press, 332 pp.

Buddington, A. F. 1959. Granite emplacement with special reference to North America. *Geol. Soc. Amer. Bull., 70,* 671–747.

Chappell, B. W., and A. J. R. White. 1974. Two contrasting granite types. *Pacif. Geol., 8,* 73–174.

Coleman, R. G. 1977. *Ophiolites.* New York: Springer-Verlag, 229 pp.

Cox, K. G., J. D. Bell, and R. J. Pankhurst. 1979. *The Interpretation of Igneous Rocks.* London: George Allen and Unwin, 450 pp.

Dewey, J. F., and J. M. Bird. 1970. Mountain belts and the new global tectonics. *Jour. Geophys. Res., 75,* 2625–2647.

Dewey, J. F., and J. M. Bird. 1971. Origin and emplacement of the ophiolite suite: Appalachian ophiolites in Newfoundland. *Jour. Geophys. Res., 76,* 3179–3206.

Dickinson, W. R., and T. Hatherton. 1967. Andesitic volcanism and seismicity around the Pacific. *Science, 157,* 801–803.

Ernst, W. G. 1976. *Petrologic Phase Equilibria.* San Francisco: W. H. Freeman and Company, 333 pp.

Fenner, C. N. 1929. The crystallization of basalts. *Amer. Jour. Sci., 18,* 225–253.

Gill, J. B. 1970. Geochemistry of Vitu Leon, Fiji and its evolution as an island arc. *Contr. Mineral. Petrology, 27,* 179–203.

Jakes, P., and A. J. White. 1969. Structure of the Melanesian Arcs and correlation with distribution of magma types. *Tectonophysics, 8,* 223–236.

Keith, S. B. 1978. Paleosubduction geometries inferred from Cretaceous and Tertiary magmatic patterns in southwestern North America. *Geology, 6,* 516–521.

Larsen, E. S., Jr., 1948. Batholith and associated rocks of Corona, Elsinore, and San Luis Rey Quadranges, southern California. Geol. Soc. Amer. Mem. 29, 182 pp.

Mitchell, A. H., and H. G. Reading. 1969. Continental margins, geosynclines, and ocean floor spreading. *Jour. Geol., 77,* 629–646.

Miyashiro, A. 1972. Metamorphism and related magmatism in plate tectonics. *Amer. Jour. Sci., 272,* 629–656.

Moore, J. G. 1959. The quartz-diorite boundary line in the western United States. *Jour. Geol., 67,* 198–210.

Peacock, M. A. 1931. Classification of igneous rock series. *Jour. Geol., 39,* 54–67.

Ringwood, A. E. 1974. The petrological evolution of island arc systems. *Jour. Geol. Soc., 130,* 183–204.

Ringwood, A. E. 1975. *Composition and Petrology of the Earth's Mantle.* New York: McGraw-Hill, 618 pp.

Ringwood, A. E. 1977. Petrogenesis in island arc systems. In *Island Arcs, Deep Sea Trenches and Back-Arc Basins,* Maurice Ewing Series, Vol. 1, pp. 311–324. Amer. Geophys. Union.

Wager, L. R., and W. A. Deer. 1939. Geological investigations in East Greenland III. The petrology of the Skaergaard intrusion, Kangerdlugssas, East Greenland. *Med. øm Grønland, 105,* 1–352.

White, A. J. R., B. W. Chapell, and J. R. Cleary. 1974. Geological setting and emplacement of some Australian Paleozoic batholiths and implications for intrusive mechanisms. *Pacif. Geol., 8,* 159–171.

Williams, H. 1973. Bay of Islands map area, Newfoundland. Geol. Surv. Canada, Paper 72–34, 7 pp.

Williams, H., and J. Malpas. 1972. Sheeted dikes and brecciated dike rocks within transported igneous complexes, Bay of Islands, Western Newfoundland. *Canad. Jour. Earth Sci., 9,* 1216–1229.

Wyllie, P. J. 1973. Experimental petrology and global tectonics—a preview. *Tectonophysics, 17,* 189–209.

Yoder, H. S., Jr., ed. 1979. *The Evolution of the Igneous Rocks, Fiftieth Anniversary Perspectives.* Princeton, N.J.: Princeton Univ. Press, 588 pp.

9

Continental Igneous Rocks

Igneous rocks formed after cessation of major orogenic periods in continental shield areas or in major continental fracture zones include tholeiitic basalts and their intrusive equivalents, anorthosites, peridotites, carbonatites, and alkali-rich rocks.

Of the various continental lavas, tholeiitic basalts are the most common, occasionally flooding thousands of square kilometers, as in the Deccan Plateau region of India or the Columbia River province in the United States. Tholeiitic melts of huge volume may also be emplaced at shallow depths in dike swarms and sills, as in Antarctica and Tasmania. Emplacement at greater depth permits slower cooling, with extensive fractionation leading to internal stratification; the largest intrusion of this type is the Precambrian Bushveld complex in South Africa (volume about 100,000 km^3). Other well-known examples are the Stillwater complex in Montana, the Muskox intrusion of Canada, and the Skaergaard intrusion of Greenland.

The Skaergaard intrusion, very small (500 km^3) in comparison to the Bushveld complex, is probably the world's most thoroughly investigated major igneous body. The Skaergaard intrusion is an initial rifting deposit and cannot properly be placed in the category "continental igneous rocks." However, it has many of the characteristics of large gabbroic layered intrusives and will be discussed here.

GABBROIC LAYERED INTRUSIONS

The Eocene Skaergaard intrusion has been long regarded as the classic case of differentiation of a tholeiitic basalt. For this reason we will discuss it in terms of both the standard interpretation of Wager and Brown (1967) as well as the more recent interpretation by McBirney (1975). Both approaches are illustrated in Figure 9–1; in these diagrams the intrusion is shown rotated from its present highly inclined position.

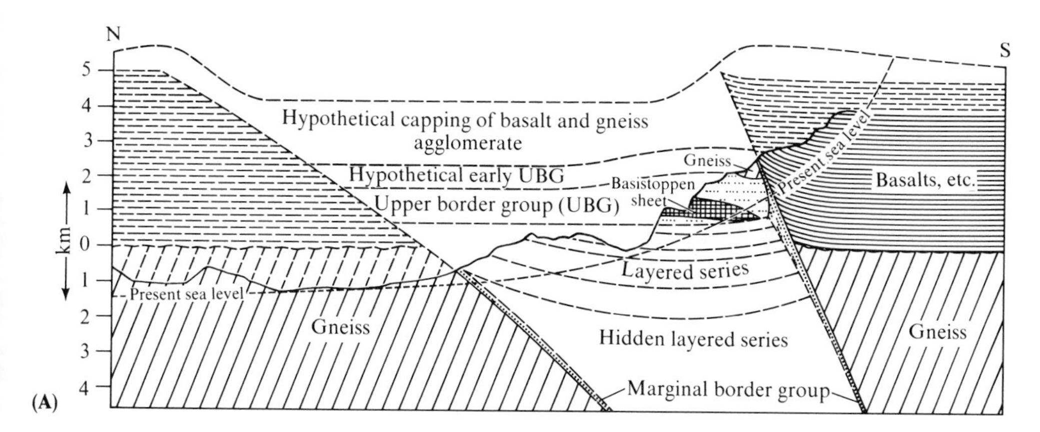

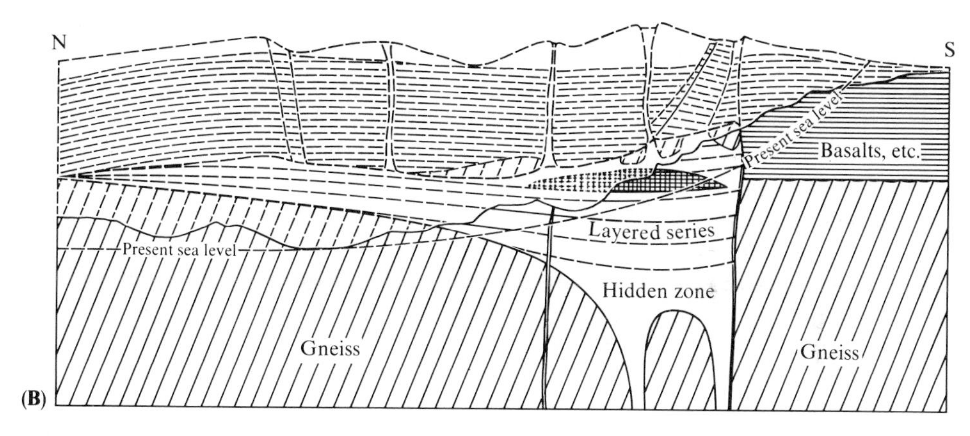

Figure 9–1
The Skaergaard intrusion in its original orientation. (**A**) The Wager and Brown interpretation. The solid irregular line indicates the present erosion surface. The intrusive is considered funnel-shaped with a large unexposed hidden zone at the base. [From *Layered Igneous Rocks* by L. R. Wager and G. M. Brown. W. H. Freeman and Company. Copyright © 1967. Fig. 8.] (**B**) The McBirney interpretation. The present level of erosion is indicated by a solid line. Much of the hidden zone has been eliminated, and the extent of the middle and lower zones has been considerably increased. [From A. R. McBirney, 1975, *Nature, 253,* Fig. 7.]

The lower portion of the intrusion is not completely exposed by erosion. Presently exposed are the contact with the country rock (the marginal border group), an upper highly differentiated portion (the upper border group), and a central layered series. It is obvious from the outcrops that there exists an unexposed lower portion, known as the hidden layered series. The size and composition of the hidden layered series are not known. Needless to say, this represents an intriguing problem, which has provoked considerable hypothesizing on the part of igneous petrologists. Interpretations of this zone have been based on a study of exposed materials and laboratory examinations of synthetic melts. Two interpretations of the size of the intrusion are shown in Figure 9–1.

Examination of the exposed portions of the intrusion has indicated a general compositional pattern. The edge is a fine-grained mafic chilled border, originally interpreted as representing the original melt composition. The lower exposed portion consists of ultramafic compositions; the central area is composed of two-pyroxene gabbro, and the upper portion is feldspathic gabbro or iron-rich diorite. It is obvious from rock type alone that this body is highly differentiated. Additional strong evidence comes from two different types of layering. One of these, called *rhythmic layering* (see Figure 9–2), shows up as a repetitious lithologic stratification; strongly developed layers of diverse mineral composition are present throughout much of the body. As these layers are composed of different proportions of light and dark minerals, they form a conspicuous pattern in outcrop. A second type of compositional change within the intrusion is called *cryptic layering;* this is usually only determined by microscopic examination or chemical analyses. It consists of a relatively continuous vertical series of chemical (and to lesser extent mineralogical) changes in the composition of minerals that display solid solution.

Intrusion of the tholeiitic melt first resulted in the formation of a chilled margin against the colder country rock. Although this fine-grained rock has been taken as representing parental melt composition, it is likely that the original melt may have assimilated some country rock and undergone secondary reactions, and perhaps contained olivine crystals that were concentrated toward the center of the chamber during intrusion; concentration of crystalline materials into the faster-moving central portions of an intruding magma has been called *flow differentiation.*

After emplacement, crystallization occurred in the cooler upper portion and adjacent to the chilled border zones. Convective motion as well as gravity settling along the cooler edges resulted in the downward motion of crystals to the chamber floor (see Figure 9–3). As a result of continued crystal accumulation, the level of the floor slowly rose, leaving the residual melt in the upper portion of the chamber. Crystal settling has thus resulted in an accumulation of crystals that is oldest at the base and youngest at the top.

The rocks formed by crystals that have settled to the magma floor are called *cumulates* (see Figure 9–4). Cumulates form from the loosely packed cumulus crystals and the interprecipitate liquid (similar to pore water in an unconsolidated sediment). This liquid, when crystallized, may enlarge the cumulus crystals or form phases different from the cumulus crystals (either as a result of differences in liquid–crystal composition or through interaction with the melt above).

The cryptic layering developed in most large mafic intrusives consists of a continuous upward change in composition of mineral solid solutions in the direction of lower liquidus temperatures. The only abrupt break that occurs is the occasional elimination of a high-temperature phase, and a simultaneous or later entrance of a lower-

Figure 9–2
Conspicuous differences in light and dark mineral content of bands within the Skaergaard intrusion. [From L. R. Wager, 1968, in H. H. Hess and A. Poldervaart (eds.), *Basalts,* vol. 2 (New York: Wiley Interscience), Plate 5.]

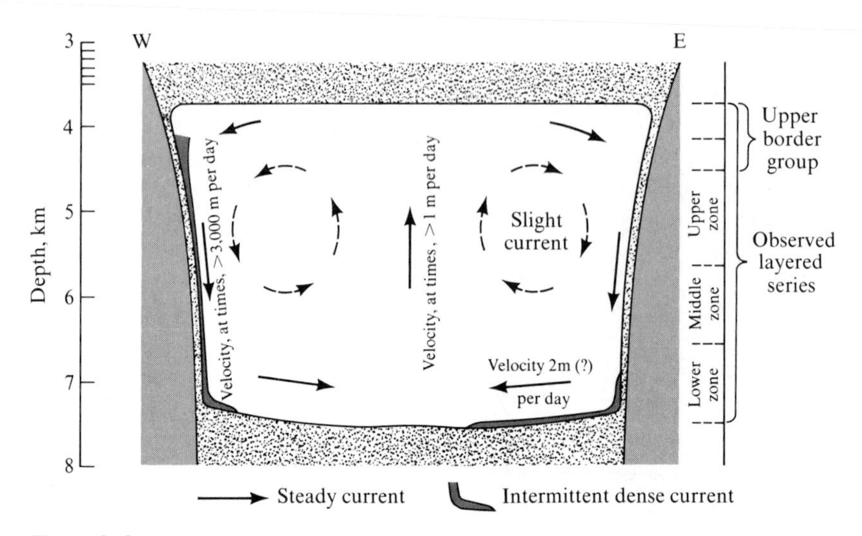

Figure 9–3
Schematic sketch of the Skaergaard intrusion showing two possible types of convective movement: a slow continuous circular or elliptical convective current with a velocity of about 2 m/day, and intermittent dense high-velocity currents that descend along the walls and undercut the slow current, depositing layers of crystals on the floor. [From L. R. Wager, 1968, in H. H. Hess and A. Poldervaart (eds.), *Basalts,* vol. 2 (New York: Wiley Interscience), Fig. 16.]

Figure 9–4 (*facing page*)
Cumulate textures. (**A**) Plagioclase orthocumulate. Early calcium-rich cumulate plagioclase is originally in a melt of different composition. The melt has crystallized and no reaction with plagioclase has occurred, resulting in a zoned plagioclase rim. (**B**) Plagioclase mesocumulate. An intermediate situation between plagioclase orthocumulate and adcumulate growth. Considerable unzoned adcumulate growth has occurred, but the outer portion of the plagioclase rim shows orthocumulate growth with zoning. Other phases have also crystallized from the melt. (**C**) Plagioclase adcumulate. Plagioclase cumulate in a melt of different composition has grown by acquisition of material from an external source. The plagioclase forms unzoned crystals as the interstitial liquid is forced out. (**D**) Polymineralic adcumulate. The three phases have grown by adcumulate growth and are all unzoned. (**E**) Olivine heteradcumulate. The cumulate olivine grains show no evidence of growth. The surrounding melt has nucleated and undergone adcumulate growth, forming large essentially unzoned poikilitic pyroxene and plagioclase. (**F**) Olivine crescumulate. Cumulate olivine has grown upward into the liquid above, changing composition slightly to correspond to the surrounding melt. All olivine shown is in optical continuity. The surrounding plagioclase and pyroxene are also essentially unzoned and grown as heteradcumulate. [From L. R. Wager, G. M. Brown, and W. J. Wadsworth, 1960, *Jour. Petrology 1,* Figs. 1, 2.]

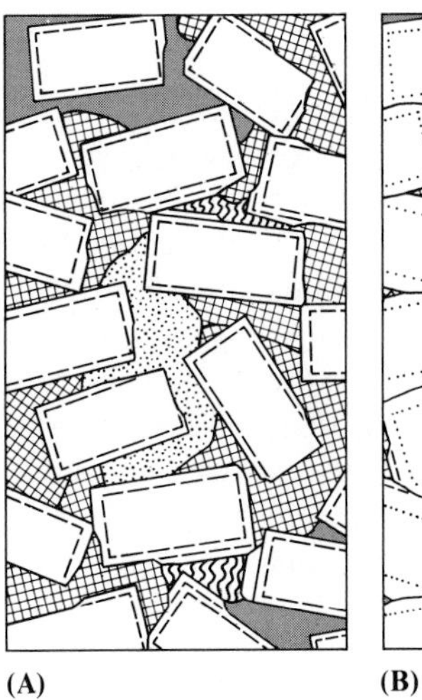

(A)

(B)

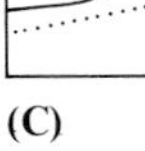

(C)

PLAGIOCLASE
Boundary of the cumulus crystals (labradorite) diagrammatically shown by the innermost rectangle. The limits of medium- and low-temperature zones, where developed, shown outside the cumulus crystal boundaries.

PLAGIOCLASE
Boundary of the cumulus crystals (labradorite) shown by the dotted line. Outside is adcumulate growth of plagioclase of similar composition. In places beyond the broken lines, lower temperature zones are shown.

PLAGIOCLASE
The cumulus part of the crystal is shown within the dotted line. This has been enlarged by growth of more plagioclase of the same composition, which fills the interstices.

Pyroxene
Olivine
Iron ore
Poikilitic crystals, zoned (but this is not indicated)

Quartz and orthoclase, locally the final residuum

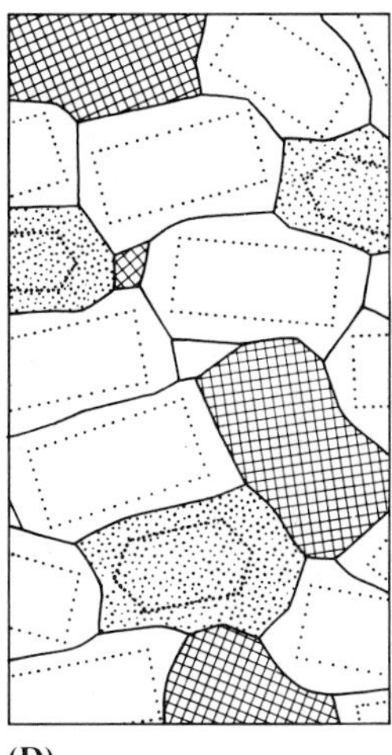

(D)

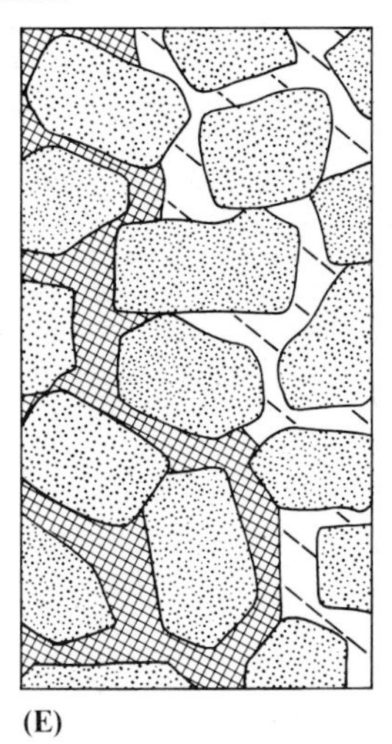

(E)

(F)

temperature phase. An example is the elimination of magnesian olivine and later appearance of orthopyroxene or ferrous olivine. This break in olivine precipitation defines the middle zone of the Skaergaard intrusive. The vertical trend of cumulates in the cryptic layering is shown in the 2,500 m of the layered series (see Table 9–1 and Figure 9–5). The plagioclase becomes more sodic upward. Magnesian olivine becomes more iron-rich, is not present in the middle zone, and again recurs in the upper zone, richer in iron. Calcium-poor (pigeonite) and calcium-rich (augite) pyroxenes occur together over most of the zones, becoming progressively enriched in iron. In the upper border groups, located above the layered series, the cryptic layering compositional trends are abruptly reversed, indicating that these rocks have grown downward from the chamber roof. In other large mafic intrusives, such as the Bushveld, reversal of the cryptic layering trend does not occur.

Rhythmic layering commonly shows gravity stratification; the heavier olivine or pyroxene may increase in amount to the base of the layer while feldspar increases upward. The base of the individual layers is sharp within the lower and middle zones of the Skaergaard, but becomes increasingly diffuse in the upper zone. Evidence that these layers solidified upward is given by cross bedding and scour structures, similar to those found in sedimentary strata. In spite of the vertical mineralogical variation shown by the many rhythmic layers, compositional changes associated with cryptic layering remain consistent throughout the 2,500 meters of the layered zones in the Skaergaard intrusive.

Table 9–1

Compositions of Coexisting Solid Solution Cumulate Crystals Within the Layered Series of the Skaergaard Intrusion

Zone	Height above lowest exposed level, m	Plagioclase An	Olivine Fo	Augite Ca	Augite Mg	Augite Fe	Pigeonite Fe
	2,500	30	0	43	0	57	
Upper	2,000	38	24	40	21	39	
(Ferrodiorites)	1,800	40	33	35	33	32	Absent
	1,580	45	40	35	37	28	46
Middle	1,000	51	Absent	37	39	24	n.d.[a]
	800	51	53		n.d.[a]		n.d.[a]
Lower	600	56	57	38	41	21	35
(Gabbro)	500	58	59		n.d.[a]		Present
	300	62	63		Present		
	100	66	67				

[a]n.d. = not determined.

Source: After *Layered Igneous Rocks* by L. R. Wager and G. M. Brown. W. H. Freeman and Company. Copyright © 1967.

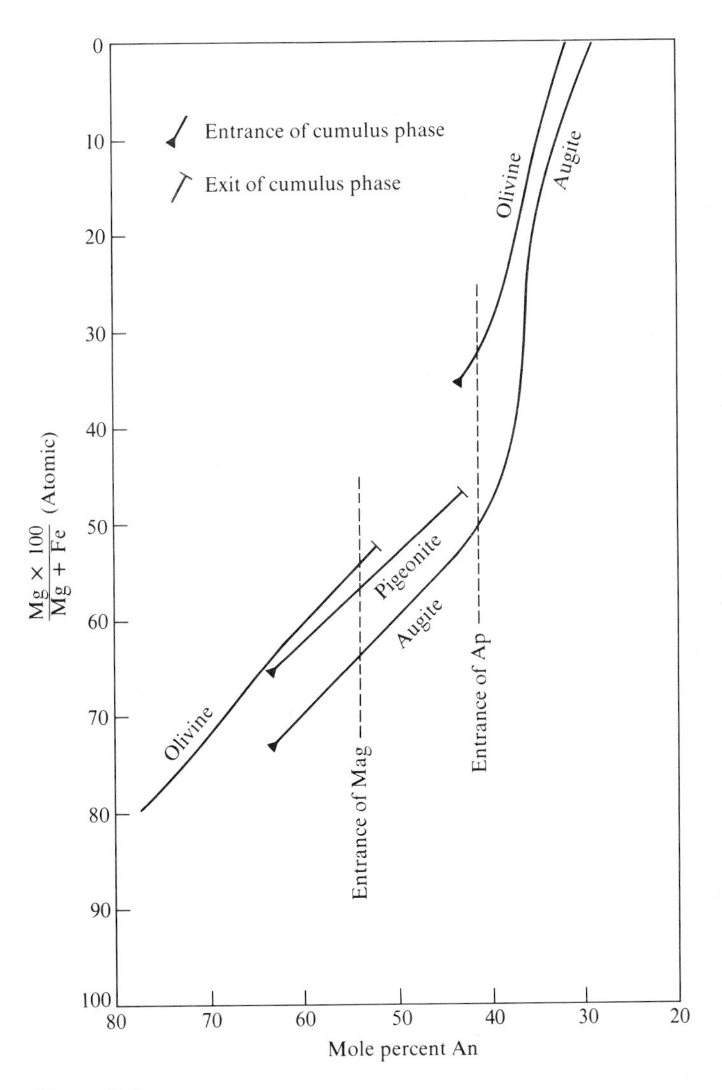

Figure 9-5
Composition of the major cumulus phases of the Skaergaard intrusion plotted against plagioclase composition (which becomes more sodic upwards). Mag = magnetite, Ap = apatite. [From L. R. Wager, 1968, in H. H. Hess and A. Poldervaart (eds.), *Basalts,* vol. 2 (New York: Wiley Interscience), Fig. 10.]

Cryptic layering provides a measure of the effectiveness of crystal fractionation processes—a verification of many of the processes suggested by Bowen (1928). The progressive change of liquid composition during differentiation (the liquid line of descent) has been determined for the Skaergaard intrusive by Wager (1960), assuming that the chilled border zone is representative of the parental magma composition, and that the final silicic differentiates are a result of fractionation rather than contamination by country rock assimilation. The computed liquid fractions (the Skaergaard trend) are shown in Figure 9–6A from the assumed base of the intrusive through the upper layered series for total oxides versus weight percent crystallized melt. These data are given in terms of FMA in Figure 8–8. Fractionation results in gradual decrease of MgO and increase in FeO (as a result of early crystallization of Mg-rich olivines and pyroxenes). Gradual increase of Na_2O and decrease of CaO are a result of the initial crystallization of high-calcium plagioclase. The gradual decrease in MgO and CaO results in a high content of FeO and alkalies in the final residual liquid. It is interesting to note that the SiO_2 content does not rise significantly until the liquid is almost completely consumed. Liquids comparable to those in the andesite–dacite–rhyolite series are only achieved when the original melt is about 98% solid, strongly indicating that significant amounts of granitic magma are not likely to arise from similar fractionation of tholeiitic magmas.

The fractionation trend of Wager has been seriously questioned by McBirney (1975), whose liquid composition trends (in terms of SiO_2 and FeO) are given in Figure 9–6B. McBirney has established that the chilled margin is in fact not representative of the original magma composition. Knowledge of the temperatures of crystallization and oxygen content of the various layered series, along with partial fusion experiments of various layered rocks, has revealed that the change of liquid compositions is quite different from that proposed by Wager. Furthermore, strong evidence of liquid immiscibility exists in the upper zone rocks (between a melt rich in iron and phosphorous and low in silica, versus one rich in silica and alkalies and low in iron). This study (still in progress) has already led to interpretations significantly different from the approach of Wager and others in terms of liquid fractionation trends, amounts of the melt present, shape of the intrusion, and the mechanism of crystallization.

ANORTHOSITES

Anorthositic massifs are unique, inasmuch as most were probably emplaced during the Precambrian—most between 1100 and 1700 m.y.b.p., with some Archaen types (pre-2500 m.y.b.p.). This perhaps implies a unique event or events that cannot be related to more recent geologic situations.

Most anorthosites are found within Precambrian shield areas. Within the Grenville tectonic province of eastern Canada anorthosites comprise about one-fifth of the

---

Figure 9–6 (*facing page*)
(**A**) Weight percentages of oxides in successive liquid fractions as a result of differentiation of the Skaergaard pluton. [As calculated by L. R. Wager, 1960, *Jour. Petrology, 1,* Fig. 14, © Oxford University Press.] (**B**) Comparison of the weight percentages of silica and iron oxide (solid lines) at successive stages of crystallization of the layered series, as determined experimentally by McBirney. The dashed lines are curves calculated by Wager (1960). [From A. R. McBirney, 1976, *Nature, 253,* Fig. 3.]

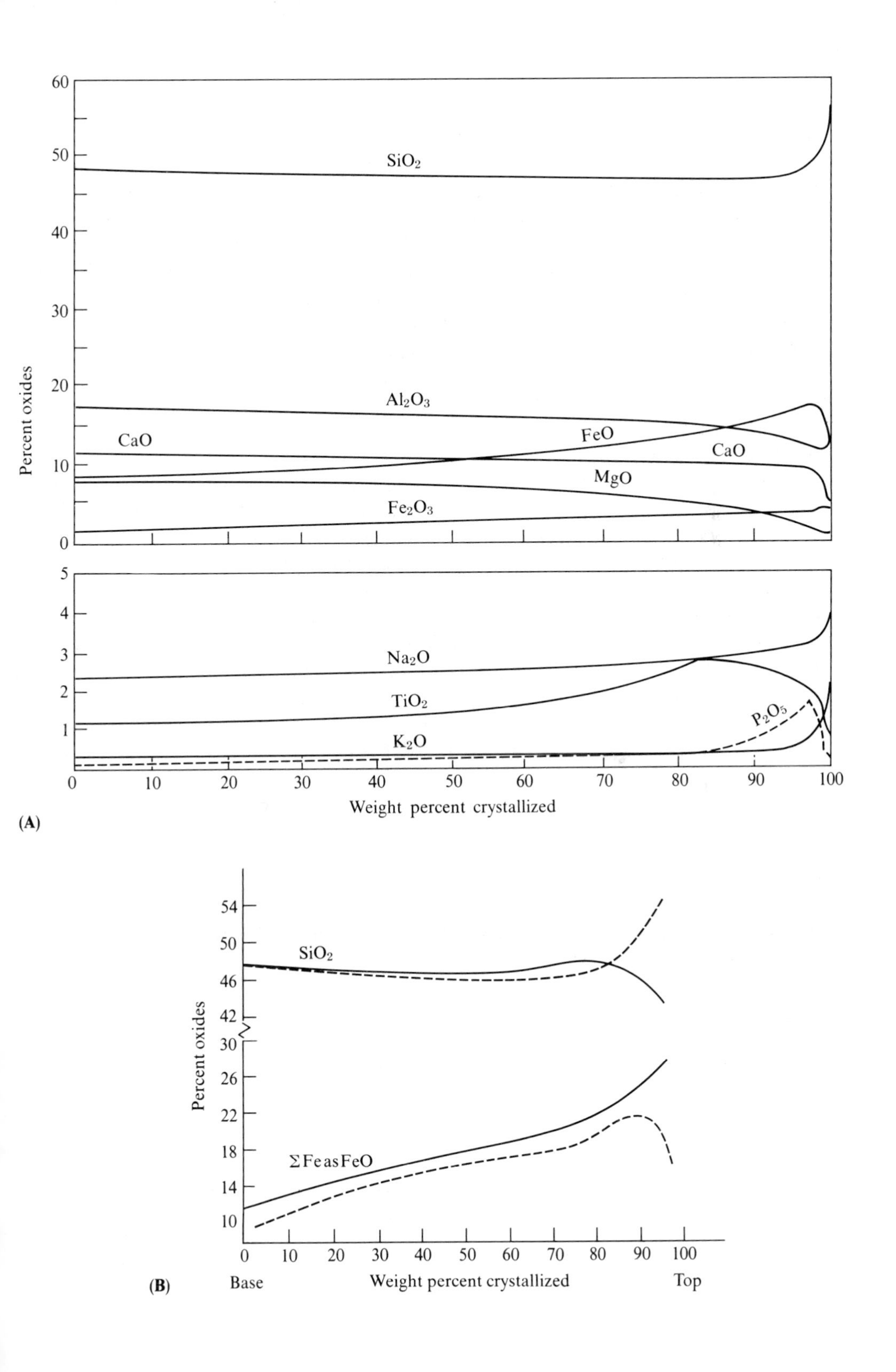

60
50
40
30
20
10
0
SiO2
Al2O3
CaO
FeO
CaO
MgO
Fe2O3
Percent oxides
5
4
3
2
1
Na2O
TiO2
P2O5
K2O
0
10
20
30
40
50
60
70
80
90
100
Weight percent crystallized
(A)
54
50
46
42
30
26
22
18
14
10
SiO2
ΣFe as FeO
Percent oxides
0
10
20
30
40
50
60
70
80
90
100
Base
Weight percent crystallized
Top
(B)

exposed bed rock. The adjacent related Adirondack Mountains in New York State contain an anorthosite body covering 200 km^2 (see Figure 9–7). These rocks, studied in detail by Buddington (1939), have been considered the model for typical anorthosite bodies, but recent work has indicated considerable gradation from this generally unlayered massif to the stratified mafic intrusives discussed above. The Adirondack and similar Norwegian massifs are lenslike or sheetlike, with an uplifted roof. The central portion is coarse plagioclase (andesine) with minor mafic minerals. Border areas are often gabbroic with up to 30% pyroxene. Associated intrusives are pyroxene-bearing varieties of granite, granodiorite, and quartz syenite. Both the anorthosite and associated igneous and metamorphic rocks are compatible with the granulite metamorphic facies.

Both magmatic and metamorphic origins have been postulated for anorthosite bodies. In light of recent data—mainly the common existence of crude cumulate layering and igneous trace-element patterns—the idea of a metamorphic origin is now largely abandoned. At present the source magma for anorthosites is viewed as one containing a high percentage of the plagioclase component, originating either through partial melting in the mantle or perhaps in the deep crust. Although various models have been postulated, thus far none is sufficiently compelling to discount all other contenders.

The development of a largely anorthositic body from differentiation of basalt has been described for an intrusion at Mineral Lake, Wisconsin (see Figure 9–8). This

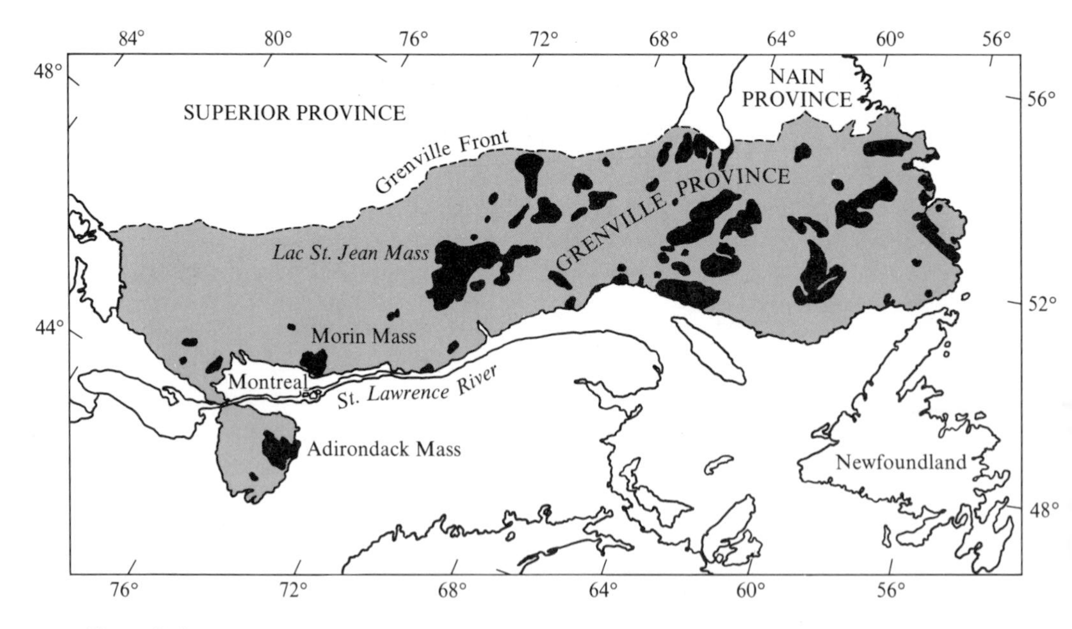

Figure 9–7
The major anorthosite plutons (in black) within the Grenville province of Canada and New York. [After J. Martignole and K. Schrijver, 1970, *Tectonophysics, 10,* Fig. 1.]

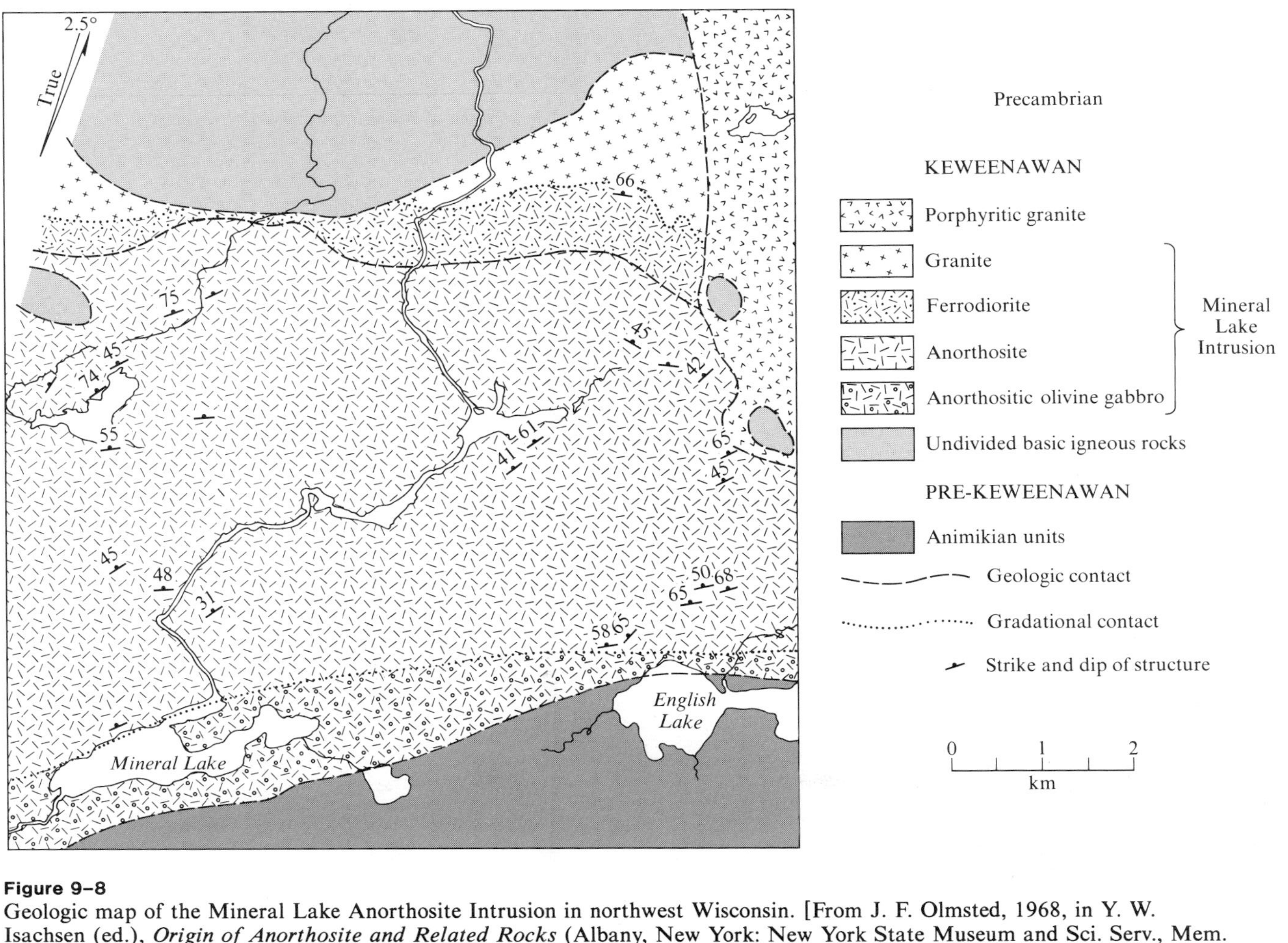

Figure 9–8
Geologic map of the Mineral Lake Anorthosite Intrusion in northwest Wisconsin. [From J. F. Olmsted, 1968, in Y. W. Isachsen (ed.), *Origin of Anorthosite and Related Rocks* (Albany, New York: New York State Museum and Sci. Serv., Mem. 18), Fig. 2.]

intrusion is of Middle Keweenawan age, has a thickness of 4,500 m, and covers an area of about 120 km^2. Injection of magma occurred along a contact between pre-Keweenawan metasediments and Middle Keweenawan rocks.

The origin of this body has been deduced (by Olmsted, 1968) from internal features such as texture, composition, and mineralogy. Let us consider the evidence.

The intrusion is a tilted differentiated sheet. The internal rock sequence grades upward from a chilled basal zone through minor ultramafic segregations, anorthositic olivine gabbro, anorthosite, ferrodiorite, to granite. Rhythmic layering is absent. Aside from a distinct change in the basal zone, mineral compositions are relatively consistent in the lower half of the intrusion. In the upper portion compositional changes show increasing ratios of Fe/Mg in pyroxene, and Na/Ca in plagioclase (see Figure 9–9). Chemical analysis of the chilled zone indicates an unusually high iron content ($FeO = 15.08\%$ by weight, and $Fe_2O_3 = 2.19\%$ by weight) as compared to typical Keweenawan basalts and the chilled border zones of intrusions such as the Skaergaard.

The relatively minor ultramafic rocks near the base are coarse-grained and rich in pyroxene and olivine. Mafic grains occur in clusters, whereas plagioclase is interstitial, suggesting that the mafic crystals were derived as cumulates from the overlying magma (see Figure 9–10).

The anorthositic olivine gabbro above is ophitic to subophitic, indicating that the pyroxenes probably crystallized from the melt around preexisting plagioclase crystals (see Figure 9–10B). The zone above is mainly anorthosite and leuco-gabbro (containing 80–90% plagioclase); pyroxene continues to be interstitial, and the plagioclase shows a strong preferred orientation (see Figure 9–10C), most of which is parallel to the edges of the intrusion.

Putting all the above information together, let us consider the possible origin. The unusually high iron content in the chilled border zone indicates that the magma was probably partially crystallized when intruded. Assuming an average Keweenawan basalt as the source magma, Olmsted (1968) has calculated that the magma was composed of about half solids (of anorthositic gabbro composition) when intruded. Wholesale contamination by iron-rich, silica-poor wall rock is rejected because of the relatively low silica content of the border zone. A partially solid character of the magma is also indicated by the well-developed preferred orientation and fractured character of much of the plagioclase. The presence of some ultramafic materials near the base indicates crystal settling.

Figure 9–9 (*facing page*)
Mineralogy of the Mineral Lake Intrusion. (**A**) Modal composition of the major minerals plotted as a function of vertical distance from the base. (**B**) Change of mineral composition as a function of vertical distance from the base of the intrusion. [From J. F. Olmsted, 1968, in Y. W. Isachsen (ed.), *Origin of Anorthosite and Related Rocks* (Albany, New York: New York State Museum and Sci. Serv., Mem. 18), Figs. 3, 11.]

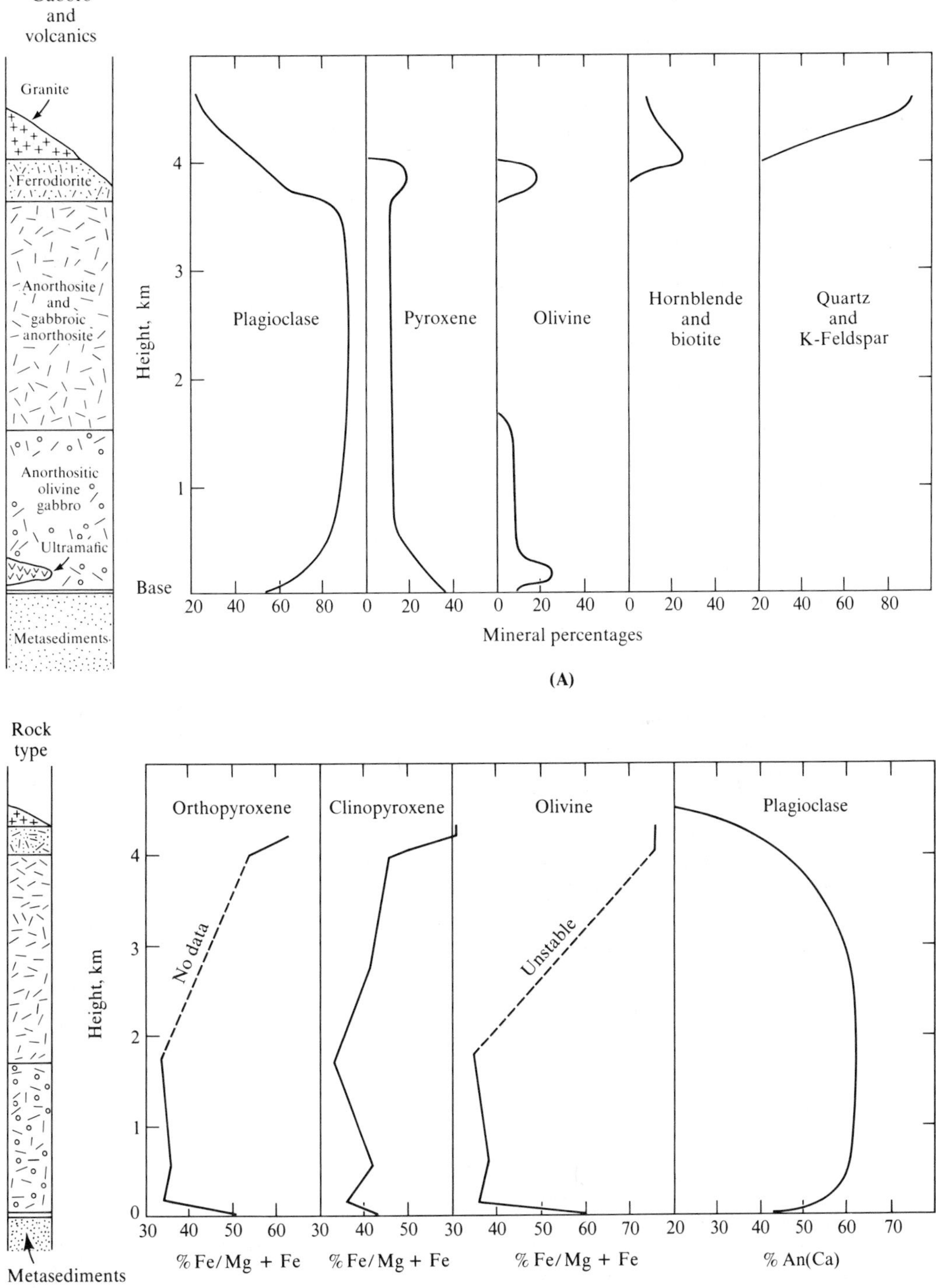
Gabbro
and
volcanics
Granite
Ferrodiorite
Anorthosite
and
gabbroic
anorthosite
Anorthositic
olivine
gabbro
Ultramafic
Metasediments
Height, km
Base
Plagioclase
Pyroxene
Olivine
Hornblende
and
biotite
Quartz
and
K-Feldspar
Mineral percentages
Rock
type
Metasediments
Orthopyroxene
Clinopyroxene
Olivine
Plagioclase
No data
Unstable
% Fe/Mg + Fe
% Fe/Mg + Fe
% Fe/Mg + Fe
% An(Ca)

(A)

(B)

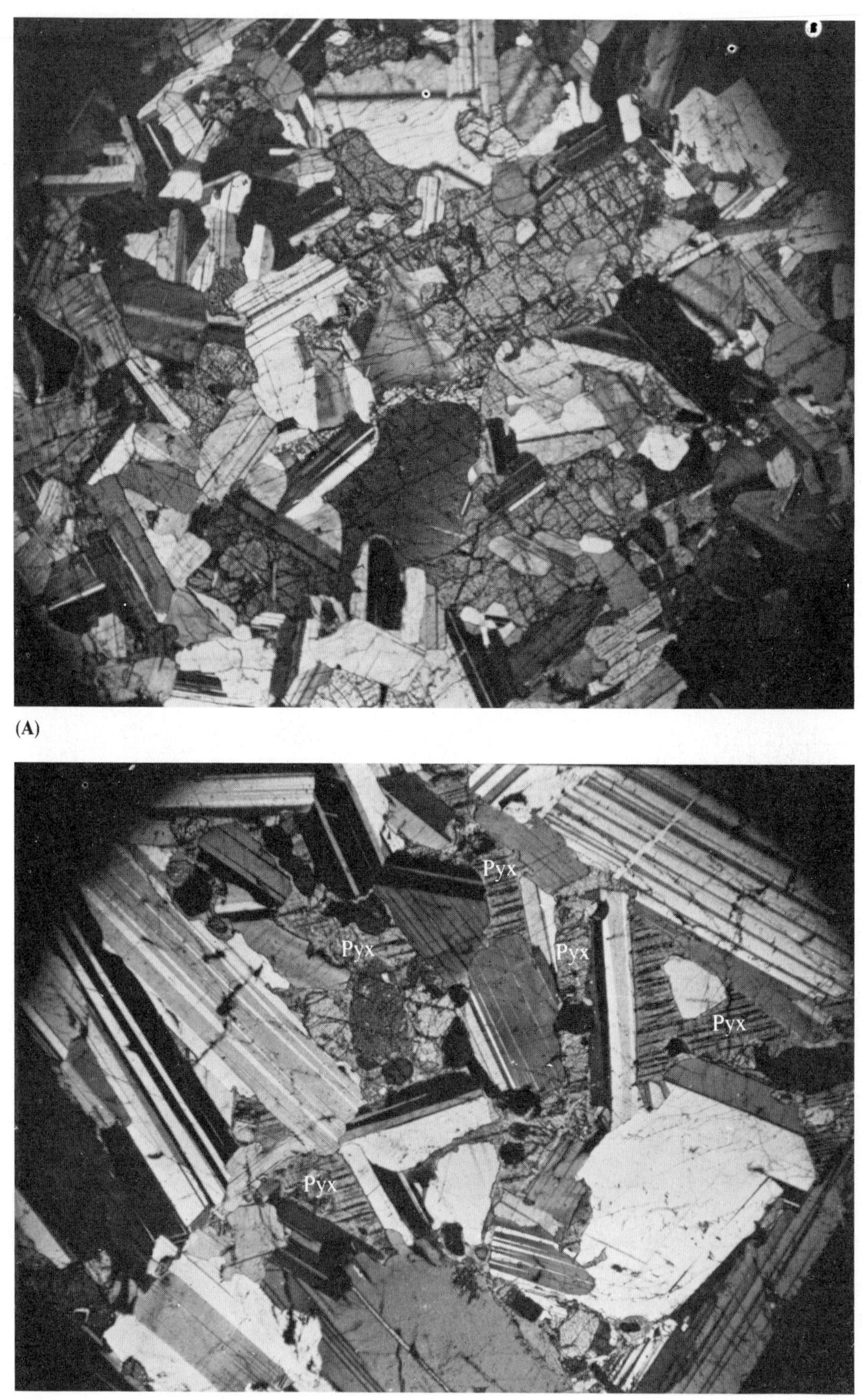
(A)
Pyx
Pyx
Pyx
Pyx
Pyx
(B)

(C)

Figure 9–10
Some textures within the Mineral Lake Intrusion. (**A**) Olivine-rich ultramafic material from zone near the base. Most plagioclase is anhedral and appears to be later than olivine. Crossed polars. Width of photo is about 3.1 mm. (**B**) Gabbroic anorthosite. Olivine (small rounded grains) and plagioclase are enclosed in a large pyroxene grain (labeled Pyx). Crossed nicols. Width of photo is about 5.5 mm. (**C**) Highly developed preferred orientation of plagioclase crystals from the central anorthositic portion of the intrusion. Crossed nicols. Width of photo is about 4.2 mm. [From J. F. Olmsted, 1968, in Y. W. Isachsen (ed.), *Origin of Anorthosite and Related Rocks* (Albany, New York: New York State Museum and Sci. Serv., Mem. 18), Figs. 6, 7, 9.]

As rhythmic banding and compositional layering are absent throughout much of the intrusion, differentiation was apparently not accomplished as a result of crystal settling after emplacement. The densities of olivine and pyroxene are known to be greater than that of a gabbroic melt, whereas plagioclase has been thought to be similar to it. It was suggested that during intrusion of the partially crystalline magma, the heavier mafic minerals settled to the base of the inclined magma chamber (and are now mainly not exposed), and the lighter plagioclase was intruded with the melt. This mechanism results in separation of much of the early cumulate mafic phases that would be expected normally as a result of basaltic differentiation. The probability of plagioclase being carried by the intruding magma has been enhanced by recent data on the Skaergaard intrusion (McBirney and Noyes, 1979, p. 492), which indicate that in a partially crystallized gabbroic liquid, plagioclase densities are slightly lower than that of the melt. During or following intrusion, it is also possible that much of the siliceous liquid portion of the melt was separated from the crystal-rich melt (perhaps by filter-pressing) and squeezed upward toward the roof of the magma chamber, where it crystallized as ferrodiorite and granite.

As can be seen from this example, the solution to the problem of origin of this anorthosite body required careful observation and correlation of both field and laboratory data. As is true with most petrologic problems, neither approach is independent of the other.

ALKALI BASALTS AND NEPHELINITES

Alkali basalts and nepheline-rich rocks are found within continental areas unrelated to subduction zones, but are perhaps related to initiation of rifting. The presence within these rocks of xenoliths containing high-pressure minerals (such as aluminous pyroxenes and spinel) clearly indicates that these magmas have a source within the mantle; the depth of origin has been placed within 40–100 km beneath the surface.

These rocks have been found in a variety of areas such as the East African rift zones, the Rhine Volcanic Province, the Oslo Graben in Norway, and the Basin and Range Province in the western United States. Varieties include nepheline-normative rocks, and hypersthene-normative rocks. The latter is called the shoshonite series.

The origin of these rocks has been the subject of considerable study, but as yet no unified model has emerged to account for their development. For example, it is known that, in a number of areas where differentiation has occurred, alkali olivine basalts will fractionate to form liquid residues of trachytic or phonolitic composition. Much of the fractionation is considered to occur in low-pressure environments, but other schemes require transfer of alkalies or separation of high-pressure phases within the mantle.

The composition of continental alkali-rich lavas shows greater variation than typical oceanic alkali lavas. A higher potassium content is common in continental environments; this may be correlated with larger amounts of Ba, Sr, and Rb. Leucite-bearing mafic lavas have no equivalent in oceanic environments. Although the reasons for these differences are not known, it is usually assumed to be related to reactions within the sialic crust.

CARBONATITES, KIMBERLITES, AND RELATED ROCKS

Carbonatites are often found with nepheline-rich rocks; as the name implies, they are composed mainly of carbonates—usually calcite, dolomite, and/or ankerite. The principal associated minerals are apatite, magnetite, and phlogopite, with minor amounts of sodic pyroxenes and amphiboles. Accessory phases are fluorite, perovskite, pyrochlore, monazite, and barite. Part of the interest in carbonatites stems from the fact that they often form the host rocks for rich concentrations of niobium, rare earth elements, and thorium.

The igneous origin of carbonatite has been argued for a long period of time, but the observations of Dawson (1962, 1966) of active carbonatite volcanism in the crater of Oldoinyo Lengai in East Africa ended the debate. It is now known and accepted that carbonatites can be intrusive or extrusive. Examples of carbonatite dikes, sills, sheets, pipes, stocks, flows, and pyroclastics have been described. In spite of their widespread occurrence in both place and time, carbonatites and their related rocks remain relatively uncommon rocks. Two environments favor their occurrence: a platform type associated with ultramafic to alkali rocks (peridotites and nepheline syenite) and a geosynclinal type associated with alkalic rocks such as nepheline syenite. Many carbonatites have been related to rift valleys on continents such as the African rifts, the Rhine graben, and the Midland Valley in Scotland.

A generalized geological map of a typical carbonatite and its related rocks is shown in Figure 9–11. The general pattern is that of a group of more or less concentric,

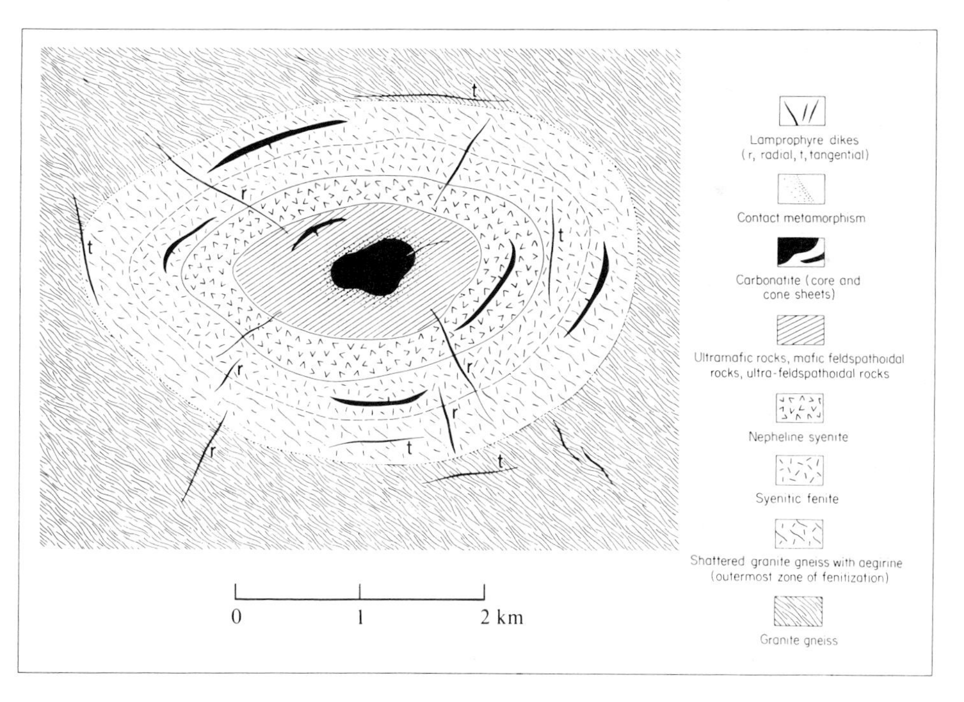

Figure 9–11
Idealized geological map of an alkalic-ultramafic carbonatite complex. Carbonatites form a central core and cone sheets. Lamprophyre dikes are radial and tangential. The complex is surrounded by a fenitic halo and shattered metamorphic rocks. [From E. W. Heinrich, 1966, *The Geology of Carbonatites* (Chicago: Rand McNally), Fig. 2–5.]

essentially vertical rings; some complexes may be elongated on fracture zones. The whole complex is usually quite small—on the order of a few kilometers in diameter. An exception to this is the Gulinsky complex in northern Siberia, which measures 30 by 50 km. Within these complexes carbonatites make up less than 10% of the total, the balance being composed of various alkalic rocks. Ring complexes always form discordant contacts with country rock. Doming, brecciation, and alteration of wall rock is common. The intrusive sequence is usually a complex one of multiple injection and alteration. The primary injections are often those of an intermediate ring—usually a phonolite or nepheline syenite. Later injections usually result in alteration of earlier injections, and an intense alkali metasomatism (fenitization). The last intrusion of the group is usually the carbonatite core, perhaps with cone sheets in the earlier rocks, or radial or concentric lamprophyre dikes. *Lamprophyres* are mafic dikes with a high percentage of mafic phases; mafic minerals may be present in phenocrysts and groundmass, but salic minerals are limited to the groundmass. Tuffs, breccias, and mafic alkaline and carbonatite flows may be present in volcanic occurrences. The association, both regionally and locally, of carbonatites with alkali-rich, silica-poor igneous rocks implies a strong genetic relationship.

Kimberlites are inequigranular alkalic peridotites that contain rounded and corroded phenocrysts of olivine, phlogopite, Mg-ilmenite, and pyrope in a fine-grained matrix of olivine, phlogopite, serpentine, perovskite, spinels, calcite, and/or dolomite. Kimberlites, occurring mainly in nonorogenic continental platforms, have become famous as the host rock for diamonds. Experimental determination of the diamond–graphite transition curve has demonstrated that this rock must have a mantle rather than crustal origin (see Figure 9–12). Associated silicate minerals, including the high-pressure SiO_2 phase coesite, indicate that this rock originated at a depth of about 100 km. Although some kimberlites may be massive, contain few inclusions, and possess the chilled borders of the usual relatively gentle process of intrusion, most kimberlites clearly indicate an explosive vertical ascent. Typical kimberlites are found in small vertical pipes that taper downward into fissures. The amount of included fragments may range up to 90%; these fragments may consist of either fragmented kimberlite or country rocks. The cementing material may be either massive kimberlite or hydrothermal vein minerals. Inclusions appear to represent a sampling of most rocks that have been penetrated by the kimberlite. These consist of ultrabasic mantle rocks that may be either mantle xenoliths or magmatic segregations (often called cognate xenoliths), high-grade metamorphic or igneous rocks from basement complexes, or a wide variety of crustal rocks; some of the latter, of considerable size, are referred to as "floating reefs" in mining terminology. The inclusions, although giving no indications of elevated temperatures, are often rounded and polished, and may be at a lower level than their original stratigraphic position.

Emplacement of kimberlite is probably initiated by the slow rise of a gas-charged crystal mush into tension fractures at a depth of about 2 km. If followed by an instantaneous drop in pressure (initiated by a fracture to the surface), rapid vesiculation and crystallization occur. Crystallizing magma is driven rapidly upward. En-

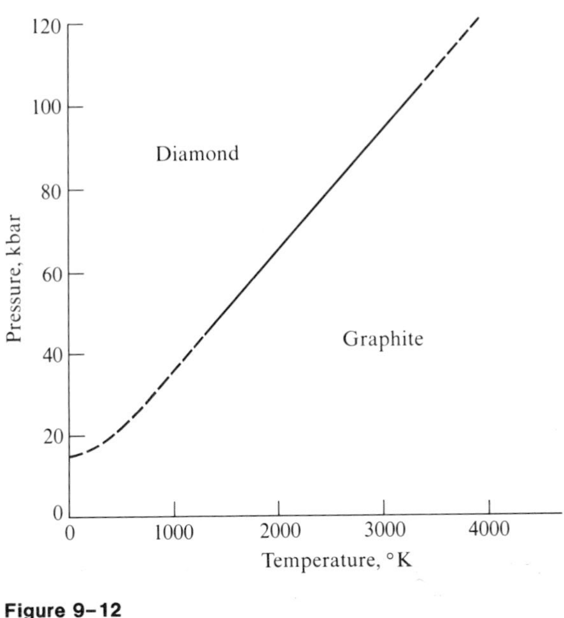

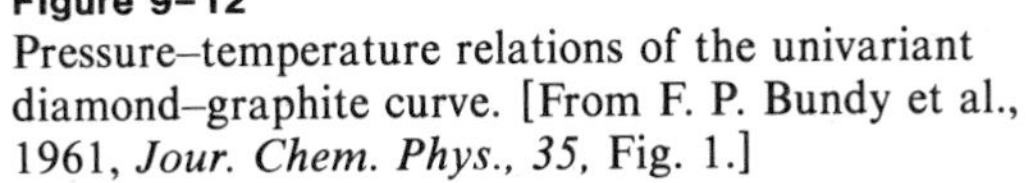
Figure 9–12
Pressure–temperature relations of the univariant diamond–graphite curve. [From F. P. Bundy et al., 1961, *Jour. Chem. Phys.*, *35*, Fig. 1.]

largement of the vent occurs as the fluidized crystallizing and fragmented kimberlite moves upward, blasting its way to the surface. Smaller fragments of dislodged country rock are moved upward by the semisolid streaming mass, while larger fragments sink gently downward. The intruded mass may be affected later by hydrothermal alteration or introduction of magma. Settling of the mass may continue for extended periods, as verified by distortion of later sediments.

Explosive volcanism of this type can be visualized by a schematic phase diagram as shown in Figure 9–13 for a binary system that can be called *X*-Gas, with *X* being any anhydrous silicate and the gas being some mixture of CO_2 and H_2O. In part A, the melt, at high pressures, is shown at point *a*. This is the minimum melting point for a gas-saturated melt. Removal of heat from this system would cause crystallization of melt *a* at this temperature to form crystals *b* and a gas phase of composition *c*. If, on the other hand, a drop in pressure occurs, the positions of the various lines on the diagram would shift. The system is shown at lower pressures in part B; point *a* is at the same location as in part A, but the minimum liquid freezing temperature is now higher. Point *a* is now located in the crystal–gas field. If the pressure drop is rapid, the system adjusts by simultaneous rapid crystallization and vesiculation. The mass streams upward to areas of lower pressure.

It has been commonly assumed that kimberlites and carbonatites are genetically related. This has been based upon the presence of primary magmatic carbonate in

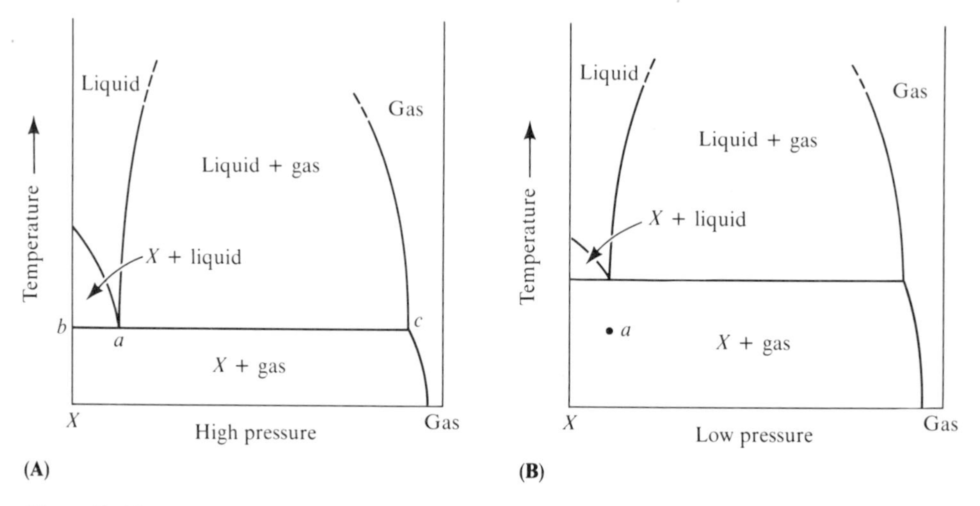

Figure 9–13
Hypothetical binary system *X*–Gas; *X* is any anhydrous solid and Gas is a mixture of CO_2 and H_2O. (**A**) At high pressures a batch *a* is at the minimum melting point of the gas-saturated melt. Cooling would initially produce crystals *b* and gas *c*. (**B**) If the pressure were decreased instantaneously such that the diagram changed as shown, the batch *a* would be instantly within the field of *X* + Gas. Rapid vesiculation and crystallization would occur with explosive volcanism.

both rock types and the assumption that kimberlites are found associated with nepheline syenite–ijolite–carbonatite complexes. These criteria have been demonstrated as invalid by Mitchell (1979) on the basis of mineralogical criteria. For example, the carbonatite-rich portions of kimberlites were shown to contain little to no magnetite, no Nb-bearing minerals, some Mg-ilmenite, and rarely Al_2O_3-poor pyroxenes; those portions of carbonatite complexes that somewhat resemble kimberlites contain Ti-rich magnetite, Nb-rich minerals, Mn-ilmenite, and Al_2O_3-rich pyroxenes. It now appears that kimberlitic magmas can differentiate to a carbonate-rich silica-poor residue that is quite different from the kimberlitelike rocks associated with the carbonatite complexes. This has important economic consequences in that one would not expect to find Nb-rich minerals associated with kimberlites or diamond deposits associated with carbonatite complexes.

SUMMARY

Of the various nonorogenic igneous continental rocks, tholeiitic basalts and their intrusive equivalents are volumetrically the most significant. Large intrusions form the most outstanding examples of the effect of fractional crystallization. Intrusions such as the Skaergaard in Greenland show both rhythmic and cryptic layering, with compositional changes grading from a tholeiitic basalt to an iron-rich final differentiate that shows little to no silica enrichment. This is in sharp contrast to the calc-alkaline differentiation trend, which shows strong enrichment in alkalies and silica, with simultaneous decrease in iron. High-temperature experimentation on synthetic melts has suggested the conditions for both differentiation processes.

Anorthosites, limited in occurrence to the Precambrian, are of igneous origin. The particular methods of origin are probably varied, and no universally accepted theory accounts for their presence. Possible origins include partial melting in the mantle or gravitational differentiation from a basaltic magma.

Alkali basalts and nephelinites are known to have originated within the mantle at depths between 40 and 100 km. Fractionation of alkali-rich melts probably occurs within crustal environments.

Carbonatites are extremely uncommon, but occur in a variety of different circumstances, such as rift valley, platform, and geosynclinal environments. Genetic associations with alkali-rich silica-poor rocks are probable.

Kimberlites have a mantle origin, as established by the presence of minerals stable only under conditions of high pressure, such as diamonds. Violent emplacement is indicated by the presence of abundant (and abraded) xenoliths brought up from both crustal and mantle levels. In spite of the presence of primary carbonate phases in both kimberlites and carbonatites, significant differences in mineralogy appear to indicate that these rock types are unrelated in origin.

FURTHER READING

Bhattacharji, S. 1967. Mechanics of flow differentiation in ultramafic and mafic sills. *Jour. Geol., 75,* 101–112.

Bowen, N. L. 1928. *The Evolution of the Igneous Rocks.* Princeton, N.J.: Princeton Univ. Press, 332 pp.

Buddington, A. F. 1939. *Adirondack Igneous Rocks and Their Metamorphism.* Geol. Soc. Amer. Mem. 7, 354 pp.

Carmichael, I. S. E., F. J. Turner, and J. Verhoogen. 1974. *Igneous Petrology.* New York: McGraw-Hill, 739 pp.

Dawson, J. B. 1960. A comparative study of the geology and petrography of the kimberlites of the Basutoland province. Ph.D. thesis, University of Leeds.

Dawson, J. B. 1962. The geology of Oldoinyo Lengai. *Bull. Volcan., 24,* 155–168, 349–387.

Dawson, J. B. 1966. Oldoinyo Lengai—an active volcano with sodium carbonatite lava flows. In O. F. Tuttle and J. Gittins (eds.), *Carbonatites,* p. 155. New York: Wiley Interscience.

Dawson, J. B. 1967. Geochemistry and origin of kimberlites. In P. J. Wyllie (ed.), *Ultramafic and Related Rocks,* pp. 269–278. New York: John Wiley and Sons, Inc.

Garson, M. S., and W. C. Smith. 1958. *Chilwa Island.* Nyasaland Geol. Surv. Mem. 1.

Green, T. H. 1969. High-pressure experimental studies on the origin of anorthosites. *Canad. Jour. Earth Sci., 6,* 427–440.

Harris, P. G., and E. A. K. Middlemost. 1970. The evolution of kimberlites. *Lithos, 3,* 77–88.

Heinrich, E. W. 1966. *The Geology of Carbonatites.* Chicago: Rand McNally, 555 pp.

Martignole, J., and K. Schrijver. 1970. The level of anorthosites and its tectonic pattern. *Tectonophysics, 10,* 403–409.

McBirney, A. R. 1975. Differentiation of the Skaergaard Intrusion. *Nature, 253,* pp. 691–694.

McBirney, A. R., and R. M. Noyes. 1979. Crystallization and layering of the Skaergaard intrusion. *Jour. Petrology, 20,* 487–554.

Mitchell, R. H. 1979. The alleged kimberlite–carbonatite relationship: additional contrary mineralogical evidence. *Amer. Jour. Sci., 279,* 570–589.

Olmsted, J. F. 1968. Petrology of the Mineral Lake Intrusion, northwestern Wisconsin. In Y. W. Isachsen (ed.), *Origin of Anorthosite and Related Rocks,* pp. 149–162. Albany, New York: New York State Museum and Science Service, Mem. 18, 466 pp.

Wager, L. R., and G. M. Brown. 1967. Layered Igneous Rocks. San Francisco: W. H. Freeman and Company, 588 pp.

Wager, L. R. 1960. The major element variation of the layered series of the Skaergaard Intrusion. *Jour. Petrology, 1,* 364–398.

Wager, L. R. 1968. Rhythmic and cryptic layering in mafic and ultramafic plutons. In H. H. Hess, and A. Poldervaart (eds.), *Basalts,* Vol. 2, pp. 573–622. New York: Wiley Interscience.

Wyllie, P. J., and W. L. Huang. 1975. Peridotite, kimberlite, and carbonatite explained in the system $CaO–MgO–SiO_2–CO_2$. *Geology, 3,* 621–624.

Yoder, H. S., Jr. 1965. Diopside–anorthite–water at five and ten kilobars and its bearing on explosive volcanism. *Carnegie Inst. Wash. Year Book 64,* pp. 82–89.

Yoder, H. S., Jr. 1968. Experimental studies bearing on the origin of anorthosite. In Y. W. Isachsen (ed.), *Origin of Anorthosite and Related Rocks,* pp. 13–22. Albany, New York: New York State Museum and Science Service, Mem. 18, 466 pp.

II

SEDIMENTARY ROCKS

10

The Occurrence of Sedimentary Rocks

As we noted in the Introduction, sedimentary rocks cover 66% of the continental surfaces and probably most of the ocean floor as well. The basic reason for this wide areal extent is the chemical instability of igneous and metamorphic rocks under atmospheric conditions. Rocks and minerals are in equilibrium only under the set of physical and chemical conditions in which they formed; under different conditions they will tend to react to reach a new equilibrium state. Igneous and metamorphic rocks form at temperatures and pressures much higher than those at the earth's surface, and in an environment containing less water, less oxygen, less carbon dioxide, and no organic influences. It is to be expected that such rocks will be unstable and undergo chemical and physical changes when brought to the surface by tectonic, erosional, or isostatic forces. These changes constitute the process we call weathering. Based on Le Chatelier's Principle, we would expect the products of this chemical change to contain more water, more oxygen, more carbon dioxide, and more organic matter than before the change. This expectation is fulfilled (see Chapter 11).

Sedimentary rocks consist almost entirely ($> 95\%$) of three types: sandstones, mudrocks, and carbonate rocks. Sand is defined as fragmental sediment between 2 mm and 0.062 mm (62 μm) in diameter; mud, as fragmental sediment smaller than 0.062 mm.* Carbonate rocks are composed largely of $CaCO_3$ (calcite or aragonite) or $CaMg(CO_3)_2$ (dolomite), with other carbonates being rare (siderite, magnesite, and so on). As expected, there are transitional rocks that do not fit neatly within the three pigeonholes. For example, coquina is a fragmental rock composed of sand-size fragments of fossil shells. It is both a sandstone and a limestone; it is usually included in the limestones. How should we classify a rock composed of subequal amounts of clay and microcrystalline carbonate material (marl)? There are no perfect answers to such questions, only generally accepted compromises; sometimes, not even these.

*Mud is further subdivided into silt (62–4 μm) and clay (<4 μm).

The most abundant sedimentary rocks are the mudrocks, which form 65% of all sedimentary rocks. A moment's reflection about the mineralogy of igneous and metamorphic rocks suggests why this is so. Igneous and metamorphic rocks are composed of approximately 20% quartz and 80% other silicate minerals; only the quartz is chemically stable under most surface conditions. The other minerals are unstable when exposed at the surface and are altered to a variety of substances, but mostly to clay minerals (see Chapter 11). Clay minerals are mud size; hence, mudrocks are the dominant sedimentary rock. The quartz in crystalline rocks is very resistant to chemical attack and occurs chemically unchanged in both mudrocks and sandstones. According to many thousands of analyses by X-ray, polarizing microscope, and chemical techniques, mudrocks and sandstones have the average detrital mineral compositions shown in Table 10–1. A weighted average shows that the detrital sediment in the sedimentary column consists of 45% clay minerals, 40% quartz, 6% feldspar, 5% undisaggregated rock fragments, and 4% others. About 85% is either clay minerals or quartz, the most stable minerals under surface conditions. Clearly, weathering has been a very effective process through geologic time.

Although the areal extent of sedimentary rocks is great, their thicknesses are not. In part, the thinness of sedimentary cover on the continents results from the definition of the word sedimentary. In some areas the base of the sedimentary column is well defined. For example, in mid-continental North America the oldest sediments contain fossils and look in every way like normal mudrocks, sandstones, or carbonate rocks. Directly below them lie granites, gneisses, or schists. The boundary between sedimentary and nonsedimentary can be drawn on the outcrop using a pencil. Similarly, in the deep ocean basins, sediment such as globigerinid ooze, brown clay, or chert is underlain by basalt, again with a very well-defined contact.

In many areas, however, the contact between sedimentary and nonsedimentary is gradational, and different geologists would draw the pencil line at stratigraphic positions several hundred meters or more apart. With increased burial depth and resultant increased temperature and pressure, sediment that has largely equilibrated with atmospheric conditions is subjected to an environment so different that it must adjust. But different minerals have different limits of stability; they do not all change at the same depth or at the same rate once the change has begun. For example, some clay minerals are known to grow and/or recrystallize to new clay minerals at a depth of about 3,000 m and a temperature of 80°C. Chert, which is microcrystalline, may recrystallize and coarsen at 150°C; and feldspar may be dissolved by circulating groundwaters at any depth from the surface downward. Where are we to draw the line between a sedimentary rock and a rock that has been so severely changed by conditions accompanying burial that we believe the term metamorphic rock is more appropriate? Standard practice is that each investigator is sovereign in setting the dividing line for his rocks. One person's hard shale is another person's slate; a metaquartzite to one investigator may still be a quartz-cemented quartz sandstone to another. In some geographic regions these differences in terminology can seriously handicap communication.

Table 10–1
Average Detrital Mineral Composition of Mudrocks and Sandstones

Mudrocks, %	Mineral composition	Sandstones, %
60	Clay minerals	5
30	Quartz	65
4	Feldspar	10–15
< 5	Rock fragments	15
3	Carbonate	< 1
< 3	Organic matter, hematite, and other minerals	< 1

The average thickness of sedimentary rocks on the continents is about 1,800 m but is quite variable, ranging from zero over extensive areas such as the Canadian Shield to more than 20,000 m in some basinal areas such as the Louisiana–Texas Gulf coastal region (see Figure 10–1). The maximum possible thickness is determined by the geothermal gradient in the area, fluid chemistry, and the chemical reactivity of the detrital particles. The average temperature at a given depth can vary greatly among geographic–tectonic areas. For example, at a depth of 10,000 m under Pittsburgh it is about 150°C; under New Orleans, 200°C; under Las Vegas, 260°C; under Los Angeles, more than 300°C. The mineral content of the sandstones in the sedimentary section can vary from those composed entirely of quartz grains, which are relatively resistant to destruction or recrystallization, to sandstones composed mostly of calcic plagioclase grains and basaltic rock fragments, which alter chemically at very low temperatures. The basaltic fragments are altered to clay minerals, micas, and zeolite minerals; the plagioclase grains to zeolites and epidote.

TYPES OF SEDIMENTARY ROCKS

The major types of sedimentary rocks are mudrocks (65%), sandstones (20–25%), and carbonate rocks (10–15%), with all others totaling less than 5%. In later chapters we will discuss each of the abundant types (and some of the minor ones) in some detail; in this section we will introduce only their general characteristics.

Mudrocks

Mud particles are less than 62 μm in size and mostly less than 5 μm in size. Because of this, the particles are easily kept in suspension by even the weakest of currents and can settle and accumulate only in still waters. Many such environments exist, both in

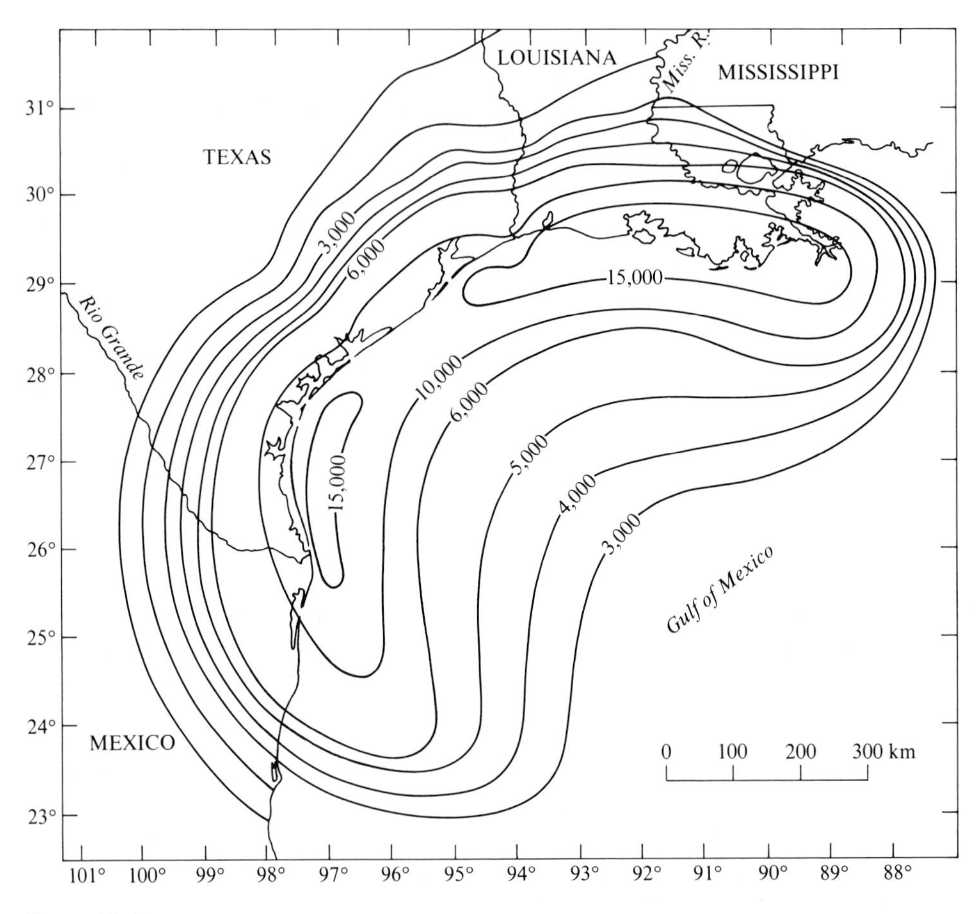

Figure 10–1
Isopach (thickness in meters) map of Cenozoic sediments in the Gulf Coast basin. (Note irregular contour interval.) Approximately 7,000 m of Mesozoic sedimentary rocks underlie the Cenozoic rocks of the delta region. [From G. C. Hardin, Jr., 1962, *Geology of the Gulf Coast and Central Texas* (Houston Geol. Soc.), Fig. 1.]

nonmarine and marine settings; for example, floodplains, deltas, and lakes on the continents; lagoons, deltas, and areas below wave base in the marine environment.

The thickest accumulations of mudrocks occur in geosynclinal settings, with mudrock thicknesses ranging up to at least 2,000 m in the central Appalachians of Pennsylvania and in the Ouachita Geosyncline in Arkansas. When interbedded with sandstones in the geosynclines, the mudrocks commonly form the bulk of the accumulation; for example, 56% of a 7,000-m Tertiary section in Indonesia and 61% of a 3,000-m Carboniferous section in the Anadarko Basin of western Oklahoma.

In contrast to the apparent siltiness of thick accumulations of geosynclinal mudrocks, those deposited in areas such as lakes, abyssal areas far from land, or shallow marine areas below wave base are exceptionally fine grained—a mixture of clay minerals, organic matter, and quartz. An excellent example is the Chattanooga Black Shale of Late Devonian age in Tennessee and surrounding states. This unit has a thickness of only 10 m and is a blanket deposit extending over tens of thousands of square kilometers. It is believed to have been deposited in an epicontinental sea in water depths of less than 30 m and is composed of about 20–25% quartz, 25–30% clay and mica, 10% alkali feldspar, 10–15% pyrite, 15–20% organic matter, and 5% miscellaneous constituents. The quartz grains are nearly all less than 15μm in size and are distributed as thin laminae within the black organic matter and clay mineral mixture. Petrologically identical sediment can accumulate also in the deep ocean, as is presently occurring in the numerous small fault basins off the coast of Southern California. We will examine the basic controls of mudrock occurrence and composition in more detail in Chapter 12.

Sandstones

Sand accumulates in areas characterized by relatively high kinetic energies; that is, environments of moving fluids. Examples of these environments include desert dunes, beaches, marine sand bars, river channels, and alluvial and submarine fans. Some sands are deposited on the shallow sea floor and are subsequently carried down to great depth in the sediment–water mixtures called turbidity currents. Many common sites of sand accumulation are elongate, such as beaches and rivers, but in the geologic record the sands deposited in these environments are commonly sheetlike in character. This difference results from the displacement of the depositional site through time; for example, a beach migrates inland during a marine transgression, resulting in a slight increase in the thickness of the sand body but an extreme increase in its width. It is, however, possible for a sand-dominated beach–dune complex to exist at the same geographic locality for a long period of time; this could occur in the tectonic setting of a slowly subsiding basin, resulting in pure sand deposits hundreds of meters thick with relatively narrow areal extents. The Cambro-Ordovician quartz sands of the western United States may be an example of this.

As is the case with mudrocks, the environment of deposition commonly can be related to mineral composition. Sands deposited in loci of highest kinetic energy, such as beaches and desert dunes, tend to be more quartz-rich than the sand bars of sluggish rivers. This occurs because of the relative ease of breakage and elimination of cleavable minerals such as feldspar or foliated fragments such as shale or schist. It is, however, not a good idea to base an environmental interpretation on detrital mineral composition. Some fluvial sands contain more than 90% quartz and many modern beaches contain high percentages of feldspar and rock particles. Mineral composition of sandstones is not a good environmental indicator, as we will discuss further in Chapter 13.

Carbonate Rocks

Modern carbonate sediments are composed almost entirely of the hard parts of marine organisms, and there is every reason to believe this has been true of carbonates since Cambrian time, at least. Because of the great chemical reactivity of calcium carbonate, however, most carbonate particles are recrystallized sometime after deposition, so that their organic origin is not always evident. This is particularly true of the microcrystalline particles that form the bulk of ancient limestones.

Because they are organic in origin, the abundance of carbonate particles is tied to the occurrence of the phytoplankton at the base of the food chain; and the phytoplankton, in turn, are tied to the depth of penetration of light into sea water. No light means no photosynthesis and no phytoplankton. The depth of penetration of light in sea water is shown in Figure 10–2, and it is apparent that below a depth of 50 m or less there will be a sharp decline in the abundance of living organisms. In fact, most organisms live within 10 m of the surface. It is worth noting, however, that the carbonate material generated in these shallow waters need not be deposited there. Planktonic, carbonate-shelled organisms such as *Globigerina* may settle to the deeper ocean floor to accumulate. Few of these deep sea carbonates will appear in the stratigraphic record because the shells dissolve in the cold waters of the deep ocean. Many of those that survive dissolution will be carried into a trench adjacent to a continental block and be subducted into the mantle. Also, carbonate-shelled planktonic organisms did not evolve until Mesozoic time. As a result, most carbonate rocks we see in the stratigraphic record are of shallow marine origin.

Sand-size fragments of carbonate-shelled organisms are not difficult to recognize and identify in thin sections of limestones, at least to the level of phylum and class. Sometimes even the genus and species can be specified and rather detailed reconstructions of depositional environments are possible using these and other data. For example, analyses of the relative amounts of the different isotopes of oxygen present in the calcite of unrecrystallized shell material can be used to determine the temperature of the water in which the organism lived (see Chapter 15). Even the types of amino acids that existed in the tissue of the organism can be estimated from the mineralogy of the shell. Living organisms are very sensitive to their environment and, because of this, limestones can be gold mines of information about the shallow marine waters of the geologic past.

DEPOSITIONAL BASINS, GEOSYNCLINES, AND PLATE TECTONICS

Based on stratigraphic data accumulated over the past hundred years, it is clear that the majority of preserved sedimentary rocks are marine, although no one has ventured to estimate the exact proportions of marine and nonmarine strata. There are several reasons for the dominance of marine sediments.

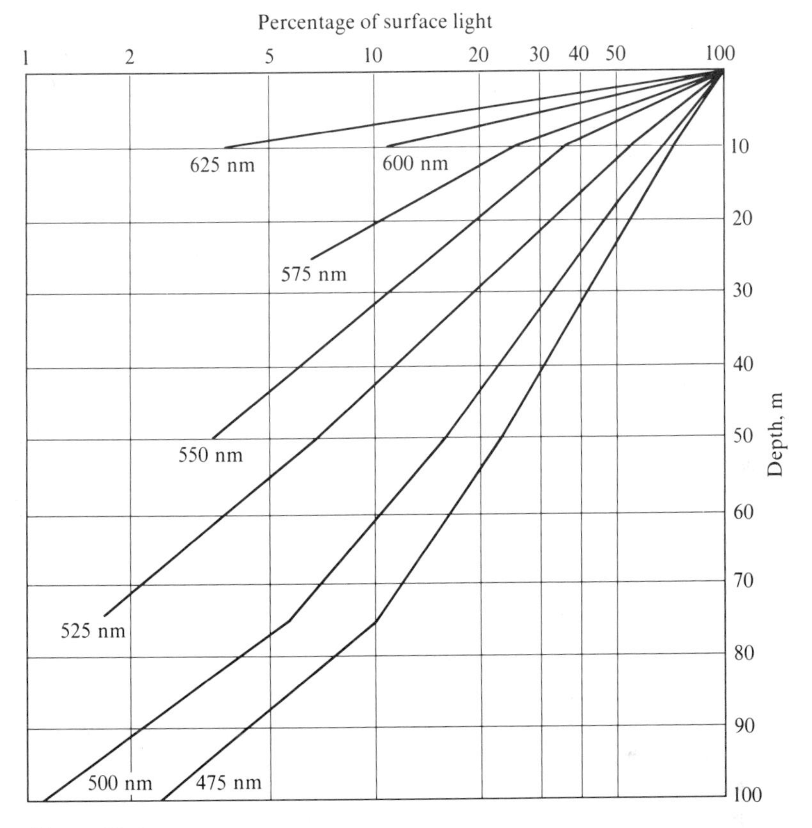

Figure 10–2
Depth to which light of various wavelengths will penetrate into clear ocean water. In coastal waters, penetration is commonly only 1–10% of the penetration in clear water. (< 400 nm = ultraviolet, 400–500 nm = blue, 500–600 nm = green, 600–700 nm = red, > 700 nm = infrared.) Plants can photosynthesize in a wide variety of wavelengths and intensities. For example, red algae prefer red light.

1. The light sialic material that forms a large proportion of continental masses is limited in volume, with the result that continental areas constitute only about 30% of the earth's surface. The marine areas are sediment traps that cover 70% of the earth's surface.

2. Because of the movement and subduction of oceanic crust and mantle at the edges of some continental blocks, topographically depressed areas (trenches) will exist at some continental margins adjacent to raised orogenic areas.

3. There has been a pronounced tendency through time for broad areas of the continental blocks to be depressed, resulting in widespread incursions of the ocean

into the interiors of the continents (epicontinental seas). The resulting shallow water sediments are laterally extensive on the craton, although they usually are thin. The tectonic explanation of these broad downwarps of the continents is not clear but may be related to phase changes caused by variations in heat flow in the asthenosphere, to eustatic changes in sea level associated with crustal rifting in the deep ocean basins, and/or to downbowing of cratons associated with convergent plate margins. Epicontinental seas can also be produced by melting of polar ice caps because of climatic change.

4. Continental deposits are, by definition, formed above sea level and hence are subject to removal should the rate of accumulation fall behind the erosion rate. It is no accident that stratigraphic sections on the continental blocks contain many unconformities, while deeper marine or oceanic sections are more complete.

Prior to the advent of the theory of plate tectonics in the 1960s, the ruling idea used to explain the thickest accumulations of marine sediments was the geosynclinal theory. Geosynclines were thought of as areally extensive, usually linear, structural depressions located most commonly at the edges of continents. In North America the type locality of the geosyncline was the Appalachian trend along the eastern margin of the United States and extending northeastward into the maritime provinces of Canada.

The idealized geosyncline was divided into two parallel segments, termed miogeosyncline and eugeosyncline. In general, miogeosynclinal terranes are characterized by clear-cut depositional contacts with underlying continental basement and a stratigraphic section whose marine sediments all were deposited in shallow water (fossils) and lack interstratified volcanic rocks or volcanic detritus. The sediments grade landward into continental, cratonic sediments rich in quartz. As a first approximation, the miogeosyncline can be interpreted as a thick accumulation of strata on the margin of a continent.

In contrast, eugeosynclinal terranes are characterized by equivocal contact relationships with continental basement, an abundance of volcanic rocks and detritus within the sedimentary sequence, and a conspicuous absence of evidence of shallow water deposition. Eugeosynclinal strata apparently form in the deeper part of an ocean basin adjacent to the shallow water miogeosynclinal strata, and in an area of active volcanic activity. In many cases of geosynclinal deposits, the boundary between the mio- and eugeosyncline (if exposed) is marked by a tectonically chaotic mixture of very large fragments of older sedimentary and crystalline rocks, some several kilometers in length, set in a muddy matrix. This intensely disturbed sedimentary rock unit is called a *mêlange*. In the Appalachian region, the miogeosynclinal rocks are exposed in the Valley and Ridge Province, and the eugeosynclinal rocks to the east in New England and the maritime provinces of Canada.

As indicated in Figure 10–3, dating of Precambrian rocks indicates that the continental block has grown during geologic time by the accretion of successive geosynclinal deposits. Figure 10–4 shows the extent of accretion onto the Precambrian

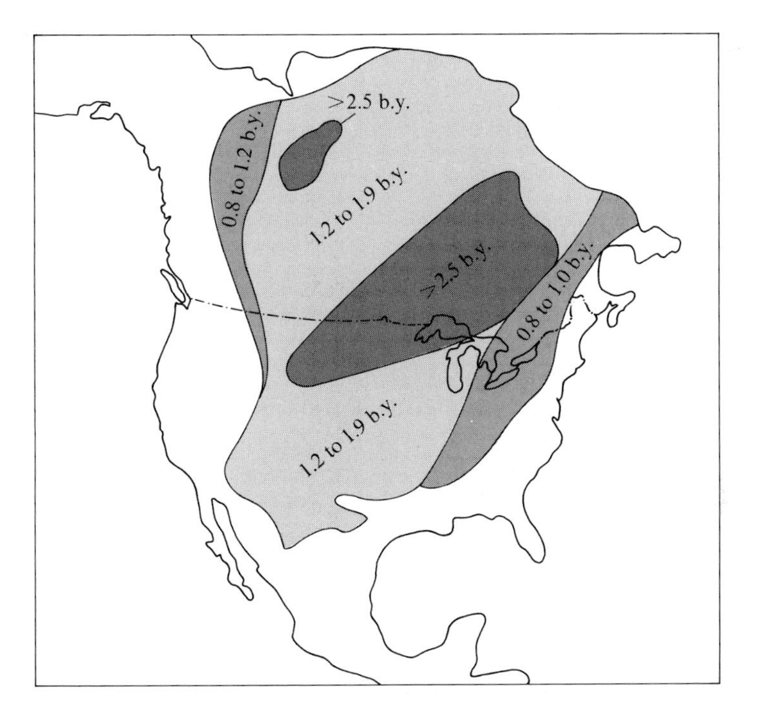

Figure 10–3
Generalized map of the ages of Precambrian rocks in North America illustrating continental nuclei and apparent enlargement of the nuclei through time. [From R. W. Ojakangas and D. G. Darby, 1976, *The Earth Past and Present* (New York: McGraw-Hill), Fig. 7–4.]

nucleus of the United States during Phanerozoic time. Also shown is the position of the United States with respect to latitude during each geologic period, and it is apparent that the area was located within 30° of the equator during the entire Paleozoic Era. This implies a continual tropical to subtropical climate, and climate is an important control of the mineral composition of sediments.

Modern theories concerned with the origin of large sedimentary basins and their accumulations of sediment center on plate movements. According to the theory of plate tectonics, the settings of basins can be described with reference to three fundamental factors: (1) the type of lithosphere that serves as substratum for the basin (oceanic, transitional, or continental); (2) the proximity of the basin to a plate margin; and (3) the types of plate junctions nearest to the basin. The term geosyncline is rapidly passing into history but still may be useful to indicate a geographically extensive area in which the accumulation of sediment is much thicker than in surrounding areas.

Figure 10–4
Present outline of conterminous United States showing accretion to the Precambrian craton during Phanerozoic time. Also shown is the current best estimate of the location of the conterminous United States in relation to the equator during each Phanerozoic period. During the Cretaceous and Tertiary the United States was located more than 2,000 km north of the equator because of a consistent drift northward of the North American plate (1,100 km equals about 10° of latitude).

Plate tectonics as a framework of thought precludes the possibility of a neat catalog of basin types. The key factors in basin evolution are the types of plate interactions and settings, but the order in which they may be arranged in space and time is variable within wide limits. A great variety of developmental schemes can be accommodated within the theory of plate tectonics. The evolution of sedimentary basins is incidental to the formation and consumption of lithosphere. The major perturbations of a stable and level earth's surface are related to the opening of oceanic basins accompanied by the rifting and fragmentation of continental blocks, and to the closing of oceanic basins accompanied by the collision and assembly of continental blocks. For this reason, the principal trends of basin evolution can be classified on the basis of their location within the continuing interplay between tectonics and sedimentation.

1. Oceanic basins underlain by oceanic lithosphere.
2. Rifted continental margins along the contact between oceanic and continental lithosphere (Atlantic-type margin).
3. Arc-trench systems where oceanic lithosphere is consumed beneath island arcs or continental margins (Island arc or Andean-type margin).
4. Suture belts where continental blocks are juxtaposed by crustal collision.
5. Intracontinental basins in the interior of continental blocks.

Oceanic Basins

The principal settings of oceanic facies controlled by tectonic relations are (see Figure 10–5): (1) bathymetric highs of ridge crests at divergent plate junctions where the layered igneous ophiolite succession is formed along the trends of the spreading centers (see Chapter 8); (2) where the oceanic substratum gradually subsides as it cools in moving away from spreading centers; and (3) deep basins beneath which the thermal contraction of the lithosphere is essentially complete. The pelagic sediment that covers the igneous portion of the ophiolite sequence has a stratigraphy and facies relationship that reflects changing water depths. Near the upper part of the ridge the

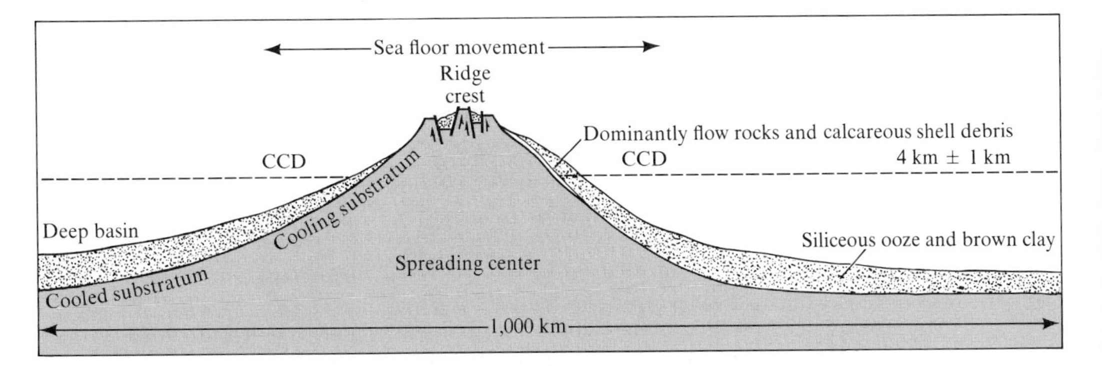

Figure 10–5
Idealized cross section of an oceanic spreading center showing accumulations of ponded and peripheral pelagic sediment and volcanic rocks in extensional fault basins at the ridge crest and on the flanks above the carbonate compensation depth (CCD), and siliceous ooze and brown clay overlying the mafic rocks to the sides of the spreading center. Turbidites may be intercalated with the siliceous ooze and brown clay near continental edges. Total sediment thickness far from continental margins may be a few hundred meters.

water depth commonly is less than about 4 km, so that pelagic shells of calcium carbonate can accumulate; below this depth, called the *carbonate compensation depth,* the degree of undersaturation of the water with respect to calcite or aragonite is so great that shell accumulation is not possible (see Chapter 15). Lower on the rise flanks and in the deep basins, siliceous shells accumulate. If the depositional site is sufficiently near a landmass, continental sediment may be carried into the deep ocean basin by *turbidity currents* (see Chapter 13) and may be interbedded with the siliceous oceanic deposits.

Rifted Continental Margins

Rifted continental margins form in pairs when a divergent plate junction forms within a continent (see Figure 10–6), and form singly when magmatic arcs are rifted

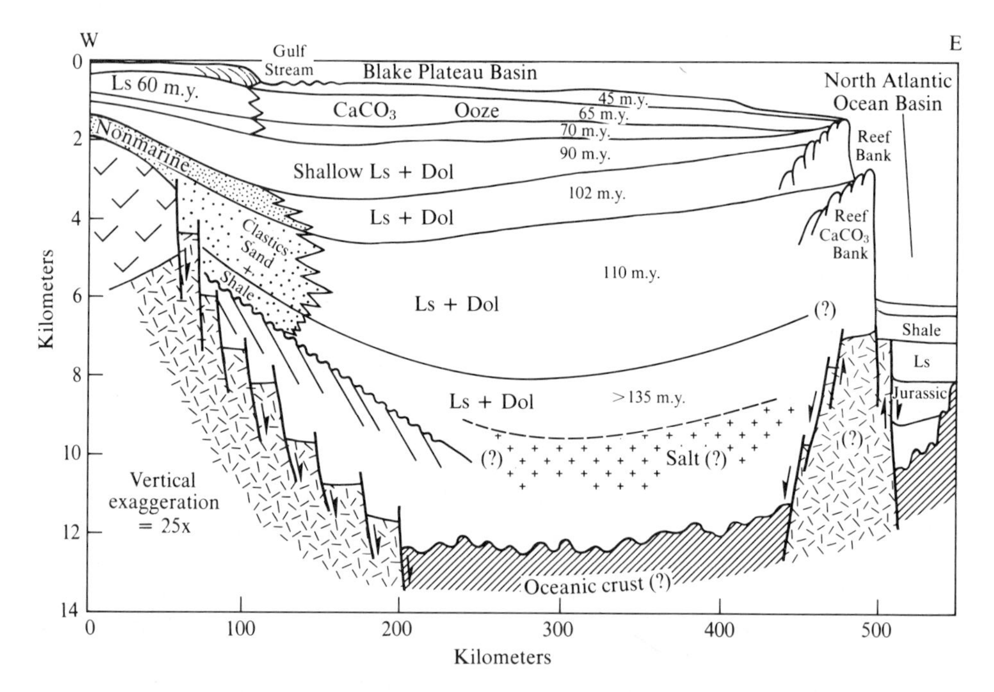

Figure 10–6
Diagrammatic cross section of the Blake Plateau east of Florida revealing the underlying sediment accumulation to be a highly faulted graben of Triassic age, now 12 km deep. Formation of the graben preceded the opening of the North Atlantic Ocean by 30 million years. Note the salt deposit at the base of the section. Most of the Gulf of Mexico Basin also has salt immediately above the Paleozoic basement rocks. Analogous graben of Triassic age occur in North Africa, the other side of the rifted Paleozoic landmass. Ls = limestone; Dol = dolomite. [From R. E. Sheridan, 1976, *Sedimentary Basins of Continental Margins and Cratons* (New York: Elsevier), Fig. 2.]

away from the margins of continental blocks by spreading behind the arcs (for example, the Sea of Japan). In the case of simple continental separation, each rifted continental margin causes the juxtaposition of a high-standing continental block (a sediment source) against a newly formed oceanic basin (a sediment sink). The resulting sediment accumulation forms a characteristic prism spanning the interface between continental and oceanic crust. The prism may thus contain strata of both miogeosynclinal and eugeosynclinal affinities. The eugeosynclinal facies normally grade imperceptibly into pelagic oceanic facies in the basinward direction.

The geometry of plate tectonics requires most continental separations to proceed as wedgelike openings, rather than as instantaneous separations along the whole length of rift belts. As a result, the prisms of sediment along rifted margins vary in age, although the sedimentologic facies in each prism may be very similar.

The first stage of rifting is thermal arching, typically associated with the extrusion of lavas rich in sodium and potassium. The balance between the rate of accumulation of such volcanics and the rate of erosion of the thermal arches they crown is uncertain, but when erosion predominates, uplifted terranes of granitic basement are prominent as sediment sources.

When sufficient crustal extension affects the arched region, rift valleys begin to form as graben (see Figure 10–7). Probably these develop first within the domal uplifts, but later they extend as an essentially continuous branching network along the full trend of the rift belt. In the rift valleys, continental redbeds are interbedded with volcanics that continue to erupt through the growing system of crustal fractures. Broad regions to either side of the eventual zone of rupture between the separating continents can be affected by the extensional faulting. For example, the Triassic basins of the Appalachian region, which are filled with richly feldspathic nonmarine sediment and volcanic flow rocks, lie as much as 250–500 km inland from the present continental slope; the slope can be taken as roughly marking the line of Jurassic continental separation.

As continued crustal separation induces subsidence along the zone of incipient continental rupture, the floors of the main rift valleys become partially or intermittently flooded to form proto-oceanic gulfs. Restricted conditions in these basins, which probably are still rimmed by uplifts that block delivery of clastic sediment from the interior of the continent, promote the deposition of evaporites in suitable climates. For example, thicknesses of 5–7.5 km of evaporites are present in the subsurface beneath parts of the Red Sea. Extensive evaporites several thousand meters thick are known also from coastal basins on both sides of the Atlantic Ocean, where they apparently are correlative and represent dismembered portions of the same elongate trend of evaporite basins (see Figure 10–8).

Subsequent evolution of the rifted region is marked by the formation of a wedge of marine and nonmarine strata built upward to form an isostatically balanced continental terrace. The terrace develops on continental crust and extends to the slope break at the shelf edge, from which the continental slope leads down to deep water

Figure 10–7
Map of the eastern margin of the United States and Canada showing the location of known Triassic fault basins (graben) that originated about 30 million years before the North Atlantic Ocean came into existence along the site of the Mid-Atlantic Ridge. Similar graben occur in Triassic rocks of northwestern Africa. [From F. B. Van Houten, 1977, *Amer. Assoc. Petroleum Geol., 61,* Fig. 1.]

where the continental rise of turbidites accumulate along the edge of oceanic crust. The distance from the thin edge of the shelf sediment wedge to the top of the slope is perhaps 100 to 250 km. Examples of such sediment wedges are the lower Paleozoic carbonate–shale sections in the Appalachian and Cordilleran miogeosynclines.

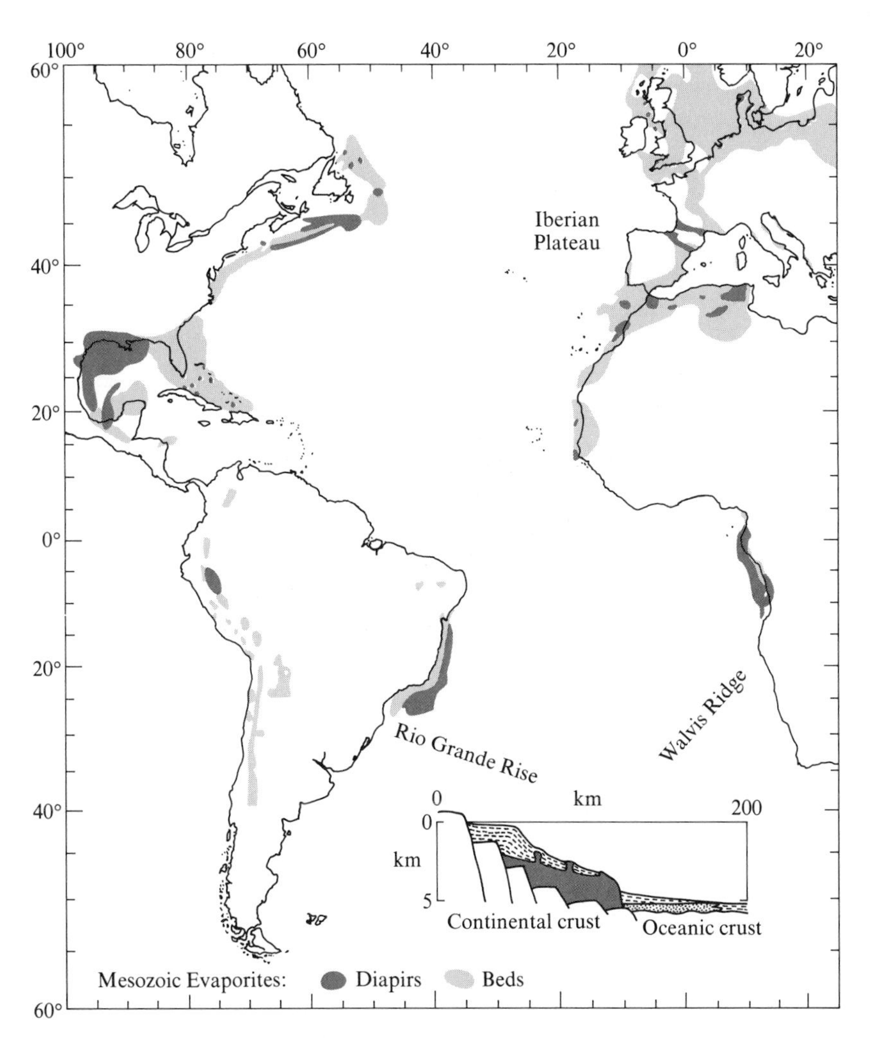

Figure 10-8
Distribution of Mesozoic evaporites and their relationship to continental margins of the Atlantic Ocean. [From K. O. Emery, 1977, Amer. Assoc. Petroleum Geol. Short Course Notes No. 5, Fig. 7.]

Arc-Trench Systems

Arc-trench systems are the characteristic geologic expression of convergent plate junctions and are the depositional sites of most of the eugeosynclinal suite of sedimentary rocks. The system consists of five morpho-tectonic elements, each of which may accumulate a different sedimentary assemblage (see Figure 10–9). The elements are:

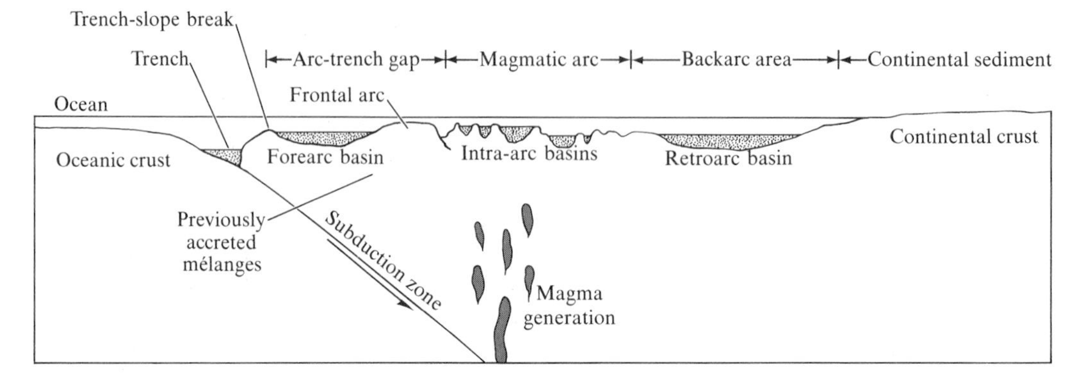

Figure 10–9
Generalized sketch of an arc-trench system along a convergent continental margin showing spatial relationships and nomenclature of plate tectonics and related sedimentary basins. Marine sediment accumulations in the basins are stippled.

1. The *trench,* a bathymetric deep floored by oceanic crust.

2. The *subduction zone* beneath the inner wall of the trench and the trench-slope break marking the top of the inner wall.

3. The *arc-trench gap,* a belt within which a *forearc* (or *frontal arc*) *basin* may occur between the trench slope break and the magmatic arc.

4. The *magmatic arc,* within which *intra-arc basins* may occur.

5. The *backarc area,* within which may lie either an *interarc basin* floored by oceanic crust and separated from the rear of the arc by a normal fault system, or a *retroarc* (or *backarc*) *basin* floored by continental basement and separated from the rear of the arc by a thrust fault system.

In each of these morphotectonic settings, sedimentation, volcanism, and plutonic intrusions occur contemporaneously, although not necessarily at the same site.

On the trench floor, variable thicknesses of turbidites are ponded above the oceanic sediment layers carried into the trench from the open ocean floor. Transport by turbidity currents within a trench is mainly longitudinal along the trench axis, although the initial entry of sediment into the trench may occur along the inner wall as well as from the ends of the trench. The volume of sediment within the trench at a given time reflects the balance between the rate of supply and the rate of plate consumption into the subduction zone.

Immediately landward of the top of the trench, within the arc-trench gap, lie forearc basins, which overlie older, deformed orogenic belts or perhaps oceanic or transitional crustal material. Forearc basins receive sediment mainly from the extensive nearby arc structures, where not only volcanic rocks but also plutonic and metamorphic rocks exposed by uplift and erosion may serve as sources. Sources may also include local uplands along the trench-slope break or within the arc-trench gap itself. There may be little transfer of sediment into the subduction zone from the forearc basins; frequently, they seem to completely override the subduction zone. The Great Valley sequence of California is an example of a forearc basin deposit.

By inference from the bathymetry of modern forearc basins and from the sedimentology of older sequences that appear to have been deposited in similar settings, forearc basins may contain a variety of facies. Shelf and deltaic or terrestrial sediments, as well as turbidites, may occur in different examples. The local bathymetry is controlled by the elevation of the trench-slope break, the rate of sediment delivery to the forearc basin, and the rate of basin subsidence acting in combination. Various facies patterns can occur in different basins.

The sedimentary strata in modern intra-arc basins include distinctive turbidite aprons of volcanic debris shed backwards from the rear sides of frontal arcs. These turbidite wedges rest almost directly on the igneous oceanic crust with few or no intervening pelagic deposits present. Away from the intra-arc spreading centers, sedimentation varies markedly. Where an intra-arc basin is bounded on the side away from the arc-trench system by a submerged remnant arc, no effective source of clastic sediment is present and oceanic pelagic deposits accumulate. If sections of remnant arc remain elevated above sea level, a broad oceanic region is formed in which the only thick sedimentary accumulations are turbidite wedges stranded behind each submerged remnant arc. If a rifted continental margin bounds the basin, nonvolcanic turbidite deposits may enter the basin and mix with arc-derived sediments.

The sedimentary record of retroarc basins includes fluvial, deltaic, and marine strata as much as 5 km thick deposited in terrestrial lowlands and epicontinental seas along elongate cratonic belts between continental margin arcs and cratons. Sediment dispersal into and across retroarc basins is both from highlands on the side toward the magmatic arc and from the craton toward the continental side. The Sea of Japan is a modern example of an extensional retroarc basin. Ancient examples of compressional (thrust-faulted) retroarc basins are the Upper Cretaceous basins of the interior and Rocky Mountain region of North America.

Suture Belts

Suture belts contain deformed examples of all the various types of sedimentary sequences discussed in connection with oceanic basins, rifted-margin prisms, and arc-trench systems. In addition, the collision process can give rise to sedimentary basins located immediately above the suture zone. The basins are generated by depression of the continental block by partial subduction. The sediment fill is clastic debris of continental origin, characteristically wedges of fluvial and deltaic strata.

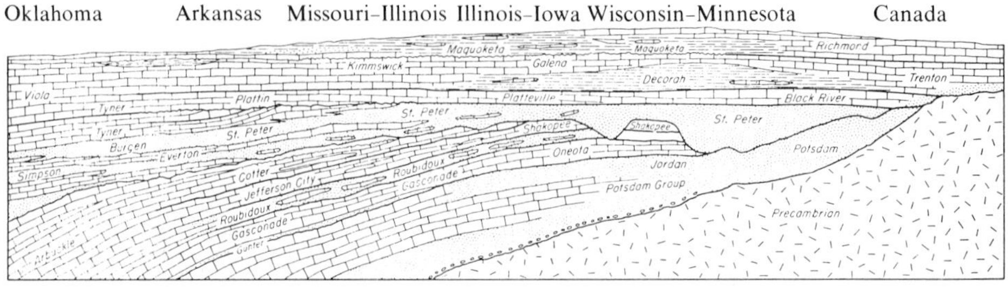

Figure 10–10
Generalized cross section of cratonic lower Paleozoic rocks in mid-continental North America. The carbonate rocks are shallow water, fossiliferous limestones and dolomites containing many small reefs; the sandstones are composed almost entirely of fine- to medium-grained quartz grains cemented by quartz and calcite. [From P. E. Potter and W. A. Pryor, 1961, *Geol. Soc. Amer. Bull., 72,* Fig. 13.]

Intracontinental Basins

Basins within the continental craton are difficult to explain in terms of activities at plate margins. Examples of such basins include the Michigan Basin, with a sediment accumulation 3,000 m thicker than its geographic surroundings; and the Williston Basin in North Dakota and Montana, with 4,500 m more than its surroundings. The basins may have resulted from aborted continental rifting, local cooling in the asthenosphere, downbowing of the crust near convergent plate boundaries, or from causes not now recognized.

The term craton is generally used to refer to tectonically passive parts of a continent, typically formed of lower to middle Precambrian igneous rocks and metamorphosed sedimentary rocks and overlain by essentially flat-lying upper Precambrian or younger sedimentary rocks. Cratons consist of geosynclinal deposits accreted to the original lower Precambrian continental nucleii. A craton tends to be an area of positive relief; portions of it are generally exposed even during times of maximum continental submergence. Cratonic sedimentary rocks are thin but laterally extensive, and contain many unconformities (see Figure 10–10). Both nonmarine and shallow marine deposits can occur, and the sandstones are highly quartzose because of repeated and intense abrasion. Some of these well-rounded and uniformly sized quartz sands may be carried by river systems long distances into bordering mobile belts.

SUMMARY

Sedimentary rocks are composed almost entirely of mudrocks (65%), sandstones (20–25%), and carbonate rocks (10–15%). Clay minerals and detrital quartz grains form about 85% of the mineral grains in these rocks. The thickness of sedimentary rocks on the continents ranges from zero over extensive areas such as the Canadian and Siberian Shields to more than 20,000 m in the deepest parts of some geosynclinal areas; the lower limit is set by the local geothermal gradient and the susceptibility of the minerals to recrystallization.

Mudrocks are composed of 60% clay minerals that, because of their small grain size, can accumulate only in areas of low kinetic energy. Sandstones dominate in areas of high kinetic energy such as beaches and desert dunes. The occurrence of carbonate rocks is controlled primarily by the depth of penetration of light into the sea, so that most carbonate rocks accumulate within a few tens of meters of the sea surface.

The location and size of depositional basins are controlled by continental drift and plate tectonics. Five distinct areas of structurally controlled accumulation of sediments can be recognized: ocean basins, rifted continental margins, arc-trench systems, suture belts, and intracontinental basins. Numerous examples of each type are known from both modern and ancient examples. The mineral composition of the sediments in each type of basin is determined by its location with respect to a continental margin, the nature of the underlying crustal material, and the types of plate junctures nearest to the basin.

FURTHER READING

Bathurst, R. G. C. 1975. *Carbonate Sediments and their Diagenesis,* 2d Ed. New York: Elsevier, 658 pp.

Burk, C. A., and C. L. Drake, eds. 1974. *The Geology of Continental Margins.* New York: Springer-Verlag, 1,009 pp. (Collection of 71 articles by various authors covering all aspects of the origin and development of continental margins.)

Burke, K. 1976. Development of graben associated with the initial ruptures of the Atlantic Ocean. *Tectonophysics, 36,* 93–112.

Conant, L. C., and V. E. Swanson. 1961. Chattanooga Shale and Related Rocks of Central Tennessee and Nearby Areas. U.S. Geol. Surv. Prof. Paper 357, 91 pp.

Dickinson, W. R. 1974. Plate tectonics and sedimentation. In *Tectonics and Sedimentation,* pp. 1–27. Soc. Economic Paleontologists and Mineralogists Spec. Pub. No. 22.

Hsü, K. J. 1973. The odyssey of a geosyncline. In R. N. Ginsburg (ed.), *Evolving Concepts in Sedimentology,* pp. 66–92. Baltimore: The Johns Hopkins Univ. Press.

Pettijohn, F. J., P. E. Potter, and R. Siever. 1973. *Sand and Sandstone.* New York: Springer-Verlag, 618 pp.

Scotese, C., et al. 1979. Paleozoic base maps. *Jour. Geology, 87,* 217–277.

Shaw, D. B., and C. E. Weaver. 1965. The mineralogical composition of shales. *Jour. Sedimentary Petrology, 35,* 213–222.

Sloss, L. L., and R. C. Speed. 1974. Relationships of cratonic and continental margin tectonic episodes. In *Tectonics and Sedimentation,* pp. 98–119. Soc. Economic Paleontologists and Mineralogists Spec. Pub. No. 22.

11

The Formation of Sediment

Sediment originates at the earth's surface because of the chemical and, to a lesser extent, mechanical instability of igneous and metamorphic rocks under atmospheric conditions. The variety of new substances produced by the chemical alteration depends on both the surface conditions (amount of water, temperature, availability of gaseous oxygen, and so on) and the chemical composition of the minerals being altered. For example, calcite cannot be produced by the alteration of orthoclase because there is no calcium in orthoclase. If calcite is to form, a calcium-bearing mineral such as plagioclase or hornblende must occur in the rock to supply calcium ions. In most igneous and metamorphic rocks a variety of minerals of greatly differing chemical composition is present, so that a variety of sedimentary minerals is produced at most geographic sites.

In addition to new minerals, all sedimentary accumulations contain greater or lesser amounts of organic matter. Although living matter forms under atmospheric conditions, it is composed of highly organized constituents that can be maintained only during the life of the organism, so that sediments contain carbonaceous material in all stages of decomposition (weathering). In both modern and ancient sediments the entire range of altered tissue is found, from nearly intact tissue through partially decomposed parts of the plant structure to free amino acids and finally to essentially pure free carbon (anthracite coal or graphite). Plants first colonized the land surface during the Silurian Period and no doubt were preceded by bacteria. It is fair to say that no sedimentary rock in the existing stratigraphic column has formed free of organic influences, although the degree of influence probably increased greatly about 420 million years ago.

CHEMICAL WEATHERING

Inorganic chemical weathering on the earth has not occurred since the evolution of land plants about 420 million years ago. The influence of the organic acids produced during the growth and decay of living tissues dominates the weathering process, and organic acids have chemical properties not possessed by inorganic acids, an important one being the ability to chelate metal ions (see below). This ability greatly increases the solubility of minerals. Unfortunately, the interaction between organic acids and minerals is very poorly understood and, as a result, the standard discussion of chemical weathering emphasizes inorganic reactions.

Granite

If we examine a new roadcut in granite in a humid temperate area, one thing we notice immediately is the freshness of the outcrop face. The nearly vertical surface has a microjagged relief, and a view through a hand lens reveals the sharp edges and corners of the individual crystals. Cleavages in minerals such as feldspars and hornblende are shiny and clear; magnetite is coal black and unstained; and the concentric, semicircular ridges of the conchoidal fracture of quartz are clearly visible. There is no evidence of life on the outcrop surface.

A return to the same cliff face a few months or years later reveals a different scene. The near-vertical face is partly covered by vegetation. A mixture of brown and black stain has spread over the outcrop. Examination of the surface using a hand lens reveals no sharp, jagged corners on crystals, but instead softer and duller crystal surfaces. Fragments of granite lying on the ground must be broken with a rock hammer to obtain a surface clean and clear enough to see cleavages or the true colors of the minerals. The wavy fracture surfaces of quartz crystals now lack the sharp edges of the original conchoidal fractures. Brownish-red stain is everywhere. What has caused these changes?

Some clues are provided by thin sections of the altered granite. The most prominent sites of chemical alteration occur along crystal boundaries, twin composition surfaces, cleavage planes, and fractures in the rock. Apparently water has percolated along cracks and weakly bonded surfaces, leaving a trail of clay minerals and iron oxides in its wake. Some minerals appear more altered than others. Hornblende, biotite, and plagioclase are in worse condition than orthoclase and muscovite crystals; quartz shows few ill effects. More detailed study of this outcrop and others in which a variety of mineral types is present reveals that a consistent ordering is present in the degree of chemical alteration (see Figure 11–1). It is apparent that the sequence is exactly the same as the sequence of crystallization of minerals from a basaltic magma—Bowen's reaction series. In weathering, however, augite does not turn into hornblende, nor hornblende into biotite. As we will see, all three are converted during weathering into the same materials; only the rates of conversion differ.

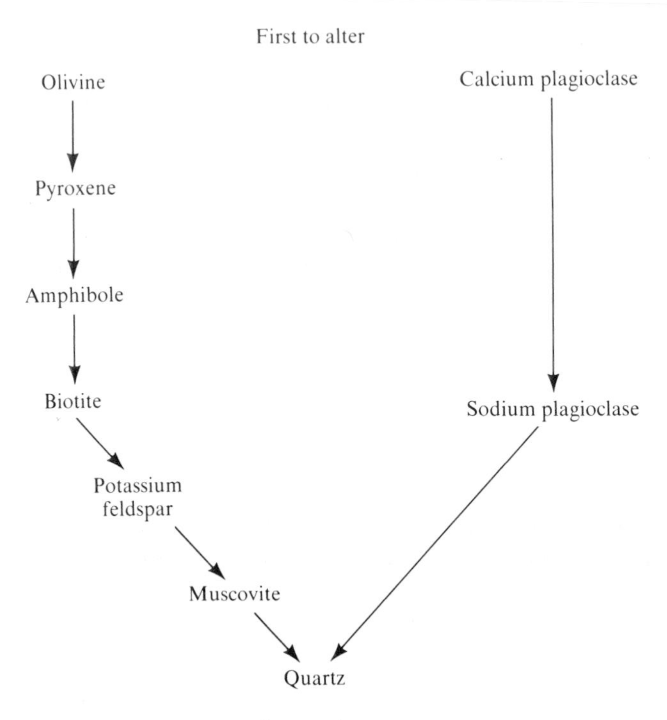

Figure 11–1
Relative rates of chemical alteration during weathering based on field and laboratory data.

The earliest fairly thorough field study of mineralogic changes in granitoid rock during weathering was made by Goldich in 1938. Goldich studied mineralogic and chemical changes in a granitic gneiss, two diabases, and an amphibolite by examining both the unaltered rocks and their weathering products. Based on his observations he established the weathering stability series shown in Figure 11–1, and these results have since been duplicated in many other weathering profiles. Goldich also determined the relative rates of loss of elements from the rocks he studied. Most rapid losses were recorded for sodium and calcium; then potassium and magnesium; weathered residues were relatively enriched in water, titanium, aluminum, and silicon. The explanation for the relative mobility of these elements is the type of bond they form with oxygen in the mineral crystal structure. Only the titanium, silicon, and aluminum form bonds with oxygen that are dominantly covalent in character. The other four elements form bonds that are predominantly ionic; and ionic bonds are more easily broken by the force of the dipolar molecules of which water is composed. (The elements in evaporite minerals also are joined by ionic bonds, which is the reason for their great solubilities.)

Goldich observed that the orthoclase crystals contained a potassium-rich phyllosilicate mineral (sericite or illite) and inferred that it had formed by alteration of the orthoclase. We can write this reaction as follows:

$$\underset{\text{Orthoclase}}{3KAlSi_3O_8} + 2H^+ + 12H_2O \rightarrow \underset{\text{Muscovite}}{KAl_3Si_3O_{10}(OH)_2} + \underset{\text{Soluble silica}}{6H_4SiO_4} + 2K^+$$

(Sericite and illite are not chemically well-defined substances but are approximately the composition of muscovite with slightly less potassium.) Goldich also noted that the most abundant clay mineral formed from the orthoclase was kaolinite. This can be produced by continued alteration of the muscovite:

$$\underset{\text{Muscovite}}{2KAl_3Si_3O_{10}(OH)_2} + 2H^+ + 3H_2O \rightarrow \underset{\text{Kaolinite}}{3Al_2Si_2O_5(OH)_4} + 2K^+$$

It appears, then, that the weathering of potassium feldspar produces a clay mineral of some kind, silica in solution, and potassium ions. We can write the reaction for the weathering of plagioclase feldspar as

$$\text{Albite} + H_2O + H^+ \rightarrow \text{Sodium montmorillonite} + H_4SiO_4 + Na^+$$
$$\text{Anorthite} + H_2O + H^+ \rightarrow \text{Calcium montmorillonite} + H_4SiO_4 + Ca^{2+}$$

In fact, both field observations and laboratory experiments reveal that all the abundant minerals (except quartz) in igneous rocks alter during weathering to (1) phyllosilicate minerals, (2) silica in solution, and (3) alkali and alkaline earth cations (see Table 11–1). This common pattern is violated only by iron released from ferromagnesian minerals. The iron in ferromagnesian minerals is present mostly in the ferrous form, but on release from the crystal structure it oxidizes immediately to the very insoluble ferric ion, which hydrates and precipitates in place as $Fe(OH)_3$. For example:

$$\underset{\text{Augite}}{(Ca, Na)(Mg, Fe, Al)(Si, Al)_2O_6} + H^+ + H_2O \rightarrow$$
$$\text{calcium montmorillonite} + H_4SiO_4 + Ca^{2+} + Na^+ + Fe(OH)_3$$

This brown, amorphous substance subsequently dehydrates to the red mineral hematite, Fe_2O_3:

$$2Fe(OH)_3 \rightarrow Fe_2O_3 + 3H_2O$$

Thus, the ubiquitous occurrence of ferromagnesian minerals in crystalline rocks accounts for the very common red color of sedimentary rocks and soils. Table 11–1 summarizes the products generated by weathering of the common silicate minerals.

Table 11–1
Summary of Weathering Reactions of the Common Minerals in Igneous Rocks

Input		Output	
Mineral	Others	Phyllosilicate or clay mineral	Others
Potassium feldspar	H_2O, H^+	Illite, sericite (muscovite)	Dissolved silica Potassium ions
Sodium feldspar	H_2O, H^+	Sodium montmorillonite	Dissolved silica Sodium ions
Calcium feldspar	H_2O, H^+	Calcium montmorillonite	Dissolved silica Calcium ions
Pyroxenes Amphiboles Biotite	H_2O, H^+	Calcium–sodium montmorillonite	Dissolved silica Calcium ions Sodium ions Magnesium ions Ferric hydroxide (precipitate)
Olivine	H_2O, H^+	Serpentine (antigorite + chrysotile)	Magnesium ions Ferric hydroxide (precipitate)

The Effect of Vegetation

It is reasonable to assume that rock covered by vegetation will alter at a faster rate than bare rock because, although both are subjected to intermittent rainfall and temperature change, the rock with plant cover is also leached by organic acids. Until recently, however, there had been no attempt to quantify this presumed difference in rate. As a result, Jackson and Keller (1970) went to Hawaii and conducted a field and laboratory study of the weathering of lava flows of known age, parts of which were colonized by the lichen *Stereocaulon vulcani* and parts of which were bare. They found that the weathering crust (soil) on the colonized rock was 10–100 times thicker than on the bare rock. Chemical analysis revealed that the lichen-free soil contained 5–10 times more silicon than the soil formed by the lichens, indicating a marked loss of this element from the soil as a result of the plant growth.

This loss of silicon is explained by the unique ability of some organic compounds to form very stable chelate complexes with cations. A chelate compound is one in which organic ring structures are coordinated by a cation so that the cation is held (sequestered) by more than one chemical bond and is thus removed from direct contact with the soil water. The water does not "see" the silicon (just as it would not "see" a copper ion in the middle of a sphalerite crystal) and hence does not become saturated with respect to SiO_2; the water mass can hold more silica because the silicon in the organic structure is not contributing toward saturating the solution. Another example of natural chelation is that of cupric ion by the amino acid glycine (see Figure 11–2). Many of the organic compounds dissolved in soil waters and ground waters have this ability to chelate. A large proportion of some economically important elements (such as copper and manganese) is transported in streams in chelate compounds, to be precipitated when the organic matter decomposes. Primitive plants such as lichens are particularly adept at snatching metal ions out of bare rock, although seed plants are able to perform this feat as well. Thin skins of lichen commonly can be seen occurring as the first colonists on the floors of granite quarries. Fry (1927) shows a series of remarkable photomicrographs of lichens "making a meal" out of rock particles.

Basalt

There were several reasons why Jackson and Keller chose Hawaiian basalts as the rocks to examine to determine rates of chemical weathering. One possibility is that Hawaii is a pleasant place to do field work; so why not? Another possibility is that the geologic setting is particularly suitable. In Hawaii, (1) the dates of eruptions during historic times and hence the dates of initiation of chemical weathering are known precisely; (2) the compositions of the flows are essentially the same; (3) annual rainfall is very high, about 1–10 m/yr; (4) basalt alters rapidly compared to granite, so that a measurable thickness of soil is developed on flows only 40–60 years old. The reason for the similarity in composition of the flows is their derivation from the same magmatic source on the sea floor (more than 10,000 m below the ocean surface) within very brief periods of time. But what is the reason for the rapid alteration compared to granite? Several explanations can be invoked.

1. Calcic plagioclase and pyroxenes contain less silicon than alkali feldspars and biotite, so fewer of the strong silicon–oxygen bonds are present to ward off chemical attack.

2. Chemical attack occurs along surfaces. Crystals in basalt are smaller than those in granite; thus, they have higher surface/volume ratios.

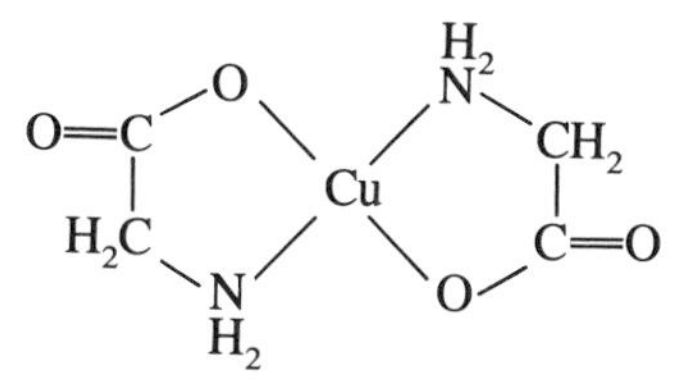

Figure 11–2
Chelation of cupric ion by two molecules of glycine, $H_2N \cdot CH_2 \cdot COOH$.

3. Volcanic flow rocks chill rapidly on leaving the site of eruption and, as a result, commonly have a glassy groundmass. Glass is amorphous and amorphous substances alter much faster than crystalline ones because the ions and ionic groupings in them are disorganized and hence bonded less strongly.

Submarine Alteration

We normally think of weathering as occurring, by definition, at the air–rock interface. However, essentially the same process can occur at the sea floor at the water–rock interface, where it is usually termed submarine alteration. When it occurs near rift zones in the ocean basins the alteration is called hydrothermal in recognition of the fact that the liquids and gases issuing from the rifts are at temperatures of several hundred degrees Centigrade. There has been only one study of undersea weathering of a granitoid rock, a granodiorite at the continental margin off the coast of Southern California. It was found that the alteration of the rock was very similar to the alteration that would have occurred in contact with the atmosphere. The alteration products were iron oxide, sericite, montmorillonite, and kaolinite, as would be expected in a rock composed of quartz, orthoclase, plagioclase, and ferromagnesian minerals. An assemblage of plants and animals was growing attached to the undersea rock wall, and it probably played a part in the alteration process.

Because of the existence of spreading centers and fracture zones throughout the world ocean, basalt either is exposed at the sea floor or lies at a shallow depth beneath the water–sediment interface over much of the ocean floor. In either case, the basalt is subject to chemical alteration by circulating sea water. The most abundant new mineral produced is a potassium montmorillonite. The montmorillonite is the clay mineral expected from the weathering of a plagioclase–augite rock; the potassium is supplied by the ocean water. Also produced are small amounts of analcite ($NaAlSi_2O_6 \cdot H_2O$) and phillipsite $(\frac{1}{2}K, Na, Ca)_3Al_3Si_5O_{16} \cdot 6H_2O$. Both of these minerals seem reasonable as results of the interaction between basalt and sea water.

Certain areas of the sea floor are notably hotter than others, presumably because of the presence of magma at shallow depth below the water–rock (sediment) interface. The sea floor rocks and the water in them are warmed, which results in the movement of hot water through the fractured rocks or sediments above. Cold sea water is drawn down into the sediment to establish a convective pattern, as in a pot of water heating on a stove. The hot water reacts strongly with the sea-floor basalt and takes into solution a variety of minor elements present in the rock. The rising hot water is cooled and diluted by the addition of cold sea water, resulting in the precipitation of elements such as copper, nickel, cadmium, and others as sulfides in fractures. Some economic mineral deposits, such as the copper deposits in Cyprus, may have formed in this manner.

Metalliferous deposits of economic value formed in association with ocean-floor spreading centers and fracture zones need not be confined to vein occurrences. An excellent example of blanket-type deposits is provided by the deposits forming today on the floor of the Red Sea. The minerals in the Red Sea are disseminated as sulfides in a widespread sediment blanket. As the rising solutions cool in contact with sea water, they spread along the sea floor, combine with sulfide ions either from the brine itself or from bacterial reduction of the sulfate in sea water, and are deposited in the enclosing detrital or carbonate sediments. Such deposits are known from the Miocene of Japan and are suspected as the explanation of metalliferous deposits in other areas, as well.

The manganese nodules that cover some parts of the floor of the Pacific Ocean result from enrichment of oceanic bottom water in manganese from basalts and rift zone fluids. The nodules contain an abundance of other valuable metals as well, and numerous proposals for mining the nodules have been made in recent years.

CLAY MINERALS

Clay minerals are the most common mineral group in sedimentary rocks, totaling about 45% by weight or volume. They are very small particles, commonly less than 1 μm in size, and are the major constituent in clay-size sediment. In addition, they are the only abundant minerals in detrital rocks that are not inherited from igneous or metamorphic parents; the variety of clay minerals formed reflects the minerals weathered to produce them, but the clay mineral structure is generated in the sedimentary environment.

The mechanism by which the phyllosilicate layer structure of clays is produced from the tectosilicate structure of a feldspar or the inosilicate structure of an amphibole or pyroxene is unknown. The scale of the transformation is too small to be traced by thin section studies and can be followed only by using an electron microscope in combination with detailed chemical analysis. But even using these techniques there are great difficulties. Microchemical analysis of a particle less than 1 μm in size is not easy, even using the most sophisticated modern instrumentation. Figure 11–3 shows the appearance of kaolinite flakes produced on a feldspar surface; but there are no apparent intermediate steps. There is either a feldspar or a layer-structure clay, with a sharp boundary between them. However, studies of the transformation using advanced techniques have begun only recently, so the next few years may witness important new insights into the process of clay mineral formation.

Clay Mineral Structure

The major clay minerals are kaolinite, montmorillonite, and illite. Nearly all clay-bearing sedimentary rocks contain more than one type of clay mineral, and this fact, coupled with their very small size and similar optical properties, requires the use of X-ray diffraction techniques to identify them. These techniques are standardized and

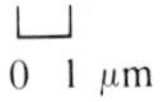

Figure 11–3
Contact of orthoclase crystal and sheaves of kaolinite in a weathering horizon of the Butler Hill granite, Missouri. Photo taken using a scanning electron microscope at a magnification of 3,000X. [From W. D. Keller, 1978, *Geology, 6,* Fig. 1.]

provide unequivocal identification of the various varieties of clay minerals. In clay mixtures, however, the estimation of relative percentages of each clay mineral is only semiquantitative. Estimates to within 10% of the amount present are acceptable for most purposes.

The crystal structure of clay minerals is similar to that of micas, sheeted-layer structures with strong bonding (covalent) within each sheet and among the sheets, but weak bonding (hydrogen bonds and van der Waals bonds) between the adjacent two- and three-sheet layers. The weak bonding between layers permits not only the excellent cleavage of clays but also adsorption of metallic cations and organic substances on clay mineral surfaces. This latter factor is important in chemical reactions that occur in weathering horizons.

The clay mineral structure contains two types of sheets. One is composed of tetrahedrally coordinated Si–O and Al–O groups in which three of the four oxygen atoms of each group are shared with adjoining groups. The cations in this layer are at least 50% silicon atoms; the proportion of aluminum is greatest in illite and least in kaolinite. The second type of sheet is composed of Al–OH or Mg–OH groups in octahedral coordination, with ferrous iron sometimes substituting for magnesium. A single flake of kaolinite clay consists of one Si–O tetrahedral sheet and one Al–O octahedral sheet (see Figure 11–4) with essentially no cation substitution in either sheet. All other clays (and micas) are composed of three sheets, an octahedral sheet sandwiched between two tetrahedral sheets (see Figure 11–5). There is abundant substitution within all three sheets: aluminum for silicon in tetrahedral sheets; magnesium and ferrous iron for aluminum in the octahedral sheet. These substitutions cause charge imbalances within the sheets (Al^{+3} vs. Si^{+4}; Mg^{+2} and Fe^{+2} vs. Al^{+3}) that are rebalanced by adsorption of metallic cations on the surfaces of each clay flake. The cations adsorbed are those available in soil waters—the potassium that was released from orthoclase, and the sodium and calcium that were released from plagioclase and ferromagnesian minerals. Illite prefers potassium; montmorillonite prefers sodium and calcium. These relationships are summarized in Table 11–2.

Clay Mineralogy and Climate

The relationship between the composition of the mineral being weathered and the composition of the initial clay mineral produced is clear. Orthoclase is a potassic

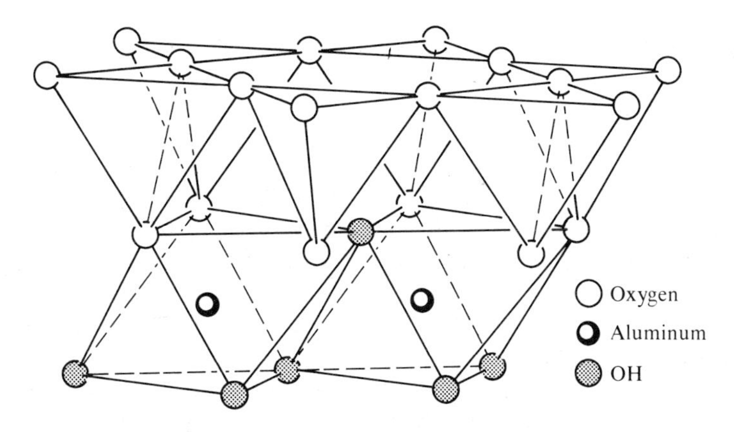

Figure 11–4
Diagrammatic sketch of the structure of kaolinite with a tetrahedral layer bonded on one side to an octahedral layer. [From C. S. Hurlbut, Jr., and C. Klein, 1977, *Manual of Mineralogy,* 19th Ed. (New York: John Wiley and Sons), Fig. 10.54.]

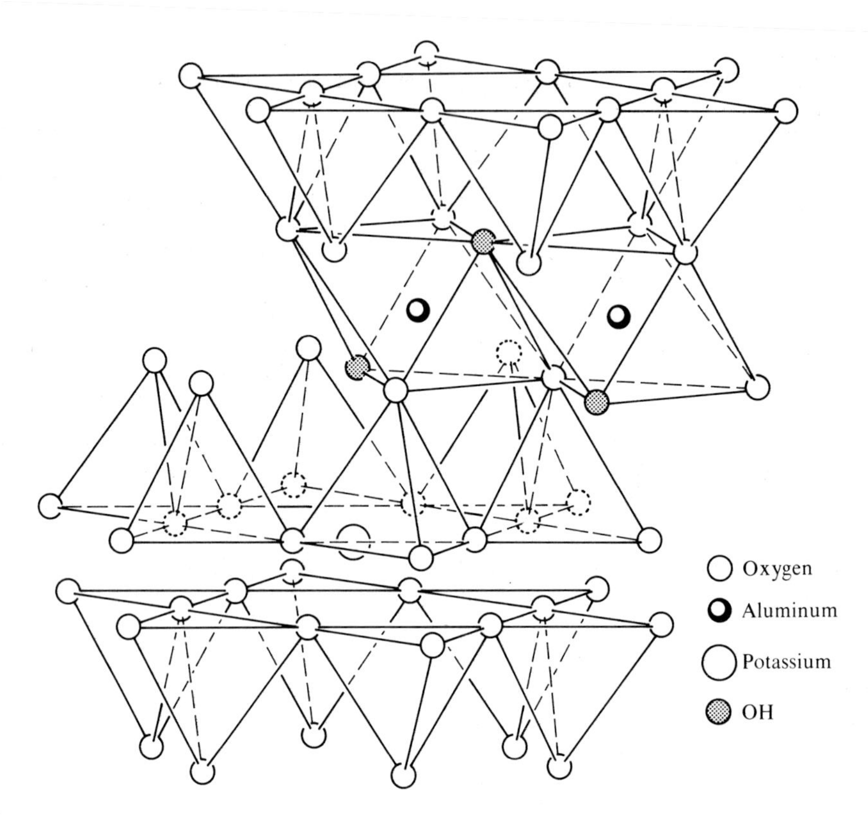

Figure 11-5
Diagrammatic sketch of the muscovite (illite) structure and the position of the required interlayer cation. [From C. S. Hurlbut, Jr., and C. Klein, 1977, *Manual of Mineralogy,* 19th Ed. (New York: John Wiley and Sons), Fig. 10.56.]

mineral and produces a potassic clay (illite). Plagioclase, amphiboles, and pyroxenes are sodic and calcic and produce sodic or calcic clay (montmorillonite). But what can we expect to occur as the initially formed clay minerals are attacked by the meteoric waters passing over them after they form? Will they be transformed into different clay minerals? What will happen if the intensity of weathering increases because of a climatic change; for example, increased temperature caused by changes in solar radiation, as during recent interglacial epochs; or increased temperature resulting from drifting of the continent from higher to lower latitudes, as during the Permo-Triassic breakup of the supercontinent Pangaea?

We have already observed that the initially formed illite or montmorillonite continues to lose potassium, sodium, or calcium as weathering proceeds. The reason this occurs is that the K^+, Na^+, and Ca^{2+}, initially released from orthoclase, plagioclase, and ferromagnesian minerals, are carried away in solution. Le Chatelier's Principle

Table 11–2
The Abundant Phyllosilicates in Rocks and Weathering Horizons Showing Their Idealized Compositions[a]

Name	Chemical formula	Comment
Two-layer structure		
Kaolinite	$Al_2Si_2O_5(OH)_4$	Almost no substitution
Antigorite	$Mg_3Si_2O_5(OH)_4$—platy	Forms from serpentine—no Al is present
Chrysotile	$Mg_3Si_2O_5(OH)_4$—fibrous	Forms from serpentine—no Al is present
Three-layer structure		
Pyrophyllite	$Al_2Si_4O_{10}(OH)_4$	Almost no substitution
Montmorillonite	$Al_2Si_4O_{10}(OH)_2 \cdot xH_2O$	Mg may partly replace Al; interlayer Na and Ca present
Muscovite (illite)	$KAl_2(AlSi_3O_{10})(OH)_2$	In illite, Mg, Fe partly replace octahedral Al; interlayer K present
Talc	$Mg_3Si_4O_{10}(OH)_2$	
Vermiculite	$Mg_3Si_4O_{10}(OH)_2 \cdot xH_2O$	
Phlogopite	$KMg_3(AlSi_3O_{10})(OH)_2$	
Biotite	$K(Mg,Fe)_3(AlSi_3O_{10})(OH)_2$	
Chlorite	$Mg_5Al(AlSi_3O_{10})(OH)_8$	

[a]The Si_2O_5 or Si_4O_{10} grouping ($\pm$Al substituting for Si) is the tetrahedral layer. The other cations and hydroxyl are the octahedral layer.

tells us that if the products of a reaction are removed, more of them will be produced. Eventually, a clay mineral containing only aluminum and silicon as cations will remain—kaolinite. The weathering paths of the primary silicate minerals are shown in Figure 11–6. It is clear that average weathering residues should contain more montmorillonite than illite, because the total volume of plagioclase plus ferromagnesian minerals in crystalline rocks is greater than the volume of orthoclase plus perthite. In a warm humid climate with good drainage, such as might occur in South Carolina or Louisiana, kaolinite should be the most abundant clay mineral produced.

Laterite

Suppose the intensity of weathering becomes extreme, as in the Amazon River Basin or southeast Asia—extremely high rainfall and good drainage, coupled with very

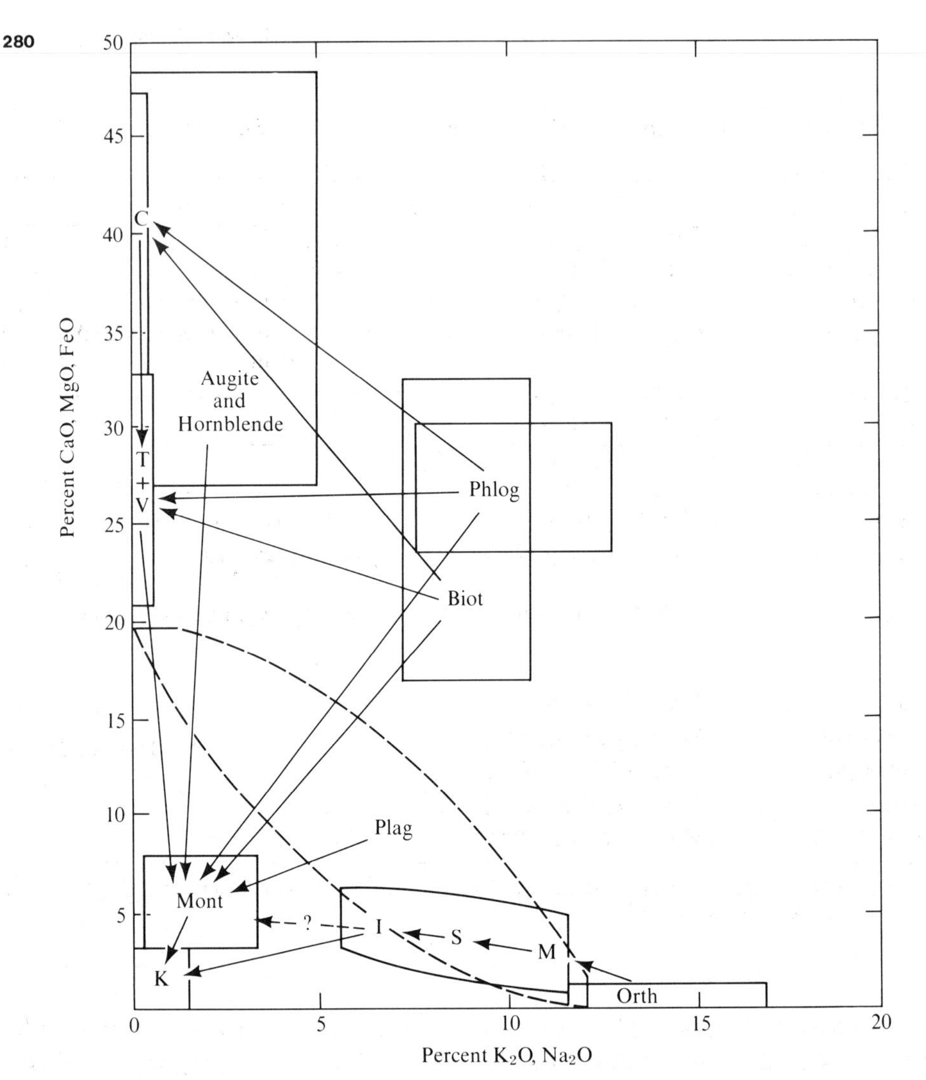

Figure 11-6
Proportions of alkali and alkaline earth oxides in the abundant primary silicate minerals and clay minerals. Biot = biotite; C = chlorite; I = illite; K = kaolinite; M = muscovite; Mont = montmorillonite; Orth = orthoclase plus perthite; Phlog = phlogopite; Plag = plagioclase; S = sericite; T = talc; V = vermiculite. Observed and suspected paths of weathering are shown by solid and dashed arrows.

high mean annual temperature. Under these conditions, even the kaolinite is unstable, and silica is leached from the clay mineral to leave an aluminous residue of amorphous and crystalline $Al(OH)_3$, the mineral gibbsite. Such a soil residue is called either aluminous laterite or bauxite, and is the world's major source of aluminum. The major economic deposits are located in Jamaica, Surinam, Guinea, and Australia. The Australian deposits are residual from a warmer and more humid climate during Tertiary times.

The behavior of iron during weathering is complicated by the fact that it can occur in the sedimentary environment in more than one state of oxidation. Most iron in ferromagnesian minerals is in the reduced state because of a general deficiency of gaseous oxygen in the environments of formation of igneous and metamorphic rocks. In an aerated weathering environment the iron is converted into the oxidized form to precipitate as ferric hydroxide. Once precipitated, it cannot be reduced easily and accumulates. Such iron-rich crusts are best developed on basalts because they contain three times as much iron as granites (12% vs. 4%); soils developed on basalt in tropical climates are ferruginous laterites and are commonly tens of meters thick. These residual soils also contain aluminum and titanium oxides and hydroxides, the other superstable weathering residues. A complete gradation in composition occurs between aluminous and ferruginous laterites.

MECHANICAL WEATHERING

Mechanical weathering is of trivial importance compared to chemical weathering because of the extraordinary dissolving power of even the smallest amount of dipolar water molecules. Even in the driest of deserts there is a significant amount of water in the air (that is, the humidity is not zero) and, in the cool predawn hours, this moisture condenses on the desert sand. Field observations reveal much flaking of fallen Egyptian monuments near the air–sand interface, and very little at the air–monument interface. The ancient inscriptions are decomposed and illegible on the monument surface partly buried in the sand, and legible on the upper surface of the monument. Thin sections of crystalline rocks exposed in desert areas always show the formation of films of iron oxide and/or clay along crystal boundaries or mineral cleavages. Desert sand originates primarily because of chemical decomposition, not mechanical abrasion by wind-transported older sediment.

The most important mechanical weathering process is frost wedging, which is visible on the upper parts of mountainous areas in wet, temperate climates. When water freezes it expands about 9% in volume, generating a force more than adequate to spring loose flat slabs of rock. Surfaces at higher elevations of the Rocky Mountains and Sierra Nevada are often covered by these slabs. The water seeps into cracks during the warm daylight hours and freezes at night, loosening the slabs. This process is repeated each day for much of the year and is very effective.

The importance of mechanical weathering is that it makes many small fragments from a single larger one, greatly increasing the surface area of the rock mass. Chemical activity operates fastest on exposed surfaces, so the night-time activity of the H_2O molecules paves the way for the more destructive activities of the liquid H_2O molecules the following day.

SUMMARY

Sediment is formed because of the instability of igneous and metamorphic rocks at the earth's surface. The crystalline rocks alter because they formed at temperatures between 200°C and 1000°C, at fairly high pressures, and generally in the presence of relatively small amounts of water and oxygen compared to the amounts available at the surface. The presence of living organisms at the surface not only accelerates the formation of sediment, but also makes possible chemical reactions not possible in purely inorganic systems.

The residue of the chemical alteration process under surface conditions is enriched in ferric iron oxide, in hydrated substances, and in aluminosilicate minerals depleted in alkali and alkaline earth cations. The most abundant aluminosilicate mineral group in sediments is clay, which forms about 45% of all minerals in sedimentary materials. The type of clay that is most abundant depends during early stages of weathering on the mineral composition of the parent rock, but in later stages depends entirely on the climate. Illite and montmorillonite appear initially, to be succeeded in humid temperate climates by kaolinite. In humid tropical areas, gibbsite and ferric iron oxides form the final residue of the original igneous or metamorphic rock.

FURTHER READING

Baas Becking, L. G. M., I. R. Kaplan, and D. Moore. 1960. Limits of the natural environment in terms of pH and oxidation-reduction potentials. *Jour. Geology, 68,* 243–284.

Degens, E. T., and D. A. Ross, eds. 1969. *Hot Brines and Recent Heavy Metal Deposits in the Red Sea.* New York: Springer-Verlag, 600 pp.

Ehlmann, A. J. 1968. Clay mineralogy of weathered products of river sediments, Puerto Rico. *Jour. Sedimentary Petrology, 38,* 885–894.

Fry, E. J. 1927. The mechanical action of crustaceous lichens on substrata of shale, schist, gneiss, limestone, and obsidian. *Ann. Botany, 41,* 437–460.

Goldich, S. S. 1938. A study in rock weathering. *Jour. Geology, 46,* 17–58.

Huang, W. H., and W. C. Kiang. 1972. Laboratory dissolution of plagioclase feldspars in water and organic acids at room temperature. *Amer. Miner., 57,* 1849–1859.

Jackson, T. A., and W. D. Keller. 1970. A comparative study of the role of lichens and "inorganic" processes in the chemical weathering of recent Hawaiian lava flows. *Amer. Jour. Science, 269,* 446–466.

Kerr, R. A. 1978. Seawater and the ocean crust: the hot and cold of it. *Science, 200,* 1138–1141, 1187.

Syers, J. K., and I. K. Iskandar. 1973. Pedogenic significance of lichens. In V. Ahmsdjian and M. E. Hale (eds.), *The Lichens,* pp. 225–248. New York: Academic Press.

12

Mudrocks

Although they form approximately two-thirds of the stratigraphic column, mudrocks are poorly understood and inadequately studied. Few sedimentary petrologists have chosen to study mudrocks because of several difficulties.

1. Mudrocks are composed mostly of clay minerals that absorb water easily and in large amounts, so that mudrocks become plastic and flow downslope readily. Hence they form valleys rather than hills. Cliffs formed of mudrock are not common.

2. Mudrocks are extremely fine-grained. In outcrop there is little that can be described other than color and the presence or absence of fissility. In thin section many of the rock constituents cannot be resolved because of small size, intermixing of the different clay minerals, and the common occurrence of opaque hematite stain or organic matter.

3. It is almost impossible to distinguish between quartz and feldspar in such small grains unless the feldspar is twinned. But because large feldspar grains tend to weather and break along twin composition surfaces, most silt-size feldspar grains are untwinned.

4. Because of their small size and sheet structure, clay minerals are easily and frequently altered after deposition. The original clays have very often been recrystallized and/or changed into clays of a different chemical composition.

Is there any hope? Can we salvage two-thirds of the stratigraphic column from the scrap heap of petrology? Recent advances in analytical techniques suggest the answer to these questions is "yes." Scanning electron microscopy permits magnifications of at least 50,000×, in contrast to the polarizing microscope with its limitation of about 500×. New wet chemical techniques permit the isolation of quartz and feldspar from the mass of clay minerals so that the feldspars can be studied petrographically. The development of the electron microprobe in about 1950 and its greatly increased use by sedimentary petrologists during the past few years have revolutionized our ability to make microchemical analyses of tiny grains in sediments. It is now possible to make quantitative chemical analyses of areas only 1 μm in diameter, the size of some individual clay flakes.

In addition, there is now a much-increased interest in the organic matter in mudrocks on the part of the major oil companies. How are the organic tissues of microscopic organisms converted into petroleum and natural gas? It seems likely that the next decade may well be "the age of mudrocks" in terms of the amount of study they receive and the increase in our understanding of these enigmatic rocks.

FIELD OBSERVATIONS

As is the case for all rocks, an adequate field description should include information about the texture, structure, and mineral composition. It is harder to do this for mudrocks than for sandstones or carbonate rocks because of the fine grain size of mudrocks.

Textures

Texture is defined as the size, shape, and arrangement of the grains or crystals in a rock. Individual grains cannot always be seen in a mudrock using only a 10× hand lens, but despite this handicap it still is possible to make a semiquantitative estimate of the ratio of silt to clay. The method is to nibble a bit of the rock between the teeth to determine whether the rock is gritty (see Figure 12–1). Mudrocks are formed almost entirely of quartz and clay. Quartz is gritty; clay is slimy. If you sense no abrasion of your teeth, the rock contains more than two-thirds clay minerals; and clay minerals are the bulk of the clay-size particles. If grit is sensed, there are between about two-thirds and one-third clay minerals. If there are less than one-third clay minerals, enough quartz silt is present to be seen using the 10× hand lens. Alternatively, if the sample is wet, rubbing with the thumb will produce a shiny surface on clay-rich materials, and a dull surface on silt-rich material.

The shape of the grains in a mudrock cannot be determined in outcrop because of their small size. This is not a problem, however, because the quartz grains in mudrocks are always quite angular. Particles with diameters less than about 60 μm travel almost entirely in suspension, being small enough to be suspended even by weak currents. This means that they are not abraded by impacts with other grains and will be as angular after transport as when they began as fragments from a crystalline rock. Clay minerals, on the other hand, have a shape determined by their crystal structure. Most commonly, they are shaped like a sheet of paper, although this cannot be seen using a 10× lens.

Figure 12–1
Geologist hungry for knowledge determining the grain-size distribution of a sample of Hennessey Formation mudrock (Permian), central Oklahoma. [Photo courtesy S. Bock.]

Structures

The major sedimentary structures visible in outcrop are fissility and lamination. *Fissility* is a property of a mudrock that causes it to break along thinly spaced planes parallel to bedding and to the orientation of the sheetlike clay flakes. The existence of fissility depends on many factors, only one of which is the abundance of clay minerals. Mudrocks with identical percentages of clay can differ greatly in fissility because of differences in the perfection of orientation of the clay flakes. In outcrop the reasons for the lack of parallelism of clays cannot be determined, but observations made in modern muddy environments suggest several possible explanations.

Clay minerals, because of their structural and chemical imperfections, have charged surfaces. As a result, when they enter saline water after stream transport they interact with ions in the sea and clump into aggregates of randomly oriented clay flakes. This process is termed *flocculation.* Both the number of flakes in a floccule and the stable size of the floccules are variable. Clearly, a mudrock formed of clay lumps rather than neatly oriented individual flakes will not be fissile.

Bottom-dwelling organisms in the depositional environment also affect the development of fissility. As they scavenge through the fresh mud for organic matter they swallow great amounts of clay and, when the clay passes through the alimentary canal of the organism, it is formed into aggregates: more clumps. Also, as the organism burrows through the mud it destroys any clay flake parallelism formed during settling of the flakes to the bottom. Studies of modern mud environments suggest that the rate of burrowing and mealtime clump formation is frequently greater than the rate of deposition of the mud particles. Perhaps it is more meaningful to ask how mudrocks can ever be fissile, rather than to ask why many are not. Fissility in some mudrocks may be produced during diagenesis of the clay minerals.

Lamination refers to parallel layering within a bed. By definition, a bed is thicker than 1 cm; a lamina, thinner than 1 cm (see Figure 12–2). Lamination can have many origins that are related both to variations in current strength during deposition of the layer and to changes in composition of the sediment deposited. For example, a bed of mudrock may contain laminae of black organic matter; or zones of green color within a dominantly red unit; or microplacering of quartz silt within an otherwise clayey mudrock. Each of these features can give unique information about oxidation/reduction reactions at the site of deposition; or changes in these conditions after burial; or variations in current strength with time during deposition of the bed. All departures from randomness and homogeneity should be recorded at an outcrop. Perhaps their meaning might be unclear at the moment, but the truth may emerge on reflection in the laboratory or office.

Color

The colors of mudrocks fall almost entirely into two groups: gray to black, and red–brown–green. The gray–black shades reflect the presence of free carbonaceous material, which in turn reflects deposition in an oxygen-deficient or reducing environment. In well-oxygenated water the concentration of gaseous oxygen is 10^{-4} mol/L. To obtain conditions sufficiently reducing for the black free carbon to accumulate requires a decrease to about 10^{-6} mol/L. This implies a lack of circulation of aerated water in the depositional environment, preventing complete oxidation of the organic tissues to carbon dioxide plus water. It is important to note that lack of circulation is not related to depth of water. Most of the ocean floor at depths of many thousands of meters is kept well oxygenated by the cold bottom currents that originate at the surface in polar regions and circulate throughout the oceans. There are many shallow areas of the ocean floor that are stagnant, such as the lagoons between the Texas Gulf Coast and the offshore barrier bars. Some of the geographically most extensive black mudrocks are known to have been deposited in water only a few tens of meters in depth; for example, the Chattanooga Shale noted earlier in Chapter 10.

The best-studied example of a large, modern stagnant environment is the Black Sea, which has an average depth of 3,700 m but no measurable oxygen in the lower

Figure 12–2
Laminated oil shale from the Mahogany ledge in the lower Piceance Creek section, Rio Blanco County, Colorado. Dark lines on the straight-edge indicate sampling zones; numbers are cumulative sample numbers in a larger investigation. [From D. A. Brobst, U. S. Geol. Surv. Prof. Paper 803, Fig. 11.]

3,500 m. Circulation of oxygen-rich water from the Mediterranean Sea is prevented by a rock barrier at the southwest end of the Black Sea (Bosporus) that rises to within 40 m of the surface.

The red–brown–green color grouping reflects the presence or absence of ferric oxide (red) or hydroxide (brown) as colloidal particles among the clay mineral flakes. Hematite (Fe_2O_3) is red; goethite [$FeO(OH)$] is brown. Only a few percent of hematite are sufficient to give a deep red color to a mudrock. If these compounds are absent, the true green color of most clay minerals shows through; illite, chlorite, and biotite are green. In many red mudrocks, ovoid or tubular green spots are present, reflecting reduction of ferric iron adjacent to a plant root or other bit of organic matter and removal of the ferrous ions in groundwater.

Table 12–1
Classification of Mudrocks

Ideal size definition	Field criteria	Fissile mudrock	Nonfissile mudrock
> ⅔ silt	Abundant silt visible with hand lens	Silt–shale	Siltstone
> ⅓ < ⅔ silt	Feels gritty when chewed	Mud–shale	Mudstone
> ⅔ clay	Feels smooth when chewed	Clay–shale	Claystone

Source: H. Blatt et al., 1980, *Origin of Sedimentary Rocks,* 2d Ed. (Englewood Cliffs, N.J.: Prentice Hall), Table 11–1.

Nomenclature

There is no uniform usage of the many terms that refer to detrital sediment finer than sand size. Such terms as shale, argillite, argillaceous, and clayey mean different things to different people. The terminology we will adopt (see Table 12–1) is simple, usable in field work, and consistent with whatever detailed laboratory studies may follow the outcrop descriptions.

LABORATORY STUDIES

Laboratory investigations of mudrocks tend to concentrate on mineralogy and chemistry rather than texture, probably because *some* textural observations can be made in the field but almost no mineralogic determinations are possible without sophisticated laboratory instruments. The most valuable techniques for studying mudrocks are the following:

1. *X-ray diffraction.* This method is used for mineral identifications and is particularly useful when the particles are so small (for example, clays) that the petrographic microscope is not effective. The sample is powdered, mounted on a glass slide, and bombarded with X-rays. The X-rays are diffracted by planes of atoms in the crystal structure and a pattern is produced on a paper chart (see Figure 12–3). The chart (diffractogram) is a plot of diffraction angle versus intensity of diffracted radiation and reveals the interplanar spacings of the mineral and, in turn, its crystal structure. This is the best method for identifying the various types of clays in a rock. It is sometimes used in conjunction with differential thermal analysis (DTA), a technique in which the sample is heated in a furnace to determine the temperatures at which water or carbon dioxide is released. Different temperatures are characteristic of different minerals.

2. *Sodium bisulfate fusion.* The purpose of this technique is to isolate the quartz and feldspar grains from the mass of clay minerals and other substances in the mud-

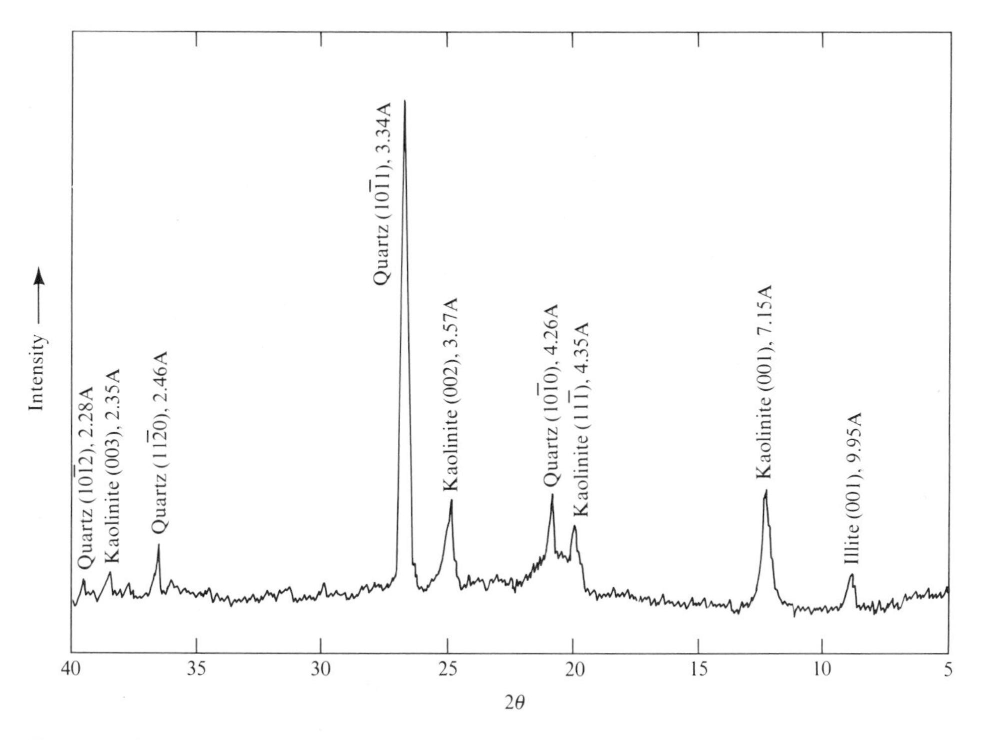

Figure 12–3
X-ray diffraction pattern of an unoriented mixture of ⅓ quartz, ⅓ kaolinite, and ⅓ illite. Differences in peak height among the three minerals result from the combined effects of differences in crystal structures and orientation of grains on the glass slide. Labels show the lattice plane that produces the particular peak and the distance between these lattice planes. θ is the reflection angle at which the X-ray wavelengths are in phase.

rock. Pea-size fragments of the sample are fused over a Bunsen burner in sodium bisulfate. The only materials that survive are quartz and feldspars, which are almost completely unaffected in either composition or grain size. These grains can then be analyzed petrographically.

3. *Electron microprobe.* This instrument provides a means of determining the chemical composition of very small volumes at the surface of polished thin sections or grain mounts. An electron beam is focused on the area of interest, which can be as small as 1 μm in diameter. The impact of the beam on the sample causes the emission of X-rays whose wavelengths are characteristic of the elements present in the area hit by the beam. The intensity of the X-rays reveals the concentration of the element (see Figure 12–4). This technique is sufficiently sensitive to determine concentrations of trace elements as well as of major elements in the sample.

4. *Scanning electron microscope (SEM).* This instrument is used for both textural and mineralogic determinations. A piece of the rock perhaps a centimeter in

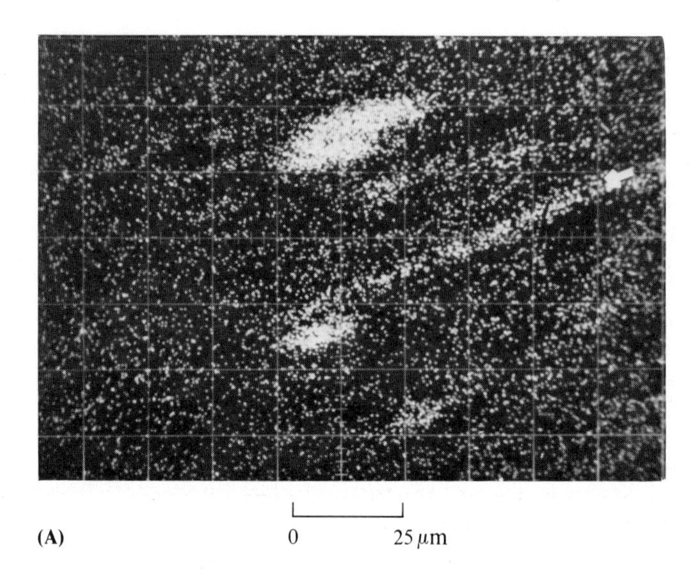

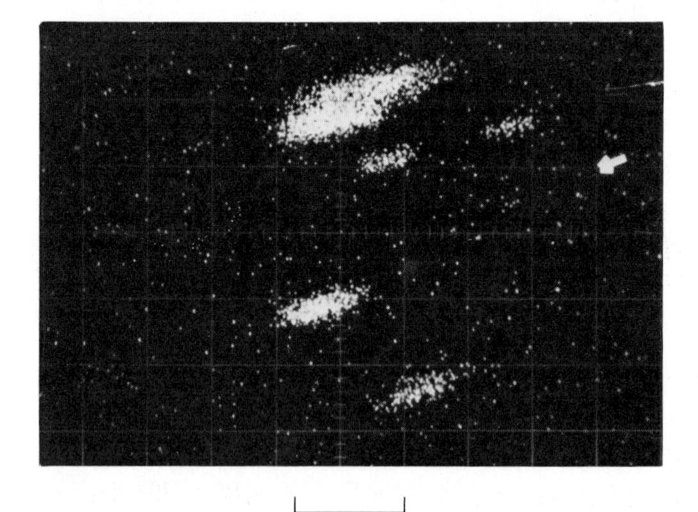

Figure 12–4
Electron microprobe scans of the Levis shale, Ordovician, Quebec, Canada. This patch of micro- to cryptocrystalline pyrite is not visible using only a polarizing microscope. The arrows point to a layer of iron-rich chlorite. (**A**) Distribution of iron (white dots). (**B**) Distribution of sulfur. [From R. Siever and M. Kastner, 1972, *Jour. Sedimentary Petrology, 42,* Fig. 1.]

diameter is coated under vacuum with a gold–palladium mixture. The coated specimen is then bombarded by electrons, which are scattered by the gold–palladium coating to produce the detailed topography of the fragment. Magnifications of 50,000× with excellent resolution and great depth of field are easily obtained (see Figure 12–5), and enlargements of 100,000× are possible with somewhat diminished but quite usable resolution. X-ray attachments to the SEM are commonly used and permit at least semiquantitative element analyses of the sample.

5. *Polarizing microscope.* Although it is possible in theory with perfect imaging and monochromatic light to resolve grains as small as 0.2 μm with a polarizing microscope, in practice it is not possible to achieve resolution better than about 1 μm. This is about the average size of the particles in a mudrock, so in thin section much of the rock appears as an irresolvable birefringent mass of clay minerals. Most mudrocks are mixtures of different clay species, and their optical properties are sufficiently similar to make distinctions among them impossible unless individual flakes can be isolated. The most appropriate use of the light microscope in mudrock studies is to study textural features that can be seen in thin section but not in outcrop, such as small-scale cross bedding, structures produced by organic burrowing, microdesiccation features, or other structures that give insights into the history of the rock.

6. *Radiography of rock slabs.* Many mudrocks appear structureless in outcrop (or in the small chip of rock present in thin section) but may contain sedimentary structures too subtle to be visible with the naked eye. These can be made visible by the use of X-ray radiography (see Figure 12–6). In this technique, a slab of mudrock 15 cm long, 10 cm wide, and 0.5 cm thick is cut perpendicular to bedding and placed directly on X-ray film. A photograph is taken using either a medical, dental, or industrial X-ray unit. Previously obscure features such as root tubules, organic burrows, and slump features, and subtle laminations are clearly visible because textural and mineralogic variations in a rock affect the penetration of the X-rays.

Mineral Composition

Clay Minerals

The most abundant mineral group is the clays—principally illite, montmorillonite, and kaolinite—a result of the great chemical stability of a sheeted Si–O, Al–OH crystal structure (see Chapter 11). Although the abundance of clays in the stratigraphic column is not surprising, the relative abundance of the three main types is. Available data, based largely on X-ray diffraction studies during the past 50 years, indicate that the relative abundances change markedly with the age of the mudrock (see Figure 12–7). Those of Cenozoic age contain about twice as much montmorillonite as illite, as would be expected because of the greater abundance in crystalline rocks of plagioclase and ferromagnesian minerals compared to potassium feldspar

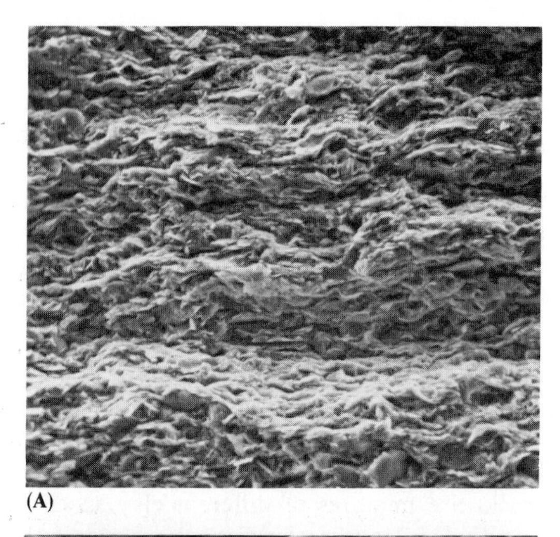

(A)

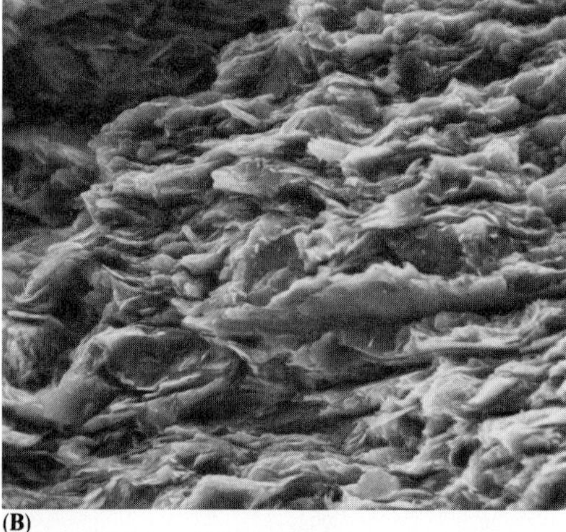

(B)

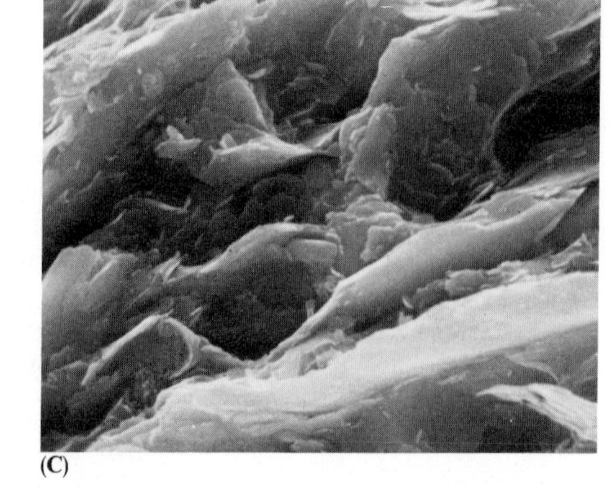

(C)

Figure 12–5
Scanning electron micrographs of the Pritchard Shale, north of Las Cruces, New Mexico. The fissility clearly results from parallelism of clay mineral flakes. (**A**) 300X. (**B**) 1,000X. (**C**) 5,000X. [Photos courtesy K. P. Helmold.]

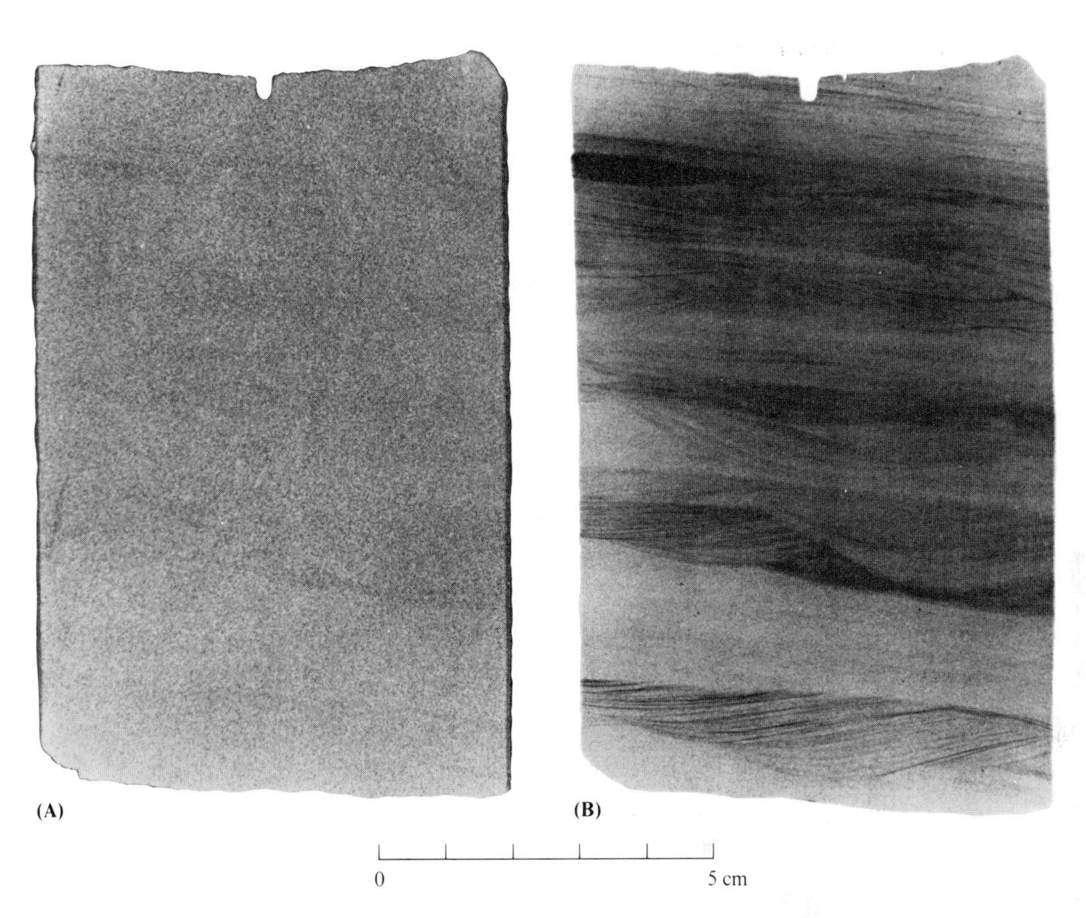

Figure 12–6
(**A**) Polished slice of core and (**B**) positive print of X-radiograph of Berea Sandstone (Mississippian), Illinois. Only vague banding is visible on the polished slab, but X-radiation reveals an apparent dip of 10°, scour and fill, and cross bedding. [From W. K. Hamblin, 1965, *Kansas Geol. Surv. Bull., 175,* Part 1, Fig. 9.]

(see Figure 12–7). Kaolinite is a poor third in abundance because of the limited latitudinal range of the hot, wet climate required for its formation. The bulk of the earth's land area is in the cool, temperate climatic belt that favors the stability of montmorillonite and illite.

As we go back in time, however, it is clear that relative abundances change, with illite becoming overwhelmingly dominant in Paleozoic mudrocks, forming as much as 80% of all clay minerals. Several explanations for this change have been suggested.

1. Continual increase in basaltic volcanic activity through Phanerozoic time, thus increasing the relative abundance on the earth's surface of plagioclase and ferromagnesian minerals, the parent materials of montmorillonite. This would imply, however, that progressively lesser amounts of granite and gneiss (which are rich in potassium feldspar) have been exposed through geologic time. Field evidence does not

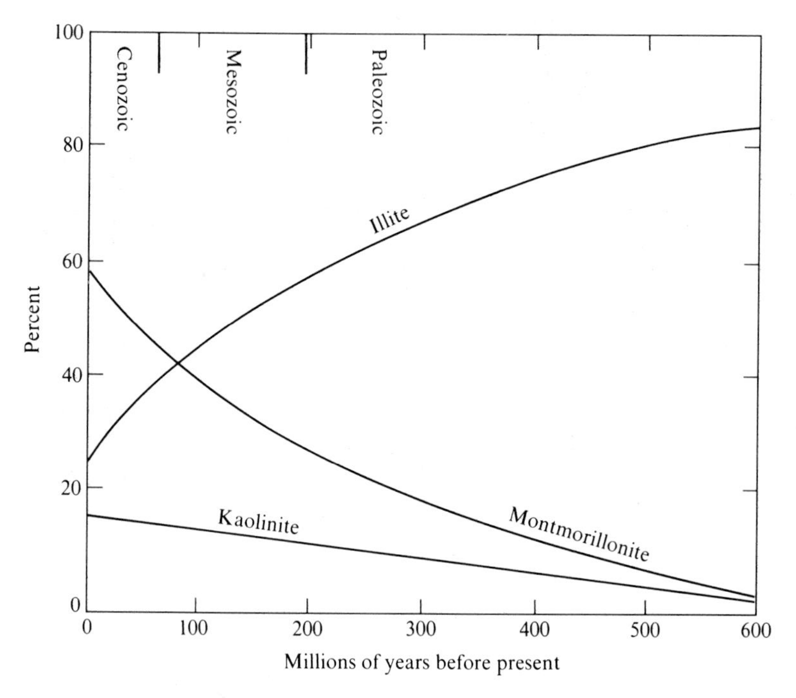

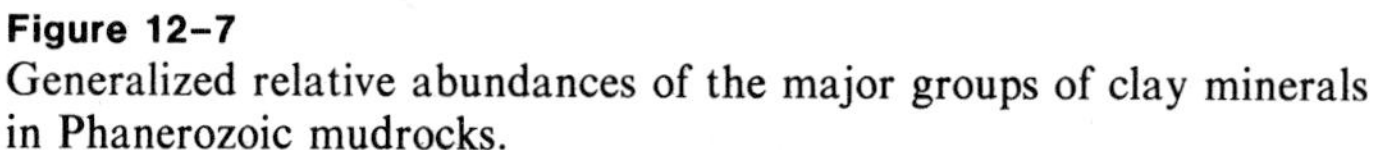

Figure 12–7
Generalized relative abundances of the major groups of clay minerals in Phanerozoic mudrocks.

support this hypothesis. The percentage of sand-size feldspar, which must originate largely in granites and gneisses, is at least as great in Mesozoic and Cenozoic sandstones as in older ones. This suggests, if anything, an *increasing* abundance of silicic crystalline rocks with time rather than a decrease.

2. Change in the biologic controls of chemical weathering. Plant life first colonized the land surface during the Silurian Period, about 420 million years ago. The great abundance of coal beds of Carboniferous age suggests that plants were widespread 220 million years ago. It has been suggested that this spread of land plants is responsible for the change in clay mineralogy through time. Plants require at least as much potassium ion for their nutrition as they do calcium and sodium combined. Perhaps the increase in the amount of plant life on the land surface has resulted in an increasing removal of potassium from developing soils, making it less possible for illite to form from the weathering of potassic feldspar. Present data concerning the illite/montmorillonite ratio through time are inadequate to determine whether the ratio changed sharply in the late Paleozoic. Also, no one knows whether the spread of plants over the land surface required 200 million years, 20 million years, or perhaps only 2 million years.

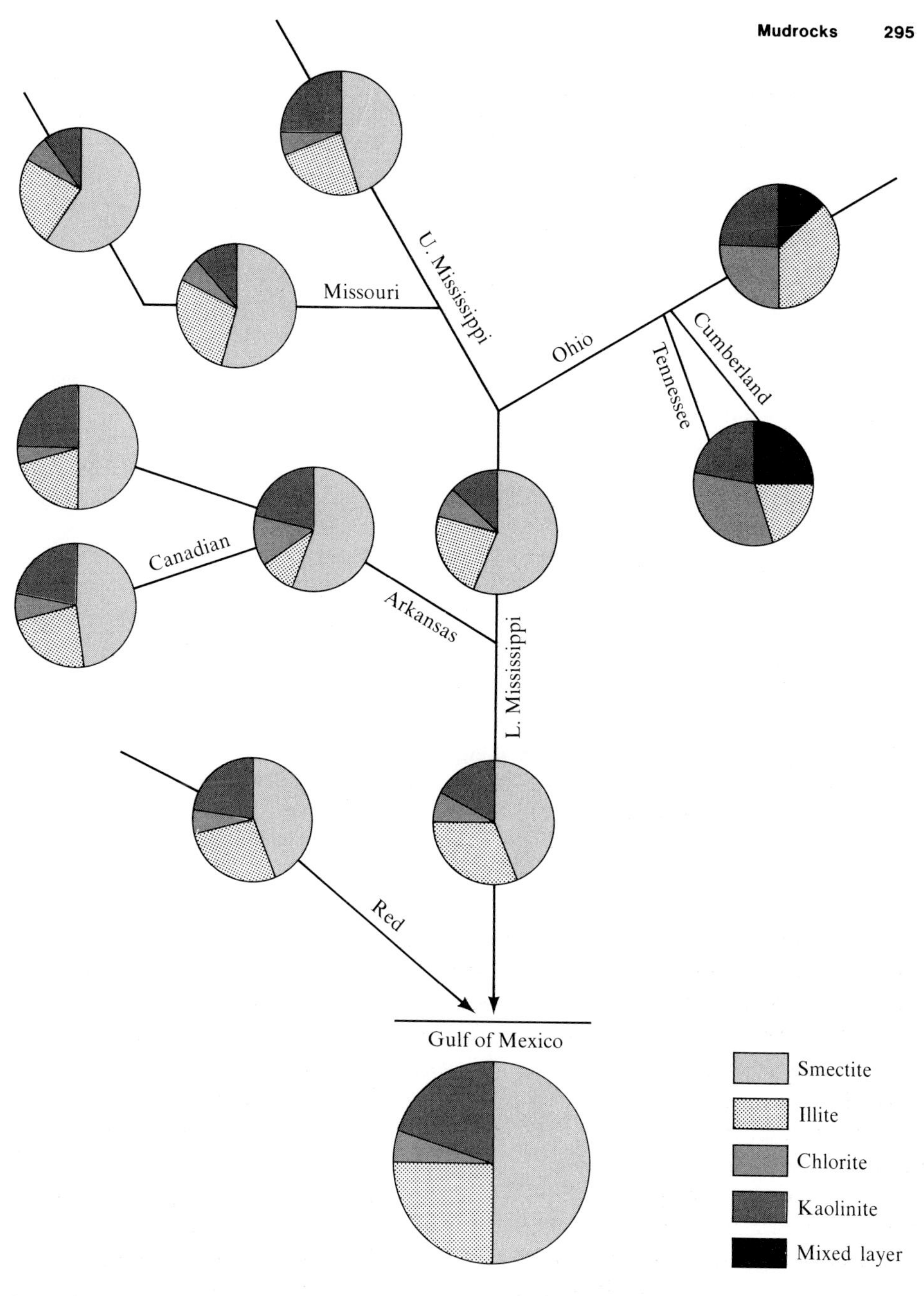

Figure 12–8
Pie diagrams showing the relative amounts of the different clay minerals in the Mississippi River, its major tributaries, and the Gulf of Mexico. Smectite is the group name for expandable clay minerals. Montmorillonite is the most abundant member of the group. [From P. E. Potter et al., 1975, *Bull. Centre Rech. Pau—SNPA, 9,* Fig. 13.]

3. The change through time toward increasing illite abundance could also be caused by diagenetic processes. During the past 20 years, many data have accumulated that suggest postdepositional processes as the major cause of the increase in the illite/montmorillonite ratio with time. In the United States, the studies have concentrated on changes in clay mineral composition in upper Tertiary sediments in the Gulf Coast basin. Similar results have been obtained from other areas of the world, such as the Niger Delta in West Africa and the Rhine Graben area of West Germany. In each area, there occurs an abundance of clay minerals composed of alternating illite and montmorillonite layers. Such crystals are most abundant at shallow depth, but as sample depth increases the montmorillonitic layers in the crystals are converted into illite layers (see Figure 12–9). Suggested sources of the required potassium ions include orthoclase grains in the mudrocks and potassium ions in the connate waters trapped in the rocks. The conversion to illite is largely complete at a depth of 4,000 m and a temperature of approximately 100°C. The chemical change can be idealized by the reaction

$$\text{Montmorillonite} + Al^{3+} + K^{+} \rightarrow \text{Illite} + Si^{4+} + Mg^{2+} + Fe^{2+} + Na^{+} + Ca^{2+}$$

The released silica may crystallize as chert in the mudrocks; the calcium may precipitate in calcite, dolomite, and ankerite; the magnesium in dolomite; the iron in ankerite or hematite. The fate of the sodium is unknown and it may leave the mudrock, escaping upward to the sea floor.

Quartz

Based on studies using both X-ray techniques and sodium bisulfate fusion, the average amount of quartz plus chert in mudrocks is 30% ± 3%. More than 95% of this is single crystals of quartz. The mean size of this sediment is about 6ϕ (15 μm) and is ⅛ fine and very fine sand, 6/8 silt size, and ⅛ clay size; all grains are angular. The amount of quartz in a mudrock is correlated with the grain size of the quartz, with a lesser amount of quartz implying a smaller mean size of the grains (see Figure 12–10). This relationship seems quite reasonable from the viewpoint of sediment transport processes—the very weak currents that allow low-density floccules of silt and clay size to settle cannot have transported relatively high-density grains such as quartz to the site of deposition.

The origin of the quartz and chert is varied. Some of it results from chipping of larger quartz grains during transport in streams or by wind, or during pounding by the surf on beaches and barrier bars. Some of it is quartz released as silt-size grains from slates, phyllites, and low-grade schists. And some of it may be secondary, having crystallized as very fine quartz or chert from the silica released during illitization of interlayered illite/montmorillonite after deposition of mud. No method presently exists to distinguish among these alternatives.

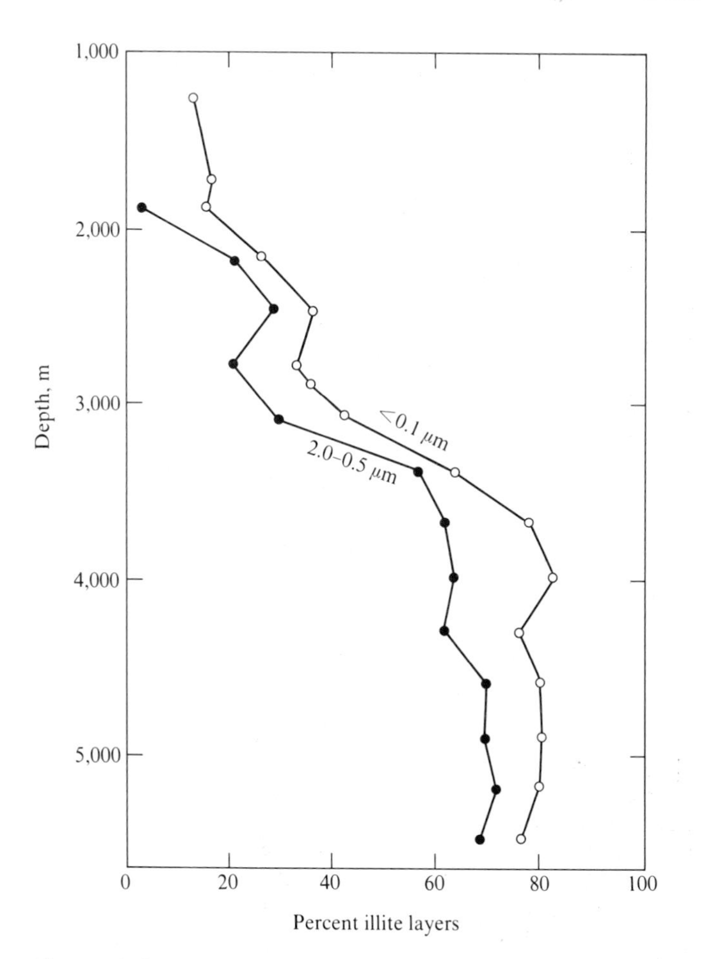

Figure 12-9
Proportion of illite layers in interlayered illite–montmorillonite clay as a function of depth in Oligocene–Miocene sediment in a Gulf Coast borehole. The conversion to illite is faster in finer-grained (hence more reactive) clay. [From J. Hower et al., 1976, *Geol. Soc. Amer. Bull., 87,* Fig. 3.]

Feldspar

Feldspar grains average about 5% ± 2% of the average mudrock. In order to study the mineralogy of these grains, they must first be isolated and concentrated using the sodium bisulfate fusion technique. This destroys the part of the mudrock that is neither quartz nor feldspar, so that the feldspar percentage is increased. It may then be examined using a variety of techniques, such as the polarizing microscope or electron microprobe. Such studies are only now beginning.

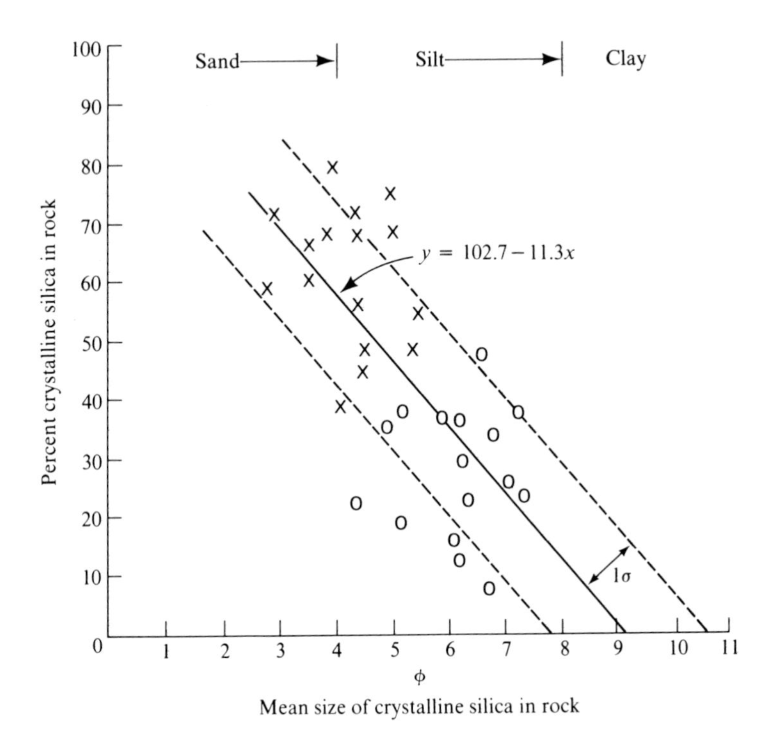

Figure 12–10
Correlation between the percentage of crystalline silica and the mean size of crystalline silica in interbedded mudrocks and sandstones. O = mudrock; X = sandstone; correlation coefficient = −0.68. The solid line is the line of best fit to the data points; $\pm 1\ \sigma$ lines enclose ⅔ of the points. $\phi = -\log_2$ mm. [From H. Blatt and D. J. Schultz, 1976, *Sedimentology, 23,* Fig. 1.]

Carbonate Minerals

Carbonate material is almost as abundant in mudrocks as feldspars but even more poorly understood. Presumably, most of it is calcite, rather than dolomite, siderite, or other mineral, but this assumption is based entirely on the fact that calcite is the dominant carbonate in sandstones. There is some evidence that dolomite becomes more abundant in older mudrocks, as is true of carbonate rocks in general (see Chapter 15).

The origin of the carbonate in mudrocks may be organic, inorganic, or a combination of the two. Perhaps it is particulate material derived from breakdown of shell material. Perhaps it is an inorganic precipitate formed from brines produced during diagenesis. No one has attempted to resolve these uncertainties.

Other Substances

Other materials commonly reported to occur in mudrocks include organic matter, pyrite, and hematite. Of these, the organic matter is probably most abundant and certainly is the most important from either an academic or economic point of view. Mudrocks contain about 95% of the organic matter in sedimentary rocks, although the amount in the average mudrock is probably less than 1%. The black matter in these rocks is much darker in color and more carbonaceous than living organic matter because of the processes that formed it. These processes depend fundamentally on the way the living tissue decomposed after death. If gaseous oxygen is present the tissue is changed into $CO_2 + H_2O$, leaving no residue. The CO_2 either is dissolved in soil water or escapes into the atmosphere. If gaseous oxygen is absent the tissue is inefficiently and only partially decomposed, leaving a residue of black substances. These organic residues are the raw material from which petroleum is produced by natural processes. That is, mudrocks are the source rocks in which our major energy resource is formed; subsequently, the oil is squeezed into the sandstones and carbonate rocks where we find it.

The environmental conditions that cause the reduction of the organic matter toward free carbon also cause reduction of other substances. For example, ferric iron is changed to ferrous form and sulfur is reduced from a valence of +6 in sulfate ion (SO_4^{2-}) to −2 (S^{2-}). These two reduced species then combine to produce amorphous FeS and FeS_2, pyrite. Many black mudrocks contain pyrite.

Hematite is a very common pigmenting material in mudrocks; possibly it is more abundant in mudrocks than in sandstones. As we noted previously, iron atoms in ferromagnesian minerals exist mostly in the reduced state, but oxidize immediately on contact with the atmosphere. The substance formed initially is brown goethite.

The granitic igneous rocks, gneisses, and schists that contain ferromagnesian minerals form deep in the crust and reach the surface through tectonic and isostatic uplift and erosion. Hence, they occur initially in highland areas, mountain ranges such as the Sierra Nevada or ancestral Rockies. Precipitation in the mountains releases the iron atoms so they may oxidize. Some of the iron ions combine with hydroxyl ions to form colloidal goethite; some are adsorbed onto clay mineral surfaces to balance charge deficiencies; some are adsorbed onto organic matter in the soil; some are adsorbed onto the amorphous materials that form about 20% of the soil solids. The oxidized iron atoms are then transported to lowland areas where they are deposited. Colloidal material, clay materials, and organic matter are hydraulically equivalent in that they all are very fine grained and will therefore settle together when water velocity decreases. In interbedded mudrock–sandstone sequences it is common to find the mudrocks redder than the sandstones.

Some iron atoms may be added to those that arrive at the depositional site. After deposition, iron may be leached from the octahedral layer of clay mineral structures or from detrital ferromagnesian minerals to contribute additional intensity to the red coloration of the fine-grained sediment (see Chapter 14).

Bentonites

Bentonite is an important variety of mudrock defined by the type of material from which it formed. It is composed almost entirely of montmorillonite and colloidal silica produced as the alteration products of glassy volcanic debris, generally a tuff or volcanic ash. In a pure bentonite the only other minerals present are clearly volcanic in origin, such as euhedral zircon and/or brown biotite, or relict glass shards (fragments). The montmorillonite is normally the calcic variety, reflecting the fact that the parent material is basaltic or andesitic. Sodic and potassic bentonites are known, however, and contain, in addition to the essential montmorillonite, euhedral sanidine and quartz grains with the high-temperature crystal habit (β-quartz). Bentonites are frequently interbedded with tuffs containing nonvolcanic constituents.

A bed of bentonite is the result of either a single eruption or of several eruptions within a very brief period, perhaps a few years. As a result, bentonite beds are normally less than 50 cm thick. Stratigraphic and sedimentologic studies have demonstrated regularities in the distribution patterns of these beds consistent with their mode of origin. The thickness of the bed and the size of the unaltered fragments in it decrease logarithmically with distance from the volcanic vent. These fragments also tend to be graded within the bed, with heavier ones near the base and lighter ones near the top. From a stratigraphic point of view, the important fact about bentonite beds is that they define a "time line" in the geologic section, or as close to an infinitely thin synchronous surface as it is possible to get in geologic materials. From an economic point of view, bentonites are important as bases for drilling mud in oil exploration. Unfortunately, the swelling characteristic that makes them desirable in these muds causes them to form very unstable slopes in outcrop. Construction projects on any mudrocks that contain montmorillonite require great care. From a general academic point of view, bentonites are important because they are among the rare mudrocks that provide a clear description of the rocks from which they formed.

Mudrocks and Source Areas

One of the tasks of sedimentary petrology is to determine paleogeography and provenance. Did the drainage basin from which the sediments were derived contain outcrops of igneous and metamorphic rocks? Or were only older sedimentary rocks exposed? If crystalline rocks were exposed, of what types were they and in what proportions? The aim is to construct paleogeologic maps for earlier periods that are as accurate and detailed as geologic maps of present outcrop patterns.

The chief clue to upstream geology is the mineral composition of the downstream sediment accumulation. If the sediment contains kyanite it means that high-grade aluminous schists were exposed in the drainage basin. Similarly, the presence of labradorite implies gabbro or amphibolite; strongly abraded quartz sand means older sandstone was present.

How can an examination of the mineral composition of mudrocks contribute toward paleogeographic reconstruction? The major constituent of mudrocks is clay minerals. But, as we have seen, the intensity of weathering and diagenetic alteration are at least as important as determinants of clay mineralogy in a mudrock, as are the crystalline rocks from which the clays developed. Kaolinite can be formed from any aluminosilicate mineral. Both kaolinite and montmorillonite are often changed to illite after burial. Because of this, the character of the original crystalline rocks probably is determined better by minerals other than clays. If we are to obtain source area (provenance) information from mudrocks, we will have to concentrate on the other one-third of the mudrock, the part composed of quartz, feldspars, and miscellaneous accessory minerals. Unfortunately, however, only one such study has yet been published, and it was concerned only with the average size distribution of quartz grains in mudrocks. The usefulness of mudrocks for provenance studies is unknown.

ANCIENT MUDROCKS

There have been very few studies of the mineral composition and petrology of ancient mudrocks, and even fewer that have tried to relate the petrology of a specific formation to depositional environment. Two of these few are the investigations by Scotford (1965) and by Weiss et al. (1965) of shales of Cincinnatian age (Upper Ordovician) in Ohio, Indiana, and Kentucky.

The Cincinnatian Series in the area studied is about 200 m thick and crops out in an oval-shaped band 40 km wide around the Jessamine Dome on the Cincinnati Arch (see Figure 12–11). It consists of shales and fossiliferous limestones deposited in one of the many shallow epeiric seas that covered parts of North America during Paleozoic time. The mudrocks range from 50% to more than 90% of the Series.

Nearly all the siltstone and shale beds are laminated and often interfinger with either the edges or the surfaces of the interbedded limestones. Whole valves or bits of skeletal material scattered on a parting plane in shale may increase in amount laterally into a distinct bed of limestone, with the lamination in the argillaceous beds bending to conform to the shape of the limestone layers. Apparently the carbonate sediment was lithified before the muds, and the resulting limestone beds served as rigid units during mud compaction. Some of the silty mudrocks are ripple-marked or cross-laminated, but not the fissile units (shales). Presumably this indicates that the clay mineral floccules were too small in size (< 10–$20\ \mu m$) to permit this type of current-generated structure to develop.

The mudrocks are medium- to fine-grained silts with less than 3% sand. No variation in grain size is present either laterally or vertically within the Series as a whole, but size variation sometimes occurs from the middle to the boundaries of individual mudrock beds; coarser sizes tend to occur near the contacts with adjacent limestone beds. Perhaps this reflects a correlation between increased current strength and an increased abundance of carbonate-secreting organisms on the shallow sea floor during Cincinnatian time.

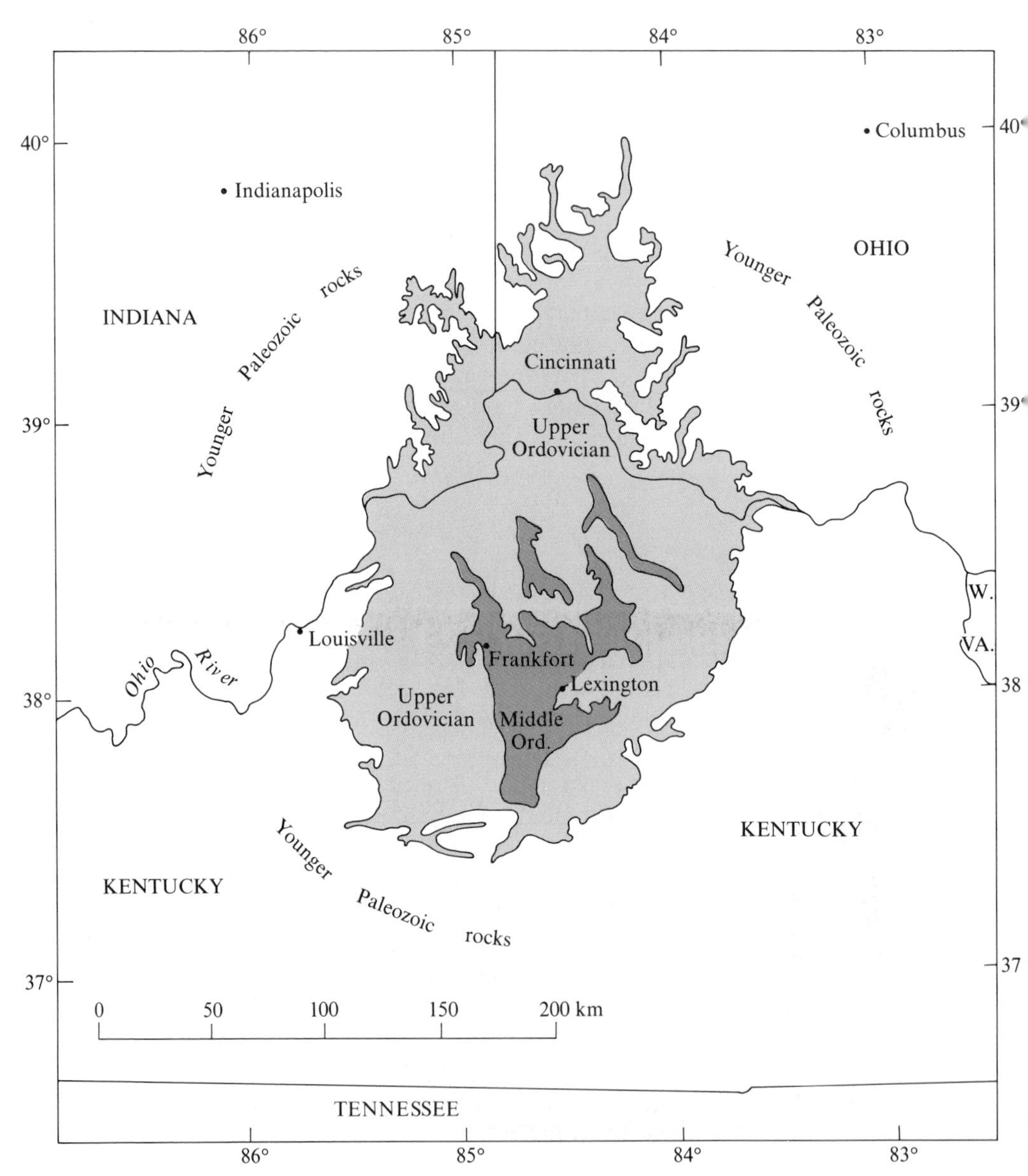

Figure 12–11
Geologic map of Jessamine Dome area, showing Cincinnatian (dark color) outcrop pattern.

Most of the mudrocks contain some calcite in the form of both shell fragments and microcrystalline calcite of uncertain origin. The bulk of the mudrocks, however, is clay minerals (72% of the noncarbonate fraction), mostly illite but with lesser amounts of chlorite. Angular, chemically etched, silt-size quartz forms almost all the rest of the insoluble fraction; trace constituents include orthoclase, plagioclase, biotite, and a few stable detrital heavy minerals such as garnet and zircon. Diagenetic pyrite occurs in some mudrocks. The uniformity and simplicity of mineral composition and fine grain size throughout the Cincinnatian mudrocks indicates a uniformity of source terrane, specifically an extensive area with very low relief and no nearby significant exposures of crystalline rocks. Based on present knowledge of early Paleozoic paleogeography and tectonics in the mid-continental area, the nearest crystalline rocks were those on the Canadian Shield in eastern Canada.

Many different types of mineralogic and element analyses were made of the Cincinnatian mudrocks and most showed no trends with distance from the axis of the Cincinnati Arch. Environmental conditions apparently were static. Statistical treatment of the data provided no evidence for the existence of the north–south trending Arch in Ordovician time. Detailed study of one member of the Series, however, did reveal numerous trends suggesting the presence of an east–west topographic high. For example, there were consistent north-to-south increases in the percentages of sand and silt, decrease in the amount of clay, and a linear increase in the silt/clay ratio. The variations were interpreted to result from changes in depositional environment during Clarksville time. The increase in detrital grain size indicates that current strengths were greater toward the south. The percentage of calcite in the mudrocks also increases toward the south, suggesting shoaling toward an area of more abundant organic growth, particularly of carbonate-shelled organisms.

It is clear that analyses of the nonclay fraction of mudrocks deposited in epeiric seas can yield important paleogeographic information. The early stages of growth of an underwater structural or topographic feature can be detected many millions of years before the structure is emergent and sheds coarser clastic debris.

MODERN MUD ENVIRONMENTS

As we noted earlier, many environments exist in which muds can accumulate. These environments can be nonmarine or marine, shallow or deep, aerated or stagnant. Our goal in the remainder of this chapter is to describe a few muddy depositional settings and interpret the mineral composition of the muds in terms of source area, climate, and water circulation.

Santa Clara River

There have been few studies of the total mineral composition of river muds, although there have been numerous investigations of clay mineral compositions in various fluvial environments. One exception is a study of soil and river mud in the Santa Clara drainage basin, located about 80 km northwest of Los Angeles. In order to establish a genetic sequence of mineralogic changes, samples were collected from the bedrock, which is approximately 75% sedimentary and 25% igneous plus metamorphic; from the *in situ* soil; from the riverbed; from suspended stream sediment; and from continental shelf sediment in the area where the river empties into the Pacific Ocean.

The drainage basin is 105 km in length from west to east and 60 km in maximum width; basin area is about 4,100 km^2. The stream gradient is rather steep in the mountainous headwaters, averaging 30 m/km, but is only 3 m/km on the coastal plain. The steepest tributaries have gradients of 60 m/km. Annual rainfall averages 60 cm in the headwaters and 40 cm on the coastal plain. Mean annual temperature is 10°C.

In the Santa Clara drainage basin the mud is derived almost entirely from disaggregation of the poorly developed soils and the sedimentary rocks. Less than 10% is obtained directly from the igneous and metamorphic rocks in the basin. In mud coarser than about 15 μm, the mineral composition clearly reflects the source rock from which it came. Rocks exposed in the upper 20 km of the basin are nearly all igneous and metamorphic, about two-thirds granite plus gneiss and about one-third anorthosite; small outcrops of schist also occur in this area. The ratio of plagioclase to quartz reflects these facts (see Figure 12–12). It is consistently greater than 3:1 in the

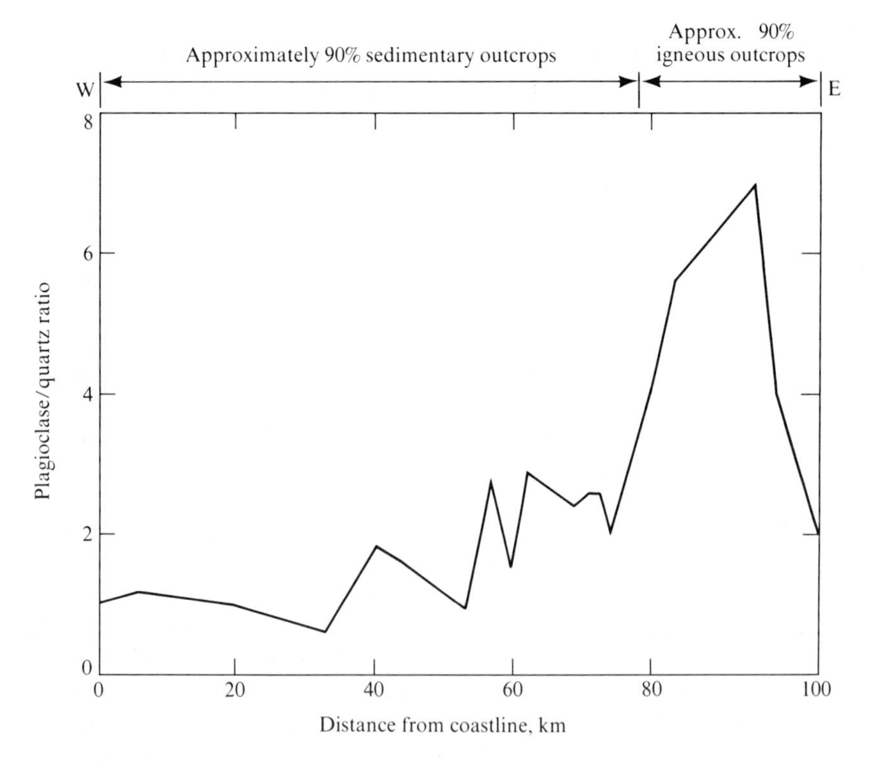

Figure 12–12
Relationship between distance of transport and the ratio of plagioclase to quartz in the silt fraction of sediment in the Santa Clara River drainage basin.

upper 20 km and reaches a value of 7:1 in the heart of the crystalline rock area. Farther downstream, where outcrops are largely friable Tertiary sandstones and mudrocks, dilution rapidly reduces the plagioclase/quartz ratio to 1:1. Abrasion during transport may be a factor in the loss of plagioclase feldspar downstream, but the nearly instantaneous decrease in the ratio as the stream leaves the area of crystalline rocks suggests that the change in source is responsible rather than abrasion. Also, silt grains travel in suspension rather than by traction, and few shattering impacts with other grains are possible. Without impacts during transport there can be little mechanical reduction in grain size.

In the accessory mineral suite of the 62–15 μm size fraction, blue-green hornblende, actinolite, and sphene occur, from the metamorphic rocks; and rounded zircon, apatite, and epidote, from the older sedimentary rocks. Ilmenite is the only distinctive mineral from the igneous outcrops, particularly from the anorthosite.

In the size fraction finer than about 15 μm clay minerals begin to dominate the detrital mineral fraction. These minerals occur most commonly in aggregates rather than as separate clay flakes. Probably the aggregated structure was produced in the soil during weathering, but it is possible that part of the aggregation (flocculation) was generated during stream transport. As would be expected from the wide variety of source rocks in the drainage basin, the aggregates consist of montmorillonite, illite, kaolinite, and mixed layer montmorillonite–illite; that is, all varieties of clay minerals are represented.

Eastern Mediterranean Sea

The Mediterranean Sea is a nearly landlocked basin enclosed on the north and east by Europe and the Middle East, on the south by Africa, and on the west by the juncture of Africa and Europe at the Strait of Gibraltar. The floor of the basin, where not covered by sediment, is oceanic basalt. The Mediterranean is an oceanic basin that formed in late Tertiary time immediately south of the former position of the Tethys seaway. Most of the Tethys was destroyed at the close of the Mesozoic Era in the plate collisions that gave birth to the Alps, Himalayas, and other mountain ranges that stretch in a west–east band across southern Europe, Turkey, Iran, Afghanistan, northern India, and southwestern China.

Although the Mediterranean Sea is nearly landlocked, its location on the northern edge of the low-latitude arid belt prevents it from becoming stagnant like the neighboring Black Sea (see below). Evaporation at the surface of the Mediterranean Sea causes the oxygen-rich surface water to become more saline, denser, and to sink. The dense water flows westward out of the basin along the sea floor through the Strait at Gibraltar, which has a sill depth of 320 m. Less saline Atlantic Ocean water flows in along the surface above the dense Mediterranean water to replenish the basin water mass. Fresh, light water also is supplied from southern Europe and the shallower, aerated part of the Black Sea, through the Bosporus and Dardanelles.

The bathymetry of the Mediterranean area divides the basin into a western and an eastern section. The eastern section (see Figure 12–13) has a maximum water depth greater than 4,000 m and an average depth of about 2,000 m. The distribution of bottom sediments, sands, and muds depends on bottom morphology. In areas with broad shelves, sands predominate; where the shelves are narrow, so are the belts of sand accumulation, and muds occur not far from the shoreline. The distribution of sediment grain sizes is also affected by climatic conditions. In the relatively humid zone of the eastern Mediterranean (Lebanon, Israel) mud production is high and sands are restricted to depths shallower than 50–100 m; muds cover most of the continental slope. In the more arid zone west of the Nile Delta sands are prominent to outer shelf depths of 100–250 m before being overwhelmed by muds. These observations are consistent with the facts that chemical weathering produces mud and that chemical weathering is minimal in arid regions.

The Nile River has been the dominant supplier of sediment to the eastern Mediterranean basin. (This ceased to be true when the Aswan High Dam was completed in 1970, 900 km upstream from the mouth of the Nile.) Although much of the sandy material is derived from the Arabian Desert, the muds originated largely in the humid-tropical upper reaches of the Nile, where chemical alteration of Cenozoic basaltic volcanic rocks resulted in a steady stream of montmorillonitic and kaolinitic clays. As would be expected, the Nile muds are markedly enriched in iron (7.8%) and titanium (1.4%). Many small globules of ferric hydrate and acicular rutile are present. These distinctive mineralogic characteristics enable the Nile River muds to be traced as they enter the eastern Mediterranean basin and are distributed by the currents.

Circulation of surface water in the eastern basin is counterclockwise, and the effect of this on the muds that poured from the Nile Delta is clearly shown by the distribution of montmorillonite (see Figure 12–14). It dominates the clay mineral suite along the entire eastern Mediterranean coastline, from the Nile to southwestern Turkey.

Chlorite and illite, on the other hand, are generated in soils of high-latitude areas with less intense conditions of chemical weathering. Such conditions occur on the northern side of the eastern Mediterranean basin and are clearly reflected in the marine muds west of the Turkish coastline, in the waters off the coast of southern Greece and Italy (see Figure 12–15). The control of clay mineral assemblages by source rock, climate in the source area, and current pattern in the marine depositional basin are exceptionally clear in the eastern Mediterranean. The same controls are almost certainly present in the nonclay mineral fraction of the muds, but this fraction has not been studied.

Black Sea

The best-studied example of a modern depositional basin dominated by reducing conditions is the Black Sea, which was formerly a part of the Mesozoic Tethys

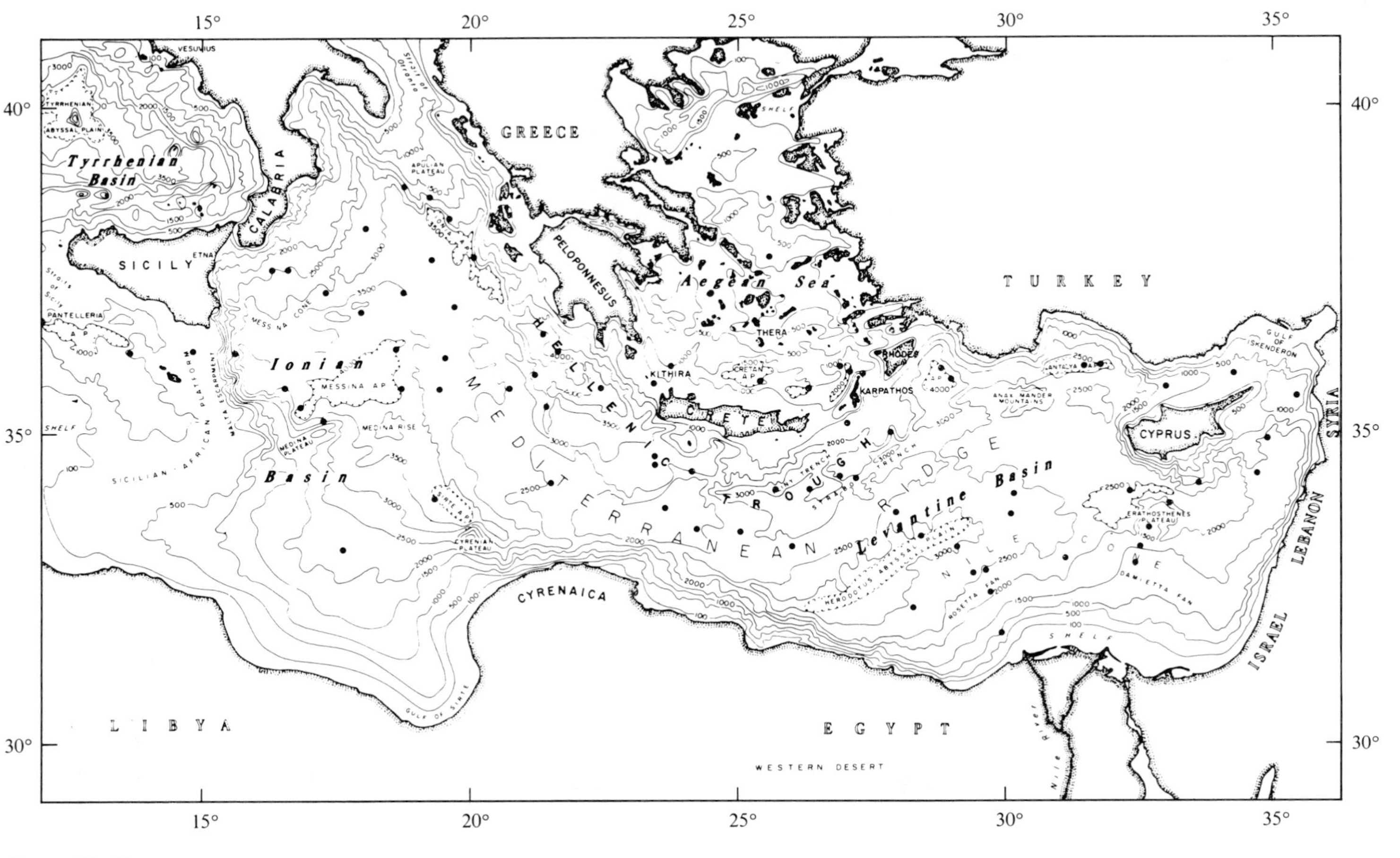

Figure 12–13
Bathymetry in the eastern Mediterranean basin. Contours in meters. The dots are data points for Figure 12–14. [From K. Venkatarathnam and W. B. F. Ryan, 1971, *Marine Geol., 11,* Fig. 1.]

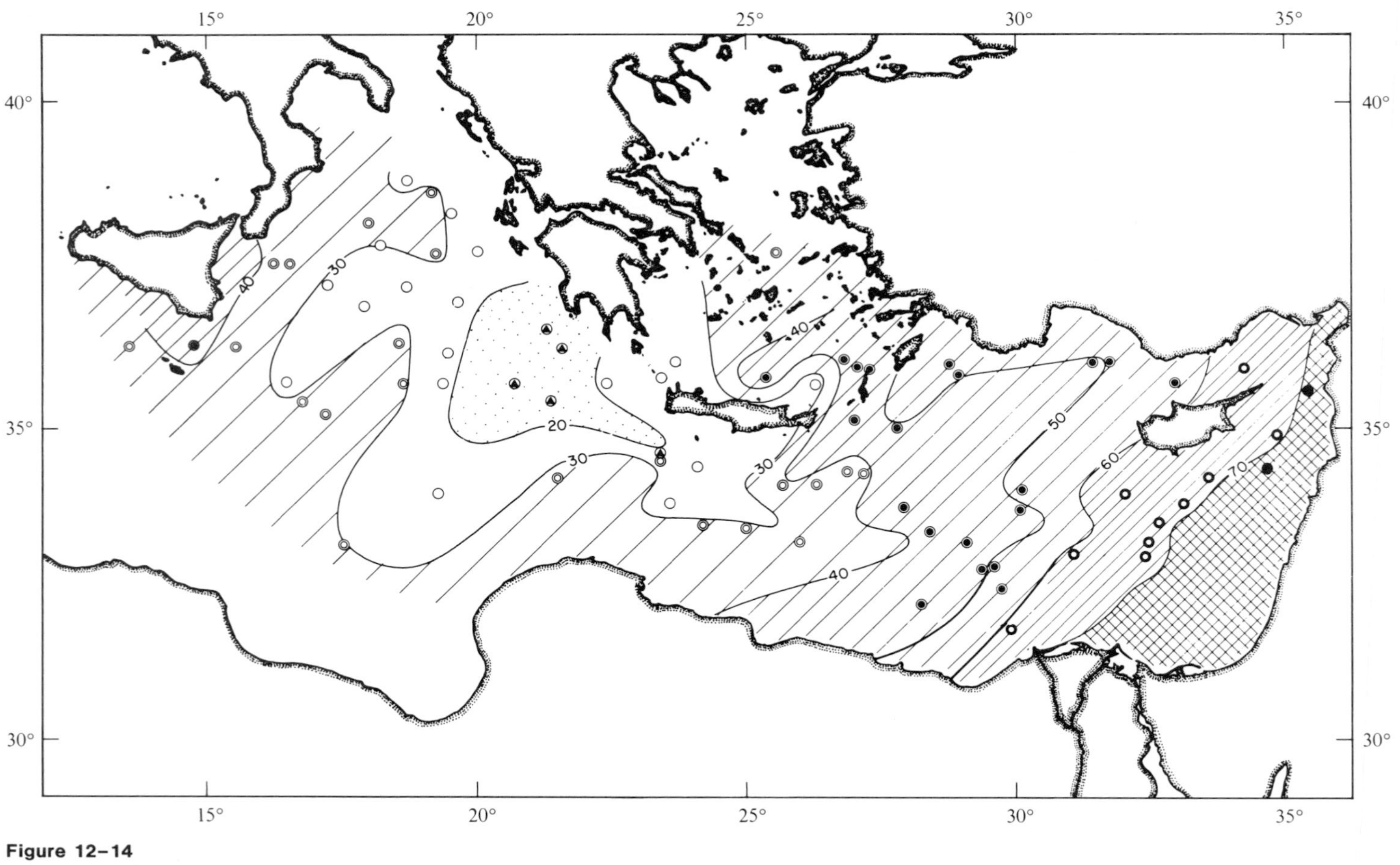

Figure 12–14
Percentage of montmorillonite in the < 2 μm fraction of the clay mineral assemblage of the eastern Mediterranean basin. The circles are sampling sites. [From K. Venkatarathnam and W. B. F. Ryan, 1971, *Marine Geol., 11,* Fig. 3.]

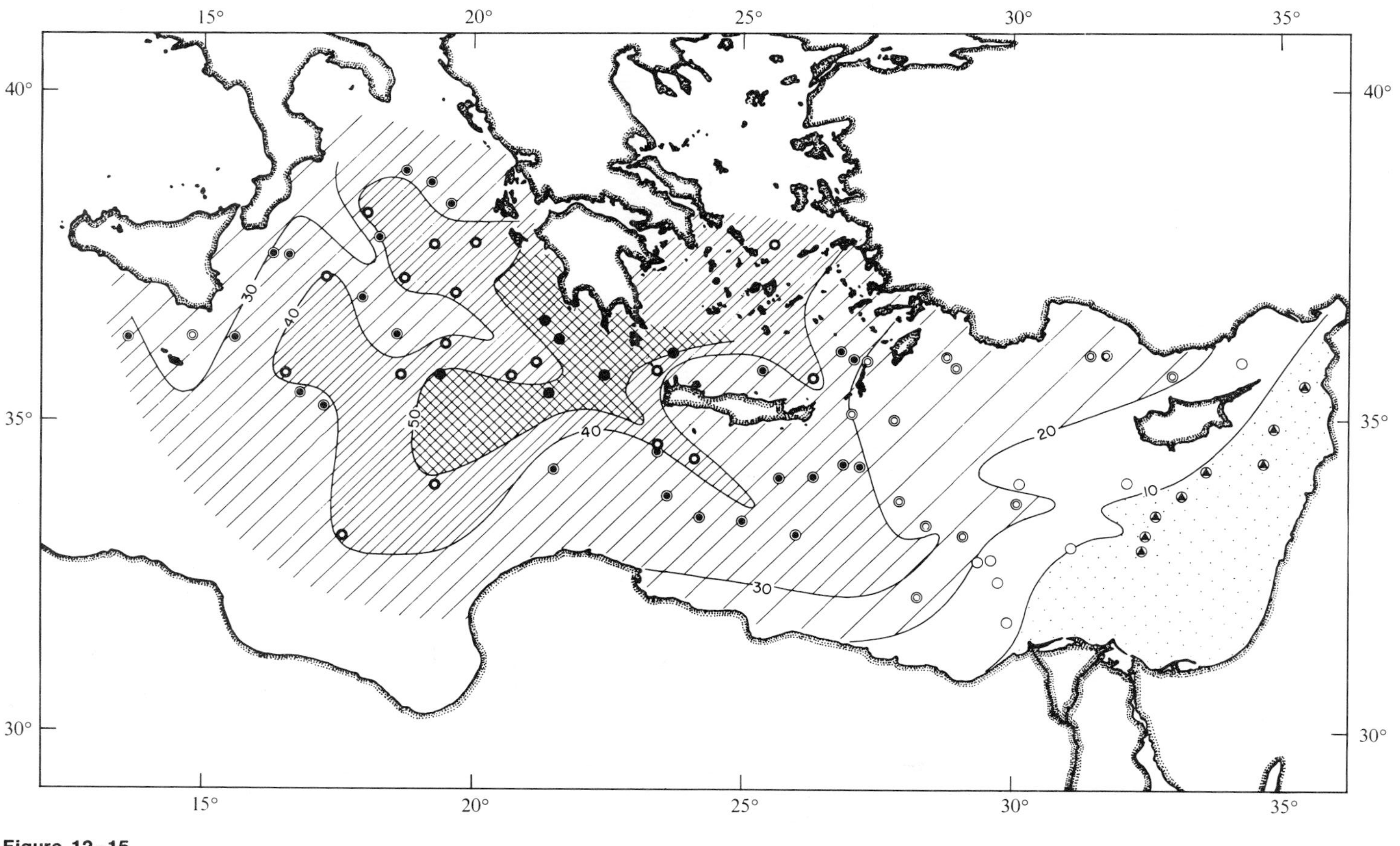

Figure 12–15
Percentage of illite in the $< 2\ \mu m$ fraction of the clay mineral assemblage of the eastern Mediterranean basin. The circles are sampling sites. [From K. Venkatarathnam and W. B. F. Ryan, 1971, *Marine Geol.*, *11*, Fig. 5.]

seaway. Drilling in the Black Sea has not penetrated the basement but, based on regional geology and seismic data, the floor underlying the basin is believed to be formed of metamorphosed sedimentary rocks. Sediments above the basement are 8–14 km in thickness and range from Cretaceous to Holocene in age.

Submarine slopes are quite steep in most nearshore areas, normally 4–6° and in places 12–14° (see Figure 12–16). Gentle slopes exist only along the Ukranian and Roumanian coasts on the northwest side of the basin. Average water depth is 1,200 m, but depths as great as 2,260 m are known.

The water mass in the Black Sea is dominated by reducing conditions that resulted from the isolation of its water from the Mediterranean Sea about 7,000 years ago. At that time, circulation through the Bosporus Strait at the southwest end of the Black Sea became very restricted as the undersea Bosporus ridge rose tectonically to within 40 m of the water surface. The ridge prevented renewal of Black Sea bottom water by oxygen-rich Mediterranean Sea water. Below a depth of 50–70 m in the Black Sea, the oxygen content of the water decreases steadily, and below 200 m there is no measurable amount of oxygen dissolved in the water. From 200 m downward no life can exist (see Figure 12–17).

The deep water sediments deposited during the last 25,000 years in the Black Sea are distinguished by three sedimentary units (see Figure 12–18). The upper unit is a microlaminated layer about 30 cm thick consisting mainly of the remains of the coccolithophore *Emiliania huxleyi.* This alga lives in the upper, aerated part of the water mass. The microlaminations consist of alternating light layers that contain at least 40% white coccolith remains, and darker layers that contain fewer coccoliths and significantly more black organic matter. Each lamina is only 0.1–0.2 mm thick and required 30–60 years to accumulate, based on radiocarbon dating. The base of this unit has been dated at 3,000 years before the present.

Underlying the upper unit with sharp discontinuity is an organic-rich mud about 40 cm thick. It is vaguely laminated by light-colored skeleta of coccoliths or dinoflagellates and dark organic matter that is so well preserved that details of the plant cellular structures can be studied using the electron microscope. The organic matter in this unit is sometimes as much as 50% of the dry sediment weight. The 40 cm of sediment in this unit accumulated during the period 7,000–3,000 years before the present. The accumulation rate of 40 cm/4,000 years is the same as that of the upper unit, whose formation required 3,000 years for 30 cm of sediment.

Underlying the algal remains with sharp discontinuity is a unit of uncertain thickness whose base is thought to be about 25,000 years old. It is microlaminated as are the two upper units but contains less than 1% organic carbon. The dark laminae are formed by concentrations of unstable ferrous sulfides such as mackinawite and greigite that quickly oxidize when the sediment cores are opened. In places, sandy and sometimes graded layers are present in this unit, indicating occasional slumping of sediment from shallow areas on the fringes of the Black Sea basin.

The three sedimentary units can be correlated over most of the Black Sea. Individual laminae, some as thin as 1 mm, can be correlated from one end of the basin to the other, a distance of about 1,000 km. This indicates extremely uniform depositional conditions during the past 25,000 years. Mud forms at least 90% of the inorganic fraction of the units almost everywhere, and the bulk of the mud is clay over at least half the basin floor. Organic carbon is an important component of bottom sediment everywhere, averaging about 2% with values of 4–5% being common (see Figure 12–19). The distribution of the clay mineral species (see Figure 12–20) reflects the source rocks and climate surrounding the basin; illite and chlorite from silicic granites, gneisses, and schists in the western Caucasus and in the Danube River drainage basin; noticeable amounts of kaolinite from sedimentary rocks in the Danube River drainage; and montmorillonite from Eocene volcanic rocks along the northern Turkish coast. Quartz and feldspars are relatively minor constituents of Black Sea sediment because of the extremely fine-grained character of the sediment (see Figure 12–21). As in most modern muds and ancient mudrocks, quartz is considerably more abundant than feldspars, except along the northern Turkish coast where the mafic volcanics that supply the montmorillonite clay also supply abundant feldspar. In this nearshore area it is common to find feldspar more abundant than quartz.

Organic matter has very high adsorptive capacities for trace elements, a fact long recognized by agricultural experts. Because of this, we would anticipate a correlation between the abundances of organic matter and the various minor elements in Black Sea muds. This is indeed the case (see Figure 12–22). Ancient mudrocks are also notable for their rich content of trace elements. In phosphatic mudrocks, for example, the value of the uranium obtained as a byproduct of industrial phosphate processing may exceed the value of the phosphate.

Atlantic Ocean

The largest areally definable marine basins are the major ocean basins—the Pacific, Atlantic, and Indian. During the past 15 years, since the advent of plate tectonics, interest in the sedimentology and petrology of oceanic sediments has increased significantly. Our understanding of both silicate and carbonate materials has improved as a result of the newer investigations, but our concern here will be with the silicate materials, particularly the muds. We will focus our attention on the Atlantic basin, but analogous descriptions can be given as well of sediments in the other oceans.

Clay minerals are the dominant mineral group on the floor of the Atlantic Ocean, and all of the major clay minerals are present: gibbsite, kaolinite, montmorillonite, illite, and chlorite. This is what we would anticipate because of the very large size of the drainage area supplying the sediment and the wide range of rock types being weathered. Further, we would expect to find a close relationship between the climate on the adjacent land surface and the type of clay supplied to the basin. Chlorite, which is most susceptible to destruction, should be most common where chemical

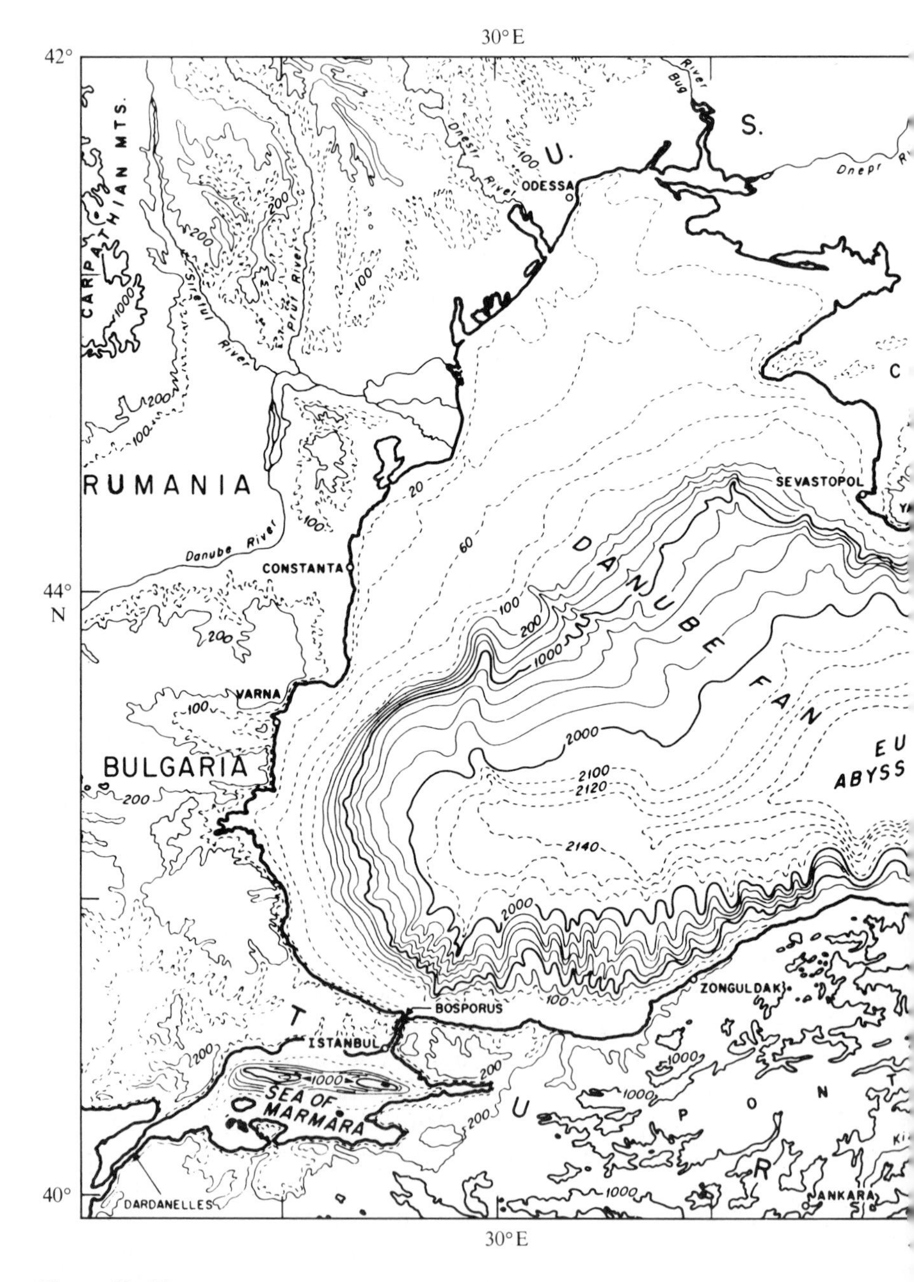

Figure 12–16
Bathymetric chart of the Black Sea. Contours in meters. Note change in contour interval at 200 m and 2,000 m. [From D. A. Ross et al., 1974, Amer. Assoc. Petroleum Geol. Mem. 20, Fig. 2.]

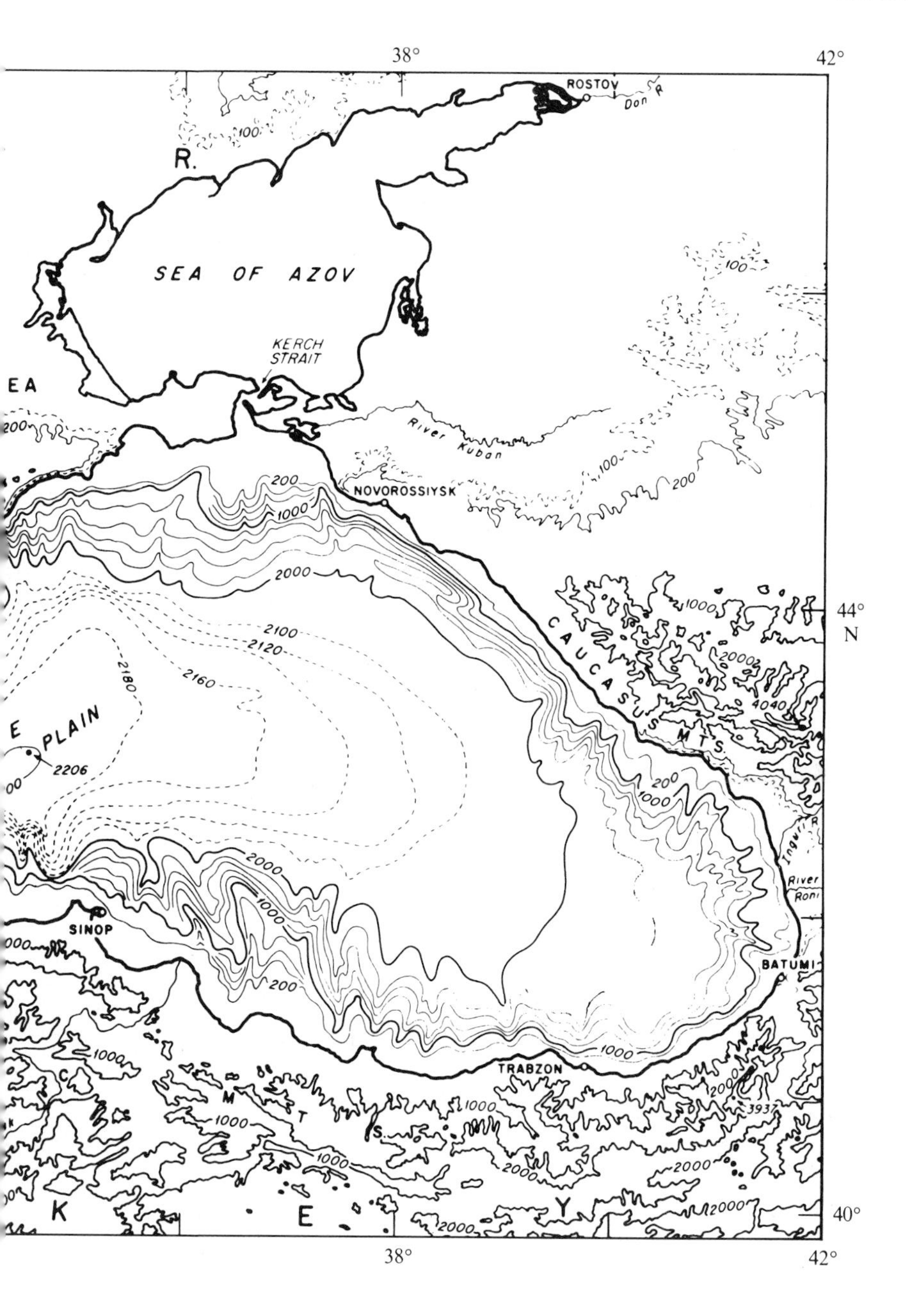
38°
42°
ROSTOV
Don R
SEA OF AZOV
KERCH STRAIT
River Kuban
NOVOROSSIYSK
CAUCASUS MTS.
44° N
PLAIN
2206
2100
2120
2160
2180
2000
1000
200
100
4040
Inguri R
River Rioni
SINOP
BATUMI
TRABZON
3937
40°

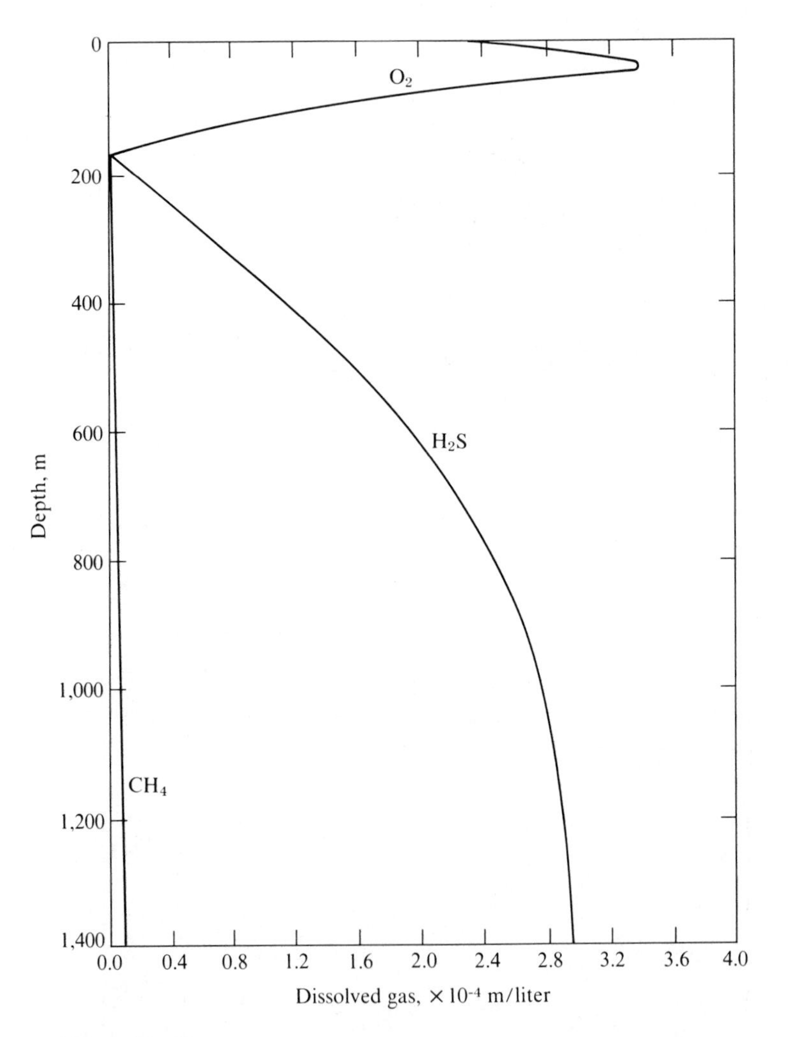

Figure 12–17
Typical profile of dissolved gas concentrations in the Black Sea. The oxygen maximum at 30 m results from the metabolism of phytoplankton.

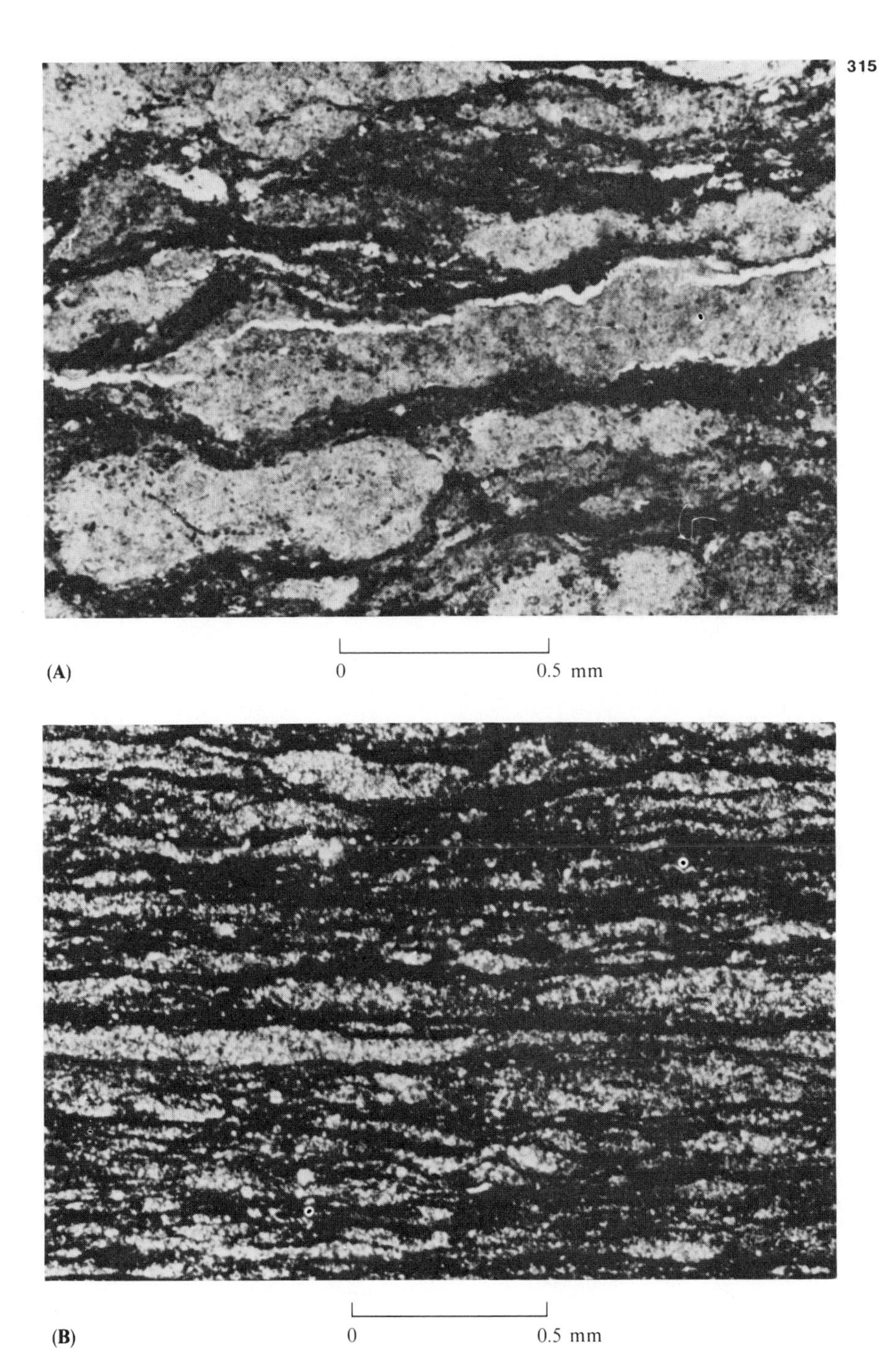

Figure 12–18
Photomicrographs of typical sediments from the Black Sea. (**A**) Upper unit, coccolith mud and interstratified carbonate layers composed of coccoliths and terrigenous material. (**B**) Middle unit, layers rich in organic matter alternating with terrigenous silt layers.

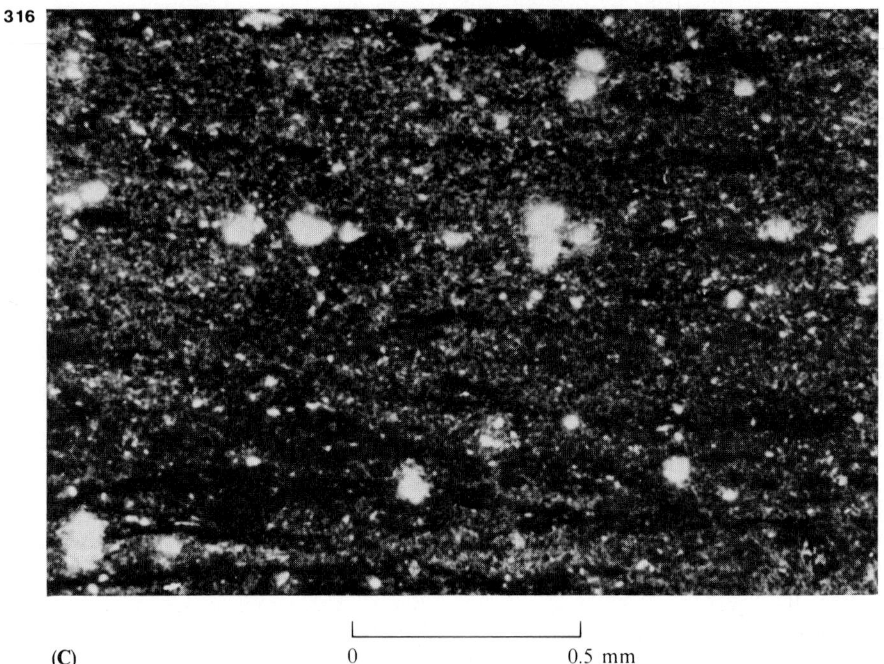

(C)

Figure 12–18 *(continued)*
(**C**) Lower unit, silt with thin layers of organic material. [From G. Müller and P. Stoffers, 1974, Amer. Assoc. Petroleum Geol. Mem. 20, Fig. 5.]

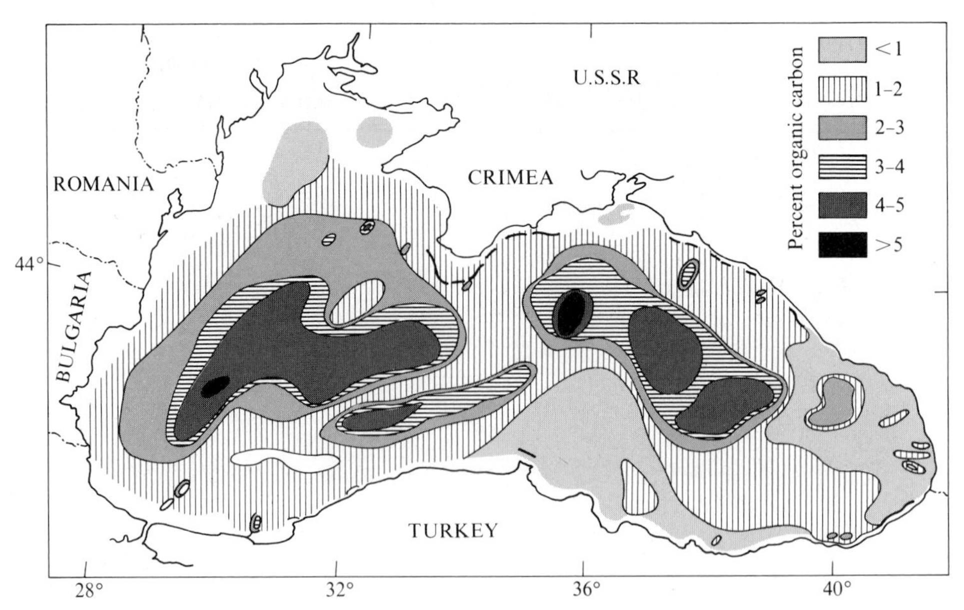

Figure 12–19
Content of organic carbon in modern Black Sea sediments. Heavy black line indicates areas not covered by modern sediments. [From K. M. Shimkus and E. S. Trimonis, 1974, Amer. Assoc. Petroleum Geol. Mem. 20, Fig. 11.]

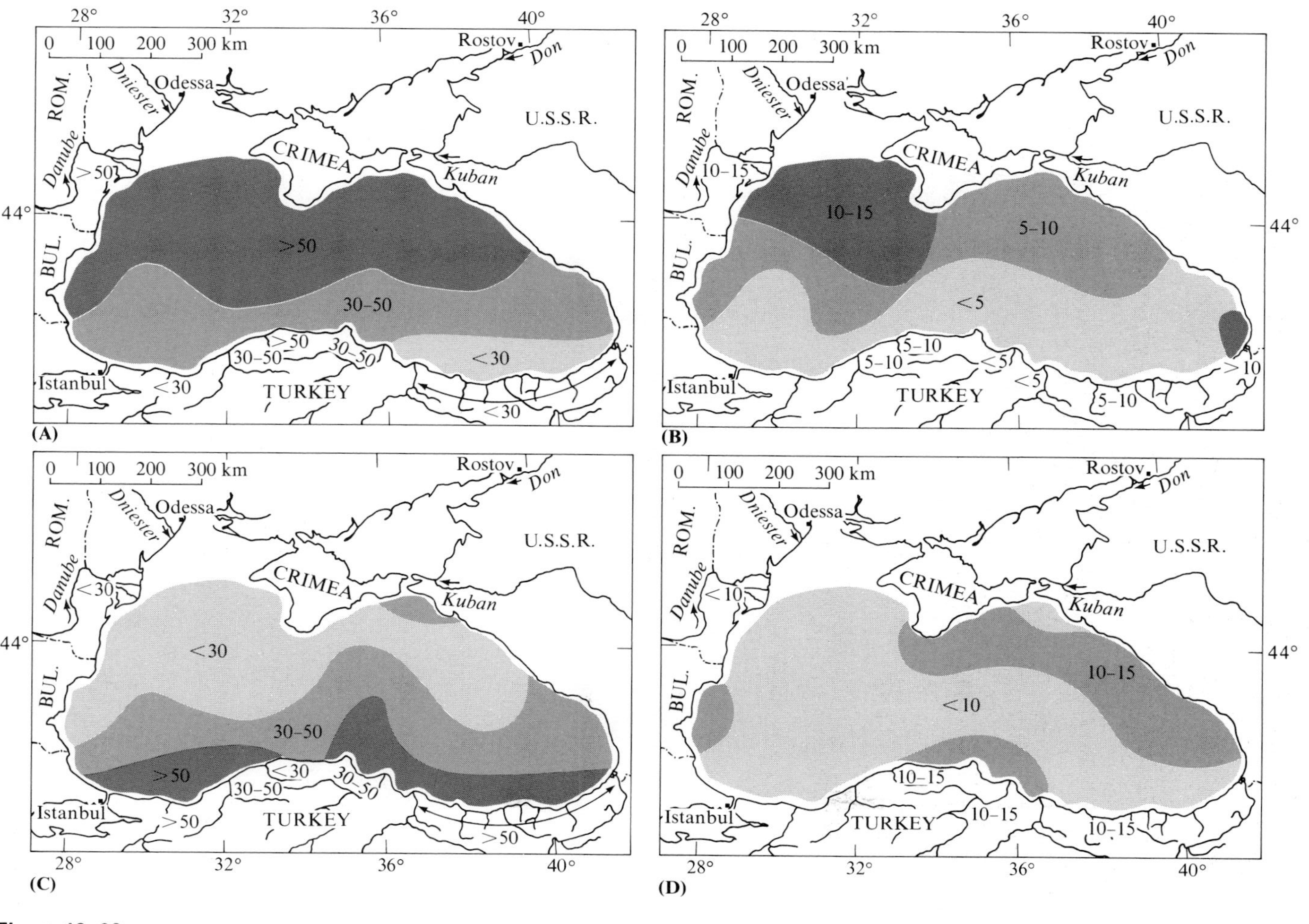

Figure 12–20
Clay-mineral composition of fraction $< 2\ \mu m$, in percent. **(A)** Illite. **(B)** Kaolinite. **(C)** Montmorillonite. **(D)** Chlorite. [From G. Müller and P. Stoffers, 1974, Amer. Assoc. Petroleum Geol. Mem. 20, Fig. 17.]

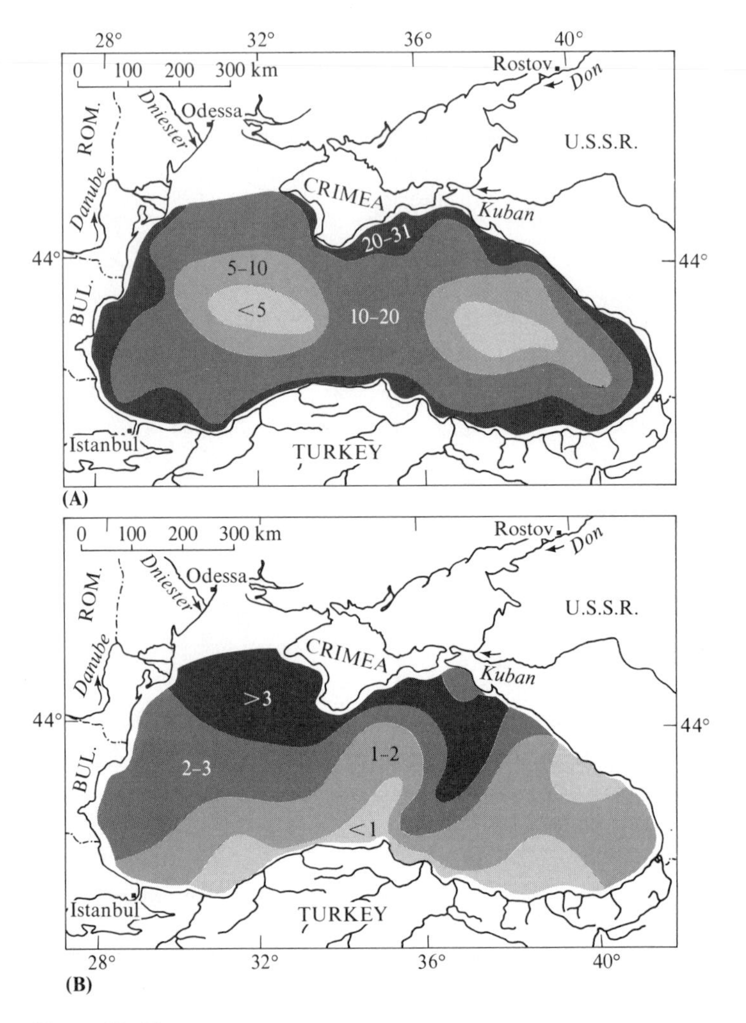

Figure 12–21
Quartz and feldspars in upper unit of Black Sea bottom sediment. (**A**) Percent quartz. (**B**) Quartz/feldspar ratio. [From G. Müller and P. Stoffers, 1974, Amer. Assoc. Petroleum Geol. Mem. 20, Fig. 12.]

weathering is least effective, in polar regions where water is perpetually frozen and therefore chemically inactive. This expectation is fulfilled. In the South Atlantic, for example, the abundance of chlorite among the clay minerals increases from less than 5% in equatorial waters to an average of 10% at lat. 30° S and to 20% at lat. 50° S (see Figure 12–23). The same trend toward increasing amounts of chlorite is present north of the equator.

The reverse gradient in abundance is shown by kaolinite (see Figure 12–24). In low latitudes it forms 25–50% or more of the clay mineral assemblage but decreases to

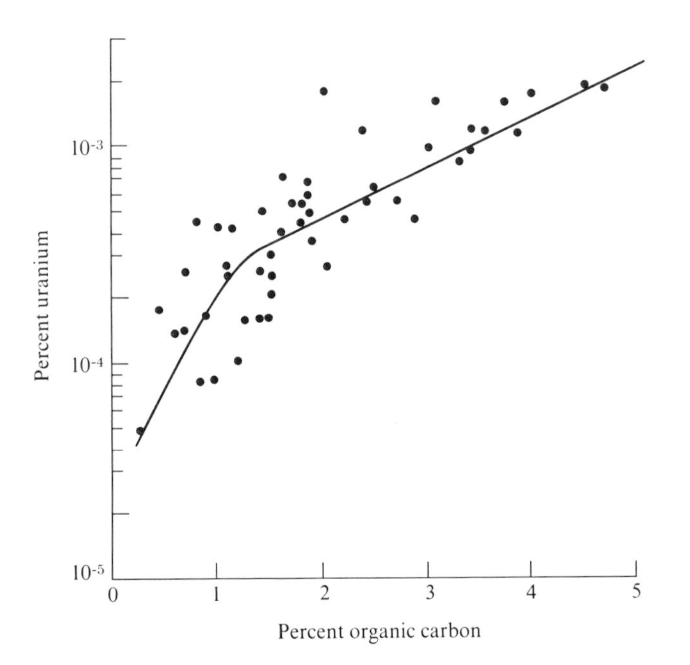

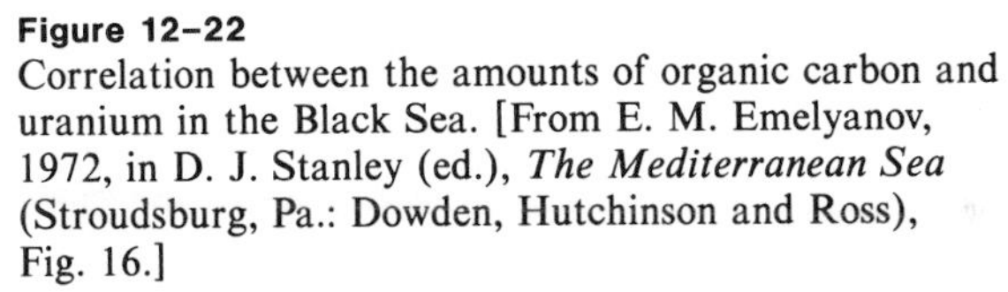
Figure 12–22
Correlation between the amounts of organic carbon and uranium in the Black Sea. [From E. M. Emelyanov, 1972, in D. J. Stanley (ed.), *The Mediterranean Sea* (Stroudsburg, Pa.: Dowden, Hutchinson and Ross), Fig. 16.]

less than 5% at lat. 50° north and south of the equator. An even more intense gradient is shown by gibbsite.

The climatic control of mineral composition shown by clay mineral distributions in the Atlantic Ocean basin is duplicated by the nonclay sediment fraction. An illustration is the amount of the chemically unstable amphibole group in the mud fraction (see Figure 12–25). The amphibole/illite ratio varies systematically from 0.05 near the equator to greater than unity in polar regions, paralleling the behavior of chlorite. Similarly, feldspar is rare in equatorial Atlantic muds but abundant in polar muds.

The clear control of detrital mineral composition in the oceans by the climate on adjacent land surfaces reflects not only the importance of climate but also reflects the unimportance of ocean water as a modifying influence. Ocean water is a very concentrated solution of ions, containing 35,000 ppm of dissolved solids in contrast to only 120 ppm in river water. Because of this difference, ocean water has the chemical potential to convert cation-poor clays such as gibbsite and kaolinite to cation-rich clays such as illite and chlorite. The fact that there is no evidence for these changes in

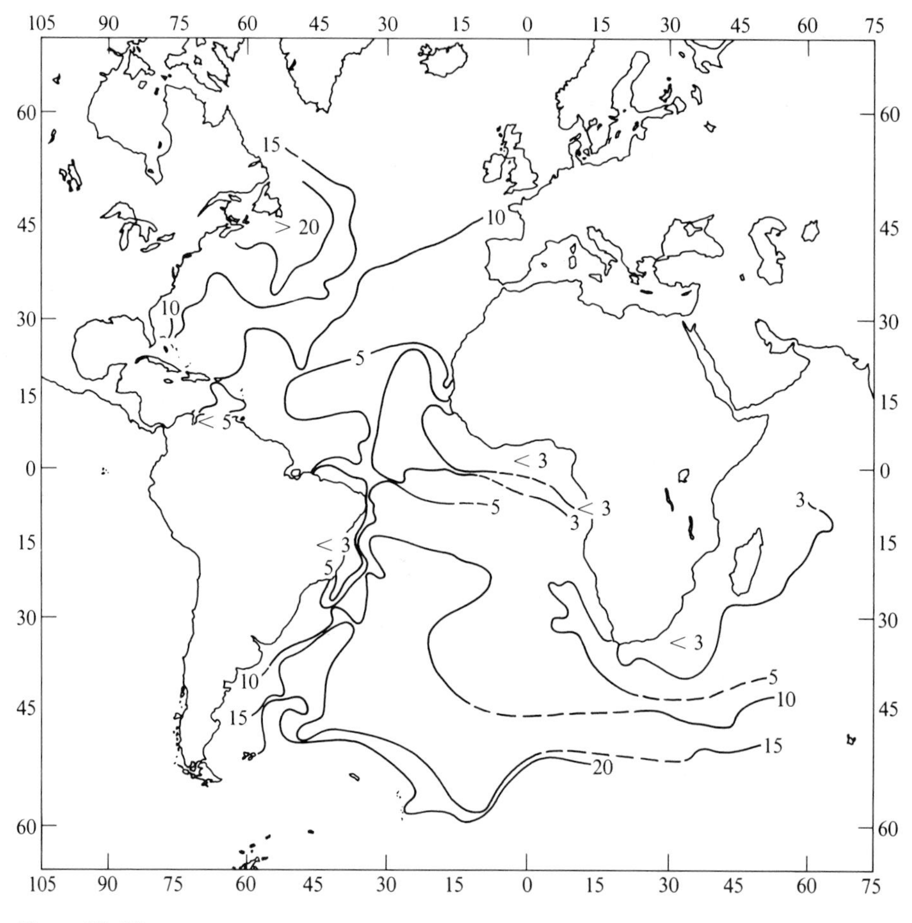

Figure 12–23
Percentage of chlorite in the < 2 μm size fraction of Atlantic Ocean basin sediment. [From P. E. Biscaye, 1965, *Geol. Soc. Amer. Bull., 76,* Fig. 5.]

the pattern of clay distribution in the ocean indicates that the process of change is too slow to be noticeable in time periods of a few million years. Many long cores of sediment have been retrieved from the ocean basins during the past decade of drilling by Joint Oceanographic Institutions for Deep Earth Sampling (JOIDES) cruises. They, too, find no evidence for change due to reaction of detrital clay with ocean water. Apparently, such change requires higher temperatures to be effective. Field evidence suggests that temperatures of at least 50°C are needed. Temperatures at the ocean floor are about 2°C.

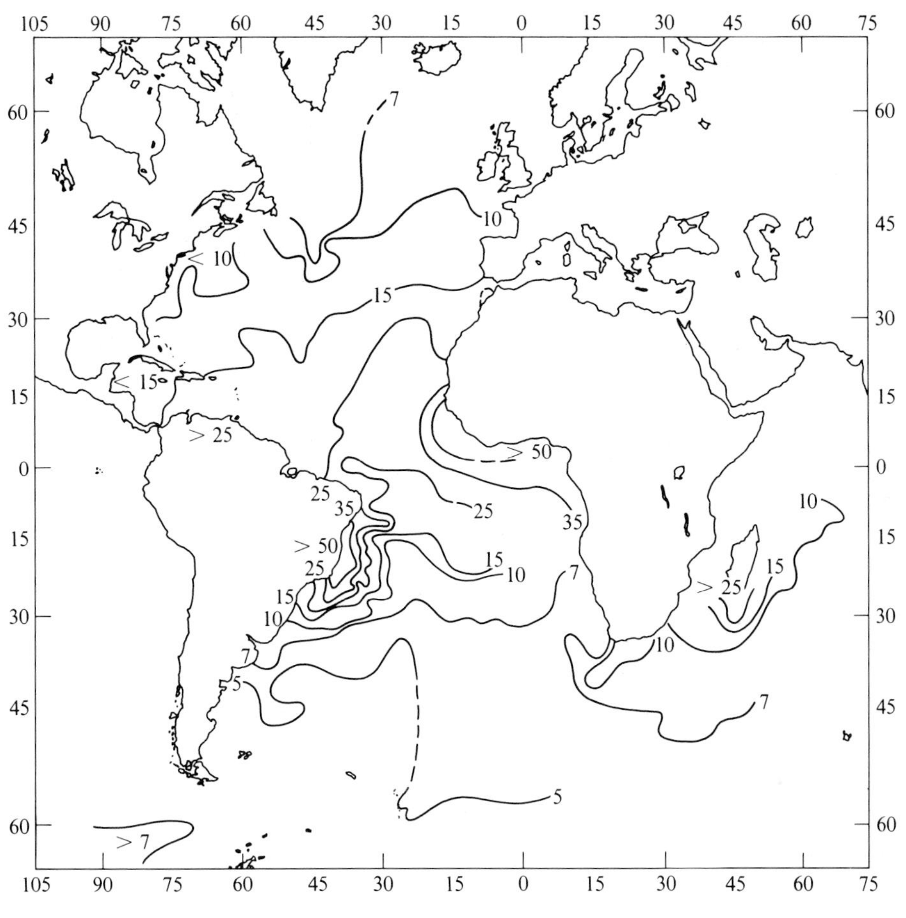

Figure 12–24
Percentage of kaolinite in the $< 2\ \mu m$ size fraction of Atlantic Ocean basin sediment. [From P. E. Biscaye, 1965, *Geol. Soc. Amer. Bull., 76,* Fig. 4.]

SUMMARY

Mudrocks are the most abundant type of sedimentary rock, although their abundance in outcrop may be less than that of sandstones. Mudrocks are not well understood because their aphanitic texture and the difficulty of disaggregating them make them hard to study.

The relative proportions of fissile and nonfissile mudrocks is not known. Fissility can be produced during deposition, but the rapid rate at which burrowing organisms on the sea floor destroy clay flake parallelism suggests that a large proportion of the fissility seen in ancient mudrocks has been produced during diagenesis.

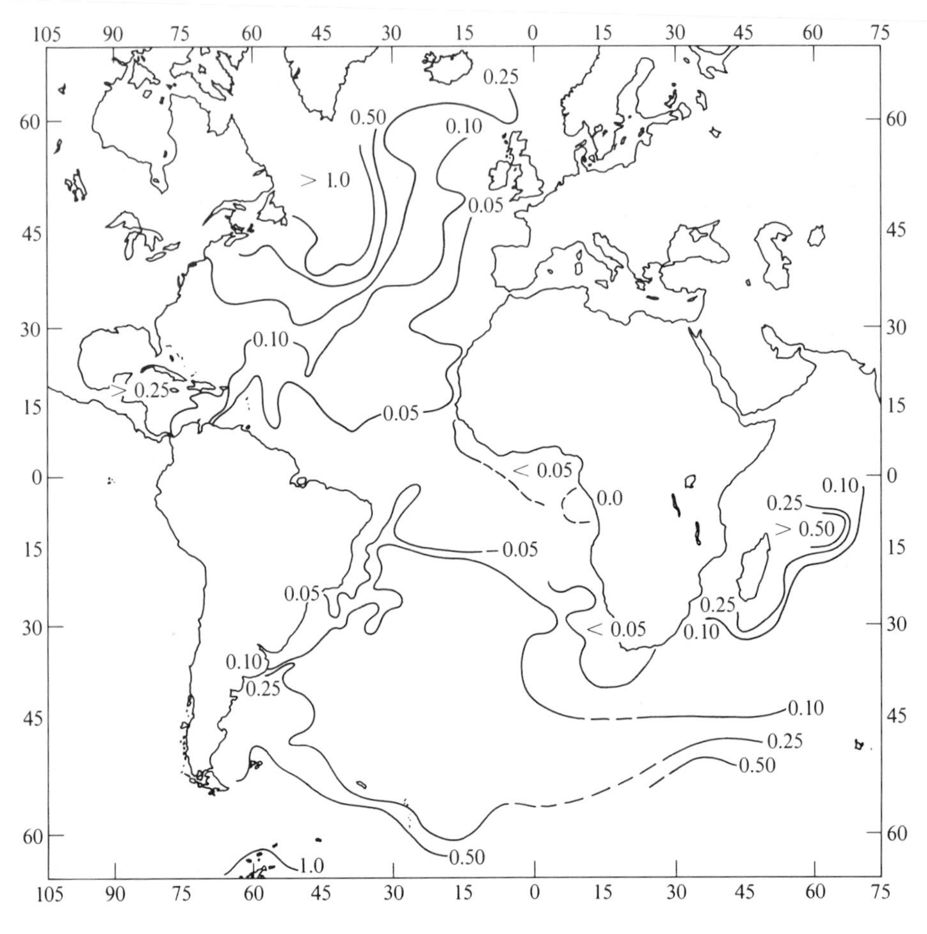

Figure 12–25
Ratio of amphibole/illite in the 2–20 μm size fraction of Atlantic Ocean basin sediment. [From P. E. Biscaye, 1965, *Geol. Soc. Amer. Bull., 76,* Fig. 14.]

Mudrocks contain most of the earth's buried organic matter, so that gray to black mudrocks are common. Black organic matter can be preserved only in an oxygen-deficient environment, and in such an environment advanced forms of life such as burrowing organisms cannot exist. As a result, black mudrocks are more likely to be fissile than red ones, which form and are preserved under oxygenated conditions.

Clay minerals form 60% of muds and mudrocks. Illite forms about 25% of modern clay mineral suites but increases in abundance with increasing age to perhaps 80% in lower Paleozoic mudrocks. Both field observations and geochemical experiments indicate that the cause of the change is the progressive change of montmorillonite and mixed layer montmorillonite/illite clay to pure illite during diagenesis. The reaction is initiated at temperatures as low as 50°C but is not completed until temperatures above 200°C are attained. Detrital silt-size quartz forms most of the remainder of the typical mudrock.

Mudrocks can be used in provenance studies of modern muds, but because the character of the clay mineral suite changes during diagenesis, only the nonclay fraction can be used in provenance studies of ancient mudrocks. Quartz, feldspar, and accessory minerals present in mudrocks can be used for this purpose.

FURTHER READING

Biscaye, P. E. 1965. Mineralogy and sedimentation of recent deep-sea clay in the Atlantic Ocean and adjacent seas and oceans. *Geol. Soc. Amer. Bull., 76,* 803–832.

Blatt, H., and D. J. Schultz. 1976. Size distribution of quartz in mudrocks. *Sedimentology, 23,* 857–866.

Degens, E. T., and D. A. Ross, eds. 1974. The Black Sea—Geology, Chemistry, and Biology. Tulsa, Oklahoma: Amer. Assoc. Petroleum Geol. Memoir No. 20, 633 pp.

Fan, P.-F. 1976. Recent silts in the Santa Clara River drainage basin, southern California: a mineralogical investigation of their origin and evolution. *Jour. Sedimentary Petrology, 46,* 802–812.

McBride, E. F. 1974. Significance of color in red, green, purple, olive, brown, and gray beds of Difunta Group, northeastern Mexico. *Jour. Sedimentary Petrology, 44,* 760–773.

Scotford, D. M. 1965. Petrology of the Cincinnatian Series shales and environmental implications. *Geol. Soc. Amer. Bull., 76,* 193–222.

Shaw, D. B., and C. E. Weaver. 1965. The mineralogical composition of shales. *Jour. Sedimentary Petrology, 35,* 213–222.

Slaughter, M., and J. W. Earley. 1965. Mineralogy and geological significance of the Mowry Bentonites, Wyoming. Geol. Soc. Amer. Spec. Paper 83, 116 pp.

Stanley, D. J., ed. 1972. *The Mediterranean Sea: A Natural Sedimentation Laboratory.* Stroudsbury, Pa.: Dowden, Hutchinson, and Ross, 765 pp.

Tourtelot, H. A. 1979. Black shale—its deposition and diagenesis. *Clays and Clay Minerals, 27,* 313–321.

Weiss, M. P., W. R. Edwards, C. E. Norman, and E. R. Sharp. 1965. *The American Upper Ordovician standard. VII. Stratigraphy and petrology of the Cynthiana and Eden Formations of the Ohio Valley.* Geol. Soc. Amer. Spec. Paper 81, 76 pp.

13

Sandstones and Conglomerates

Conglomerates and sandstones form 20–25% of the stratigraphic column, and the grains in them are coarse enough to be seen easily, identified mineralogically, and described texturally using a 10× hand lens. In addition, sandstone beds commonly display obvious internal structures such as cross bedding, ripple marks, or size grading that can be diagnostic of a specific environment of deposition. Furthermore, they supply about half of the world's production of petroleum and natural gas. As a result, sandstones have been studied intensively for more than a hundred years. The proportion of books and articles in professional journals that deal with sandstones is far greater than the 20–25% we might anticipate.

FIELD OBSERVATIONS

Textures

A wide variety of textural and structural features can be observed and described from outcrops of sandstones. Perhaps the most fundamental is grain size (see Table 13–1), both the average size and the range of sizes in the rock. It is difficult, at least initially, to calibrate your eyes precisely enough to distinguish among 0.5 mm, 0.25 mm, and 0.125 mm. Therefore, it is useful to make a simple size comparator for field use; two pieces of thin cardboard, a hole puncher, transparent tape, and some laboratory-sized sand grains are all that is needed (see Figure 13–1). With this aid it is not difficult to estimate average grain size and the spread of sizes in the rock. The spread of sizes is normally given as the range (in ϕ units) that includes approximately two-thirds of all the grains. This range is twice the *standard deviation.* For example, if the mean size is 2.0 ϕ and two-thirds of the grains have sizes between 1.5 ϕ and 2.5 ϕ, then the

Table 13–1
The Udden–Wentworth Grain Size Scale for Clastic Sediments[a]

	Name	Millimeters	Micrometers	ϕ
		4,096		−12
GRAVEL	Boulder			
		256		−8
	Cobble			
		64		−6
	Pebble			
		4		−2
	Granule			
		2	—	−1
SAND	Very coarse sand			
		1		0
	Coarse sand			
		0.5	500	1
	Medium sand			
		0.25	250	2
	Fine sand			
		0.125	125	3
	Very fine sand			
		0.062	62	4
MUD	Coarse silt			
		0.031	31	5
	Medium silt			
		0.016	16	6
	Fine silt			
		0.008	8	7
	Very fine silt			
		0.004	4	8
	Clay			
		↓	↓	↓

[a]As devised by J. A. Udden (1898) and C. K. Wentworth (1924). The ϕ scale (Krumbein, 1934) was devised to facilitate statistical manipulation of grain-size data and is commonly used. $\phi = -\log_2$ mm.

standard deviation is 0.5 ϕ. Standard deviation is the accepted measure of the *sorting* of the sediment (see Figure 13–2), and sorting values tend to differ among different sedimentary environments. We will return to this point later in the chapter.

After the average grain size and variety of sizes have been estimated, the sediment must be named to make communication easier. As we noted above, there is general agreement concerning a scale of grain size and the degree of sorting. There is, unfortunately, no agreement among geologists concerning a method of naming mixtures of different sizes. For example, analysis of typical usage reveals that field geologists are prone to call a deposit conglomerate even if gravel forms less than 50% of the rock. The presence of only 10% or 20% pebbles or cobbles creates a dominating impression

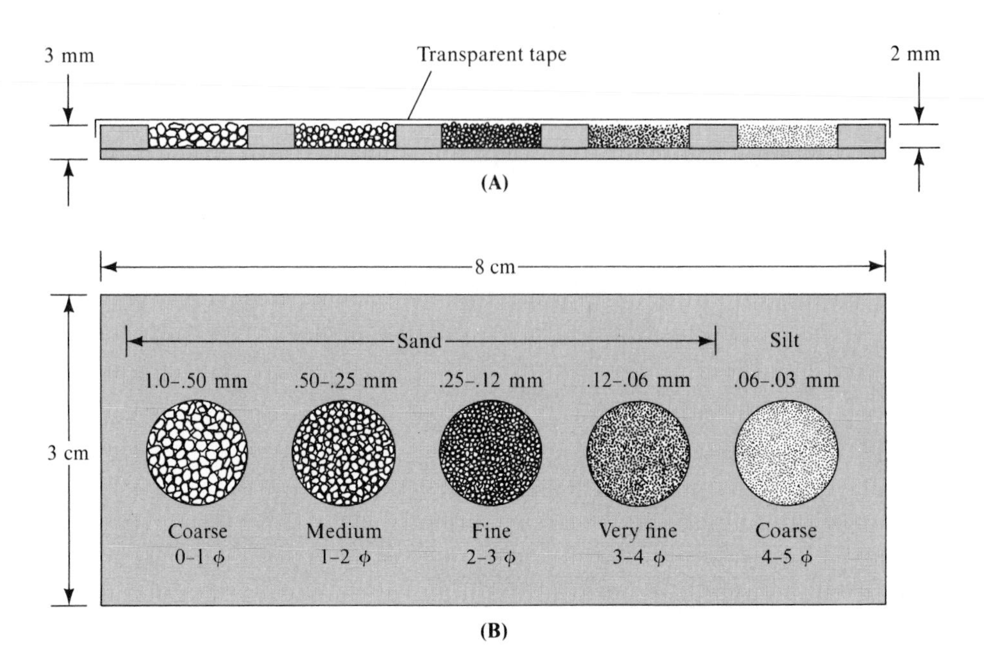

Figure 13–1
Sketch showing the construction of a simple grain size comparator for field use. (**A**) Side view. (**B**) Top view.

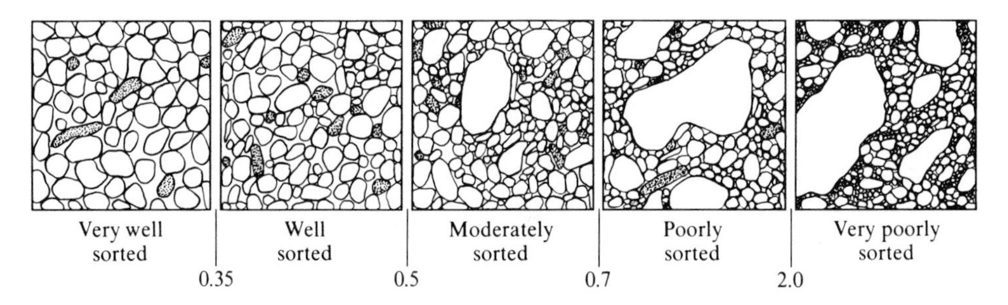

Figure 13–2
Classification of degrees of sorting as seen through a square hand lens. Silt- and clay-size sediment are indicated by the fine stipple. The values of standard deviation that divide each class of sorting are also shown. [From R. R. Compton, 1962, *Manual of Field Geology* (New York: John Wiley and Sons), Fig. 12–1.]

when observing an outcrop. Nevertheless, it clearly is desirable to have a uniform usage. One system in common use is shown in Figure 13–3.

The *shapes of grains* in sandstones and conglomerates vary widely from those shaped like spheres to those that approximate a disc or a rod. Two aspects of grain shape can be seen either in outcrop or in a hand specimen: *sphericity* and *roundness* (lack of sharp corners). Sphericity is defined by the relative equidimensionality of

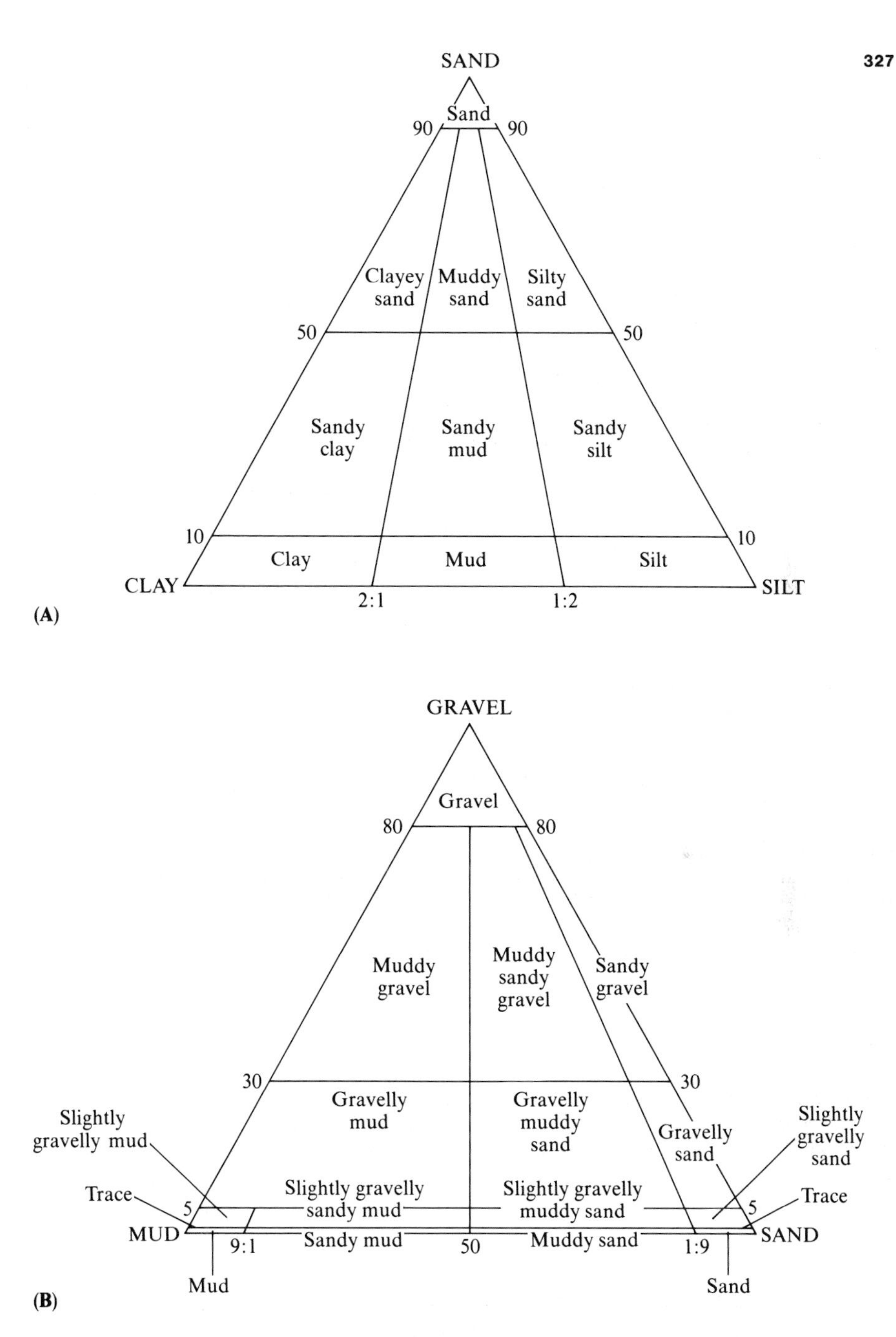

Figure 13–3
Triangular classification of grain sizes in detrital rocks. If no gravel is present, triangle A is used; if gravel is present, triangle B. Note the emphasis given to even a trace amount of gravel. [From R. L. Folk, 1954, *Jour. Geology, 62,* Fig. 1.]

three mutually perpendicular axes through the grain. If the axes are approximately equal the grain is spherical; if two of the axes are noticeably longer than the third the grain is disc-shaped or platy; if two axes are noticeably shorter than the third the grain is rod-shaped or elongate. Pebbles formed of relatively mechanically isotropic material such as quartz tend to be either rods (in rivers) or discs (on beaches), although shape overlap is common between the two environments. Strongly nonisotropic pebbles have a sphericity controlled largely by the structure of the fragment. For example, phyllite fragments, because of their foliation, will be platy irrespective of either their mechanism of transport or environment of deposition.

Grain *roundness* is determined by grain size, hardness, and environments of transportation and deposition (see Figure 13–4). Coarser grains tend to be rounder because they have more impacts with other grains and impact with more inertia during their sedimentary transport. If they travel in a stream or in the sea, larger grains spend more time being abraded during rolling along the bottom, while silt-size grains are being rapidly transported in suspension. If they are wind blown, larger grains have their sharp corners chipped off by wind-driven impacts with other grains. Silt grains are kept suspended by desert winds and settle only when they leave the desert area and enter vegetated areas where winds are less intense. Such wind-deposited silts are known as *loess*.

Cobbles and pebbles are transported almost entirely by rolling or sliding, either in the upper reaches of rivers, in braided streams in lowland areas near the base of a highland area, or on beaches. For this reason they are nearly always well rounded. It is so uncommon to find angular gravel in an ancient rock that a rock containing an abundance of such fragments is given a special name, ***breccia***. Breccias can occur at

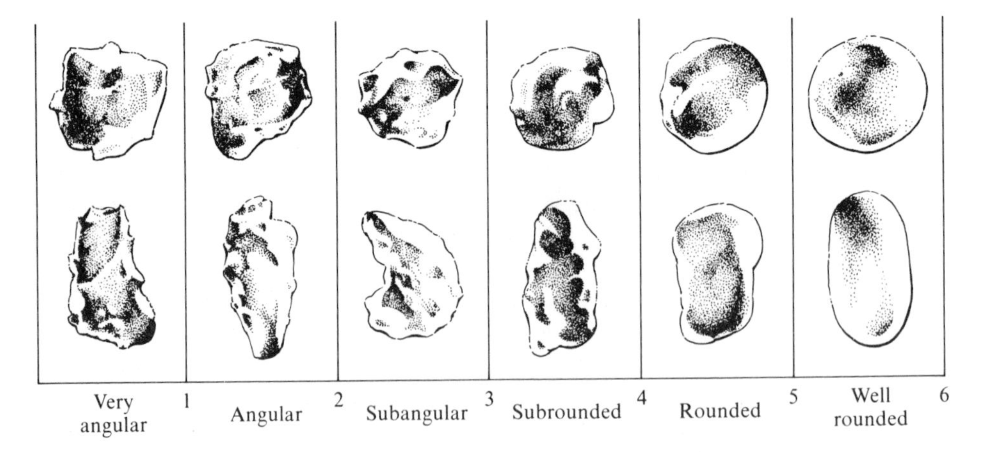

Figure 13–4
Terminology for degree of rounding of detrital grains using a hand lens. The numbers assigned to each roundness class permit calculation of mean roundness and standard deviation. [After M. C. Powers, 1953, *Jour. Sed. Petrology, 23,* Fig. 1.]

the base of a source rock (talus breccia); in a glacial deposit where the gravel was transported without abrasion by being frozen in a mass of ice; in a collapse structure such as a sinkhole; or as a result of a major structural disturbance (tectonic breccia, fault breccia).

Color

The *color* of a rock is an obvious characteristic that should be described in the field because it can sometimes be used effectively for tracing sedimentary units. Sandstones and conglomerates occur in a variety of colors, ranging from uncolored to black. Several charts for describing rock colors are available; the two most widely used are those published by the Geological Society of America and by the Munsell Color Company. The Munsell chart was designed for use with soils but is equally usable with rocks. Both charts consist of coded "paint chips" about 15 mm square mounted on heavy paper and covering the range of natural colors.

The common colors of rocks are red, brown, and yellow, which result typically from the presence of ferric oxide cement (hematite); and gray to black, which reflects the presence of free carbon (organic matter). Colorless rocks, such as quartz-cemented quartz sandstones, contain neither ferric oxide nor free carbon. Local variations in the color of a rock can result from weathering phenomena; for example, a bit of organic matter in a sandstone can react chemically with surrounding hematite cement to reduce the ferric iron to the ferrous state so that the red color vanishes. Conversely, weathering at the outcrop of a colorless rock can release ferrous iron from minerals such as hornblende and biotite, oxidize the iron, and produce hematite.

Distance from Source

Many studies have been made of the effects of topography, climate, and other variables on the generation of detrital sediment. It is clear from these studies that relief is the most important factor; rate of erosion increases very rapidly with increasing relief (see Figure 13–5). Thus, most detrital particles originate in highland areas such as the Alps, North American Rocky Mountains, or Himalayas.

Because of the great relief, stream gradients are relatively high in mountainous regions, the ability of the stream to transport large particles (*stream competence*) is correspondingly high, and coarse particles can be moved downstream. As the stream leaves the mountains and the stream valley widens, however, competence decreases and coarse particles can no longer be moved. Many studies of conglomerates have used this concept to infer distance of transport from a mountain front (see Figure 13–6). It is also possible to apply it to sandstones, although as grain size decreases it becomes easier for local variations in stream character to obscure the larger-scale trend toward size decrease in the downstream direction; that is, the scatter of the data points increases.

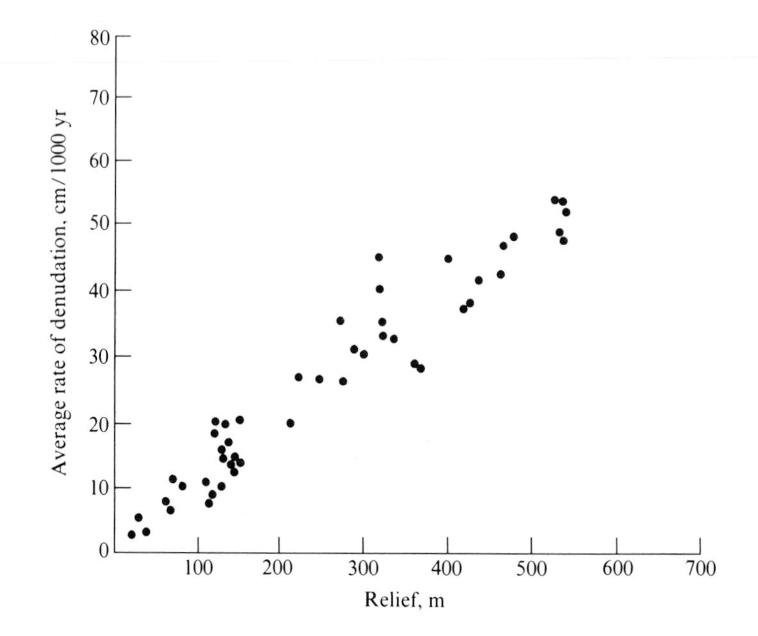

Figure 13–5
Average rate of denudation as a function of relief on an andesitic strato-volcano of Holocene age in northeastern Papua. [From B. P. Ruxton and I. Mcdougall, 1967, *Amer. Jour. Sci., 265,* Fig. 5A.]

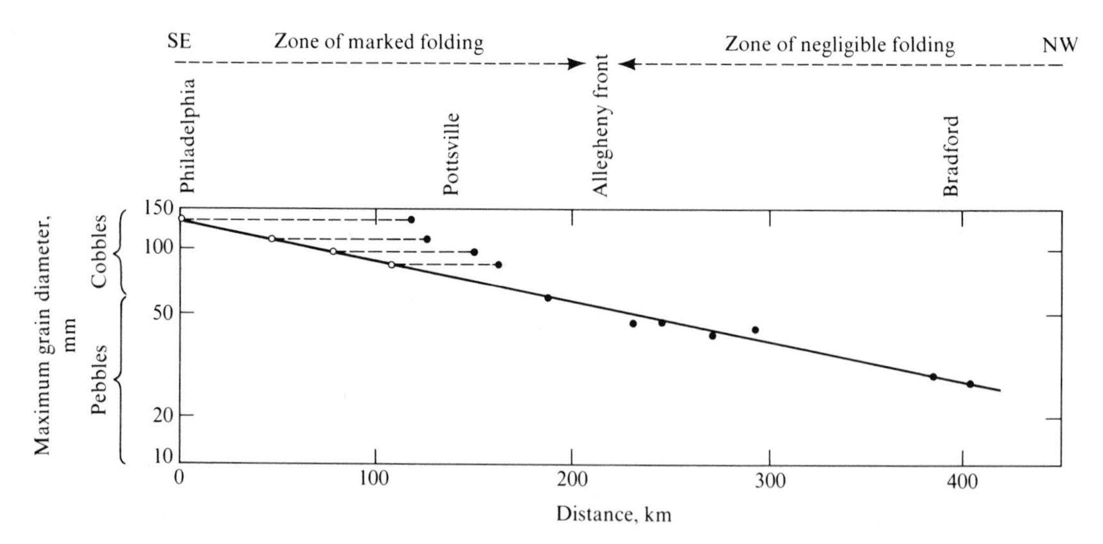

Figure 13–6
Relationship between distance of stream transport and maximum grain size in Pottsville conglomerates (Pennsylvanian), Pennsylvania. Solid circles, observed data; open circles, presumed prefolding position of the beds based on extrapolation from observed data points. Source of the gravel is thought to be in the vicinity of Philadelphia, based on existing structural and deformational patterns. [From B. R. Pelletier, 1958, *Geol. Soc. Amer. Bull., 69,* Fig. 18.]

Figure 13–7
Turbidite bed in the Cloridorme Formation (Ordovician), Quebec, showing large changes in thickness due to a combination of scour and load casting. Most laminae in the shale below the sandstone bed are not truncated, but simply thin below the thickest parts of the sandstone bed, indicating that the irregular lower contact is caused more by loading than by scour. [Photo courtesy G. V. Middleton.]

Structures

Structures are larger-scale features than textures and can be formed in many ways, such as by changes in depositional conditions, by erosion, or by diagenetic accentuation of an originally gradational boundary (see Table 13–2). Detailed description and interpretation of all these structures requires an understanding of the interaction between sediment and moving fluids, which, in turn, must be preceded by an introduction to fluid mechanics. Such an approach is beyond the scope of this book. Instead, we will describe the appearance of some of the most common and useful structures and indicate the types of information that can be obtained from them. Photographs of eight structures listed in Table 13–2 are shown in Figures 13–7 to 13–14; detailed discussions of them can be found in most recent sedimentology texts.

The most common sedimentary structure is *bedding,* the subplanar discontinuity that separates adjacent layers of rock. At the outcrop, we describe both its external geometry and internal character as indicated in Table 13–2. The distinction between bedding and lamination is nongenetic (see Table 13–3). When measurements of the thicknesses of large numbers of beds are made in a stratigraphic unit, it is common to

Table 13–2
Classification of Primary Sedimentary Structures

Bedding, external form
1. Beds *equal* or *subequal* in thickness; beds laterally uniform in thickness; beds continuous
2. Beds *unequal* in thickness; beds laterally uniform in thickness; beds continuous
3. Beds *unequal* in thickness; beds laterally variable in thickness; beds continuous
4. Beds *unequal* in thickness; beds laterally variable in thickness; beds discontinuous
Bedding, internal organization and structure
1. Massive (structureless)
2. Laminated (horizontally laminated; crosslaminated)
3. Graded
4. Imbricated and other oriented internal fabrics
5. Growth structures (stromatolites, etc.)
Bedding plane markings and irregularities
1. On base of bed:
a. Load structures (load casts)
b. Current structures (scour marks and tool marks)
c. Organic markings (ichnofossils)
2. Within the bed:
a. Parting lineation
b. Organic markings
3. On top of the bed:
a. Ripple marks
b. Erosional marks (rill marks; current crescents)
c. Pits and small impressions (bubble and rain prints)
d. Mud cracks, mud-crack casts, ice-crystal casts, salt-crystal casts
e. Organic markings (ichnofossils)
Bedding deformed by penecontemporaneous processes
1. Founder and load structures (ball-and-pillow structures, load casts)
2. Convolute bedding
3. Slump structures (folds, faults, and breccias)
4. Injection structures (sandstone dikes, etc.)
5. Organic structures (burrows, "churned" beds, etc.)

Source: F. J. Pettijohn, P. E. Potter, and R. Siever, 1973, *Sand and Sandstone* (New York: Springer-Verlag), Table 4–2.

Table 13-3
Terminology for Distinguishing Between Bedding and Lamination

Term	Criterion
Very thickly bedded	Thicker than 1 m
Thickly bedded	30–100 cm
Medium bedded	10–30 cm
Thinly bedded	3–10 cm
Very thinly bedded	1–3 cm
Thickly laminated	0.3–1 cm
Thinly laminated	Thinner than 0.3 cm

Source: R. L. Ingram, 1954, *Geol. Soc. Amer. Bull., 65,* Table 2.

Figure 13-8
Groove casts (a), prod casts (b), and brush marks (c). Current from bottom to top. Gardeau Formation, Devonian, near Dansville, New York. [From F. J. Pettijohn and P. E. Potter, 1964, *Atlas and Glossary of Primary Sedimentary Structures* (New York: Springer-Verlag), Plate 66.]

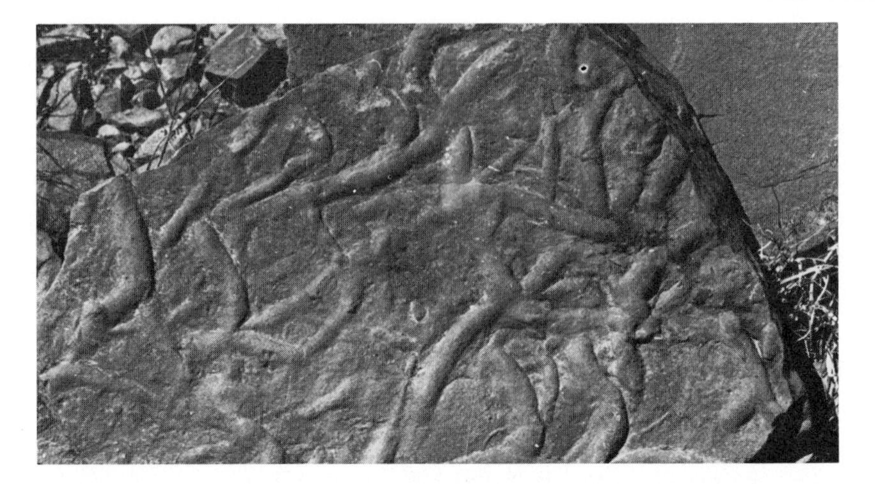

Figure 13–9
Casts of trace fossils of uncertain origin on a bedding plane, Haymond Formation (Pennsylvanian), near Marathon, Texas. Photo courtesy F. J. Pettijohn. Part of the head of a geologic pick in lower left of photo shows scale. [From F. J. Pettijohn and P. E. Potter, 1964, *Atlas and Glossary of Primary Sedimentary Structures* (New York: Springer-Verlag), Plate 70A.]

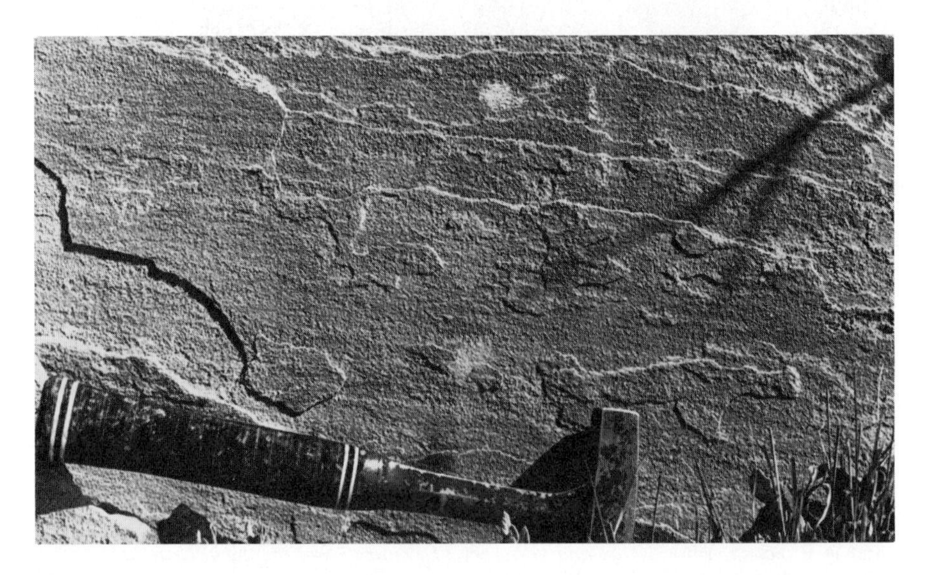

Figure 13–10
Parting lineation in fine-grained sandstone, Haymond Formation (Pennsylvanian), Pecos Co., Texas. Fluvial current flowed parallel to the hammer handle. [From F. J. Pettijohn and P. E. Potter, 1964, *Atlas and Glossary of Primary Sedimentary Structures* (New York: Springer-Verlag), Plate 76A.]

Figure 13-11
Dish structures (concave upward) and vertical fluid-escape pipes (pale), a syndepositional feature formed by turbidity currents. Cap Enrage Formation (Cambro-Ordovician), Quebec, Canada. [From R. G. Walker, 1978, *Amer. Assoc. Petroleum Geol., 62,* Fig. 7.]

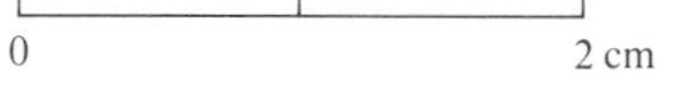

Figure 13-12
Casts of halite crystals in siltstone, Moenkopi Formation (Triassic), New Cameron, Arizona. [From F. J. Pettijohn and P. E. Potter, 1964, *Atlas and Glossary of Primary Sedimentary Structures* (New York: Springer-Verlag), Plate 98A.]

Figure 13–13
Polished slabs of siltstone in the Martinsburg Formation (Ordovician), Pennsylvania, showing convolute lamination. [From E. F. McBride, 1962, *Jour. Sed. Petrology, 32,* Fig. 10.]

find that the thicknesses cumulate as a straight line when plotted on semilog graph paper (see Figure 13–15). The relationship stems from the interaction between the average current velocity and the size of sediment being transported and deposited, but the exact cause is not understood.

A bedding surface testifies to a change of some sort during deposition of the sediment, but often the nature of the change is uncertain. Many of these changes represent *diastems,* brief periods of time during which no sediment was deposited, none was eroded, and no rock record exists at that location. When sedimentation resumed, some factor had changed so that a discontinuity was created (see Figure 13–16).

Bedding surfaces commonly are generated by contrasting textures (see Figure 13–17); for example, the contrast between the very coarse texture of a conglomerate and the finer texture of an overlying sandstone. What mechanism produced the boundary between the coarser- and finer-grained layers? Was it a rapid decrease in

Figure 13–14
Highly contorted folds resulting from slumping of semilithified beds soon after deposition (penecontemporaneous deformation). Note the finely bedded undisturbed beds both above and below the disturbed zone. Cerro Toro Formation (Late Cretaceous), southern Chile. Width of photo is 55 m. [From R. D. Winn, Jr., and R. H. Dott, Jr., 1979, *Sedimentology, 26,* Fig. 4.]

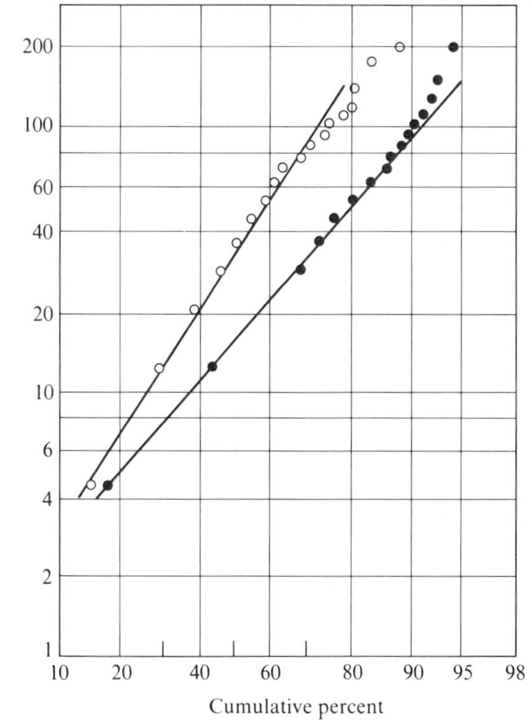

Figure 13–15
Illustration of lognormal distribution of bed thicknesses of turbidite sandstone, Cretaceous, Chile; based on measurements of 271 beds. [From K. M. Scott, 1963, *Amer. Assoc. Petroleum Geol., 50,* Fig. 13.]

Figure 13–16
Thinly interbedded sandstone and shale overlain by very thickly bedded uniform sandstone, Moehave Formation (Triassic), Utah. The cause of the radical change in bedding characteristics is unknown. [From F. J. Pettijohn and P. E. Potter, 1964, *Atlas and Glossary of Primary Sedimentary Structures* (New York: Springer-Verlag), Plate 25.]

stream current velocity; or a lateral shift in channel position; or the transgression of an offshore marine sand bar over a gravel beach? The causes of bedding surfaces are many, and often they are not understood. Their existence, however, means that *something* must have changed during the interval of time represented by the infinitely thin surface; perhaps a change in water chemistry caused incipient cementation of the earlier sediment; or perhaps there was a change in water temperature resulting from a change in solar radiation; or perhaps a change in stream current velocity occurred

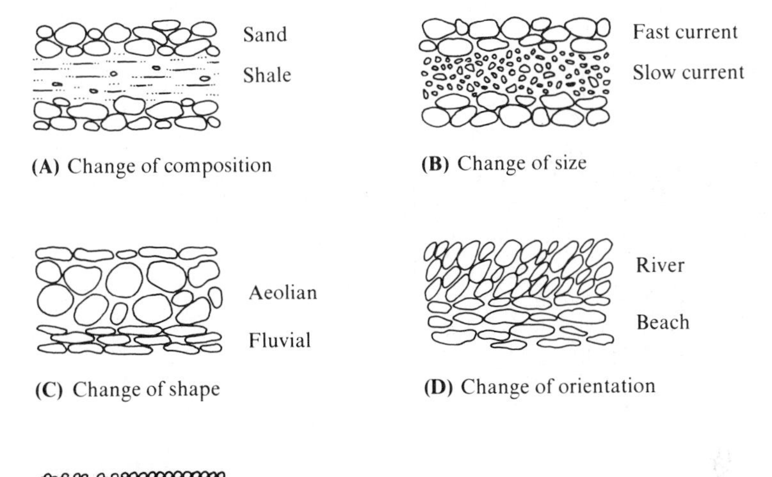

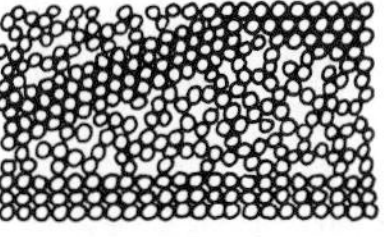

(E) Change of packing

Figure 13–17
Idealized examples of some types of textural changes that are sensed by our eyes as bedding. [From J. C. Griffiths, 1961, *Jour. Geology, 69,* Fig. 3.]

as a result of a spring flood on April 7th of the early Middle Devonian. It is worth remembering that most of the time since the earth was formed is represented in the geologic record by bedding surfaces, not by rocks. Hence, the bedding surfaces we see in outcrops deserve careful examination. Their commonness should not make them less noteworthy.

Graded bedding. Graded beds are layers of detrital sediment marked by a gradation in grain size from base to top, normally coarser at the base and finer at the top (see Figure 13–18). They range in thickness from less than a centimeter to several meters; typically, thicker graded units show more changes in grain size between base and top than do thinner units. Most graded detrital rocks are deposited by ***turbidity currents,*** coherent mixtures of gravel, sand, mud, and water that move downslope from shallow depth in lakes or in the sea toward the basin floor. As the turbidity current slows toward the foot of the continental slope or lake center, the coarser grains in the sediment–water mixture are deposited first to form the base of the graded bed, with the finer grains of silt and clay following in sequence. As the mixture continues outward on the submarine alluvial fan on the deep basin floor somewhat depleted in coarser grains, a lateral change develops until at the farthest edge of the bed only the finest sediment remains. The thickness of the bed decreases

Figure 13–18
Outcrop showing graded bedding with very coarse sand at the base and muddy sediment at the top; total bed thickness is about 6 cm. Minnitaki Group (Archean), Ontario, Canada. [From F. J. Pettijohn et al., 1973, *Sand and Sandstone* (New York: Springer-Verlag), Fig. 4–8 (photo courtesy R. G. Walker).]

in the basinward direction as well, as a lesser and lesser amount of sediment remains to be deposited. Individual graded beds often can be traced laterally in the field for several kilometers in a cross section of an ancient submarine fan (see Figure 13–19).

Destruction of bedding by bioturbation. Many types of mobile organisms live on the sea floor, and the bulk of them feed by burrowing through freshly deposited sediment looking for edible organic matter. This activity destroys bedding and lamination. Burrows are recognized in cross section by the contrast between the texture or mineral composition of the burrow-fill and that of the host rock. If burrows are sufficiently abundant, only vestiges of the original bedding may remain; in extreme cases complete homogenization is produced within thick sand units (see Figure 13–20).

Directional structures. A large variety of structures can indicate direction of current flow in sandstone beds. Some of them occur within beds, such as imbrication or cross bedding; others occur on the surfaces of the beds, such as ripple marks or flutes. Measurement of the orientation of such structures in fluvial or turbidite sediments defines the *paleoslope* or regional depositional dip. For example, the paleoslope in the

Figure 13–19
Cross section of a Miocene submarine fan showing long, even beds of sand deposited by sporadic turbidity currents interbedded with pelagic shale. Valley of Santerno River, northern Italy. [From F. J. Pettijohn et al., 1973, *Sand and Sandstone* (New York: Springer-Verlag), Fig. 11–38 (photo courtesy P. E. Potter).]

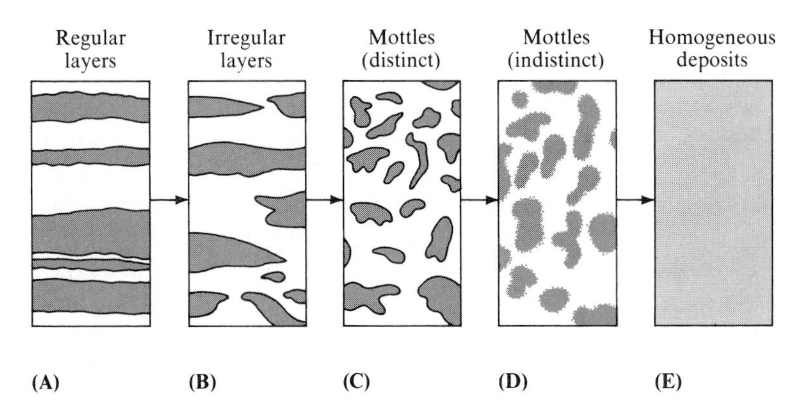

Figure 13–20
Diagrammatic sketch showing progressive destruction of bedding by burrowing organisms. It is also possible for burrowing within a homogeneous sand to result in a mottled sand when the burrows are filled with material not originally present in the sand (arrows). [From D. G. Moore and P. C. Scruton, 1957, *Amer. Assoc. Petroleum Geol., 41,* Fig. 12.]

Figure 13–21
Imbrication in Jurassic conglomerate, southwest Oregon. The scale is parallel to the bedding and imbrication is clearly visible just under the scale. Current flowed from left to right. Imbrication generally is more pronounced in modern sediments than in ancient rocks because of the effect of compaction during burial. [Photo courtesy R. G. Walker.]

mid-continental United States since the Cretaceous Period has been toward the south; that is, the drainage direction for the past 125 million years has been toward the Gulf of Mexico. The roots of the Mississippi River run deep.

Imbrication refers to a stacking of grains with their flat surfaces at an angle to the major bedding plane. It is seen most clearly in gravels and conglomerates (see Figure 13–21) but is present in sands as well, as has been demonstrated by examination of thin sections cut perpendicular to bedding but parallel to the direction of current flow as defined by other criteria. In conglomerates the imbrication is developed best where the pebbles and cobbles are in direct contact (that is, not much sand is present) and the gravel particles are platy in shape. The gravel dips upstream because this is the stable position; gravel oriented in the opposite direction tends to be overturned by the moving water.

Cross bedding is perhaps the most familiar of sedimentary structures because it is so common in sand and silt deposits (see Figure 13–22); it is rare in conglomerates. A vast literature exists concerning its origin and interpretation. High-angle cross bedding is produced largely by avalanching of sand or coarse silt down the lee (downstream) slope of wavelike structures such as ripples and dunes. Therefore, cross beds dip downcurrent, opposite to the direction of imbrication. The initial angle of dip is controlled by the angle of repose of loose sediment in air or water (about 35–45°), but in ancient rocks the dip angle is more likely to be 15–20° because of compaction following burial. Cross bedding at a low angle ($< 10°$) is produced during deposition by accretion on sloping surfaces, such as on the downstream slope of wavelike structures or on the face of a beach. In tide-dominated environments, where the current direction is constant for six hours before reversing, a distinctive herringbone pattern can be produced (see Figure 13–23).

Figure 13–22
Well-developed cross bedding in siltstone, Pierce Canyon redbeds (Permian), Eddy County, New Mexico. Fluvial current flowed from right to left. [From J. D. Vine, 1963, *U.S. Geol. Surv. Bull., 1141-B,* Fig. 7.]

Figure 13–23
Herringbone cross bedding in Winchell Creek member, Great Meadows Formation (Ordovician), New York State, underlain by planar lamination and overlain by ripple forms. Bar = 15 cm. [From S. J. Mazzullo, 1978, *Jour. Sed. Petrology, 48,* Fig. 8.]

Ripple marks are formed as the first structure on a bedding surface when the current becomes fast enough to transport sand. They are usually asymmetrical in cross section, with the downstream slope steeper than the upstream slope (see Figure 13–24). This type of ripple is formed by frictional stresses between the moving fluid and the sandy bed when there is continued movement of the fluid in the same general direction. The fluid may be air or water. Desert wind currents, stream currents, marine longshore currents, turbidity currents, or deep sea geostrophic currents all produce rippled sand surfaces. Hence, ripples occur on sand dunes, stream bottoms, and at all depths in the sea. In some cases there may be differences in either scale or form of the ripples in different environments because of variations in the fluid dynamics of ripple formation in different environments. Oscillatory movement of fluid (back and forth movement rapidly repeated) such as in some lakes produces symmetrical ripples that cannot be used to indicate the downstream direction.

Flute casts are asymmetrical structures found on the soles (bottoms) of some sandstone beds (see Figure 13–25), particularly in environments where turbidity currents have been common. For this reason, flute casts have become associated with deep water deposits, although turbidity currents are not restricted to the deep ocean basins. The scooplike structure of the flute is produced by erosion of freshly deposited mud on an underwater slope. A turbidity current flows over the mud and turbulent

Figure 13–24
Cuspate ripple marks on thin-bedded sandstone, Caseyville Formation (Pennsylvanian), Illinois. Current flowed from left to right. [From F. J. Pettijohn and P. E. Potter, 1964, *Atlas and Glossary of Primary Sedimentary Structures* (New York: Springer-Verlag), Plate 84A.]

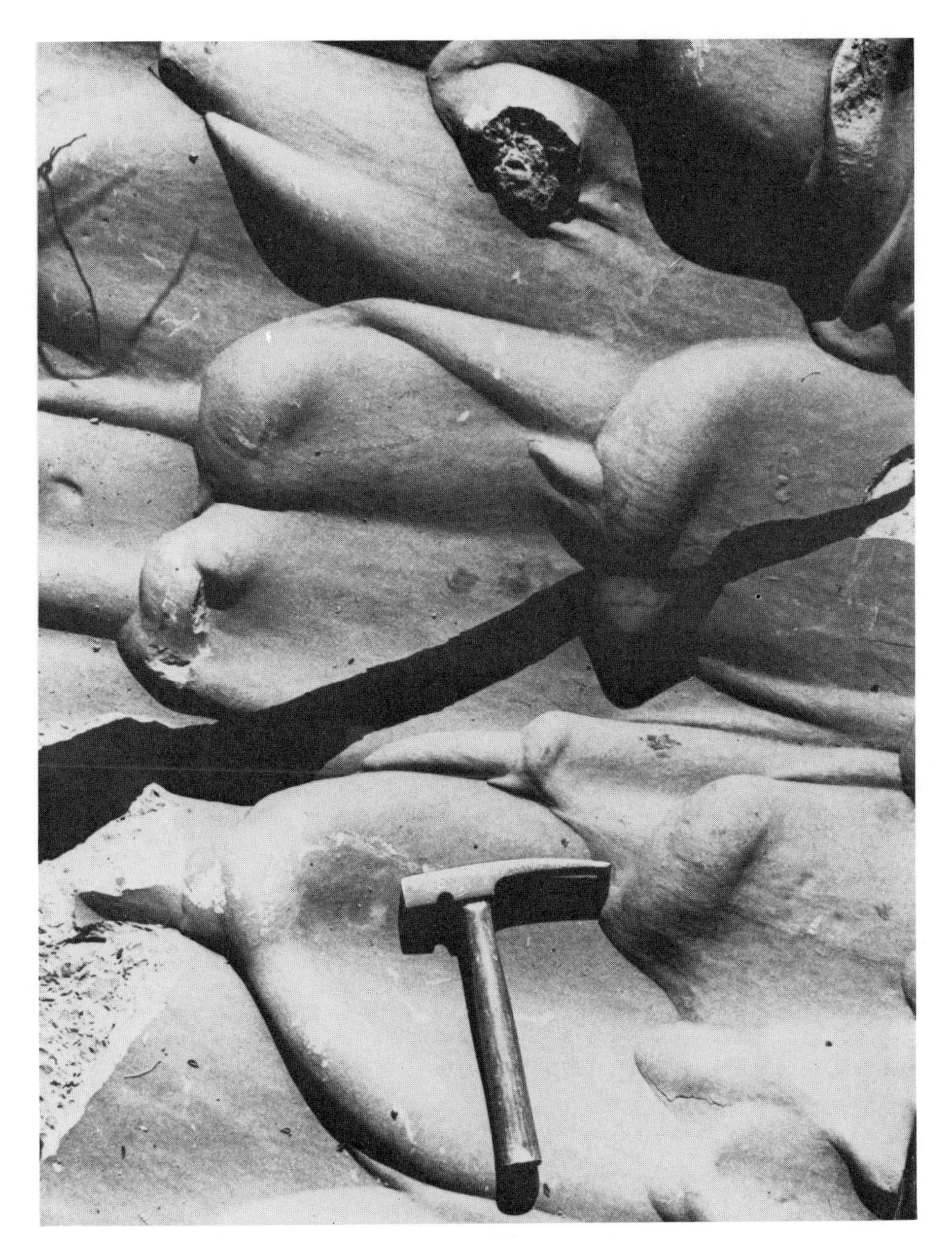

Figure 13–25
Large, closely spaced and sometimes overlapping flute casts on the base of a sandstone bed, Burnett Co., Texas. Current flowed from left to right. [From F. J. Pettijohn and P. E. Potter, 1964, *Atlas and Glossary of Primary Sedimentary Structures* (New York: Springer-Verlag), Plate 58.]

eddies form at the current–mud interface at the head of the current, resulting in the erosion of a spoon-shaped gob of mud with the deepest part of the cup upcurrent (see Figure 13–26). As the tail of the turbidity current passes over the cup, fine-grained sand is deposited in it. This sand is subsequently seen as a downward projection on the base of the sandstone that overlies the mud in the stratigraphic record. In outcrop, normally the mud (now mudrock) is weathered away to reveal the oriented flutes.

A large number of other sedimentary structures have been described from outcrops of sandstones and conglomerates, and most of them can be used to interpret some aspect of depositional or diagenetic conditions. For example, in conglomerates the long axes of gravel particles can have a preferred orientation either parallel to or transverse to the paleoflow direction. Parallel orientation is produced by rapid mass transport processes, such as characterize turbidity currents; transverse orientation occurs when the gravel particles are transported by the rolling action typical of rivers and beaches.

Loose sediment settles into the cracks of a mud-cracked surface, preserving the mud cracks in positive relief (see Figure 13–27). This occurs almost exclusively in continental rocks. Gypsum and halite crystals formed in muds in an evaporative environment may be preserved as casts on the same sole.

In most underwater environments, various burrowing organisms leave traces of their presence in the sediment (see Figure 13–28), and the character of these structures indicates depth of water. For example, deep vertical burrows several tens of centimeters in length are common in the littoral zone because the organisms want to

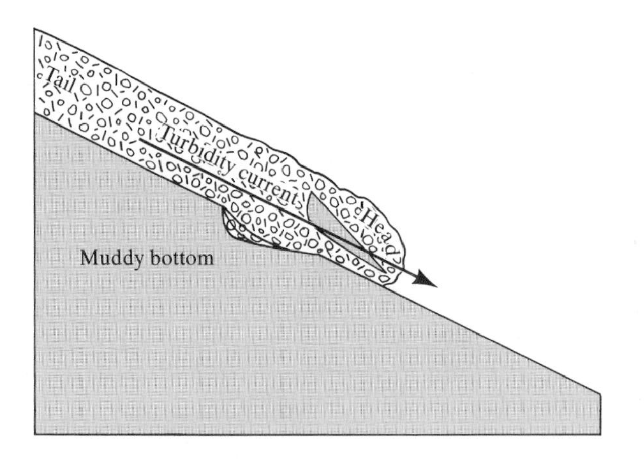

Figure 13–26
Schematic cross section illustrating the formation of a flute by bottom scour from the head of a turbidity current on the sea floor. Cross-bedded sand from the tail of the current has begun to fill in the flute. The current is normally many orders of magnitude larger than the flute.

Figure 13–27
Infilled mud cracks in mudstone of the Thurso Sandstone (Devonian), Scotland. Some of the infillings in the lower left of the photo stand out in positive relief as the less resistant mud is removed by weathering processes. [From F. J. Pettijohn and P. E. Potter, 1964, *Atlas and Glossary of Primary Sedimentary Structures* (New York: Springer-Verlag), Plate 96B.]

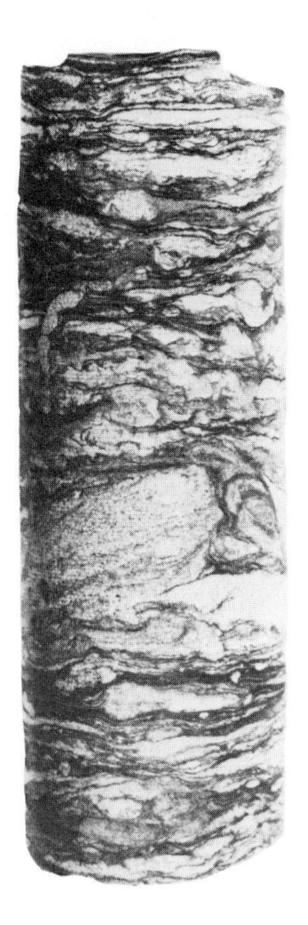

Figure 13–28
Bedding of siltstone (light) and organic-rich shale (dark) disrupted by burrowing of benthonic marine organisms. Tradewater Formation (Pennsylvanian), Kentucky. Diameter of core is 6 cm. [From F. J. Pettijohn et al., 1973, *Sand and Sandstone* (New York: Springer-Verlag), Fig. 4–21 (photo courtesy P. E. Potter).]

avoid the large daily and seasonal changes in temperature and salinity that characterize very shallow water. In deeper water, surface grazers are dominant and leave various types of traces on the bedding surfaces as they hunt for organic matter.

In the field the general rule to follow is that anything that departs from isotropism and homogeneity has the potential to supply useful information about the origin of the sediment. Careful observation and recording in either notebook or on film often proves invaluable in the office. Every little squiggle has a meaning all its own.

Mineral Composition

The main control on our ability to identify and interpret mineralogic variation in outcrop is grain size. The coarser sizes of detrital fragments are particularly valuable because they are most often pieces of the aggregated source rock itself rather than individual mineral grains such as orthoclase or single crystals of quartz that might have originated in a wide variety of rock types. In addition, most accumulations of gravel are located much nearer the source terrane than sand deposits. As a result they have undergone less modification since being released from their parent rock, and the paleogeologic interpretation of pebbles is more reliable. Pebbles can be formed from almost any type of rock—vein quartz, chert, granite, basalt, gneiss, and so on—and the proportions of the different types can form the basis of a detailed paleogeologic map.

Accurate identification of sand-size mineral grains is difficult using only a 10× hand lens. Quartz is easy to spot, as is pink potassium feldspar; but chert can look like rhyolite, schist looks like shale, and those little black things you see through the hand lens may be basalt, magnetite, or black chert. Considerable experience is needed in order to achieve a reasonable degree of precision in the identification of sand-size particles in hand specimens. Most of these mineralogic interpretations typically are postponed until thin sections and a polarizing microscope are available. Areal changes in mineral and rock fragment composition are very important in paleogeographic work, however, and as much should be done in the field setting as possible. The question of where to go next for a meaningful rock sample can often be answered by mineralogic trends pieced together from individual observations made at scattered outcrops. Time in the field is too valuable to be wasted wandering around wondering where to go next.

Summary of Field Observations

As we have noted, many characteristics of sandstone sequences are best seen in the field. For this reason, they should be described while you are in the field at the outcrop. Back in the laboratory, it is difficult to recall whether bedding contacts were sharp or gradational; whether the graded beds you observed were abundant or exceptional; whether imbrication was present; whether the conglomerates (which were too coarse to sample) contained a variety of types of rock fragments or only quartz pebbles; or whether the conglomerates were more poorly cemented than the interbedded sandstones (which you did sample). Each of these observations has the potential to contribute toward the interpretation of the rock sequence—its provenance, depositional environment, and diagenetic history.

Table 13–4 is a checklist of features that might be observed and described in the field. Many of them will not occur at any particular outcrop but most will be present somewhere in most stratigraphic sections.

LABORATORY STUDIES

Although some textural and mineralogic features of sandstones can be determined at the outcrop, an adequate evaluation requires laboratory study. Details of texture can reveal whether the transporting medium was a glacier, a stream, or the wind. Accurate mineralogic work can reveal whether the source rocks were greenschist, rhyolite, gabbro, or granite. In addition, it is possible using a polarizing microscope and other instruments, such as an electron microprobe and mass spectrometer, to establish the types of cements, the order in which they formed, and whether they were precipitated from fresh or saline waters. In many cases the inferences made are quite dependable. In other cases they are imprecise and speculative and there is as much art involved as science. An interpretation of the history of a group of sand grains is not a sacred document but simply a best guess based on data available at the time. And the quality of the guess varies not only with the available data but also with the caliber of the geologist doing the work.

Textures

Texture in rocks can be considered both as an aggregate property and as a property possessed by individual grains. For example, grain size distribution, sorting, and permeability are aggregate properties; grain orientation and shape are properties of individual particles.

Size distribution and sorting are estimated in the field and determined more accurately in the laboratory using either sieves, pipette, or settling tube. The rock specimen is disaggregated using a 10% solution of cold or hot hydrochloric acid to dissolve calcite, dolomite, and hematite cements; silica cement cannot be dissolved without also dissolving the detrital grains, so the sizes of sand grains in silica-cemented rocks must be determined by measurements of grain lengths as seen in thin section. The resulting size distribution is plotted on one of several types of graph paper and the average size and sorting are calculated. Size and sorting can be used to infer current strength and depositional environment.

Table 13–4
Checklist of Features of Detrital Rocks to Be Described in the Field

General characteristics
1. Extent of outcrop: height and width
2. Relative abundance of conglomerate, sandstone, and mudrock
3. Degree of intermixing: Is the lower one-third mudrock, the middle one-third conglomerate, and the upper one-third sandstone? Or are the rock types intimately intermixed?
4. If intermixed, is there a characteristic sequence, such as conglomerate units normally followed by sandstones?
Sedimentary structures
1. Thickness and lateral extent of beds
2. Nature of contacts with beds above and below: sharp or gradational
3. Character of bedding surfaces:
a. Planar, wavey, or irregular; relation to lithology
b. Ripple marks: symmetry, orientation, height, distance between crests
c. Parting lineation: orientation
d. Load casts
e. Flute casts: scale and orientation
f. Groove casts: description and orientation
g. Organic tracks and trails: description
h. Mud cracks, raindrop impressions, or other features
i. Cut-and-fill structures
4. Internal character of beds:
a. Laminations: cause (e.g., organic matter, grain size change), thickness, continuity; fissility
b. Organic burrows: abundance, type, relation to lithology
c. Cross bedding: scale, orientation, angle of dip
d. Graded bedding: thickness of graded unit, frequency, grain size variation
e. Convolute bedding, intraformational breccias
f. Orientations of fossils (e.g., preferred orientation of elongate shells; brachiopod shells mainly concave up or down) or pebbles
g. Imbrication: orientation

Central to the use of texture to infer depositional environment is the concept of *textural maturity*. The principle involved is that different aspects of sediment texture are modified at different rates and the rates are determined by environment. For example, suppose we fill a dump truck with a mixture of mud and sand and empty it into a river. The first textural change that occurs will be the separation of mud from sand; within a few minutes the mud will be on its way downstream. The separation of the various grades of sand takes a little while longer, perhaps a few hours or days, and the rounding of the sand grains during transport requires still longer periods of time.

Table 13-4 (*continued*)

Sedimentary textures
1. Mean grain size and sorting
2. Clay content
3. Shape of gravel particles: rod, disc, sphere
4. Rounding of grains: relation to grain size
5. Color: relation to rock type and grain size
Mineral composition
1. Percentages of quartz, feldspar, and rock fragments; variation with grain size
2. Degree of weathering: e.g., K-feldspar mostly altered to white kaolinite but twinned plagioclase is fresh
Diagenetic features
1. Degree of lithification
2. Type of cement: quartz, calcite, hematite, or other
3. Porosity: Does water sink into a hand specimen?

A soft fragment such as mica schist will round in perhaps a few weeks; a feldspar grain in perhaps a few years; a sand-size quartz grain will not round at all during stream transport. The sequential changes of clay removal, sand grain sorting, and sand grain rounding are termed stages of textural maturity (see Figure 13–29). The rate at which the textural maturity increases depends on the kinetic energy characteristic of the environment.

Kinetic energy determines not only the rate of textural modification but the limiting degree of modification as well. As we noted, a stream does not have the capability to round a sand-size quartz grain because the amount and intensity of grain-to-grain impacts are inadequate to do the job. In a sand dune or on a beach, however, impacts are more frequent and more intense so that rounding of quartz sand can occur, although even in these high-energy environments the process requires millions of years. The general relationship between sedimentary environment and textural maturity is shown in Figure 13–30. It is important to keep in mind that the relationship shown refers to the time of transport of the sediment and takes no account of textural changes that can occur during diagenesis. As we shall see, the degree of apparent rounding can be decreased by the addition of quartz to the detrital grain after burial (quartz overgrowths), and the clay content of the sand can be increased by diagenetic conversion of unstable grains into clay minerals. The possibility of such postdepositional changes must be considered before the meaning of textural maturity of a sandstone can be determined.

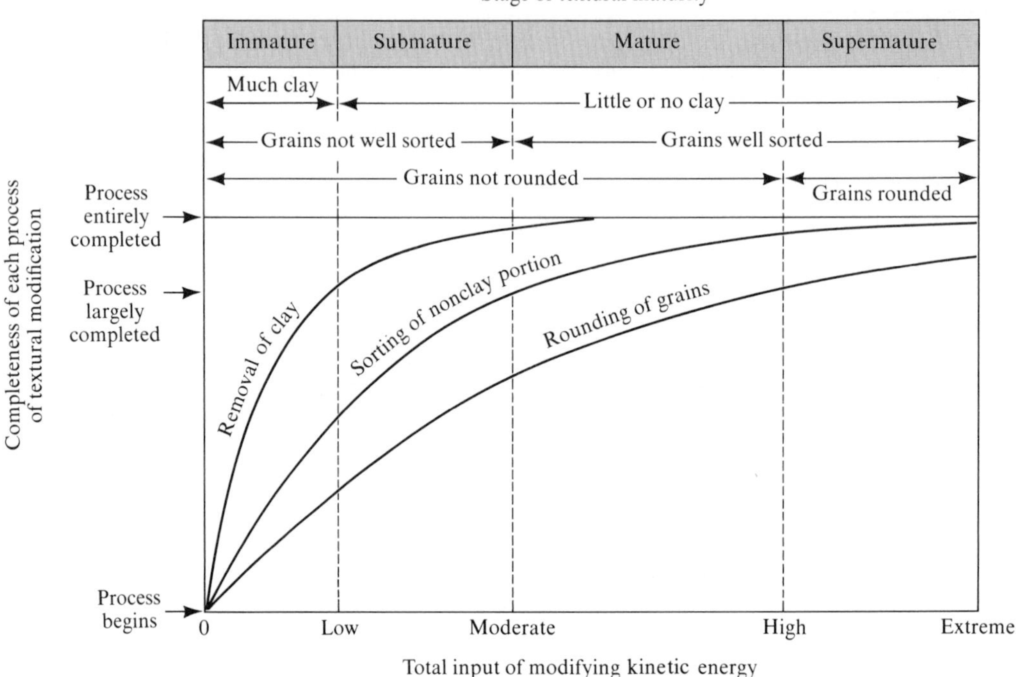

Figure 13-29
Textural maturity of sands as a function of the input of kinetic energy. [From R. L. Folk, 1951, *Jour. Sed. Petrology, 21,* Fig. 1.]

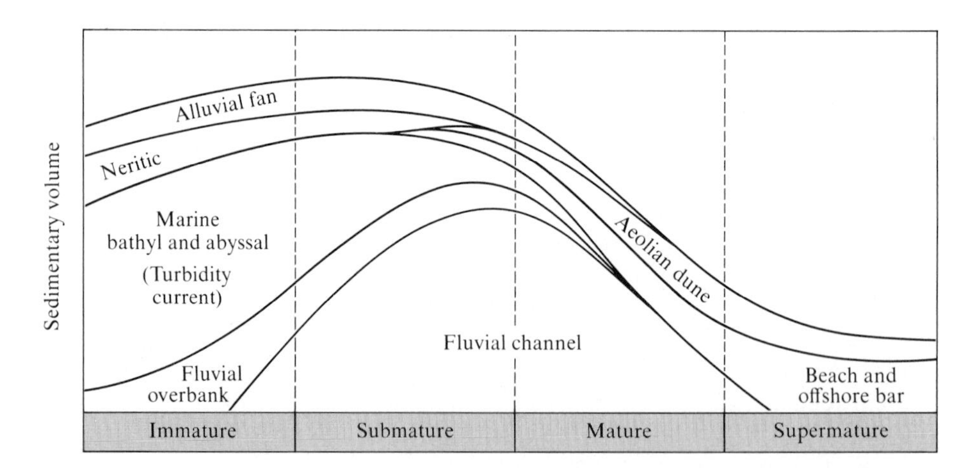

Figure 13-30
Relationship among sedimentary volumes, environments of deposition, and textural maturity. The diagram is qualitative; adequate numerical data do not exist.

Porosity and Permeability

The initial porosity (void volume/total rock volume) and permeability (ability to transmit a fluid) of a sediment are directly related to textural maturity. The presence of clay decreases porosity because the clay minerals are only 1/100–1/1000 the size of sand grains and, as a result, lodge in the spaces between the sand. The clays also decrease permeability because of the very large surface/volume ratio of platy clay flakes. Attractive forces exist between the mineral surfaces and the moving fluids (water, petroleum, or natural gas), and the more surface area the fluids must pass over, the slower their movement. Growth of clay minerals after burial is a common phenomenon in many sandstones (see Chapter 14).

Sorting of the sand grains also affects initial porosity and permeability. Although the smaller grains of sand are closer to the size of their fellow travelers than are the clay particles, they will tend to fill in the empty spaces between the larger grains. Poorer sorting (lower textural maturity) implies lesser porosity. Permeability is lowered as well by poorer sorting, although not as much as when clay minerals are present.

The roundness of sand grains has only a small effect on porosity and permeability. Increased roundness allows the grains to be packed closer together and therefore decreases porosity and permeability.

Mineral Composition

The mineral composition of sandstones is studied to determine two things about the history of the rock: (1) the character of the source rocks from which the detrital grains were derived (provenance), and (2) the diagenetic sequence of events following deposition of the grains, particularly the chemical events. We will discuss provenance determinations in this chapter and defer discussion of diagenetic events to Chapter 14.

Quartz

About two-thirds of the detrital fraction of the average sandstone is quartz; because of this fact, an enormous literature exists concerning its physical and chemical character in detrital rocks. Quartz is an extraordinarily stable mineral in the sedimentary environment because of its hardness, lack of good cleavage, and lack of metallic cations. But great stability is a two-edged sword as far as provenance studies are concerned; its sedimentary history leaves few readable marks on a quartz grain. Trying to obtain information from a quartz grain is like trying to get blood from a turnip.

Quartz grains in sediments are usually grouped as either polycrystalline or monocrystalline (see Figure 13–31). A polycrystalline grain consists of more than one crystal; that is, it is a rock fragment composed entirely of quartz. Polycrystalline quartz grains are less stable than single crystals and are preferentially destroyed by

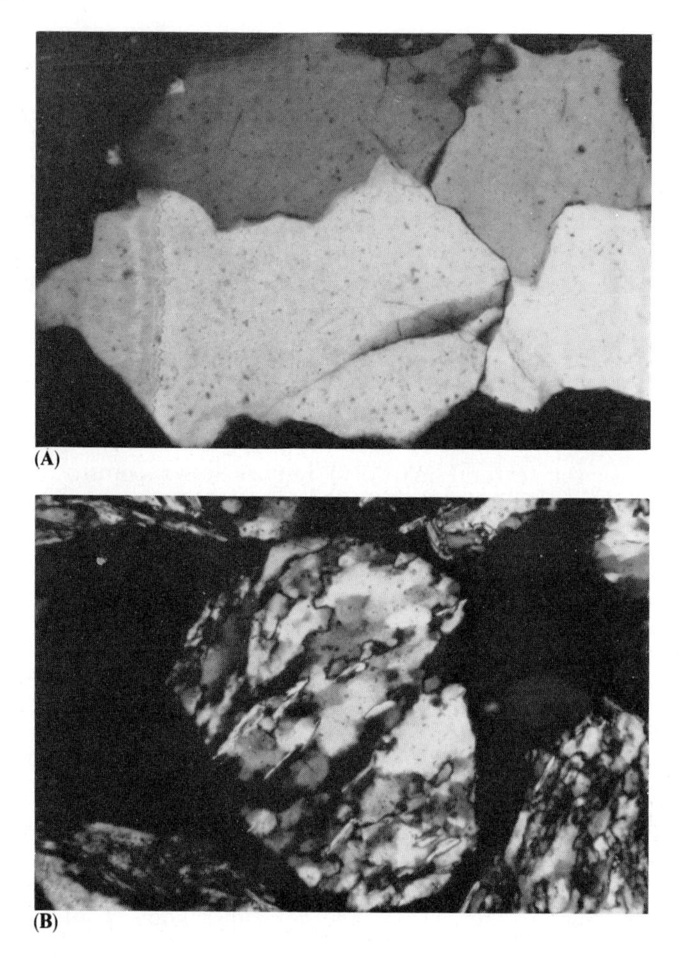

Figure 13–31
Photomicrographs of coarse sand-size quartz grains (crossed nicols) from (**A**) an igneous rock and (**B**) a metamorphic rock. The granitic quartz is composed of fewer crystals of more equant shape and with little or no intercrystalline suturing. The metamorphic grain consists of perhaps ten times as many crystals, which are stretched and elongate and have intensely sutured contacts. These grains are extreme examples of polycrystalline grains from the two classes of crystalline rocks.

sedimentary processes, largely by chemical attack. As the percentage of quartz in a sandstone increases, the percentage of the quartz that is polycrystalline decreases. Sandstones composed entirely of quartz typically contain no polycrystalline quartz.

The average granite is coarser grained than the average metamorphic rock (such as schists) so that individual quartz crystals in a granite are larger. As a result, a sand-size polycrystalline quartz grain from a granite will contain fewer quartz crystals than a sand-size polycrystalline grain from a metamorphic rock. Also, many metamorphic rocks crystallize in a nonhydrostatic stress field (shearing stress) so that the shape of the crystals in the polycrystalline detrital fragment is elongate. Igneous rocks crystallize in a hydrostatic stress field, giving rise to crystals that are equant in shape.

Because siliceous igneous rocks are relatively coarse-grained, coarse-grained single crystals of quartz are derived mostly from granitoid igneous rocks. Fine-grained single crystals in a sediment are not necessarily derived mostly from metamorphic rocks, however. They can be produced by abrasion of coarser quartz grains of any origin. Quartz crystals typically are plastically deformed by stresses during or after their crystallization in the igneous or metamorphic environment, and the deformation produces the phenomenon known as undulatory extinction. These deformed crystals are less stable than undeformed ones and are preferentially destroyed by sedimentary processes, as are the polycrystalline quartz grains. Pure quartz sands average about 50% apparently undeformed grains, in contrast to the 15% typical of igneous and metamorphic rocks.

Feldspar

Feldspars form 60% of the average crystalline rock but only 10–15% of the average sandstone because of the instability of feldspars in the sedimentary environment. In a tectonic setting characterized by mountain building, such as in Colorado during the Pennsylvanian Period or in Southern California during the late Tertiary, erosion and burial are rapid and the resultant sands may contain 50% feldspar. In a quiescent setting such as during the early Paleozoic Era in central North America, the sediment produced is reworked repeatedly by waves and currents as the shallow epicontinental seas advance and retreat so that almost no feldspar survives.

The variety of feldspar (see Figure 13–32) that is most abundant in a sandstone also depends to an important degree on the tectonic setting in which the sandstone forms. Plagioclase will dominate when erosion and burial are rapid, as along convergent margins. Potassic feldspars (orthoclase and microcline) dominate the feldspar suite of sandstones formed in cratonic settings. In a setting such as Southern California the plagioclase/K-feldspar ratio may be 2:1, with abundant unstable calcic feldspar; in the Cambrian rocks of the central United States the ratio may be 1:5 with only very sodic plagioclase present, even though there is no difference in composition between the granites in Southern California and those in the mid-continent. These relationships between the abundance and composition of feldspars in sandstones are summarized in Figure 13–33.

The proportion of total feldspar in a sandstone also is affected by the grain size distribution in the rock. Detrital feldspars, because of their relative instability, are normally finer grained than associated quartz grains. As a result it is not uncommon to find, for example, that the medium sand fraction contains 90% quartz and 10% feldspar; in the fine sand fraction, 80% and 20%; and in the coarse silt fraction, 70% and 30%. The percentage of feldspar then decreases to about 5% in the fine silt

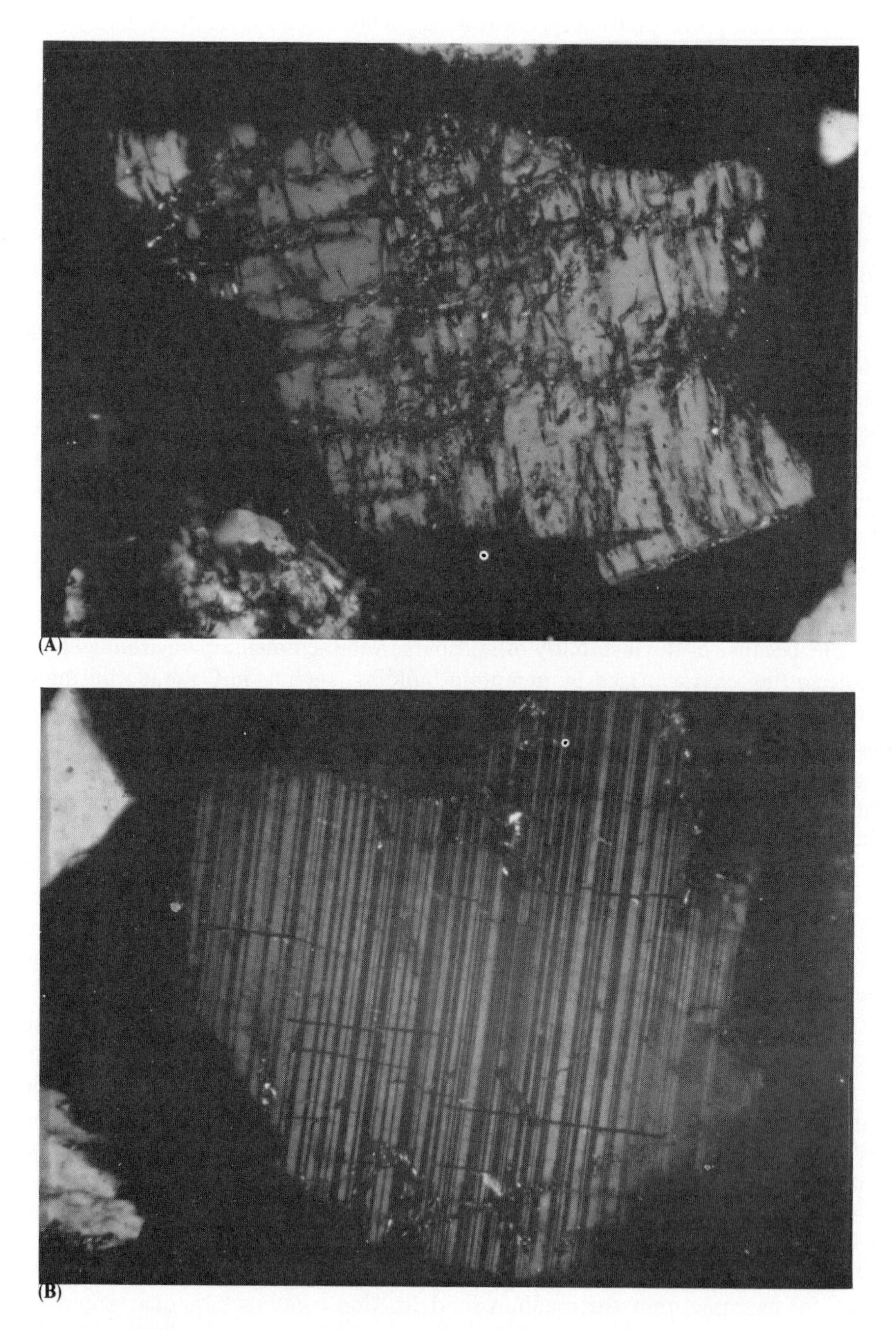

Figure 13–32
Photomicrographs of medium sand-size feldspar grains (crossed nicols). (**A**) Orthoclase with alteration parallel to right-angle cleavages. (**B**) Plagioclase showing polysynthetic albite twinning.

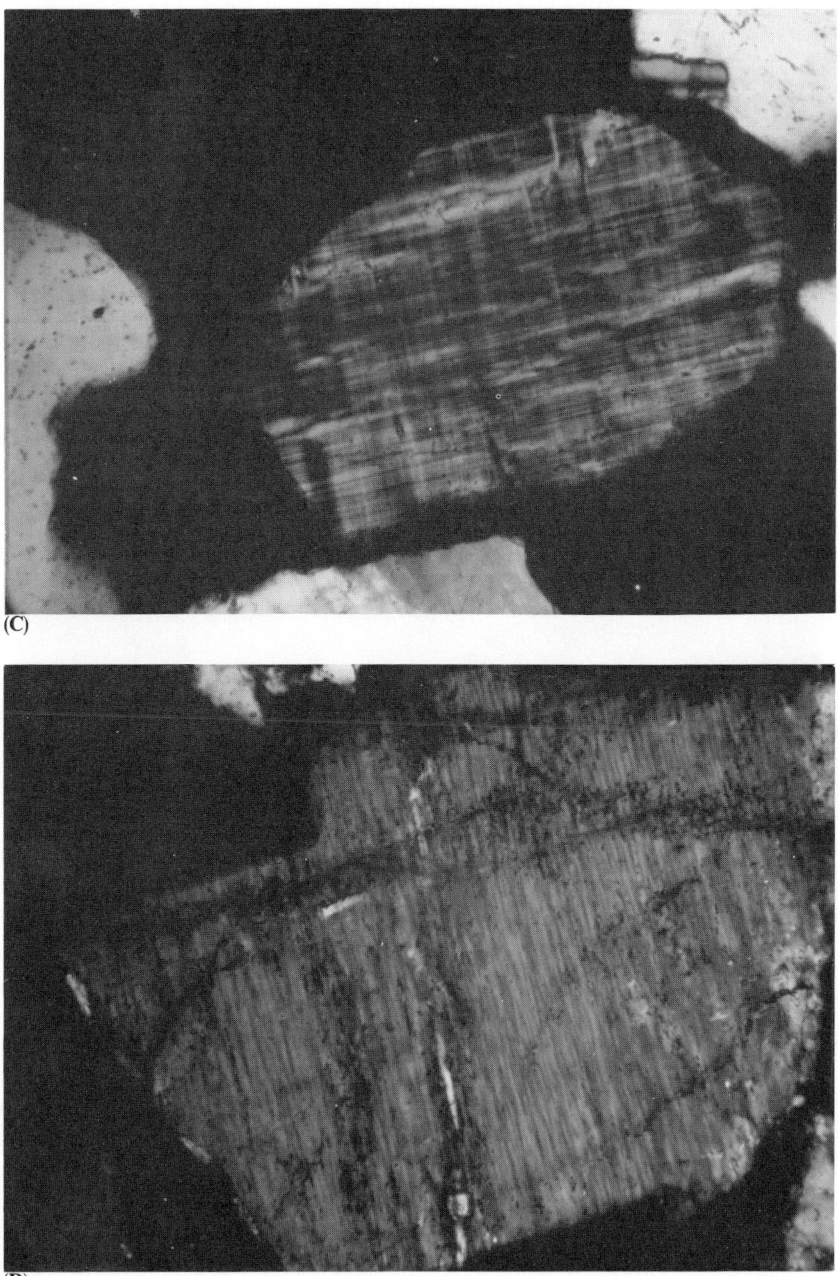

(C) Microcline showing characteristic grid-twinning. (D) Microperthite showing spindle-shaped exsolution lamellae of sodic and potassic feldspar.

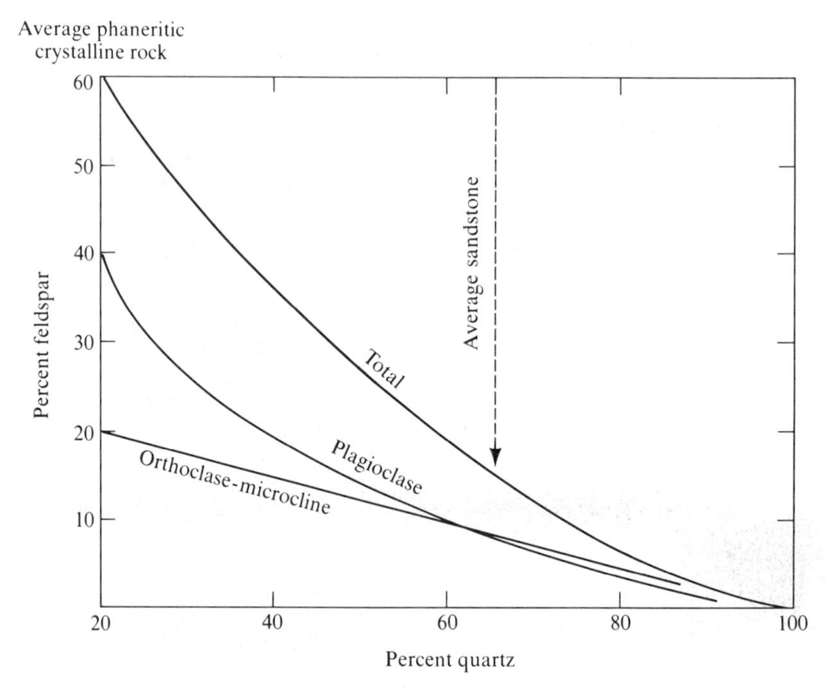

Figure 13–33
Generalized relationship between abundance and composition of feldspars in sandstones. The percentage of detrital grains not accounted for by quartz and feldspar is the percentage of unstable lithic fragments plus clay minerals.

fraction as the rapidly increasing surface/volume ratio of the smaller grains causes a rapid increase in the dissolution rate of the feldspars. The average mudrock contains about 5% feldspar. The clay-size fraction contains almost no feldspar.

Unstable Lithic Fragments

Pieces of undisintegrated source rock (see Figure 13–34) form about 15% of the average sandstone but can supply much more than 15% of the provenance information about the rock. Unlike quartz or feldspar grains, a piece of basalt or mica schist in a sandstone is unequivocal evidence of the nature of the source rock. These fragments indicate not only whether the source rock was igneous or metamorphic, but also can reveal such things as the silica content of the magma, its rate of crystallization, or the character of the premetamorphic sedimentary rocks from which the metamorphic rock was formed.

Although any type of rock fragment can be found in a sandstone, some types are much more common than others. The factors that determine which will occur are:

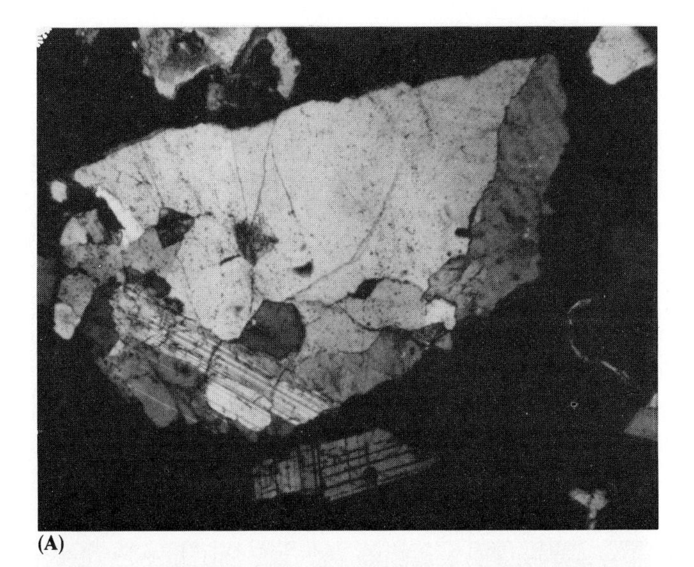
(A)

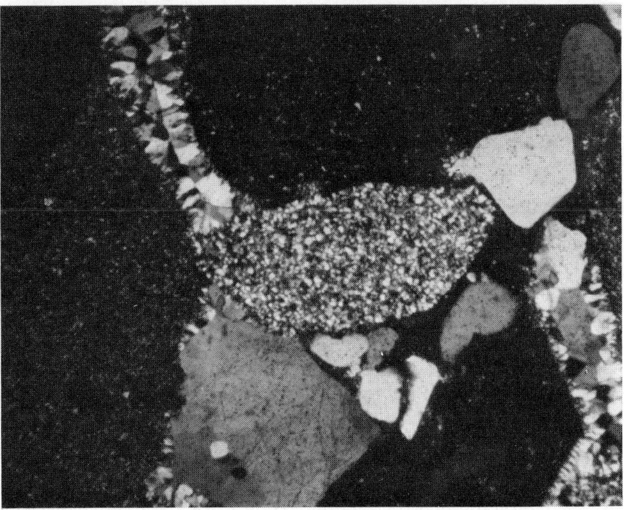
(B)

Figure 13–34
Photomicrographs of coarse sand-size lithic fragments (crossed nicols), showing characteristic appearances. (**A**) Granite fragment composed of a large quartz crystal, an orthoclase crystal (right side, dark gray), and a twinned plagioclase crystal (lower left). (**B**) Chert fragment (center) surrounded by three very finely microcrystalline chert fragments (black with gray speckles) and several quartz fragments (white and gray). Elongated crystals of quartz cement have grown from chert grain surfaces to bind the grains.

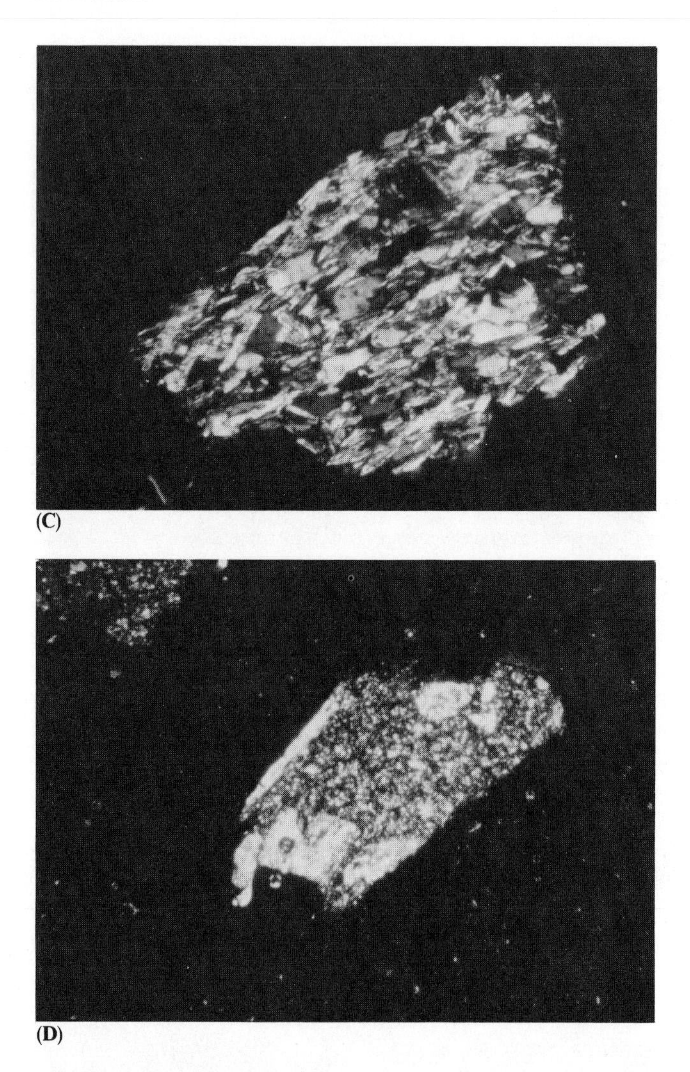

(C)

(D)

Figure 13–34 (*continued*)
(**C**) Mica–quartz schist fragment containing a few opaque mineral crystals. (**D**) Volcanic rock fragment containing three large and altered K-feldspar crystals (one of which is euhedral) and a large mica flake (NW edge of grain) set in a groundmass of felsitic crystals of low birefringence, probably quartz and K-feldspar.

1. Areal abundance in the drainage basin.

2. Location in the drainage basin, whether in the highland or lowland areas.

3. Susceptibility of the rock fragments to chemical and mechanical destruction by sedimentary processes.

4. Size of the crystals within the fragments.

Obviously, the greater the areal extent of the source rock the better the chance to find pieces of it downstream, and we have already noted the greater rates of erosion from areas of high relief.

Factors 3 and 4 determine the *survival potential* of the fragment, and their importance can be illustrated by consideration of mudrock sources. Mudrocks cover two-thirds of the continental surfaces and probably most of the mudrock is shale. Therefore we might anticipate that most undisaggregated lithic fragments in sandstone would be fragments of shale. But this is the opposite of what is found; shale fragments are quite uncommon in ancient rocks. The explanation is that shale fragments are practically untransportable because they are very soft and also split rapidly along fissility surfaces. They have a low survival potential. Extension of this principle leads to the expectation that fragments of gabbro will be poorly represented because of their chemical instability relative to granite; fragments of older sandstones will be rare because of the easy breakage during transport of the common cements calcite and hematite. Most fragments of older sandstone that survive will be either pieces of quartz cemented quartz sandstone or chert fragments.

The crystal size within the rock fragment determines the minimum size of fragment necessary for the fragment to exist. For example, a fragment of granite cannot occur in a fine-grained sandstone because fragments tend to break along crystal boundaries and the crystals in a granite are coarser than fine sand. Fragments such as rhyolite or chert, however, can occur with equal ease in sand of any size. Neglect of this factor can lead to erroneous paleogeologic inferences. It is clear that the interpretation of upstream paleogeology from sandstone petrology is not straightforward, even when the sandstone contains pieces of the source rock itself.

Accessory Minerals

The accessory minerals in sandstone include all detrital minerals except quartz and feldspar, although micas are typically excluded from the accessory group because of their anomalous behavior during transport. Any mineral that occurs in crystalline rocks can occur in sandstones. The relative amounts of the minerals in the sediment depend on the abundance of the mineral in the source rock, its survival potential, and its specific gravity. Because of the wide range in specific gravities of the common accessory minerals, there often is significant segregation among them during transport. The range in specific gravity among the common accessories is 3.0–5.2. In contrast, the range among quartz and feldspars is only 2.56–2.76.

Excepting micas, no common detrital minerals occur with specific gravities in the range 2.8–3.0, and the usual method for separating quartz plus feldspar from the accessory minerals is based on this fact. The loose sediment (or disaggregated sandstone) is dropped into a liquid with a specific gravity in the 2.8–3.0 range, with the result that the quartz and feldspar float while the accessories sink. For this reason the accessory minerals are termed *heavy minerals*. The heavy minerals typically form less than 1% of a sandstone.

Most species of heavies in sandstones originate in metamorphic rocks. This is the case because metamorphic rocks form in a much wider range of *TP* conditions than

Table 13–5
Common Accessory Minerals in Sandstones and the Types of Crystalline Rocks in Which They Usually Originate

Igneous rocks	Metamorphic rocks	Indeterminate[a]
Aegerine	Actinolite	Enstatite
Augite	Andalusite	Hornblende
Chromite	Chloritoid	Hypersthene
Ilmenite	Cordierite	Magnetite
Olivine	Diopside	Sphene
Topaz	Epidote	Tourmaline
	Garnet	Zircon
	Glaucophane	
	Kyanite	
	Jadeite	
	Rutile	
	Sillimanite	
	Staurolite	
	Tremolite	
	Wollastonite	

[a]Minerals in the Indeterminate group are common in both igneous and metamorphic rocks.

igneous rocks, permitting a larger number of species to crystallize. For the purposes of provenance interpretation it is useful to group accessory minerals by the type of crystalline rock in which they usually form (see Table 13–5); unfortunately, many of the more common accessories in sandstones, such as zircon, tourmaline, and magnetite, form in both metamorphic and igneous rocks.

RECYCLING OF GRAINS

Our consideration of sand mineralogy thus far has concentrated on two questions: (1) What are the abundant minerals (and rock fragments) in sandstones? (2) Did these minerals originate in metamorphic or in igneous rocks? That is, we have been concerned with the ultimate sources of the grains. However, two-thirds of the continental surface is covered by sedimentary rocks, not by metamorphic and igneous rocks. If we are to construct an accurate paleogeographic map for the Devonian or Jurassic Periods we will have to determine which of the sand grains are coming directly from igneous or metamorphic sources and which are being released from older sedimentary rocks. We must distinguish between *ultimate* sources and *proxi-*

mate sources. Perhaps the quartz or garnet grain in our Jurassic sandstone emanated last from a Triassic sandstone, and before that resided in a Permian mudrock and an Ordovician conglomerate since being released from a Proterozoic gneiss.

Four approaches are currently accepted for separating ultimate from proximate sources.

1. The percentage of quartz among the detrital grains. The principle involved is that repeated reworking over long periods of time is required to remove completely all the feldspars and lithic fragments from an assemblage of sand grains. Therefore, if a sandstone is composed entirely of quartz, the grains probably were derived from older sandstones rather than directly from an igneous or metamorphic rock.

2. The percentage of superstable accessory minerals in the heavy mineral assemblage. The principle is the same as for quartz. The most resistant heavies are zircon, tourmaline, and rutile, so that the "ZTR index" is a commonly used criterion of the importance of recycling.

3. The degree of rounding of the quartz grains. It requires repeated abrasion over long periods of time in a beach or dune environment to produce a well-rounded quartz grain from the angular grains released from crystalline rocks. Therefore, an assemblage of well-rounded grains indicates not only environments of deposition but recycling as well. Most pure quartz sands in the geologic column consist almost entirely of well-rounded grains.

4. The presence of abraded secondary growths on quartz grains. It is common to find secondary quartz deposited from underground waters onto the surfaces of detrital quartz grains. Subsequently, the rock may be disaggregated and the enlarged quartz grains released and abraded. The abraded overgrowths can be seen in thin sections of the later sandstone deposit that includes the overgrown grains and constitute excellent evidence of recycling (see Figure 13–35). Unfortunately, this criterion for recycled grains is useful only with quartz grains and is uncommon even on quartz.

It is apparent that the key mineral for evaluating the abundance of older sedimentary rocks in a drainage basin is quartz. Its percentage and grain shape are easily determined by petrographic studies; these variables serve as cornerstones of drainage basin analysis. Several indices of recycling have been used to make quantitative evaluations of the proportion of older sediments that existed upstream; for example, the ratio of quartz plus chert to feldspar plus rock fragments; or the ratio of monocrystalline quartz to the more easily destroyed polycrystalline quartz grains. Other useful indices can be devised for particular circumstances.

An example of the complexities in paleogeologic interpretation caused by recycling is provided by a regional study of Phanerozoic sandstones of the upper Mississippi Valley conducted by Potter and Pryor (1961). Their integrated interpretation of petrology, cross bedding, major unconformities, and facies maps led them to the provenance interpretation shown in Figure 13–36.

Figure 13–35
Photomicrograph of a medium sand-size detrital quartz grain (crossed nicols), showing multiple rounded overgrowths, Weber Sandstone (Pennsylvanian–Permian), Utah. The border of the detrital part of the grain is marked by the inner oval ring of water-filled vacuoles. [From I. E. Odom et al., 1976, *Jour. Sed. Petrology, 46,* Fig. 9.]

CLASSIFICATION OF SANDSTONES

Texture and mineral composition are the aspects of a sandstone that give the most insight into the genesis of the rock. For this reason, it makes sense to devise a classification scheme around these two variables, and a large number of such schemes has been published. As we did for mudrocks, we will use a scheme that can be applied in field studies and will only be refined rather than completely changed by subsequent laboratory studies. The easiest mineralogic separation to make in the field is among quartz, feldspar, and rock fragments. These can form the poles of a classification triangle (see Figure 13–37), and the inside of the triangle can be subdivided in any convenient manner. The triangle shown is one of many in common use among professional sandstone petrologists.

The texture of the sandstone is indicated by the use of the concept of textural

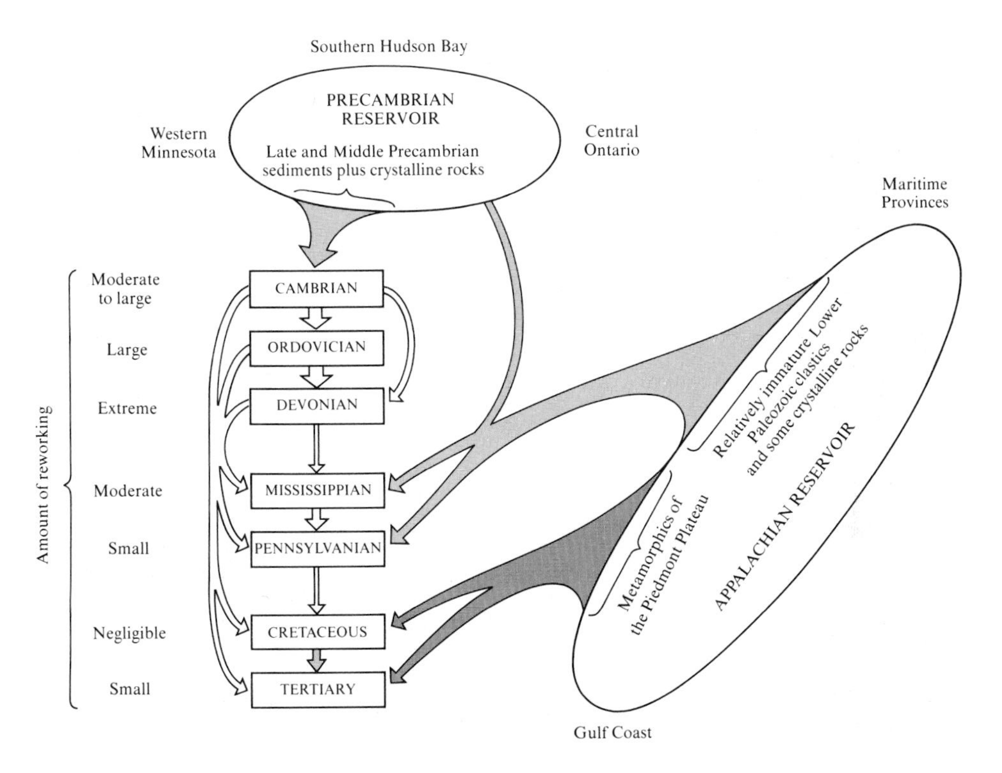

Figure 13-36
Dispersal centers, source relations, and recycling sequence of Phanerozoic clastic rocks in the upper Mississippi Valley. [From P. E. Potter and W. A. Pryor, 1961, *Geol. Soc. Amer. Bull., 72,* Fig. 14.]

maturity. It also is useful to indicate in the rock name the type of cement that holds the grains together. Examples of this system of nomenclature are:

1. Supermature, medium-grained, quartz-cemented quartzarenite.
2. Immature, fine-grained, clay-cemented arkose.
3. Submature, pebbly, calcite-cemented lithic conglomerate.
4. Mature, coarse-grained, calcite-cemented subarkose.

ANCIENT SANDSTONES

Cratonic Sandstones

As an example of a comprehensive study of an ancient cratonic sandstone, we can examine the study of the Fort Union Formation (Paleocene) in Wyoming by Courdin

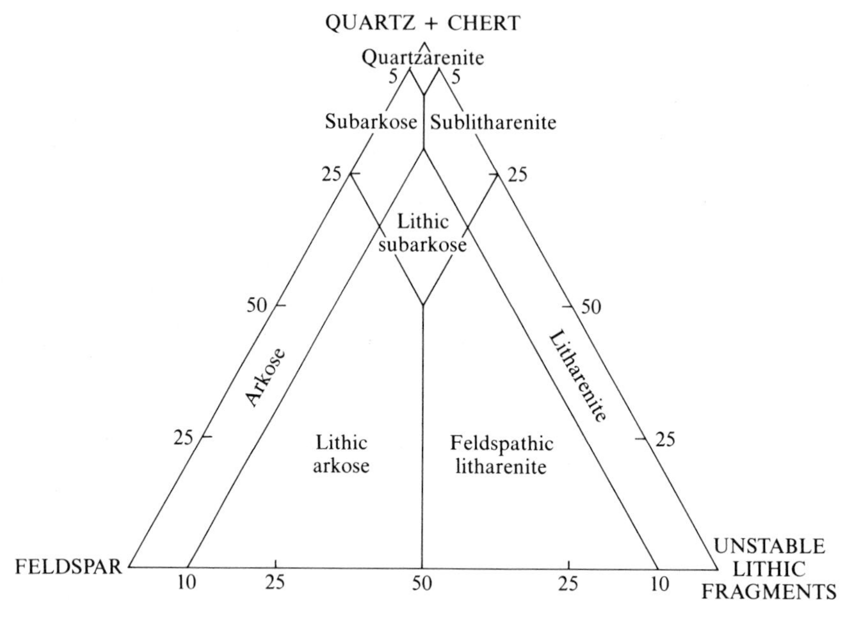

Figure 13–37
One of the many mineralogical classifications of sandstones in common use. [From E. F. McBride, 1963, *Jour. Sed. Petrology, 33,* Fig. 1.]

and Hubert (1969). The Fort Union is a series of conglomerates, sandstones, and mudrocks exposed in the Wind River Basin of west-central Wyoming (see Figure 13–38). It outcrops in narrow, discontinuous belts that trend parallel to the axis of the Basin. Along the western and southern margins its thickness varies from 60 to 300 m, depending largely on the extent of erosion prior to deposition of the overlying sediments (lower Eocene). The Fort Union thickens northward and eastward toward the axis of the Basin and locally exceeds 2,400 m in the subsurface (see Figure 13–39).

Stratigraphy

The lower part of the formation is characterized by white to gray, fine- to coarse-grained, massive to cross-bedded sandstone, interbedded with dark gray to black shale, claystone, siltstone, and brown carbonaceous shale. Abundant thin brown-weathering ironstone beds are present in most places, and lenticular coal beds occur locally. In places much coarse-grained sandstone and conglomerate is present, and some of these units have upwardly concave bases that cut across the bedding of the underlying deposits. Terrestrial plant fossils have been found in some of the finer-grained sandstones and mudrocks.

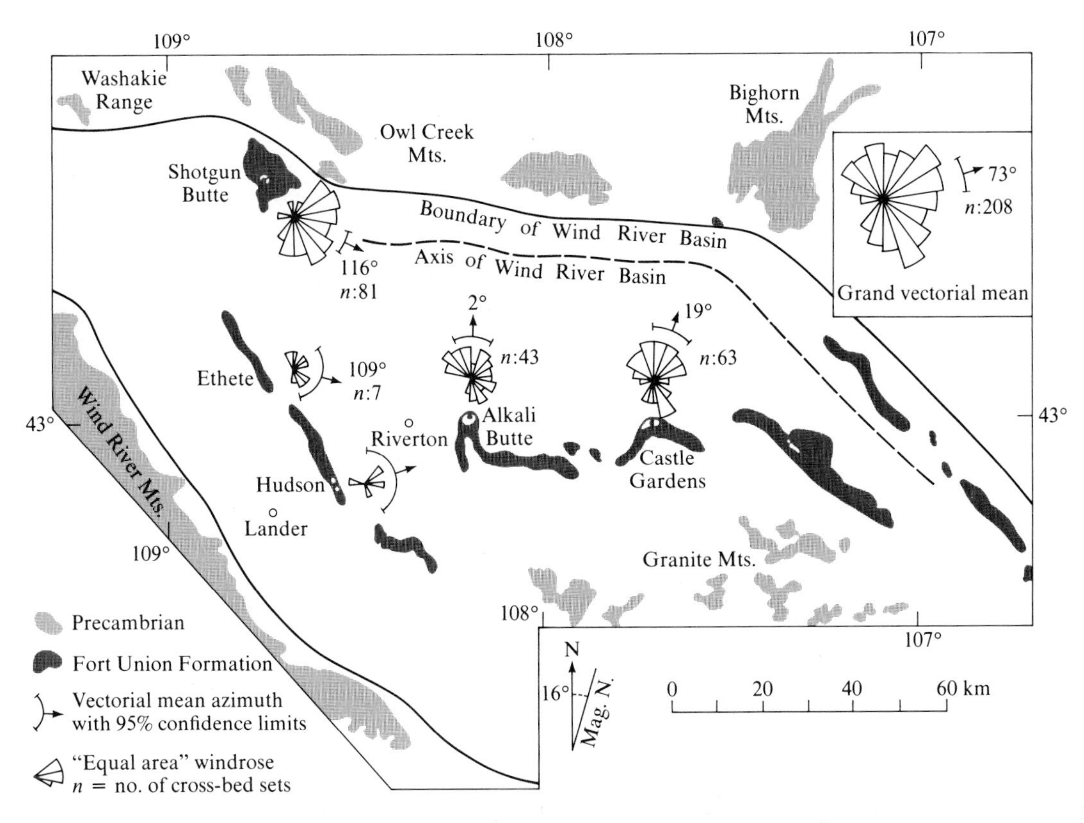

Figure 13–38
Outcrop map of Fort Union Formation in west-central Wyoming showing results of measurements of the directions of cross bedding at each outcrop. [From J. L. Courdin and J. F. Hubert, 1969, *21st Field Conf., Wyoming Geol. Assoc.*, Fig. 1.]

Overlying the lower Fort Union Formation is the Waltman Shale Member of the unit, 200 m thick at its type locality in the eastern part of the Wind River Basin but absent in most areas because of the erosion that preceded the deposition of the overlying unconformable Wind River Formation. In the subsurface the Waltman is up to 1,000 m thick. The shale unit is a homogeneous dark brown to black silty micaceous shale; the minute muscovite mica flakes are easily seen using a hand lens. In some areas, the member contains several sandstone units that are 8–9 m thick. Some of this sandstone is conglomeratic, having abundant pebbles of black chert and scattered cobbles of white granite as much as 15 cm in diameter in a coarse-grained arkosic sandstone matrix.

In two locations in the Wind River Basin, the Waltman Member is overlain by the Shotgun Member, 860 m thick at one locality and 490 m thick at the other. The unit

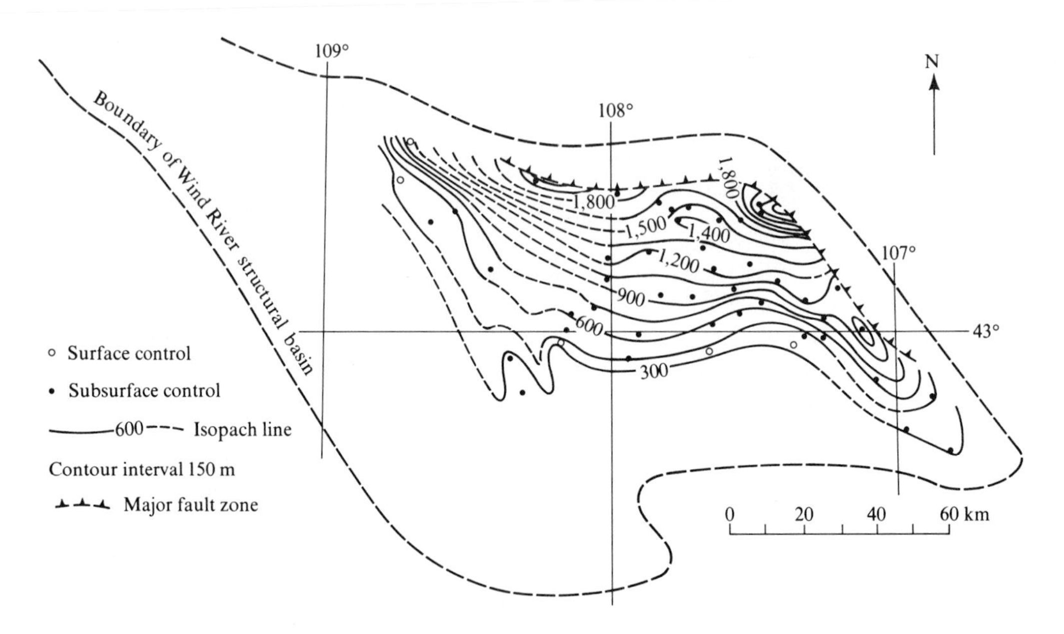

Figure 13–39
Isopach (thickness) map of the Fort Union Formation. [From J. L. Courdin and J. F. Hubert, 1969, *21st Field Conf., Wyoming Geol. Assoc.*, Fig. 2.]

is more widespread in the subsurface. It consists mostly of varicolored mudstones of different types with occasional thin, fine-grained sandstones.

Petrology

In the field, Courdin and Hubert noted the character of the contacts between beds (parallel or cross-cutting), measured a total of 1,960 m of stratigraphic column in the formation at five locations, measured the orientation of 200 cross beds at these sites, and determined the lithology of 941 pebbles coarser than 5 mm in diameter. The results of the cross-bedding measurements are shown in Figure 13–38. The average direction of sediment transport was to the ENE, but a great deal of scatter is present in the measurements; this is interpreted to reflect meandering streams. The paleoslope is clear, however, as almost no cross beds dip toward the WSW. Standard deviations of the fine-grained channel sands vary between 0.4 ϕ and 0.7 ϕ, normal values for fine-grained fluvial sands. The mudrocks are silty and were interpreted to be overbank and floodplain deposits.

The pebble count revealed that 87% are sedimentary types, all from pre-Paleocene formations that still outcrop in the Wind River Basin. Chert is the most common type of pebble (76%). Varieties of chert present include black phosphatic chert traceable to the Permian Park City Formation of Permian age; gray-banded chert from the Madison Limestone of Mississippian age; and "bird's-eye" agate chert from the Morrison Formation (Jurassic). Most of the remaining 11% pebbles of sedimentary origin are composed of quartzarenite fragments from the Tensleep Formation (Pennsylvanian); a few (3%) are siliceous shale pebbles of the Mowry Formation (Cretaceous). About 13% of all pebbles are fragments of the Precambrian granite exposed at the border of the basin.

In the laboratory, thin sections and petrographic analyses were made of 49 randomly selected sandstones from the Fort Union Formation. The thin sections were cut perpendicular to lamination to avoid the possibility of slicing along a lamina with an unrepresentative mineral composition. In each slide, 300 detrital grains coarser than 0.03 mm were identified along linear traverses on the slide. Twenty-three sandstones were then disaggregated by dissolving their hematite and calcite cements in dilute cold HCl; the freed grains were recovered and a heavy liquid was used to separate the quartz and feldspar from the heavy accessory minerals. The heavies were then mounted in epoxy on a glass slide and 200 nonopaque mineral grains were identified. During counting, features of the grains such as degree of rounding, euhedral character, and color were noted.

The light mineral fraction consists of 77.5% quartz, 11.6% chert, 6.0% feldspar, and 4.9% rock fragments. Nearly all of the quartz (94%) is angular to subrounded single crystals that are undiagnostic of either ultimate or proximate source. Probably most were derived from older sandstone units that fringe the basin, but there is no way to be certain. A few of the monocrystalline quartz grains are well rounded and, therefore, certainly were not derived directly from the granitic rocks in the area. The chert grains are of the same types found in the pebble count and occasionally contain ghosts of replaced fossils and other carbonate particles that reveal the chert to have formed by replacement of older limestones in the pre-Paleocene source rocks. The black chert sand grains contain phosphatic inclusions, reflecting their origin in the chert member of the Permian Park City phosphate formation.

About 95% of the feldspar grains in thin section are alkali feldspar. The frequency of occurrence of feldspars in the Fort Union is perthite > orthoclase > microcline > plagioclase. The relative abundance of perthite, which is very easily decomposed and disintegrated, indicates that the feldspar probably was derived directly from nearby granitic rocks rather than being of recycled origin. Perthite has a low survival potential.

The rock fragments are 55% sedimentary (shale and siltstone), 35% metamorphic (schist), and 10% igneous (granite). The fact that rock fragments form only 4.9% of the detrital grains suggests that the dominant sources for the Fort Union Formation were older sediments rather than igneous and metamorphic rocks. The dominance of easily disaggregated particles such as shale and schist testify once again to the nearness of the source rocks of the formation.

The suite of accessory minerals is dominated by zircon (72%), garnet (8%), tourmaline (6%), and rutile (4%). Other minerals present include sphene, zoisite, apatite, staurolite, and anatase. The dominance of ZTR and garnet (which ranks just behind ZTR in stability in the sedimentary environment) supports the idea obtained from the light minerals that recycled sediments were the dominant source of the Fort Union.

Diagenesis

The sequence of precipitation of the major cements is iron oxides, silica, and carbonates. Limonite and hematite were deposited as coatings on the grains and as pore-filling cements. Secondary growths of quartz (overgrowths) were precipitated from migrating pore solutions on top of many iron-oxide-stained detrital quartz grains. Coarse, clear calcite is the dominant pore-filling cement in Fort Union sandstones. Occasionally, dolomite and siderite euhedra occur. The carbonate cements commonly corrode and replace quartz overgrowths, indicating that they formed later than the silica cement.

Other diagenetic growths present include feldspar overgrowths, tourmaline overgrowths, and euhedral crystals of anatase, kaolinite, and pyrite.

Summary

During Paleocene time, the granitic mountains that border the Wind River Basin were rising highlands that enclosed most of the subsiding basin. Fluvial, deltaic, lacustrine, and paludal (swamp) sediments of the Fort Union Formation accumulated in the basin. The cross-bedded sandstones were deposited as point bars on the inner slopes of meander bends of the shifting streams. The average grain sizes and standard deviations of these sands are typical of such deposits. Cross-bedding azimuths show that the rivers flowed basinward from the southern and western margins.

The initial detritus eroded from the Mesozoic and Paleozoic sedimentary rocks that overlie the crystalline basement consisted dominantly of quartz, chert, sandstone–siltstone fragments, and round ZTR grains. Further uplift and local unroofing of the granitic cores of the uplifts led to local influxes of granitic detritus as revealed by changes in detrital petrology at Castle Gardens, Alkali Butte, and Hudson, where a transition is observed from cherty litharenite and sublitharenite in the lower parts of the sections to subarkose, lithic arkose, and arkose in the upper parts. The Precambrian cores of the Washakie, Owl Creek, and northern end of the Wind River Mountains were exposed only locally during the Paleocene.

Sandstones at Convergent Plate Margins

Sandstones deposited on the craton are characterized typically by very high percentages of quartz and chert, 85–90% in the example of the Fort Union Formation. The mineral composition of these sandstones is dominated by resistant detrital minerals that have survived the repeated and vigorous destructive processes that typify a stable tectonic setting.

The opposite extreme of tectonic activity and detrital mineral composition occurs in the vicinity of the convergence of an oceanic plate and a continental plate, such as has existed near the west coast of North America for approximately the past 200 million years. Sandstones deposited in forearc basins, between the oceanic trench and the volcanic island arc, have a mineral composition characterized by an abundance of basaltic volcanic rock fragments, calcic plagioclase grains, and noticeable amounts of accessory minerals such as olivine or augite. These sediments usually have not suffered much transport and the bulk of them have been deposited in marine waters below wave base. In addition, the convergent plate margins have some areas of higher than average heat flow. The combination of high heat flow and an abundance of detrital particles that are chemically very unstable at low temperatures results in pervasive diagenetic alteration of the grains at relatively shallow depths of burial. The diagenetic production of clay matrix from the detrital grains and the crystallization of zeolite minerals are common phenomena in plate-margin sandstones.

As an example of plate-margin sandstones, we can consider the Mesozoic lithic sandstones in central Oregon studied by Dickinson et al. (1979). The sandstones occur in inliers surrounded mostly by Cenozoic plateau basalts of the Columbia River Sequence (see Figure 13–40A) and are similar in aspect to the sediments in the eugeosynclinal Fraser Belt exposed to the north in British Columbia and to the south in Northern California. The most extensive outcrop of Mesozoic sandstones in central Oregon is the John Day inlier, approximately 2,500 km^2 in extent. Based on previous studies, the position of the inlier within the eugeosynclinal belt is shown in Figure 13–40B.

Stratigraphy

The Mesozoic sequence of the John Day inlier includes about 15,000 m of dominantly clastic strata that were deposited between mid-Triassic and mid-Cretaceous times. Most sandstones are turbidites, but facies changes are intricate in detail and shelf deposits occur locally as lateral equivalents of nearby turbidite successions. At least six unconformities break the stratigraphic sequence, which has been subdivided into about 25 formally named stratigraphic units. Many of these are quite thin and have limited lateral extent.

Petrology

As is typical in eugeosynclinal rocks, both depositional texture and mineral composition are obscured in many samples by intense alteration of framework grains, growth of diagenetic matrix, or replacement by secondary carbonates. Dickinson et al. (1979) concentrated their petrographic studies on sandstones whose texture and limited degree of alteration permitted positive identification of the framework grains.

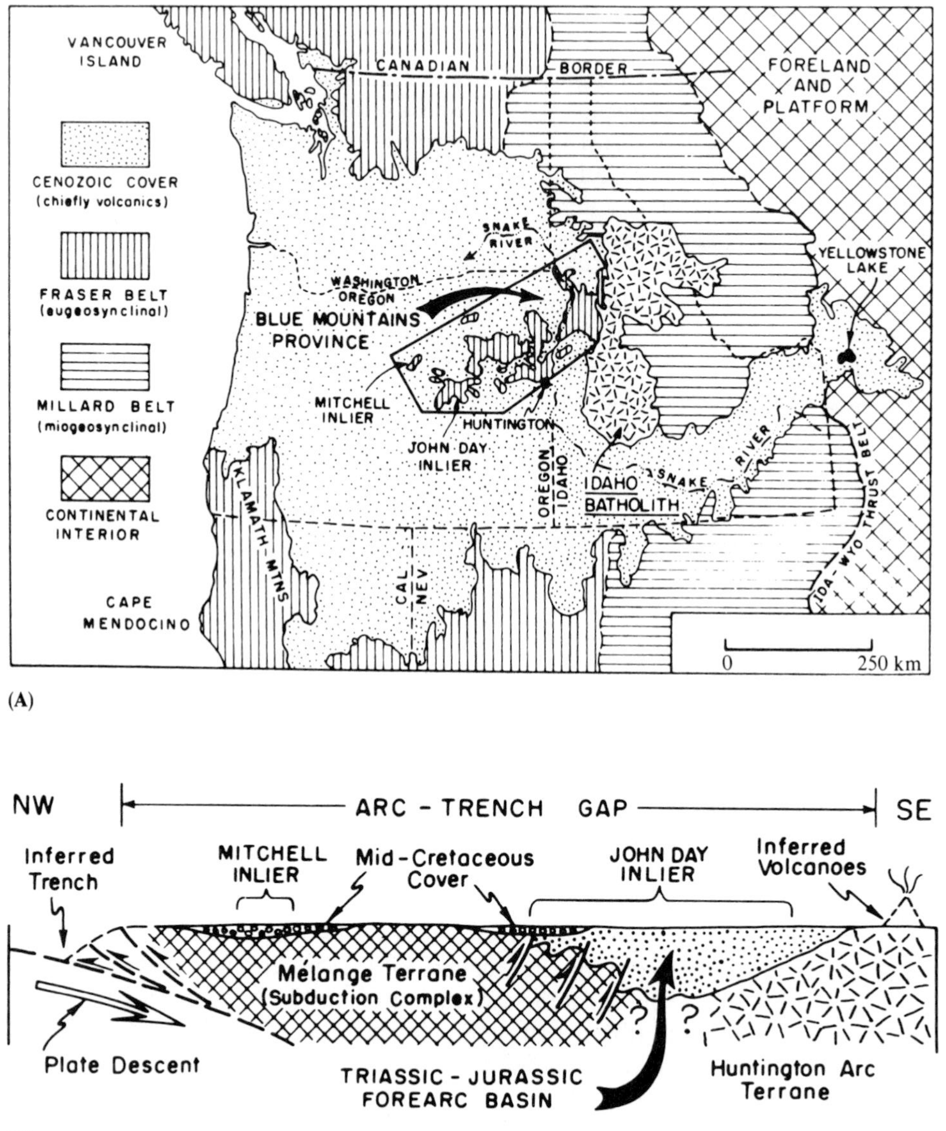

Figure 13–40
(**A**) Location of John Day and Mitchell inliers in relation to regional orogenic terranes.
(**B**) Schematic restored profile across pre-Tertiary terrane of central Oregon showing inferred position of the John Day inlier within a mid-Mesozoic arc-trench gap. [From W. R. Dickinson et al., 1979, *Jour. Sed. Petrology, 49,* Figs. 1, 2.]

Table 13–6
Average Modal Compositions of John Day Inlier Sandstones Derived from Each of the Three Distinct Provenance Groups[a]

	Mélange-derived sandstones	Volcaniclastic sandstones	Mixed-provenance sandstones
Interstitial matrix and cement	7	7	9
Monocrystalline mineral grains			
Quartz grains	5	1	13
Plagioclase grains	12	30	21
K-feldspar grains	1	1	4
Polycrystalline quartz fragments			
Chert	38	1	5
Polycrystalline quartz	6	0	1
(Meta-) Volcanic rock fragments			
Microxenolithic grains	18	56	32
Felsite grains	6	3	10
Microgranular hypabyssal	3	0	1
(Meta-) Sedimentary rock fragments			
Argillite grains	1	2	7
Shale/slate grains	5	0	6
Quartz–mica tectonite	4	0	0
Clinopyroxene grains	0	3	0
Mica flakes	0	0	1
Miscellaneous framework grains	1	2	0

[a]Compositions total 100% excluding matrix and cement.
Source: W. R. Dickinson et al., 1979, *Jour. Sed. Petrology, 49.*

In each of the 34 thin sections examined, 400 grains were identified, noting not only whether the grain was quartz, feldspar, or a rock fragment, but also the type of feldspar (identified by selective staining for K-feldspar and plagioclase) and internal structure of the quartz and lithic fragments.

Based on the results of the point counts, it was possible to group the 34 thin sections into three provenance categories (see Table 13–6).

1. Eleven samples whose grains were derived entirely from the upper Paleozoic to lowermost Mesozoic mélange terrane.

2. Nine samples of volcaniclastic turbidites derived mainly from contemporaneous Jurassic volcanic eruptions nearby.

3. Fourteen samples of lithic sandstones derived from mixed provenances exposing varied rock types not limited to volcaniclastic or mélange sources. Even in this

provenance category, however, the most abundant grain types are volcanic rock fragments.

The source rocks of the mélange-derived sandstones are mainly chert, greenstone, and altered felsite, and slate or phyllite within the mélange terrane. Feldspar grains evidently were derived chiefly from the phenocrysts in the volcanic components of the mélange, for the feldspar content is highest in the samples that are richest in volcanic rock fragments. The very low percentages of quartz and K-feldspar indicate no direct plutonic or basement source of these minerals was present (for example, granite). Mélange-derived sandstones are most common in the lower part of the John Day inlier.

The volcaniclastic sandstone group has the clearest provenance of the three groups. About 60% of the framework grains are volcanic rock fragments, mainly microxenolithic, and 30% are volcanic feldspar grains, almost entirely plagioclase of phenocrystic origin. As we would expect from a provenance of mafic volcanic rocks, quartz and chert are nearly absent and clinopyroxene grains are uncommonly abundant, forming 3% of the framework of the sandstones. The rocks with volcaniclastic provenance are most prominent in the middle section of the John Day inlier.

The sandstones of mixed provenance have compositions intermediate between those of the other two groups, plus additions from a sedimentary or metasedimentary terrane. Volcanic rock fragments, which total 27% in the mélange-derived sandstones and 59% in the volcaniclastic sandstones, are 43% in the sandstones of mixed provenance. Chert plus sedimentary rock fragments are 48% in the mélange-derived sandstones, 3% in those of volcaniclastic provenance, and an intermediate 18% in sandstones of mixed parentage. The sandstones of mixed parentage dominate in the upper parts of the John Day inlier sequence, as would be expected. The provenance is interpreted to have been an eroded arc terrane in which some of the folded and intruded substratum beneath the volcanoes had begun to undergo erosion.

Diagenesis

The samples studied by Dickinson et al. were specifically chosen to be largely free of diagenetic effects, but it is clear that all grades of alteration are present. In the most affected sandstones, pervasive diagenesis has largely obliterated the detrital mineralogy so that the identification of the detrital grains is uncertain. Probable volcanic lithic fragments appear as roundish aggregates of clay and fine mica whose boundaries can sometimes be inferred and usually are indistinct.

In the 34 samples that were examined, however, the amount of silt- and clay-size matrix is less than 5% in all three groups of sandstones, indicating very little conversion of the unstable mafic volcanic fragments into clay. The sandstones of volcaniclastic and mixed provenance do, however, contain 3–5% of clearly authigenic green chlorite as a cementing material. The chlorite rims the framework grains and is oriented normal to their surfaces, which would not be possible if the chlorite were detrital in origin.

Summary

Lithic sandstones that were deposited within a Mesozoic forearc basin in central Oregon include mélange-derived, volcaniclastic, and mixed-provenance suites in which relative proportions of feldspar grains, chert grains, volcanic–metavolcanic rock fragments, and sedimentary–metasedimentary rock fragments are different and diagnostic. The stratigraphic variations in sandstone petrology provide evidence about the tectonic evolution of the uplifted mélange ridge and the igneous arc terrane that flanked the forearc basin on opposite sides during the mid-Mesozoic.

Although petrologic relationships in eugeosynclinal terranes may be obscured in many samples by diagenetic alteration, it generally is possible to obtain samples whose depositional character is well preserved.

MODERN SAND ENVIRONMENTS

The chief reason petrologists study the composition of the detrital fraction of ancient sandstones is to decipher paleogeology and paleogeography. But before we can do this we must accumulate a data base from modern stream studies. How rapidly do schist fragments disintegrate during stream transport? How rapidly is the percentage of basalt fragments lowered by the input of quartz sand released from friable sandstones surrounding the area of outcrop of the basalt? Until we know the answers to questions such as these it will not be possible to interpret adequately the mineral composition of ancient sands.

With the important exception of turbidity current deposits, sands accumulate in environments of high kinetic energy, such as the high-velocity sections of stream channels or alluvial fans, marine beaches or sand bars, and desert dunes. Therefore our goal in the remainder of this chapter will be to describe and interpret variations in the mineral composition of sands in modern environments, both nonmarine and marine. The present is the key to the past.

Elk Creek, South Dakota

Elk Creek is located in the northern part of the Black Hills, a partially unroofed asymmetrical dome approximately 150 km long and 80 km wide (see Figure 13–41). The domed area rises 800 m above the surrounding plain. The core of the uplift is composed predominantly of fine-grained Precambrian schist, minor amphibolite, and lesser amounts of granitic gneiss. In the northern Black Hills where Elk Creek is located, numerous shallow intrusives of Tertiary age cut the schist and its overlying Paleozoic sedimentary rocks. Precambrian granite is exposed in the southern part of the Black Hills but none crops out as far north as the drainage basin of Elk Creek.

A study was conducted to determine the rate of decrease of the schist fragments in Elk Creek sediment with increasing distance from their areas of outcrop. Samples

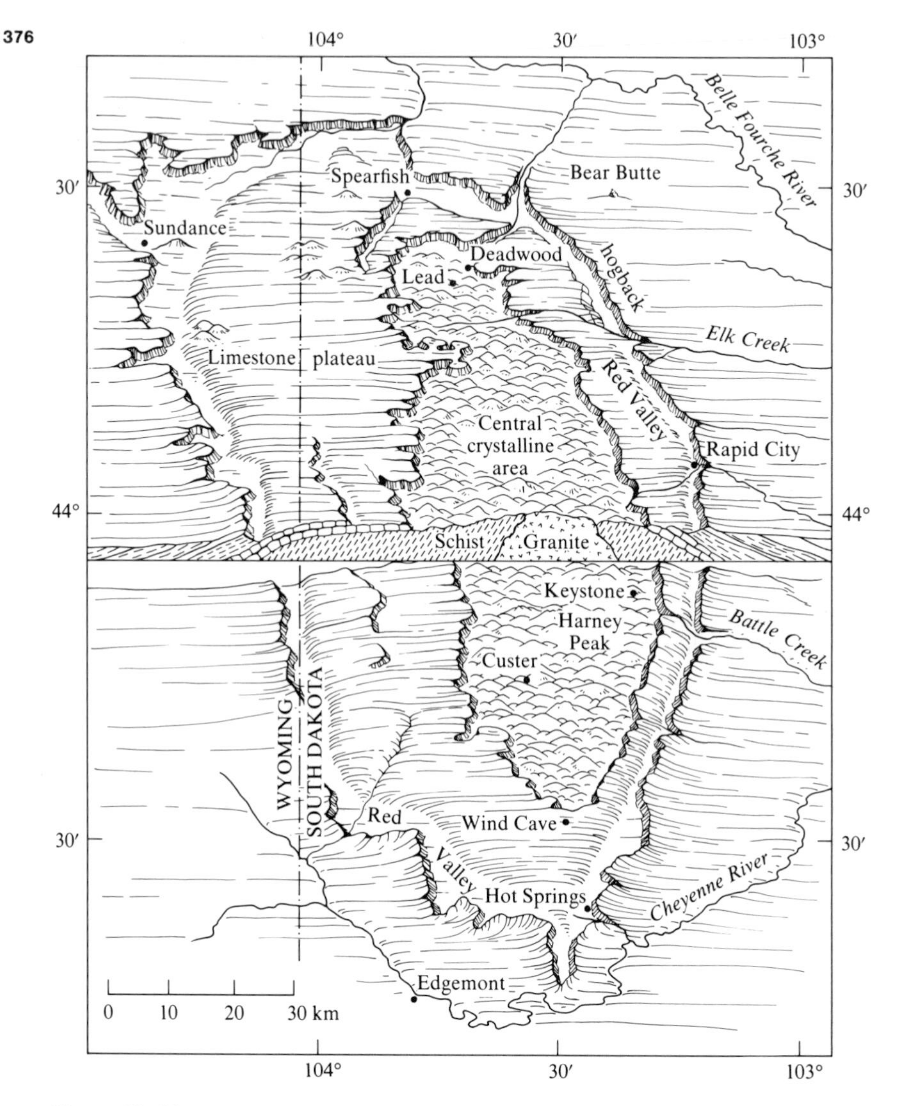

Figure 13–41
Bird's-eye view by Canada geese of the physiography and general geology of the Black Hills, as seen while flying north. [From K. L. Cameron and H. Blatt, 1971, *Jour. Sed. Petrology, 41,* Fig. 1.]

were taken both from the residual soil developed in the source area and from the sand in the stream at arbitrarily selected distances downstream. Schist is exposed over 45% of the drainage area in the headwaters of the stream, and the sand fraction of the residual soil on the schists consists almost entirely of schist fragments. Nevertheless, after less than 0.5 km of stream transport, schist fragments form only 30% of the very coarse and coarse sand sizes and less than 20% of the two finer sand sizes. After

21 km of stream transport (13 airline km) schist fragments have decreased to less than 2% of the sediment (see Figure 13–42). It was concluded that schist fragments have a low survival potential because of mechanical instability. Apparently, soil-forming processes weaken the bonds between adjacent mica flakes sufficiently so that the schist fragments disaggregate almost immediately on entering the high gradient part of the stream (20 m/km). Schist fragments are nearly untransportable in this semiarid area. This suggests that ancient sandstones that contain large amounts of schist sand grains must be located either within a few kilometers of the source area or else adjacent to a source area with much greater topographic relief than is present in the drainage basin of Elk Creek.

Vogelsberg Area, West Germany

An attempt was made in West Germany to investigate the loss of volcanic fragments during stream transport. In the State of Hessen there occurs a nearly circular mass of Tertiary basaltic rocks about 250 km^2 in area. The volcanic outcrop has a maximum relief of 400 m and stream gradient in the upland area of 25–30 m/km; maximum relief is only half of that in the Black Hills but the stream gradient is slightly greater. Samples of stream sand were collected both within the area of volcanic outcrop and outward for several tens of kilometers and the proportion of volcanic fragments determined. In the coarse sand fraction the percentage of volcanic debris decreases to 5% or less within 10–20 km from the basalt outcrop; in the fine sand fraction the decrease to 5% occurred within only 5 km (see Figure 13–43). The more rapid loss of

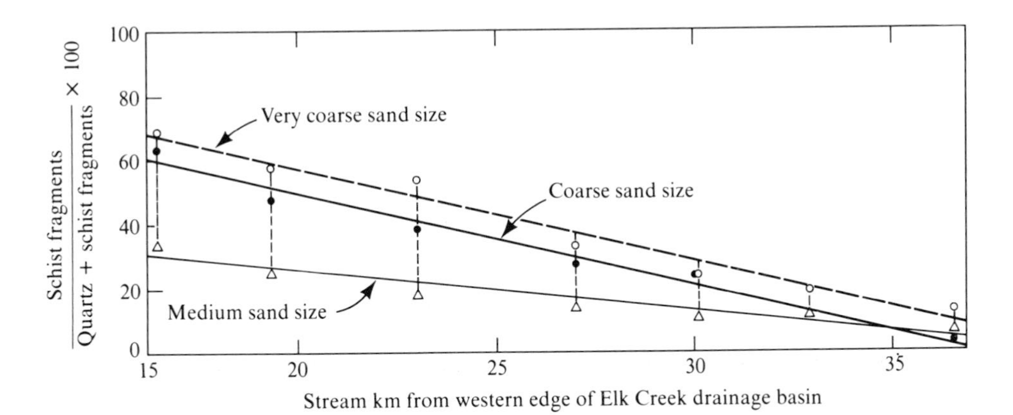

Figure 13–42
Decrease in abundance of schist fragments with distance of transport in Elk Creek, expressed as the ratio of schist to quartz plus schist in the three coarsest sand fractions. [From K. L. Cameron and H. Blatt, 1971, *Jour. Sed. Petrology, 41,* Fig. 10.]

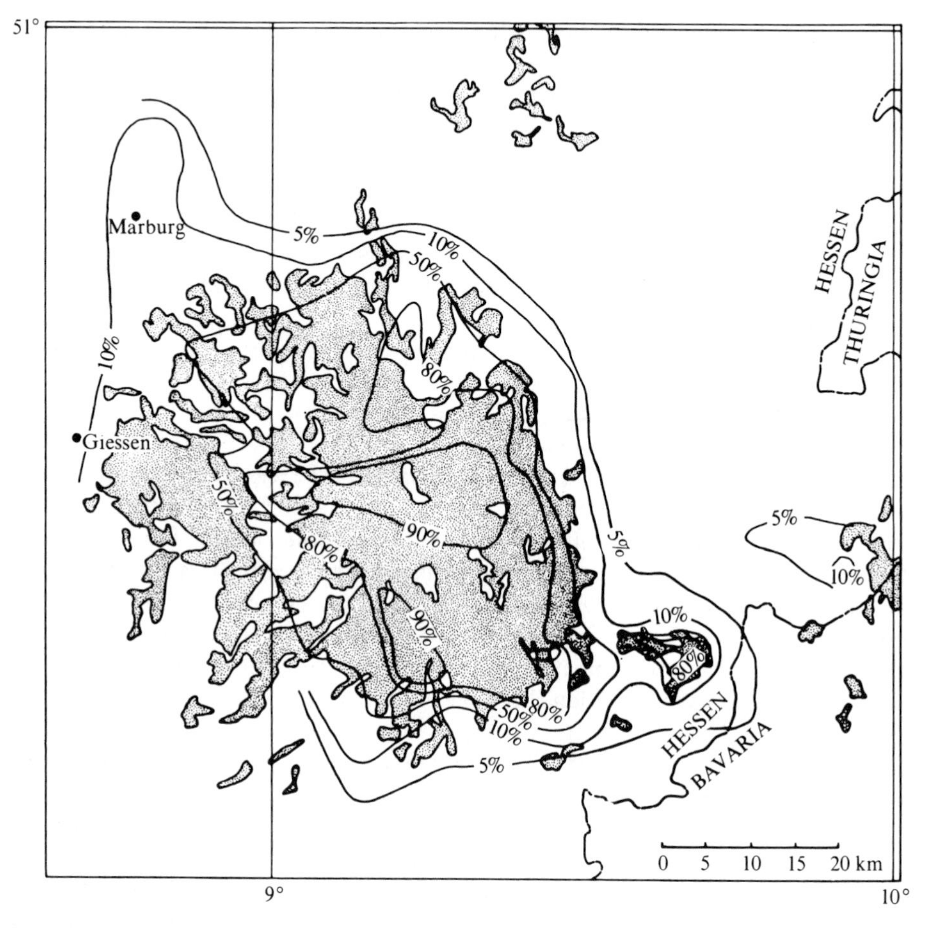

Figure 13–43
Map of outcrops of volcanic rocks in central Hessen. Isopleths are the percentage of volcanic detritus in the fine sand fraction of stream sediment. (Note irregular contour interval.) [From H. Blatt, 1978, *Geol. Rundschau, 67,* Fig. 3.]

volcanic debris from the fine sand fraction was attributed to greater dilution by the fine-grained friable quartz sandstone that surrounds the circular basalt mass.

In general, then, the results of the Black Hills study and the German study were similar. In areas of crystalline rock outcrop a few hundred square kilometers in area and with topographic relief of a few hundred meters, debris from the outcrop disappears very rapidly. After a distance of stream transport of 20 km, less than 5% of the stream sediment will consist of crystalline rock debris. In the Black Hills, the disappearance results from the mechanical instability of foliated fragments; in the German study the disappearance results from intense dilution by sand released from surrounding sandstones. Clearly, areas of outcrop of crystalline rocks that are only a few hundred square kilometers in extent will be very difficult to detect in the drainage basin of an ancient sandstone.

Gulf of Mexico

The Gulf of Mexico is a large marine basin of uncertain tectonic origin in a humid subtropical climatic region. It receives detrital sediment from several directions and from a large variety of different types of source rocks, ranging in character from the polycyclic Phanerozoic debris of the mid-continental United States to the late Tertiary volcanic rocks of eastern Mexico. In addition, there is extensive development of carbonate reefs at shallow water sites where the influx of detrital sediment is not large. Figure 13–44 illustrates depositional provinces in the Gulf based on sand distribution patterns defined by differences in the distribution of heavy minerals. Similar boundaries could be drawn based on the distribution pattern of clay minerals, but the boundaries would not be as clearly defined because there are not as many distinctly different clay minerals as there are heavy minerals. It might also be possible to define depositional provinces in the Gulf based on feldspars (which form 20% of Mississippi River sediment) and rock fragments (which form 10% of Mississippi River sediment and a much larger part of the sediment entering the Gulf from the mountains of eastern Mexico), but this has not yet been done.

The nonopaque heavy mineral suite of the Eastern Gulf Province is characterized by large percentages of kyanite plus staurolite (32%) derived from the Cretaceous and younger sedimentary mantle of the Appalachians. The minerals were derived ultimately from the high-grade metamorphic rocks in that region.

The Mississippi Province is identified by the presence of abundant augite (23%) but also contains much hornblende (40%), epidote (16%), and garnet (9%). This assemblage has been derived from glacial deposits in the upper part of the Mississippi River drainage.

The Central Texas Province contains as much hornblende (58%), epidote (17%), and garnet (7%) as the Mississippi Province but contains almost no augite (3%). The source of this association is believed to be largely the Colorado River of Texas with subordinate contributions from the Mississippi and Rio Grande Rivers.

The Rio Grande Province contains as much epidote (15%) as the other provinces but contains 23% hornblende (much more than in the Eastern Gulf but much less than the Mississippi and Central Texas Provinces), 24% augite (the same as the Mississippi but eight times more than the Eastern Gulf and Central Texas Provinces), and 7% brown hornblende derived from the basaltic rocks that are abundant in the Rio Grande drainage basin. Brown hornblende is nearly absent in the other three provinces (0–2%).

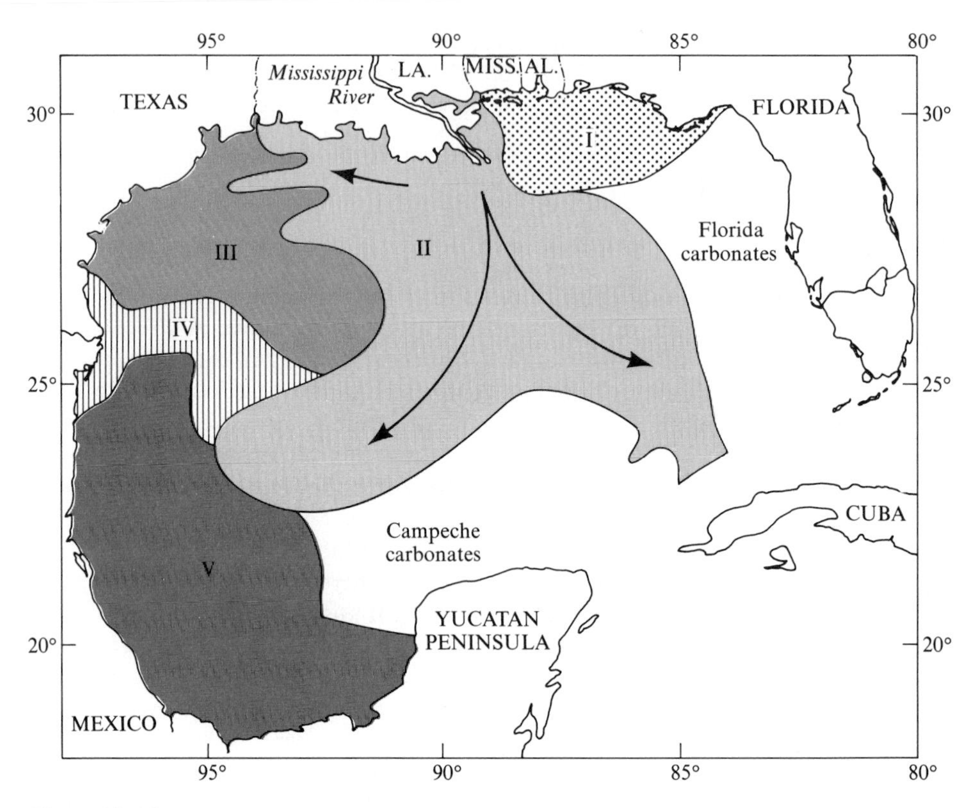

Figure 13–44
Major heavy mineral provinces, areal distribution of Mississippi River sediment, and principal sediment dispersal directions in the Gulf of Mexico. I = Eastern Gulf Province; II = Mississippi Province; III = Central Texas Province; IV = Rio Grande Province; V = Mexican Province. [From D. K. Davies and W. R. Moore, 1970, *Jour. Sed. Petrology, 40,* Fig. 1.]

There have been few studies of the Mexican Province in the southwestern part of the Gulf of Mexico, but it appears to be similar in heavy mineral composition to the Rio Grande Province with the exception that the Mexican Province lacks brown hornblende.

Sea of Japan

Very few sedimentary petrologists have attempted to characterize modern sediment distributions by variations in feldspar content or chemical composition. This results from the relatively time-consuming nature of feldspar studies. When dealing with lithic fragments and heavy minerals a large number of easily identifiable types is

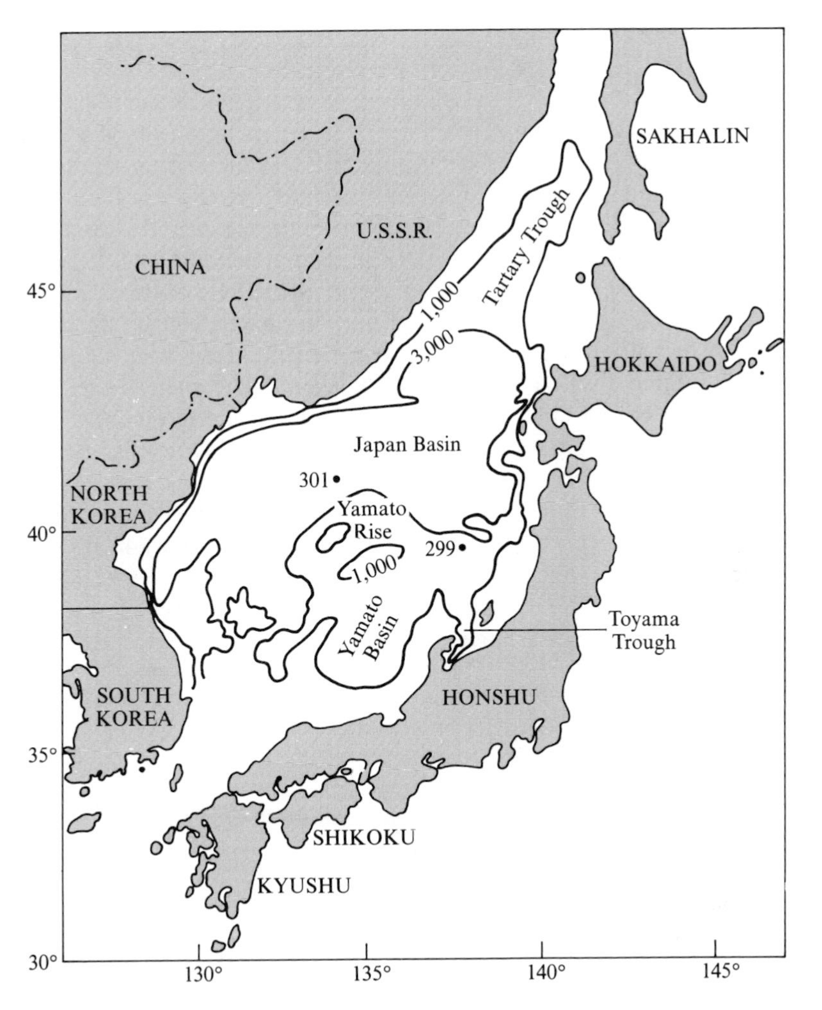

Figure 13–45
Location of sample sites 299 and 301 in relation to physiographic features in the Sea of Japan. Bathymetric contours in meters. [From J. C. Ingle, Jr., et al., 1975, *Initial Reports of the Deep Sea Drilling Project, 31,* Fig. 2.]

present. There exist many varieties of igneous, metamorphic, and sedimentary lithic fragments and as many species of heavy minerals. But there are few easily identifiable types of feldspars (orthoclase, microcline, plagioclase, and very rarely microperthite or sanidine), and determination of the compositions of individual plagioclase grains is usually not attempted.

A notable exception is a recently published study of the provenance of turbidite sediments from the water–sediment interface to a depth of about 500 m in the Sea of Japan (see Figure 13–45), a backarc sedimentary basin. At site 299, 33 samples were taken from 8 cores; at site 301, 8 samples from 4 cores. The sediments are dominantly silts, so it was more convenient to examine the feldspars in the 74–34 μm size rather than in the fine sand size only. Results for the 41 samples from the two sites are summarized in Figure 13–46, and it is clear that two source areas for the sediments can be distinguished. The feldspars at site 299 contain significantly lesser amounts of orthoclase and greater amounts of plagioclase than at site 301 and, in addition, site 299 feldspars are more calcic than those at site 301. There also is a lower quartz/feldspar ratio in site 299 sediment. These data led to the conclusion that site 299 receives most or all of its sediment from the Japanese volcanic arc via the north–south trending Toyama Trough (see Figure 13–45), whereas site 301 receives detritus from both the arc and the granitic Asian mainland to the west.

Samples from the two sites also show numerous stratigraphic changes in mineralogy through the 500-m length of the cores from each site (see Figure 13–47). It is clear that large changes in mineralogy occur over very short stratigraphic intervals; for example, the increase in calcic feldspar ($n > 1.54$) from 24% to 44% in the interval between 44.53 m and 44.98 m depth. So sharp a change within 0.45 m presumably reflects a sampling of different turbidity current deposits, the more calcic feldspars reflecting a greater sediment contribution from the island arc.

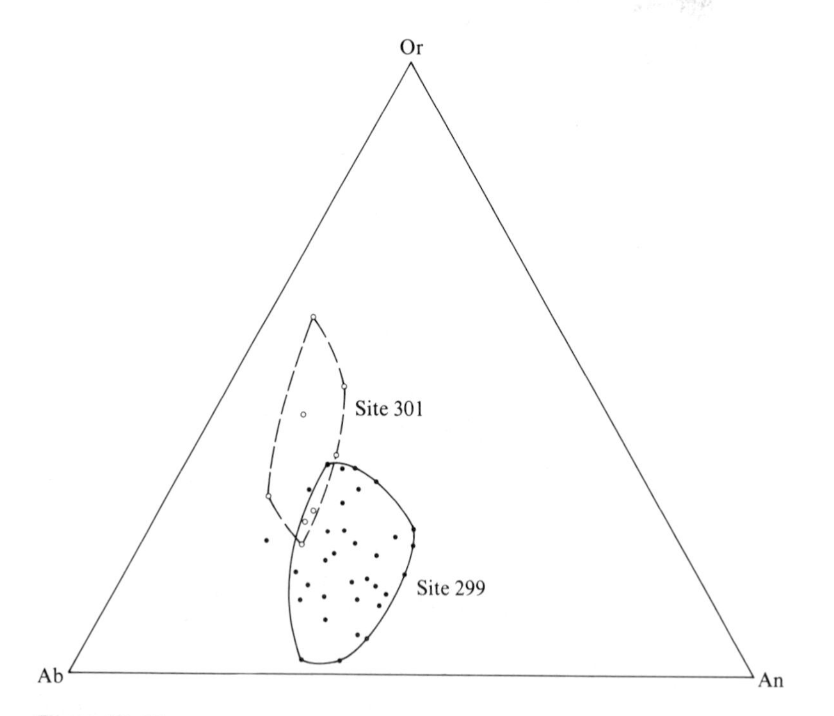

Figure 13–46
Compositions of unaltered feldspars in samples from sites 299 and 301 in the Sea of Japan. Each data point indicates the relative amounts of orthoclase and plagioclase as well as the composition of the plagioclase. [From D. F. Sibley and K. J. Pentony, 1978, *Jour. Sed. Petrology, 48,* Fig. 2.]

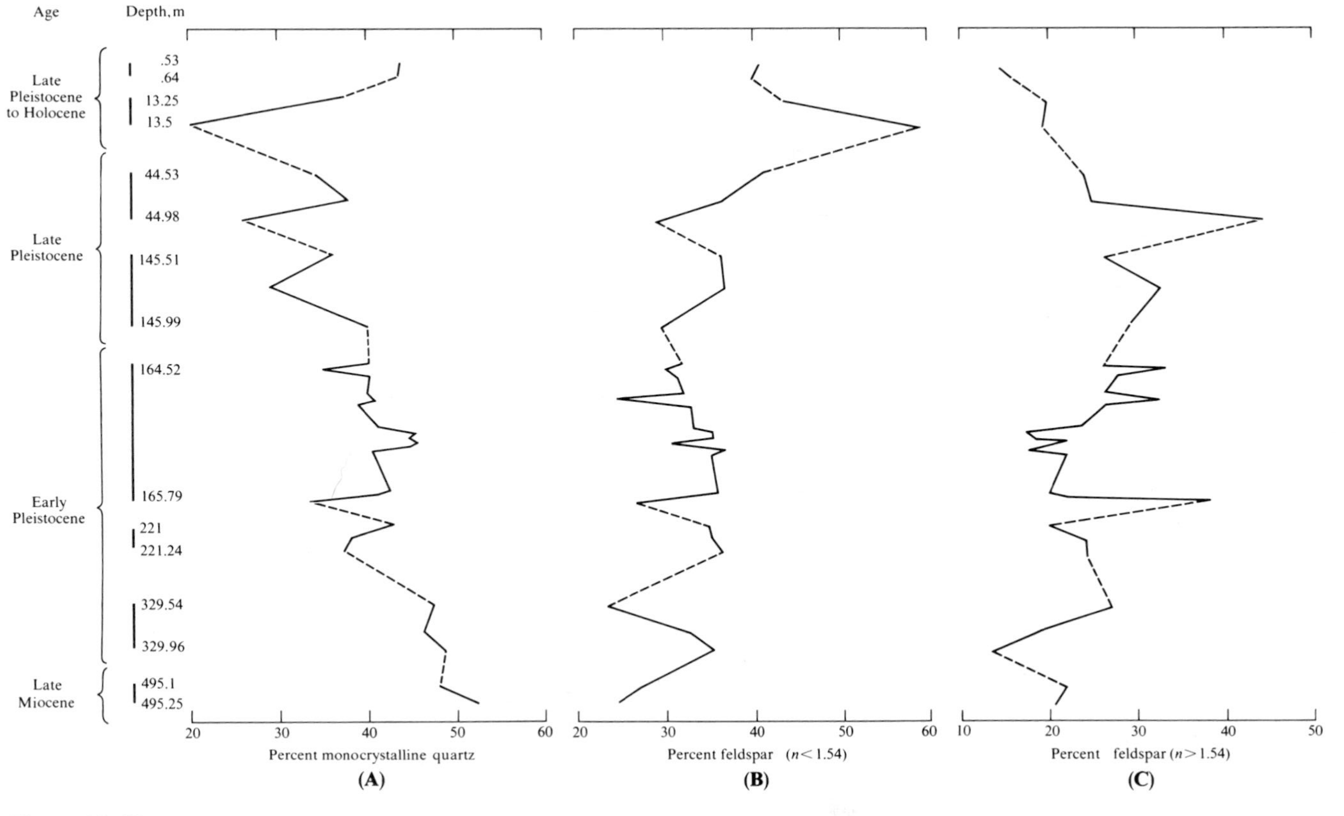

Figure 13–47
Stratigraphic changes in quartz and feldspars at site 299 based on petrographic distinctions. [From D. F. Sibley and K. J. Pentony, 1978, *Jour. Sed. Petrology, 48,* Fig. 3.]

SUMMARY

The essential features of conglomerates and sandstones that must be described and interpreted are texture, structure, and mineral composition. Textural features supply clues to distance of transport, mechanism of transport, and depositional environment. Sedimentary structures can indicate the fluid mechanics of the depositional process, direction of fluid flow, and give indications of water depth through changes in the type of ichnofossil assemblage (tracks, trails, burrows of bottom-dwelling organisms).

The mineral composition of the detrital rock is the best indication of the types of source rocks that supplied sediment to the rock being studied. Most provenance investigations require a sound understanding of igneous and metamorphic petrogenesis to be successful because mineralogic variations in sandstones frequently are quite subtle. Quartz grains may be plastically deformed or not, monocrystalline or polycrystalline, and polycrystalline grains may show a variety of types of internal structures characteristic of an igneous or metamorphic parentage. Feldspar grains may be orthoclase, microcline, sanidine, a perthitic intergrowth, or plagioclase of variable composition. Different feldspar types and compositions are characteristic of different source terranes. Lithic fragments are the least equivocal grains with respect to their origin but, even with these grains, problems of interpretation exist. Minor accessory minerals are valuable sources of information, but many of the ones most abundant in sandstones occur in several types of crystalline source rocks.

The mineral composition of sandstones can be used to infer the tectonic setting in which the rocks formed. Cratonic sands are rich in quartz and chert because of repeated reworking in shallow water environments of high kinetic energy. Sandstones formed along divergent continental margins are somewhat less rich in quartz and chert and typically contain 20–30% of foliated sedimentary or metamorphic lithic fragments. In marginal rift zones such as the Triassic rocks in New England or the Pennsylvanian rocks of southern and southwestern Oklahoma, richly feldspathic sands are formed (dominantly potassic feldspars) as granitic rocks are exposed by tensional forces. Along convergent continental margins, volcanic lithic fragments tend to be abundant and are accompanied by a feldspar suite with a high proportion of fairly calcic (An-content greater than 30%) plagioclase grains. Within a convergent margin setting, many smaller tectonic units are present, and it may be possible to distinguish among them mineralogically.

FURTHER READING

Blatt, H. 1978. Sediment dispersal from Vogelsberg basalt, Hessen, West Germany. *Geologische Rundschau, 67,* 1009–1015.

Courdin, J. L., and J. F. Hubert. 1969. Sedimentology and mineralogical differentiation of sandstones in the Fort Union Formation (Paleocene), Wind River Basin, Wyoming. *Wyoming Geol. Assoc. Guidebook, 21st Field Conf.,* 29–38.

Davies, D. K., and W. R. Moore. 1970. Dispersal of Mississippi sediment in the Gulf of Mexico. *Jour. Sedimentary Petrology, 40,* 339–353.

Dickinson, W. R. 1970. Interpreting detrital modes of graywacke and arkose. *Jour. Sedimentary Petrology, 40,* 695–707.

Dickinson, W. R., K. P. Helmold, and J. A. Stein. 1979. Mesozoic lithic sandstones in central Oregon. *Jour. Sedimentary Petrology, 49,* 501–516.

Graham, S. A., R. V. Ingersoll, and W. R. Dickinson. 1976. Common provenance for lithic grains in Carboniferous sandstones from Ouachita Mountains and Black Warrior Basin. *Jour. Sedimentary Petrology, 46,* 620–632.

Pettijohn, F. J., and P. E. Potter. 1964. *Atlas and Glossary of Primary Sedimentary Structures.* New York: Springer-Verlag, 370 pp.

Pettijohn, F. J., P. E. Potter, and R. Siever. 1973. *Sand and Sandstone.* New York: Springer-Verlag, 618 pp.

Potter, P. E., and F. J. Pettijohn. 1977. *Paleocurrents and Basin Analysis,* 2d Ed. New York: Springer-Verlag, 460 pp.

Potter, P. E., and W. A. Pryor. 1961. Dispersal centers of Paleozoic and later clastics of the Upper Mississippi Valley and adjacent areas. *Geol. Soc. Amer. Bull., 72,* 1195–1250.

Russell, R. D. 1937. Mineral composition of Mississippi River sands. *Geol. Soc. Amer. Bull., 48,* 1307–1348.

Sibley, D. F., and K. J. Pentony. 1978. Provenance variation in turbidite sediments, Sea of Japan. *Jour. Sedimentary Petrology, 48,* 1241–1248. (See also 1975, *Initial Reports of the Deep Sea Drilling Project, 31,* 507–514.)

14

The Diagenesis of Sandstones

Diagenesis is defined as all the physical, chemical, and biological changes that a sediment is subjected to after the grains are deposited but before they are metamorphosed. Some of these changes occur at the water–sediment interface and are termed *halmyrolysis,* but the bulk of diagenetic activity takes place after burial. Examples of halmyrolytic activity include such things as the alteration of biotite flakes into the mineral glauconite on the sea floor; or the destruction of clam shells by boring sponges as the shells lie on the sea floor. During deep burial the main diagenetic processes are compaction and lithification.

COMPACTION

Field Observations

Everyone who has ever stepped into a patch of solid-looking mud in a field or along a roadside is immediately aware that mud is a very compactable sediment. It is clear that freshly deposited mud contains a great deal of water that is easily squeezed out by a relatively small amount of pressure; indeed, the "mud" is mostly water rather than solid mineral matter. Measurements of modern muds reveal that a mixture of mud and water on the sea floor is typically at least 60% water, and values as high as 80% have been recorded—the "sediment" has a "porosity" of up to 80%! The water is easily squeezed out because the clay minerals in the mud are ductile (flexible) and platy and thus can be compacted very tightly at relatively low pressures.

A marked contrast in compactability is found by stepping on a beach composed of quartz sand. The crunching sound produced reflects the friction of grain against grain as the pressure causes an increase in tightness of packing. Very little water spews out of the sand around your feet. Quartz grains are subspherical to ellipsoidal in shape, relatively closely packed when deposited, and are not ductile at sedimentary pressures and temperatures. Hence, the thickness of the beach sand accumulation is not decreased appreciably by sun-worshipers and joggers applying as much as several bars of pressure to its upper surface many times each day.

The differential compaction between sands and muds can be seen in ancient rocks as well, and a particularly graphic example on a small scale is shown by the crumpled small sandstone dike in Figure 14–1. A sandstone dike is formed by an intrusion of loose sand under pressure into mud and is tabular in shape, the sedimentary analog of an igneous dike. Sandstone dikes typically occur in swarms and are produced by the cracking of an impermeable mud seal above an overpressured sand layer. The seal formed before much water was squeezed from the sand with the result that the sand grains did not touch and did not support the full overburden load. The load was partly supported by the water. When stresses cracked the seal a sand–water mixture squeezed upward to form a sandstone dike, which subsequently was lithified. In the example shown, the dike formed before there was much compaction of the mud layers so that when they did compact the still unlithified tabular sand body was crumpled.

Figure 14–1
Part of a subsurface core composed of thinly layered quartz sand (white) and organic-rich mud (dark), illustrating several contorted sandstone dikes, Dakota Sandstone (Cretaceous), North Dakota. [From J. W. Shelton, 1962, *Jour. Sed. Petrology, 32,* Fig. 2.]

By graphically stretching out the crumpled structure to its original tabular shape the original thickness of the mud layers was determined to be between 2.0 and 3.4 times greater than at present; that is, the amount of compaction was between 50% and 70%, in the range of what we would expect in a slightly silty mud.

The pressure exerted on a layer of sedimentary rock at depth is equal to ρgh (density $\times$ gravitational acceleration $\times$ height of the column), and if we assume that the sediment is permeable from its burial depth to the surface, fluid pressure is equal to the weight of a column of water of that height, approximately 10 bar/100 m. In some areas of rapid sedimentation, such as the Gulf Coast, many impermeable clay seals are formed in the stratigraphic column. In such cases the sediment pile is not permeable from certain layers of sand up to the surface; water in the sand supports some of the weight of the overburden (both rock and water), and the pressure of the water in the sand layer is greater than 10 bar/100 m. Fluid pressures may be as high as 25 bar/100 m, 2.5 times the pressure exerted by a column of water at that depth. The upper limit is set by the density of the overlying column of rock.

Laboratory Studies

Examination of thin sections of sandstones reveals several effects of compaction. A freshly deposited quartz sand has a porosity of about 45% ± 5% depending on sorting and grain angularity. A thin section of a well-lithified quartz sand, however, reveals that the percentage of original pore space (now filled with secondary cement) is perhaps 30% ± 5%, and there is no evidence that the detrital grains have been dissolved or otherwise altered since deposition. The difference between the original 45% porosity and the 30% paleoporosity that was available to be filled with cement must have been caused by compaction. The weight of overburden reduced the amount of empty space in the sand by about one-third of the amount originally present.

Examination of thin sections of sandstones that contain ductile lithic fragments or clay supplies additional evidence of compaction. Sand-size pieces of mud and fragments of micaceous rocks such as shale, slate, phyllite, and schist are deformed. Their shapes are not the platey ones they had when deposited and they bend around the more rigid quartz and feldspar grains and are squeezed into pore spaces (see Figure 14–2). In highly compacted lithic sandstones porosity can be reduced from the original 45% to nearly zero simply by the squeezing of ductile grains into pores (see Figure 14–3). In such cases the original thickness of the sand must have been reduced by nearly 50%!

CEMENTATION

How is an assemblage of loose detrital fragments converted into a solid mass that resists a sharp whack with a hammer? How is the process called lithification accom-

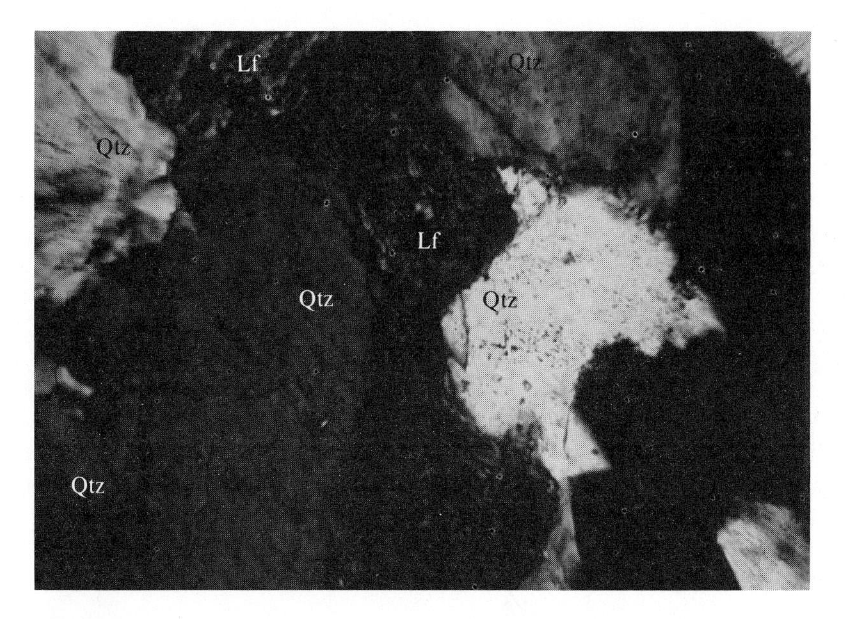

Figure 14–2
Photomicrograph (crossed nicols) of a micaceous lithic fragment (Lf) plastically deformed by rigid quartz grains (Qtz) in a compacted coarse-grained lithic sandstone. Kingston, New York. [From H. Blatt, 1966, *12th Ann. Conf., Wyoming Geol. Assoc.*, Fig. 7.]

plished? Can compaction alone turn a sand into a sandstone? To answer such questions we must go to the laboratory; field work cannot supply the answers.

The question of whether compaction alone can produce lithification can be answered by laboratory experiments. We can take a cylindrical metal container open at one end, half fill it with a sediment–water mixture, insert a piston into the open end of the cylinder, and apply pressure on the piston to compact the sediment. The base of the cylinder (or the piston) is microperforated to permit escape of water from the cylinder during the experiment.

Our first experiment is with a quartz–water mixture and the results are negative. We apply pressures up to the limit of sedimentary conditions to no avail. When the piston is removed the sand pours out as easily as it entered. We next repeat the experiment using 80% quartz and 20% schist fragments or mud and find that although we have not produced a rock from the loose grains, we have produced some aggregates. Quartz grains have become indented into the mud or schist fragments to create a partial lithification, pieces of "rock" a few grain diameters in size. Apparently the presence of ductile fragments can lead to lithification by compaction. We decide to go all the way and run an experiment using 100% mud fragments and schist fragments plus water. The result confirms our inference; when the cylinder is inverted at the end of the experiment a cylindrical piece of artificial rock falls out.

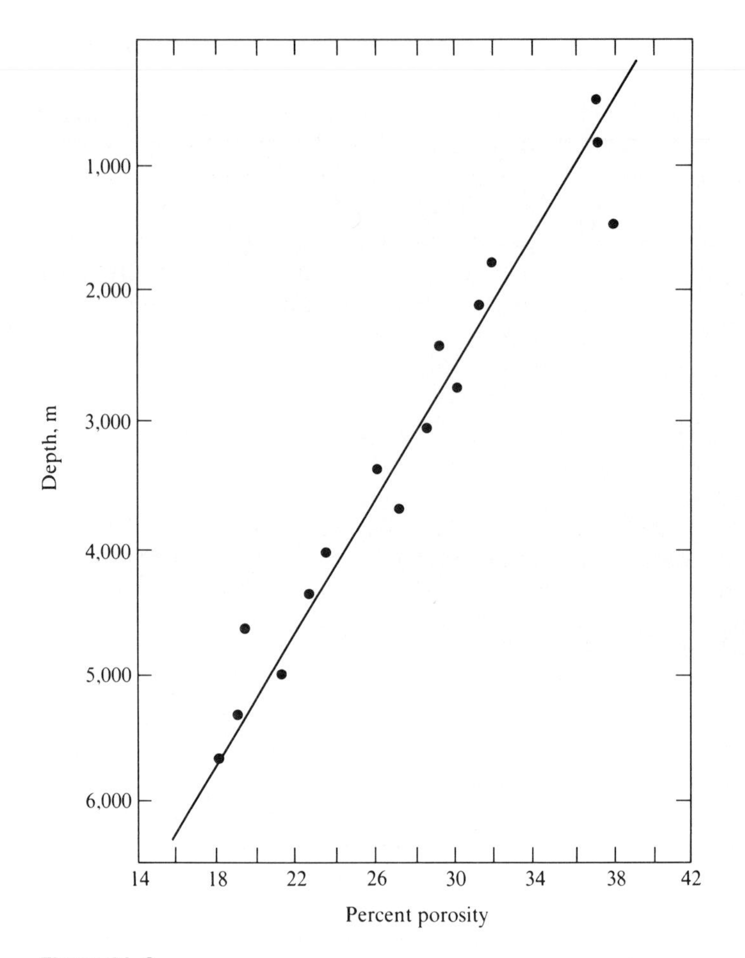

Figure 14–3
The effect of the presence of ductile rock fragments on the compaction of uncemented Tertiary sands in the subsurface of south Louisiana. Compaction is reflected by the decrease in average porosity with depth; based on 17,367 cores averaged for each 1,000 feet. [From H. Blatt, 1979, Soc. Econ. Paleon. & Miner. Spec. Pub. 26, Fig. 7.]

This result raises additional questions. Why does the presence of ductile fragments help to lithify a compacted sediment? To answer this we make thin sections of the partially lithified fragments and fully lithified artificial rock to examine features too small to be visible otherwise. The polygranular fragments from the experiment with 80% quartz look exactly like the completely lithified "rock." It appears that indentation of grain against grain is sufficient to cause lithification; the reason the experiment using 100% quartz failed to produce lithification is that quartz is too rigid to permit adjacent grains to indent under the conditions of our experiment. The quartz grains are in contact only at points when deposited (poured into the cylinder) and compression alone cannot increase the area of grain-to-grain contact. When ductile fragments are in the sediment, however, the area of intergrain contact *can* be increased by pressure, leading to lithification. The success of lithification by compaction is correlated directly with the content of ductile material. Mud will be converted into mudrock by a relatively small thickness of overburden; lithic sandstones will require a much greater thickness and perhaps even the intense squeezing caused by deformation and tectonic activity long after burial. It probably is not possible to lithify a pure quartz aggregate by compaction alone except under localized conditions such as immediately adjacent to a fault surface where stresses are unusually intense and the quartz grains are granulated.

Pore-Filling Cements

The average sandstone contains only 15% lithic fragments, and many of these fragments are not ductile; for example, rhyolite, gneiss, and granite. And the amount of detrital clay in the average sandstone probably is less than 5%. Intimate grain-to-grain contact in most sandstones is achieved largely by the introduction of chemical precipitates—"cements." The growth of new mineral matter into the depositional pores creates the intimate contact needed for lithification. The degree of lithification will depend on the amount of cement-to-grain contact produced. If only a small amount of secondary mineral matter is precipitated in the pores, the rock can be disaggregated into individual grains by finger pressure; such rocks are termed *friable*. An increased amount of pore filling produces a rigid but still porous sandstone, the type of sandstone in which much of the world's petroleum, natural gas, and uranium are located. In the extreme, all porosity is lost and the sandstone is truly "hard as a rock."

Note that the exact nature of the pore-filling precipitate is of secondary importance in the lithification process. The intimate contact of quartz cement with a detrital quartz grain produces a stronger bond than the contact of calcite cement with the quartz grain, but both types of pore filling cause the rigidity that converts a loose pile of sand into a rock.

Field Observations

The occurrence and type of cementation can sometimes be related to other aspects of the sandstone, such as its detrital mineral composition, environment of deposition, or stratigraphic position. Each of these three features can be either determined with certainty or estimated with some precision at the outcrop. The most common cements are quartz and calcite, with hematite and clay minerals running a poor third and fourth.

1. *Quartz*. Cementation by quartz is confined almost entirely to sandstones whose detrital composition is 90–100% quartz. The reverse is not true, however. Not all pure quartz sandstones are cemented by quartz; calcite and hematite also are

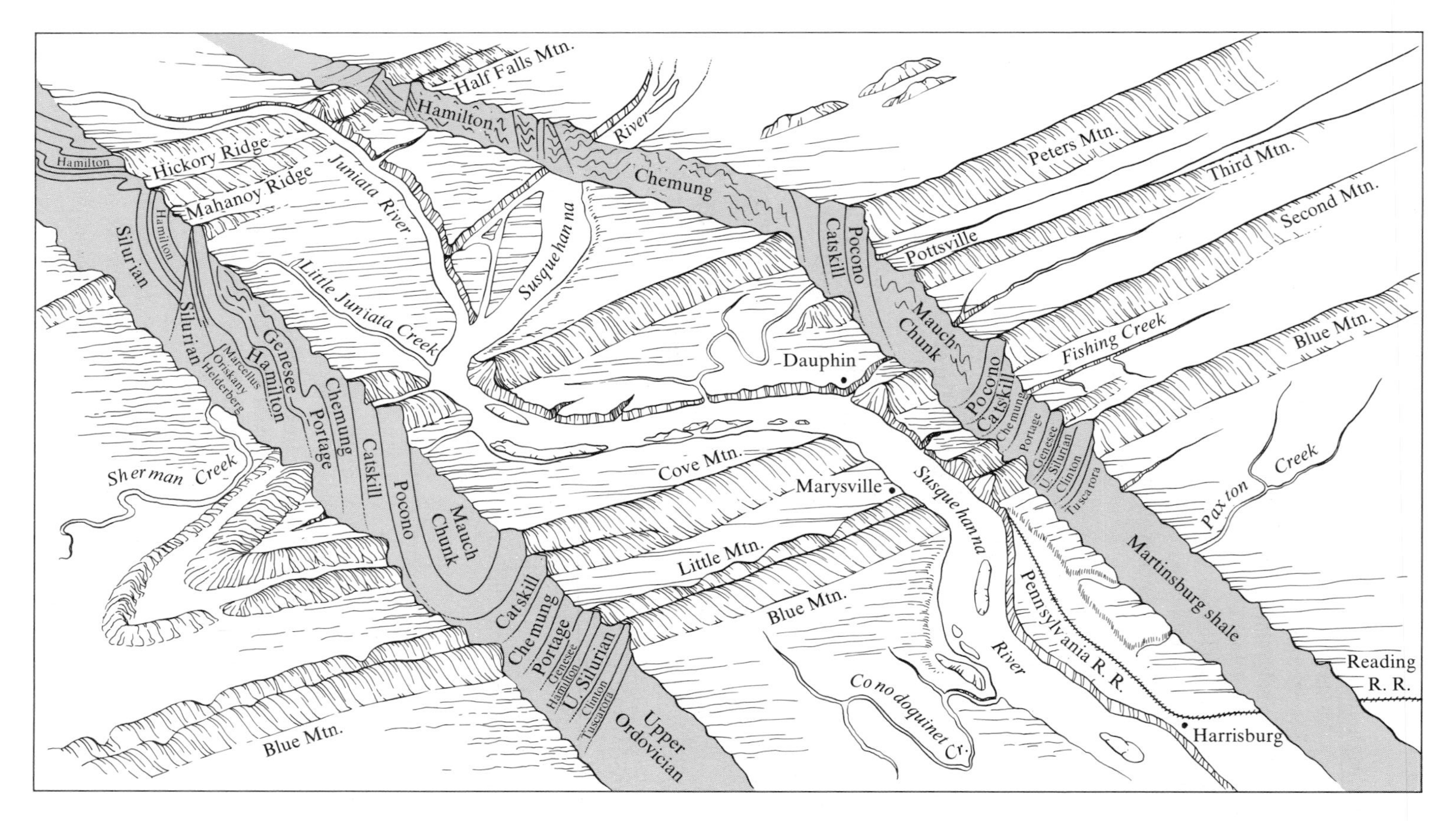

Figure 14–4
Idealized bird's-eye view looking north (parallel to Susquehanna River) from near Harrisburg, Pennsylvania, with the cut-away of the earth's surface showing the close control of topography by the lithology and differential resistance to erosion of early Paleozoic bedrocks. Structures are steepened by vertical exaggeration. Width of diagram is about 20 km. The Pocono and Tuscarora Formations are quartz-cemented quartz sandstones, and the other ridges are underlain by well-lithified, hematite-cemented, coarse-grained sandstones and conglomerates. Topographically low areas are underlain by mudrocks. [From G. W. Stose et al., 1933, *XVI Internat. Geol. Cong., Guidebook 10,* Plate 5.]

common cements in these rocks. Pure quartz sandstones are most common in high-energy depositional environments and, therefore, quartz cement occurs most commonly in ancient beaches, marine bars, desert dunes, and some fluvial sand bars. It is rare in deposits such as alluvial fans and turbidite sandstones.

Quartz-cemented quartzarenites are very resistant to weathering and typically stand out in relief at an outcrop. On a more regional scale they are ridge formers; for example, in the central Appalachians, where they define clearly the plunging anticlines and synclines that form the structural pattern in the area (see Figure 14–4).

2. *Calcite.* Calcite cement occurs commonly in sandstones of all depositional environments and mineral compositions. At the outcrop it is easily identified by its immediate reaction with dilute hydrochloric acid, a plastic bottle of which should always be carried in the field. Carbonate-cemented sandstones may grade laterally into sandy limestones and pure limestones, an occurrence that suggests original environmental conditions as responsible for the cement, rather than diagenetic conditions.

Cementation of sandstones requires some initial depositional permeability to permit the migration of saturated waters through the sand, and the cementation pattern at the outcrop can reflect this fact. Sometimes the scale of variation in original permeability is large, as in the case where a well-sorted and clay-free fluvial sand bar is both overlain and underlain by muddy overbank sands. Sometimes the scale of variation is very small, on the order of a centimeter or less (see Figure 14–5), and the explanation is obscure. There is no apparent difference in the laminae other than the fact that one is cemented and the other is friable or unconsolidated.

Calcite cement can also form *concretions,* locally cemented areas within a more friable layer of sandstone. The concretions can occur in a wide range of shapes (see Figure 14–6) ranging from spherical to very irregular, probably in response to permeability variations within the sandy unit. The longest dimension of irregularly shaped concretions is nearly always in the plane of the bedding, the surface of greatest permeability. The irregular shapes of concretions are evidence of the intricacies of permeability variation within sands and sandstones.

The relationship between sedimentary structures in the concretion and those in the enclosing sandstone can reveal the time of formation of the concretion. If laminations within the sandstone pass undeflected through the concretion, the concretion must have formed after the sand was compacted. If laminae curve around the concretion (see Figure 14–7), the concretion must have formed very soon after deposition of the sand, probably before it was buried more than a few meters.

Calcite is several orders of magnitude more soluble than quartz, hematite, or clay minerals, so its distribution as a cement in sandstone beds tends to be patchy. The sandstone may grade within a few meters laterally from well lithified to friable for no apparent reason. Possibly the water from which the calcite precipitated simply became depleted in calcium and/or carbonate ions as it moved through the sand layer. It also is possible, however, that the calcite cement is being dissolved away by acidic rain and soil waters at the present outcrop. Such an occurrence was documented in the Bethel Sandstone in southern Indiana and northwestern Kentucky.

Figure 14–5
Outcrop face cross section of a partially lithified modern desert sand dune near Ashquelon, Israel. Some cross bedding is defined by differential cementation to the left of the lower part of the meter stick. Lithified laminae contain 40–55% $CaCO_3$; uncemented laminae, 4–10%. [From D. H. Yaalon, 1967, *Jour. Sed. Petrology, 37,* Fig. 3.]

Figure 14–6
Concretion of hematite from the Garber Sandstone (Permian), central Oklahoma, showing control of shape by permeability parallel to bedding in the sandstone.

Figure 14–7
Concretions of dolomite in the Monterey Formation (Miocene), California. Bedding passes without deflection through the middle of the concretions but enclosing beds bulge around them, indicating that the concretion formed while the enclosing sediments were still unlithified. [From D. L. Durham, 1974, U. S. Geol. Surv. Prof. Paper 819, Fig. 29.]

The Bethel is a sinuous Mississippian shallow water channel sand (see Figure 14–8) composed of well-sorted, fine-grained quartzarenite. The channel is 225 km long, 800–1,300 m wide, and as much as 80 m thick. In the subsurface, cementation is well developed and calcite forms nearly 30% of the cement (see Figure 14–9). Other cements are quartz, illite, kaolinite, and pyrite. At the surface the Bethel contains no calcite and only minor amounts of other types of cement. Porosities and permeabilities increase upwards as the degree of lithification decreases. Pyrite,

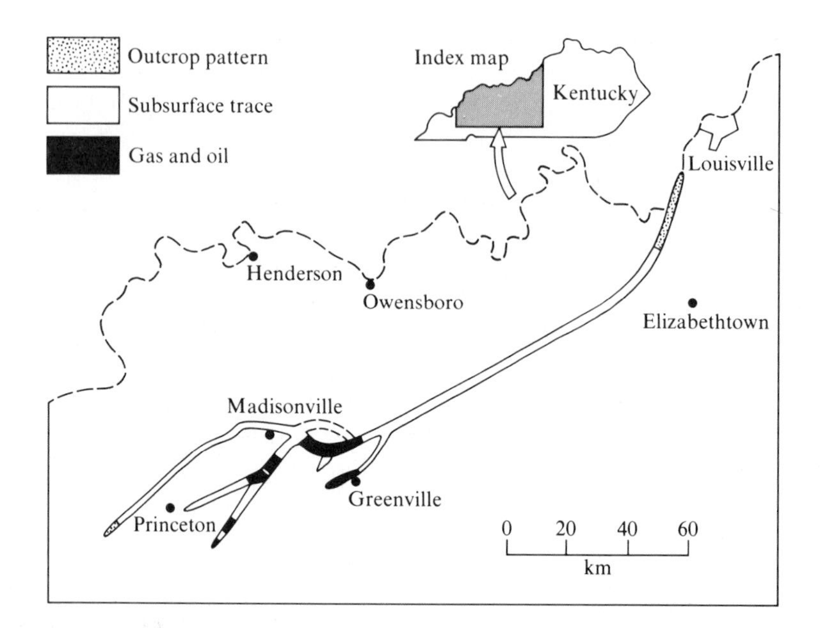

Figure 14–8
Bethel Sandstone distributary system in western Kentucky, showing locations of oil and gas accumulations. Sediment was transported from the northeast to the southwest. Dashed lines indicate probable channel. [From D. W. Reynolds and J. K. Vincent, 1967, Ky. Geol. Surv. Ser. 10, Spec. Pub. No. 14, Fig. 3.]

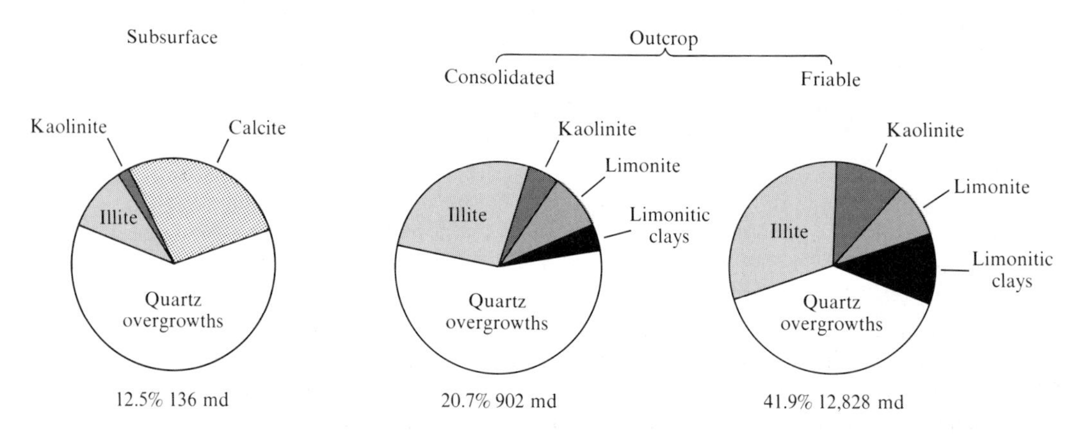

Figure 14–9
Pie diagrams showing relative abundances of cementing agents in Bethel Sandstone in the subsurface and in outcrop. Below each pie are corresponding average porosities (in percent) and permeabilities (in millidarcies), which reflect decreasing amounts of total cement toward the surface. [From Sedimentation Seminar, 1969, Ky. Geol. Surv. Rept. Inves. 11, Fig. 21.]

present as a minor cement in the subsurface, is absent because it was converted to limonite at the surface. Laboratory studies reveal that the illite/kaolinite ratio decreases toward the surface because of leaching of illite and growth of kaolinite, and the percentage of feldspar decreases from 2% to 0.2% because of near-surface leaching.

3. *Hematite.* Hematite cement in sandstones is easily identified in the field by its red color, which is present with only 1% or less Fe_2O_3. Hematite is an intense pigmenting material. The iron atoms in hematite are derived mostly from weathering and leaching of the common iron-rich minerals such as magnetite, ilmenite, biotite, and hornblende, less commonly from augite and olivine, and secondarily from iron-bearing clay minerals.

In some sedimentary sequences different intensities of red coloration can be seen, reflecting the fact that most hematite forms after a sediment is deposited, and some of these occurrences have been studied in detail by Walker (1967). One of them is located in the hot, arid Sonoran Desert of northeastern Baja California, Mexico (see Figure 14–10A). Mean annual rainfall is only 10 cm/yr at present, and the presence of evaporites and a flora of semiarid character in Pliocene sediments in the area indicates no major change for the past 5–10 million years. Throughout this period alluvial fans derived from the crystalline highlands of the peninsular ranges have coalesced to form extensive bahadas, and these grade eastward into intertidal muds and salt flats that border the Gulf of California (see Figure 14–10B). Normal faulting parallel to the mountain front has produced numerous alluvial fan cross sections, some more than 300 m in height. It is in these Holocene–Pliocene stratigraphic sections that the origin of the hematite can be seen.

The recent alluvium occurs as a veneer of sheetflood deposits on the surfaces of active fans and is rarely more than a meter thick. It is composed of gray arkosic detritus, chemically unaltered fragmental granitic debris from the plutons that form the backbone of Baja California. This Holocene alluvium is conspicuously rich in chemically unstable ferromagnesian minerals, particularly hornblende and biotite. Underlying Pleistocene fan sediment has undergone mild chemical alterations that have produced distinctive characteristics: some degree of lithification and a reddish color.

Four major Pleistocene soils can be identified in the alluvium. The younger soils are yellowish but have a distinct reddish hue. Successively older soils become progressively more red. The intense red color of older soils is as red as some of the reddest ancient red beds. Given the evidence of constant climate during the time of deposition of the alluvial fans, and given the observation that the fan sediment changes in color from gray to red with increasing age, it is apparent that the red color has been produced after burial. Ferric oxide has formed during early diagenesis. We can even catch this process in the act by careful field work (see Figure 14–11). Apparently iron-bearing minerals are being leached by the infrequent rains that percolate through the sediment with the result that the ferrous iron in the minerals is released, oxidized, and dispersed throughout the sediment. Subsequent laboratory studies have elaborated the deductions made at the outcrop. For example, the modern alluvium

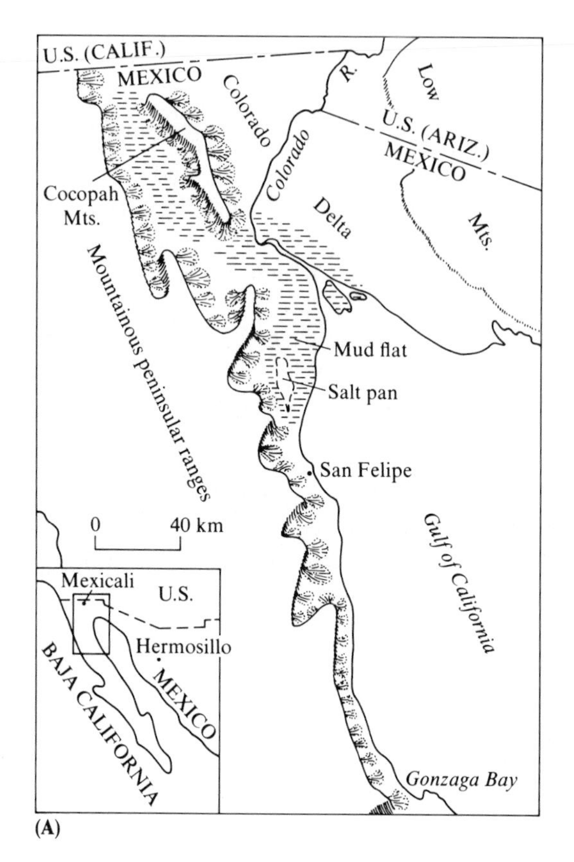

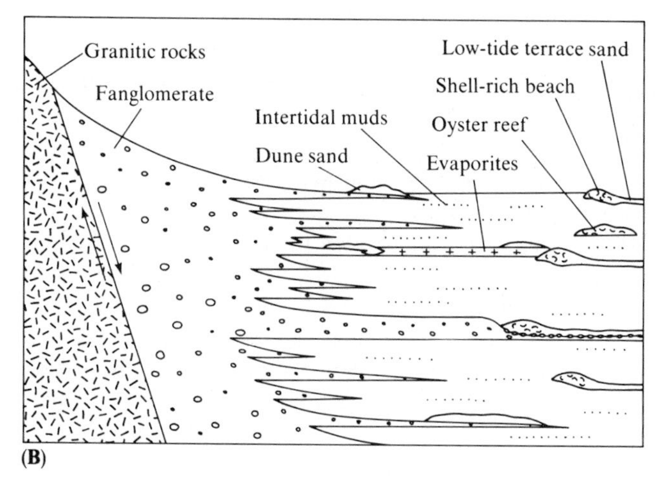

Figure 14–10
(A) Map of northeastern Baja California showing the topographic and environmental setting in which hematite cement is forming. **(B)** Diagrammatic cross section showing facies relationships of Holocene, Pleistocene, and Pliocene sediments, northeastern Baja California, Mexico. [From T. R. Walker, 1967, *Geol. Soc. Amer. Bull., 78,* Figs. 1, 2.]

Figure 14–11
(**A**) Outcrop of gray mudrock and sandstone at crest of hill, underlain by uniformly red mudstone and sandstone that contains hematite development "caught in the act." (**B**) Close-up of hand specimen showing the red halo spreading downward from a zone of concentration of ferromagnesian minerals. [From T. R. Walker, 1967, *Geol. Soc. Amer. Bull., 78,* Plate 2.]

contains 7.3% heavy minerals, of which 61% is angular to subangular hornblende. In the Pliocene red fanglomerates there is only 4.8% heavy minerals, of which only 35% is hornblende; and the hornblende grains in the Pliocene deposits have been etched by percolating rain waters into exotic cockscomb forms (see Figure 14–12). In advanced stages of alteration, the hematite stain is concentrated in an iron-rich montmorillonite formed by the intrastratal alteration of the hornblende.

It may be that the red coloration in many of the upper Paleozoic red bed–evaporite sequences in the western United States has formed in this way (see Figure 14–13). A significant piece of data that supports this hypothesis is the fact that these ancient intensely red alluvial fan deposits contain no hornblende, despite the fact that their source rocks, the Front Range granites, contain abundant hornblende, as do the modern stream sands that drain the granites.

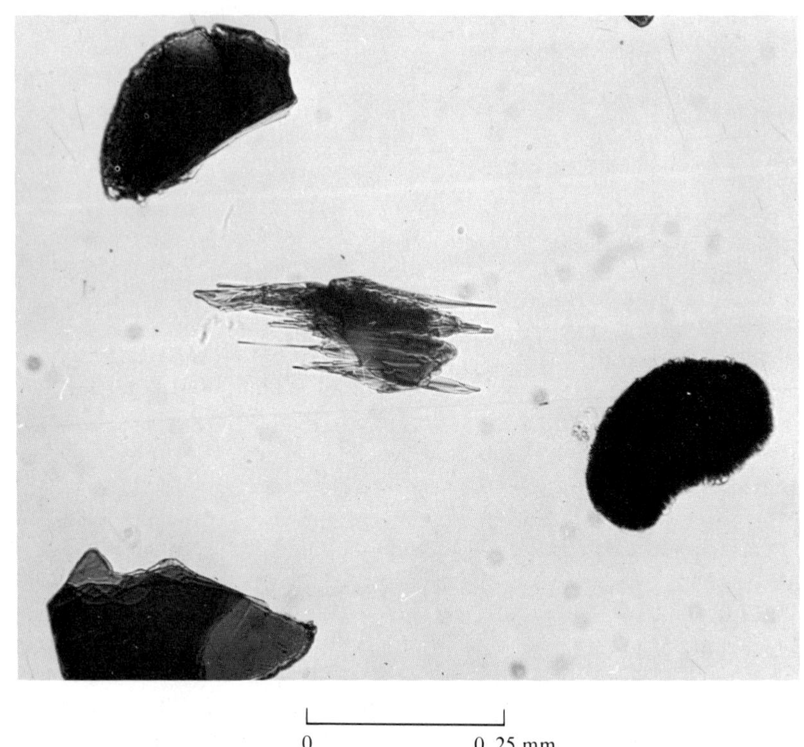

0 0.25 mm

Figure 14–12
Hornblende grain (center) etched after deposition in an alluvial fan deposit, Pliocene, Baja California. The spikelike shape must be postdepositional because the delicate shape could not survive transportation. Biotite grains at upper and lower left; black opaque grain at right. [From T. R. Walker, 1967, *Geol. Soc. Amer. Bull., 78,* Plate 1.]

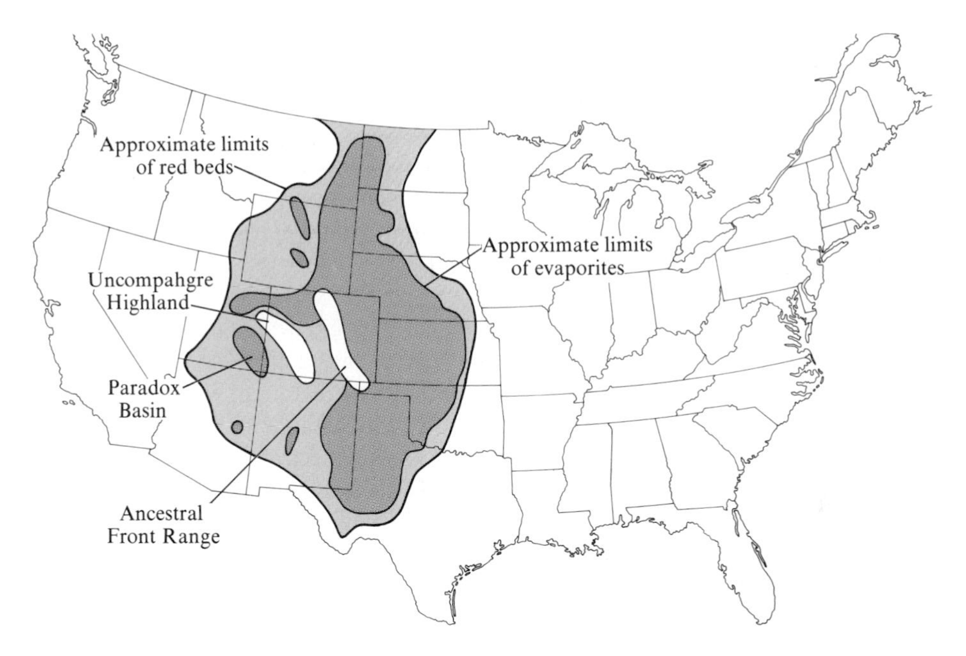

Figure 14–13
Areal distribution of red beds and evaporites of late Paleozoic age in the western interior United States. [From T. R. Walker, 1967, *Geol. Soc. Amer. Bull., 78,* Fig. 5.]

4. *Clay minerals.* Individual clay minerals are, of course, too small to be seen using a 10× hand lens, but it sometimes is possible to detect their presence in a hand specimen. If present in amounts of 10% or more, clay minerals will coat the sand grains to the extent that the boundaries of the sand appear fuzzy or indistinct. In many texturally immature sandstones, however, clay is essentially impossible to detect at the outcrop.

The best way to recognize the presence of clay in a hand specimen is by inferring its presence from the absence of other cementing agents in a lithified sandstone. We know that quartz cement is prominent only in pure quartz sandstones and is easy to see in such rocks using a hand lens. Calcite cement is identified by its reaction with acid. (Break off a cubic centimeter of rock and drop it into a small porcelain dish with the acid. The extent of disaggregation reveals the importance of calcite cement.) Iron oxide cement is identified by its red, brown, or yellow (limonite) color. Most hematite-cemented sandstones also contain some clay, but laboratory studies are required to be sure. The reason for the association between hematite and clay is twofold. If the hematite forms during early diagenesis by alteration of hornblende, biotite, and so on, montmorillonitic clay is commonly generated (see Chapter 11). If the hematite entered the depositional environment as coatings on other particles, such as organic matter or clay minerals, it entered as particles hydraulically equivalent to clay so that no separation during deposition was possible between the hematite and its fellow travelers.

Clay pore-filling in sandstones can be either detrital or authigenic; that is, the clay may have been carried in by the current that carried the other particles in the rock or the clay may have grown in the sediment after deposition of the sand grains. At the outcrop it is not possible to distinguish between these two possibilities, although a well-sorted quartzarenite is rather unlikely to contain detrital clay. Laboratory studies are required.

Laboratory Studies

Laboratory studies of lithification depend both on examination of existing rocks and on geochemical experiments in which we try to duplicate the features seen in the rocks. For example, in many thin sections and scanning electron photomicrographs of pure quartz sandstones we observe that some of the detrital grains have rhombohedral or prismatic crystal outlines (see Figure 14–14), an apparent result of growth of quartz from a groundwater solution after burial. What are the chemical conditions in

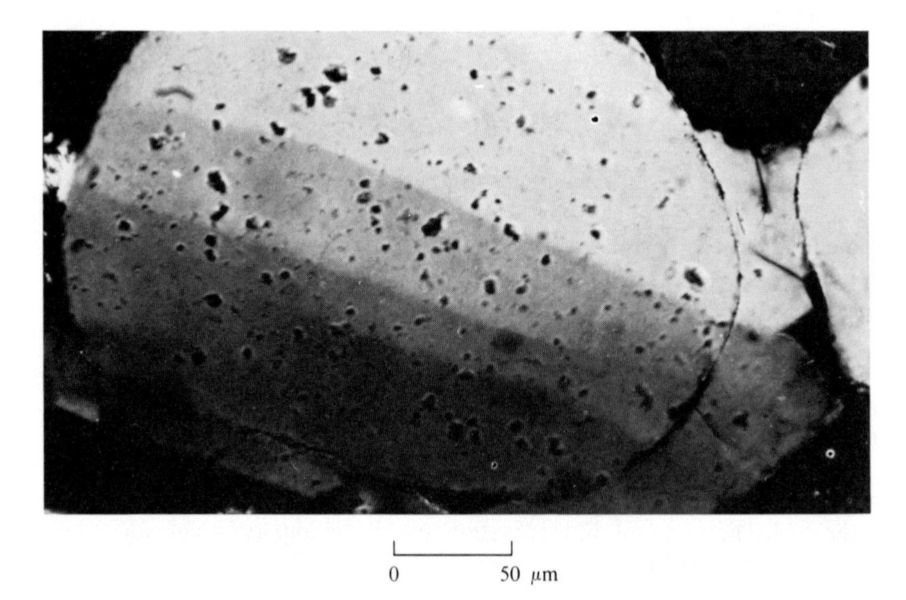

Figure 14–14
Thin section photomicrograph (crossed nicols) of detrital quartz grain with rhombohedral and prismatic overgrowths. The outline of the detrital grain is marked by a discontinuous coating of "dirt." Bands of undulatory extinction pass through both detrital grain and overgrowth. [From E. D. Pittman, 1972, *Jour. Sed. Petrology, 42,* Fig. 6A.]

Figure 14–15
Photomicrograph (crossed nicols) of an assemblage of very coarse sand-size volcanic rock fragments and gray quartz grains. Arrows indicate location of grain boundaries, made indistinct by partial conversion of volcanic rock fragments to clay by circulating pore waters. Triassic, New Zealand.

which this can occur? What is the solubility of quartz in water? In sandstones rich in volcanic rock fragments, such as many graywackes, we see in thin section that many of the grains have indistinct outlines and grade imperceptibly into a mushy-looking clay matrix of montmorillonite (see Figure 14–15). It appears that the detrital grains are in the process of decomposing. Under what conditions can such a reaction occur? These questions can be answered only by conducting geochemical experiments.

Experiments reveal that increased temperature is, in general, more important than increased pressure as a factor in lithification. There are several reasons for this:

1. The range of pressures associated with diagenesis (as contrasted to metamorphism) is insufficient to deform rigid materials such as quartz, feldspar, and most

nonfoliated rock fragments at room temperature. Increased temperature decreases the rigidity of materials so they can deform plastically.

2. Most materials are more soluble at higher temperatures. For example, the solubility of opal is 120 ppm at 20°C and 500 ppm at 100°C.

3. Some chemical reactions will not occur until an energy barrier has been overcome—the activation energy. Increased temperature can supply the energy needed to overcome this barrier.

The increase in temperature with burial depth is a consequence of the transfer of thermal energy from the interior of the earth to the surface where it is dissipated. Different geothermal gradients are observed, depending on the overall thermal conductivity of the rocks, regional heat flow, and subsurface water movement. The world average gradient is approximately 25°C/km, but gradients as low as 5°C/km and as high as 90°C/km have been measured. The lower values occur on the stable cratons; the higher values on the continental side of convergent oceanic plate–continental plate margins. For example, in the Central Tertiary Basin of Sumatra the gradient averages 68°C/km.

The boundary between diagenesis and metamorphism is conventionally taken as either 150°C or 200°C, but the changes that we associate with metamorphism, such as the development of foliation or the recrystallization of quartz, occur over a range of temperatures. The changes are gradational, as are all natural processes. No specific temperature is adequate as the point at which diagenesis becomes metamorphism. Based on the range in geothermal gradients that have been measured, sedimentary rocks may change into metamorphic rocks at depths between 3 km and 40 km, with an average depth for the change at 6–8 km.

Our concern in this section is the types of minerals that may form in noticeable amounts during diagenesis and the conditions in which these changes occur.

1. *Quartz.* As we noted previously, significant amounts of pore-filling secondary quartz are essentially restricted to pure quartz sandstones. Such sandstones occur mostly on cratons and divergent plate margins; they are unknown near convergent margins. The secondary quartz typically forms at temperatures less than 100°C. The quartz *overgrowths* are precipitated as coatings on the detrital quartz grains and in crystallographic continuity with them (see Figure 14–14). The growths take root at several points on the detrital grains. Assuming the pore waters remain saturated with respect to quartz (6 ppm at 20°C $\rightarrow$ 63 ppm at 100°C), the numerous rhombohedra soon coalesce into a single large one. If growth continues, the planar faces from adjacent grains will abut somewhere in the pore space. The planar faces that persist as long as the quartz grows into a void space are lost when the overgrowths abut, and the overgrowth–overgrowth contacts seen in thin section are crystallographically irrational compromise boundaries.

It also is possible for a quartzarenite to be lithified without the introduction of pore-filling of any kind. As we noted earlier in this chapter, quartz grains are subequant in shape, are not ductile, and when deposited are supported by their neighbors only at local points of grain-to-grain contact. In laboratory experiments the area occupied by these points cannot be increased sufficiently to permit adhesion (lithification) to develop. At depths of several thousand meters however, and over long periods of time, it seems that migrating pore waters interact with the quartz grains at the points of contact, dissolving quartz at these points so that the grains can settle into each other under overburden pressure, greatly enlarging their areas of contact. These enlarged quartz–quartz contact surfaces can be fairly flat in thin section or highly interdigitated. In either case it can be impossible to distinguish between these *pressure solution* surfaces and overgrowth–overgrowth contacts using standard thin section petrography. That is, it may not be possible without special techniques to determine whether the quartzarenite was lithified by the precipitation of secondary quartz in the pore spaces or was lithified by pressure solution without the introduction of secondary material.

The special technique that resolves this problem is *luminescence petrography.* In this technique, a thin section is bombarded by electrons causing certains parts of minerals to luminesce. The parts that luminesce are those that either contain "activator elements" as trace impurities (commonly transition elements or rare earth elements) or contain certain types or amounts of crystal defects. In the case of quartz, detrital grains nearly always luminesce but secondary growths do not (see Figure 14–16). Presumably this difference in response to electron bombardment can be related to the temperatures of formation of the quartz: above 300°C for the detrital grains but less than 150°C for the secondary growths.

2. *Calcite.* Coarsely crystalline calcite cement is common in all three types of sandstones—quartzarenites, arkoses, and litharenites—and in all tectonic settings. Calcite cement can form more rapidly and be dissolved more rapidly than quartz cement and, as a result, euhedral growths of calcite cement are rarely seen in sandstones. The typical morphology is an anhedral mosaic of cement crystals 10 μm or larger in size with compromise crystallographic boundaries like those of quartz cement.

Although partial scalenohedra or rhombohedra of calcite cement are seen only rarely in sandstones, there is evidence that, like quartz, the calcite cement commonly grows into pore space spasmodically. Periods of growth are interrupted by periods of nondeposition. Sometimes the composition of the pore waters is changed somewhat during these hiatuses, and this can be seen using luminescence petrography. As we noted earlier, transition elements are activators of luminescence. Iron is a relatively abundant transition element and if the change in composition of the pore fluid includes an increase in the amount of iron, the iron can be included in trace amounts in the growing calcite cement crystals, causing a change in the luminescence characteristics. As is the case with quartz cement, the durations of the hiatuses are unknown, as are the causes of the changes in composition of the pore waters. Perhaps they resulted from structural or topographic changes in the area where water is entering the formation; perhaps they were caused by mixing of different formation waters as a

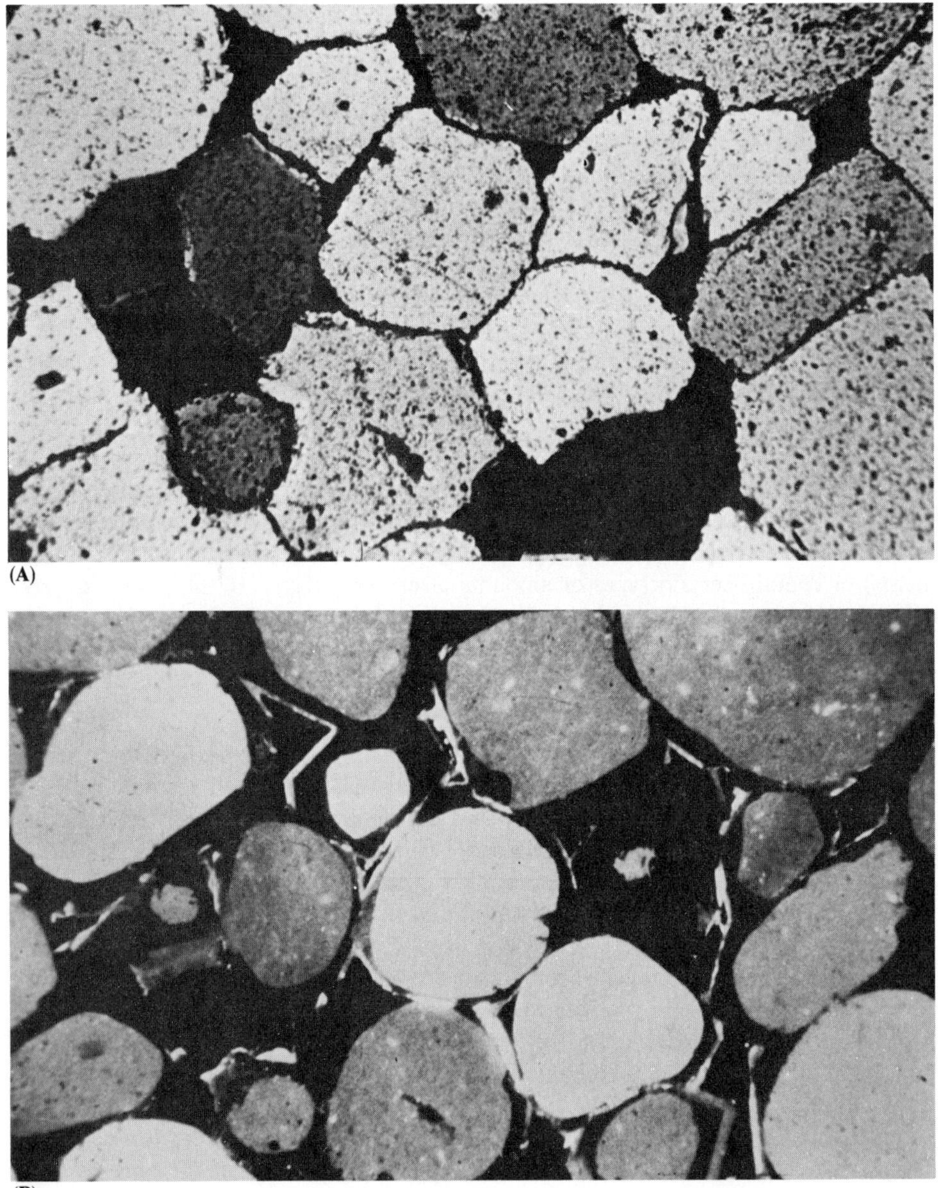

Figure 14–16
Photomicrographs (uncrossed nicols) of medium sand-size quartz grains in the Hoing Sandstone (Devonian), Illinois, (**A**) under crossed nicols and (**B**) under cathodoluminescence. Apparent pressure solution contacts are clearly seen in luminescence petrography to be formed of secondary quartz abutting in original pore space. [From H. Blatt, 1979, Soc. Econ. Paleon. & Miner. Spec. Pub. 26, Fig. 9.]

result of faulting in the depositional basin; or perhaps they resulted from a change in content of dissolved oxygen in the pore waters that caused a chemical reduction of ferric iron in an adjacent hematite-cemented unit.

The sizes of the calcite crystals that form the cement mosaic range over several orders of magnitude, from about 10 μm to several centimeters. When the sizes of the crystals exceed the sizes of the detrital grains, the cement crystals can enclose several detrital grains within them. If a cement crystal does not abut against others and is able to express its preferred crystal habit, a *sand crystal* is formed, a scalenohedron of calcite containing many thousands of detrital grains (see Figure 14–17). The reason some cement crystals grow to such exceptionally large sizes is unknown. They seem to be confined to well-sorted sandstones, however, possibly because of the high permeabilities of well-sorted sands.

Figure 14–17
Sand crystals from a Miocene fluvial sandstone exposed in the Badlands, South Dakota. The crystal outlines are those of scalenohedra of calcite; about ten calcite crystals are present in the aggregate. The salt-and-pepper appearance is caused by the innumerable quartz crystals included within each scalenohedron. The fully developed crystal oriented NNE–SSW in the left half of the aggregate is about 10 cm long. [From F. J. Pettijohn, 1975, *Sedimentary Rocks,* 3d Ed. (New York: Harper & Row), Fig. 12–3.]

Red matrix
Relic hornblende
Dissolution voids
(A)
0
0.25 mm
Authigenic adularia
Dissolution void
Relic hornblende
(B)
0
50 μm

Figure 14–18
(**A**) Thin section (crossed nicols) and (**B**) SEM photomicrographs of an intensely etched and partly dissolved hornblende grain in sandstone. Hayner Ranch red beds (Miocene), north of Las Cruces, New Mexico. [From T. R. Walker, 1976, *The Continental Permian in Central, West, and South Europe* (Dordrecht, Holland: D. Reidel), Figs. 12, 13.]

---

3. *Hematite.* Hematite pore-filling occurs as crystals about 1 μm in size that cannot be resolved using a petrographic microscope. It appears in thin section as an irresolvable opaque area that is red in reflected light (yellow for limonite) and typically is thin or absent at grain contacts, reflecting its postcompaction origin.

Earlier in this chapter we discussed the field evidence indicating that hematite cement in sandstones is diagenetic rather than depositional. We noted the reddish halo surrounding concentrations of accessory minerals in partially red sandstones. Detailed further documentation of the diagenetic origin of hematite can be made in the laboratory. Examination of iron-bearing accessory minerals in thin section reveals them to be intensely etched (see Figure 14–18) and reduced in size; study of the sandstone using the SEM reveals beautifully euhedral crystals of hematite growing in pore spaces (see Figure 14–19). In areas of the thin section where montmorillonite clay has formed in place from the altered hornblende grain, the clay is found by the electron microprobe to be iron-rich, with the amount of iron decreasing from more than 20% in the hornblende grain to less than 3% at the outer fringe of the clay as a result of leaching (see Figure 14–20). Hematite can form whenever iron-bearing minerals are present in a sandstone and oxidizing conditions exist.

4. *Clay minerals.* Laboratory studies of pore-filling clay cements are typically also concerned with the problem of *matrix*. The matrix of a detrital sediment is defined as the finer-grained material in the rock. In a conglomerate the sand grains are the matrix; in a sandstone the matrix is dominantly clay minerals. Sandstones that contain matrix normally have two distinct peaks in their grain-size distribution; one in the sand size, the second in the clay size.

In sandstones whose detrital sand grains are all quartz, any clay minerals that grow during diagenesis can be recognized and identified easily either in thin section or with the SEM because of their euhedral shape (see Figure 14–21). This also is true in most arkoses. In lithic sandstones, however, the origin of clay matrix is less clear, and in many rocks we cannot be certain whether the clay is detrital, secondary, or perhaps simply crushed micaceous lithic fragments. The problem is particularly severe in deeply buried sandstones deposited in basins at convergent plate margins, such as along the west coast of North America and in southern Europe. In thin section we find an irresolvable clay mush that not only fills pore space completely but seems to grade imperceptibly into the detrital grains themselves. Much of the clay appears to have formed from chemical reaction between detrital volcanic fragments, phyllo-

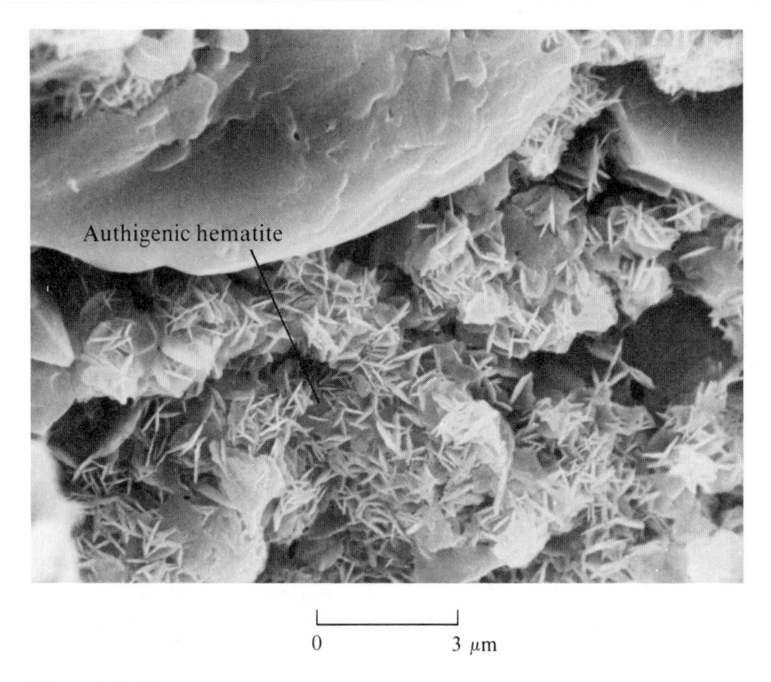

Figure 14–19
SEM photomicrograph of bladed authigenic hematite crystals grown as clusters of rosettes, Moenkopi Formation (Triassic), Cameron, Arizona. Large grain in upper part of photo is rounded detrital quartz grain with small overgrowths; larger, clearly euhedral overgrowths project from unseen quartz grains into the void. The rosettes are grown on surfaces of detrital and secondary quartz. [From T. R. Walker, 1976, *The Continental Permian in Central, West, and South Europe* (Dordrecht, Holland: D. Reidel), Fig. 25B.]

silicate fragments, and the pore waters; but some of the mush may be detrital. Such matrix-rich sandstones are characteristic of convergent plate margins although not restricted to them.

INTRASTRATAL SOLUTION

In our discussion of weathering in Chapter 11 we noted that alteration of mineral grains in igneous and metamorphic rocks was not restricted to the air–rock interface.

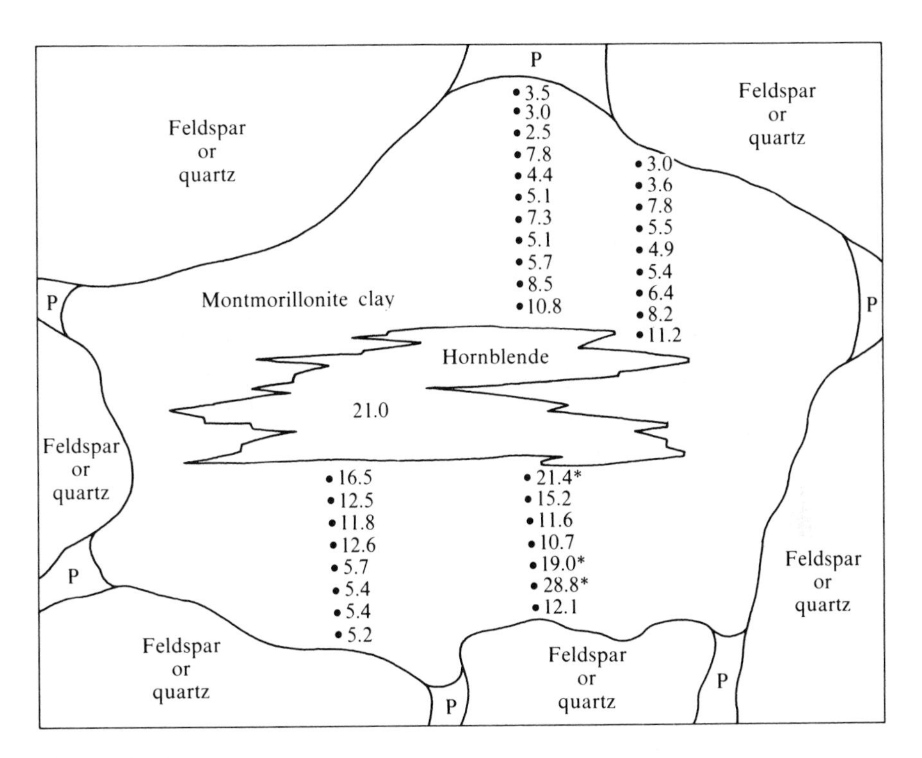

Figure 14–20
Results of electron microprobe analyses of spots 4 μm in diameter in and around an altered hornblende grain 0.4 mm in length in an arkosic red bed in an alluvial fan of Pliocene age, Sonoran Desert, northeastern Baja California, Mexico. The numbers are percentages of Fe_2O_3; high values within the clay (asterisks) are local areas of hematite stain or micro-islands of hornblende. P = pore. [From T. R. Walker et al., 1967, *Geol. Soc. Amer. Bull., 78,* Fig. 2A.]

It also occurs below sea level at continental margins and at the sea floor. The only requirement for alteration to occur is the presence of unstable minerals and water. It comes as no surprise, therefore, to find that water–sediment interaction continues after burial and can result in the partial or complete destruction of certain mineral species. Most susceptible are those formed at very high temperatures and with low silicon–oxygen ratios, such as olivine, hypersthene, or augite. These minerals are extremely rare in pre-Tertiary sandstones. Preferential loss of some types of minerals results in preferential enrichment of others in the residue, such as zircon, tourmaline, and rutile. Paleozoic sandstones characteristically have high proportions of ZTR in their accessory mineral suite.

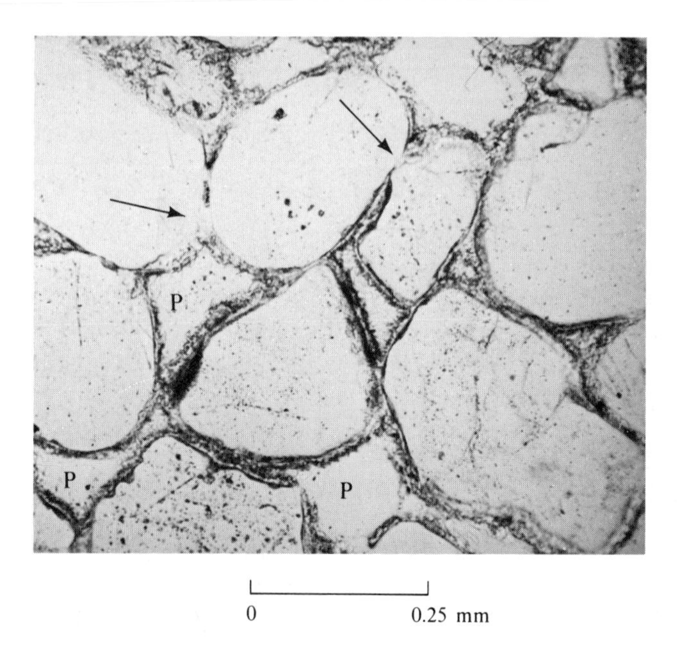

Figure 14–21
Thin section photomicrograph (crossed nicols) of Spiro Sand (Pennsylvanian), Oklahoma, showing authigenic coatings of chlorite fibers or plates growing normal to the surfaces of detrital quartz grains. Coatings are thin or absent at grain-to-grain contacts (arrows). Abundant porosity remains in this gas-producing unit. [From E. D. Pittman and D. N. Lumsden, 1968, *Jour. Sed. Petrology, 38,* Fig. 1.]

EXAMPLES OF DIAGENETIC INVESTIGATIONS

Most studies of sandstone diagenesis are made as a part of more comprehensive investigations aimed at unraveling the history of the unit. Because of this, diagenetic studies vary greatly in scope and detail. We will describe two to illustrate approaches that have been used.

Burial Depth and Permeability: Ventura Oil Field, California

Permeability of reservoir beds in oil fields generally decreases in progressively deeper producing zones. But the decrease may be due to several factors, such as finer grain size, poorer sorting, increased compaction, or increased precipitation of chemical cements. Each of these factors carries different implications with regard to the prospects for future oil production. For example, grain size and sorting are controlled largely by depositional environment, such as position on a delta or submarine fan. It is possible that completing a well a few hundred meters to one side of an existing hole might result in improved sorting and better production. If the decrease in permeability is due to increased compaction of the sand because of increased squashing of ductile rock fragments, we can probably forget about the possibility of significant production at greater depth but might consider broadening the scope of our exploration at shallower depths. If the decrease in permeability results from precipitation of cements in pore spaces, we need to consider the type of cement that is causing our problem. Can we expect the amount of cement to increase continually with depth until all pores are filled; or is the cement likely to be redissolved as depth increases? In some areas the amount of carbon dioxide increases with depth (as a result of decarbonation reactions, and so on), resulting in the solution of earlier-formed calcite cement and regeneration of pore space. These are the kinds of questions that can be solved by a well-planned study of diagenesis.

Hsü (1977), a geologist then employed by the Shell Oil Company, wanted to determine the reasons for the decrease of permeability (and oil-producing potential) with depth in the Ventura Basin oil field, Southern California. The Ventura oil field is located on the crest of the Ventura anticline and produces from several horizons in a Pliocene section of sands, silts, and shales more than 3,000 m thick. Regional geology and the sedimentologic character of cores from producing horizons are consistent with the interpretation that the sands were deposited by turbidity currents as linear bodies trending east–northeast in a submarine fan complex, much like the setting off the coast of Southern California today (see Figure 14–22). Current indicators within the sands reveal an east–west flow direction.

The Pliocene oil-producing formations of the Ventura field have been subdivided on the basis of depth into nine producing zones of varying thickness. The shallowest, zone 1, is 800 m thick; the deepest, zone 9, is 300 m thick. Figure 14–23 shows the range of permeabilities measured on many samples in cores from five of the zones; it is clear that a sequential decrease exists with increasing depth. To determine the reason for the decrease, Hsü made large numbers of sieve analyses of disaggregated samples from each zone and examined many thin sections of core samples. The results of these studies revealed that although there were large differences in grain size and sorting among individual sandstone units, there was no difference among the nine zones. Further, there were no large or systematic differences among the zones in mineral composition, detrital grain shape, or grain angularity. Only one factor differed among the nine zones, the degree of closeness of the grains as seen in thin section (packing) (see Figure 14–24).

Based on studies of modern sediment accumulations, the closeness of packing of sand grains in these sands at the time of deposition is known. By using this information Hsü was able to quantify the amount of compaction with depth. As grains get closer together the sizes of the pores decrease, so that it is possible to correlate the

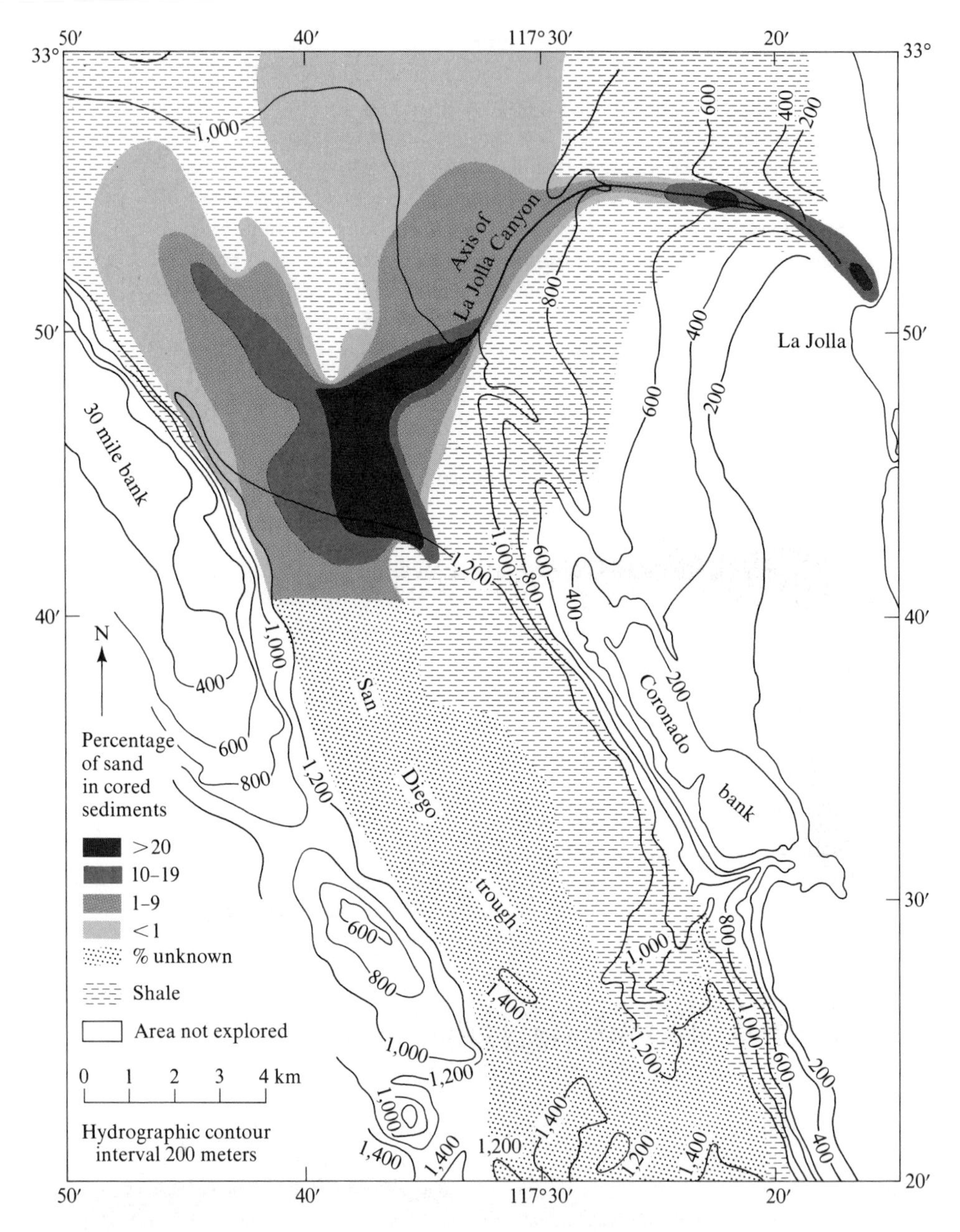

Figure 14–22
Lithologic and bathymetric map of the sea floor off La Jolla, Southern California, illustrating the formation of a modern submarine fan by sediment funneling down La Jolla Canyon. [From K. J. Hsü, 1977, *Amer. Assoc. Petroleum Geol., 61,* Fig. 12.]

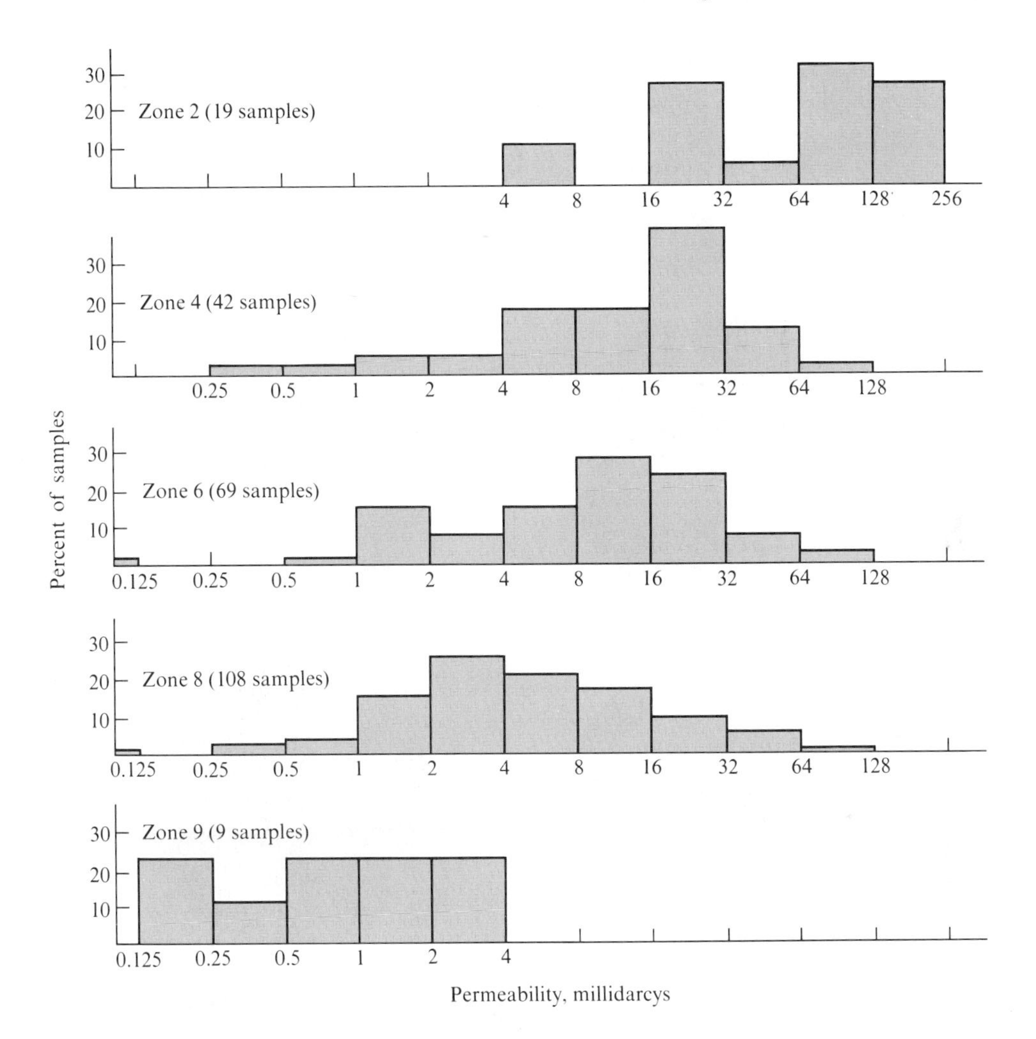

Figure 14–23
Range of permeabilities in sandstones in five of the nine oil production-depth zones, Ventura field. [From K. J. Hsü, 1977, *Amer. Assoc. Petroleum Geol., 61,* Fig. 3.]

measured permeabilities with the degree of compaction. The correlation is excellent in the producing zones of the Ventura oil field. Intense compaction is present in all zones and even in zone 2, 1,000–1,400 m depth, the permeability averages only 4.5% of depositional permeability (which was 1,300 md). Values decrease to 1.2% in zone 4; 0.5% in zone 6; 0.23% in zone 8; and 0.09% in zone 9. On the basis of this decrease, he predicted that permeabilities 200 m below the base of zone 9 would be only 0.02% of depositional permeability, too low for economically useful oil production.

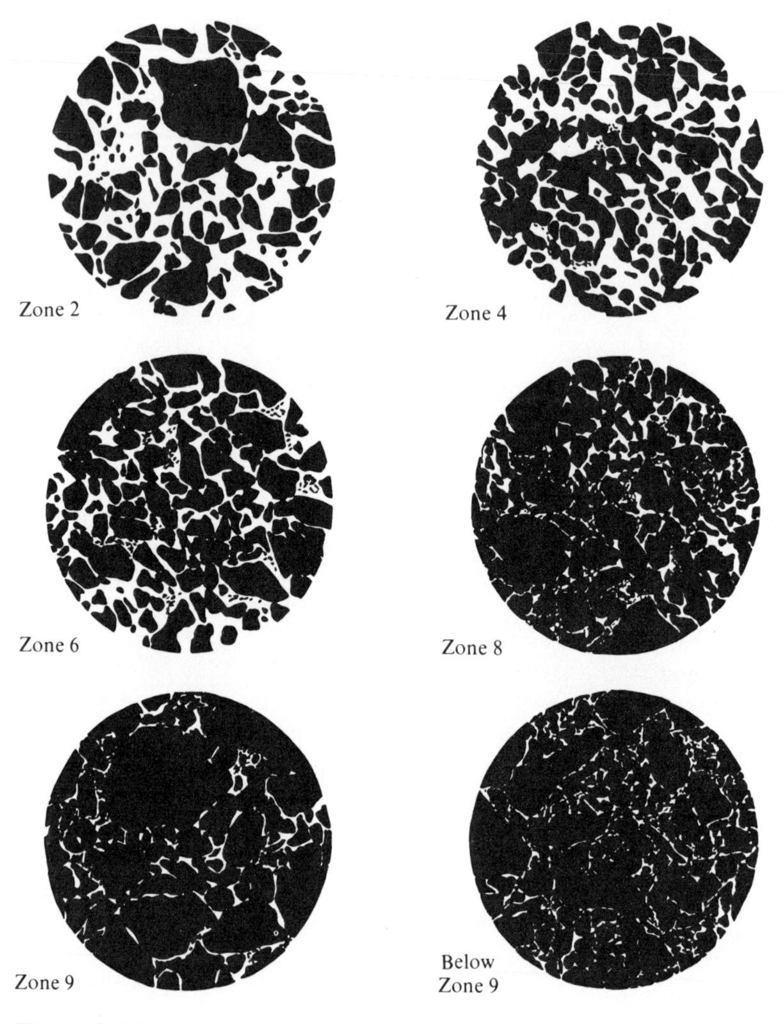

Figure 14–24
Tracings of outlines of the framework grains as seen in typical thin sections of sandstones from the Ventura oil field. Field diameters are 1 mm. The effect of increased depth on porosity and permeability is clearly shown. [From K. J. Hsü, 1977, *Amer. Assoc. Petroleum Geol., 61,* Fig. 6B.]

Chemical Diagenesis: Morrison Formation, New Mexico

In the Ventura oil field there was little evidence of significant chemical diagenesis. Detrital grains were not dissolved and no new minerals were precipitated by the fluids that migrated through the sands after burial. In part, the apparent absence of chemical diagenesis is due to the young age of the sandstones, 5–10 million years. In part, it probably results from a lack of significant change in the composition of the migrating pore fluids during this period. In part, it results from the fact that many of the sandstones were impregnated with oil very early in their diagenetic history. If water is not in contact with the grains there would be no precipitation of new mineral matter in the pores. In older sandstones, however, extensive chemical interaction is common between detrital grains and the waters that have passed through the sediment during the tens or hundreds of millions of years since it was buried.

The criteria used to distinguish materials produced by diagenetic processes include (1) delicate external and internal morphologies that preclude sedimentary transport, (2) spatial relationships of detrital or diagenetic components indicating an origin that postdates deposition or an earlier diagenetic stage, (3) compositions that differ radically from similar materials of detrital origin, and (4) textures that are unlikely to have been produced by depositional processes.

The primary criterion used to determine the relative timing of diagenetic alterations is based on the assumption that pores fill inward. If a sequence of materials lines a pore, that material closest to the detrital grains is assumed to be the oldest, and the material in the center of the pore to be the youngest. It is also assumed that a cement will tend to fill or line all types of available pores (with the exception of cements requiring a nucleus of similar structure or composition upon which they can develop). Thus if cavities in partially dissolved grains are not lined or filled with cementing agents present in the rock, these cavities postdate the cements.

A good example of the complexity of these chemical interactions is provided by Flesch and Wilson (1974) in their investigation of outcrops of the Morrison Formation (Jurassic) in northwestern New Mexico. The study was a comprehensive one concerned with environments of deposition and provenance of the detrital grains, as well as diagenetic aspects, but our focus here will be principally on diagenesis.

The Morrison Formation in this area consists of about 55% fine- to coarse-grained white to yellowish sandstone and 45% montmorillonitic claystone of various colors. Depositional environments were interpreted using the geometry of the sandstone units, sedimentary structures, and lithologies. Based on these field data, four members were recognized. The first (lowest) and third members are braided stream sands composed of more than 95% sandstone; the second and upper members are dominantly claystone and were interpreted as meandering stream deposits. The detrital mineral composition of the Morrison Formation sandstones is 55–80% quartz, 12–36% feldspar (subequal amounts of K-feldspar and plagioclase), 5–15% lithic fragments (almost all igneous), and 1–2% chert. These framework grains form only 67–77% of the whole rock, the rest being 9–31% clay plus chemical cement and 1–20% pore space.

Many chemical changes have affected the Morrison Formation in the area studied. The most common involve cementation by grain coatings of montmorillonite, interlayered montmorillonite–illite, chlorite, and chalcedony; overgrowths on quartz and feldspar grains; pore fillings of calcite and kaolinite; and dissolution of detrital feldspars. Less common alterations include the diagenetic formation of gibbsite, pyrite, possibly anatase, and replacement of iron-rich materials by hematite and limonite. Some of the diagenetic features are illustrated in Figure 14–25.

It is clear from these changes that there have been many different types of pore waters passing through the formation during the 160 million years or so since it was deposited by Jurassic streams. For example, secondary growths of feldspar, calcite, montmorillonite, and illite require basic solutions rich in dissolved cations such as calcium and potassium. The formation of kaolinite and gibbsite and the dissolution of detrital feldspars requires acidic, cation-poor solutions. Replacement of iron-rich minerals by hematite requires oxidizing waters (waters with a high content of dissolved oxygen). Precipitation of pyrite, on the other hand, requires the absence of dissolved oxygen in pore waters. The chalcedony cement may have crystallized from an opaline predecessor, and the precipitation of opal from a pore solution requires an unusually high concentration of dissolved silica in pore waters, at least 120 ppm. Normal near-surface waters average only 13 ppm. Quartz cement, on the other hand, forms from dilute solutions that contain only 6–10 ppm dissolved silica and crystallizes directly from solution with no opal or chalcedony precursor.

Any attempt at a complete analysis of the origin of all the cements, pore fillings, replacements, and dissolutions would take us deeply into physical chemistry and the general topic of sedimentary geochemistry. We will not pursue that type of analysis. We can, however, construct a chronologic sequence of the chemical changes seen in the Morrison, based on textural evidence in thin sections and SEM observations. The result is shown in Figure 14–26, with the approximate timing of events based on structural events known from earlier studies of regional geology.

SUMMARY

Diagenetic processes in detrital rocks include some of the most complex phenomena in geology. They involve both physical stresses and extreme variations in chemical composition of natural waters, and the way in which these variables interact with the mineral and rock particles in a sediment is further compounded by the length of geologic time. Diagenetic studies, however, are rewarding in proportion to their complexity. The effort and background knowledge required to resolve most diagenetic complexities are great, but so are the rewards in terms of better understanding of the history of the rock.

The amount of compaction of a freshly deposited sand depends on five factors: clay content, sorting, percentage of ductile fragments, angularity of the nonductile grains, and burial depth or tectonic stresses. Compaction is a more effective process in sandstones that contain abundant ductile lithic fragments such as schist, phyllite, slate, shale, mica flakes, or fragments of floodplain and sea-floor mud. With increased depth of burial, these fragments are squeezed into adjacent depositional pore spaces, thinning the stratigraphic section and diminishing the porosity and permeability of the sandstone.

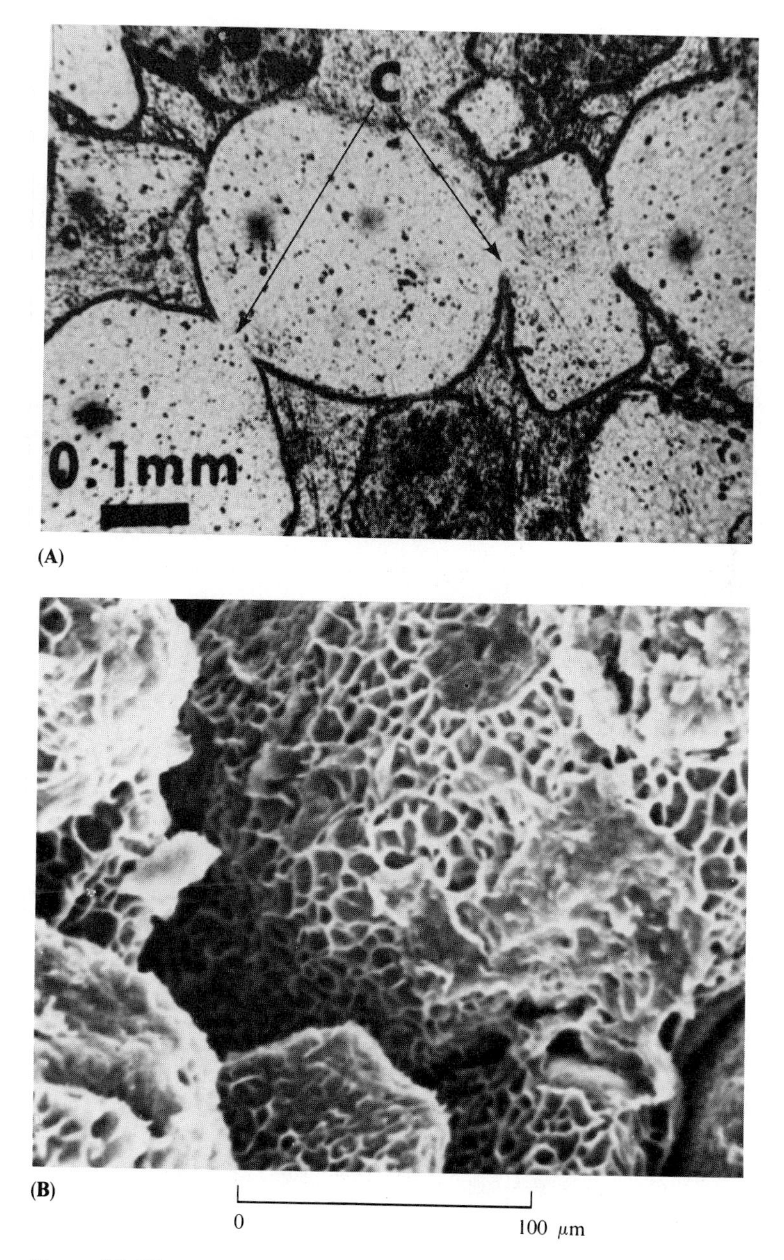

Figure 14–25
Photomicrographs (crossed nicols) of some diagenetic features seen in sandstones of the Morrison Formation (Jurassic), Sandoval Co., New Mexico. Photos courtesy C. T. Siemers. **(A)** Thin section showing clay growths into pore spaces from the surfaces of detrital grains. Note absence of clay growth at locations where grains are in contact (C); the flow of pore waters was insufficient to form a visible precipitate at these sites. **(B)** The coatings of clay (montmorillonite) on the grains in part A as seen using an SEM. The clays are again seen to have grown normal to grain surfaces; a honeycomb appearance is produced.

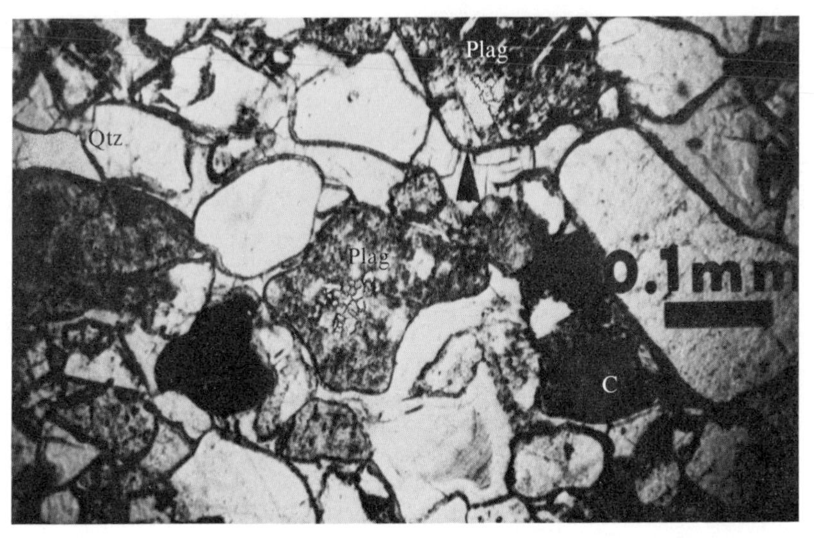

(C)

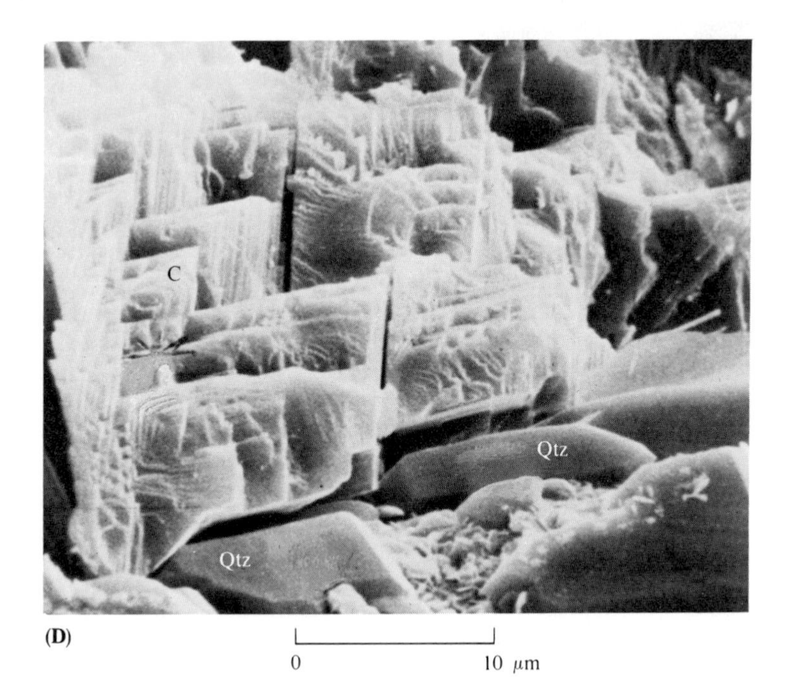

(D)

Figure 14–25 (*continued*)
(**C**) Thin section of subarkosic sandstone showing chalcedony cement (light color) with fibers grown normal to grain surfaces (Qtz); partly altered plagioclase grains (Plag); and a chert grain (C). (**D**) Scanning electron micrograph showing rhombs of calcite cement (C) covering quartz overgrowths (Qtz), which, in turn, partially coat euhedral flakes of authigenic chlorite (between Qtz and Qtz at lower right of photo). [From G. A. Flesch and M. D. Wilson, 1974, *25th Field Conf., New Mexico Geol. Soc.,* Figs. 3A, 4A, 3D, 4E.]

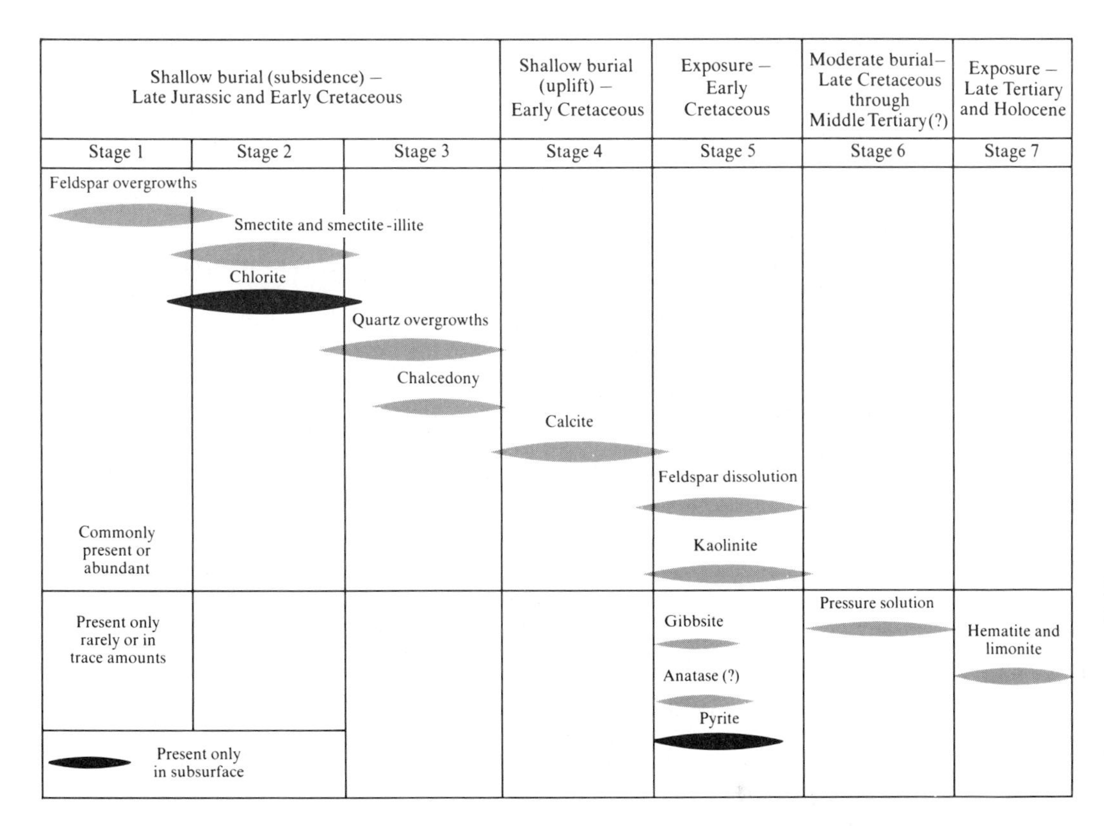

Figure 14–26
Types of chemical alterations and their timing in the Morrison Formation sandstones, northwestern New Mexico, based on thin section and SEM studies. [From G. A. Flesch and M. D. Wilson, 1974, *25th Field Conf., New Mexico Geol. Soc.*, Fig. 7.]

The growth of secondary minerals in a sandstone requires that the pore fluid be oversaturated with respect to the mineral being precipitated. It also is necessary for very large amounts of pore fluid to pass through the rock over long periods of geologic time. Quartz and calcite probably form the bulk of authigenic mineral growths in sandstones, but others, such as hematite and clay minerals, can dominate in some sandstones. Most hematite seems to be produced very rapidly in modern semiarid areas by leaching and oxidation of ferrous iron from ferromagnesian minerals. However, continued destruction of ferromagnesian minerals much later in the geologic history of the rock can cause hematite to be formed throughout the life of a sandstone.

FURTHER READING

Flesch, G. A., and M. D. Wilson. 1974. Petrography of the Morrison Formation (Jurassic) sandstone of the Ojito Spring Quadrangle, Sandoval County, New Mexico. *New Mexico Geol. Soc. Guidebook, 25th Field Conf.,* 197–210.

Hsü, K. J. 1977. Studies of the Ventura Field, California, II: lithology, compaction, and permeability of sands. *Amer. Assoc. Petroleum Geol. Bull., 61,* 169–191.

Journal of the Geological Society (London). 1978. *135,* Pt. I, 1–156. (An issue composed of 14 articles plus discussions of the state of the art in studies of sandstone diagenesis.)

Schluger, P. R., ed. 1979. *Diagenesis as it Affects Clastic Reservoirs.* Soc. Economic Paleontologists and Mineralogists Spec. Pub. No. 26, 443 pp.

Scholle, P. A. 1979. *A Color Illustrated Guide to Constituents, Textures, Cements, and Porosities of Sandstones and Related Rocks.* Amer. Assoc. Petroleum Geol. Memoir No. 28, 201 pp.

Sedimentation Seminar. 1969. Bethel Sandstone (Mississippian) of western Kentucky and south-central Indiana, a submarine-channel fill. *Kentucky Geol. Surv. Report of Investigations No. 11,* pp. 7–24.

Shelton, J. W. 1962. Shale compaction in a section of Cretaceous Dakota Sandstone, northwestern North Dakota. *Jour. Sedimentary Petrology, 32,* 873–877.

Siever, R. 1979. Plate-tectonic controls on diagenesis. *Jour. Geology, 87,* 127–155.

Walker, T. R. 1967. Formation of red beds in modern and ancient deserts. *Geol. Soc. Amer. Bull., 78,* 353–368.

Walker, T. R., P. H. Ribbe, and R. M. Honea. 1967. Geochemistry of hornblende alteration in Pliocene red beds, Baja California, Mexico. *Geol. Soc. Amer. Bull., 78,* 1055–1060.

Wilson, M. D., and E. D. Pittman. 1977. Authigenic clays in sandstones. Recognition and influence on reservoir properties and paleoenvironmental analysis. *Jour. Sedimentary Petrology, 47,* 3–31.

15

Limestones and Dolomites

Limestones and dolomites form 10–15% of the sedimentary column and are nearly always quite pure. Impurities total less than 5% of the rock, and typically are confined to clay minerals, fine sand-size and coarse silt-size quartz grains, and very finely granular to powdery quartz of uncertain origin. Because of the essentially monomineralic nature of limestones and dolomites, their study is largely a study of textures and structures, supplemented by geochemistry. Because the minerals calcite and dolomite are very soluble at near-surface conditions, carbonate rocks recrystallize easily and frequently, obliterating many of the diagnostic textures and structures needed to interpret the origin of the rocks. As a result, the study of carbonate rocks is, at least initially, more difficult than sandstones. Experience is even more necessary for proper analysis of limestones and dolomites than it is for sandstones.

FIELD OBSERVATIONS OF LIMESTONES

Carbonate rocks are recognized in outcrop by their reaction to a few drops of dilute hydrochloric acid, their softness, and the typically interlocking texture of their crystals. Limestone is distinguished from dolomite by the fact that dolomite does not react visibly to dilute hydrochloric acid unless powdered. Also, dolomite commonly weathers with a dull brownish yellow cast (buff color) because it usually contains some ferrous iron as a substitute for magnesium in the crystal structure. The iron is released from the carbonate during weathering and oxidizes to cause the color.

The relative proportions of calcite, dolomite, and quartz (or other silicate minerals) in a carbonate rock can be estimated by etching the rock surface at the outcrop. Hydrochloric acid is dripped onto a clean flat surface of a hand specimen until the calcite is dissolved in sufficient amounts so that the less soluble materials stand out in

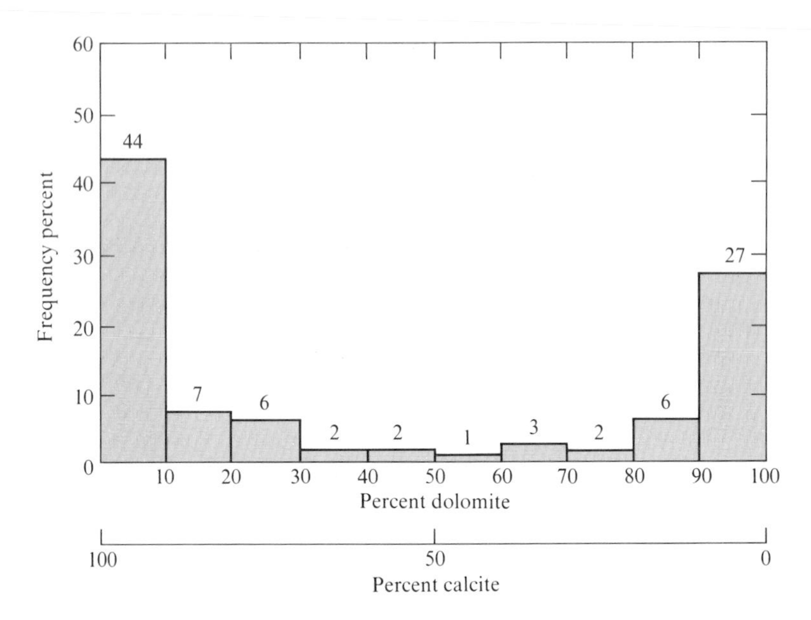

Figure 15–1
Computed percentages of calcite and dolomite for 1,148 modal analyses of North American carbonate rocks. [From E. Steidtmann, 1917, *Geol. Soc. Amer. Bull., 28,* Fig. 1.]

relief. Quartz is identified by its translucency and glassy appearance, chert by its hardness (if the grains are large enough to be scratched and examined using a 10× lens), and dolomite by its white color, softness, and crystal outlines. Calcite only rarely occurs as scalenohedra or rhombohedra, but dolomite often occurs as rhombs scattered in the limestone. The etching technique also reveals the distribution in the rock of these relatively insoluble constituents. For example, chert may be located along bedding planes; fossils may have been converted into chert but the bulk of the rock may be dolomite; the limestone–dolomite contact in the outcrop may be at an angle to the bedding of the rocks, proving a replacement origin for the dolomite.

Extensive field studies during the past 60 years in many parts of the world have documented the fact that carbonate beds tend to be either all calcite or all dolomite (see Figure 15–1). Apparently, whatever the conditions are that produce limestones and dolomites, a marked tendency exists to form one or the other but not subequal mixtures of the two. If we assume the limestones and dolomites to be primary precipitates from surface waters, what inferences might we make about the chemical composition of these waters? If we believe the dolomite is of replacement origin, what can we say about the difference in composition between the surface waters and the diagenetic waters from which the dolomite precipitated? These are questions that are raised by field observations but can be answered fully only by a combination of field and laboratory studies. We will examine the problems of the origin of limestone and dolomite later in this chapter.

Textures

The textures of limestones are extremely variable because of the complex origins of carbonate rocks. Limestones can have textures that are identical to those of detrital rocks (grain rounding, sorting, and so on; see Figure 15–2A) or to those of chemical precipitates (equigranular, interlocking crystals, "porphyritic," and so on; see Figure 15–2B). Many carbonate rocks display both types of textures. In addition, they commonly have biologically produced textures characteristic of the growth habits of living organisms, such as algae or corals (see Figure 15–2C). Some of these biologic textures are so intricate they defy adequate written description (see Figure 15–2D). In these cases a labeled photograph is indeed worth a thousand words.

Most limestones can be described by their gravel- and sand-size particles, the calcium carbonate mud matrix, and the coarsely crystalline calcite cement.

Allochemical Particles

Allochems are the coarse silt-, sand-, and gravel-size carbonate particles that form the framework in mechanically deposited limestones. They are the equivalent of the quartz, feldspar, and lithic fragments in sandstones. Four types of allochems are common: fossils, peloids, ooliths, and limeclasts.

1. *Fossils* are the solid carbonate remains of living organisms—the hard parts. They are identified in the field by their biologically determined shapes, such as the concavo-convex outline of a clam, the leafy shape of a bryozoan, or the segmented tubular shape of a crinoid stem (see Figure 15–3A). Because they are identified by their shape, however, they become harder to recognize if they were broken into small pieces before burial, and this is a common phenomenon in the shallow marine environment where most limestones were deposited. Both high current velocities and predators can cause the fragmentation. The minimal size required for recognition depends on the microstructure of the organism. For example, a fragment of a coral 10 mm in diameter might be indistinguishable from a bryozoan or an algal fragment; a similar-size piece of crinoid column would be easily identifiable. An echinoid fragment of this size probably would not be recognized as a fossil fragment at all because echinoderm skeletons disaggregate into sand-size crystals of calcite that look like cement crystals in limestone hand specimens.

2. *Peloids* are round, ellipsoidal aggregates of microcrystalline (aphanitic) calcium carbonate that lack internal structure (see Figure 15–3B). Most are fecal pellets and are mostly of coarse silt size to very fine sand size, presumably reflecting the uniform anal dimensions of the organisms that produce them. Marine worms, for

Figure 15–2
(**A**) Well-rounded, poorly sorted pebbles of algal micritic limestone in a matrix of black (organic matter) microcrystalline limestone. The algal nature of the pebbles is visible only in thin section. Cool Creek Limestone (Ordovician), Arbuckle Mountains, Oklahoma. (**B**) Laminated limestone composed of interlocking microcrystalline calcite crystals a few micrometers in diameter. The laminations indicate the original clastic nature of the crystals that has been obliterated during recrystallization from aragonite to calcite. Kindblade Formation (Ordovician), Arbuckle Mountains, Oklahoma.

(**C**) Limestone composed of fragments of calcified algal mat showing the platy limestone fragments characteristic of such algal mats. Also visible are original void spaces now filled with coarser translucent calcite (sparry calcite), the dark gray areas roughly parallel to the bedding. McLish Limestone (Ordovician), Arbuckle Mountains, Oklahoma. (**D**) Polished slab of reef rock composed of the coelenterate (?) *Tubiphytes* (white), sponges and bryozoa (light gray), and the alga *Archaeolithoporella* (laminated white crusts). Cements are shades of gray bands. Capitan reef (Permian), New Mexico. [Photo courtesy J. A. Babcock.]

(A)

(B)

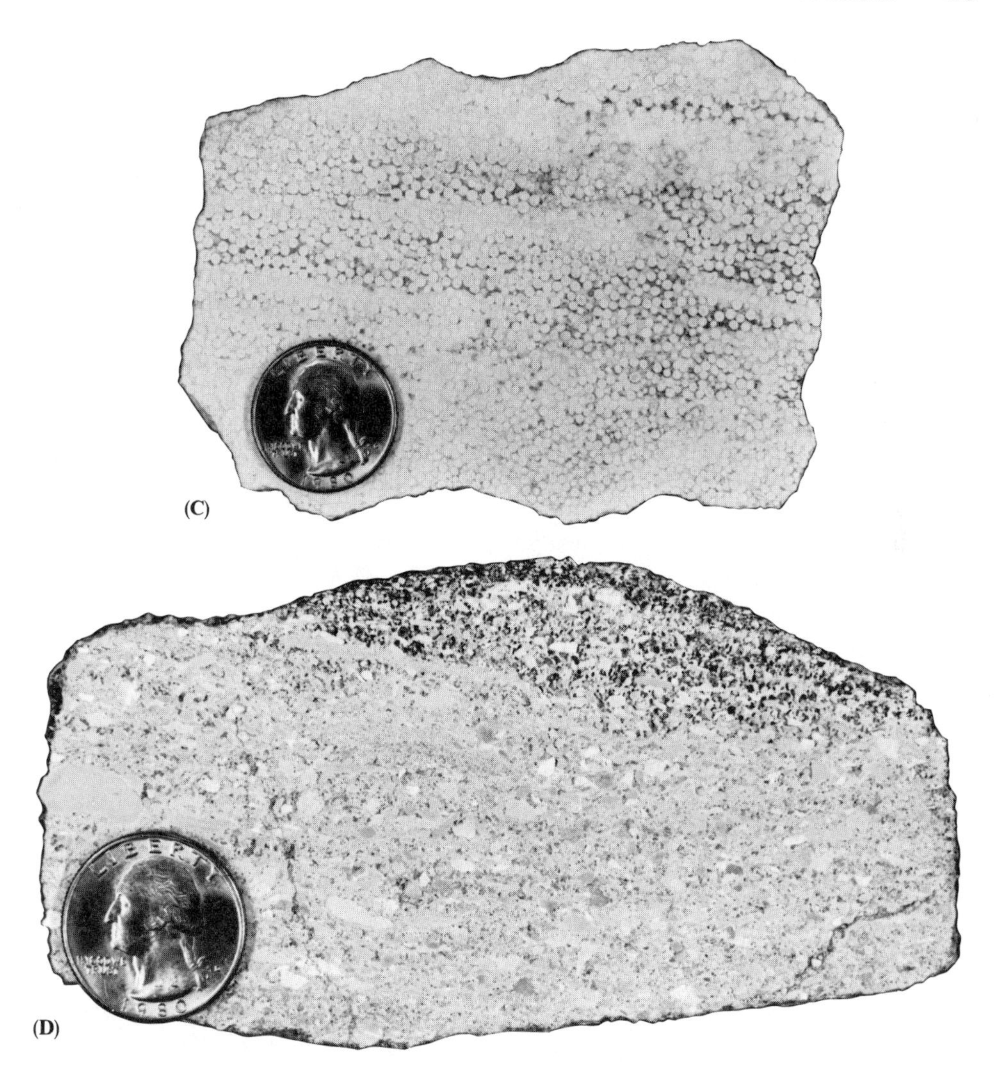

Figure 15-3
(**A**) Large, unbroken fossils in a matrix of microcrystalline calcite, about natural size. Easily distinguishable are brachiopods and crinoid stems. Probable gastropod at upper right; dark objects may be fish plates. Note extremely poor sorting, indicating quiet water deposition, as does the micritic nature of the matrix. [Photo courtesy F. J. Pettijohn.] (**B**) Peloids (structureless micritic intraclasts) at center left of photo. Also visible are a stylolite seam and many large algal (?) fragments and intraclasts and a pelloidal intraclast (top center). Arbuckle Formation (Ordovician), southern Oklahoma. (**C**) Ooliths cemented by translucent sparry calcite. Upper part of slab shows horizontal bedding; at lower right the oolitic layers are cross-bedded, dipping to the left. Chimney Hill Limestone (Silurian), Arbuckle Mountains, Oklahoma. (**D**) Micritic intraclasts cemented by sparry calcite and oriented with long dimensions parallel to bedding. Dark-colored layer (hematitic) of finer-grained intraclastic debris overlies the coarser layer. West Spring Creek Formation (Ordovician), Arbuckle Mountains, Oklahoma.

example, burrow through the carbonate sediments on the shallow sea floor swallowing anything that is small enough to ingest that contains nourishing organic matter. The ingested sediment passes through the alimentary tract of the organism and is emitted finally as the microcrystalline aggregates we call pellets. Peloids may also be produced by diagenetic alteration of other types of allochems.

3. *Ooliths* are nearly spherical, polycrystalline carbonate particles of sand size that have a concentric or radial internal structure (see Figure 15–3C). These particles always have a nucleus of some sort, such as a quartz grain or fossil fragment, around which the oolitic coating has formed. If the nucleus is sufficiently large it can be seen in hand specimens of the rock. The outer rind of the oolith is chemically precipitated from agitated water and, therefore, the presence of ooliths is evidence that the particle has been transported by strong currents. Oolitic limestones commonly are cross bedded for this reason. Pseudo-ooliths are sometimes formed by algal borings and alteration of the outer part of a shell fragment or other allochemical particle.

4. *Limeclasts* are fragments of earlier-formed limestone. They may originate in a number of ways. Most are ***intraclasts,*** pieces of penecontemporaneous lithified carbonate rock from within the basin of deposition (see Figure 15–3D). Perhaps they are pieces of semiconsolidated carbonate mud torn from the sea floor by a winter storm on January 3, 100 million B.C. Perhaps they are aggregates of pellets (*grapestone*) that stuck together because of mucilaginous organic coatings or were cemented together. Perhaps they are fragments produced by drying and cracking of intertidal mud from carbonate sediment on the margin of the basin. Any of these origins qualifies the particles as intraclasts.

Limeclasts also exist as fragments that were carried into the basin of deposition from the surrounding area; for example, a piece of Mississippian fossiliferous limestone in a Cretaceous limestone. Most limestones do not contain such fragments, and those that do typically also contain other evidence of externally derived (terrigenous) detritus, such as noticeable percentages of detrital quartz, feldspar, or silicate lithic fragments. Commonly the limestone that contains such fragments was deposited in a tectonically active area such as the Alpine region during the closing of the Tethys seaway or in the Marathon region of west Texas during its Pennsylvanian orogenic episodes.

Orthochemical Particles

Orthochems are the calcium carbonate matrix and cement that bind the allochems to lithify the sediment. Orthochemical material is of two types, microcrystalline $CaCO_3$ and coarsely crystalline cement.

1. *Microcrystalline calcite* (micrite) is the most abundant type of calcium carbonate material found in limestones. In hand specimens it is dull, opaque, and aphanitic like a piece of rhyolite or chert and can vary in color from white to black,

depending on impurities. It may be present in small amounts between the allochemical grains or form most of the rock with the allochems dispersed through the rock like plums in a pudding.

Nearly all microcrystalline carbonate sediment originates as disarticulated algal material. From the viewpoint of a carbonate geologist, the algal world can be divided into two groups, those that have hard parts (calcareous algae) and those that do not (seaweed and others). Some of the algae have an internal structure formed of needles of aragonite a few micrometers in length connected by organic tissue; when the organism dies the organic tissue decomposes, releasing the needles onto the intertidal or shallow sea floor (see Figure 15–4). It is these aragonite needles, subsequently recrystallized to equant blocks of calcite 1–4 μm in size, that we see in the fine-grained limestones in the stratigraphic column.

2. *Coarsely crystalline calcite cement* (sparry calcite) in limestones has the same origin as the calcite cement in sandstones. It is formed of interlocking coarse silt- to sand-size crystals of calcite precipitated from a moving pore fluid that was supersaturated with respect to calcium carbonate. This cement is termed *sparry calcite.* Typically, sparry calcite and microcrystalline calcite matrix do not occur together; if one is abundant, the other normally is absent.

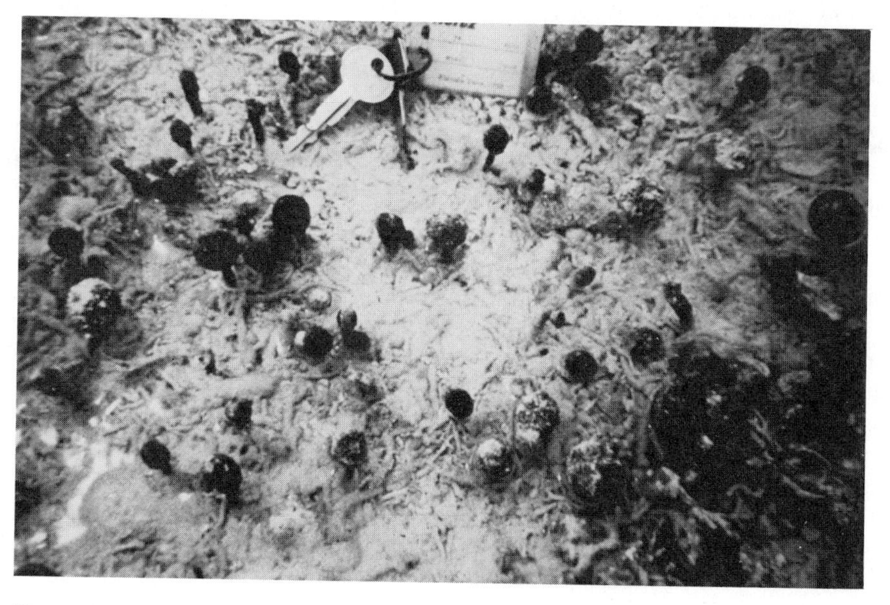

Figure 15–4
A mini-forest of Penicillus plants (shaped like asparagus tips) surrounded by aragonite needles produced by decay of their ancestors. Assorted organic debris litters the "forest floor." Depth of water is a few meters. [Photo courtesy R. N. Ginsburg.]

Grain Size, Sorting, and Rounding

The interpretation of these textural features in limestones is more difficult than in sandstones, largely because of the biologic character of fossils and pellets. For example, the fossils in a limestone can be whole ostracods of a particular species that were buried in the carbonate mud in which they lived. The fact that these allochems are of a certain size and are very well sorted is not directly related to current strengths in the depositional environment. Both mean size and sorting are biologic in origin rather than hydrodynamic.

We noted earlier the biologically determined excellent sorting of fecal pellets. Rounding can be similarly biologically determined. Crinoid columns and fecal pellets are always round, irrespective of whether the depositional environment is of high or low kinetic energy. The variety of shapes and sizes of biologic particles makes "hydraulic parity" very difficult to determine.

The energy level of the environment of carbonate deposition is evaluated mostly from the presence or absence of calcium carbonate mud. It is assumed that aragonite needle-producers are ubiquitous, so that microcrystalline ooze is always available in carbonate environments. Therefore, if a limestone lacks these aphanitic particles it means current strengths were high enough to remove them. If the limestone is rich in microcrystalline carbonate, we interpret the depositional environment as having been of low kinetic energy. It must be remembered, however, that the source area of most carbonate particles is very close to the site of final deposition; allochems may be produced rapidly and in large numbers, thus forming a rock with many allochems and little mud in a low-energy environment.

Other factors used to evaluate the energy level of the environment are: (1) evidence of mechanical abrasion during transport, (2) presence of ooliths, and (3) current structures such as cross bedding.

Noncarbonate Mineralogy

The average limestone contains only 5% noncarbonate material, of which nearly all is quartz, clay minerals, and chert. The quartz and clay minerals are silicates derived from outside the basin of deposition (as are a small percentage of carbonate limeclasts) and are considered *terrigenous detritus*. They reflect the presence of source areas somewhere within the regional drainage, with the grain size of the terrigenous sediment reflecting the velocities of currents as they enter the depositional basin. The reason such detritus is so uncommon in limestone-depositing basins is the biologic origin of nearly all carbonate particles.

An influx of silicate detritus implies an influx of fresh water. Most carbonate-secreting organisms live in very shallow water and cannot tolerate a significant change in salinity. In addition, their reproductive rate is inadequate to keep up with the rate of influx of mud characteristic of most streams, so they are killed off rather quickly. For example, limestones are not forming today near the Mississippi Delta or for hundreds of kilometers to its west where the Mississippi detritus is spread. Simi-

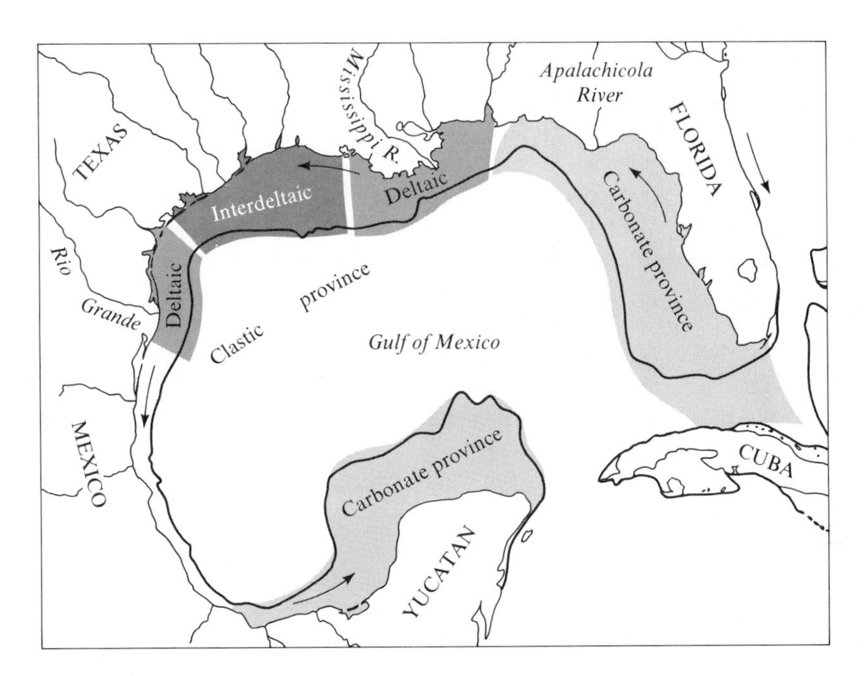

Figure 15–5
Modern sediment types in the nearshore area of the Gulf of Mexico. Modern carbonate deposition occurs along the west coast of Florida and along the north side of the Yucatan Peninsula. Both Florida and the Yucatan are underlain by Tertiary carbonates. The east coast of Florida receives quartz and clay from the southern Appalachians; the central Gulf Coast, from the mid-continental United States; the east coast of Mexico, from the Sierra Madre Oriental. Arrows indicate direction of nearshore current flow and sediment transport. Heavy line is the shelf boundary. [From C. E. B. Conybeare, 1979, *Lithostratigraphic Analysis of Sedimentary Basins* (New York: Academic Press), Fig. 4–1.]

larly, carbonates are absent along the northeastern coast of Mexico because of sand and mud originating in the Sierra Madre Oriental. Limestone reefs are abundant, however, along the west coast of Florida beyond the reach of Mississippi detritus, and along the southeast coast of Mexico and the Yucatan Peninsula, which are unaffected by drainage from mountainous areas (see Figure 15–5).

Chert in carbonate rocks is of two origins. A minor amount is extrabasinal, with the same source as other terrigenous particles. Nearly all chert in carbonates is intrabasinal and forms by crystallization of the amorphous silica from shells of siliceous organisms. The bulk of the amorphous silica is secreted by siliceous sponges (Cambrian–present), Radiolaria (Ordovician–present), and diatoms (Jurassic–present), with the diatoms doing about 80% of the secreting at present. These organisms remove dissolved silica from sea water (diatoms also live in fresh waters) and precipitate it as an amorphous, solid support for their soft tissues. When they die,

Figure 15–6
Chert nodules and lenses developed along surfaces parallel to bedding, Onondaga Limestone (Middle Devonian), Albany, New York. Length of ruler at lower left is 15 cm. [Photo courtesy R. C. Lindholm.]

they sink to the sea floor, and some of the silica shells are buried in the carbonate sediment, dissolved, and later crystallized as microcrystalline quartz (chert). The shells of diatoms and radiolaria are mostly of silt size and, therefore, tend to accumulate in quiet-water areas with muddy sediment.

The intrabasinal chert in limestone occurs either as very finely crystalline to powdery quartz or as nodules centimeters to meters in length (see Figure 15–6). When in the nodular form it is concentrated along visible bedding planes, presumably reflecting migration paths of the silica dissolved from the siliceous shells. The reason the dissolved silica migrates to the centers of crystallization we see as nodules is not known.

Classification

Classification of limestones is based on textural variations almost exclusively because of the lack of mineralogic variations. But despite the constant mineralogy of ancient carbonates, a useful classification scheme can be constructed by analogy with sandstones. The allochemical grains, calcium carbonate mud, and sparry calcite cement are the analogs of sand grains, clay matrix, and cement in sandstones. Figure 15–7 demonstrates the similarity of the classification schemes for sandstones and limestones.

In constructing descriptive names for limestones, the key terms are modified so that fossil becomes *bio-*, peloid becomes *pel-*, oolith is shortened to *oo-*, intraclast is *intra-*, microcrystalline calcite is *mic-*, and sparry is *spar-*. Thus we have biosparites,

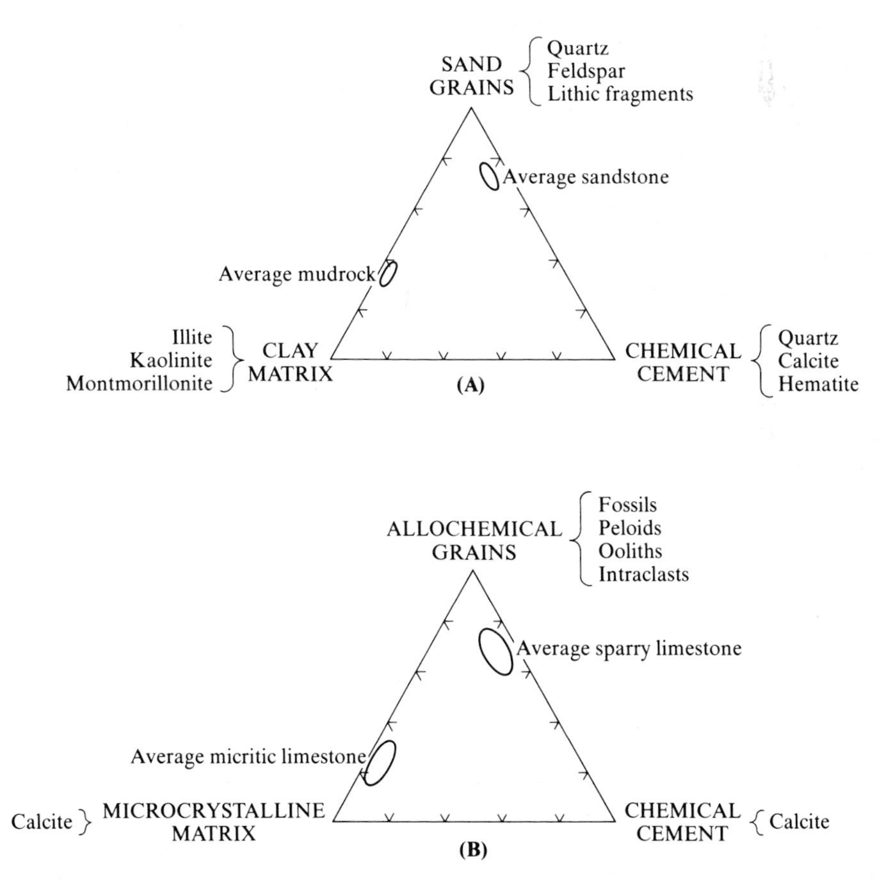

Figure 15–7
Triangles showing the analogy between the components of (**A**) sandstones and (**B**) limestones.

Table 15–1
Representative Compositions of Limestones and Their Best Descriptive Names

Composition of limestone		
Allochems	Orthochems	Appropriate name
70% Pelecypods 30% Ooliths	Sparry cement	Oolitic pelecypod biosparite
80% Ooliths 5% Fossil fragments 15% Glauconite	Sparry cement	Glauconitic oosparite[a]
60% Pellets 30% Fossil fragments 10% Intraclasts	Microcrystalline carbonate matrix	Fossiliferous pelmicrite
70% Intraclasts (gravel size) 25% Trilobites 5% Pellets	Sparry cement	Trilobite intrasparudite[b]
40% Crinoid fragments 40% Brachiopods 20% Clay minerals	Microcrystalline Carbonate matrix	Clayey crinoid-brachiopod biomicrite

[a]If terrigenous detritus forms 10% or more of the sand- and gravel-size debris, it is added to the rock name as a modifier.
[b]If the allochems are of gravel size (> 2 mm), *rudite* is added to the rock name.

pelmicrites, and oosparites. The main part of the name is based on the major allochem and orthochem; appropriate modifiers may precede the main part of the name. Table 15–1 illustrates typical limestone compositions and their descriptive names. As in all classifications, rocks occur in nature that are difficult to place in pigeonholes, but the concept behind the nomenclatural scheme is sound and leads to better communication.

Structures

A wide variety of sedimentary structures occur in limestones. Some of the structures reflect biologic origin (for example, reefs); some, current origin (for example, cross bedding); and others, diagenetic origin (for example, stylolites).

Bedding and Lamination

These structures are found in carbonate rocks of any origin and any composition. As is true of sandstones, the origin of bedding and lamination in many carbonates is obscure. Consider bedding in micrites, for example. The micrite is sensibly identical on both sides of the bedding plane (even in thin section) and no concentrations of organic matter, clay minerals, or other materials are evident. Perhaps there was incipient cementation or recrystallization of the carbonate ooze before the next lamina of ooze was deposited. But why? What characteristic of the sea bottom water or exposure to air on a tidal flat was responsible for the very rapid lithification? Such questions can rarely be answered.

Sometimes the origin of bedding and lamination is clear. This is true where there is a change in size or type of allochem, a change in the ratio of micrite to allochems, or where an obvious surface of subaerial exposure is present.

Laminated carbonate sediment can be produced by blue-green algal mats that lack hard parts. The soft-bodied algae grow as filamentous mats in the intertidal zone. The filaments are mucilaginous and as they are repeatedly swept over by wave-generated currents and tides they trap and bind microcrystalline carbonate particles in the water, resulting in the formation of laminated layers consisting of a mixture of organic tissue and micrite (see Figure 15–8). These structures are called *stromatolites* and are common in rocks of all ages, from Precambrian to the present. In pre-Holocene stromatolites the organic tissue has decomposed and only laminated carbonate sediment remains. In many deposits the layers appear to have accumulated with a bulbous upper surface, a growth topography caused by erosive effects on the accumulating algal mat.

Stromatactis Structure

This peculiar structure (see Figure 15–9) is common in micritic reef knolls and consists of layers or irregular masses of coarsely crystalline calcite within an otherwise homogeneous micrite. The features are 1–5 mm thick, can be 10 cm or more in length, and are oriented parallel to the bedding of the limestone. The base of the structure is flat and commonly is floored with laminated micritic sediment, but the bulk of the structure is sparry calcite. The origin of these structures is uncertain, but they seem to form during very early diagenesis. Cavities are partially filled by micritic sediment, and then the upper part of the cavity is filled by the coarse spar that typically forms the bulk of the stromatactis structure. They are useful for defining the bedding in otherwise uniform micritic mounds.

Dune Forms, Ripples, and Cross Bedding

Allochemical limestones can display the same variety of current structures as sandstones, although they typically are not as evident at first glance as in sandstones. As with sandstones, the most common are dunes, ripples, and cross bedding. Many marine limestones are cross bedded, with the thickness of cross-bedded sets ranging up to several meters. Commonly, the azimuths of the cross bedding in marine limestones

(A)

(B)

Figure 15–8

(**A**) Subtidal stromatolites in water 2 m deep, Shark Bay, Western Australia. Elongation of mounds is in direction of wave movement. The mounds are built by colloform mat growth and are soft on top, increasingly lithified with depth into the mound. [From P. E. Playford and A. E. Cockbain, 1976, *Stromatolites* (New York: Elsevier), Fig. 9B.] (**B**) Plan view of an eroded surface of a limestone showing cross sections of algal stromatolites, Hoyt Limestone (Cambrian), Saratoga Springs, New York. The white scale is 10 cm in length. Comparison with part A shows that for stromatolites, the present is a good key to the past. [From J. L. Wilson, 1975, *Carbonate Facies in Geologic History* (New York: Springer-Verlag), Plate 14A.]

Figure 15–9
Stromatactis structure in a micritic mud mound, Ireland. The lighter-colored filling of the structures is sparry calcite. Length of specimen is 15 cm. [From H. Blatt et al., 1980, *Origin of Sedimentary Rocks,* 2d Ed. (Englewood Cliffs, N.J.: Prentice-Hall) Fig. 14–3 (photo courtesy R. C. Murray).]

show a bimodal distribution (see Figure 15–10), an expression of reversing tidal currents like that described earlier from intertidal sandstones. Two opposed directions of cross bedding are a more common feature of limestones than of sandstones because a larger proportion of most preserved limestones were formed in the intertidal zone because of biologic controls. Sandstones form in a much greater variety of environments.

Some carbonates are transported and deposited by turbidity currents and show graded bedding, flute casts, and many of the other sedimentary features characteristic of sandstone beds formed by this mechanism. Limeclasts of extrabasinal origin are common in such limestones.

Mounds and Reefs

Sandstones and mudrocks generally are deposited in topographically low areas. This is not as true for limestones. Because the carbonate sediment is mostly produced in place by living organisms, local sites may build thick accumulations. This can result in a topographic feature that strongly influences surrounding sedimentation. The locally thick limestone section may consist of fossils that have grown attached to each

Figure 15–10
Bimodal cross bedding reflecting two opposing directions of water movement (opposing tidal currents) in sparry limestones of the Kansas City Group (Upper Pennsylvanian), eastern Kansas. Upper cross beds dip to right; those below dip to left. [From W. K. Hamblin, 1969, *Kansas Geol. Surv. Bull., 194,* Pt. 2, Fig. 4A.]

other to form a wave-resistant structure called a *reef.* Numerous modern examples of reefs occur along the coasts of Florida, Australia, and elsewhere, and around the edges of mid-ocean volcanoes. The internal structure of these limestone accumulations is extremely complex, reflecting the biologic shapes and ecologies of the organisms involved. The deposit lacks the usual two-constituent fabric of framework and cement that characterizes allochemical limestones. Often, whole, undisturbed fossils are abundant and are interwoven with micrite that settled into crevices among the growing organisms. Ooliths and intraclasts are rare within the reef itself, but intraclasts are abundant in the talus deposits that fringe the reef. The reef core typically is massive and unbedded, although exceptions exist where laminations of coralline algae define the orientation of the air–water interface.

In Paleozoic times, framework-building organisms were less common than in modern seas, and many of the topographic constructional highs in carbonate areas were formed by accumulations of micrite. These highs are called *lime-mud mounds* (see Figure 15–11). Many of these mounds may have originated as accumulations of sand-size pellets swept together by current activity, as occurs today in some areas of Florida Bay. Perhaps there was also a biologic control in some cases. Sometimes the structureless micritic core of a mud mound is capped by micrite that contains organisms capable of trapping fine carbonate sediment, such as sponges, algae, or bryozoa.

Figure 15–11
Aerial photograph of lime-mud mounds in Florida Bay. The mounds are several hundred meters long and are composed of micrite pellets swept together by currents, partially lithified, and further held together by the roots of the mangrove trees that form the bulk of the vegetation on the islands. The mounds are only slightly above mean sea level. [Photo courtesy R. N. Ginsburg.]

Reefs and Paleoclimate

Limestones are composed of calcite and are essentially organic remains; that is, nearly all limestones are coquinas. Biologists tell us that marine organisms reproduce more rapidly in warm waters. Chemists tell us that calcite is less soluble in warm water than in cold water. From this information, we would anticipate that modern carbonates and reefs should be located near the equator, and this is indeed the case (see Figure 15–12A). Nearly all reefs are located within 30° of the equator where the temperature of the surface water is at least 20°C.

This suggests that the locations of ancient reefs might reflect paleolatitudes of ancient land masses. At present, only about 30% of the ancient reefs are located within 30° of the equator (see Figure 15–12B); but when the present latitudes are replaced by the paleolatitudes as determined from paleomagnetic measurements (see Figure 15–12C), the proportion rises to 90%.

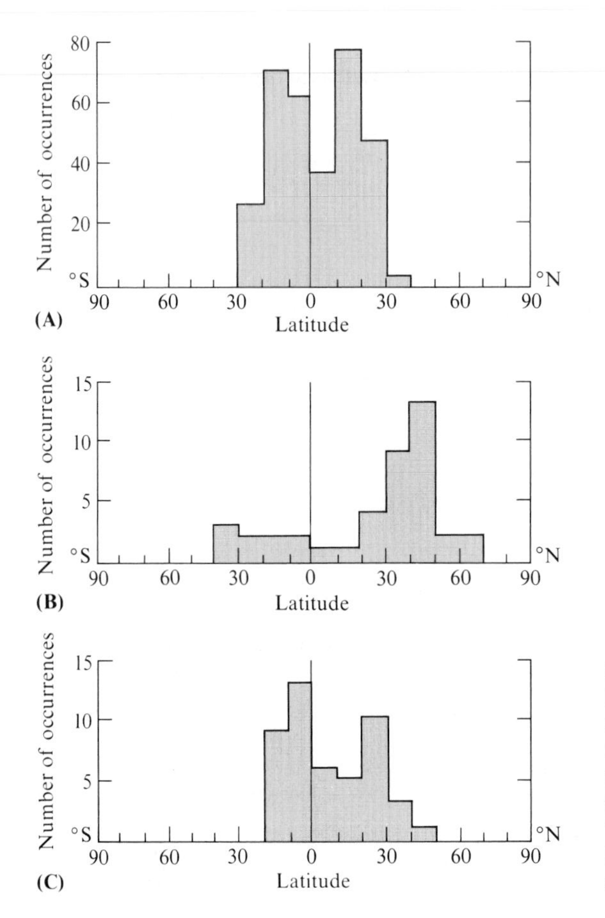

Figure 15-12
Latitudinal histograms for organic reefs. (**A**) Present latitude of modern reefs. (**B**) Present latitude of ancient reefs. (**C**) Paleolatitude of ancient reefs. [From B. F. Windley, 1977, *The Evolving Continents* (New York: John Wiley and Sons), Fig. 10.1.]

Diagenesis

Limestones, because of their solubility, undergo extensive modification after deposition. For example, the micritic sediment is deposited as aragonite needles but is calcite in all ancient rocks. Many allochems were aragonite originally but they, too, are calcite in pre-Pleistocene limestones. In hand specimen, however, it is not possible to distinguish aragonite from calcite, so analysis of these changes must await laboratory studies.

Diagenetic changes that can be observed in the field are cementation by sparry calcite and replacement of limestone by other minerals, particularly by chert and dolomite. We have previously considered chert and the source of the silica from which it forms. Chert nodules appear to form by replacing areas of limestone, but the reason this occurs is not clear. Most explanations appeal to a change in the hydrogen ion concentration of pore waters passing through the limestone. At pH values above 7 or 8, silica is more soluble than calcite; at values below 7 the reverse is true. Therefore, a lowering of the pH in pore waters could result in the simultaneous dissolution of calcite and precipitation of quartz (chert).

Replacement of limestone by dolomite is proved by several field observations.

1. The dolomite–limestone contact frequently is at an angle to bedding.

2. Fossils in the limestone are composed of dolomite rather than calcium carbonate, and there is no evidence that any group of organisms has ever formed their shells of dolomite.

3. Rhombs of dolomite cut through allochems (see Figure 15–13); that is, the allochems are partly calcite and partly dolomite. There are, however, many beds of pure dolomite that contain no allochems, and it is not possible to determine from field observations whether the bed was originally a limestone.

Figure 15–13
Photomicrograph (crossed nicols) of two dolomite rhombs cutting across and replacing micritic oolitic coatings around fossil fragments. Ghosts of the micrite are clearly visible in three areas within the dolomite rhomb at the left of the photo; also in the upper half of the rhomb at the right side of the photo. The sparry crinoid fragment in the upper right corner of the photo has a laminated coating, indicating an algal origin. The large dolomite rhomb is about 100 μm in diameter. [Photo courtesy R. C. Murray.]

Stylolites

A *stylolite* is an irregular surface within a bed and is characterized by mutual interpenetration of the two sides, the columns, pits, and teethlike projections on one side fitting into their counterparts on the other (see Figure 15–14). The seam is made visible by a concentration of insoluble constituents, such as clay or organic matter.

Stylolites can be seen cutting across allochems such as fossils, with the upper or lower part of the fossils apparently dissolved away. Hence, the seams are a diagenetic

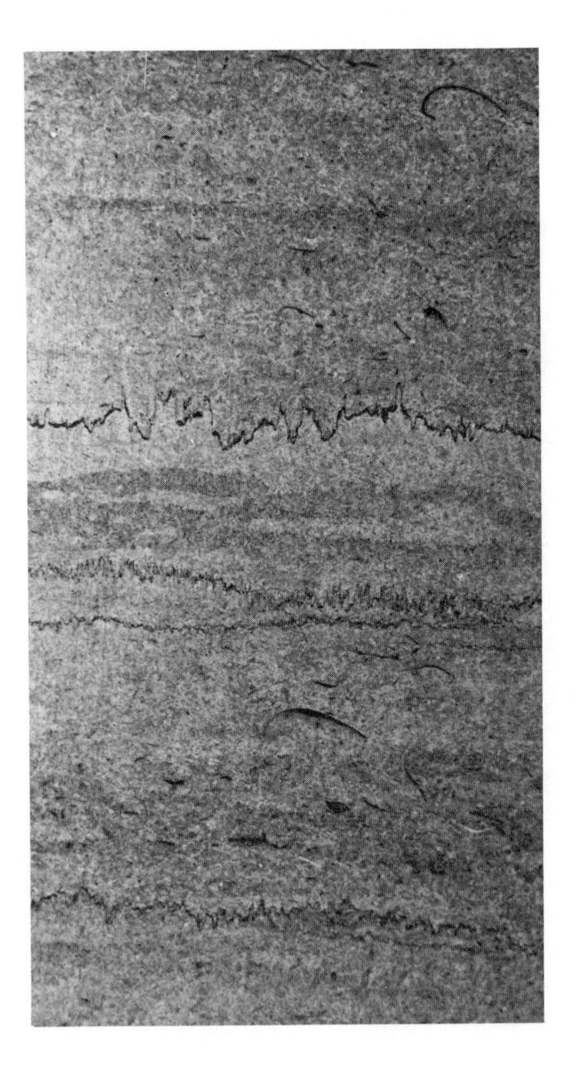

Figure 15–14
Stylolites in a Mississippian brachiopod–crinoid biosparite used as a dividing panel between toilet stools in the Oklahoma Memorial Union, Norman, Oklahoma. The stylolite seams are formed of material less soluble than the limestone, a mixture of clay and carbonaceous material in this example.

feature. The orientation of the seams is nearly always approximately parallel to bedding, implying that the dissolution was caused by the same type of interaction between overburden stress and pore waters that forms the pressure solution surfaces in some quartzarenites. The only difference between the two is that the stylolites have greater lateral continuity (often several meters in length) and are marked by material of a different composition than the main rock mass (noncarbonate sediment). Stylolites are very common structures in limestones because of the relatively great solubility of carbonates, and their abundance testifies to the huge thicknesses of carbonate sediment that can be removed by stylolite formation. By comparing the percentage of clay in the bulk limestone to the percentage in some stylolite seams it has been inferred that up to at least 40% of the original carbonate sediment has been dissolved.

LABORATORY STUDIES OF LIMESTONES

The prime tool used in laboratory studies of limestones is the standard polarizing microscope, but in recent years increasing use has been made of luminescence petrography, the SEM, and the electron microprobe. These newer tools have provided data and insights not previously obtainable.

Allochemical Particles

1. *Fossils.* The outer form and internal microstructure of the hard parts of organisms are extremely complex, but several picture books are available that illustrate some of the appearances (Horowitz and Potter, 1971; Scholle, 1978). Figure 15–15 shows a few typical examples. Proper identification of some types of fragments poses no difficulty, such as the spiral structure of a snail or the multiple segments of a crinoid stem. But many identifications are more speculative, such as distinguishing some corals from bryozoans; or crinoid plates from echinoid fragments. The only way to learn the appearance of the major types of fossils is to examine a variety of fossiliferous limestones with one eye peering down a polarizing microscope and the other eye examining the various picture books until the appropriate photomicrograph is discovered. Most of the time a suitable match is found, but it is common to find problematica for which one person's guess is almost as valid as anyone else's, particularly when the fossil fragments are small. Proper identification of fossils is worth the effort, however, because of the ecologic information that can be obtained. For example, some organisms like a higher salinity than others; some prefer colder waters than others; and a repeated assemblage of whole shells of several species can establish a useful biofacies for stratigraphic correlation.

Some types of shells in ancient limestones, such as those of brachiopods and ostracods, are almost never recrystallized. They display the same details of internal structure present in modern members of the two groups. Other ancient shells, such as

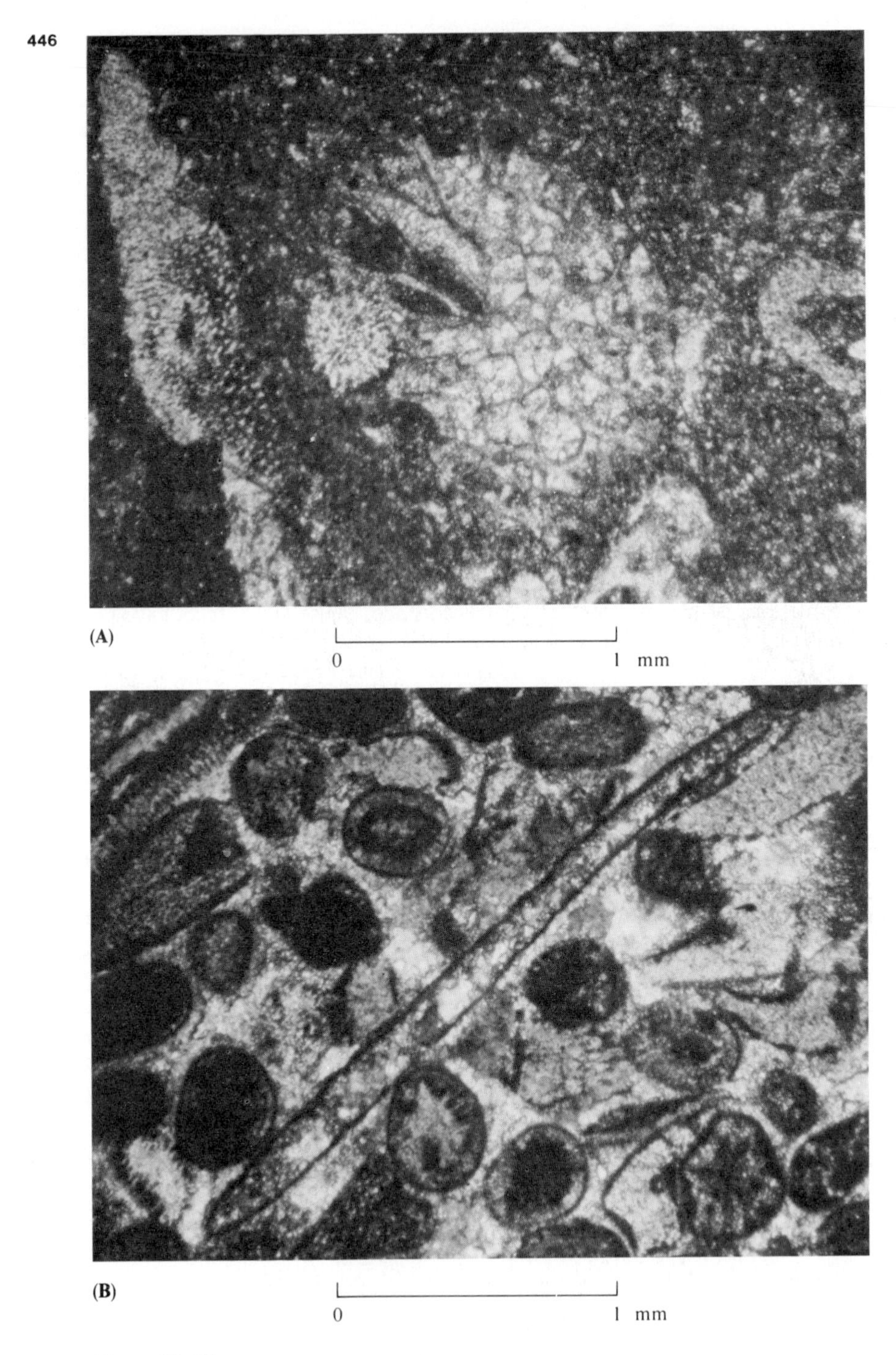

Figure 15–15
Characteristic appearances of carbonate-shelled fossils in thin section (crossed nicols). (**A**) Bryozoan frond (center) and assorted echinoderm fragments in a micrite matrix. (**B**) Long pelecypod fragment with thin micritic coating formed by boring algae, surrounded by ooliths with nuclei of echinoderm fragments and cemented by sparry calcite. The pelecypod shell has been replaced by an equigranular sparry crystal mosaic, as have nearly all ancient pelecypod shells.

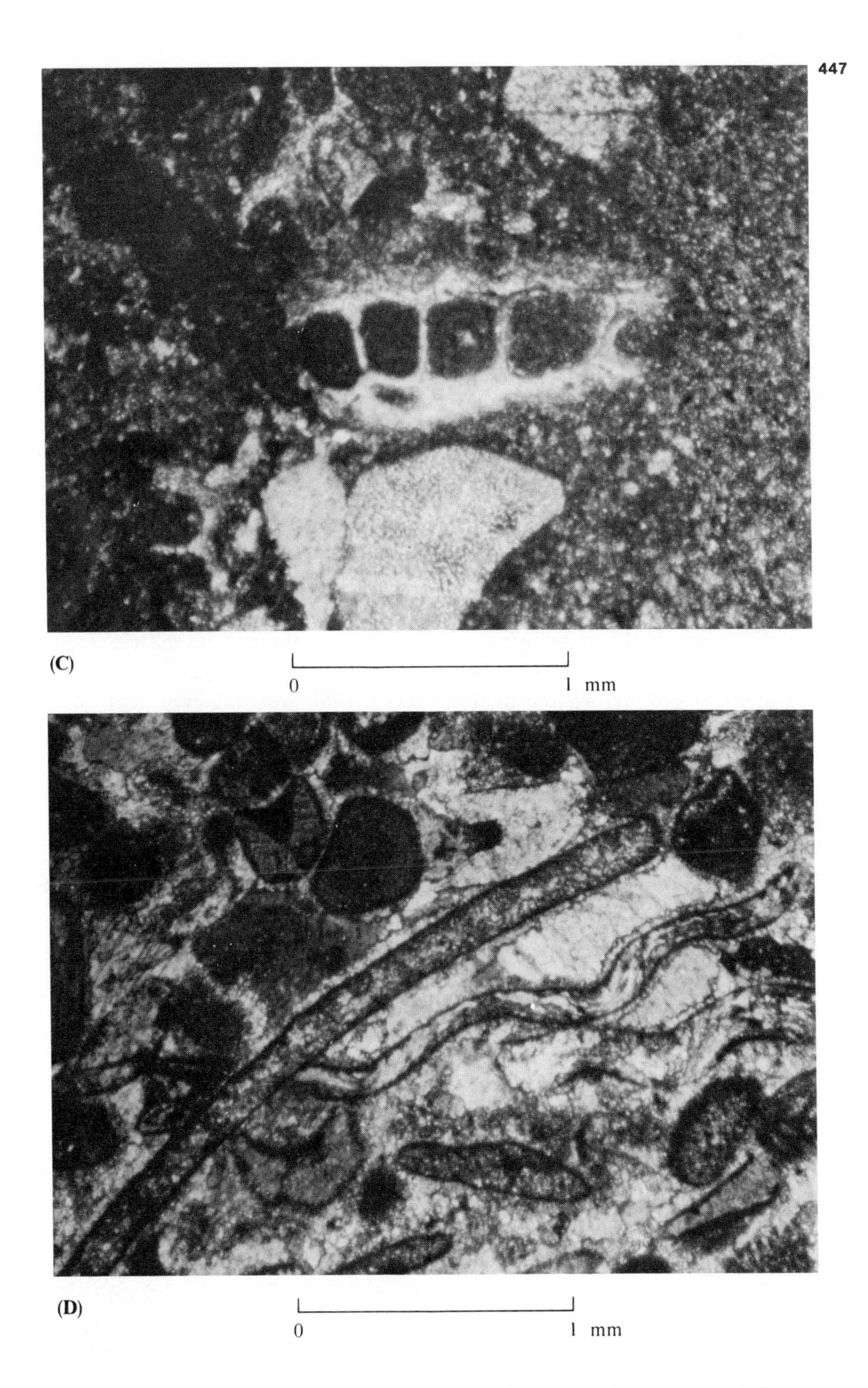

(C) An encrusting foraminifera with chambers filled with micrite surrounded by echinoderm fragments and set in a matrix of fossil hash and micrite. **(D)** Sinusoidal brachiopod showing its fibrous structure parallel to the length of the shell. Also present are a long pelecypod fragment and numerous crinoid fragments coated with micrite generated by boring algae, set in sparry calcite cement.

those of pelecypods, are nearly always recrystallized. The inside of their shells consists of an equigranular mosaic of calcite despite the fact that all modern species have complex aragonitic shells. Analyses of such variations has led to the discovery that shells composed of relatively pure calcite (with no magnesium substituting for calcium in the crystal structure) are very resistant to recrystallization; those formed of less pure calcite crystals are commonly recrystallized; and those that are aragonite during the life of the organism are nearly always recrystallized in ancient rocks. This has permitted petrographers to establish the mineral composition of shells of extinct organisms, such as trilobites. Trilobite hard parts nearly always show internal structure, indicating that they have always been formed of relatively pure calcite.

2. *Peloids.* Fecal pellets seem to be more common in sparry-cemented limestones than in those with a micritic matrix, but the reason for this is uncertain. It makes no ecologic sense to conclude that an organism will scavenge for food in both agitated and quiet waters but move to agitated waters to deposit its waste products. A more reasonable explanation is that fecal pellets are composed of structureless micrite that can be distinguished from micrite matrix only by the outlines of the pellets. When deposited, the pellets are ductile, and if they are compacted before they are rigidified they will be squashed and be indistinguishable from the micrite matrix. Hence, pelsparites are seen more commonly then pelmicrites. Many micritic limestones contain vague globular outlines visible in thin section that probably represent pellets that have partly lost their identities through diagenesis.

3. *Ooliths.* Ooliths are known from limestones of all ages and are easy to recognize in thin section. Their internal structure can be radial, concentric, or equigranular calcite mosaic. Modern ooliths are composed of aragonite needles that, in most cases, are oriented tangentially to the oolith surface. In ancient limestones the aragonite has been replaced by calcite with the destruction of much of the original oolithic fabric.

4. *Limeclasts.* Limeclasts in thin section can have any internal structure that can occur in carbonate rocks. Thus, there are biomicrite intraclasts, oosparite intraclasts, and intraclasts composed of dolomitized micrite (dolomicrite).

Orthochemical Particles

1. *Micrite.* In thin sections of ancient limestones, micrite appears as subequant blocks of polyhedral calcite about 1–4 μm in diameter. As we noted earlier, microcrystalline carbonate in modern carbonate environments is composed largely of aragonite. Therefore, if we assume that the present is an accurate key to the past for this material, we must conclude that ancient micrites are thoroughly recrystallized rocks.

2. *Sparry calcite.* Sparry calcite occurs mostly as intergranular cement among allochemical particles in the same manner as it occurs among quartz, feldspar, and lithic fragments in sandstones. Spar is most abundant in current-sorted limestones in which permeabilities were high and is less common in limestones that contain micrite,

quite analogous to sandstones where well-sorted deposits are cemented by chemical precipitates but clayey sandstones are not.

Calcite spar occurs also as a recrystallization product of chemically less stable materials. Fossils that were originally aragonite or magnesium-rich calcite appear in ancient limestones as sparry calcite. Even micrite can be further recrystallized to coarsely crystalline spar, as apparently has occurred in many Paleozoic limestones in the Great Basin of the western United States. These rocks consist entirely of coarsely crystalline spar and may be indistinguishable in thin section from some metamorphic marbles.

Cementation of carbonate particles by sparry calcite can occur intermittently (see Figure 15–16) and over long periods of time, as was shown for sandstones by luminescence petrography. At present there is no way known to obtain absolute dates for these episodes of cementation, but probably periods on the order of tens of millions of years can be required.

3. *Microsparry calcite.* Typical micrite crystals that form the matrix in many limestones have diameters of 1–4 μm. Sparry calcite pore-filling has crystal sizes coarser than 20 μm. In some limestones that appear micritic in hand specimen, however, the "micrite" crystals are seen in thin section to be microsparry calcite in the size range 5–15 μm (see Figure 15–17). Except for its crystal size, the microspar has a distribution pattern similar to that of micrite. It can form the entire rock (obviously impossible if it were a cement); it is most common in rocks that contain few allochems as contrasted to current-sorted limestones; and it occurs as fringes around allochems within a micrite matrix. Based on these observations microspar is interpreted as a product of re-recrystallization of micritic calcite.

SITES OF CALCIUM CARBONATE DEPOSITION

As we have observed, carbonate sediments are, with the exception of ooliths, collections of the hard parts of organisms in various states of disaggregation. We also have noted that the formation of calcium carbonate is favored by shallow waters and warm temperatures. These facts determine the sites of accumulation of carbonate sediments and, therefore, the sites of origin of limestone (and dolomite). Three types of settings are generally recognized:

1. *Epeiric seas.* Epeiric seas can be defined as those arms of the ocean that spread over broad areas of the central parts of continents, such as the seas that existed on the North American craton during much of the Paleozoic Era. These seas covered tens of thousands of square kilometers, and when the continents were located in low latitudes, thin but extensive clastic limestones were formed in them. Based on the ubiquitous occurrence and great abundance of fossils in these limestones, and on the sedimentary structures in the rocks, it is probable that the maximum depth of water in the epeiric seas did not exceed 30 m. The slope of the sea floor must have

(A)
0
0.6 mm
(B)
0
0.6 mm

Figure 15–16 (*facing page*)
Crinoidal biosparite from the Lake Valley Limestone (Mississippian), New Mexico. (**A**) A normal thin section view (uncrossed nicols) shows a crinoid plate (left center) surrounded by an apparently uniform, one-stage growth of calcite cement. (**B**) Approximately the same area viewed using cathodoluminescence shows at least five generations of cement that can be correlated from sample to sample and related to a variety of tectonic and erosional events. [From P. A. Scholle, 1978, Amer. Assoc. Petroleum Geol. Mem. 27, p. 230 (photo courtesy W. J. Myers).]

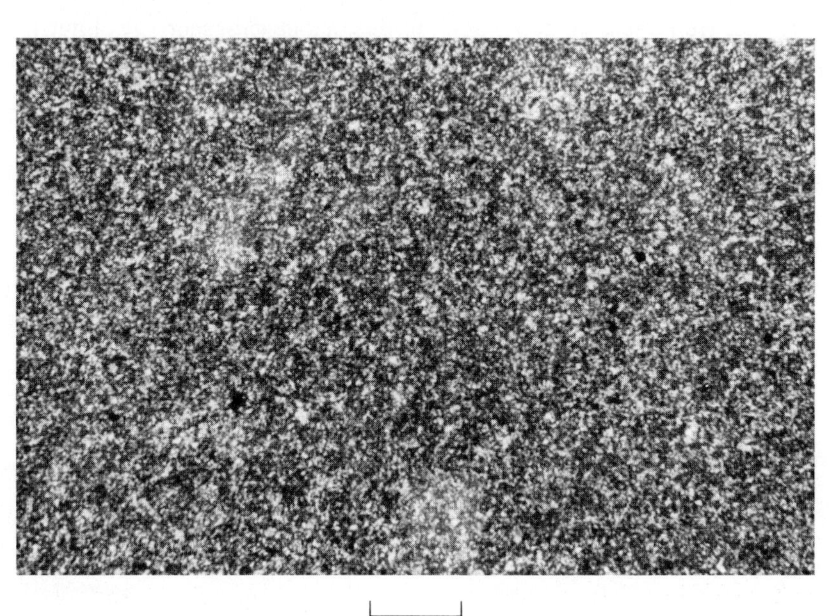

0 0.4 mm

Figure 15–17
Photomicrograph (crossed nicols) of Solenhofen Limestone (Jurassic), West Germany. In hand specimen this rock is so finely crystalline that individual crystals cannot be resolved and it appears to be a homogeneous micrite. The thin section reveals that much microspar is present (the clear translucent areas) among the micrite crystals (the darker opaque areas). The micrite seems vaguely peloidal in places. [From P. A. Scholle, 1978, Amer. Assoc. Petroleum Geol. Mem. 27, p. 175.]

been about 2 cm/km, an extraordinarily low gradient. For comparison, we might note that the present average slope of the world's continental shelves is 125 cm/km. Examples of limestones formed in such epeiric seas include the Ordovician and Mississippian rocks of the mid-continental United States.

The shallowness of epeiric seas over wide areas damps out lunar tidal effects that would create turbulence and mixing of waters. These seas also are thought to have been too shallow to permit extensive wave action. Long-period swells propagated from a deep sea basin would have been damped at the shallow margins of an epeiric sea. Within the epeiric sea, the only mechanism for agitating the water would be local winds. This scenario of relatively restricted water agitation and lack of appreciable influx of water from the deep ocean basin results in a general shortage of nutrients, so reef growth would be unlikely. Hence, epeiric seas are dominated by particulate carbonate sediment rather than by massive reef growth.

The carbonate sediments in the central parts of these seas tend to be micrites. Toward the margins of the basin sand-size fossils become more abundant, and peloidal muds occur in the more saline, shallower water near the basin edge. In the tidal flat and supratidal environment (see Figure 15–18), dolomite and gypsum typically are formed.

2. *Shelf margins.* A growing reef has an exceptionally large need for nutrients, and the formation of carbonate structures is favored by agitation of the waters.

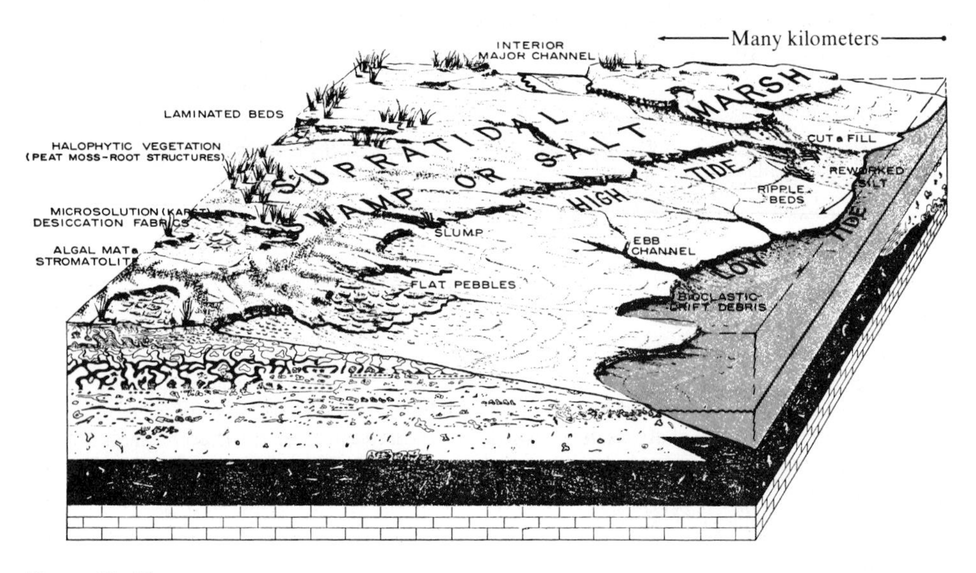

Figure 15–18
Idealized view of an epeiric sea shoreline showing the appearance of the tidal flat and supratidal environment of low-energy carbonate deposition. Vertical and horizontal scales unequal. [From P. O. Roehl, 1967, *Amer. Assoc. Petroleum Geol. Bull., 51,* Fig. 6.]

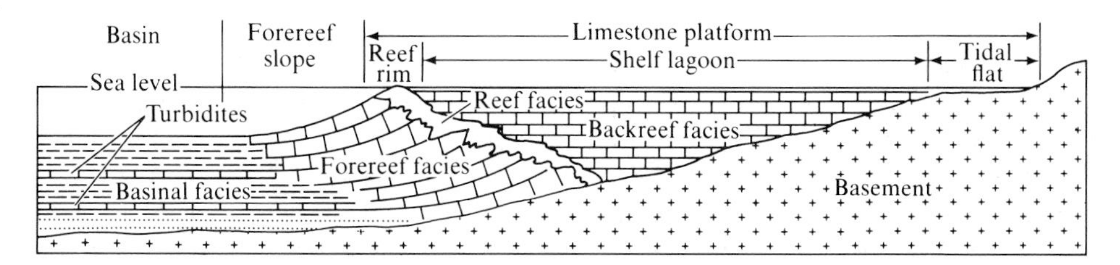

Figure 15–19
Schematic cross section illustrating principal environments of deposition near a reef. Vertical growth of the reef was made possible by a gradual rise in sea level. Probably only 10% of the reef complex consists of in-place living reef organisms, the remainder being debris broken from the living reef (mostly on the seaward side of the reef) and the remains of carbonate-shelled organisms that have found a home in the environment created by the reef growth. [From F. J. Pettijohn, 1975, *Sedimentary Rocks,* 3d Ed. (New York: Harper & Row), Fig. 10–2.]

(Agitation drives off carbon dioxide dissolved in the water and causes the pH to increase.) It follows, therefore, that the most favorable site for reef development is a location where (a) cool water is being warmed and agitated so that CO_2 is driven off, and (b) the cold water is rising from relatively deep areas of the ocean because deep waters are rich in nutrients. The place where such conditions are common is at a fairly sharp break in slope, such as at the edge of the continental shelf or around the edge of a mid-ocean volcano. Modern examples of such sites include the reef tract along the southern tip of Florida, the Great Barrier Reef along the east coast of Australia, and the circular reef growths that partially surround many volcanoes in the Pacific Ocean in the lower latitudes. Figure 15–19 shows an idealized sketch of a reef and its associated facies.

3. *Deep sea.* Carbonate sediment is found in bathyl and abyssal depths in the modern ocean basins, and limestones formed in such depths are known from rocks of Mesozoic and Cenozoic age. The Holocene deposits consist of the shells of planktonic organisms, particularly of the foraminifera *Globigerina,* but also including hard parts of the algal family Coccolithophoridae and the planktonic microscopic mollusks called pteropods. About 48% of the ocean floor is presently covered by sediment in which the remains of these organisms form at least one-third of the particles.

The calcium carbonate shells in these deep sea deposits were formed in the upper few meters of the sea, as are the shells and rigid reef structures we considered previously. But the organisms whose hard parts accumulate in shallow waters are bottom-living forms (such as clams, crinoids, bryozoa); those in deep water are chambered, floating forms.

The areal extent of calcium carbonate accumulations in the deep sea is limited by two factors. The first of these is the same as is present for shallow water accumulations—sea surface temperatures. Most surface sea water in low latitudes is saturated

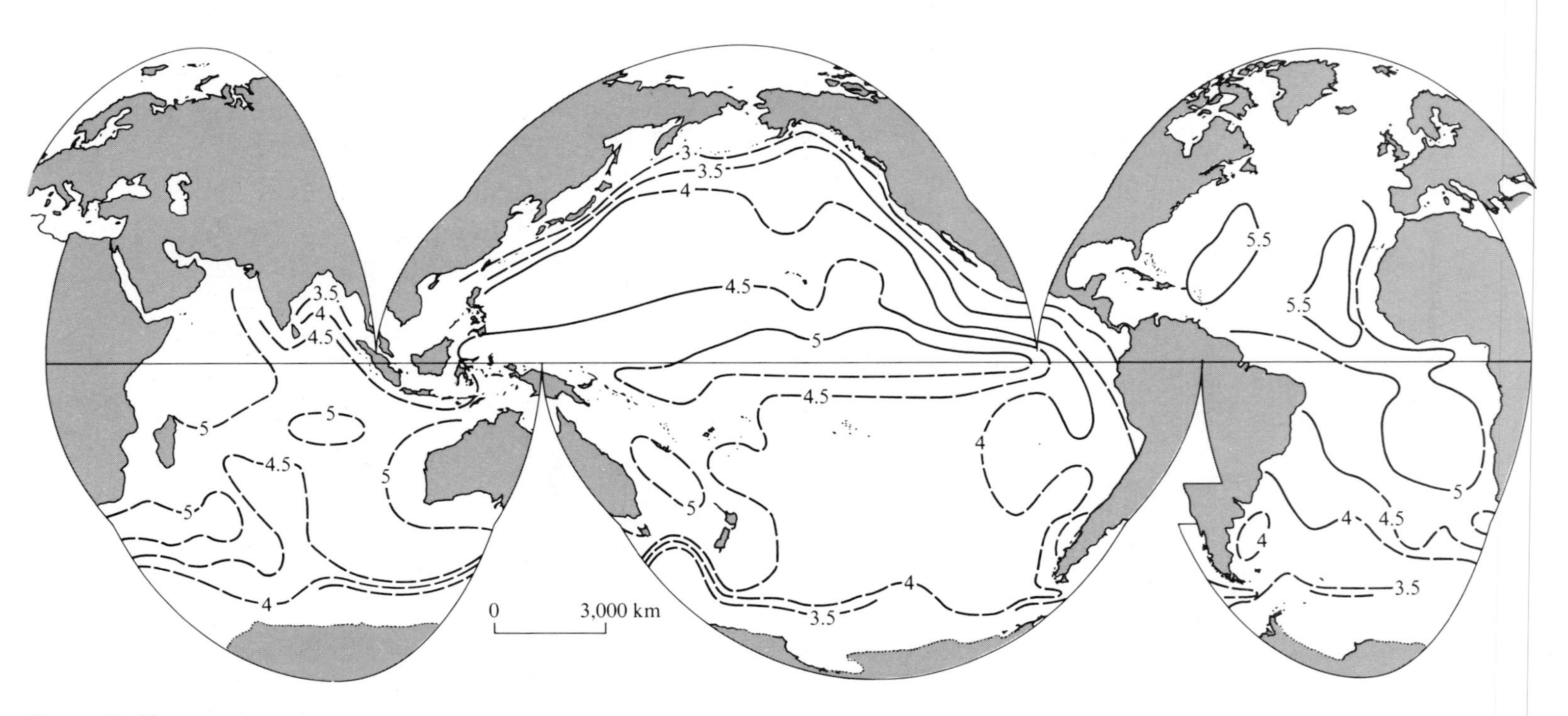

Figure 15–20
Depth in kilometers of the calcium carbonate compensation surface, defined by interpolating the local facies boundaries between calcareous sediments and sediments containing less than a few percent of carbonate. Dashed lines are based on fewer than 20 control samples. Depths are in km. [From W. H. Berger and E. L. Winterer, 1974, Internat. Assoc. Sedimentologists Spec. Pub. No. 1, Fig. 1.]

with respect to calcite and aragonite so that it is not difficult for marine organisms to remove calcium and carbonate ions from the water and form it into these minerals as skeletal material. *Globigerina* and pteropods are more common in tropical waters than in higher latitudes.

The second limitation on the occurrence of deep sea carbonates is depth. Sea water is colder at depth than at the surface and colder waters contain more dissolved carbon dioxide than warm waters. The increased carbon dioxide causes an increase in carbonic acid (H_2CO_3) in the water and results in dissolution of the calcitic and aragonitic shells as they settle toward the sea floor after death of the organism. Few shells survive below about 5,000 m. The increased hydrostatic pressure at depth also increases the solubility of calcium carbonate, but this effect is of less importance than the temperature–CO_2 effect. The depth below which no calcium carbonate accumulates is called the *carbonate compensation depth,* about 4,000 m in equatorial regions but rising gradually toward the sea surface at higher latitudes because of the lower temperature of surface sea water (see Figure 15–20).

Deep sea limestones in ancient rocks are identified in outcrop chiefly by the stratigraphic and tectonic settings in which they occur. Obviously, it requires rather severe tectonic activity to cause bathyal and abyssal sediments to be raised upward several thousand meters and incorporated into the continental mass. Deep sea micrites are known from the Franciscan Formation (Jurassic) in California, and from several units in the Alps, both locations where obduction of oceanic crust has occurred, apparently with the incorporation of some of the sea-floor deposits (the carbonate sediment) onto the continental plate. In these limestones, the most abundant fossils typically are the silt-size plates of planktonic algae called coccoliths, seen using an electron microscope (see Figure 15–21). As with the *Globigerina,* the plates settled to the deep sea floor following the death of the organism. Carbonate-secreting planktonic organisms did not evolve until the Mesozoic so that there are no Paleozoic or Precambrian "microbiomicrites," and pre-Mesozoic deep sea limestones consist of sand- and gravel-size limeclasts carried into deep oceanic waters by turbidity currents. The source of the detritus was an actively rising highland, such as the Alpine area created by the continental plate collisions that closed the Tethys seaway through Europe and Asia.

MODERN REEF ENVIRONMENTS

The only modern locations that qualify as epeiric seas are the Baltic Sea and Hudson Bay. Both of these sites are located above lat. 50° N and, as a result, contain no Holocene carbonate sediments. The present cannot be used as a key to the past for epeiric sea limestones. Reefs are abundant in modern seas, however, and their study by geologists over the past 30 years has provided a good understanding of the ecologic and mechanical factors that govern the formation of reefal limestones. Examples of modern extensive reef development include the Great Barrier Reef along the

Figure 15–21
Transmission electron micrograph of Laytonville Limestone (Cretaceous), Franciscan Formation, Sonoma County, California, a coccolith coquina showing whole segmented oval plates and a hash of separate segments. Length of chipped plate in left center is 6 μm. [Photo courtesy R. E. Garrison.]

northeastern coast of Australia, the Yucatan Peninsula of southern Mexico, the Bahama Platform (a rifted continental fragment) 250 km northeast of Cuba, and the reef tract along the southern tip of Florida. These areas are all located where the sea surface temperature during the coldest winter months is at least 24°C, between about 25° north and south of the equator.

Southern Florida Reef Tract

The reef complex located at the southern tip of Florida is the only areally extensive carbonate depositional environment along the coast of the United States. It was one of the first modern carbonate environments to be studied in detail (Ginsburg, 1956), and the insights it provided still guide much present-day thinking about the biologic and environmental controls of limestone formation.

The South Florida reef tract is located in a subtropical area at lat. 25° N, bordered on the northwest by the mangrove swamps of the Everglades and on the southeast by the warm Gulf Stream waters of the Atlantic Ocean (see Figure 15–22). The area can be considered as composed of two parts, the linear reef front that is washed by the warm Gulf waters and the backreef area located in Florida Bay. The two areas are distinguished by different water circulation patterns, salinities, flora and fauna, and types of carbonate particles.

The Reef

On the main reef the annual range in temperature of surface waters is between 15°C and 32°C; salinity, between 35,000 and 37,000 ppm. This relatively small range in salinity is characteristic of marine areas that contain abundant living organisms. Most organisms are very sensitive to salinity and cannot survive large changes. The calcareous flora of the reef is dominated by two families of algae, the green Codiaceae and the red Corallinaceae. The fauna is extremely varied but is dominated by corals, mollusks, echinoderms, foraminifera, worms, bryozoans, and crustaceans.

The reef is conveniently divided into three parts, the forereef, main reef, and backreef, which grades westward into Florida Bay. The forereef is the area in front of the growing reef below the depth of effective light penetration, between 50 and 100 m. The sediment in this part of the reef setting is composed of poorly sorted carbonate gravel and sand derived from disintegration of the growing reef, a result of pounding by surface waves and the boring activities of marine predators. It is a submarine talus deposit. In its lower parts there may be some submarine cementation by fibrous calcium carbonate; at the upper part there can be cementation by algal growth where the forereef grades into the shallow, actively growing heart of the reef complex.

The main reef consists of a series of elongate living reefs and rocky shoals separated by areas of deeper waters that are usually floored with ripple-marked carbonate sand and gravel. As shown in Figure 15–22, the reefs with distinct topographic relief rise from water shallower than 20 m, and more commonly less than 15 m, reflecting the depth below which more than half the visible light has been absorbed by the sea water. The primary structural element of the main reef mass is the moosehorn coral, *Acropora palmata* (see Figure 15–23), which grows at a rate of about 2 cm/yr. Sand- and gravel-size detritus broken from the reef accumulates among the coral branches, and the entire mass is stabilized by a covering of the hydrozoan *Millepora alcicornis* and encrusting algae. Behind the wall-like masses that take the full force of the waves (which generate the forereef talus), many other genera of corals grow and their broken fragments accumulate. In summary, the actively growing part of the reef is an intricately complex mixture of *in situ* coral–algal growth (termed *biolithite* in ancient rocks) and its debris.

The backreef area is floored either by sediment with or without a thick cover of marine grass or by patch reefs. The marine grass is the species *Thalassia testudinum*

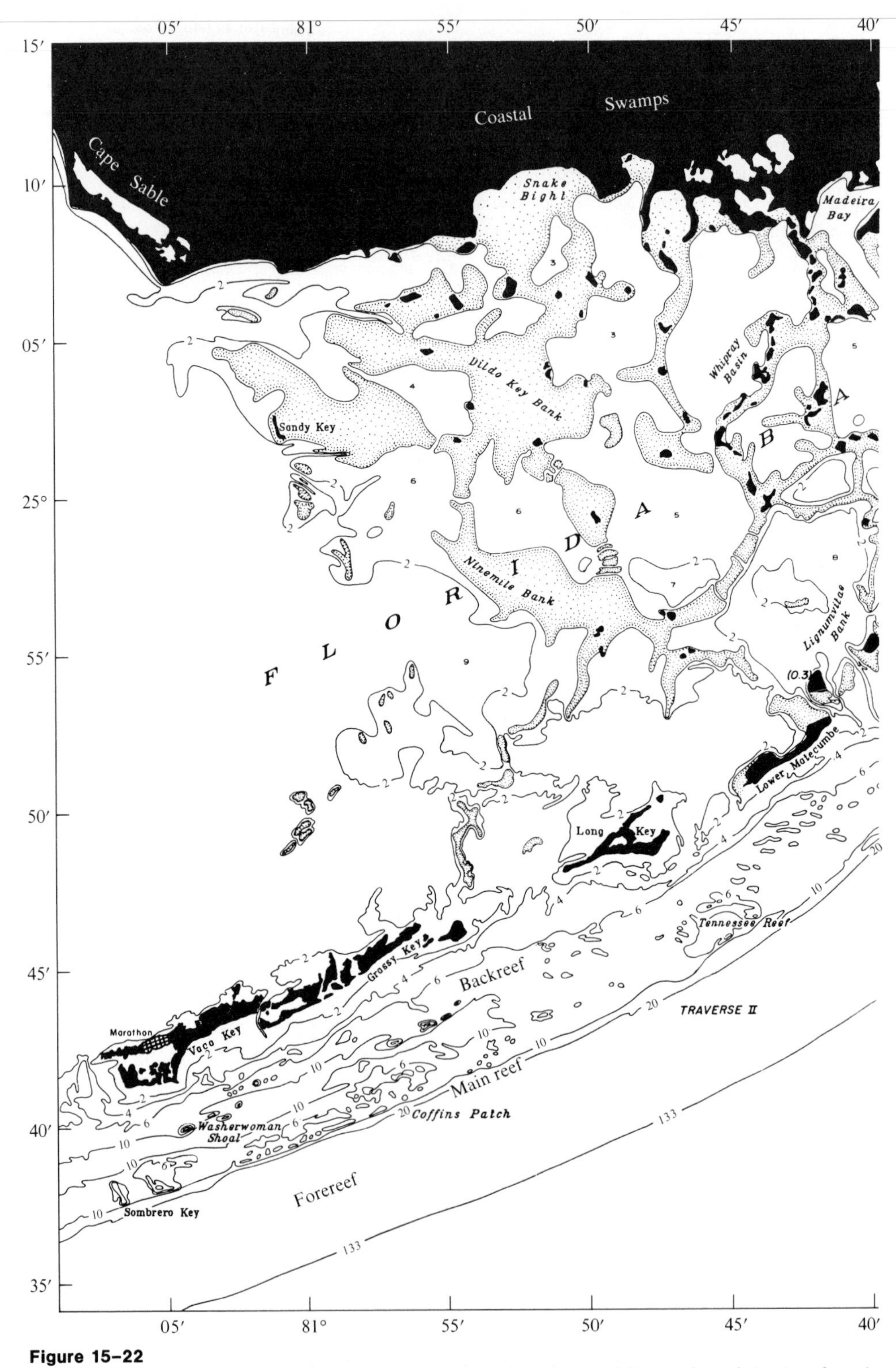

Figure 15–22
Bathymetric chart of South Florida reef tract showing locations of forereef, reef, backreef, and

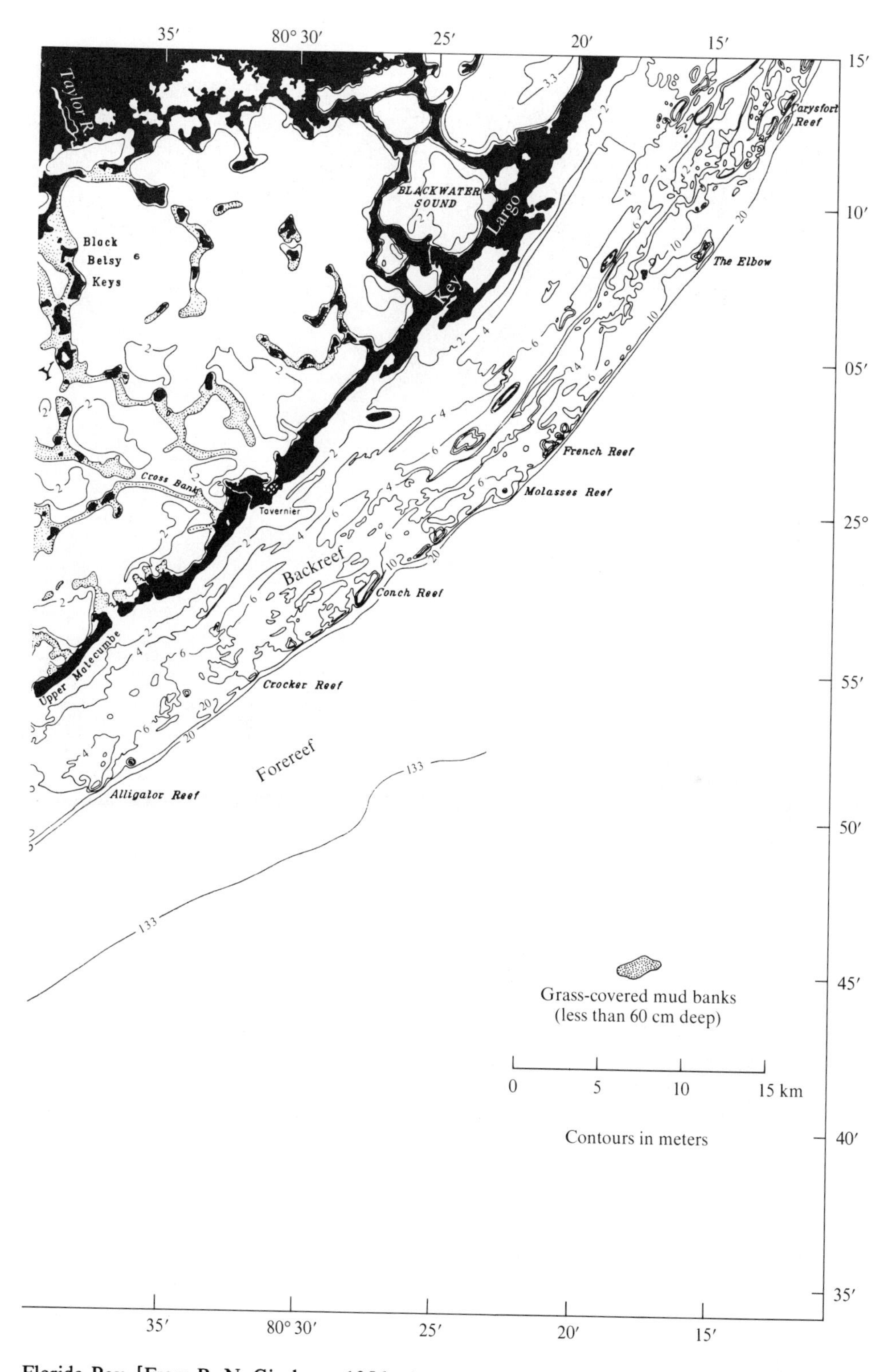

Florida Bay. [From R. N. Ginsburg, 1956, *Amer. Assoc. Petroleum Geol. Bull., 40,* Fig. 5.]

Figure 15–23
Branching colonies of *Acropora palmata* in upper half of photo, the wavy bladed hydrozoan *Millepora complanata* in the foreground, and two varieties of soft-bodied Gorgonian corals (lower right and left center). Depth is 3 m; width of photo 2 m. Key Largo, Florida. [Photo courtesy R. C. Murray.]

("turtle grass"; see Figure 15–24). The bafflelike carpet of grass provides a protected habitat that is inhabited by a variety of calcareous algal and coral genera. The prominent algae are *Halimeda, Penicillus,* and *Goniolithon; Porites* is the abundant coral genus. The grass baffles also trap microcrystalline carbonate sediment suspended in the water that is continuously washing over the area.

Florida Bay

Surface water temperatures in Florida Bay behind the reef have a range similar to that over the reef, but salinities are much more variable because of seasonal

Figure 15–24
Sharp blades of *Thalassia* grass, a solitary *Penicillus* (top center; also two clusters in upper one-third of photo), and hairy filaments of *Batophora*. Berry Islands, Bahamas. Depth 1.5 m. Blades of *Thalassia* are about 15 cm long. [From R. G. C. Bathurst, 1975, *Carbonate Sediments and Their Diagenesis* (Amsterdam: Elsevier Scientific Publishing Co.), Fig. 153.]

freshwater runoff from mainland Florida. As a result, the flora and fauna are more limited in both diversity and abundance. Most of the Bay is floored by micrite trapped by the *Thalassia* grass, and carbonate mud mounds rise to the sea surface in many areas (see Figure 15–22). The relatively small amount of sand-size sediment in the Bay is composed of 80% molluscan fragments with almost all the remainder being foraminiferal debris. Algal pieces form only 1% of the sand. The striking difference between the sediment in the reef tract and that in Florida Bay is shown in Figure 15–25.

An Ancient Analog: Lower Cretaceous of Mexico and Texas

An apparent ancient analog of the south Florida reef occurs in the El Abra reef complex, which crops out west of Tampico, Mexico, and in the Edwards Formation in

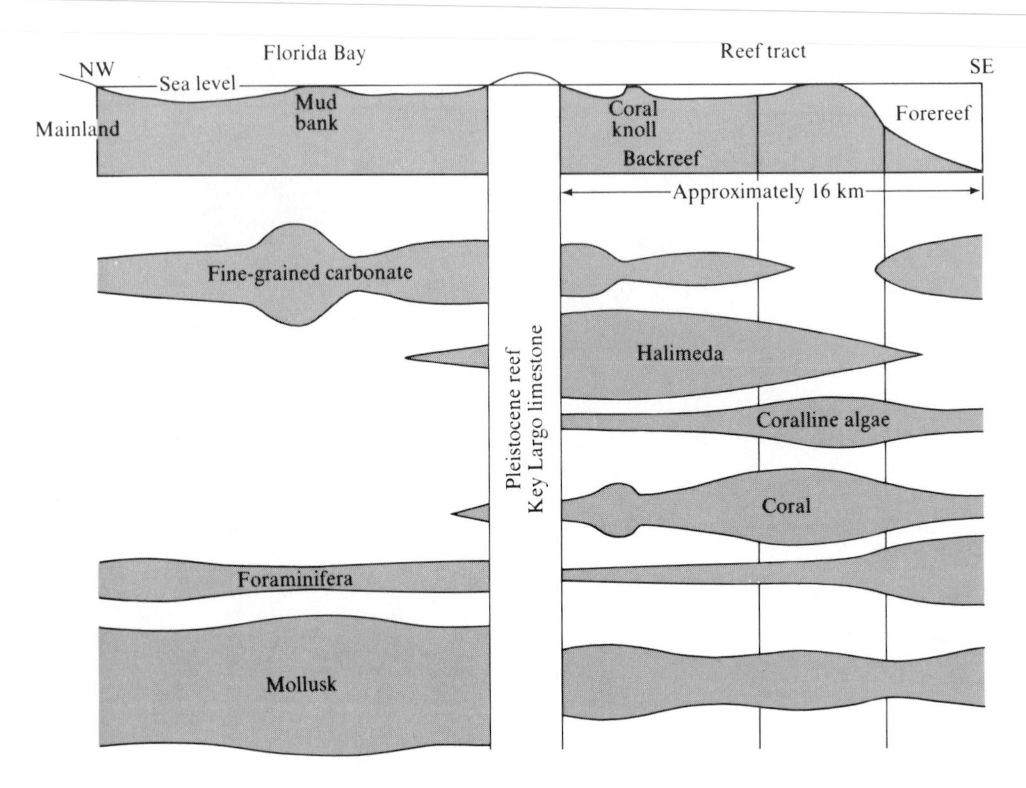

Figure 15–25
Generalized variation in relative abundance of grain types in south Florida reef tract sediment. [From R. N. Ginsburg, 1956, *Amer. Assoc. Petroleum Geol. Bull., 40,* Fig. 1.]

the subsurface of southeast Texas (see Figure 15–26). Analysis of the El Abra Formation indicates that there are four major facies present. Each facies is composed of several types of particles (as is each of the Florida facies) but can be conveniently referred to by its dominant constituent. The stratigraphic arrangement of the facies is shown in Figure 15–27, which is a cross section through quarry outcrops near the southern end of the reef complex. Because of the very large scale of the cross section, the El Abra reef seems more complex than the modern Florida tract, but in fact it is not. Our discussion of the modern reef complex considered only the facies pattern existing at the present sea floor surface and ignored the complexities that would appear in a cross section. The complexities in the third dimension are caused by changes in the pattern of reef growth, current patterns, and shoreline position through time. The period of time represented by the 40–50 m thickness of limestone in the cross section is about 10 million years, plenty of time for many shifts in sediment distribution pattern to occur.

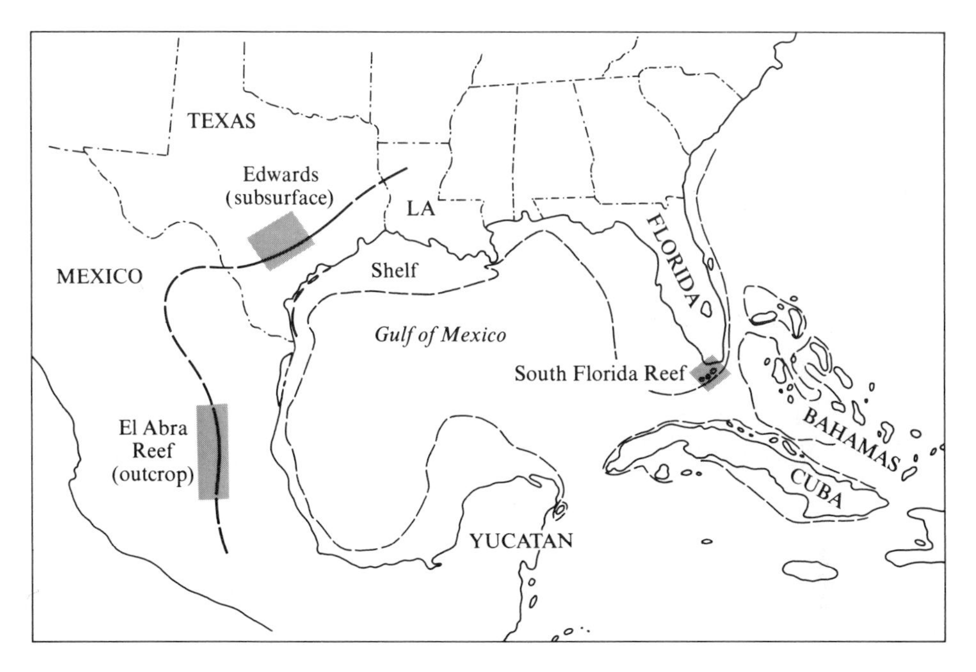

Figure 15–26
Map showing locations of El Abra reef and Edwards Formation reef (Cretaceous), and the modern reef of south Florida. Heavy dashed line indicates trend of reef. Light dashes show shelf boundary. [After L. S. Griffith et al., 1969, Soc. Econ. Paleon. & Miner. Spec. Pub. 14, Fig. 2.]

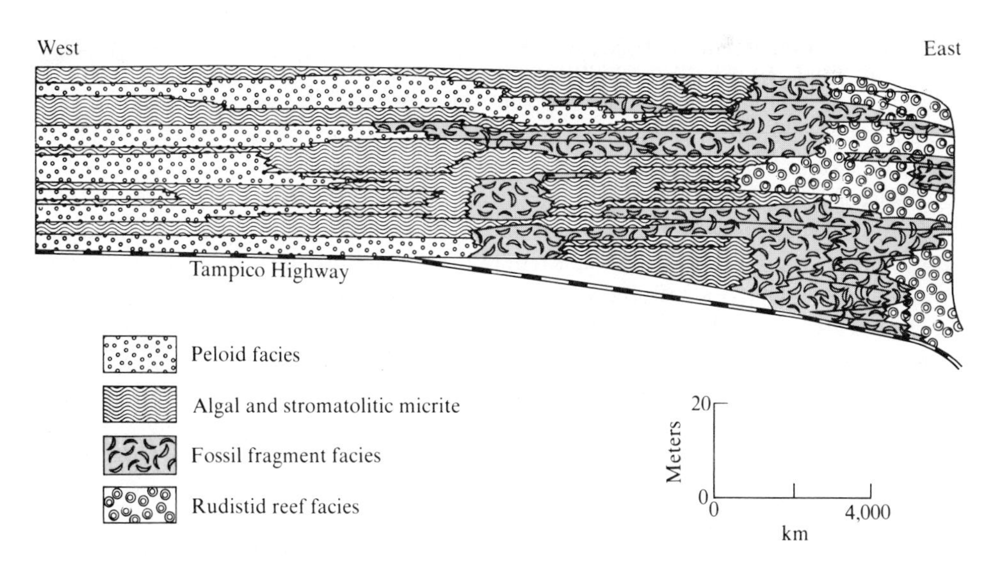

Figure 15–27
Cross section showing the distribution of the four major facies of the El Abra Limestone. [Generalized from L. S. Griffith et al., 1969, Soc. Econ. Paleon. & Miner. Spec. Pub. 14, Fig. 8.]

Figure 15–28
Colony of upper Cretaceous rudistids (*Hippurties mexicana*) in a limestone bed of the Cuautla Formation near San Lucas, Mexico. [From K. Segerstrom, 1962, *U.S. Geol. Surv. Bull., 1104-C,* Fig. 39.]

In the El Abra reef, the reef framework builder is an extinct and aberrant type of attached pelecypod called a rudistid (see Figure 15–28), which was a major reef-former during Cretaceous time. The skeletal sand facies to the west of the organic reef is massive, shows little bedding, and is composed of detritus derived from the physical and biological destruction of the rudist reef. To the west, the skeletal sand changes abruptly into stromatolites, algal micrite, and peloids (see Figure 15–27), the same type of sediment seen in Florida Bay.

The reef facies in the Edwards Formation (Lower Cretaceous) of southeast Texas were described from well cuttings and cores obtained during exploration for oil and gas. The distribution of facies reveals a depositional pattern identical to that observed in the El Abra outcrops with the addition of dolomite and anhydrite facies shoreward of the backreef area. The dolomite–evaporite facies represents a supratidal zone (see discussion of ancient dolomite, below) in which calcium sulfate precipitated and dolomite then replaced the original limestone.

DOLOMITE

Dolomite occurs in rocks of all ages but is noticeably more abundant in older rocks. It is rare in modern carbonate environments, forms about one-fourth of Paleozoic carbonate rocks, and about three-fourths of Precambrian ones.

The field characteristics of dolomites are the same irrespective of geologic age. In some occurrences, dolomite beds a meter or more in thickness are interbedded with limestone beds of similar thickness. In other outcrops, the dolomite–limestone contact is irregular or cuts across bedding planes at a high angle. Dolomite units typically grade laterally into either limestones or evaporites, particularly gypsum or anhydrite.

There is both local and regional evidence indicating that dolomite commonly forms nearer the shoreline than limestone and on the shelf rather than within the basin. In ancient reef environments where the reef is growing at the edge of a continental mass, the seaward front of the reef is typically limestone but the landward, backreef part has been replaced by dolomite. On a regional scale and over extensive stratigraphic intervals, the same tendency is present (see Figure 15–29). When we consider that the limestone was formed only a few meters below high tide the association between dolomite and extremely shallow water becomes even more striking.

Dolomite and paleolatitude. In our discussion of limestones we noted that they occurred in low latitudes, generally within 30° of the equator. Subsequently, we found that at least most dolomites are formed by geologically instantaneous replacement of these limestones; that is, replacement within a million years after the limestone was formed. Because of this, we would expect ancient dolomites to have the same latitudinal distribution as limestones and, as shown in Figure 15–30, we are not disappointed. As was true of limestone reefs (see Figure 15–12), the present locations of ancient dolomites cluster in the high latitudes but the latitudes in which they formed are almost all within 30° of the equator.

Laboratory Studies of Dolomites

In thin section, the replacement origin of many dolomites is even more evident than in the field. Coarse-grained rhombs of dolomite cut across calcitic allochems, obliterating the internal structure of the allochem. In thin sections that are entirely dolomite, faint outlines (ghosts) of previously calcitic or aragonitic fossils can be detected. Some dolomite beds consist entirely of interlocking rhombs of the mineral, except for scattered grains of detrital quartz that testify to the original particulate nature of the carbonate.

After removing from the "questionable origin" category all dolomites showing replacement phenomena either in the field or laboratory, there remains a residue of dolomites, mostly dolomicrites, in which no evidence of replacement origin can be seen. Might these be primary precipitates from a saturated solution of surface water? Thin section studies cannot answer this question; only chemical experimentation can help. Many such experiments have been conducted, with the conclusion that primary precipitation of dolomite from sea water is possible. None occurs, however, because of the slowness of the precipitation process. It seems that all dolomite forms by replacement of preexisting calcium carbonate, even though evidence for replacement is not always visible in the resulting dolomite rock.

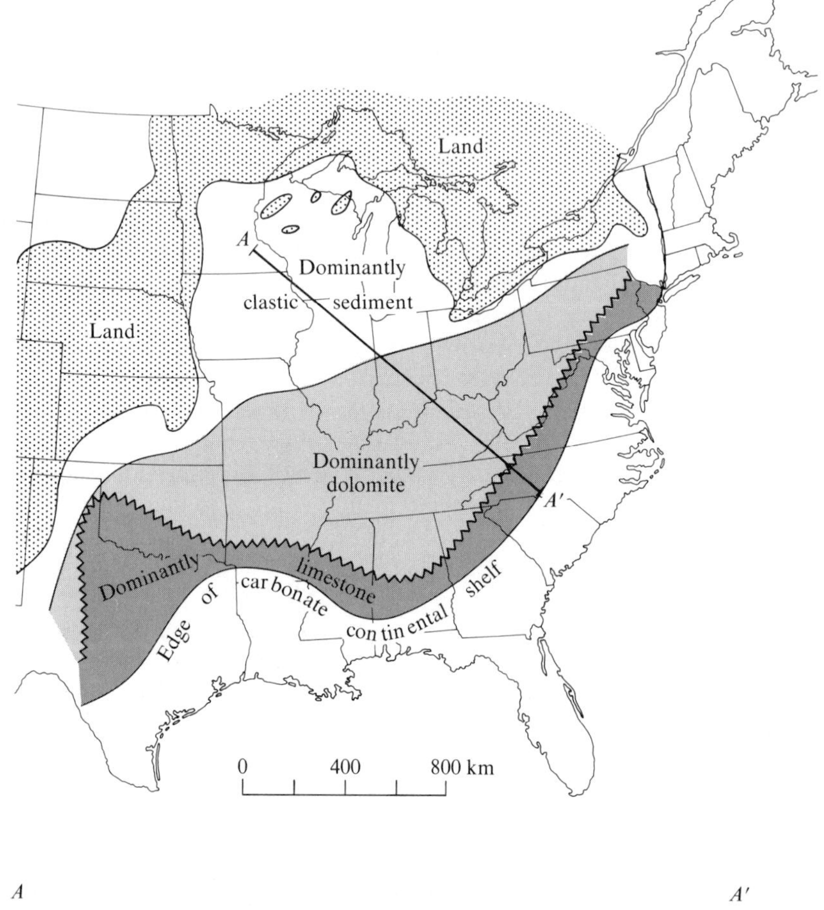

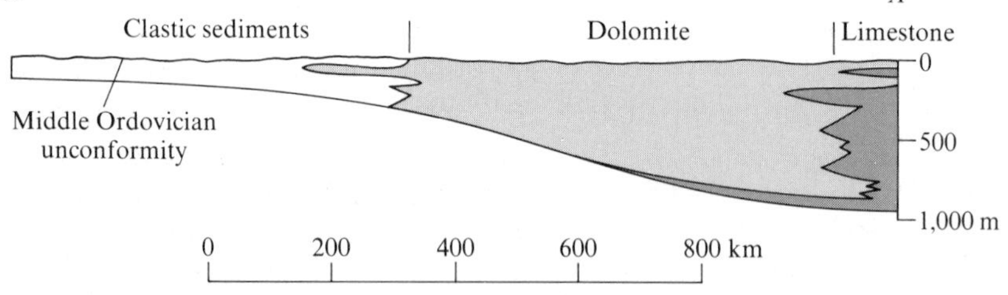

Figure 15–29
Generalized paleogeographic map and cross section of Late Cambrian and Early Ordovician time showing regional distribution of facies of the Knox Group and equivalent rocks on the cratonic shelf. [From L. D. Harris, 1973, *U.S. Geol. Surv., Jour. Research, 1,* Fig. 1.]

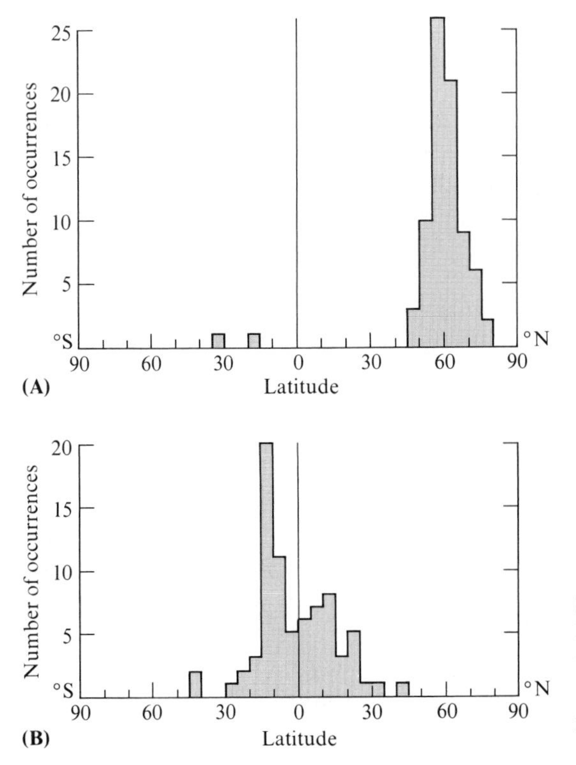

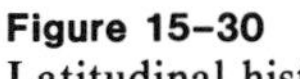
Figure 15-30
Latitudinal histograms for dolomites. (**A**) Present latitude. (**B**) Paleolatitude. [From J. C. Briden and E. Irving, 1964, *Problems in Paleoclimatology* (New York: Wiley Interscience), Fig. 18.]

Modern Dolomite

Holocene dolomite occurs only in small amounts and in very few places, the best known being on the island of Bonaire in the Netherlands Antillies, on Andros Island in the Bahamas, on the south shore of the Persian Gulf, in Deep Spring Lake, California, and in an area of ephemeral lakes called the Coorong in South Australia. These scattered areas have three things in common: (1) the dolomite is forming in exposed tidal flat sediment or in very shallow water; (2) textural relationships demonstrate the dolomite is forming by replacement of preexisting limestone; and (3) the waters from which the dolomite is forming have unusually high ratios of magnesium to calcium. The molar Mg/Ca ratios range from unity to 100, with values in the range of 10–20 being most typical. This range can be compared with the world average ratio of 0.46 in river water and 5.2 in sea water. It appears that the formation of dolomite requires enrichment of magnesium relative to calcium in the waters that flow through the pores of limestones. To produce dolomite, we need (1) water of suitable chemical composition, and (2) a mechanism to move this water through previously deposited limestone.

Studies of modern environments in which dolomitization is occurring have revealed two mechanisms by which such moving waters are produced.

1. *Evaporative reflux.* This mechanism was first suggested on the basis of field observations of the Permian reef complex in west Texas and New Mexico, where dolomite is associated with evaporites. Since that time, several areas have been found where it is thought to be operating today. The environments are in areas of limestone deposition where evaporation rates are high, such as in Saudi Arabia on the south side of the Persian Gulf. Along the shore there are broad areas of supratidal flat known locally as *sabkhas* that are subjected periodically to extensive flooding by Persian Gulf sea water, which has salinities of 40,000–65,000 ppm (normal open ocean water has a salinity of 35,000 ppm). Evaporation of this water to about 120,000 ppm causes the concentrations of calcium ion and sulfate ion to be so high that gypsum ($CaSO_4 \cdot 2H_2O$) precipitates. This has two effects. The evaporation increases the density of the remaining brine and the loss of calcium ion from the brine causes a large increase in the Mg/Ca ratio. Values above 10 are typical. The dense, magnesium-enriched brine sinks downward through the underlying limestone (see Figure 15–31), dolomitizing it within periods of a few hundred years. The replacement rate is determined by the Mg/Ca ratio and the rate at which the brine flushes through the limestone. As the positions of the shoreline and associated sabkha migrate through time, laterally extensive beds of dolomite can form.

2. *Mixing of fresh and marine waters.* The compositions of fresh water and sea water are such that it is possible by mixing approximately 95% fresh water with 5% sea water to produce a water undersaturated with respect to calcite but supersaturated with respect to dolomite. This mechanism for dolomite formation was suggested in 1971 and subsequently termed *Dorag* dolomitization. Dorag means "mixed blood" in the Persian language. For some distance landward of the shoreline the shallowly buried interface where fresh water is in contact with underlying marine water would be a dolomitizing zone (see Figure 15–32). This zone would migrate in the landward direction as the continental edge was depressed or as sea level rose; it would migrate seaward under the reverse conditions. As is the case with evaporative reflux, repeated back and forth migrations of the position of the shoreline would

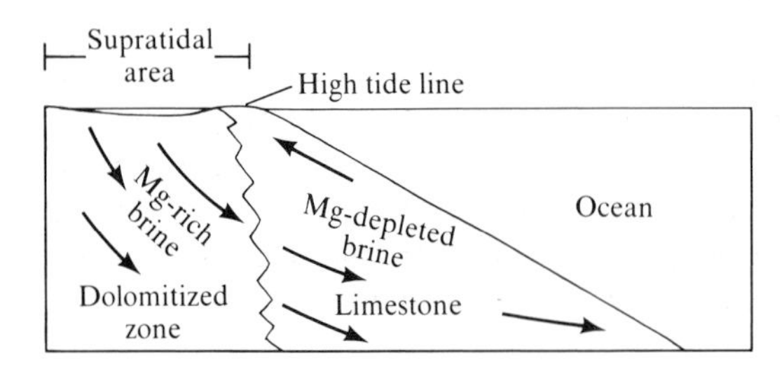

Figure 15–31
Schematic cross section showing paths of water flow in the reflux mechanism of dolomitization under a sabkha.

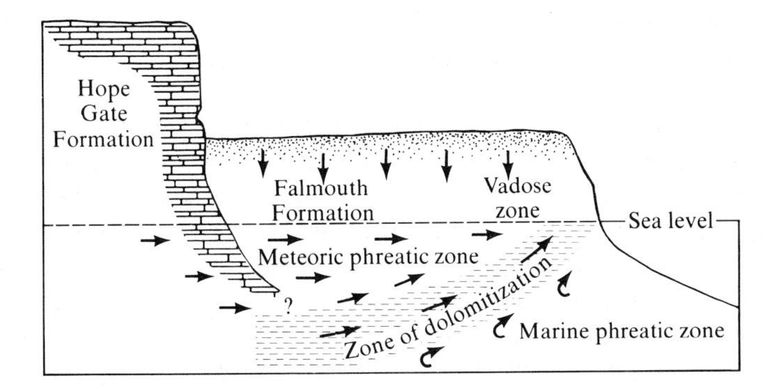

Figure 15–32
Schematic cross section showing the field relations between the Hope Gate and Falmouth Formations (Pleistocene) and groundwater zones on the north coast of Jamaica. The volume of dolomite that might be produced depends on the lateral extent of shoreline migration through time. [After L. S. Land, 1973, *Sedimentology, 20,* Fig. 1.]

produce dolomite accumulations that are both laterally extensive and stratigraphically thick. Dorag dolomitization is an attractive explanation for those dolomites that lack evidence of associated evaporite deposits. Such evidence includes not only the physical presence of the evaporites but also indirect evidence such as gypsum and halite molds and solution breccias.

SUMMARY

Limestones and dolomites typically are very pure, containing less than 5% minerals not soluble in hydrochloric acid. This purity results from the biologic origin of nearly all carbonate particles. Carbonate-shelled organisms cannot flourish unless the environment has a rather constant salinity and is visibly clear. A surrounding highland area that supplies abundant detrital sediment and fresh water will rapidly kill shallow marine, bottom-dwelling flora and fauna. The distribution of limestones also reflects present and ancient latitudes because the reproductive rates of marine organisms are higher in warmer waters.

The textures and structures of limestones are complex because these rocks can be sedimented as sand-size particles (clams, ooliths, and so on), as mud-size particles (disarticulated algae) that recrystallize easily and rapidly, or can be formed in place as a reef growth. Most limestones can be viewed as calcium carbonate equivalents of silicate rocks. The essential textural components are sand-size particles and matrix. *In situ* reef growths can have exotic textures that reflect the ecologic habits of the organisms.

Deposits of carbonate sediment in the deep sea are limited by the fact that the solubility of calcium carbonate is higher in colder water. The depth below which carbonate sediment cannot accumulate ranges between 3,000 and 5,000 m and reflects the balance between the rate of sediment production at the sea surface and the rate of dissolution of the particles as they settle toward the ocean floor after death.

Diagenetic changes in limestones are very common and pervasive. All the microcrystalline calcium carbonate particles in ancient rocks are a product of recrystallization. Many of the allochemical particles also are partly or completely recrystallized after burial. Nevertheless, it is usually possible to recognize the level of kinetic energy in the depositional environment and possibly other ecologic features as well.

Dolomite forms by replacement of preexisting limestone. The process requires a ratio of magnesium to calcium ion in the pore waters that exceeds unity, and a hydraulic mechanism for flushing large volumes of water through the pores. Several methods of accomplishing this are known to occur in shallow, nearshore marine environments.

FURTHER READING

Bathurst, R. G. C. 1975. *Carbonate Sediments and their Diagenesis,* 2d Ed. New York: Elsevier, 658 pp.

Fischer, A. G., S. Honjo, and R. E. Garrison. 1967. *Electron Micrographs of Limestones.* Princeton, N.J.: Princeton Univ. Press, 141 pp.

Ginsburg, R. N. 1956. Environmental relationships of grain size and constituent particles in some south Florida carbonate sediments. *Amer. Assoc. Petroleum Geol. Bull., 40,* 2384–2427.

Griffith, L. S., M. G. Pitcher, and G. W. Rice. 1969. Quantitative environmental analysis of a Lower Cretaceous reef complex. Soc. of Economic Paleontologists and Mineralogists Spec. Pub. No. 14, pp. 120–138.

Horowitz, A. S., and P. E. Potter. 1971. *Introductory Petrography of Fossils.* New York: Springer-Verlag, 202 pp.

Milliman, J. D. 1974. *Marine Carbonates.* New York: Springer-Verlag, 375 pp.

Pray, L. C., and R. C. Murray, eds. 1965. *Dolomitization and Limestone Diagenesis.* Soc. of Economic Paleontologists and Mineralogists Spec. Pub. No. 13, 180 pp.

Scholle, P. A. 1978. *Carbonate Rock Constituents, Textures, Cements and Porosities.* Tulsa: Amer. Assoc. Petroleum Geol. Memoir 27, 241 pp.

Stockman, K. W., R. N. Ginsburg, and E. A. Shinn. 1967. The production of lime mud by algae in south Florida. *Jour. of Sedimentary Petrology, 37,* 633–648.

Wilson, J. L. 1975. *Carbonate Facies in Geologic History.* New York: Springer-Verlag, 471 pp.

16

Other Types of Sedimentary Rocks

Although mudrocks, sandstones, and carbonate rocks form at least 95% of all sedimentary rocks, there are numerous areas on the continents where other types of sedimentary rocks are prominent. For example, about 35% of the continental area covered by sediments is underlain at some depth by evaporite beds (see Figure 16–1), rocks composed almost entirely of unusually soluble minerals such as halite. In parts of Arkansas, Oklahoma, west Texas, and California thousands of square kilometers of bedded chert deposits occur, often hundreds of meters thick. Across Wyoming and Idaho extensive bedded deposits of nearly pure apatite crop out, phosphate rock that is of great commercial interest as a source of fertilizer. Linear belts of sedimentary iron ores that supply most of the world's iron are located in the Lake Superior region, in several areas in the Appalachian Mountain belt, in Australia, and elsewhere.

Evaporites, bedded cherts, phosphate rocks, and sedimentary iron ores all form at the earth's surface in settings where neither abundant mud nor sand accumulates. Where are such environments found today and what information can these rock types give us about the climatic, topographic, and environmental conditions in which they formed? In this final chapter on sedimentary rocks we will attempt to answer these questions.

EVAPORITES

Evaporite minerals are defined as those minerals that are precipitated by the evaporation of naturally occurring waters, marine or nonmarine. The specific minerals produced depend on the composition of the original water, so there are a large number of possibilities. More than 70 evaporite minerals are known. Calcite, aragonite,

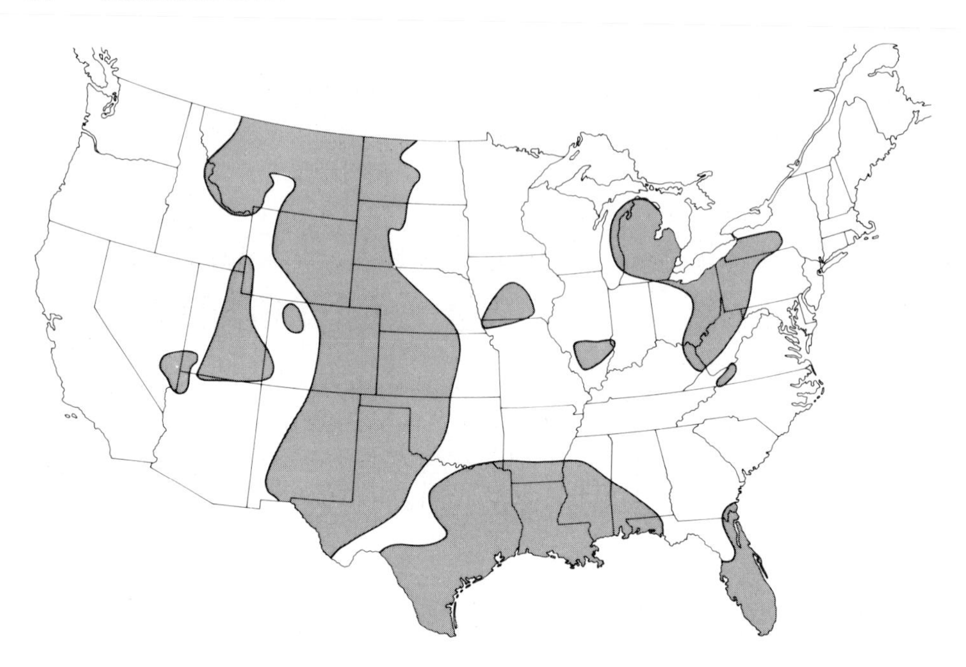

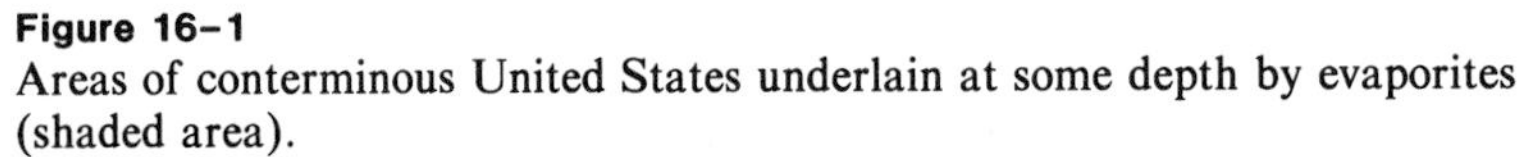

Figure 16–1
Areas of conterminous United States underlain at some depth by evaporites (shaded area).

and dolomite are not generally considered evaporite minerals even though they are common in evaporite deposits.

Evaporites are, by definition, very soluble. Therefore, they rarely occur in outcrop despite their abundance in the subsurface. Outcrops are largely confined to the more arid areas of the land surface.

In outcrop, gypsum is the most common evaporite mineral because (1) it is the least soluble mineral of the group, and (2) anhydrite ($CaSO_4$) uplifted to the surface is rapidly replaced by gypsum ($CaSO_4 \cdot 2H_2O$). In the subsurface, halite is commonly more abundant than either of the calcium sulfate minerals. Gypsum is easily identified by its normal white color and its softness, being the only common nonmicaceous mineral that can be scratched by a fingernail.

Evaporites are primary chemical precipitates and, therefore, typically are associated in the field with carbonate rocks rather than with coarse clastic sediments. Often the carbonates are dolomitic because the waters that precipitate evaporites also cause dolomitization. Commonly, fine clastic sediments interbedded with the evaporites are red because of the genetic association between continental desert climates where many evaporites form and the early diagenetic production of red pigment.

The main sedimentary structures visible in evaporite beds are dark organic laminations, which commonly exhibit intralayer crumpling (enterolithic folding; see Figure 16–2) and nodular or "chicken-wire structure" (see Figure 16–3), elongate gobs of

Figure 16–2
Laminated gypsum in the Castile Formation (Permian), New Mexico, showing diagenetic crumpling and folding. The varved laminae can be traced laterally for more than 100 km. [From F. J. Pettijohn and P. E. Potter, *Atlas and Glossary of Primary Sedimentary Structures* (New York: Springer-Verlag), Plate 111A.]

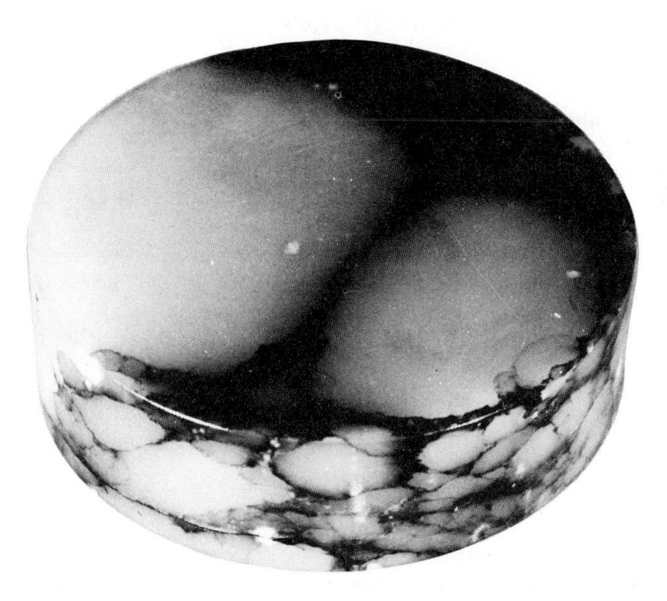

Figure 16–3
Part of a subsurface core showing well-developed nodular or chicken-wire structure in anhydrite. From Mississippian rocks in Montana. Diameter of core is 10 cm. [From H. Blatt et al., 1980, *Origin of Sedimentary Rocks,* 2d Ed. (Englewood Cliffs, N.J.: Prentice-Hall), Fig. 15–10 (photo courtesy R. C. Murray).]

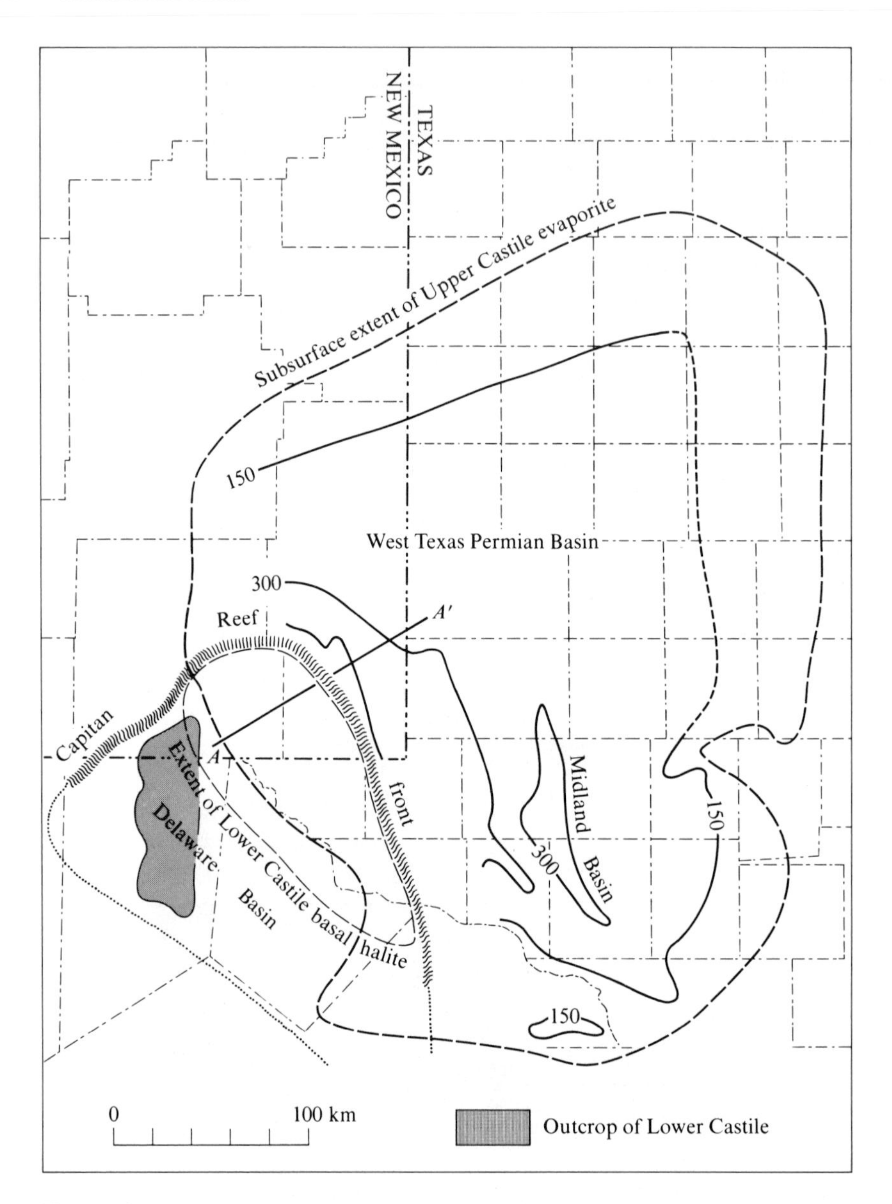

Figure 16–4
Map of west Texas and southeast New Mexico showing the location and thickness of the Upper Castile evaporite (meters). *A–A′* is the line of cross section in Figure 16–5. [After G. A. Kroenlein, 1939, *Amer. Assoc. Petroleum Geol. Bull., 23,* Fig. 1.]

anhydrite set in a darker microcrystalline anhydrite matrix. Chicken-wire structure forms as gypsum dehydrates to anhydrite and the resulting watery mush compacts with the expulsion of the water released from the gypsum. Laminated evaporites are deposited in standing bodies of water, while the chicken-wire nodules form from individual gypsum crystals that grow in soft, subareally exposed sediment.

Castile Formation (Permian): Texas and New Mexico

Perhaps the most spectacular evaporite deposit is the anhydrite–halite sequence in the Castile Formation (Upper Permian) of west Texas and southeastern New Mexico (see Figure 16–4). The Castile crops out only in the western part of the Delaware Basin (as 250 m of gypsum) but in the subsurface underlies an area more than 100,000 km^2 in extent. It has a maximum thickness of 1,200 m and a volume of approximately 25,000 km^3. The original depositional volume of the formation was even greater than at present, as evidenced by erosional contacts laterally and many anhydrite solution breccias caused by removal of halite. Near the surface the anhydrite is converted to gypsum. The ratio of halite to the calcium sulfate minerals is estimated to be about 1:1 at present.

The depositional environment of this giant evaporite deposit was a deep basin bordered by a carbonate platform that stood more than 500 m above the basin floor (see Figure 16–5). In the basin the evaporites are underlain by organically laminated

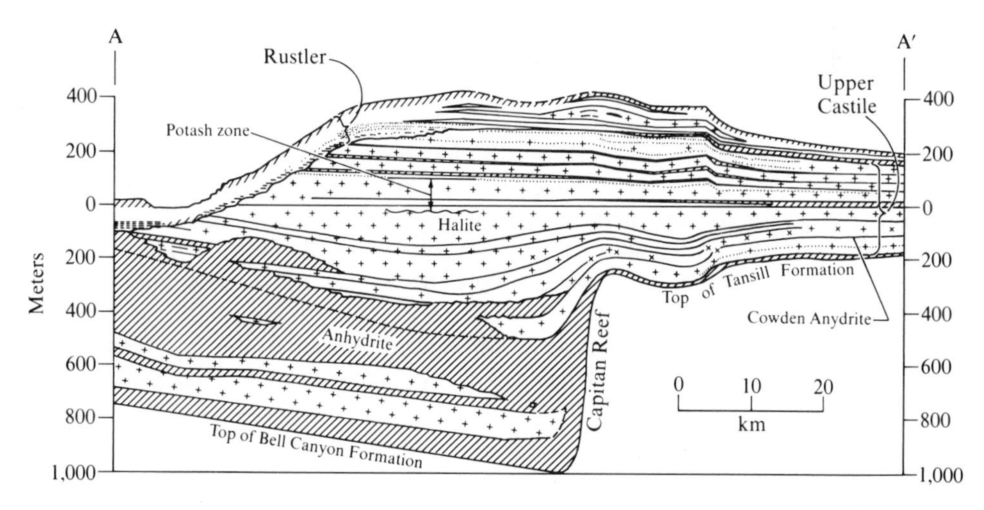

Figure 16–5
Cross section of Castile Formation normal to the reef front showing changes in evaporite composition and thickness between the Delaware Basin and backreef area. [After G. A. Kroenlein, 1939, *Amer. Assoc. Petroleum Geol. Bull., 23,* Fig. 3.]

quartz siltstone, fine sandstone, and carbonates of the Bell Canyon Formation; on the platform the beds below the evaporites are shallow water carbonates of the Tansill Formation. Above the Castile in both areas is the Rustler Formation, mostly anhydrite with lesser amounts of fine-grained red clastics, dolomite, and halite. Evaporite deposition was terminated in the region with an influx of fluvial muds and fine sands derived from the foreland of the Ouachita–Marathon orogen to the south and east.

The lower portion of the Castile is composed of alternating thin laminae of calcite and anhydrite or organic matter and anhydrite. The calcite crystals are subhedral to ovoid in outline and uniformly 25 μm in size, possibly a result of recrystallization. The anhydrite crystals range up to 100 μm in diameter and have interlocking rectangular outlines. The laminae in the calcite-laminated anhydrite average 1.6 mm in thickness and are remarkably persistent laterally, having been traced for distances up to 113 km. Each lamina is synchronous, and each couplet has been interpreted as representing an annual layer of sedimentation—a varve. A total of 260,000 varves has been measured and correlated within the basin by Anderson et al. (1972).

The Upper Castile is dominated by halite and the onset of halite precipitation seems to have been synchronous throughout the basin. The halite, however, is not as easily traceable in detail as is the anhydrite below it. The halite typically is thoroughly recrystallized and occurs in laminae up to 10 cm thick alternating with anhydrite. Some potash beds, composed chiefly of polyhalite [$Ca_2K_2Mg(SO_4)_4$], occur within the halite sequence, testifying to an extreme phase of desiccation within the basin.

Duperow Formation (Devonian): Williston Basin

The Castile Formation in west Texas and New Mexico exemplifies the first of two main types of evaporite deposits, thick accumulations of essentially pure evaporites initiated in the central parts of a desiccating standing body of water. The second type of evaporite accumulation forms at the margins of a marine basin on a sabkha, typically interbedded with dolomitic carbonate rocks (see Chapter 15). The Duperow Formation is one of several evaporite–carbonate sequences that have been studied in detail.

The Duperow is part of a great sheet of upper Devonian strata that extends beneath the Prairie Provinces of Canada southeastward from the Canadian Rockies and arctic Canada into the United States as far south as the transcontinental arch of South Dakota and Nebraska. Its maximum thickness is about 250 m in southern Saskatchewan. It thins southeastward to 150 m in northwestern North Dakota and disappears because of erosion in south-central South Dakota. The formation is composed of about 12 regular carbonate–evaporite cycles. Each cycle consists of a lower member of burrowed, fossiliferous micrite containing a normal marine shallow water

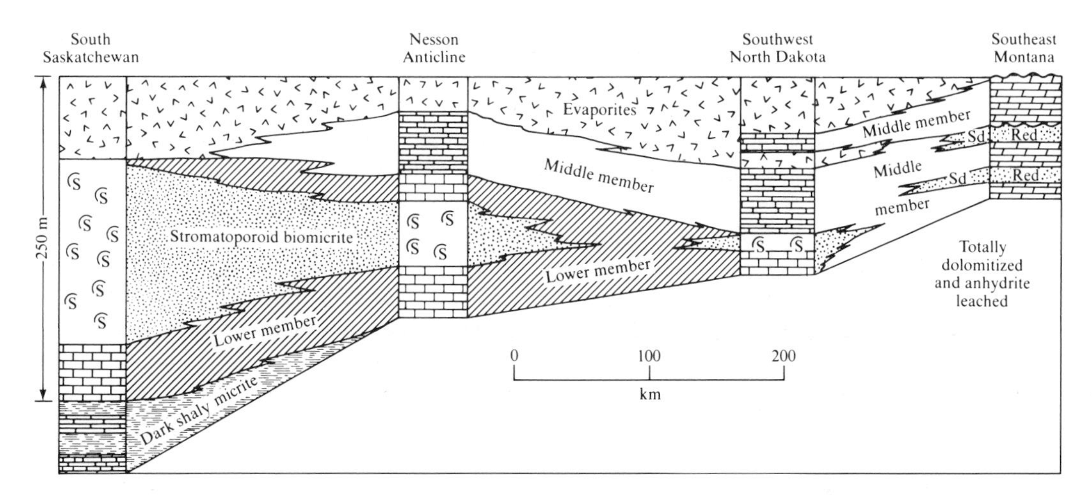

Figure 16–6
Idealized stratigraphic diagram showing interpreted Duperow facies belts from north to south across the Williston Basin. Normal marine burrowed biomicrite and stromatoporoid reef(s) grade into restricted marine laminated micrite with a microfauna (white area). The microfaunal micrite grades upward into sabkha dolomicrites and evaporites. [After J. L. Wilson, 1967, *Bull. Canadian Petroleum Geol., 15,* Fig. 5.]

fauna of brachiopods, crinoids, and stromatoporoids (a group of extinct calcareous coelenterates); a middle member of brown micrite with a restricted microfauna of ostracods interbedded with unfossiliferous peloidal beds or laminated micrite; and an upper member of bedded anhydrite with gray-green silty dolomicrite displaying intertidal and supratidal sedimentary structures. The anhydrite member typically is the thinnest of the three members (see Figure 16–6). Duperow cycles are exceedingly widespread, and constituent beds only 3–5 m thick can be traced for several hundred kilometers across the Williston Basin.

Deposition occurred in a vast backreef lagoon south of a reef belt in Alberta and stretched to a sandy shore in South Dakota and Wyoming. This very shallow basin was periodically and apparently rapidly flooded with marine water, permitting certain benthonic organisms to flourish and even patch reefs to grow at times. Gradual shallowing as sedimentation filled the basin resulted in extensive tidal flats and evaporitic sabkhas; extensive dolomitization occurred on the peripheral shelves. The time of deposition of each cycle can be estimated as 500,000 to a million or so years assuming a constant rate of sedimentation through the Late Devonian.

The bedded anhydrite of the upper member of the Duperow Formation occurs in beds from 0.3 to 3.0 m thick and can be distinguished from the blebs and nodules of replacement anhydrite by fibrous texture and ubiquitous lamination. This includes horizontal laminae about 1 mm thick, secondarily distorted laminae, "slump-and-crumple structure," and "ball-and-flow structure," discernible when the anhydrite is interlaminated with microcrystalline dolomite. In such beds, microbreccia and microconglomerate are also present, some presumably caused by early solution of the sulfate and consequent breakage of the thin layers of carbonate. Postdepositional flowage of anhydrite layers is responsible for other microconglomerates. Arched laminae, indicative of algal stromatolites and mud cracks, are also observable in the bedded anhydrites, particularly where dolomites are associated.

Rarely, the anhydrite is interlaminated or otherwise closely associated with lime mud instead of cryptocrystalline dolomite. Except where associated with cryptocrystalline dolomite, the anhydrite is without clastic impurities and, of course, is without fossils or any trace of life, unless some of the rounded grains of the microconglomerate can be interpreted as fecal pellets.

The abundance of sedimentary structures, such as mud cracks and microbreccias, is taken as evidence that the evaporites were formed on periodically exposed tidal flats.

Secondary Anhydrite

Replacement anhydrite forms 5–10% of the upper member of the Duperow. It appears as scattered, well-formed, tabular crystals, occasionally rather large (up to several millimeters across). The source of the anhydrite is plain, for the coarse blocky replacement anhydrite appears much more commonly in beds immediately above or below the thin beds of primary evaporite. This is seen best in wells in offshelf positions where thin anhydrites clearly mark off the cycles. Where bedded anhydrite and restricted environment carbonates are thicker and replace the normal limestone portions of the cycle, much more secondary anhydrite is observed throughout the cycle.

Some of the lime mudstone associated with evaporites has excellently developed bladelike brown anhydrite crystals formed in rows along bedding planes. The crystal form is such that this may represent an anhydrite replacement after gypsum. Recent supratidal deposits formed in arid conditions commonly contain such selenite crystals formed in the carbonate sediments by displacement. Most commonly, however, replacement anhydrite is seen in the coarser dolomite of the stromatoporoid–*Amphipora* facies, which seems particularly susceptible to both replacement and infilling by anhydrite. In fact, even in the middle of cycles farthest removed from bedded anhydrite layers, this rock type may be seen to have unusually high amounts of secondary anhydrite.

Anhydrite matrix with carbonate pellets and with anhydrite partly filling small vugs or even mud cracks is occasionally seen within the Duperow. This is interpreted to be the result of actual precipitation of anhydrite in previously existing void space. In the more evaporitic layers of the Duperow cycle, the instability of anhydrite has resulted in intense brecciation, slumping, and other microdisarrangement of layers. It is thus difficult to distinguish in these beds between very early anhydrite reorganization and disruption of the interlaminated carbonates, and anhydrite filling and replacement.

Origin of the Giant Evaporite Deposits

Giant evaporite deposits such as the Castile appear to have been formed in standing bodies of water that were isolated or nearly isolated from the sea by reef growth, tectonic activity, volcanic eruptions, or other cause. Therefore, the simplest model to consider for the origin of thick, pure evaporite deposits is a straightforward evaporation to dryness of a standing body of sea water. If the rate of potential evaporation exceeds the rate of inflow of water over a prolonged period of time, the concentration of dissolved salts in the water will increase and an evaporite mineral will be precipitated.

Laboratory experiments performed by numerous scientists during the past 130 years have revealed the order in which we should expect the evaporite minerals to precipitate. If we start with normal sea water, gypsum will precipitate when the water has been reduced to one-third of its original volume; halite at one-tenth the original volume. The other 70 evaporite minerals (sulfates, borates, halides, and carbonates) will not appear until the volume is less than one-twentieth of the original.

What thickness of evaporite rock might we expect from the complete evaporation of sea water? Suppose we assume a column of water 1,000 m deep with 3.5% dissolved salts. Complete evaporation would yield a nonporous precipitate of sediment with a thickness of 0.035 (1,000)/specific gravity of the mineral precipitated. The average evaporite mineral has a specific gravity of 2.2, so that the total thickness of the deposit will be 15.9 m. If we assume that all the sodium in sea water is precipitated as NaCl (specific gravity = 2.16), then Table 16–1 indicates that 41.80×2 or 83.6% of the evaporite will be halite. This is 13.5 m $[(35/2.16) \times 0.836]$. If all the calcium comes out as gypsum (specific gravity = 2.32), Table 16–1 indicates that 0.91×2 or 1.82% of the evaporite will be gypsum. This is 0.27 m $[(35/2.32) \times 0.018]$. The remaining 2.1 m of evaporite will be composed of some of the 70 or so other evaporite minerals in appropriate proportions.

Our calculation indicated that the evaporation of a standing body of water as much as 1,000 m deep can yield only 15.9 m of evaporite deposit. As we have seen in the

Table 16–1
Relative Amounts of the Most Abundant Dissolved Chemical Constituents in Sea Water

Dissolved species	Molarity	Percentage
Cl^{1-}	0.535	48.72
Na^{1+}	0.459	41.80
Mg^{2+}	0.052	4.74
SO_4^{2-}	0.028	2.50
Ca^{2+}	0.010	0.91
All others	0.014	1.33
	1.098	100.00

Castile Formation, however, the thicknesses of gypsum–halite sequences can be several hundreds of meters. Most evaporite deposits are several meters to several tens of meters thick. The amount of evaporation this requires seems staggering (think of the humidity!) but that is only because of the brief span of human life as compared to the millions of years during which uninterrupted evaporation can proceed in an isolated body of water.

The simple evaporation-to-dryness model we tested with our calculation has defects in addition to the fact that it could not supply the great thicknesses of evaporites seen in the field. For example, our model produced 50 times as much halite as gypsum because the dissolved material in sea water is mostly sodium and chloride ions. How are we to explain ratios of halite to gypsum that are greatly different from 50:1, such as the Castile Formation, in which the ratio is about 1:1? Study of ancient evaporites reveals there is no norm for this ratio. Some deposits are all gypsum; others, all halite.

The mechanism that permits this to occur is called *reflux*. The requirements for reflux are:

1. Constant influx of sea water.
2. Constant evaporation within a restricted basin.
3. Loss of the heavy brine produced, either by seepage down through the underlying sediments (subsurface reflux) or by flow of the brine out of the basin beneath the inflow of new sea water.

Under these conditions a constant salinity can be maintained and a single evaporite mineral deposited. The main factor controlling what the constant salinity will be is the rate of brine loss by reflux. Greater reflux produces lower salinity; less reflux, higher salinity.

Applying this model to the Castile Formation, we would infer that the salinity achieved during evaporation and reflux was sufficient initially to cause precipitation of calcium sulfate (lower Castile), but as reflux flow decreased, halite was precipitated. Calculations indicate that an evaporation of 290 cm of water per year would be needed to produce the observed annual increment of anhydrite.

These evaporite accumulations will be somewhat younger in age than the rocks of the basin walls and will not grade laterally into contemporaneous sedimentary facies.

If the rate of evaporation and reflux is greater than the rate of inflow, evaporative drawdown can occur. The Great Salt Lake is a familiar example, a remnant of the much larger glacial Lake Bonneville; Death Valley in southern California was the site of glacial Lake Manly, as evidenced by the present salt flat and bordering wave-cut benches marking the levels of former shorelines.

Mediterranean Sea Evaporites

During the late 1960s, seismograms were taken of several areas of the Mediterranean Sea basin, and they revealed structures resembling salt diapirs under the sea floor. In

1970, these structures were drilled and the presence of Miocene evaporites up to 3 km thick was discovered in several areas within the basin. The evaporites are both underlain and overlain by pelagic muds that were deposited in water more than 1,000 m in depth. Apparently, prior to Late Miocene time the input–output dynamics of water in the Mediterranean basin were similar to those existing today, a deep basin filled with fairly normal ocean water. Then, in the Late Miocene, plate interactions associated with the closing of the Tethys seaway isolated the Mediterranean (2.5 million km^2) to the extent that nearly all the water evaporated and evaporites precipitated. The eastern closing (Turkey, Syria, Lebanon, Israel, Egypt) still exists today but at the end of the Miocene the western end at Gibralter opened slightly to permit refilling of the basin with waters from the Atlantic Ocean, and pelagic muds were again deposited. An empty bathtub 4,000 m deep was refilled within 5 million years.

The topography of the floor of the Mediterranean Sea divides it into four basins (see Figure 16–7). The relief between the basin lip and its bottom varies between 1,000 and 2,500 m among the basins. In each basin is a sequence of Late Miocene evaporites averaging 300–500 m in thickness that includes gypsum, halite, and the more soluble salts and, in each, the areal distribution resembles a bull's eye pattern (see Figure 16–8). It looks much like the distribution in a desert playa, with the least soluble salt (gypsum) at the fringes and the more soluble salts appearing toward the

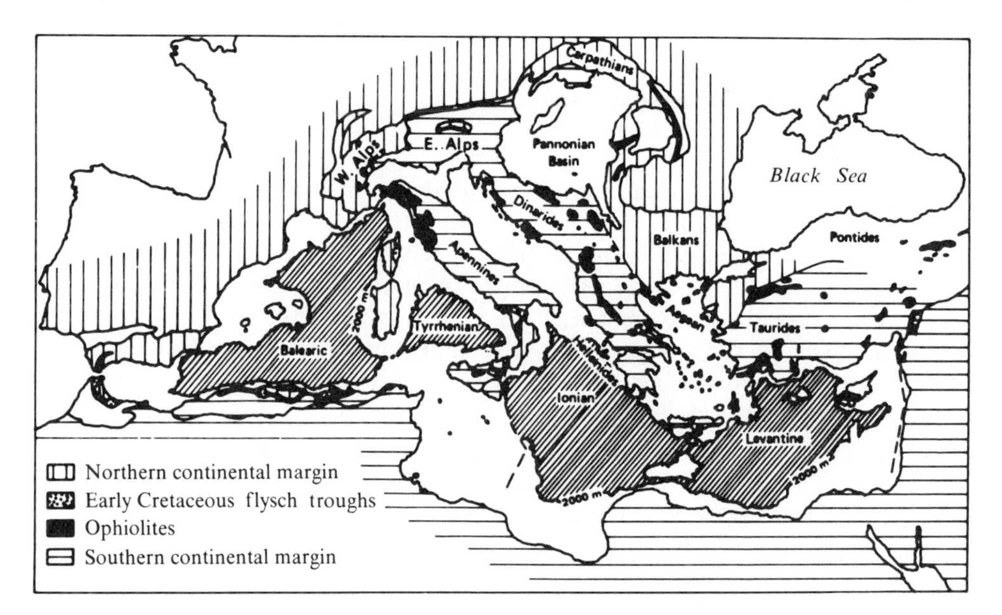

Figure 16–7
Tectonic setting of the Mediterranean basins, in which thick evaporites accumulated during Miocene time. [From K. J. Hsü et al., 1978, *Initial Reports of the Deep Sea Drilling Project, 42,* Pt. 1, Fig. 1.]

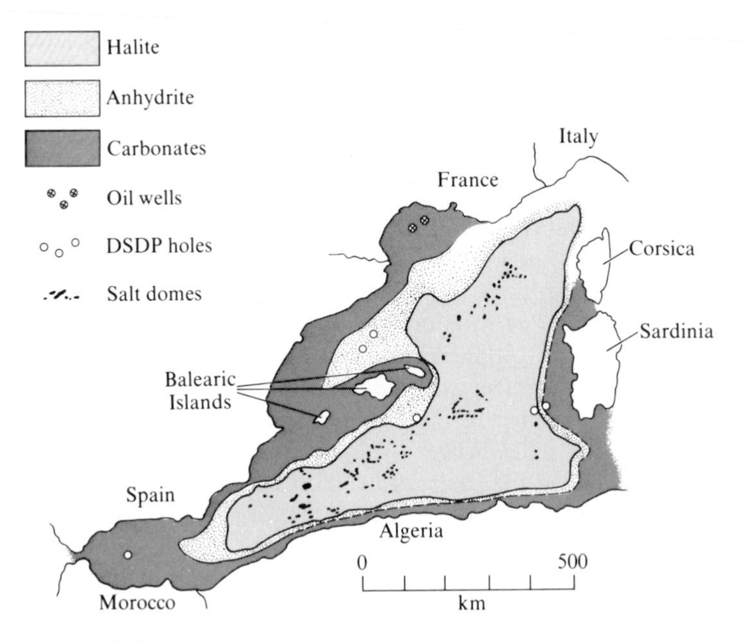

Figure 16–8
Evaporite distribution in Miocene rocks of the Balearic Basin of the Mediterranean Sea immediately east of the Strait of Gibraltar. The Balearic was one of several topographically low salt pans on the floor of a desiccated Mediterranean. Potash salts (not shown) occur in the middle of the halite area. [From K. J. Hsü, 1972, *Earth Sci. Reviews, 8,* Fig. 4.]

center as the amount of water decreased and salinity increased. Recalling that 1,000 m of sea water produces only 15.9 m of evaporite deposit, it is clear that about 30,000 m of water must have been evaporated during the Late Miocene (5 million years) to produce the 300–500 m of evaporite present in each of the four Mediterranean basins. Apparently, the small amount of inflow into each basin during the Miocene was adequate to maintain the salt concentration between that needed for gypsum to form (three times normal salinity) and total dryness. Certainly there was never water 30,000 m deep in the Mediterranean area.

Evaporites and Climate

We have seen that the formation of extensive evaporite deposits requires a relatively high rate of evaporation coupled with a low rate of precipitation. In what areas of the present earth's surface do such conditions exist? To find the answer to this question we must examine the gross circulation pattern of the atmosphere (see Figure 16–9). The intensity of solar radiation reaching the earth's surface is determined primarily

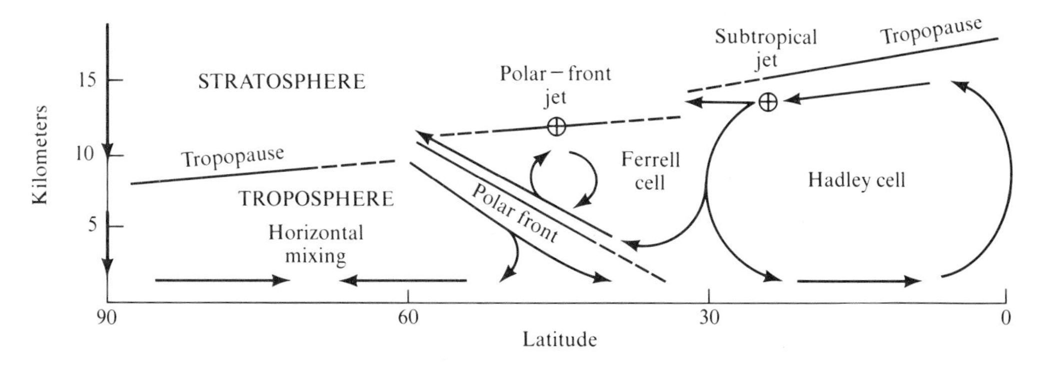

Figure 16–9
Longitudinal cross section through the earth's atmosphere showing circulation between the earth's surface and the tropopause. Surrounding 30° latitude is a belt of descending, dry air that causes the arid belt at the earth's surface.

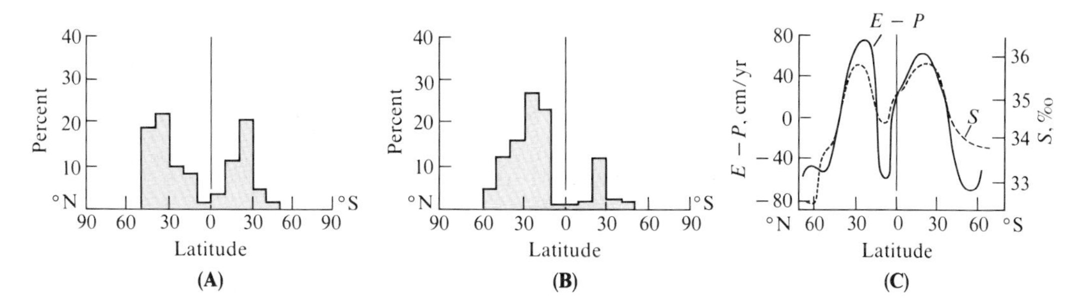

Figure 16–10
Distribution by latitude of (**A**) modern evaporites; (**B**) areas with less than 25 cm annual precipitation, excluding high-latitude deserts; (**C**) average evaporation (*E*) minus precipitation (*P*) and surface salinity (*S*) of modern oceans. Evaporite precipitation centers around 30° latitude on both sides of the equator. [From W. A. Gordon, 1975, *Jour. Geology, 83,* Fig. 1.]

by the angle of incidence of the sun's rays—the more nearly perpendicular to the earth's surface, the greater the intensity. As a result, the area around the equator is heated more than at higher latitudes so that this zone is characterized by rising air. As the air rises it cools, causing its moisture to be released and forming the rainy tropical jungles. The air continues to cool and compress as it moves away from the equator until it sinks back toward the surface at about 30° latitude. At this point it is a dry air mass because it has dropped nearly all its moisture as it cooled moving toward the poles and, as a dry air mass, it has the capability to absorb surface moisture as it descends. Thus, the belts of desert originate at 30° latitude on both sides of the equator. As shown in Figure 16–10, modern evaporite deposits center at these 30° latitude locations, as do belts of high average salinity in the oceans.

The control of evaporite occurrence by atmospheric circulation suggests that the locations of ancient evaporite deposits might reflect paleolatitudes of ancient land masses. At present, the latitudinal distribution of ancient evaporites is haphazard (see Figure 16–11A) because of differences in amount and direction of drift of the continents. But if the deposits are placed in their paleolatitudes as determined from paleomagnetic measurements (see Figure 16–11B), some semblance of order appears. In the Northern Hemisphere, a single mode appears in the 10–20° belt; in the Southern Hemisphere the pattern is still very diffuse. When the occurrences are grouped irrespective of hemisphere (see Figure 16–11C), not much improvement is obtained. Probably the lack of the perfect pattern found for modern evaporite locations results from changes in solar radiation through time. For example, in nonglacial times the intensity of solar radiation was generally greater and latitudinal belts were less clearly defined. This would cause the belts of evaporite formation to be spread out over a greater latitudinal range.

BEDDED CHERTS

Bedded cherts range in age from early Precambrian to Oligocene and occur on all continents, possibly excepting Antarctica. Their peak occurrence was during the middle Precambrian, when they may have formed about 15% of the entire stratigraphic section. In Phanerozoic rocks they are uncommon, but when they do occur they typically are several hundred meters in thickness.

Phanerozoic Marine Cherts

On a regional scale, bedded chert sequences most commonly occur with graded turbidites, ophiolites, and mélanges, in tectonic settings associated with convergent plate margins. Examples include the Franciscan Formation (Jurassic) in California, many Mesozoic occurrences in the Mediterranean–Himalayan (Tethys) region, the Ordovician cherts in Scotland, and Jurassic cherts in New Zealand. Bedded chert deposits not associated with ophiolites and mélanges include Carboniferous–Jurassic rocks in Japan, the Caballos Formation of Devonian age in Texas, and the Arkansas Novaculite of the same age in Arkansas and Oklahoma.

On an outcrop scale, stratigraphic sections of bedded cherts are composed of several types of rocks, among which the chert layers are most prominent. Individual chert layers range in thickness from a few centimeters to a meter or more (see Figure 16–12) and are composed of essentially pure microcrystalline quartz. Their color is snow-white when pure but impurities are common, giving rise to red (hematite), green (illite or chlorite), black (organic matter and manganese oxides), or other hues whose cause is uncertain. The upper and lower surfaces of chert layers can be either smooth or wavy. Both ripples and cross bedding have been observed in bedded

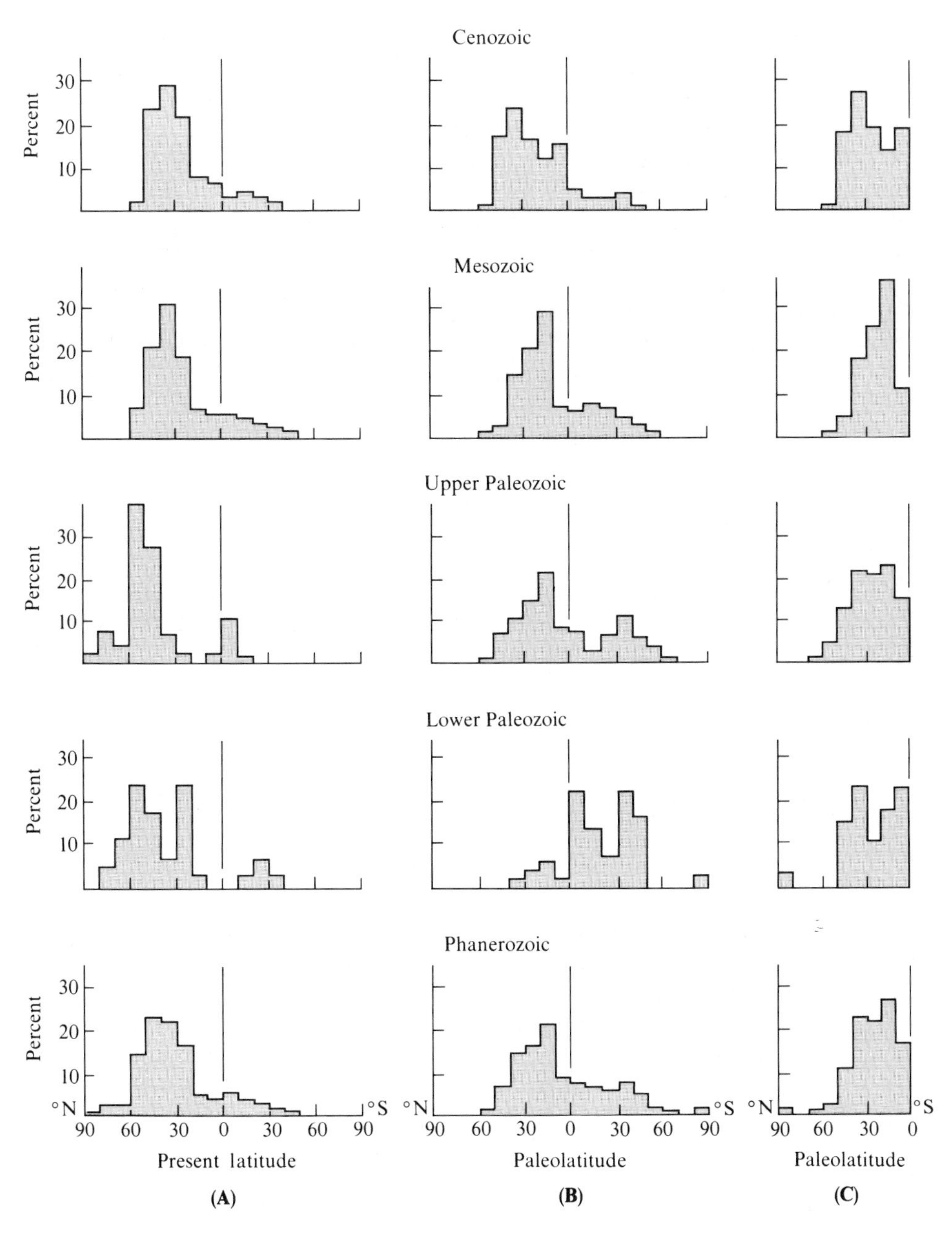

Figure 16-11
Distribution of ancient evaporites by latitude. (**A**) Present latitude. (**B**) Paleolatitude based on paleomagnetic data. (**C**) Paleolatitude without respect to hemisphere. [From W. A. Gordon, 1975, *Jour. Geology, 83,* Fig. 3.]

Figure 16–12
Outcrop of Caballos chert beds showing character of bedding, typical bed thicknesses, and minor partings of shale. [From E. F. McBride and A. Thomson, 1969, *Stratigraphy, Sedimentary Structures and Origin of Flysch and Pre-Flysch Rocks, Marathon Basin, Texas* (Dallas Geol. Soc.), Fig. 2.]

chert layers, but at most outcrops no sedimentary structures other than bedding are evident.

Interbedded sedimentary layers are typically thin, green or dark-colored shales, often siliceous and containing small amounts of pyrite. Less commonly, the interbeds are micrite, graded bioclastic limestone and fine sandstone, and fine-grained conglomerate lenses.

Caballos Formation (Devonian): West Texas

One of the best-studied bedded chert deposits is the Caballos Formation in the Marathon Basin of west Texas (see Figure 16–13) and its correlative in the Ouachita Mountains (Arkansas–Oklahoma), the Arkansas Novaculite. The two units are believed to be correlative because the total Paleozoic sequence in the two areas is very similar and both areas are believed to have been at the continental margin during Paleozoic time. It is generally accepted that the two areas are connected structurally in the subsurface through Texas.

The Caballos is a sequence of novaculite (white chert) and green, gray, brown, and tan chert beds with 10% interbeds of green and red clay shale and siliceous shale. The formation has a maximum thickness of 200 m but is generally between 60 and 150 m thick. Underlying the Caballos is the Maravillas Chert, the uppermost unit of an Ordovician succession of limestone, shale, chert, and conglomerate that has a total thickness of 600 m. The nature of the contact between the two formations is uncertain and may be unconformable. Resting conformably on the Caballos is the Tesnus Formation, a Mississippian turbidite unit about 2,000 m thick.

The depositional environment of the Caballos has traditionally been thought to be deep water because of the following evidence:

1. The cherts contain no shallow water fossils such as corals, bryozoa, or crinoids, all of which are characteristic of nearshore Paleozoic deposits. The only fossils are radiolaria, pelagic organisms characteristic of the open ocean; and siliceous sponge spicules (see Figure 16–14), whose abundance testifies to the lack of detrital sediment from a nearby land area.

2. The tectonic setting of the formation is that of a continental margin, with the thick units both above and below containing features characteristic of deposition in deep water, such as repeated turbidite sequences, graded bedding, organic burrows only along bedding planes, and the absence of exclusively shallow water sedimentary structures such as reefs, intertidal features, or mud cracks.

As is true of nearly all environments interpreted to be in deep water, the evidence is largely negative rather than positive. The conclusion regarding depth of water relies heavily on the absence of indigenous shallow water structures and fossils rather than the presence of deep water features. This is typical of such interpretations because there are few bathymetric indicators known below the photic zone where ecologic data can indicate water depth. In geologic reports the term deep water is as likely to mean 100 m as 1,000 m.

The deep water interpretation of the Caballos Formation has been challenged by Folk (1973), who has found features in parts of the formation that he interprets as evidence of deposition in a supratidal–intertidal environment. These features include stromatactis (?) structures, chert breccias he believes resulted from evaporite solution and collapse, jasperized (red chert) areas resembling silicified fossil soils, and traces of fossils that are thought to be freshwater fungi. The controversy is still unresolved. Folk's observations, however, point out once again the importance of careful and detailed field observations.

Modern Analogs of Ancient Bedded Cherts

Field evidence such as the typical association of bedded cherts with graded sandstones, mélanges, and ophiolites indicates that the cherts are marine, were associated with active tectonism, were at one stage located at a continental margin, and were probably formed in deep water. Where on the present earth do such conditions exist

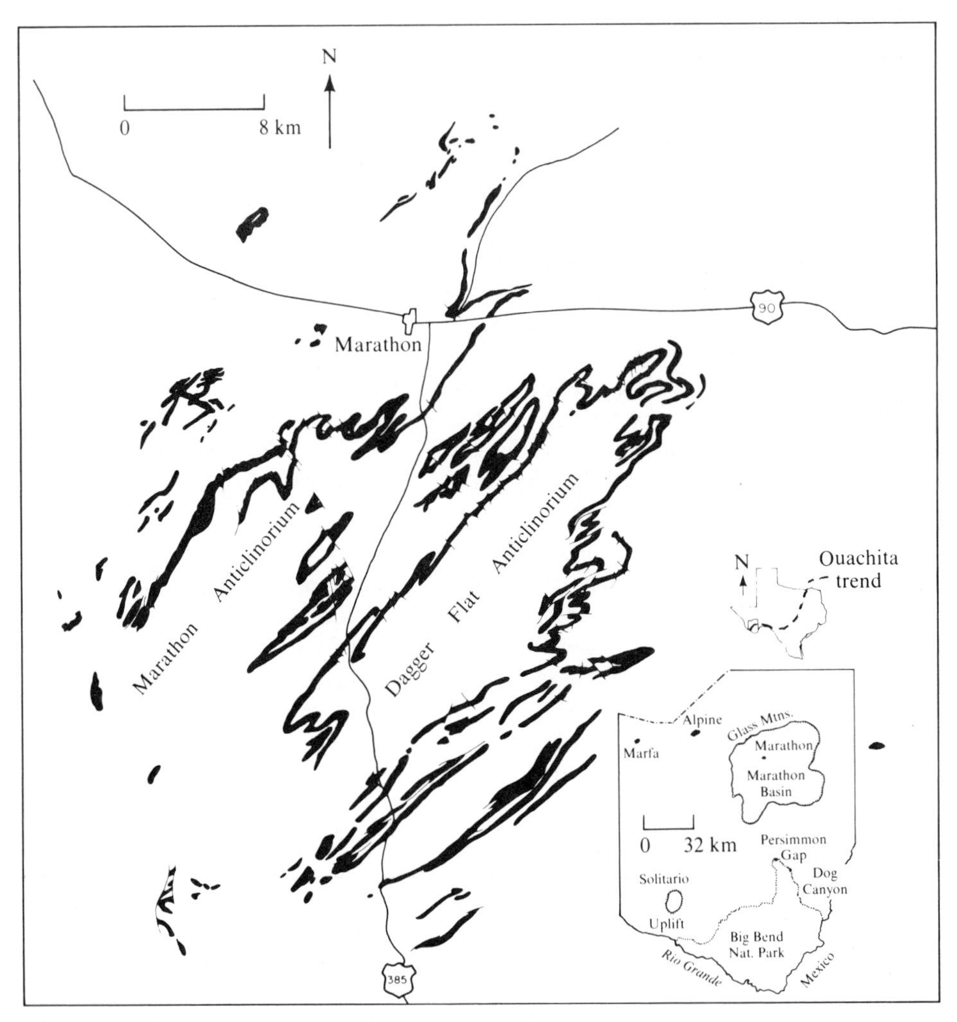

(A)

Figure 16–13
(**A**) Locality and outcrop map of the Caballos Formation. [From E. F. McBride and R. L. Folk, 1977, *Jour. Sed. Petrology, 47,* Fig. 1.]

today? The clear answer is in the ocean basin. The presence of siliceous fossils in thin sections of ancient bedded cherts suggests that the source of at least some of the silica was skeletons of organisms such as radiolaria, siliceous sponges, and diatoms. All these groups occur in marine waters, and diatoms occur in fresh waters as well.

The type and distribution of sediment currently accumulating on the deep ocean floor was not well known until relatively recently. Numerous "grab samples" from the ocean floor had been available since the globe-circling voyage of the ship *Challenger*

(**B**) Aerial view of the northeastern part of the Marathon Basin showing Caballos Formation cherts forming resistant ridges. [Cover photo of E. F. McBride and A. Thomson, 1969, *Stratigraphy, Sedimentary Structures and Origin of Flysch and Pre-Flysch Rocks, Marathon Basin, Texas* (Dallas Geol. Soc.).]

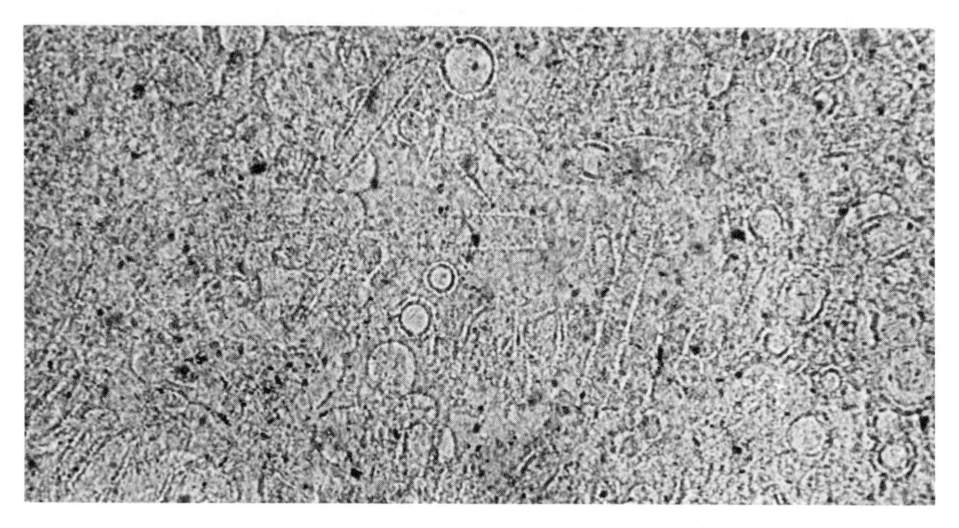

Figure 16–14
Photomicrograph of novaculite in Caballos Formation in uncrossed nicols showing ghosts of sponge spicules about 0.03 mm in diameter. [From E. F. McBride and A. Thomson, 1969, *Stratigraphy, Sedimentary Structures and Origin of Flysch and Pre-Flysch Rocks, Marathon Basin, Texas* (Dallas Geol. Soc.), Fig. 4.]

in the 1870s, but the number of samples was small and all were necessarily from the surface layer; nothing was known of the sediment more than a meter or so below the surface. Then in 1964, four of this country's major oceanographic institutions formed a working group named Joint Oceanographic Institutions for Deep Earth Sampling (JOIDES), whose objective was to construct a ship capable of drilling holes in the deep ocean floor, core the sediment and rocks encountered, and bring the cores up to the surface. The idea was sold to the U.S. government for financing and a ship, the *Glomar Challenger,* was built; it was launched in 1968. To date, the vessel has drilled in water up to 6,240 m in depth. The maximum depth penetrated below the sea floor so far is 1,741 m, and the location of the coring site can be determined using satellite fixing to within ± 1.6 km. So far, about 550,000 km of ship track have been logged in the world's oceans and seas, and about 60 thick volumes of results have been published as *Initial Reports of the Deep Sea Drilling Project* (DSDP reports). The *Glomar Challenger* continues to operate and accumulate geological, geophysical, and oceanographic data about the 70% of the earth's surface that is covered by marine waters.

The cores recovered by JOIDES commonly contain carbonate oozes, clays, and thick sequences of nearly pure siliceous sediment (which covers 14% of the ocean floor) and rock. At and near the sea floor, the sediment consists of accumulations of skeletons of diatoms and radiolaria. At depth, all stages of transition are observed from unlithified aggregates of amorphous opaline shells through various stages of recrystallization of these materials to the final product, microcrystalline quartz or chert. Beds of chert underlie many areas of the ocean floor. Apparently, the shells can be dissolved and the silica reprecipitated within a few million years after burial. Below the chert layers are often found mafic and ultramafic igneous rocks, the essential ingredients of the ophiolite suite.

Phanerozoic Nonmarine Cherts

Field geologists noted many years ago that some bedded cherts contain slump structures, intraformational breccias, and worm borings. Such observations were interpreted to indicate the presence of a silica gel during the formation of the chert bed. It was never clear, however, how such a gel could form. Gels form only by coagulation in supersaturated solutions, and all normal waters are undersaturated with respect to amorphous silica by an order of magnitude or more. Since 1965, geologic studies have found several localities where inorganic, nonmarine formation of bedded cherts is occurring. In these localities (South Australia, Oregon, east-central Africa) the pH of lake waters is raised to values exceeding 10, either by the photosynthetic activity of algae or by dissolution of nearby sodium carbonate lavas (carbonatites). At this high pH, silicate minerals are dissolved and unusually large amounts of silica can be taken into solution in the lake waters. During the rainy season the lake waters are diluted by fresh waters, the pH is lowered, and silica is precipitated as a gelatinous material on the lake floor. In the area where the lake waters are rich in sodium from the weathered lavas, the gel is not pure silica but is a sodium silicate material that, within a few hundred years, is leached of its sodium to leave a chert residue. Inorganic cherts have been identified in the Morrison Formation (Jurassic) and the Green River Formation (Eocene) and may be more common than presently believed. Remember that the only nonmarine siliceous organisms known are the diatoms and that they did not evolve until Jurassic times. Therefore, all Paleozoic nonmarine cherts may have formed by the mechanism of raising, then lowering the pH of the waters.

Precambrian Cherts

At the start of our discussion of bedded cherts we noted that such rocks are most abundant in Precambrian terranes and may have formed 15% of the middle Precambrian sedimentary column. In subsequent considerations we saw that nearly all Phanerozoic bedded cherts are marine and seem to have been formed from the skeletons of siliceous marine organisms. These two observations pose problems that are not yet resolved with regard to the origin of Precambrian bedded cherts. The diatoms did not evolve until the Jurassic; radiolarians until the Ordovician; and siliceous sponges until the earliest Cambrian. If there were indeed no siliceous organisms on earth prior to 600 million years ago, there must have been a different mechanism operating to form marine cherts.

The currently most popular suggestion is inorganic precipitation. If we assume there were no silica-shelled organisms (marine or nonmarine), what would happen to the dissolved silica weathered from rocks and carried in streams to lakes and oceans? At present it is nearly all removed by diatoms and radiolaria, which keep waters much below silica saturation. If these creatures were suddenly to vanish, sea water would become saturated with silica in a geologic instant and opal, chert, or quartz might form. Geochemical calculations, however, indicate that opal saturation would not be reached because various siliceous minerals would crystallize first and keep the level of dissolved silica below the 120 ppm needed to form opal. Further, petrographic studies of Precambrian sedimentary rocks do not reveal significant accumulations of quartz crystals. Hence, precipitation of microcrystalline quartz—chert—seems most likely, simply because it cannot be ruled out by the data as easily as can the other two possibilities.

Some students of bedded cherts suggest an alternative origin for these enigmatic Precambrian sedimentary rocks. They argue that there were indeed siliceous organisms during most of Precambrian time but that they have been largely destroyed by later diagenetic processes. Some cherts more than 2 billion years old do contain round, organic-looking siliceous structures that resemble radiolaria (see Figure 16–15). Recent laboratory experiments, however, have shown that normal coagulation into colloids of dissolved silica can produce round structures very similar in appearance to the "fossils" seen in Precambrian cherts. The controversy is as yet unresolved.

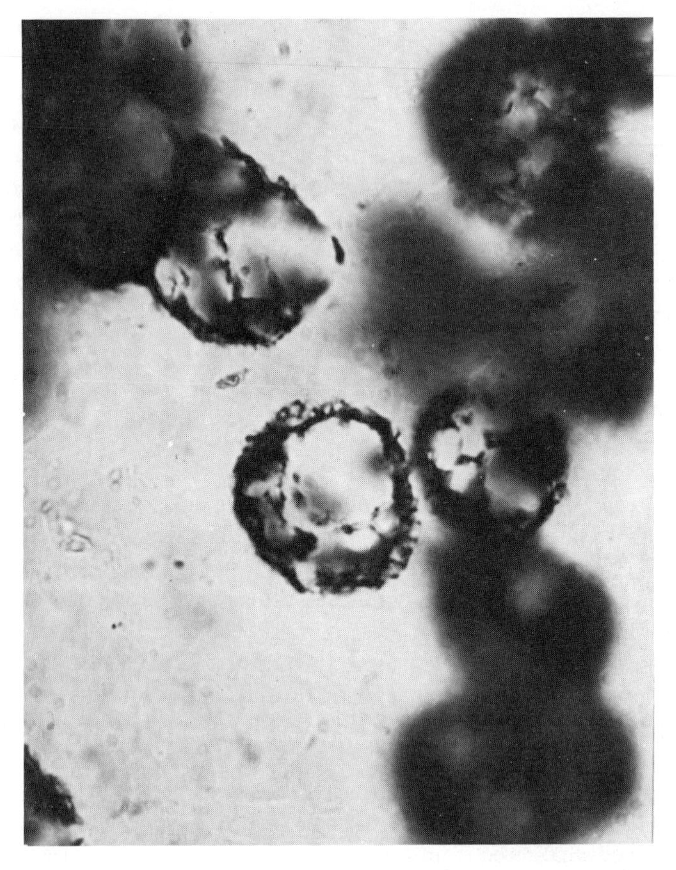

Figure 16–15
Possible recrystallized radiolarian (?) shells, Brockman Iron Formation, Hammersley Range, Western Australia. The structures are about 25 μm in diameter. [From H. Blatt et al., 1980, *Origin of Sedimentary Rocks,* 2d Ed. (Englewood Cliffs, N.J.: Prentice-Hall), Fig. 16–3 (photo courtesy G. L. LaBerge).]

BEDDED PHOSPHATE ROCKS

Bedded sedimentary phosphate rocks are defined as units that contain at least 20% P_2O_5 by laboratory chemical analysis, an enormous enrichment over the amounts in normal rocks. For example, limestones average only 0.04% P_2O_5, and mudrocks 0.17%. The phosphorous is present in ancient phosphate rocks (phosphorites) in the mineral fluorapatite, $Ca_5(PO_4)_3F$, typically with trace amounts of CO_3^{2-} substituting for PO_4^{3-} in the crystal structure. Most commonly the apatite is cryptocrystalline; phosphorites are not concentrations of detrital apatite grains.

Phosphorites occur in outcrop as black beds between 1 mm and a few meters thick, commonly interbedded with carbonates, dark-colored chert, and carbonaceous mudstone. Sandstone units are rare. The total thickness of dominantly phosphatic sequences can be several hundred meters, although normally they are much thinner. Most studies of sedimentary phosphate rocks have been made for economic purposes (fertilizer, byproduct uranium, vanadium, and rare-earth elements), so the abundance of thin phosphorite sequences or individual beds in otherwise nonphosphatic sections is not known.

The great bulk of the phosphatic material in phosphorites is in the form of pellets and nodules, with interstitial micritic or sparry phosphatic cement (see Figure 16–16). In a few thin layers the phosphatic sediment can be bioclastic, phosphatic

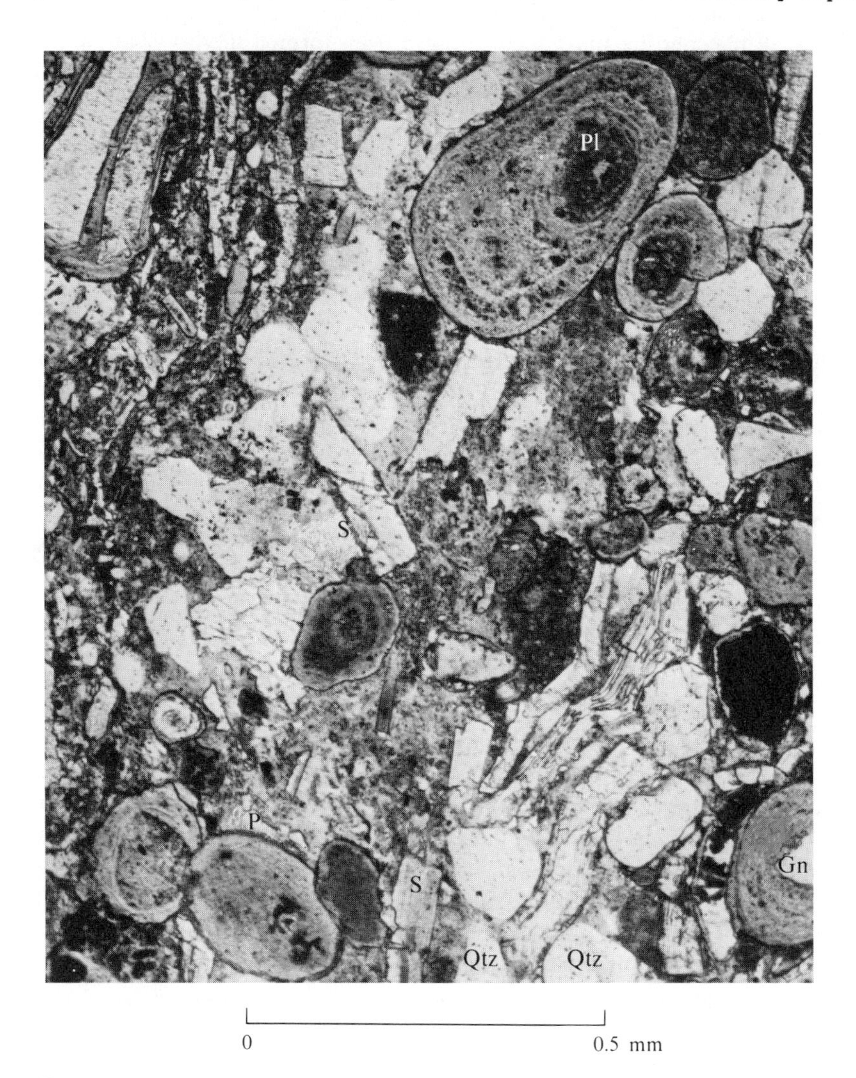

Figure 16–16
Photomicrograph (crossed nicols) of phosphorite in the Phosphoria Formation containing quartz sand grains (Qtz), skeletal fragments of collophane (S), phosphatic peloids (P), and oolitic coatings of phosphate around quartz grains (Gn) and peloids (Pl). [From E. R. Cressman and R. W. Swanson, 1964, U.S. Geol. Surv. Prof. Paper 313-C, Fig. 112A.]

brachiopods (*Lingula*), and fish scales. Sequences of phosphorite units typically are quite complex texturally, containing pyritic and highly carbonaceous mudstone, finely pelletal phosphatic mudstone, dense and structureless phosphorite, fine-grained pelletal phosphorite, oolitic phosphorite, pisolitic phosphorite, nodular phosphorite, bioclastic phosphorite, and finely crystalline argillaceous and phosphatic (pelletal) dolomite or limestone. The stratigraphic relationships, structures, and textures of phosphorite sequences and individual beds strongly resemble those of limestones, with the exception that there are no phosphorite reefs. There are no phosphatic reef-building organisms.

Nearly all phosphorite units seem to have been formed in very shallow water, as evidenced by the presence of shelf-dwelling calcareous organisms (many replaced by phosphate), remnants of phosphatized reef-building algae, ooliths, cross bedding, abrasion features on phosphoclasts, and interfingering with quartz sandstones having similar shallow water characteristics. Oolitic and bioclastic phosphorite facies dominate in the shoreward direction; carbonaceous and pyritic mudstone facies (phosphomicrite) in the basinward direction.

Except for the difference in mineral composition, apatite rather than calcite, thin sections of phosphorites look like thin sections of limestones. The terminology used to describe carbonates in the laboratory can also be used to describe phosphorites. We find phosphomicrites, phosphosparites, and varieties of both that contain variable percentages of fossils, peloids, ooliths, and phosphoclasts. Among the fossils, only the linguloid brachiopods were originally phosphatic. All other phosphatic shells in phosphorites were originally calcium carbonate or opal and were replaced by apatite during diagenesis. Phosphopeloids are structureless and somewhat elliptical in shape and range up to at least 3 cm in diameter. They generally are well sorted. Some peloids have oolitic coatings, and some peloids are aggregated into phosphograpestone lumps.

Origin of Phosphorites

Phosphorite beds occur in rocks of all ages, from Precambrian to Holocene, and on all continents. Therefore, the conditions in which they form have not been uncommon through geologic time. Nevertheless, phosphorites are peculiar rocks, consisting as they do of concentrations of a minor element (phosphorous) to more than 100 times the amount in other types of sedimentary rocks.

Where are phosphorites forming today? When we asked that question about limestones, which are similar in many respects to phosphorites, we were able to look around and see limestone reefs and calcium carbonate shells to examine. Using the principle that the present is the key to the past, we studied these materials and applied the information gained to the interpretation of the ancient limestones. In the case of phosphorites, however, this approach is inadequate. No phosphatic organisms build reefs. Only a single brachiopod group among the invertebrates in the sea secretes a phosphatic skeleton (*Lingula*), and it is not abundant. So we must use a different approach.

It is clear that phosphorites are chemical or biochemical rocks rather than terrigenous detrital rocks. We also know from the fossils they contain that nearly all phosphorites are marine. Therefore, it may be informative to determine the character of the waters in which phosphorous can be concentrated. The first person to consider this question was a Russian, Kazakov, in 1937. From oceanographic data he determined that the P_2O_5 content of marine waters is at a maximum at depths between 30 and 500 m. At depths less than 30 m the phosphorous is consumed by phytoplankton during photosynthesis. At depths greater than 500 m the content of carbon dioxide in the water is so great that the waters cannot become saturated with respect to apatite; apatite, like calcite, is soluble in acid. As we have seen, ancient phosphorites formed in fairly shallow waters (fossils, cross bedding) so that depths greater than 100 m or so are not of much interest to us in any event; at least not as far as the major phosphorite deposits are concerned.

Where in the modern ocean is phosphorite forming today? Kazakov pointed out that modern deposits are located in low latitudes on the eastern sides of marine basins where deep phosphate-rich waters are upwelling adjacent to a shallow shelf. In most locations the upwelling is caused by the trade winds blowing surface waters offshore, so that they are replaced at the surface by the deeper waters. Has this preferred location been characteristic of ancient phosphorites as well? Once again, paleomagnetic data are required so that the present locations of ancient phosphorites can be changed to the locations they occupied when they formed (see Figure 16–17). It is clear that ancient phosphorites formed in the trade wind belt in low latitudes, apparently in latitudes even lower than the Holocene average of 25°. The reason for this difference is not clear. It is clear, however, that upwelling of phosphate-rich cold bottom waters is critical to phosphorite formation. The largest and best known phosphorite deposit, the Phosphoria Formation of Permian age in the western United States, was located along the western margin of the North American continent at the time it was formed and within 5° of the Permian equator. We will consider the Phosphoria Formation in more detail a bit later in this chapter.

Metasomatism of Calcium Carbonate

As we noted earlier, the stratigraphic relationships, structures, and textures of many phosphorite deposits resemble those of limestones. In addition, some phosphorites contain originally calcareous fossils that have been phosphatized or ooliths that have been partially converted to apatite after burial. We are reminded of dolomite replacing limestone. In these cases we are certain that the apatite was not a primary precipitate but is diagenetic in origin, as is the dolomite. Unfortunately, most phosphorites show no evidence of a replacement origin.

Nevertheless, as geologists we are interested in defining the conditions in which phosphatization of calcium carbonate can occur, as we are in the case of dolomitiza-

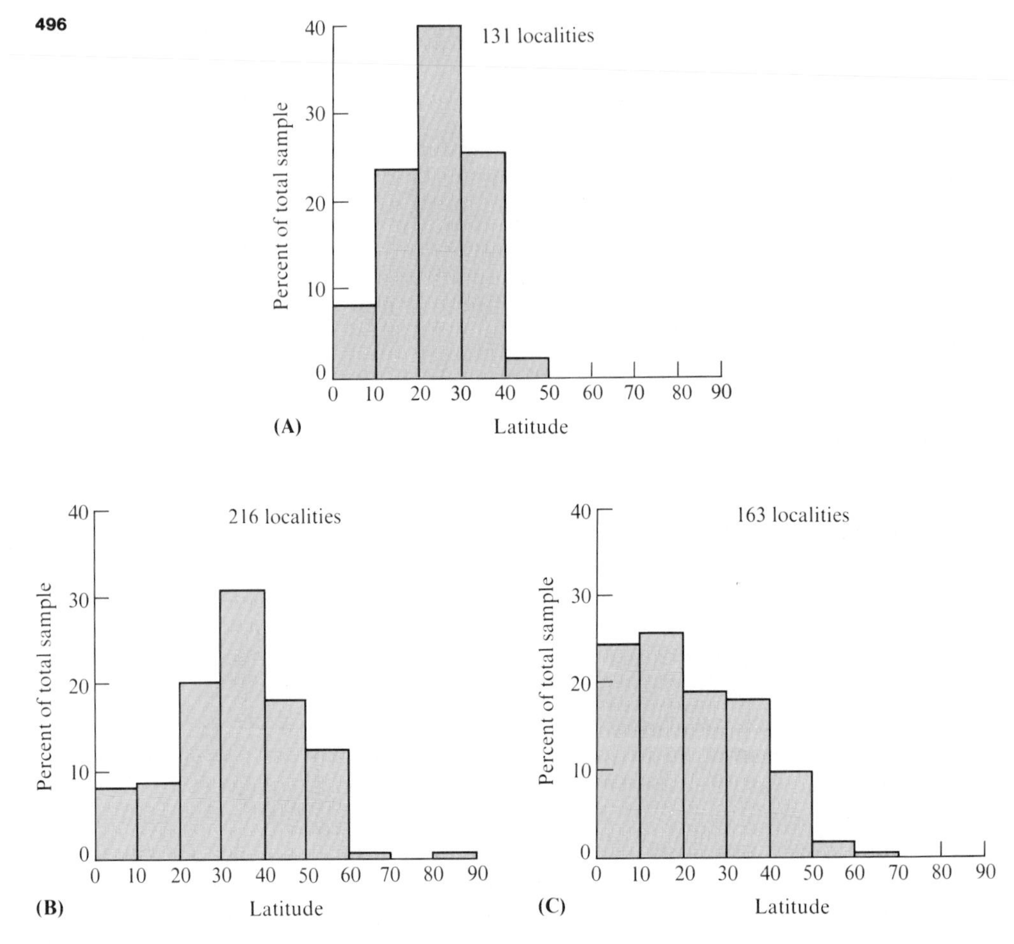

Figure 16–17
(**A**) Latitudinal distribution of Quaternary and late Tertiary phosphorite deposits.
(**B**) Present latitudes and (**C**) original latitudes of older phosphorites. [From P. J. Cook and M. W. McElhinny, 1979, *Econ. Geol., 74,* Fig. 2.]

tion. To study the phosphatization process, Ames (1959) conducted some laboratory experiments in which he passed a basic solution of sodium phosphate and sodium hydroxide through a tube packed with fragments of calcite. From time to time he sampled the composition of the solution emitted from the bottom of the tube and studied samples of the "sediment" in the tube to determine its composition. He found that the original calcite was replaced and perfectly pseudomorphed by apatite and that the apatite contained a few percent CO_3^{2-} substituting for PO_4^{3-} in its crystal structure, just as occurs in most ancient phosphorites. Ames' apatite was a hydroxyl apatite rather than a fluorapatite because he had no fluorine in his phosphatizing solution:

$$3Na_3PO_4 + NaOH + \underset{\text{Calcite}}{5CaCO_3} \rightarrow \underset{\text{Apatite}}{Ca_5(PO_4)_3OH} + 5Na_2CO_3$$

He found that phosphatization of calcite would occur in PO_4^{3-} concentrations as low as 0.1 ppm. In areas of oceanic upwelling the PO_4^{3-} content is usually 0.3–0.8 ppm, more than adequate to cause replacement of calcite by apatite. In reducing environments immediately below the sea water–sediment interface it is common to find 1.0 ppm PO_4^{3-} in solution because of the decay process of organic tissue in such an environment. It certainly seems possible that a significant amount of phosphorite could be produced by very early diagenetic replacement just below the sea floor and, just as is true in many instances of early replacement of calcite by dolomite, the replacement process cannot be detected in the resulting rock.

The Phosphoria Formation

The Phosphoria Formation (Permian) in the northwestern United States is the best studied phosphorite-bearing unit in the world. For this reason it is often used as a "type section" of phosphorite occurrence, and based on what is known of phosphorite units in other parts of the world, it is well suited for this purpose.

The Phosphoria phosphorites were deposited over an area of about 350,000 km^2 in both the platform and geosynclinal portions of the Paleozoic Cordilleran structural belt (see Figure 16–18). The regional stratigraphy of the Phosphoria reflects its structural setting not only in the great increase in thickness toward the geosyncline but also by lithologic changes. Spiculiferous chert and cherty carbonaceous mudstone dominate the stratigraphic section in the deeper parts of the depositional basin and are succeeded successively eastward by dolomitic limestones and calcareous sandstones and ultimately by continental red shales and sandstones. Highly phosphatic beds occur in both the geosynclinal and platform facies and appear to reach a maximum thickness near the "hinge line" separating the two facies.

The thickest and most extensive phosphorite unit in the Phosphoria is the Meade Peak Member, composed of phosphatic shale, mudstone, and carbonate rock. The phosphorites are mostly thin-bedded, dark-colored, well-sorted, peloidal rocks consisting of about 50% apatite (see Figure 16–19). The peloids include pellets, ooliths, nodules, pisoliths, and compound nodules. Pellets and ooliths are, by definition, smaller than 2 mm in diameter; the other peloids are larger. Linguloid brachiopods, phosphatic fish scales, and phosphatized gastropods, cephalopods, bryozoans, echinoid spines, and sponge spicules also occur in the allochemical fraction. In the nearshore area, well-aerated waters are suggested by the noncarbonaceous, cross-bedded, well-sorted spheroidal phosphorites that contain abraded fossils and lineated elliptical pellets. Contrasting with this lithology are the carbonaceous, pyritic, phosphatic mudstones, whose few fossils show no evidence of current transport. The change in lithologic character is strikingly similar to that encountered in limestones as the rocks are

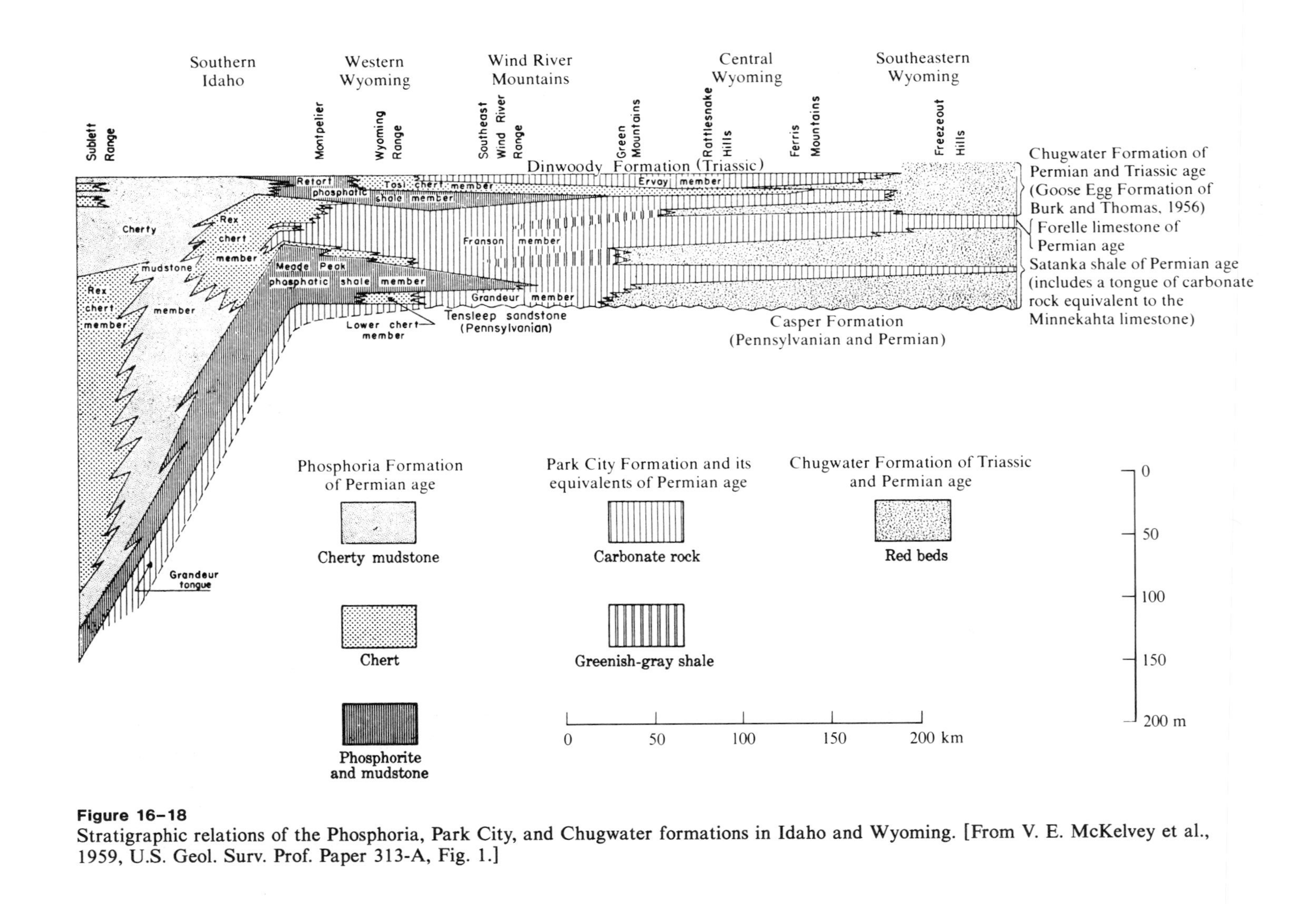

Figure 16–18
Stratigraphic relations of the Phosphoria, Park City, and Chugwater formations in Idaho and Wyoming. [From V. E. McKelvey et al., 1959, U.S. Geol. Surv. Prof. Paper 313-A, Fig. 1.]

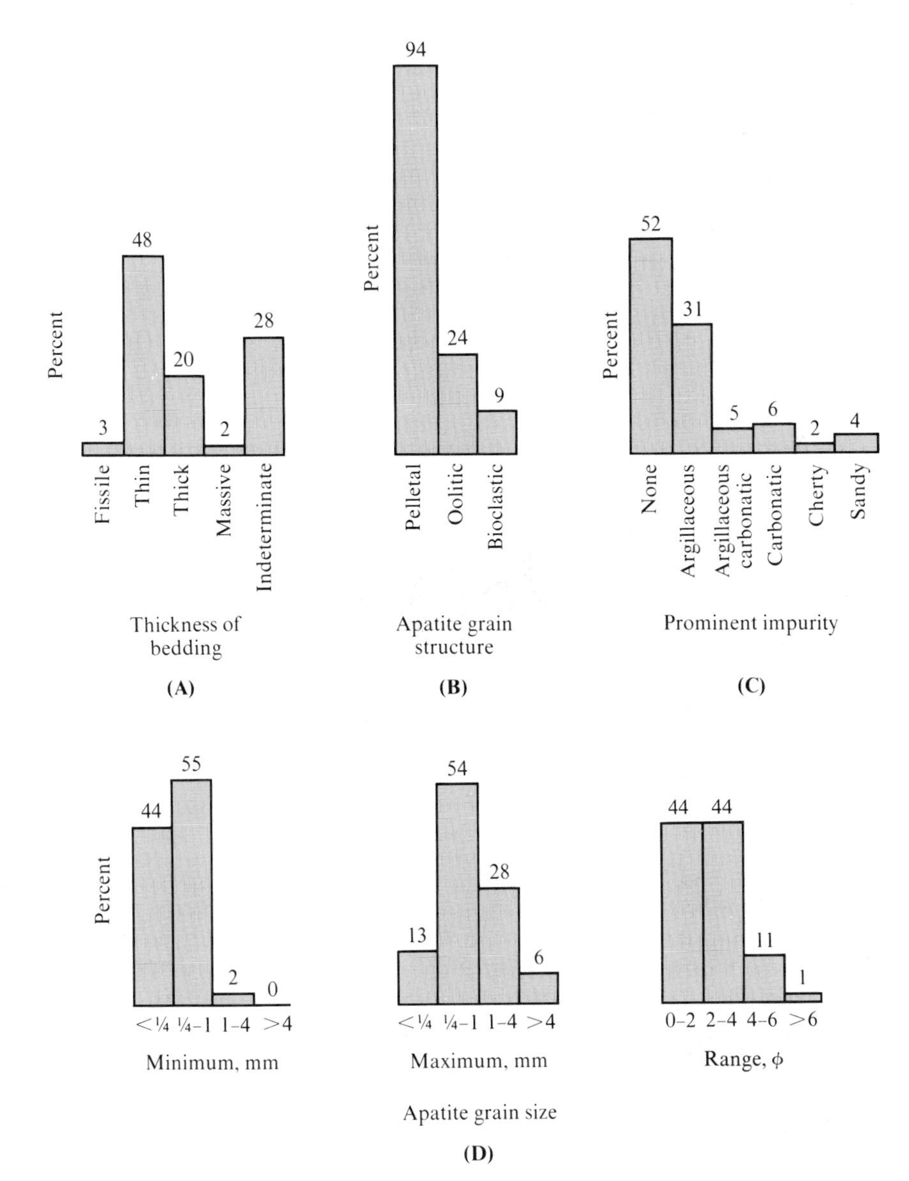

Figure 16–19
Lithic characteristics of phosphorite in the Meade Peak phosphatic shale Member. [From R. P. Sheldon, 1963, U.S. Geol. Surv. Prof. Paper 313-B, Fig. 21.]

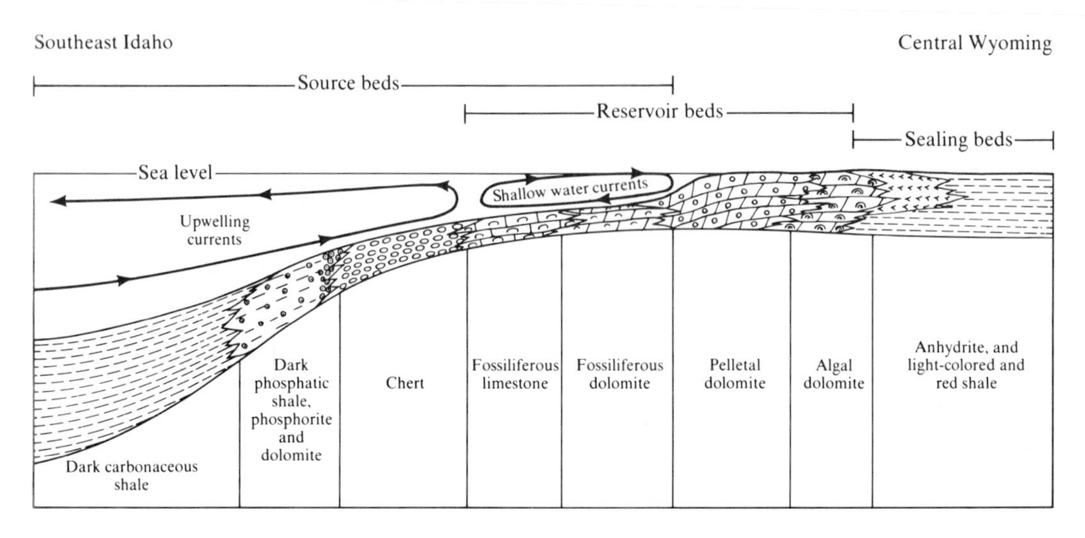

Figure 16–20
Idealized model of sedimentation in the Phosphoria sea. The Phosphoria is believed to be the source rock for major petroleum accumulations; the zones of source rock, reservoir rock, and impermeable seal are also shown. [From R. P. Sheldon, 1963, U.S. Geol. Surv. Prof. Paper 313-B, Fig. 86.]

followed from the shoreline into a micritic area offshore. An idealized model of sedimentation in the Permian Phosphoria sea is shown in Figure 16–20.

BEDDED IRON DEPOSITS

As with bedded phosphate rocks, bedded iron deposits are defined by a chemical composition rather than a mineral composition, texture, or structure. By accepted definition, the iron-rich deposits contain at least 15% Fe. This would be either 21.4% Fe_2O_3, 19.3% FeO, or some combination of the two totaling 15% Fe. These amounts are much greater than the average for mudrocks (4.8% Fe), sandstones (2.4% Fe), or limestones (0.4% Fe). All of these three rock groups can, however, grade laterally into an iron deposit. The iron deposits are termed either ironstone or iron formation; in the Lake Superior region the unweathered cherty iron formation that is mined for the metal is known as taconite.

Oolitic Iron Formations

Phanerozoic iron formations are largely oolitic and of definite shallow marine origin, as revealed by facies relationships, fossils, textures, and structures. Their areal distribution is much more restricted than that of the Precambrian banded iron formations, with individual depositional basins not exceeding 150 km in length and the thickness of the units only a few meters to a few tens of meters. The major iron-bearing mineral in the Phanerozoic deposits is goethite, but hematite and chamosite are common. Calcite and dolomite are the abundant carbonate minerals.

The most extensive and best known Phanerozoic ironstone in the United States is the Clinton Formation (Silurian), which crops out sporadically over a distance of 1,500 km along the Appalachian Mountain belt from New York State southward into Alabama. The Clinton is the local source of iron for the manufacture of steel in the southeastern United States. It is composed of oolitic hematite up to 8 m thick, iron-stained sandstone, and shale. Locally the beds are calcareous and grade into impure limestones. Fossils are abundant and have been replaced by hematite. The ooliths may be cemented by hematite or calcite, and the ooliths themselves are formed around nucleii of rounded quartz grains or fossil fragments (see Figure 16–21). In general aspect the Clinton seems identical to sediments formed in modern and ancient beach–nearshore neritic settings except that the ooliths are composed of

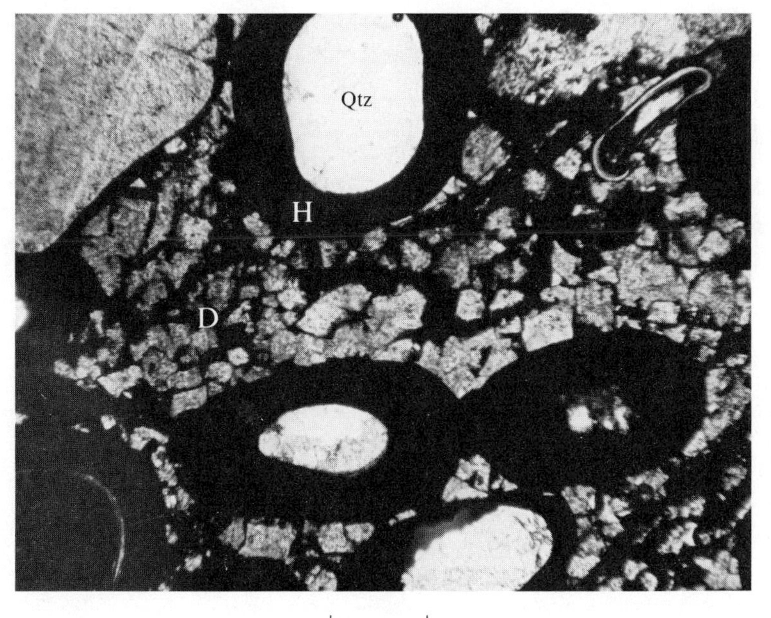

Figure 16–21
Photomicrograph (crossed nicols) of Clinton Formation (Silurian), Clinton, New York, showing detrital quartz nucleus (Qtz) surrounded by a coating of hematite (H) to form an oolith. Dolomite rhombs (D) are formed by replacement of original calcium carbonate. [From R. Schoen, 1964, *Jour. Sed. Petrology, 34,* Fig. 1.]

hematite rather than calcite or aragonite. For this reason, it is generally believed that the ooliths were formed by replacement of calcium carbonate, probably soon after deposition, which would have permitted iron-rich solutions to permeate completely the unconsolidated sediment.

Banded Iron Formations

The most abundant type of ironstone is the banded cherty iron formations of Precambrian age that are the world's major source of iron (see Figure 16–22). Typically,

Figure 16–22
Outcrop in upper part of the Negaunee Iron Formation, Marquette District, Michigan. Red (dark) bands are hematitic chert (jasper); light bands (gray) are specular hematite. [From H. Blatt et al., 1980, *Origin of Sedimentary Rocks,* 2d Ed. (Englewood Cliffs, N.J.: Prentice-Hall), Fig. 18–1 (photo courtesy H. L. James).]

these units occur in belts hundreds to thousands of kilometers long and up to several hundred kilometers wide, with thicknesses at least as great as 600 m. They contain as much chert as iron-bearing minerals. The most common and abundant iron mineral in them is hematite, but magnetite, greenalite ($FeSiO_3 \cdot nH_2O$), siderite, and locally other primary or diagenetic minerals occur. Detrital quartz, feldspar, rock fragments, or clay are extremely rare. Many of the Precambrian iron formations have undergone low-grade metamorphism so that a number of otherwise rare iron silicate minerals may be present in addition to the more usual sedimentary ones.

Most of the sedimentary structures in banded iron formations are not easily recognized in outcrop, except in rare cases where they have been accentuated by differential weathering. The interlayered cherts and microcrystalline iron-bearing minerals, such as hematite, characteristically show tabular laminations on the order of a millimeter in thickness. The laminations are caused by a variation in the content of the iron minerals or by alternation of carbonate and silicate-rich laminae. The appearance and thickness of laminations can change over brief distances laterally because of solution and reprecipitation of silica and recrystallization processes. Some of the laminae have been traced laterally over hundreds of km and may represent annual layers (varves).

The list of sedimentary structures reported to occur in banded iron formations includes many of the ones prominent in normal sandstones and limestones, such as cross bedding, graded bedding, ooliths, intraclasts, erosion channels, stromatolites, and pisoliths. For this reason, some large areas of banded iron formation have been interpreted as replaced limestones, although such an interpretation is questioned by many geologists.

Distribution and Age of Banded Iron Formations

The Precambrian banded iron formations occur in the shield areas on all continents (see Figure 16–23), and it is obvious that many of them lie close to borders of the continents or on the margins of the cratonic masses now surrounded by younger fold belts and platform sediments. It seems likely that some of the iron formations now on the borders of the different continents were originally part of a single depositional basin before being separated by continental drift. Where the general geologic setting has been studied in more detail, a comparison of iron formations on opposite continents fits fairly well with the assumption of originally coherent iron formation belts. The type and depositional environment of these iron formations with their associated rocks are similar, if not identical, and so are their ages and metamorphic histories.

Iron formations include some of the oldest rocks still preserved on the earth, about 3.8 billion years old. Those older than 2,600 million years are mostly small, lenticular bodies of various thicknesses, mostly less than a few km in length, and genetically related to volcanic rocks and clastics that appear to be deep water graywackes. Typical beds are composed of red or gray chert and interbanded magnetite- and hematite-rich layers, but other types of iron-bearing minerals also occur. Carbonaceous or

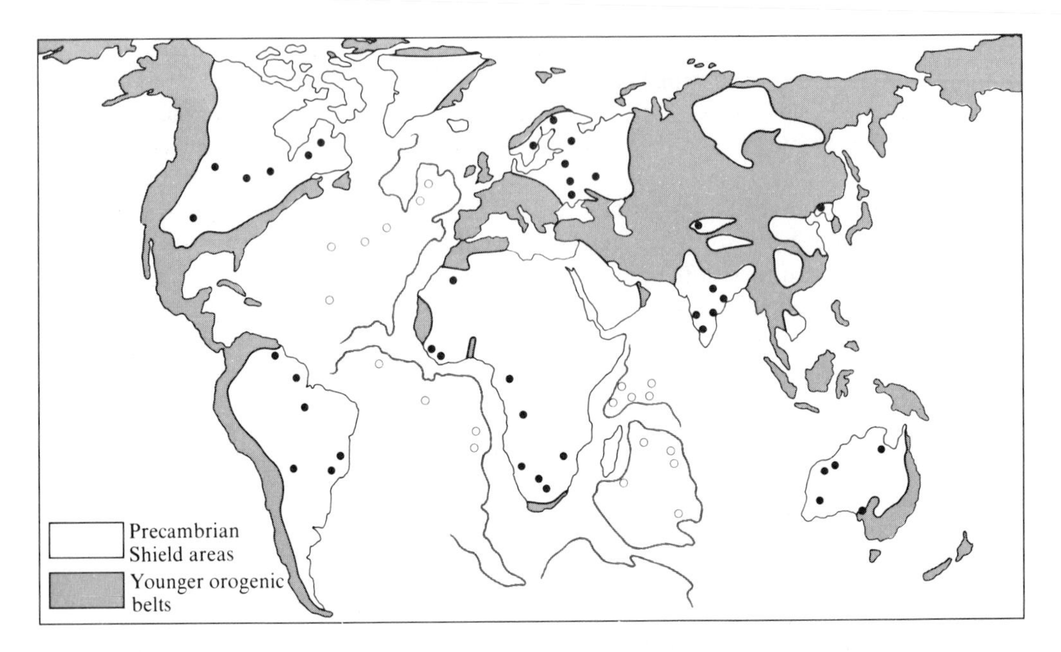

Figure 16–23
Distribution of prominent Precambrian banded iron formations (dots) and reconstruction of continental locations prior to continental drift. [From J. Eichler, 1976, *Handbook of Strata-Bound and Stratiform Ore Deposits,* vol. 7 (New York: Elsevier), Fig. 1.]

graphitic units are common. In general, this stratigraphic suite is what might be expected to form in an island arc setting, with volcanic activity supplying the iron.

Banded iron formations between 2,600 and 1,800 million years old form the bulk of the Precambrian iron-rich rocks. In contrast to the older ones, they are associated with clastic sedimentary rocks of shelf or rifted continental-margin origin: quartzites, dolomites, and mudrocks. There is no indication of volcanic association.

Banded ironstones become noticeably less common in rocks younger than 1,800 million years; the youngest is 800 million years old, the same age as the oldest metazoan fossils, the Ediacara fauna of Australia. Ironstones younger than 800 million years, the Phanerozoic ones, are very different in character from their older relatives, as we have already seen.

Origin of Banded Iron Formations

Banded cherty iron formations are very peculiar rocks for which there is no modern counterpart. None of these rocks has been formed during the past 800 million years. The possible reasons for the occurrence or absence of cherty ironstones can, however, be narrowed considerably by field observations, as pointed out by Dimroth (1976).

1. Shrinkage textures and the speckled or clotted microscopic textures typical of these rocks suggest that most have been formed from siderite and mixed gels of silica and iron oxide or iron silicate. These substances have been precipitated from sea water, lake water, or diagenetic pore solutions at a very early stage in the history of the deposit.

2. Measurements of the thicknesses of laminations, assumed to be varves, suggest that sedimentation rates of the banded cherty iron formations are similar to those of many modern nearshore marine deposits of limestones and other sediments.

3. This type of ironstone is commonly restricted to one or two formations in each Precambrian depositional basin, indicating that the supply of iron was episodic rather than continual. Therefore, banded cherty iron formations cannot be regarded as normal precipitates from Precambrian seas.

4. The rock texture is composed of interlocking crystals and contains fine laminations that are continuous laterally over distances of tens or hundreds of kilometers. Current-generated structures such as cross bedding and graded bedding are present but rare. These facts suggest that the rock formed by precipitation from a solution saturated with respect to silica and several iron compounds. However, if the speckled and clotted textures result from recrystallizated shells of Precambrian radiolaria, as some geologists believe, the interlocking texture might result from early diagenetic solution and reprecipitation of these organic materials, as occurs in modern seas. Or, if the banded ironstones were originally carbonate stromatolites, the interlocking texture might be the product of the replacement process.

5. The dominance of hematite among the iron-bearing minerals means that some free oxygen must have been present to oxidize the ferrous iron released from igneous and metamorphic source rocks. This does not necessarily mean, however, that the atmosphere contained oxygen. It is possible that ferrous iron was sufficiently abundant in solution to act as a "sink" for plant-generated oxygen for several billion years before the amount of oxygen released by photosynthesis was great enough to overwhelm the capacity of the sink. Until this point was reached, no oxygen could accumulate in the atmosphere. Iron is the third most abundant cation in the crust, and it requires a great deal of oxygen to convert the ferrous iron ions released by weathering into ferric ions.

In summary, the Precambrian banded iron formations continue to be enigmatic rocks illustrating that the present is a far from perfect key to the past.

SUMMARY

Evaporites underlie a large part of the United States, and the bulk of the deposits is composed of gypsum, anhydrite, and halite. Evaporites are formed in two types of settings: (1) large standing bodies of water in which the rate of evaporation exceeds precipitation, and (2) at the margins of a marine basin on an evaporative tidal flat, accompanied by interbedded carbonates. The first type forms relatively pure deposits of soluble salts and is younger in age than the rocks at the basin margins; the second type contains much carbonate sediment and interfingers at its margins with contemporaneous sediment.

Bedded chert deposits can be several hundred meters thick and are formed mostly in deep marine waters far from the continents. Bedded cherts are typically very pure, the only impurities being small amounts of clay, hematite, and calcium carbonate. The source of the silica in Phanerozoic chert deposits is the opaline skeletons of diatoms, radiolaria, and siliceous sponges, which crystallize to microcrystalline quartz during early diagenesis. The source of the silica in Precambrian cherts is uncertain, with some scientists relying on the possible existence of Precambrian opaline organisms and others assuming inorganic precipitation from oceans oversaturated with respect to quartz.

Bedded phosphate rocks are composed of apatite, are thought to form in the same types of sedimentary environments as carbonate rocks, and have nearly the same range of textures and structures as carbonates. The main difference between the two is the need for upwelling phosphate-rich marine waters so that oversaturation of the water with respect to apatite is achieved. Some investigators believe that many phosphorite deposits are replaced limestones, although there is little textural evidence supporting this point of view in most deposits.

Bedded iron ores are enigmatic rocks of two different types. The more abundant type is the Precambrian bedded iron ores or cherty ironstones. These rocks typically are composed mostly of chert but contain at least 15% iron in the form of hematite, magnetite, siderite, or other minerals. These minerals occur concentrated in varve-like layers within the chert. The iron-bearing minerals were precipitated from marine water, lake water, or from pore solutions during very early diagenesis. No modern counterparts of these rocks are known, so their origin remains uncertain.

The second type of ironstone is the Phanerozoic oolitic iron ores. These units are much thinner than their Precambrian relatives and much less extensive. The ooliths are composed mostly of goethite and contain a nucleus of detrital quartz or a carbonate shell fragment; the units grade laterally into shallow water limestone deposits. It is believed that the iron oxide ooliths were originally composed of calcium carbonate but were replaced by goethite during early diagenesis.

FURTHER READING

Evaporites

Anderson, R. Y., W. E. Dean, D. W. Kirland, and H. I. Snider. 1972. Permian Castile varved evaporite sequence, west Texas and New Mexico. *Geol. Soc. Amer. Bull., 83,* 59–85.

Gordon, W. A. 1975. Distribution by latitude of Phanerozoic evaporite deposits. *Jour. Geology, 83,* 671–684.

Hsü, K. J. 1973. Origin of the saline giants: a critical review after the discovery of the Mediterranean evaporite. *Earth-Science Reviews, 8,* 371–396.

Scruton, P. C. 1953. Deposition of evaporites. *Amer. Assoc. Petroleum Geol. Bull., 37,* 2498–2512.

Wilson, J. L. 1967. Carbonate–evaporite cycles in lower Duperow Formation of Williston Basin. *Bull. Canadian Petroleum Geol., 15,* 230–312.

Worsley, N., and A. Fuzesy. 1979. The potash-bearing members of the Devonian Prairie Evaporite of southeastern Saskatchewan, south of the mining area. *Economic Geol., 74,* 377–388.

Bedded Cherts

Calvert, S. E. 1974. Deposition and diagenesis of silica in marine sediments. In K. J. Hsü (ed.), *Pelagic Sediments: On Land and Under the Sea,* pp. 273–299. International Assoc. of Sedimentologists Spec. Pub. No. 1.

Folk, R. L. 1973. Evidence for peritidal deposition of Devonian Caballos Novaculite, Marathon Basin, Texas. *Amer. Assoc. Petroleum Geol. Bull., 57,* 702–725.

LaBerge, G. L. 1973. Possible biological origin of Precambrian iron-formations. *Economic Geol., 68,* 1098–1109.

McBride, E. F., and A. Thomson. 1970. *The Caballos Novaculite, Marathon Region, Texas.* Geol. Soc. Amer. Spec. Paper 122, 129 pp.

Bedded Phosphate Rocks

Ames, L. L., Jr. 1959. The genesis of carbonate apatites. *Economic Geol., 54,* 829–841.

Cook, P. J. 1976. Sedimentary phosphate deposits. In K. H. Wolf (ed.), *Handbook of Strata-Bound and Stratiform Ore Deposits,* Vol. 7, pp. 505–536. New York: Elsevier Scientific Publishing.

Cook, P. J., and M. W. McElhinny. 1979. A reevaluation of the spatial and temporal distribution of sedimentary phosphate deposits in the light of plate tectonics. *Economic Geol., 74,* 315–330.

Riggs, S. R. 1979. Petrology of the Tertiary phosphate system of Florida. *Economic Geol., 74,* 195–220.

United States Geol. Survey. 1959–1973. *Geology of the Permian rocks in the western phosphate field.* U.S. Geol. Surv. Professional Paper 313 (in parts A–F by several authors).

Bedded Iron Deposits

Dimroth, E. 1976. Aspects of the sedimentary petrology of cherty iron-formation. In K. H. Wolf (ed.), *Handbook of Strata-Bound and Stratiform Ore Deposits,* Vol. 7, pp. 203–254. New York: Elsevier.

Eichler, J. 1976. Origin of the Precambrian iron-formations. In K. H. Wolf (ed.), *Handbook of Strata-Bound and Stratiform Ore Deposits,* Vol. 7, pp. 157–202. New York: Elsevier.

James, H. L. 1966. *Chemistry of the iron-rich sedimentary rocks.* U.S. Geological Survey Professional Paper 440-W, 60 pp.

Kimberley, M. M. 1979. Origin of oolitic iron formations. *Jour. Sed. Petrology, 49,* 111–132.

III

METAMORPHIC ROCKS

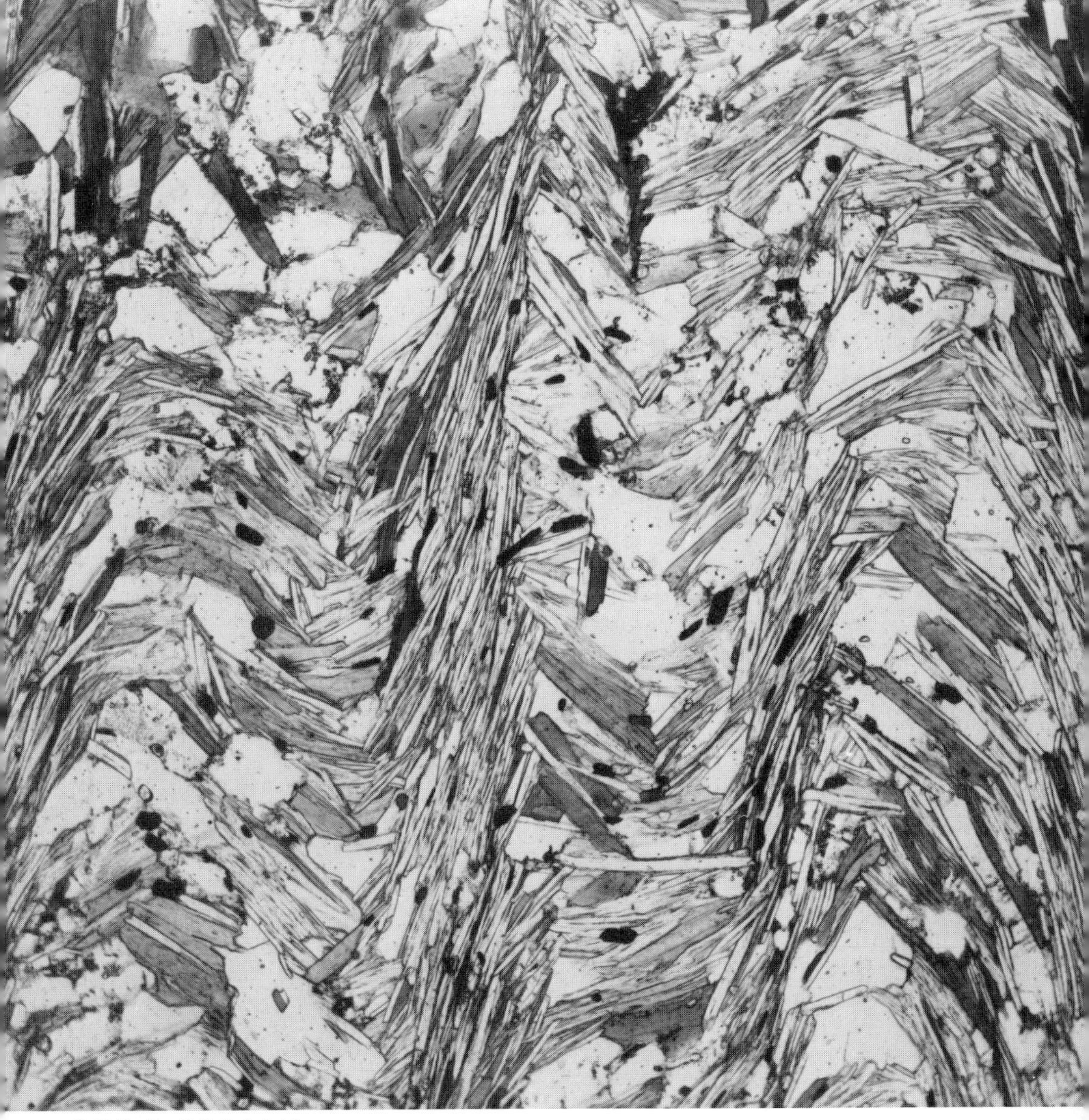

D. M. Sheridan, U. S. Geological Survey

17

The Occurrence of Metamorphic Rocks

Igneous and sedimentary rocks are formed in response to the interaction of chemical, physical, and/or biological processes and conditions both within the earth and on its surface. But the earth is a dynamic system, and once formed, rocks may be subjected to a new set of conditions that can result in extensive changes in both texture and mineralogy. If these changes occur at pressures and temperatures above those of diagenesis and below those of melting, they are referred to as metamorphism. A rock may be subjected to several environmental changes through time, which can result in a polymetamorphic rock. The principal characteristic of metamorphic changes is that they occur while the rock remains solid.

One of the favorite subjects of metamorphic petrologists is the definition of the level at which diagenesis ends and metamorphism begins. The most important cause of both diagenetic and low-grade metamorphic changes in a rock is increased temperature; increased pressure is relatively ineffective during incipient metamorphism. Compositional changes within the rock also are negligible at this stage; that is, the changes are isochemical and consist of local redistribution of elements and volatiles among the minerals that are most reactive. The common approach to drawing a boundary between diagenesis and metamorphism is to define the lower level of metamorphism as the first appearance of a mineral that does not normally form in surficial sediments, such as epidote or muscovite. Although this can be done it produces a rather sloppy limit. For example, metamorphism of shale results in the reaction of kaolinite with other constituents to produce muscovite. However, experiments have shown that this reaction does not take place at a specific temperature but occurs between 200° and 350°C depending upon the pH and potassium content of surrounding materials. Other minerals considered to form at the onset of metamorphism are laumontite, lawsonite, albite, paragonite, or pyrophyllite. Each forms at a somewhat different temperature under different conditions, but in general about 150°C or higher is required. In the subsurface, temperatures in the vicinity of 150°C are accompanied by a lithostatic pressure of about 500 bar.

The upper limit of metamorphism is taken as the point where appreciable melting of the rock occurs. Here again we have a variable, as melting temperature varies as a function of rock type, lithostatic pressure, and vapor pressure. A range of 650 to 800°C covers most situations. The upper limit of metamorphism can be defined by the presence of rocks called migmatites. These rocks display a combination of textural features, some of which appear to be igneous, and others metamorphic.

RECOGNITION OF METAMORPHIC ROCKS

A wide variety of textures, structures, and compositions is possible in metamorphic rocks, making it difficult to list one or even a few features that are diagnostic of metamorphism in all cases. It is therefore best to consider simultaneously as many different features as possible (see Table 1–1).

Many of the more obvious features in outcrops of metamorphic rocks are a result of unequal application of stress. The rocks may have undergone plastic flow, fracturing and granulation, or recrystallization. Some textures or structures in metamorphic rocks may be inherited from the premetamorphic rocks (for example, cross bedding), but most of these are eliminated during metamorphism. The application of unequal stresses, particularly when accompanied by the formation of new minerals, often causes parallelism of textural and structural features. When this is planar it is called *foliation*. When the planar structure is composed of somewhat diffuse layers or lenticles of minerals of different texture—for example, layers rich in granular minerals (such as feldspar and quartz) alternating with layers rich in minerals with tabular or prismatic habit (such as ferromagnesian minerals)—the texture is referred to as *gneissic*. If the foliation is caused by a parallel arrangement of medium- to coarse-grained platy minerals (usually mica or chlorite) it is called *schistosity*. Breakage of the rock is usually roughly parallel to the schistosity, resulting in a poorly developed rock cleavage (see Figure 17–1).

Metamorphism of fine-grained materials such as mudrocks often leads to a well-developed rock cleavage. Metamorphism of a shale results in the initial formation of a recrystallized fine-grained equivalent called a *slate*. Slates consist mainly of phyllosilicates; the platy phyllosilicates are in essentially parallel (coplanar) orientation. The type of foliation developed is called *slaty cleavage*. Slaty cleavage typically occurs at an angle to the original bedding plane of the rock (which might be distinguished by slight color or grain-size differences). The foliate features in a metamorphic rock are often not continuous in direction over long distances, but may be gently or sharply folded, or faulted, as a result of changes in the direction or intensity of the applied stress. Several foliation features may be present within the same rock, indicating a complex deformation history.

In addition to planar features, metamorphism may develop a linear fabric. *Lineation* may be developed by a preferred orientation of elongate grains, or mineral aggregates, or by the intersection of planar features, such as subparallel cleavage surfaces. Much lineation consists of a preferred orientation of elongate grains in the plane of the foliation.

Figure 17–1
Metamorphic rocks as seen in the field. **(A)** Exposure of platy gray schist near the Salmon River, Riggins Quadrangle, Idaho. The schistosity (brought out by erosion) is so predominant that it controls the angle of the hill slope. The small scale just below center is about 15 cm long. [From W. B. Hamilton, 1963, U.S. Geol. Surv. Prof. Paper 436, Fig. 24.] **(B)** Interlayered biotite and hornblende biotite gneisses (light and dark gray). The white layers are rich in potassium feldspar. The variable amount of mafic versus silicic phases in the various layers results in an unusually strong foliation in this exposure. Scale is about 18 cm long. Caldwell County, North Carolina. [From B. H. Bryant, 1970, U.S. Geol. Surv. Prof. Paper 615, Fig. 68.]

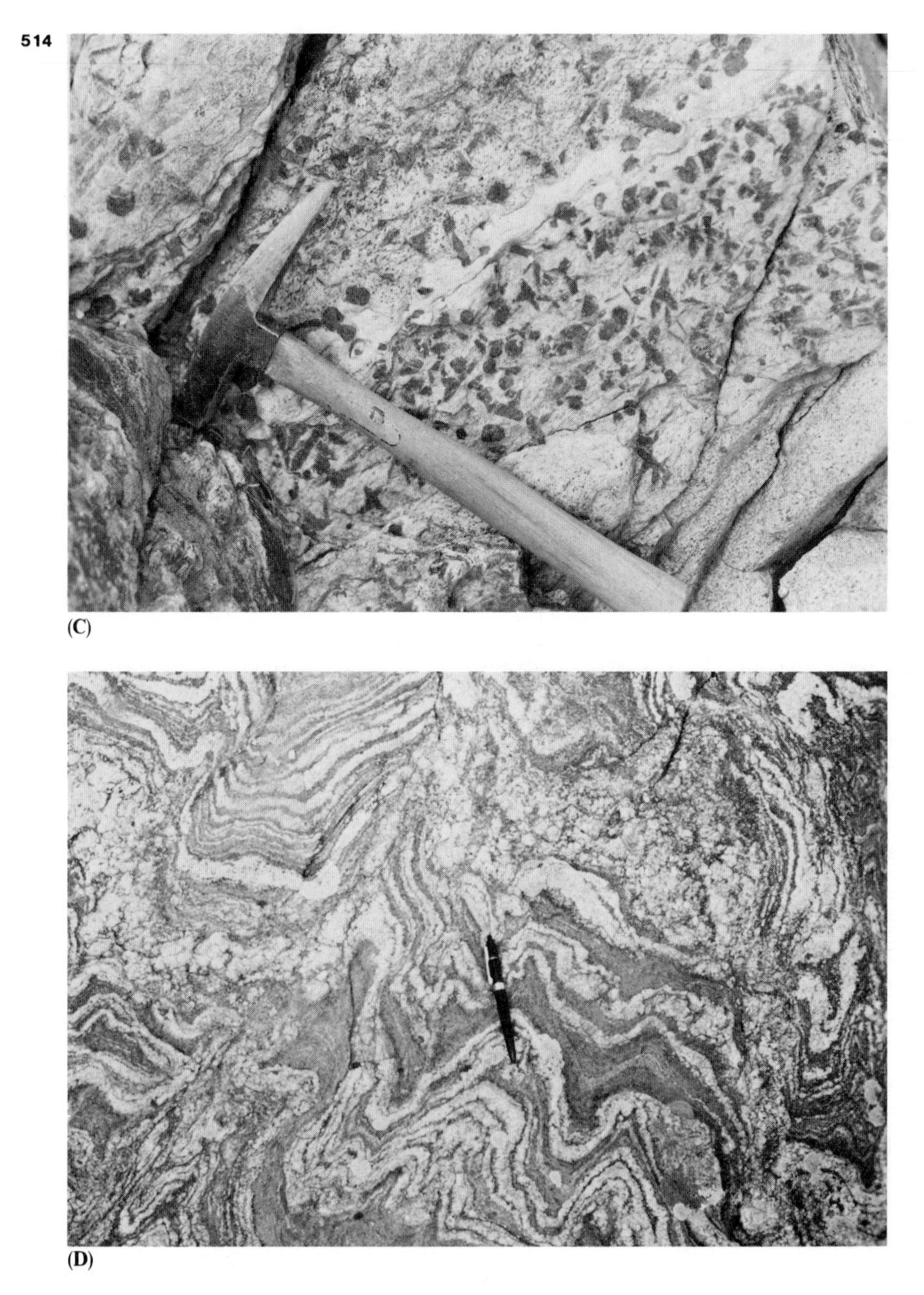

(C)

(D)

Figure 17–1 *(continued)*
(**C**) A quartz muscovite schist in Daggett County, Utah. Larger dark crystals are mainly garnet and twinned staurolite. Abundant specks near the hammer point and butt of handle are biotite. A weakly developed schistosity is evident from upper right to lower left. [From W. R. Hansen, 1965, U.S. Geol. Surv. Prof. Paper 490, Fig. 6.] (**D**) A highly contorted gneiss at the north rim of the Black Canyon of the Gunnison, Montrose County, Colorado. Intense folding of this type is found near katazonal intrusions and in Precambrian terranes. Light-colored bands are mostly feldspar and quartz; dark bands are mostly quartz and biotite. [From W. R. Hansen, 1965, *U.S. Geol. Surv. Bull., 1191,* Fig. 9.]

Textures developed during metamorphism typically are named with words that have the suffix *-blastic*. For example, a metamorphic rock composed of equigranular crystals is termed *granoblastic*. Commonly one or more minerals are present that are distinctly larger than the average; these larger crystals are called *porphyroblasts*. Porphyroblasts, in casual inspection, may be confused with phenocrysts (of igneous origin), but they usually can be distinguished by their mineralogy and the common foliate nature of the matrix. Microscopic examination of porphyroblasts often reveals that they contain numerous grains of the matrix materials, in which case they are called *poikiloblasts*. It is usually assumed that the poikiloblast has been created by growth of the larger crystal around relics of earlier minerals, but it may be that the poikiloblast has simply grown at a faster rate than the matrix minerals, and engulfed them. The included materials may indicate (by their shape, orientation, or distribution) an earlier directional feature in the rock (such as schistosity or original bedding); in this case the porphyroblast or poikiloblast is said to possess *helicitic texture*.

The growth of new minerals or recrystallization of preexisting minerals as a result of pressure and/or temperature change results in the formation of crystals that have either good, medium, or poor facial development; these are called (respectively) *idioblastic, hypidioblastic,* or *xenoblastic*. Occasionally metamorphic rocks contain aggregates of grains that have a lenticular or elliptical shape; these aggregates are called *augen* (German for "eyes"), and are commonly a result of cataclasis (fracturing, granulation, and rotation). These residual aggregates are produced within a granulated matrix. The general term for such an aggregate is *porphyroclast*.

COMMON METAMORPHIC ROCKS

After we have determined that a rock is of metamorphic origin, we must name it. Unfortunately, the procedure for naming metamorphic rocks is not as systematic as the procedures for igneous and sedimentary rocks. The names of most metamorphic rocks are based primarily on textural and structural features. The general name is often modified by a prefix indicating a conspicuous or important aspect of texture (for example, augen gneiss), one or more minerals present (for example, chlorite schist), or the name of an igneous rock of similar composition (for example, granite gneiss). A smaller number of names is based on the predominant mineral (for example, metaquartzite) or is related to the metamorphic facies to which the rock belongs (for example, granulite).

Regional metamorphism of mudrocks involving changes in both pressure and temperature results initially in recrystallization and modification of the clay minerals present. The grain sizes remain microscopic, but a new direction of preferred orientation may be developed in response to the applied stress. The resultant fine-grained rock possesses excellent rock cleavage and is called a *slate*. Continued metamorphism often results in a maintenance of the preferred orientation of platy minerals within the rock and an increase in grain size of micas and chlorites. This results in a fine-grained rock called a *phyllite,* which is similar to a slate but has a silky sheen on the cleavage surface. Close examination with a hand lens occasionally reveals tiny porphyroblasts breaking the smooth reflections of the cleavage surfaces. At higher levels of metamorphism, crystals are visible without a lens. Here we usually find distinct platy or elongate minerals in a strongly preferred orientation forming a conspicuous schistosity. The rock, now called a *schist,* can still be split into flakes or slabs. The development of porphyroblasts is common; these can often be identified as characteristic metamorphic minerals such as garnet, staurolite, or cordierite. At still higher levels of metamorphism the schistosity becomes less pronounced; the rock now contains medium to coarse bands of differing texture and mineralogy. This is gneissic banding and the rock is called a *gneiss*. The banding consists of layers relatively rich in quartz and feldspar, alternating with those containing ferromagnesian minerals (micas, pyroxenes, and amphiboles). The mineralogy is often similar to that of an igneous rock, but a distinct gneissic banding is usually taken to indicate a metamorphic origin; within the bands a moderate amount of preferred orientation is often present. Additional metamorphism could convert the gneiss into a migmatite. In this case, the light-colored bands look distinctly igneous, and ferromagnesian-rich layers retain a distinct metamorphic aspect.

Other metamorphic rock types are named only on the basis of mineral composition. A *marble* is composed almost entirely of calcite or dolomite; typically the texture is granoblastic. A *quartzite* is a granoblastic metamorphic rock consisting mainly of quartz, formed by the recrystallization of sandstone or chert. Other common types of metamorphic rocks include the following:

Amphibolite: A medium- to coarse-grained rock consisting predominantly of an amphibole (usually hornblende) and plagioclase.

Eclogite: A medium-grained rock consisting mainly of the pyroxene omphacite (a sodium- and aluminum-rich diopside) and a pyrope-rich garnet. Eclogites are chemically equivalent to basalts, but contain denser phases. Some eclogites are of igneous origin.

Granulite: An even-grained rock consisting of minerals (mainly quartz and feldspar, with lesser pyroxene and garnet) having a granoblastic texture. A weakly developed gneissic structure may consist of flat lenses of quartz and/or feldspar.

Hornfels: A fine-grained, thermally metamorphosed rock containing equidimensional grains in random orientation. Some porphyroblasts or relic phenocrysts may be present. A coarser-grained equivalent is called *granofels*.

Mylonite: A fine-grained streaky or banded rock produced by granulation or flowage of coarser rocks. A rock may be a protomylonite, mylonite, or ultramylonite, depending upon the amount of residual fragments. If the rock has schistosity surfaces of silky, recrystallized micas, it is called a *phyllonite*.

Serpentinite: A rock consisting almost entirely of minerals of the serpentine group. Accessory minerals include chlorite, talc, and carbonates. Serpentinite is derived from alteration of previously existing ferromagnesian silicate minerals, such as olivine and pyroxene.

Skarn: An impure marble containing crystals of calc-silicate minerals such as garnet, epidote, and so on. Skarns are created by compositional changes of country rock at igneous contacts.

A group of figures showing some of these textures and rock types follows. It should be understood that there is no substitute for study of a well-labeled set of metamorphic rocks and minerals.

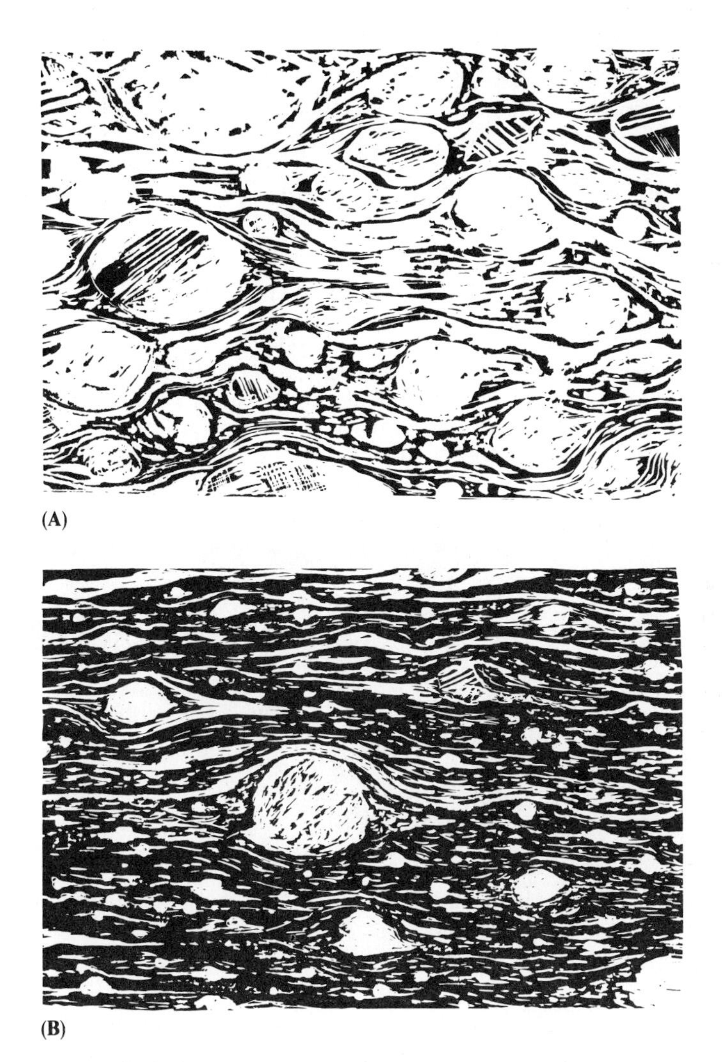

(A)

(B)

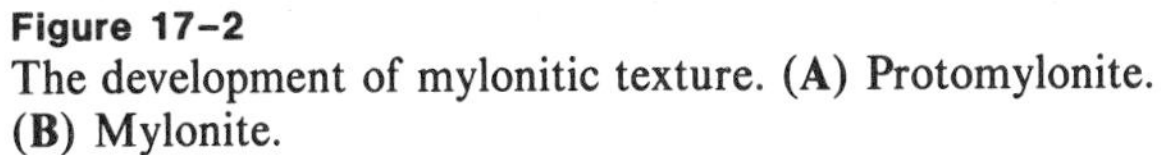

Figure 17–2
The development of mylonitic texture. (**A**) Protomylonite. (**B**) Mylonite.

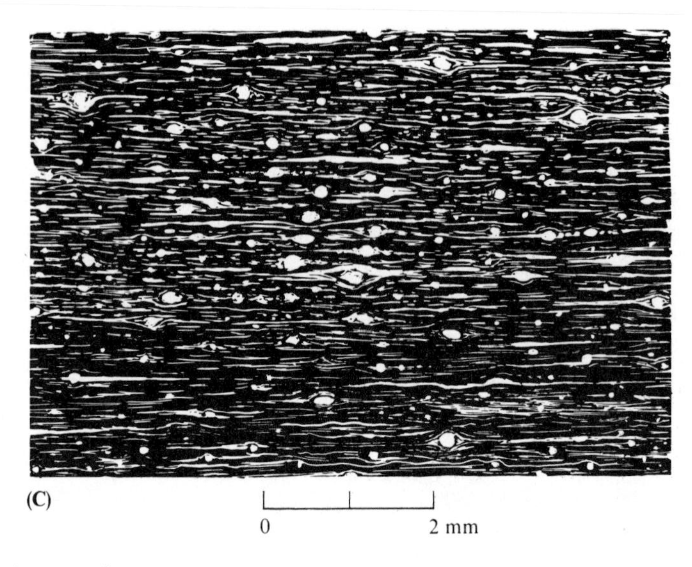

Figure 17–2 (*continued*)
(C) Ultramylonite. Many ultramylonites are considerably finer grained than that shown. The white areas represent porphyroclasts (relatively large resistant fragments of minerals or aggregates in a cataclastic rock) or trails of finer brecciated materials. Black areas represent very finely comminuted materials and dark minerals. All three represent rocks produced from the same parent—a coarse-grained granite. [From M. W. Higgins, 1971, U.S. Geol. Surv. Prof. Paper 687, Fig. 7.]

Figure 17–3
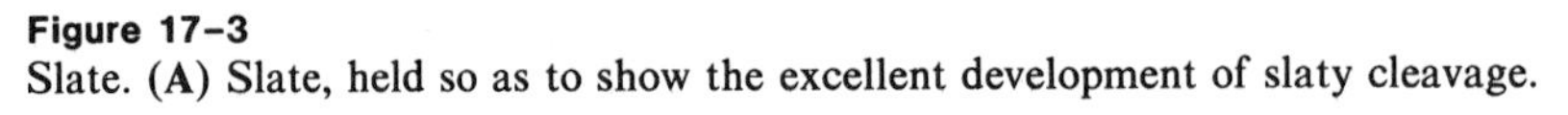
Slate. (A) Slate, held so as to show the excellent development of slaty cleavage.

(B)

(**B**) The same rock oriented to view the cleavage surface. This dark gray rock has several dark bands that represent original bedding planes in the premetamorphic mudrock. Original compositional differences are often revealed by color bands or slight grain-size differences (such as planes containing visible quartz grains). The orientation of slaty cleavage developed during metamorphism is at an angle to original sedimentary bedding surfaces.

(A)

Figure 17–4
Schists. (**A**) Augen structure is shown by the large (light) rounded feldspar in center. Well-developed schistosity is shown by biotite (dark) and quartz–feldspar foliation; this is seen best at the left side.

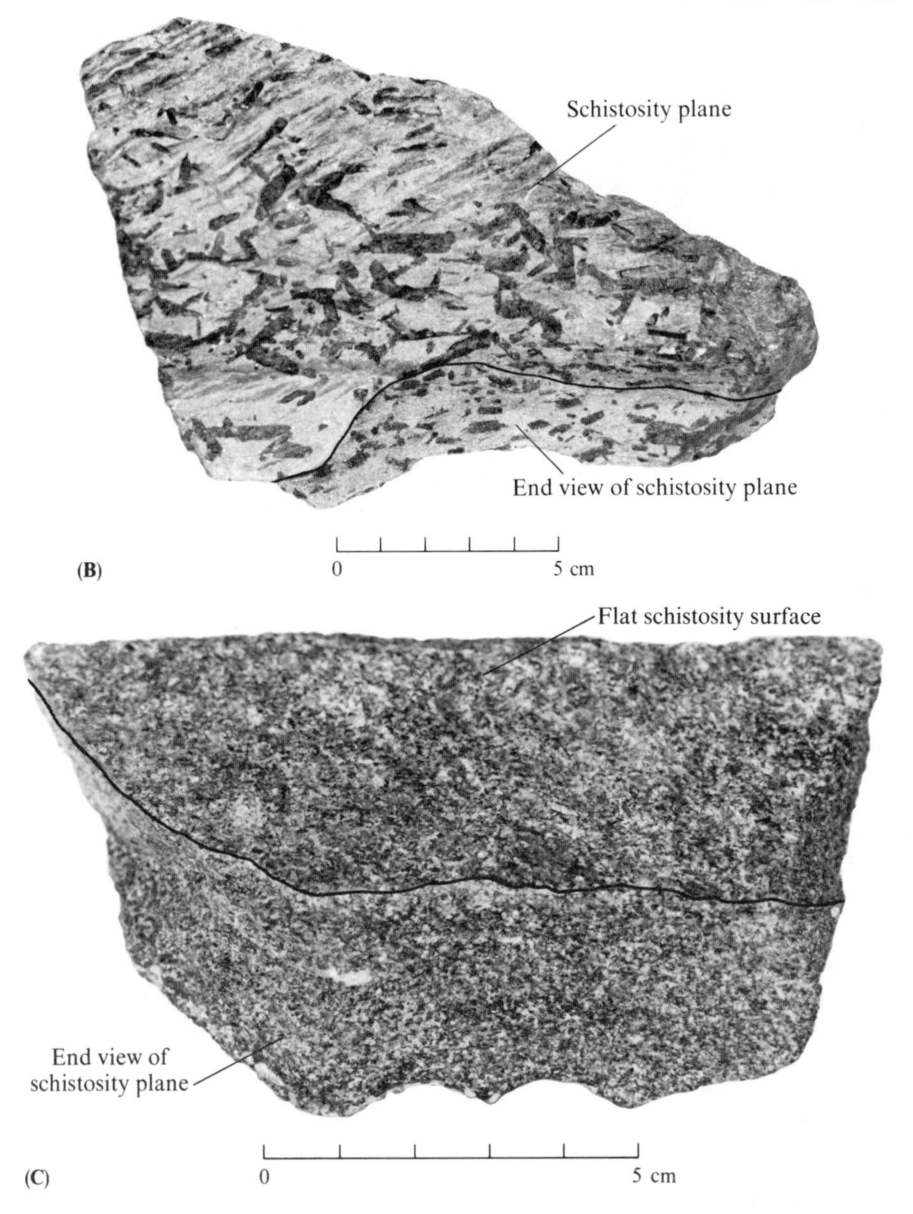

Figure 17–4 (*continued*)
(**B**) Tourmaline mica schist. The schistosity is developed by foliated muscovite. Elongated dark tourmaline porphyroblasts lie generally parallel to the schistosity. (**C**) Biotite hornblende schist. Schistosity is present, but poorly developed by dark hornblende and biotite; nevertheless there is sufficient foliation so that the major cleavage direction of the rock is essentially flat. White material consists of quartz and feldspar.

(D) Garnet schist. The photo is taken perpendicular to the schistosity plane. Large, dark, somewhat rounded masses are red-brown garnet. Elongate dark areas are biotite, and lighter areas are mainly muscovite, with lesser quartz. A definite lineation direction (vertical) is shown by the distribution of biotite. **(E)** Kyanite schist. The photo is taken perpendicular to the plane of schistosity. Pale blue (dark) porphyroblasts of kyanite with no apparent lineation generally parallel the schistosity. The lighter areas are mainly quartz and muscovite.

Figure 17–5
Gneisses. (**A**) A gneiss containing irregular dark bands of biotite and hornblende. Lighter areas are quartz and feldspar. (**B**) Biotite gneiss. The compositional banding contains biotite (dark) alternating with quartz and feldspar.

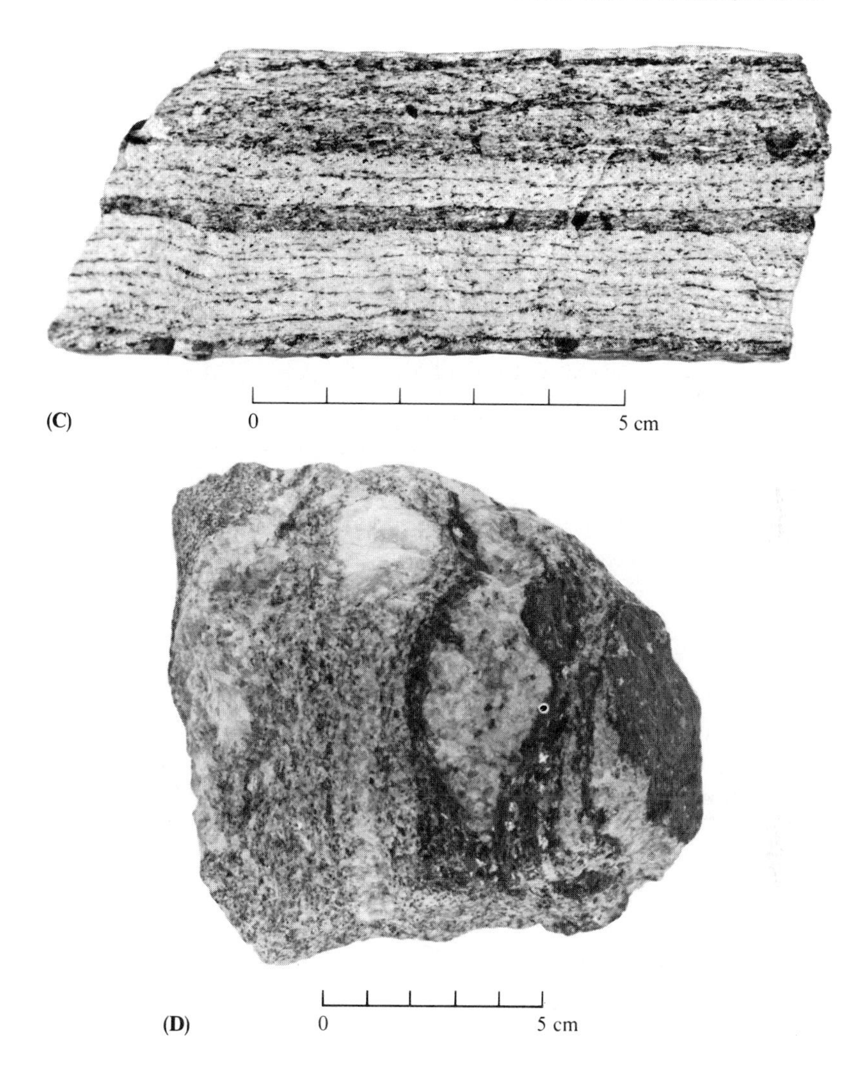

(C) A distinctly banded gneiss. Lighter layers are mainly quartz and feldspar, and darker areas are biotite and garnet. **(D)** Augen structure (light potassium feldspar) in a biotite (dark) quartz feldspar gneiss.

Figure 17–5 (*continued*)
(E) Polished slab of a granite gneiss showing segregations of darker ferromagnesian minerals and lighter quartz and feldspar. If found as the border phase of an igneous pluton the compositional banding might be taken as flow structure and the rock would be called a gneissic granite. Found within a metamorphic terrane it would probably be called a granite gneiss.

Figure 17–6
Hornfels. (**A**) This hornfels is composed of a mixture of mainly calcite (light) and green diopside (dark). No preferred orientation is present. (**B**) Talc-actinolite granofels. The actinolite consists of dark green prismatic crystals. Talc forms very pale green platy crystals.

Figure 17-6 (*continued*)
(C) This hornfels contains dark patches of red-brown garnet, green diopside, and silvery flakes of graphite. The lighter areas are calcite.

Figure 17-7
Marble. (A) This marble consists of coarse (white) masses of calcite, and (dark) brown phlogopite mica. The several very white patches (upper left) are reflections from cleavage surfaces.

Figure 17–7 (*continued*)
(**B**) Marble. The color of the calcite is orange, except near the scattered green diopside crystals, where it is white. (**C**) Chondrodite marble. The large (dark) brown crystals are chondrodite. Graphite is present as individual flakes and aggregates; the white areas are reflections from cleavage surfaces.

Figure 17-8
Polished slab of serpentinite. The rock is irregularly colored from dark green to black. Lighter areas are veins and concentrations of magnesite. Fine-grained talc is present in the lighter area at left. Tiny crystals of magnetite or chromite are commonly present in such rocks.

FIELD OBSERVATIONS OF METAMORPHIC ROCKS

Observations in the field have revealed that metamorphic rocks can be grouped into several categories as a function of their regional extent and circumstances of origin. The general pressure–temperature conditions for these groups are shown in Figure 17–9.

Regional Metamorphism

The most common type of situation is that of *regional metamorphism.* This is metamorphism that occurs over wide areas and within large orogenic belts; the intensity of metamorphism is often related to a geographic axis (or axes), and only to a limited extent to adjacent igneous activity. However, igneous activity may accompany the metamorphism. Regional metamorphism is more or less isochemical when considered over broad regions (aside from the loss of volatiles).

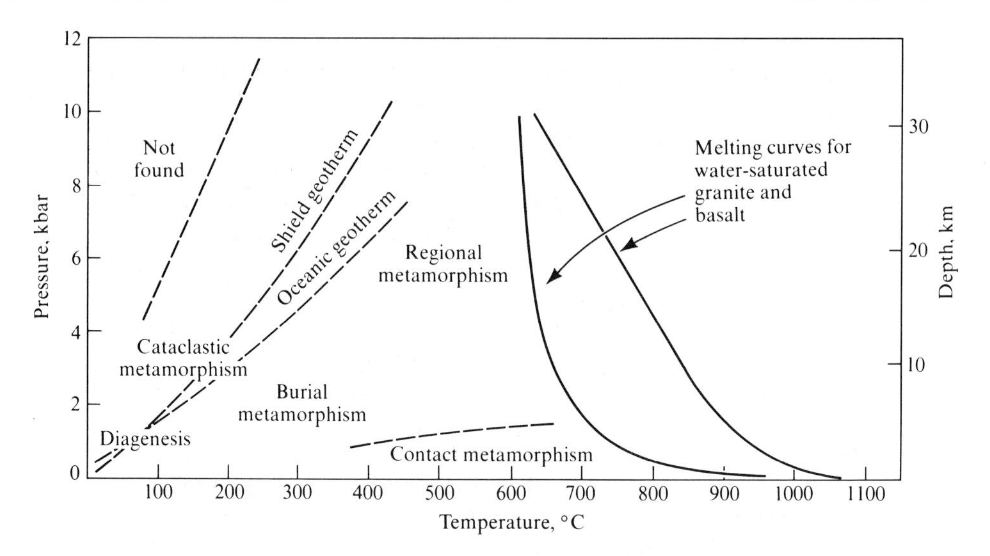

Figure 17–9
The principal types of metamorphism and their general pressure–temperature ranges. [The melting curves for water-saturated granites and basalts are from W. C. Luth, R. H. Jahns, and O. F. Tuttle, 1964, *Jour. Geophys. Res., 69,* Fig. 1; and H. S. Yoder Jr. and C. E. Tilley, 1962, *Jour. Petrology, 3,* Fig. 33.]

Regional metamorphism is found in all of the major orogenic regions of the world. Mountain-building processes are accompanied by large changes in both temperature and pressure. This is seen in the metamorphic rocks as recrystallization and the formation of new phases. As these processes generally occur during rock flowage (nonhydrostatic stress field), fabrics that have a preferred orientation of platy or elongate minerals are typical (slates, schists, and gneisses). Part of the mapping of these rocks should include directional features such as foliation, lineation, or rock cleavage. Thin sections may be made from hand specimens whose field orientation has been recorded earlier. The principal fabric directions may be determined by microscopic analysis, using oriented sections to find the preferred directions of grain elongation, crystallographic axes, and mineral cleavage. The study of both macroscopic and microscopic directional features (called petrofabrics or structural petrology) can provide a detailed picture of the deformational history of the rock.

One of the most obvious things to do in the field is to determine the rock type. As mentioned earlier, some metamorphic rocks are named on the basis of texture (such as slate, schist, and gneiss), and others on the basis of composition (such as marble and amphibolite). The texture is usually a reflection of the degree and type of metamorphism, whereas compositional differences are usually a function of the preexisting rock composition. It is generally possible to come up with a distribution of the intensity of metamorphism by putting these two factors together. Metamorphism of mudrocks, for example, results in an increase in grain size; a sequence of slate, phyllite, schist, gneiss, and migmatite in rocks of similar composition is indicative of increasing intensity of metamorphism. If these rocks are interbedded with marble it will be found that the marble shows a corresponding relationship in which original sedimentary textures are gradually eliminated in favor of a granoblastic texture; the granoblastic texture in turn shows a progressive coarsening of grain sizes, as does the metamorphosed mudrock. Quartz sandstones show a corresponding change to quartzites of increasing grain size. The general rule with increasing metamorphic intensity, regardless of original rock type, is an increase in grain size.

Along with mapping the general rock type, it is necessary to determine the rock mineralogy as completely as possible. This may be done successfully in the field if the minerals are of sufficient size. Commonly porphyroblasts are easily determined on fresh specimens with a hand lens, whereas the matrix material might require thin section examination. Determination of differences in mineralogy permits the metamorphic region to be subdivided into mineralogic zones. Thus an area of schists might be mapped as kyanite schist in one area and andalusite schist in another.

Differences in mineralogy among metamorphic rocks are a result of two factors: (1) original rock composition and (2) intensity of metamorphism. Although it is ultimately necessary to distinguish between these two factors, this is not always possible to do in the field, as subtle changes in composition might occasionally result in distinct mineralogical changes. Therefore any mineralogic changes should be recorded, and fresh specimens collected for later examination and analysis. Usually, though, compositional changes are easily recognized (as in crossing the contact between metamorphosed equivalents of mudrocks and sandstones).

Mapping of metamorphic rocks by means of changes in mineralogy has shown that certain rocks (such as quartzite) are mineralogically stable throughout the entire range of metamorphism. Others of more complex composition (such as metamorphosed mudrocks) are composed of minerals unstable during the changing conditions of metamorphism, and form a variety of new phases as a result of changes in metamorphic intensity. The changes that occur during metamorphism are of a progressive nature, in that a series of reactions occurs as a result of change in the degree of metamorphism. The mineralogy of a metamorphic rock does not make only one transition to arrive in a condition compatible with the imposed pressure–temperature conditions, but rather makes a series of changes (because extended time is usually available for constant equilibration). Arguments in favor of the progressive nature of metamorphic reactions are the following:

1. Many metamorphic rocks contain mineral relics indicative of immediately lower *PT* conditions. A wollastonite-containing rock may show partially reacted relics of calcite or quartz ($CaCO_3 + SiO_2 \rightarrow CaSiO_3 + CO_2 \uparrow$).

2. As most mineral reactions can be duplicated in laboratory experiments in days or weeks, and the *PT* rise in regional metamorphism occurs over millions of years, it

is reasonable to assume that mineralogical changes in metamorphic rocks have kept pace with the imposed *PT* conditions.

3. Directly adjacent metamorphic rocks of the same composition usually contain minerals that, although different, are characteristic of closely related *PT* environments.

The first investigator to show the progressive change of mineralogy in rocks of similar composition on a regional scale was the British geologist George Barrow (1893). He examined part of the Dalradian series in the Scottish Highlands southwest of Aberdeen (see Figure 17–10). The rocks present include a series of essentially isochemical metapelites (metamorphosed mudrocks) that could be subdivided into *mineralogical zones* containing staurolite, kyanite, and sillimanite. Later work by Barrow (1912), Tilley (1925), and Harker (1932) confirmed and extended this work over much of the Highlands. Textural changes in the pelitic rocks from slates to phyllites to mica schists to gneisses correspond to the following series of mineralogical zones, which were related to increasing temperature:

1. Chlorite zone, typically quartz–chlorite–muscovite ± albite.
2. Biotite zone, typically biotite–chlorite–muscovite–albite–quartz.
3. Garnet zone, typically quartz–muscovite–biotite–almandine–sodic plagioclase.
4. Staurolite zone, typically quartz–muscovite–biotite–almandine–staurolite–plagioclase.
5. Kyanite zone, typically quartz–muscovite–biotite–almandine–kyanite–plagioclase.
6. Sillimanite zone, typically quartz–muscovite–biotite–almandine–sillimanite–plagioclase–potash feldspar.

Each mineralogic zone is characterized by the presence of a new mineral, termed an *index mineral*. After a mineral has formed, it may persist through several higher level zones. A line drawn on a map to represent the group of exposures marking the first appearance of a new mineral with increasing metamorphic grade is known as an *isograd*. Note that the isograd is created as a result of mineralogical reactions resulting from changes of pressure and/or temperature. Some zone boundaries (as indicated by differences in mineralogy) may not be due to *PT* changes, but instead to initial differences in rock chemistry; these are compositional boundaries, not isograds. It was initially assumed that an isograd was a line of constant *P* and *T*. As it is now known that pressure and temperature can vary independently of each other, present usage favors the definition of an isograd simply as a line indicating a change in mineralogy within rocks of the same composition; nothing is assumed as to constancy of pressure and/or temperature. The isograd, mapped as a line on a map, represents the intersection of a mineralogical reaction surface with the earth's surface.

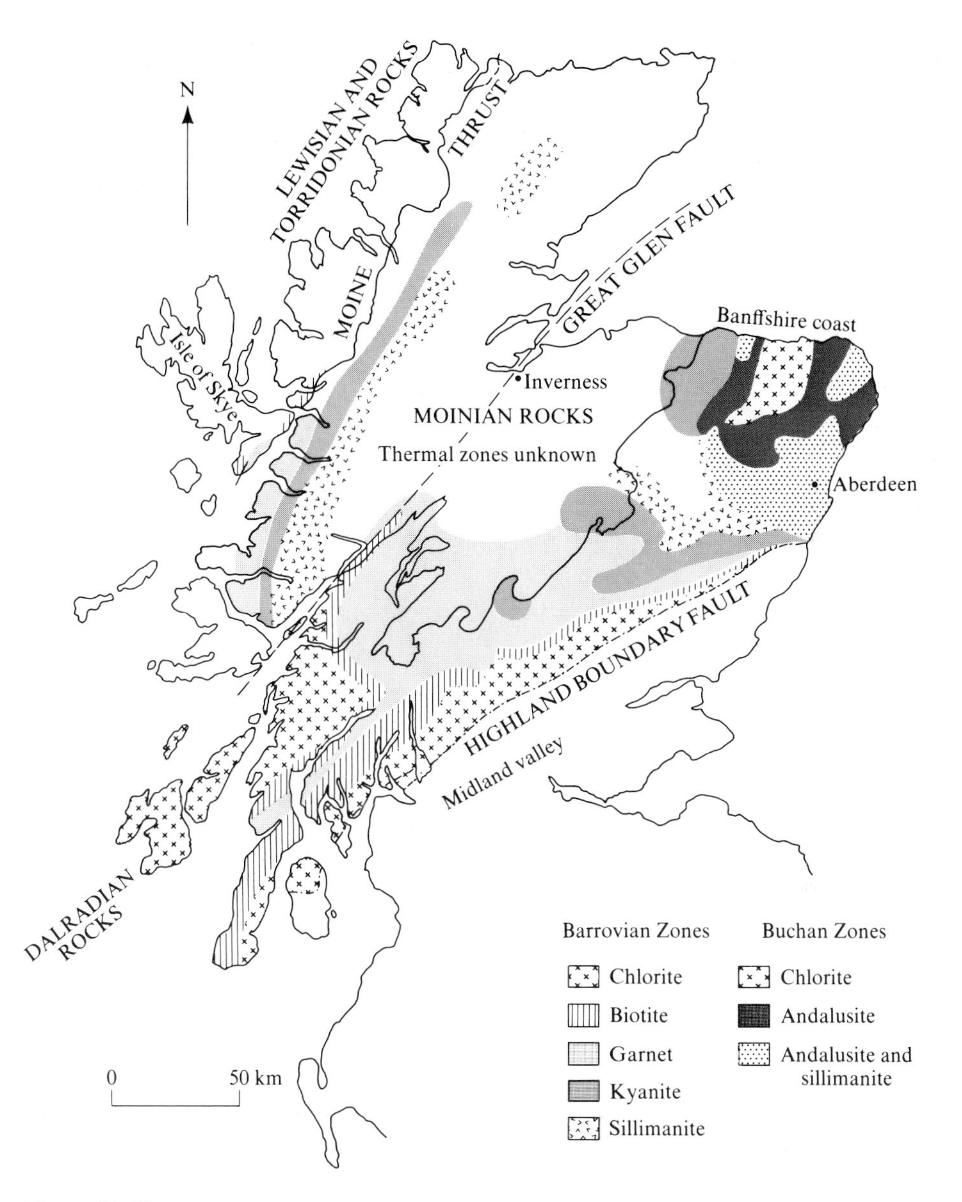

Figure 17–10
The general mineralogical zoning of Dalradian rocks in Scotland. The classic Barrovian zones are in southwestern Scotland. Separation of Dalradian and Moinian rocks is indicated by the irregularly curved solid line. The Buchan zones, characterized by relatively higher temperatures and lower pressures, are in the vicinity of Aberdeen. [From R. W. Johnson, 1963, *Geol. en Mijnbouw, 42,* Fig. 1, based mainly on earlier data.]

The mapping of mineralogical changes by isograds is usually accomplished fairly easily, because mineralogic changes often occur within a zone several meters wide. In some areas, though, difficulty is encountered because an index mineral may not appear in abundance along a closely defined line, but rather will increase in a very sporadic manner in small amounts for several kilometers below the isograd (where it might be located in large amounts). Geographic spread of the isograd is often attributed to minor changes in rock composition. In other cases, isograd mapping may be difficult due to lack of rocks of suitable composition.

Contact Metamorphism

Contact metamorphism occurs in country rock directly adjacent to igneous intrusions. It is generally a static thermal event of local extent. The metamorphic rocks that are produced lack a preferred orientation and are generally called hornfels. The maximum width of a contact zone may range from several meters to a maximum of a

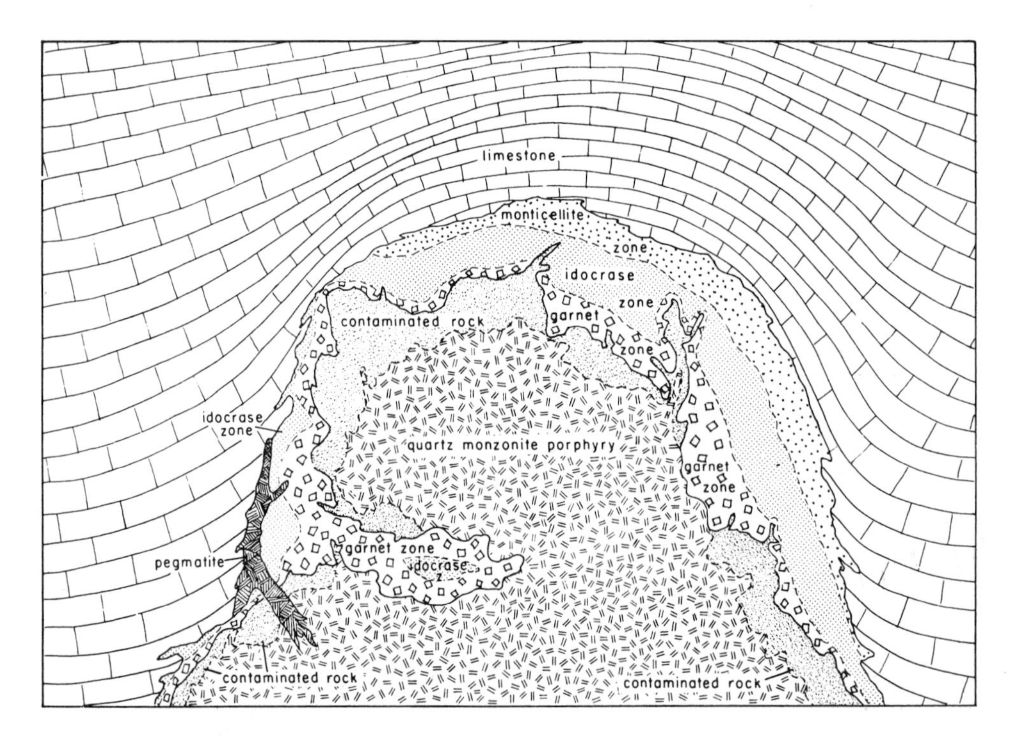

Figure 17–11
Idealized north–south cross section of an intrusion of quartz monzonite into marble at Crestmore, California. The contact aureole is up to 17 m in thickness, and is composed of three mineralogic zones. [From C. W. Burnham, 1959, *Geol. Soc. Amer. Bull., 70,* Fig. 3.]

few kilometers. The width of this zone is mainly a function of the thickness of the intrusion. Small intrusions have narrow contact zones that consist of "baked" rock; this is usually harder than the normal country rock, often with a redder color as a result of oxidation. Thicker intrusions generally have a broad thermal *aureole,* resulting in a wide zone of contact rock. Within the contact area may be several zones of differing mineralogy, which can be related to distance from the intrusion.

An additional effect of intrusion may be the change of composition of the country rock within the contact zone. The process of compositional change is referred to as *metasomatism.* Compositional changes may be caused by fluids emanating from an igneous intrusion, or from fluid migration activated within the country rock by the presence of the intrusion. Metasomatic effects are usually greatest in carbonate rocks adjacent to silicic intrusions. Rocks whose compositions have been altered by addition of new materials are called skarns or tactites. In limestone the minerals produced are usually calcium-rich silicates, and commonly include grossularite, andradite, epidote, wollastonite, or tremolite. Dolomites produce serpentine, diopside, and minerals of the chondrodite group.

A classic case of contact metamorphism is located at Crestmore, Riverside Country, California. Here a porphyritic quartz monzonite has created a contact aureole up to 150 m thick in the surrounding marble (see Figure 17–11). A series of reaction zones has formed in relation to both temperature gradient and changes in composition due to magmatic emanations. The innermost "contaminated" rock is very high in silica, and changes outward into garnet-, idocrase-, and monticellite-containing zones. Changes in composition are plotted in Figure 17–12. The major changes appear to be the result of silica, aluminum, and iron metasomatism of an initially silica-poor magnesian marble.

Other Types of Metamorphism

The most recently recognized type of metamorphism is *burial metamorphism.* Sediments and interlayered volcanic rocks deeply buried in sedimentary basins at converging plate margins may attain temperatures of several hundred degrees with high water pressure. This produces changes in the mineralogical constituents while preserving most of the original fabric; retention of much of the original fabric makes these rocks very difficult to recognize in the field. Typically, numerous zeolitic minerals are produced, which can be identified only with the polarizing microscope or X-ray diffraction techniques. The term burial metamorphism is often used as a synonym of high-grade diagenesis.

Less significant and usually very local in occurrence is dynamic or *cataclastic metamorphism,* which results from the fracturing and granulation of rocks in the vicinity of faults, overthrusts, or broadly disturbed areas. The general effect is a decrease in grain size; as little to no heat is usually present, the formation of metamorphic minerals characteristic of regional or contact metamorphism is prevented.

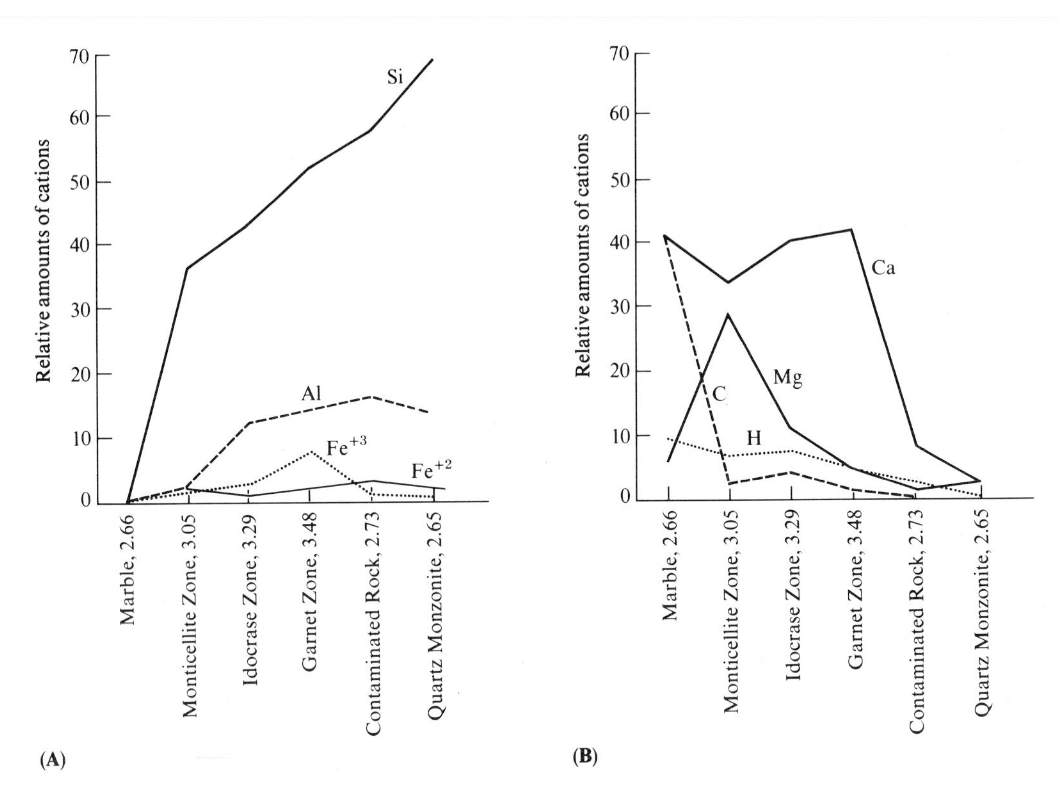

Figure 17–12
Variation in chemical composition of the major elements and specific gravity of the contact rocks at Crestmore, California. The various reaction zones are shown between the country rock marble (left) and the intrusive quartz monzonite (right). The data for the analyses are based on the assumption that a unit volume of marble remained constant during conversion into a calc-silicate rock. Part A shows those elements that have increased in amount in the marble, and Part B shows those elements that have decreased. [After C. W. Burnham, 1959, *Geol. Soc. Amer. Bull., 70,* Fig. 5.]

When the cataclasis occurs at shallow levels in the crust, a fault breccia is produced. This is commonly a fine-grained microbreccia; tiny broken fragments within a still-finer matrix may be seen with a hand lens. This fine-grained material is very susceptible to chemical alteration, which usually results in rapid erosion and topographic emphasis of the fault. In other cases, recrystallization of the fault breccia produces a resistant ridge at the fault.

At deeper levels cataclasis produces fractured and recrystallized mylonites; lens-shaped fragments of preexistent rock can often be found within a fine matrix. These fragments tend to display planar orientation. Occasionally some lineation may be present. Recrystallization of mylonites results in mechanically strong rocks that are often resistant to erosion.

Isograds

It is worth considering isograds in greater detail because the determination of isograds is one of the principal techniques used in mapping metamorphic rocks. The classic Barrovian isograds are based upon the first appearance of a mineral as a result of increased pressure and/or temperature. Isograd reactions may occur in a number of ways, such as:

1. Polymorphic inversions: kyanite (Al_2SiO_5) $\rightarrow$ sillimanite (Al_2SiO_5).
2. Dehydration reactions: $Mg(OH)_2 \rightarrow MgO + H_2O.\uparrow$
3. Decarbonation reactions: $CaCO_3 + SiO_2 \rightarrow CaSiO_3 + CO_2.\uparrow$
4. A change in the solid solution limits of one or more minerals.
5. A change in the arrangement of tie lines between coexisting phases.

Although an isograd is often regarded as a line through a series of outcrops or across a topographic map, it should be kept in mind that the line represented by the isograd is merely the surface expression of an isograd surface. Thompson (1976, 1977), while working in the Lepontine Alps in southeastern Switzerland in an area of 2,000-m relief, was able to consider the three-dimensional patterns shown by isograd surfaces and how these patterns related to pressure–temperature distribution during regional metamorphism. From this type of approach it becomes obvious why isograds are often not parallel to each other, and may occasionally cross.

In most discussions of regional metamorphism it is almost tacitly assumed that there is a fixed relationship between pressure and temperature in the sense that for each pressure there also exists a particular temperature (see Figure 17–13). *PT* gradients of this type have been estimated in many areas. It follows from these situations that the pressure and temperature gradients (as indicated by isobars and isotherms) are parallel to each other. This is shown in the block diagram of Figure 17–14A. The schematic metamorphic reaction in Figure 17–14B, showing the *PT* stability ranges for the minerals *a*, *s*, and *k*, is similar to polymorphic inversions involving the aluminum silicates. The schematic reaction shown in Figure 17–14C is similar to a typical dehydration curve of the type common in many metamorphic reactions. The dashed lines in both diagrams show a possible geothermal gradient (increase of both *P* and *T*) for a metamorphic terrane. The mineral *a* converts to *s* at $T = 2.25$ and $P = 2.25$, and *x* converts to *y* at $T = 3.75$ and $P = 3.75$. The block diagram in Figure 17–14A shows the increase of both *P* and *T* with depth during metamorphism. The isotherms and isobars are parallel. Similarly the planes (heavy lines) representing the levels of the two reactions are parallel.

If this area were tilted 45° and again eroded to a horizontal surface (see Figure 17–15), the isotherms and isobars that existed during metamorphism, as well as the reaction surfaces, would be inclined to the surface at 45°. The reaction surfaces, now seen as isograds, would be mapped as parallel lines.

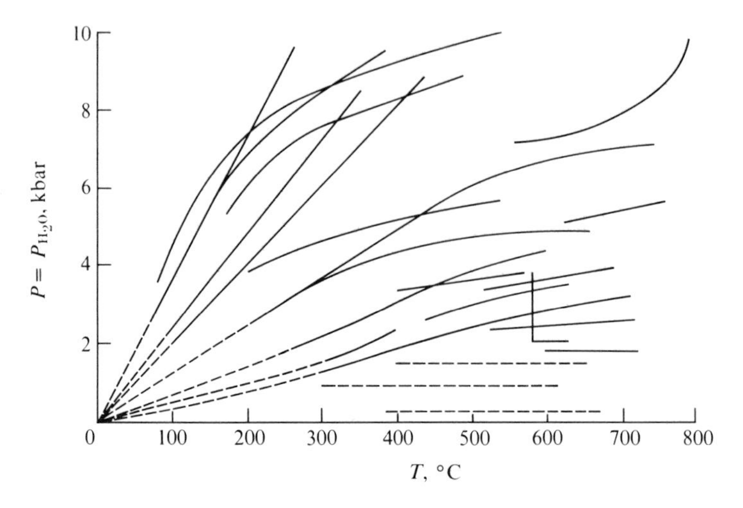

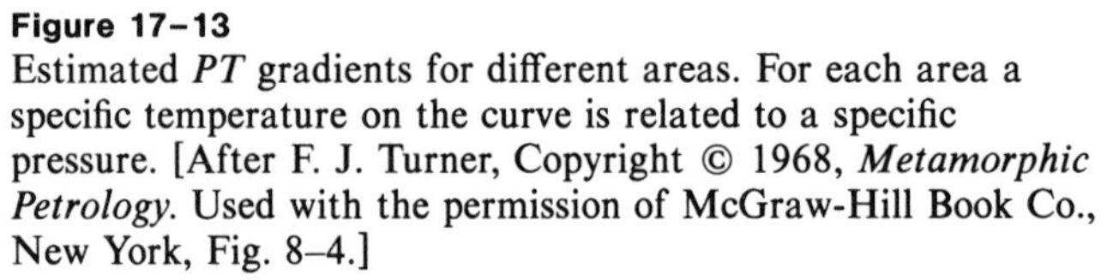

Figure 17–13
Estimated *PT* gradients for different areas. For each area a specific temperature on the curve is related to a specific pressure. [After F. J. Turner, Copyright © 1968, *Metamorphic Petrology.* Used with the permission of McGraw-Hill Book Co., New York, Fig. 8–4.]

Crossing or nonparallelism of isograds can occur when the temperature and pressure gradients are not parallel. If a metamorphic reaction were dependent upon temperature only, the isograd would parallel an isothermal surface; a reaction dependent upon pressure only would produce an isograd surface parallel to an isobaric surface. But metamorphic reactions are dependent upon both pressure and temperature; consider Figure 17–16A. We will assume the same two reactions shown in Figure 17–14. The pressure gradient is vertical, as in Figure 17–14, but the temperature gradient slopes at 45° (due to a heat source at the lower left). Here the isograd surfaces (one a plane surface and one curved) intersect at depth, but would emerge at the surface of the block as parallel lines. If the block were eroded and developed a new surface *ABCD*, as shown in Figure 17–16B, it would be obvious that the isograds were nonparallel. Note that for situations of this sort a single *PT* line cannot be drawn for the entire metamorphic region, as in Figure 17–13, because each isotherm is at a variety of pressures and vice versa. Additional discussions of this subject can be found in Thompson (1976, 1977) and Turner (1968, p. 376).

SUMMARY

Metamorphic rocks form as the result of a new set of physical and/or chemical conditions being imposed on a preexistent rock. The metamorphic rock differs signifi-

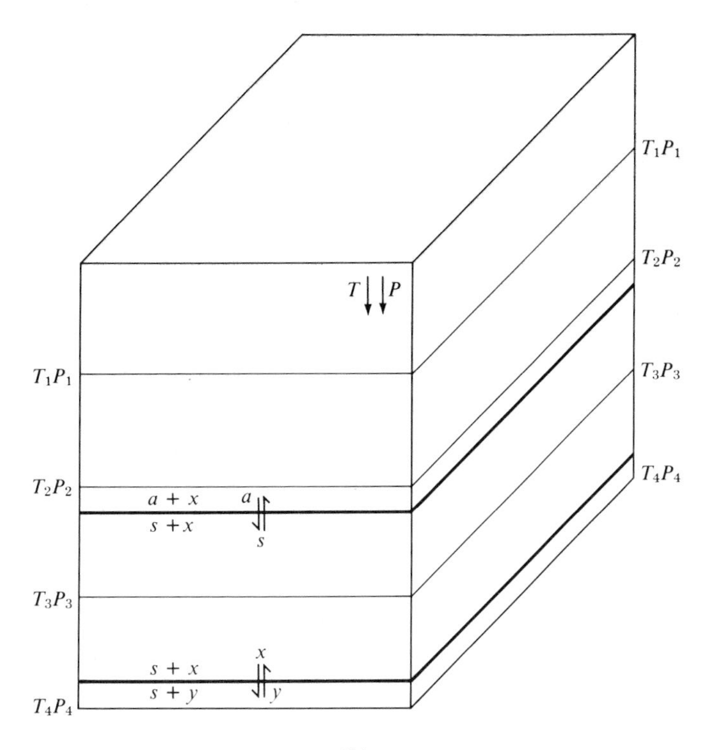

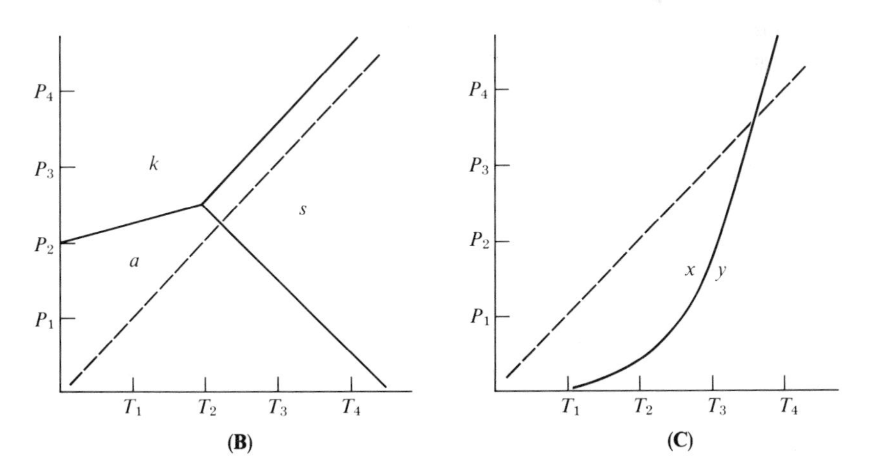

Figure 17–14
(A) Block diagram illustrating the change of pressure and temperature with depth. The top of the block is an erosion surface. Temperature increases from T_1 through T_2, T_3, and T_4. Pressure increases from P_1 through P_2, P_3, and P_4. Temperature and pressure gradients are parallel. **(B, C)** Phase diagrams. The dashed lines represent the geothermal gradient for the block in part A. The solid lines are schematic univariant curves. The locations of metamorphic reactions ($a \rightleftharpoons s$) and ($x \rightleftharpoons y$) are indicated by heavy lines in part A.

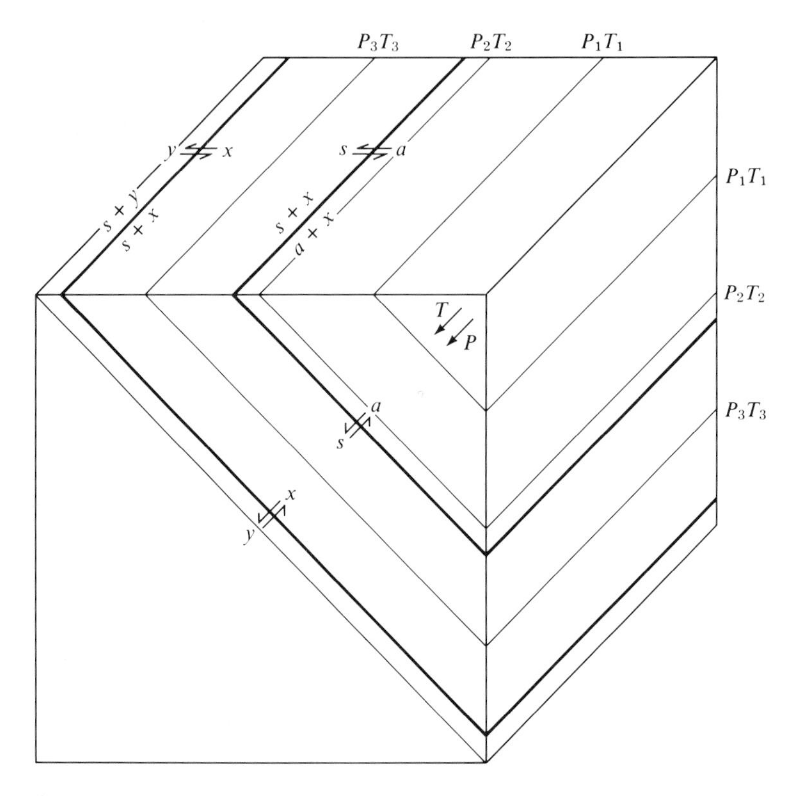

Figure 17–15
A continuation of Figure 17–14. The block has been tilted and its upper surface eroded level. The heavy lines marking the reactions $a \rightleftharpoons s$ and $x \rightleftharpoons y$ can now be mapped as isograds.

Figure 17–16 (*facing page*)
(A) The block, representing a mass of metamorphic rocks, has a vertical pressure gradient and an inclined temperature gradient. The heavy lines indicating the reactions $a \rightleftharpoons s$ and $x \rightleftharpoons y$ (from the schematic phase diagrams in Figure 17–14) are now nonparallel and cross each other at depth. Emergence of the isograd surfaces at the upper surface of the block would be visible as parallel lines. **(B)** The same block eroded to a new surface *ABCD*. The intersection of the isograds is now visible.

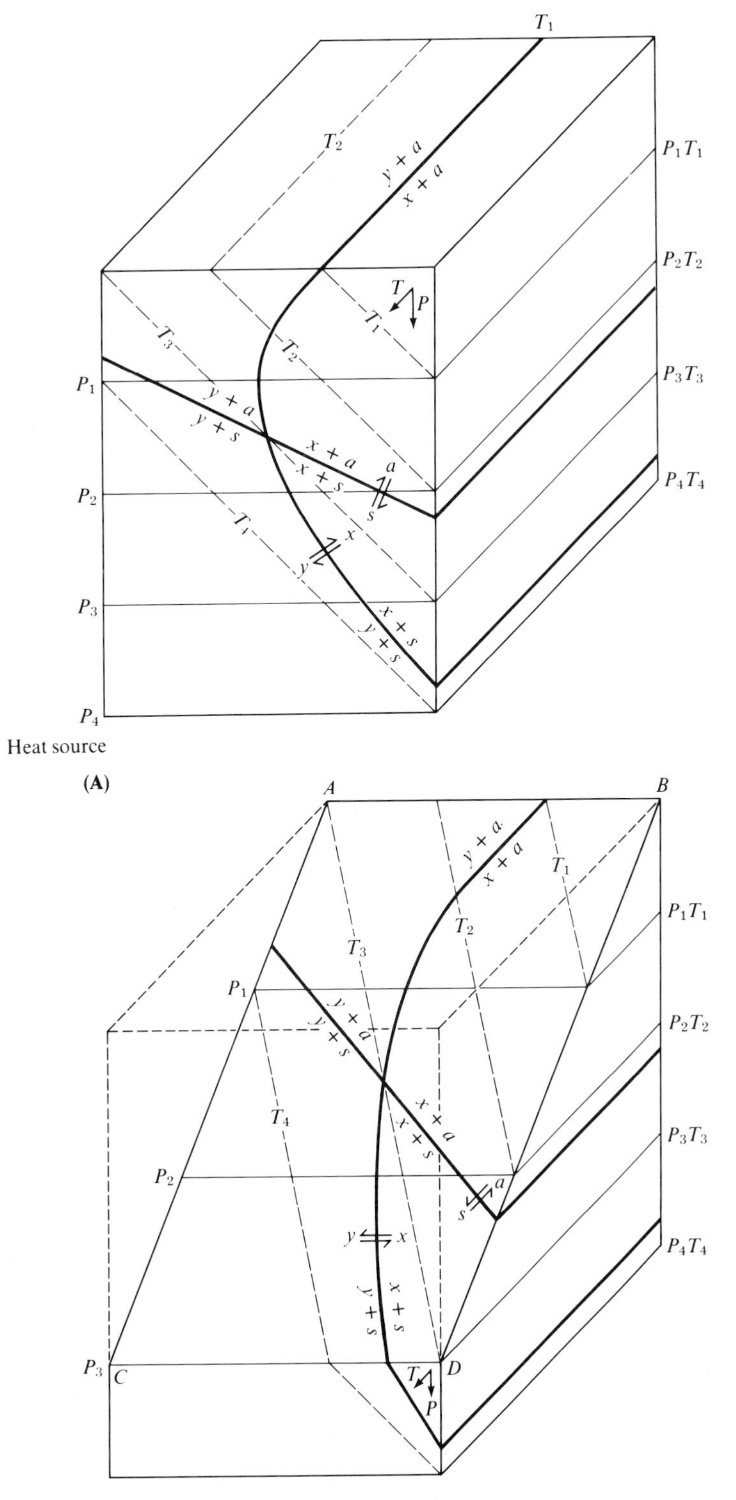

T_1
T_2
$y + a$
$x + a$
P_1T_1
P_2T_2
P_3T_3
P_4T_4
T
P
T_1
T_2
T_3
P_1
$y + a$
$y + s$
$x + a$
$x + s$
a
s
P_2
T_4
x
y
P_3
$x + s$
$y + s$
P_4
Heat source
(A)
A
B
$y + a$
$x + a$
T_1
T_2
T_3
P_1T_1
P_1
$y + a$
$y + s$
P_2T_2
T_4
$x + a$
$x + s$
P_3T_3
P_2
a
s
y
x
P_4T_4
$x + s$
$y + s$
P_3
C
D
T
P
(B)

cantly in mineralogy, texture, and (perhaps) chemistry from its precursor. Metamorphic rocks form within the pressure–temperature range between diagenesis and melting.

Most metamorphism is regional in nature and is related to major orogenic events. As these mountain-building processes generally occur during conditions of rock flowage at high temperatures and pressures, most metamorphic rocks possess a distinct anisotropic fabric (such as foliation, lineation, rock cleavage). The naming of metamorphic rocks is based to a large extent on these textural features (for example, slate, phyllite, schist, and gneiss). However, some metamorphic rock names are based upon composition (marble, serpentinite) or method of origin (skarn). Less common types of metamorphism, such as contact, burial, or cataclastic metamorphism, result in distinctive textural or mineralogical features, with corresponding differences in nomenclature.

Metamorphic rocks are mapped on the basis of texture and mineralogy. In most circumstances more or less isochemical rocks show regional changes in mineralogy and texture. These changes, broadly correlated with regional changes in pressure and temperature, result in the formation of mineralogical zones. Mineralogical zones are separated by fairly narrow regions of transition known as isograds. Isograds, mapped as lines on a geological map, are actually planar or curved mineralogical reaction surfaces that are usually inclined to the earth's surface. Isograds are parallel to each other when metamorphic pressure and temperature gradients are more or less parallel (that is, when each temperature corresponds to a particular pressure). When metamorphic pressure and temperature gradients are inclined to each other, isograd surfaces may be either nonparallel or intersecting.

FURTHER READING

Barrow, G. 1893. On an intrusion of muscovite–biotite gneiss in the east Highlands of Scotland, and its accompanying metamorphism. *Quart. Jour. Geol. Soc. London, 49,* 330–358.

Barrow, G. 1912. On the geology of lower Dee-side and the southern Highland Border. *Proc. Geol. Assoc., 23,* 274–290.

Harker, A. 1932. *Metamorphism.* London: Methuen, 362 pp.

Higgins, M. W. 1971. *Cataclastic Rocks.* U.S. Geological Survey Prof. Paper 687, 97 pp.

Mason, R. 1978. *Petrology of the Metamorphic Rocks.* Boston: George Allen and Unwin/Thomas Murby, 254 pp.

Spry, A. 1969. *Metamorphic Textures.* New York: Pergamon Press, 350 pp.

Tilley, C. E. 1925. A preliminary survey of metamorphic zones in the southern Highlands of Scotland. *Quart. Jour. Geol. Soc. London, 81,* 100–112.

Thompson, P. H. 1976. Isograd patterns and pressure–temperature distributions during regional metamorphism. *Contr. Miner. Petrol., 57,* 277–295.

Thompson, P. H. 1977. Metamorphic *P–T* distributions and the geothermal gradients calculated from geophysical data. *Geology, 5,* 520–522.

Turner, F. J. 1968. *Metamorphic Petrology.* New York: McGraw-Hill, pp. 375–380.

Williams, H., F. J. Turner, and C. M. Gilbert. 1955. *Petrography.* San Francisco: W. H. Freeman and Company, pp. 161–247.

18

Facies and Graphic Representation

THE FACIES CONCEPT

Metamorphic rocks from all parts of the world have been examined chemically and mineralogically for many years. As might be expected, it has been noted that the chemical composition of metamorphic rocks is generally quite similar to the original rocks from which they arose. A second and perhaps more profound observation has been that the number of abundant minerals present in metamorphic rocks is usually limited in number, with a common maximum of five.

The Norwegian geologist V. M. Goldschmidt (1911) examined a group of pelitic (muddy), calcareous, and sandy hornfelses in the Oslo region of Norway. These Paleozoic rocks had been metamorphosed by intrusion of silicic plutons. In spite of their wide range of chemical composition, the hornfelses of the inner contact zone tended to be mineralogically simple; that is, each of the quartz-bearing hornfelses contained four or five abundant phases. In addition, certain mineral pairs (such as anorthite and hypersthene) were consistently present, whereas chemically equivalent pairs (such as andalusite and diopside) were absent. Noting the close correlation of the mineral assemblage with the chemical composition of the rocks (as well as its simplicity), Goldschmidt inferred that these rocks had achieved thermodynamic equilibrium during the elevated pressures and temperatures of metamorphism.

Goldschmidt's observations near Oslo were confirmed by a study of contact metamorphic rocks by the Finnish geologist P. Eskola (1915) in the Orijärvi mining region of Finland. Eskola found a simple and consistent relationship between the mineralogical and chemical compositions of the metamorphic rocks. From this, he inferred that the rocks had achieved thermodynamic equilibrium under metamorphic conditions. Some of the rocks in the Orijärvi and Oslo regions that were chemically equivalent produced the same mineral assemblages. However, other chemically equivalent rocks produced a different suite of minerals. For example, chemically equivalent rocks contained potash feldspar ($KAlSi_3O_8$) and cordierite [$(Mg,Fe)_2Al_4Si_5O_{18}$] in Oslo, whereas in Orijärvi they contained biotite [$K(Mg,Fe)_3AlSi_3O_{10}(OH)_2$] and muscovite [$KAl_3Si_3O_{10}(OH)_2$]. Eskola concluded that because the chemistry of the rocks was similar, the difference in mineralogy must be a result of the differences in the intensity of metamorphism. He (correctly) decided that the Orijärvi rocks (which contain abundant hydrated phases) had formed at lower temperatures and higher pressures than those in Oslo.

Based on these observations and the fact that essentially identical metamorphic mineral assemblages are of worldwide occurrence, Eskola originated the concept of *metamorphic facies*. This is a scheme for the description of metamorphic rocks based upon their mineralogy. As stated by Eskola, "In any rock of a metamorphic formation which has arrived at a chemical equilibrium through metamorphism at constant temperature and pressure conditions, the mineral composition is controlled only by the chemical composition. We are lead to a general conception which the writer proposes to call metamorphic facies" (1915, pp. 114, 115).

The concept is descriptive. Temperatures and pressures are not implied; however, they are determined secondarily from field and laboratory investigations. Assuming that the mineralogy of the rock is controlled both by its chemical composition and by the imposed metamorphic conditions, we can consider the definition of a metamorphic facies by Ramberg: "Rocks formed or recrystallized within a certain *P*, *T*-field, limited by the stability of certain critical minerals of defined composition, belong to the same mineral facies" (1952, p. 136).

Note that the assumption here is that rocks of the same composition within a particular metamorphic facies produce the same mineralogy. The total number of minerals used to define a metamorphic facies is about 11, but any particular rock within the facies usually contains no more than 6 of these.

Although metamorphic facies were defined descriptively in terms of recurring mineral assemblages, metamorphic mineral assemblages can now be thought of as indicators of a range of pressures and temperatures within which different rock compositions yield different but consistent mineral assemblages.

Eskola's five original facies were designated mainly on the basis of the names of metamorphosed basaltic rocks, as he worked mostly with metabasaltic rocks. His original facies have been expanded to include a number of other facies, named mainly on the basis of significant mineralogy. The number of metamorphic facies varies through time as a function of whether the "splitters" or the "lumpers" happen to be in vogue. The generally accepted facies are shown in Figure 18–1 with their approximate limits of pressure and temperature as determined mainly by laboratory experimentation. It should be understood that the limits are tentative, and will be revised in response to future laboratory data. A list of typical minerals from the various facies is given in Table 18–1. Details of the various facies will be discussed in later chapters.

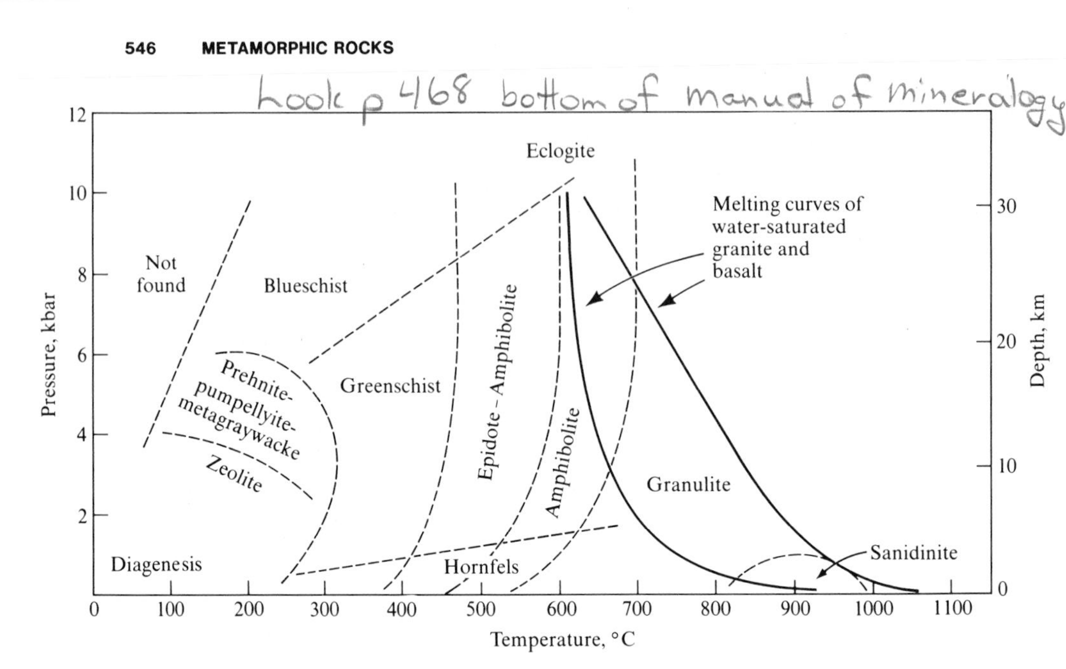

Figure 18–1
The metamorphic facies plotted as a function of pressure and temperature (from various sources). The hornfels facies is occasionally subdivided into pyroxene hornfels, hornblende hornfels, and albite–epidote hornfels. Hornfels facies rocks are distinguished from higher-pressure facies mainly on the basis of texture. Compare Figure 17–9 for the relationships between general types of metamorphism and facies. [The melting curves for water-saturated granites and basalts are from W. C. Luth, R. H. Jahns, and O. F. Tuttle, 1964, *Jour. Geophys. Res., 69,* Fig. 1; and H. S. Yoder, Jr., and C. E. Tilley, 1962, *Jour. Petrology, 3,* Fig. 33.]

EQUILIBRIUM

The discovery that the same type of rock has produced limited but consistent assemblages of minerals under metamorphic conditions led to the idea that most of the metamorphic mineral assemblages we now see were formed in equilibrium with the pressure–temperature conditions of metamorphism; that is, most metamorphic rocks contain an equilibrium mineral assemblage. Given infinite time under metamorphic conditions, no further mineralogic changes can occur. How do we know this is true?

Initial examination of metamorphic rocks quickly reveals that the textures, structures, and minerals characteristic of igneous and sedimentary environments are usually obliterated or at least partially destroyed. However, profound change alone does not indicate that a complete adjustment has been made to the conditions of metamorphism. More evidence than this is needed.

One approach is that of experimentation. By measuring various thermodynamic parameters it is possible to calculate the pressure–temperature range in which a particular mineral assemblage is most stable. This approach, however useful, is not

Table 18–1
Typical Minerals of the Major Metamorphic Facies

Facies	Protolith (precursor) rock type		
	Mafic igneous	Pelite	Calcareous
Blueschist	Glaucophane, lawsonite, pumpellyite, jadeite, chlorite	Glaucophane, lawsonite, chlorite, muscovite, quartz	Tremolite, aragonite, muscovite, glaucophane
Greenschist	Chlorite, actinolite, epidote, albite	Chlorite, muscovite, albite, quartz	Calcite, dolomite, tremolite, phlogopite, epidote, quartz
Epidote–amphibolite	Hornblende, epidote, albite, almandine garnet, quartz	Almandine garnet, chlorite, muscovite, biotite, quartz	Calcite, dolomite, epidote, plagioclase, tremolite, forsterite or quartz
Amphibolite	Hornblende, andesine, garnet, quartz	Garnet, biotite, muscovite, sillimanite, quartz	Calcite, dolomite, diopside, plagioclase, quartz or forsterite, wollastonite
Granulite	Diopside, hypersthene, garnet, intermediate plagioclase	Garnet, orthoclase, intermediate plagioclase, quartz, kyanite or sillimanite	Calcite, plagioclase, diopside, wollastonite, forsterite or quartz
Eclogite	Jadeitic pyroxene, pyropic garnet, ± kyanite	—	—
Hornfels	Diopside, hypersthene, plagioclase	Biotite, orthoclase, quartz, cordierite, andalusite	Calcite, wollastonite, grossularite

practical when we have only a thin section and microscope available. In the latter case it is necessary to search for textural and compositional evidence. One criterion involves examining the thin section to determine whether all of the principal minerals in the rock are in physical contact with each other. If they are in contact, you should then check to see whether any of the minerals show evidence of partial reaction with other minerals. Lack of reaction favors strongly the idea that the minerals existed in stable equilibrium during metamorphism. Any type of reaction relationship, such as replacement textures, reaction rims, or alteration within grains, indicates at least a partial response to changing conditions, and opens the possibility of the existence of a nonequilibrium assemblage. Another useful criterion for establishing equilibrium is the absence of compositionally zoned grains. A crystal that is not homogeneous cannot be in equilibrium with itself; in addition, the central portion of the crystal may not be compatible with other minerals outside of the grain; reaction of the core with external minerals might be prevented by the presence of a rim of different composition. These criteria are difficult to observe in some cases and can require considerable experience. An estimation of equilibrium should always be attempted when describing a metamorphic rock, as an equilibrium assemblage can be useful in determining the particular *PT* conditions to which the rock has been subjected.

It has been discovered that although most metamorphic rocks indicate textural equilibrium, detailed determinations of mineral compositions using the electron microprobe have revealed that on a small scale (the size of a hand specimen or smaller) compositional equilibrium typically has not been completely achieved. A particular mineral species may differ slightly in composition from one crystal to the next, or from center to edge. These small differences indicate that although equilibrium has been established on a large scale, it is not present on a small scale. Such minor non-equilibrium relationships do not rule out the placement of a metamorphic assemblage in a general *PT* framework.

As metamorphism is usually the result of a rise and fall of conditions of pressure and temperature, it is reasonable to ask which of the various *PT* conditions the rock has adjusted to. The answer is that the mineralogy of the rock is usually considered to be compatible with the maximum *PT* conditions. Evidence for this is present in most metamorphic terranes. In contact metamorphic situations it is common to find a series of mineralogic zones demonstrating that mineral assemblages stable at the higher temperatures are closest to the igneous intrusion. In regional metamorphism the sequence of mineralogic zones shows a stepwise change toward a central thermal axis or axes. These changes can be seen best at isograds, where textures often reveal the partial destruction of low *PT* assemblages in favor of those stable at higher pressures and temperatures. Such textural evidence indicates that metamorphic changes are *progressive* in nature; that is, the minerals present in the rock go through a series of changes as conditions of pressure and temperature are raised. Changes taking place during a rise in pressure and/or temperature are called *prograde.* They usually involve the loss of volatile substances such as H_2O and CO_2. As metamorphic conditions decline it would seem reasonable to suppose that the rock would undergo *retrograde* changes in an attempt to adjust to the lower *PT* conditions. However, H_2O and CO_2 are no longer available for recombination, so the rock cannot revert to its original mineral assemblage. Increase in grain size and decrease in porosity during prograde metamorphism also cause a decrease in the rate of retrograde reactions. Evidence of retrograde reactions is often found in metamorphic rocks, but the extent of these reactions is usually small compared to prograde reactions.

THE PHASE RULE

If we accept the idea that most metamorphic assemblages represent a generally complete mineralogical adjustment to the maximum *PT* conditions of metamorphism, then metamorphic rocks can be considered in terms of the Phase Rule (discussed in Chapter 2). As similar mineral assemblages can be found worldwide, it can be assumed that they have not formed at precisely fixed conditions of P and T, but rather over a range of PT conditions. Such a situation represents *divariant equilibrium*—that is, either P or T can vary independently and the same mineral assemblage persists. Univariant and invariant assemblages would be expected to be much less common, as the conditional restrictions are greater. Assuming divariant equilibrium, there are 2 degrees of freedom for a typical metamorphic system, and the Phase Rule, $F = c - p + 2$, can be written as $2 = c - p + 2$, or simply $p = c$. The number of phases equals the number of components. If the number of phases in the rock is larger than the number of components, then (1) a nonequilibrium assemblage is present, (2) the assemblage represents univariant or invariant equilibrium, or (3) the number of components has been chosen incorrectly. Univariant equilibrium sometimes can be determined by noting the presence of both products and reactants of a particular reaction coexisting in the rock in an apparently stable relationship [for example, a rock containing both kyanite (Al_2SiO_5) and its polymorph sillimanite (Al_2SiO_5)]. Similarly, invariancy could be indicated in a case where all three Al_2SiO_5 polymorphs are found together in stable equilibrium. However, metastable persistence of these phases outside their stability fields is common. Only a few cases have been found worldwide in which the three polymorphs are present within the same rock. In every case metastable persistence is probable. At least one of the polymorphs has been shown to be (1) in a replacement relationship to another or (2) incompatible with the pressure–temperature conditions indicated by other minerals in the rock. Many cases of the occurrence of two polymorphs together have been established as metastable equilibrium on the same bases. Divariancy is the general rule.

The relation $p = c$ suggests that we have a method of deciding whether an equilibrium assemblage is actually present. But a problem arises in deciding upon the number of chemical components in the rock. It is not sufficient to take a chemical analysis of the rock and assume that each oxide present is a component. Certain elements, because of similarities of size and charge, are able to substitute for other elements in a crystal structure. For example, some minor elements are camouflaged in the sense that they are present in minor amounts and do not result in the creation of a phase in which they are a major constituent; small amounts of chromium may be present in diopside, or traces of cadmium may be present in sphalerite. As the behavior of such minerals is essentially unchanged by these substitutions, most minor elements are not considered as components when considering equilibrium relations of the rock. On a larger scale FeO, MgO, and MnO commonly substitute for each other, and together are often taken as a single component. Finally there are certain stoichiometric associations to be taken into account; for example, at the higher levels of metamorphism, the K_2O that is present will combine with an equal molecular amount of Al_2O_3 (and $6SiO_2$) to yield potassium feldspar. Similarly, CaO and/or Na_2O will combine with an equal amount of Al_2O_3 (and silica) to produce plagioclase feldspar. Thus two of the usual six components found in the average metamorphic rock are taken as

$K_2O \cdot Al_2O_3$ and $(Ca,Na_2)O \cdot Al_2O_3$. The other four commonly chosen components are $(Si,Ti)O_2$, $(Mg,Fe,Mn)O$, H_2O, and $(Al,Fe)_2O_3$ (which utilizes the Al_2O_3 not combined in feldspars). In many pelitic rocks, FeO and MgO must be considered as separate components. Note that along with oxygen these components include Si, Ti, K, Al, Ca, Na, Mg, and Fe, which are the eight most abundant cations in the crust and constituents of most common rock-forming minerals.

Assuming divariant equilibrium and the components given above, an average metamorphic rock should consist of five or fewer minerals. A sixth fluid phase will be present during metamorphism. A maximum of five solid phases suggests that equilibrium has been attained (Goldschmidt, 1911). If the number of minerals in the rock is greater than five, either an incorrect choice of components or nonequilibrium is suggested. Note carefully that these are suggestions, and equilibrium can only be strongly indicated when it is firmly established by determination of appropriate thermodynamic parameters.

GRAPHIC REPRESENTATION

As a six-component system cannot be treated graphically, it is common in metamorphic petrology to plot rock compositions in terms of three or four components. Although this occasionally leads to oversimplification, the approach is usually worthwhile. One useful approach devised by Eskola is the *ACF* diagram. Here rock compositions can be plotted in terms of three components, called *A*, *C*, and *F*. The components are determined by taking into account both the chemical analysis and the mineral assemblage of the rock.

Certain minerals present in metamorphic rocks are stable throughout a wide range of metamorphic conditions. As these minerals are not useful in demonstrating mineralogic changes during metamorphism, their contribution to the chemical composition of the rock are eliminated from consideration, and only the remaining chemical constituents are plotted. The minerals eliminated from consideration are albite, potassium feldspar, magnetite, ilmenite, sphene, apatite, and micas. The remaining chemical constituents react to form minerals that undergo significant mineralogic changes during metamorphism.

The *ACF* components were defined by Eskola as follows:

$$A = Al_2O_3 + Fe_2O_3 - (Na_2O + K_2O)$$
$$C = CaO - 3.3\ P_2O_5$$
$$F = FeO + MgO + MnO$$

Aluminum oxide and Fe_2O_3 are grouped together because they commonly substitute for each other in crystal structures, as do FeO, MgO, and MnO. The percentages of Na_2O and K_2O are subtracted from the *A* component, as these oxides are presumed to unite with equal amounts of Al_2O_3 in the alkali feldspars. The quantity

3.3 P_2O_5 is subtracted from CaO because all of the P_2O_5 is assumed to be utilized in the mineral apatite. In apatite, $Ca_5(PO_4)_3(OH,F,Cl)$, the molecular ratio CaO/P_2O_5 is 10:3; that is, apatite contains 3.3 times as much CaO as P_2O_5. The amounts of any of the other minerals listed above (ilmenite, magnetite, and so on) are determined by point counting of the thin section, and appropriate amounts of the *A*, *C*, and *F* components are removed from the analysis. The method uses molecular proportions, which are obtained by dividing the weight percent of each oxide by its molecular weight. After the *A*, *C*, and *F* components are determined, these three numbers are recalculated on a percentage basis; thus if the *A*, *C*, and *F* components were 2, 3, and 15 respectively, this would recalculate on a percentage basis to 10%, 15%, and 75%. The rock analysis can then be plotted on a triangular graph. In addition to rock analyses, the *ACF* diagram can be used to plot minerals of interest on the basis of their specific or general composition (see Figure 18–2). The minerals muscovite and

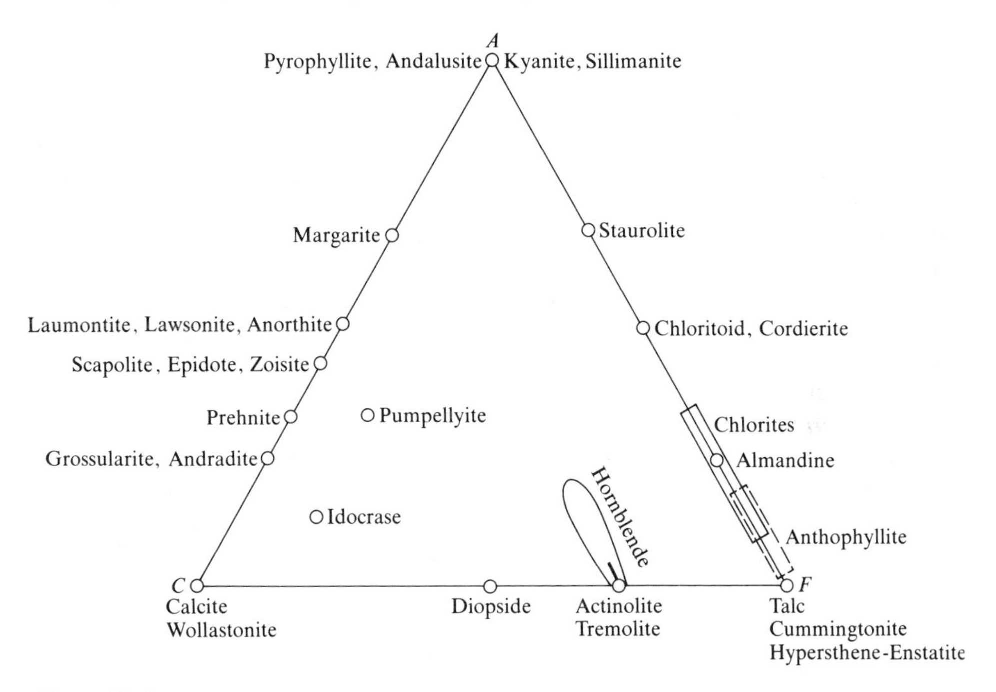

Figure 18–2
Common metamorphic minerals plotted in terms of the components *A*, *C*, and *F*. As an example of the method, consider the garnet grossularite, $Ca_3Al_2(SiO_4)_3$; this mineral contains 3CaO and $1Al_2O_3$. Hence component $C = 75$ and component $A = 25$, which places it along the left side of the diagram. Staurolite, $4FeO \cdot 9Al_2O_3 \cdot 8SiO_2 \cdot H_2O$, contains $9Al_2O_3$ and 4FeO, which yields the components $A = 69$ and $F = 31$. The mineral plots on the *AF* side of the triangle, near the *A* corner. Other minerals and rock compositions are plotted by similar techniques. Note that in this and the following two diagrams, mineral compositions may be plotted as points, lines, or regions; this is done in order to take account of compositional variability of each phase.

biotite are included on *ACF* diagrams by some petrologists. Details of the method are given by Winkler (1979) and Mason (1978).

Another commonly used system of representation is the *A'KF* diagram. *A'* is used rather than *A*, as the meaning is different than the *A* in the *ACF* system:

$$A' = Al_2O_3 + Fe_2O_3 - (Na_2O + K_2O + CaO)$$
$$K = K_2O$$
$$F = FeO + MgO + MnO$$

Many of the same type of restrictions apply with respect to subtracting the pertinent oxides of phases that are not plotted on the triangular *A'KF* diagram. Typical minerals on such a diagram are shown in Figure 18–3.

A third type of diagram, devised by Thompson (1957), has been found very useful in showing the mineralogical relationships in metapelites. This method, based on the tetrahedron K_2O–Al_2O_3–FeO–MgO, produces what is known as an *AFM* or *AKFM* diagram. This type of diagram was created to take into account the fact that in many metapelites, FeO and MgO do not act as a single component and must be treated individually.

Representation of a rock or mineral composition in terms of four components (K_2O, Al_2O_3, FeO, and MgO) requires that the plotted point lie within a three-dimensional compositional tetrahedron. As this is extremely difficult to represent on a two-dimensional sheet of paper, Thompson chose to project points within the tetrahedron to the triangular base and its extension. The technique is shown in Figure

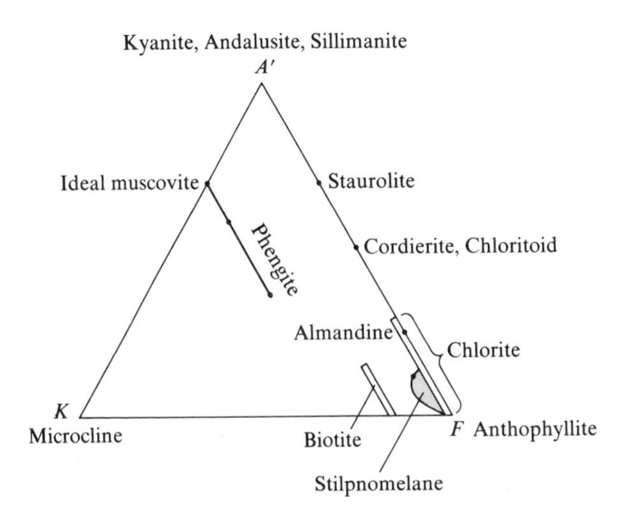

Figure 18–3
Common metamorphic minerals plotted in terms of the components *A'*, *K*, and *F* (described in text).

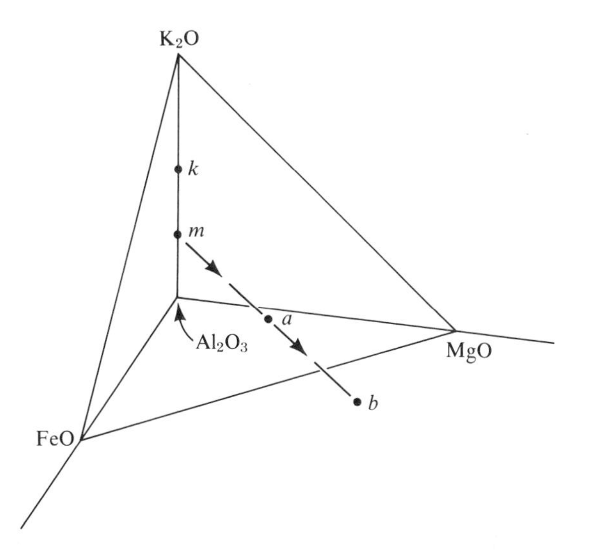

Figure 18–4
The tetrahedron K_2O–Al_2O_3–FeO–MgO with the basal plane Al_2O_3–FeO–MgO extended. Mineral or rock compositions within the tetrahedron, such as *a*, are projected from *m*, the muscovite composition point, onto the base as at *b*. Points projected to the base may fall within or outside of the tetrahedron. Point *k* is the composition point for K-feldspar.

18–4. The rock or mineral composition point *a* lies within the tetrahedron. The point from which *a* is projected is indicated as *m*; the point *m* is the composition point for the mineral muscovite. If potassium feldspar rather than muscovite is present in the rock, then the point *k* is used as the projection point. Generally no other potassium-rich, iron–magnesium-poor minerals are present. Projection of the point *a* from *m* brings it to an extension of the Al_2O_3–FeO–MgO base at *b*. The tetrahedron is then removed from the *AFM* basal triangle and all of the pertinent rock or mineral compositions are shown on the base, as in Figure 18–5. Potassium feldspar (point *k*, Figure 18–4) is the only common mineral that does not project from *m* to the *AFM* base or its extension; to cover this, potassium feldspar is indicated, when present, with arrows as in Figure 18–5. The technique of plotting a rock analysis on *ACF* and *AFM* diagrams is shown in Table 18–2.

Let us consider how these three types of diagrams may be applied to a rock such as a metapelite. One common mineral assemblage consists of kyanite, staurolite, muscovite, biotite, quartz, and plagioclase. We will make the assumption (usually justified by textural evidence) that this assemblage formed in equilibrium with the imposed metamorphic conditions. The composition points of the minerals kyanite, staurolite,

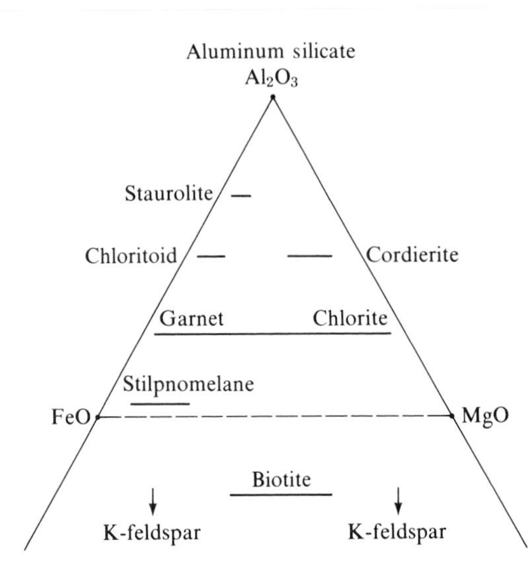

Figure 18–5
Metamorphic minerals plotted on the *AFM* base of the tetrahedron and its extension (see Figure 18–4). Compositional variation is shown by lines. K-feldspar, which does not project onto this surface, is indicated by the arrows and labels. The minerals plotted may exist in the presence of muscovite and quartz.

and plagioclase can be plotted on the *ACF* diagram (see Figure 18–6A). As these minerals are compatible (in equilibrium) with each other, we can show this by joining the three mineral composition points by heavy lines.* The rock composition, x, must fall somewhere in the compatibility triangle formed by the heavy lines. Plotted on the *AFM* diagram (see Figure 18–6B), the point x again falls within a compatibility triangle, whose corners are indicated by the composition points of kyanite, staurolite, and biotite; in fact, although this is seen as a triangle on the *AFM* base, the point x falls within a compatibility tetrahedron, as muscovite is stable with these phases, and forms the upper corner of the tetrahedron. These same four minerals can be located on an *A'KF* diagram (see Figure 18–6C); here the rock composition falls within a quadrilateral rather than a triangle.

Let us attempt to unravel the significance of these arrangements. Assuming that this rock represents a condition of divariant equilibrium, the Phase Rule ($F = c - p + 2$) reduces to $p = c$. By means of a representation such as the *ACF* plot we are making the assumption that the number of significant components can be reduced to three—those characterized by A, C, and F—such that the diagram represents a three-component system A–C–F. If the assumption is correct and $p = c$, then this three-component system should indicate assemblages of three coexisting phases under conditions of divariant equilibrium (as $p = c$ is $3 = 3$). The rock composition should

*Note that in these diagrams a rock composition is plotted in terms of the abundance of three minerals; these minerals may be located at the corners, edges, or within the reference triangle, and themselves form a subtriangle. Compositions are plotted in terms of this subtriangle. In sedimentary petrology, similar triangular graphs are used, but compositions are only plotted in terms of the three constituents that are indicated at the corners of the larger reference triangle.

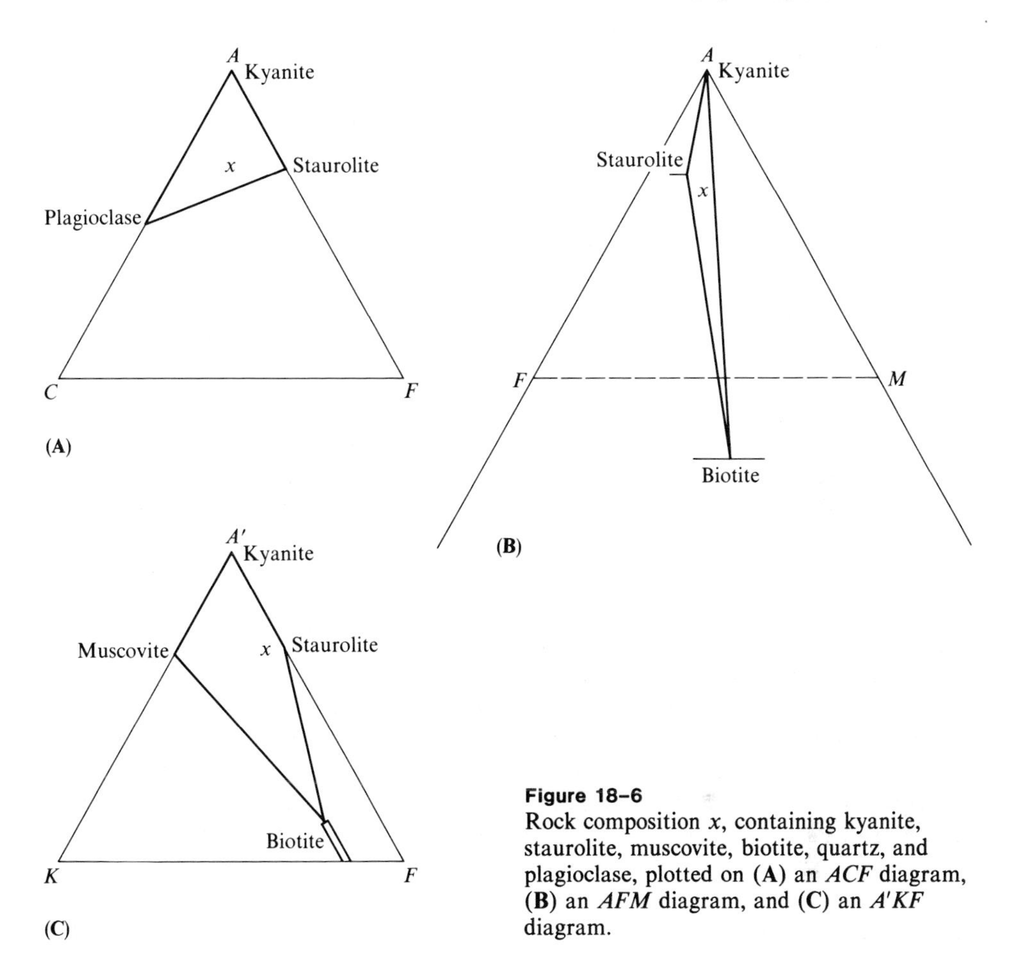

Figure 18–6
Rock composition x, containing kyanite, staurolite, muscovite, biotite, quartz, and plagioclase, plotted on (**A**) an *ACF* diagram, (**B**) an *AFM* diagram, and (**C**) an *A′KF* diagram.

fall within a compatibility triangle, whose corners consist of the composition points of three coexisting minerals. This is often the case, and for particular conditions of *P* and *T*, the *ACF* triangle can be broken into smaller three-phase compatibility triangles for various rock compositions, as seen in Figure 18–6A. The general validity of the various assumptions is indicated by the fact that it is the normal situation rather than the exception to find three of the *ACF* minerals in a rock rather than four or five. Similar relationships are true with the *AFM* and *A′KF* diagrams. Mineral compatibilities can usually be shown on these compositional triangles as three or fewer coexistent phases. However, exceptions are often found to this simplistic triangular arrangement, as is evident in the *A′KF* diagram of Figure 18–6C. Four of the minerals in this rock are kyanite, staurolite, biotite, and muscovite. No problem in representation arises in the *AFM* diagram, as the rock composition can be located within

Table 18–2
An Example of the Calculation of *ACF* and *AFM* Values from a Chemical Analysis

	Wt. %	Mol. wt.	Mol. prop.
SiO_2	65.74	60.07	1.0943
Al_2O_3	17.35	101.82	0.1703
Fe_2O_3	1.90	159.68	0.0118
FeO	3.35	71.84	0.0466
MgO	1.90	40.31	0.0471
CaO	1.25	56.07	0.0223
Na_2O	1.78	61.97	0.0287
K_2O	3.28	94.20	0.0348
MnO	0.03	70.93	0.0004
P_2O_5	0.82	141.92	0.0058
H_2O	2.01	18.01	—
	99.41		

Convert weight percent from the chemical analysis into molecular proportions by dividing the weight percent of each oxide by its molecular weight as above. It is then necessary to remove from consideration any minerals in the rock that are not plotted on the *ACF* diagram and that contain some of the *ACF* components. These minerals are determined in volume percent by means of point counts, converted to weight percent, and appropriate molar amounts subtracted from the *A*, *C*, and/or *F* components. These techniques are described in some detail by Winkler (1979), and will be eliminated for simplification. We will assume that all Na_2O and K_2O are combined in feldspars and that all P_2O_5 is present in apatite.

As each mole of K_2O and Na_2O combines with one mole of Al_2O_3 in potassium and sodium feldspar, the molecular amounts of K_2O and Na_2O are subtracted from the *A* component.

$$\begin{aligned} A &= Al_2O_3 + Fe_2O_3 - (Na_2O + K_2O) \\ &= 0.1703 + 0.0118 - (0.0287 + 0.0348) \\ &= 0.1821 - .0635 \\ &= 0.1186 \end{aligned}$$

For the *C* component it is necessary to take the mineral apatite into consideration. Every mole of P_2O_5 in apatite is combined with 3.3 moles of CaO.

$$\begin{aligned} C &= CaO - 3.3\ P_2O_5 \\ &= 0.0223 - 3.3\ (.0058) \\ &= 0.0223 - 0.0191 \\ &= 0.0032 \end{aligned}$$

No subtractions are necessary for the *F* component in this example:

$$\begin{aligned} F &= FeO + MgO + MnO \\ &= 0.0466 + 0.0471 + 0.0004 \\ &= 0.0941 \end{aligned}$$

The values of *A*, *C*, and *F* are reset to equal 100%, and the rock composition is plotted on the *ACF* triangle.

$$\begin{aligned} A &= 0.1186 \\ C &= 0.0032 \\ F &= \underline{0.0941} \\ & 0.2159 \end{aligned} \qquad \begin{aligned} A &= 54.93\% \\ C &= 1.48\% \\ F &= \underline{43.58\%} \\ & 99.99 \end{aligned}$$

Table 18-2 (*continued*)

The same rock can be plotted on an *AFM* diagram. We will assume here that both quartz and muscovite (neglected above) are present. Again, for simplification, no account is taken of other minerals that might be present, whose amounts must be determined by point counts. The rock composition point is determined in terms of molecular amounts of components *A* and *M*, where

$$A = \frac{Al_2O_3 - 3K_2O}{Al_2O_3 - 3K_2O + MgO + FeO}$$

$$= \frac{0.1703 - 3(0.0348)}{0.1703 - 3(0.0348) + 0.0471 + 0.0466}$$

$$= \frac{0.1703 - 0.1044}{0.1703 - 0.1044 + 0.0471 + 0.0466}$$

$$= \frac{.0659}{0.1596}$$

$$= 0.4129$$

$$M = \frac{MgO}{MgO + FeO}$$

$$= \frac{0.0471}{0.0471 + 0.0466}$$

$$= \frac{0.0471}{0.0937}$$

$$= 0.5027$$

Using the values $A = 0.4129$ and $M = 0.5027$, the rock composition point is plotted on the *AFM* triangle below.

Note once again that following plotting of the rock composition point, composition points of pertinent minerals present in the rock are plotted as well. This will usually show the rock composition to be located within a compatibility triangle, the corners of which represent the compositions of the pertinent minerals present.

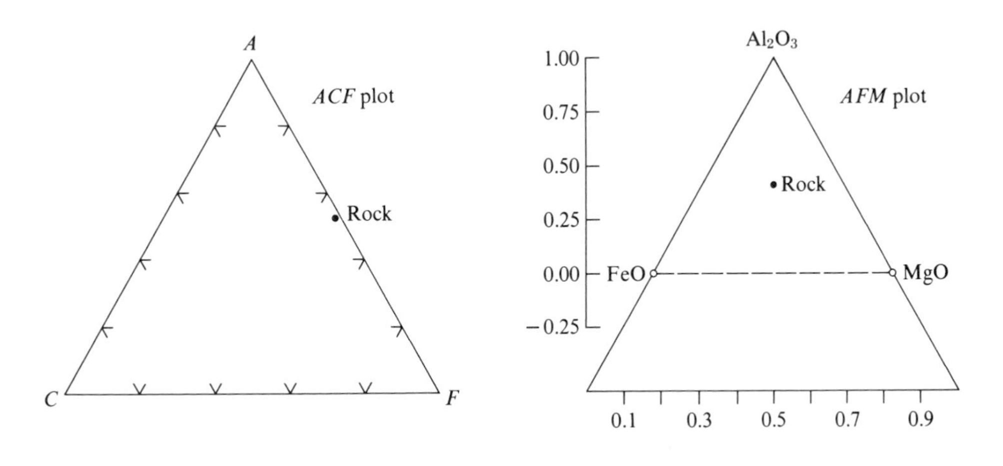

the compatibility triangle staurolite–kyanite–biotite. As muscovite (and quartz) are assumed to be present with all assemblages on the *AFM* base, no conflict arises. But the four minerals kyanite, staurolite, muscovite, and biotite can all be located in the *A'KF* diagram; this places a typical composition x within a quadrilateral rather than a triangle, which implies either (1) that equilibrium has not been achieved (as $p \neq c$), (2) that the assemblage is univariant, and perhaps contains both products and reactants for a reaction such as

$$\text{Kyanite} + \text{Biotite} + H_2O \rightleftharpoons \text{Staurolite} + \text{Muscovite}$$

or (3) that the system cannot be regarded as being three-component in terms of *A'KF*. The common occurrence of this assemblage argues against the first two possibilities. The latter possibility is correct as seen by reexamination of the *AFM* diagram, where it is obvious that all four of these phases exist together in terms of four components rather than three. It can be seen in the *AFM* diagram that biotite and staurolite contain different amounts of FeO and MgO (with more MgO in biotite than staurolite). This unequal partitioning of FeO and MgO indicates that these oxides are not acting in an equivalent manner. Therefore the two oxides cannot be grouped together and considered as a single component, as is done in the *A'KF* arrangement. Hence, relative to these minerals, the rock can only be reduced to four components, and the *A'KF* diagram cannot be used to show equilibrium relations for this rock correctly. For reasons like this, *AFM* diagrams have had much greater popularity than the *A'KF* type in recent years.

We have considered the plotting of a single mineral assemblage in the example above. However, petrologists have been determining mineral compatibilities in metamorphic rocks for many years. All of the major rock types and their constituent minerals have been determined and plotted on diagrams of the type described (as well as others for special purposes). These diagrams provide a shorthand method of describing the major mineral assemblages for each of the metamorphic facies and have proved to be an invaluable tool in analyses of metamorphic rocks. Chapter 20 discusses the major mineral assemblages in each of the metamorphic facies.

SUMMARY

Metamorphic rocks of any composition that have been formed within a limited range of pressure and temperature constitute a metamorphic facies; metamorphic rocks are determined to be within a particular facies on the basis of the presence of certain characteristic minerals.

The repetitive occurrence of particular mineral assemblages in rocks of similar composition formed under a variety of *PT* conditions suggests that the average metamorphic rock has formed under conditions of divariant equilibrium ($p = c$) at the maximum pressure and temperature to which the rock has been subjected. Other criteria generally confirm this for large-scale textural and mineralogic features.

The usual six chemical components of a metamorphic rock often can be reduced to three or four for simple graphic representation. The plotting of mineral assemblages on a triangular graph provides a shorthand method of recording equilibrium assemblages for a wide variety of rock compositions.

FURTHER READING

Eskola, P. 1915. On the relations between the chemical and mineralogical composition in the metamorphic rocks of the Orijärvi region. *Bull. Comm. Geol. Finlande, 44,* 109–145.

Goldschmidt, V. M. 1911. *Die Kontaktmetamorphose im Kristianiagebiet.* Oslo Vidensk. Skr. I, Math-Nat. Kl., No. 11.

Hyndman, D. M. 1972. *Petrology of Igneous and Metamorphic Rocks.* New York: McGraw-Hill, pp. 261–264.

Mason, R. 1978. *Petrology of the Metamorphic Rocks.* Boston: George Allen and Unwin/ Thomas Murby, pp. 44–60.

Miyashiro, A. 1973. *Metamorphism and Metamorphic Belts.* New York: John Wiley and Sons, pp. 120–135.

Ramberg, H. 1952. *The Origin of Metamorphic and Metasomatic Rocks.* Chicago: University of Chicago Press, 317 pp.

Thompson, J. B. 1957. The graphical analysis of mineral assemblages in pelitic schists. *Amer. Miner., 42,* 842–858.

Turner, F. J. 1968. *Metamorphic Petrology—Mineralogical and Field Aspects,* 1st Ed. New York: McGraw-Hill, pp. 46–82.

Turner, F. J. 1980. *Metamorphic Petrology,* 2d Ed. New York: McGraw-Hill, pp. 48–61, 181–191.

Winkler, H. G. F. 1979. *Petrogenesis of Metamorphic Rocks,* 5th Ed. New York: Springer-Verlag, pp. 28–54.

19

Controls and Processes of Metamorphism

THE CONTROLS: PRESSURE, TEMPERATURE, AND COMPOSITION

The stability relations of metamorphic minerals are best discussed by considering the variables of pressure, temperature, and composition to which they are subjected. This can be done by the use of phase diagrams and consideration of chemical equilibria.

Pressure can be of a variety of types. As metamorphism occurs at some depth, the rocks are subject to the weight of the overlying materials. Various authors refer to this as burial pressure, lithostatic pressure (P_{lith}), load pressure, or confining pressure (P_{conf}). This is a pressure that squeezes the mineral grains together. Under burial conditions this type of pressure is strictly hydrostatic in nature—that is, it can be considered to be applied equally in all directions. Under conditions of metamorphism when large masses of rock are being deformed by ductile flow, differential stress may be present. This type of stress results in foliation and folds, and is typical of regional metamorphism.

In addition to pressures from adjacent solids, a vapor or fluid phase is present. This usually consists essentially of H_2O, but also may contain O_2 or CO_2, as a function of the rock composition. In most cases the P_{H_2O} and P_{lith} are considered to be equal.

The stability of metamorphic minerals is influenced by the particular type of pressure in the system, as also is true with igneous minerals. Consider the phase relations between the three polymorphs of Al_2SiO_5 (andalusite, kyanite, and sillimanite), shown in Figure 19–1A. Andalusite is stable at fairly low pressures and temperatures, sillimanite at high temperatures, and kyanite at high pressures. These stability regions have been determined in devices that maintain high confining pressures, so the pressure coordinate on the diagram refers to lithostatic pressures. But what if the type of pressure were changed?

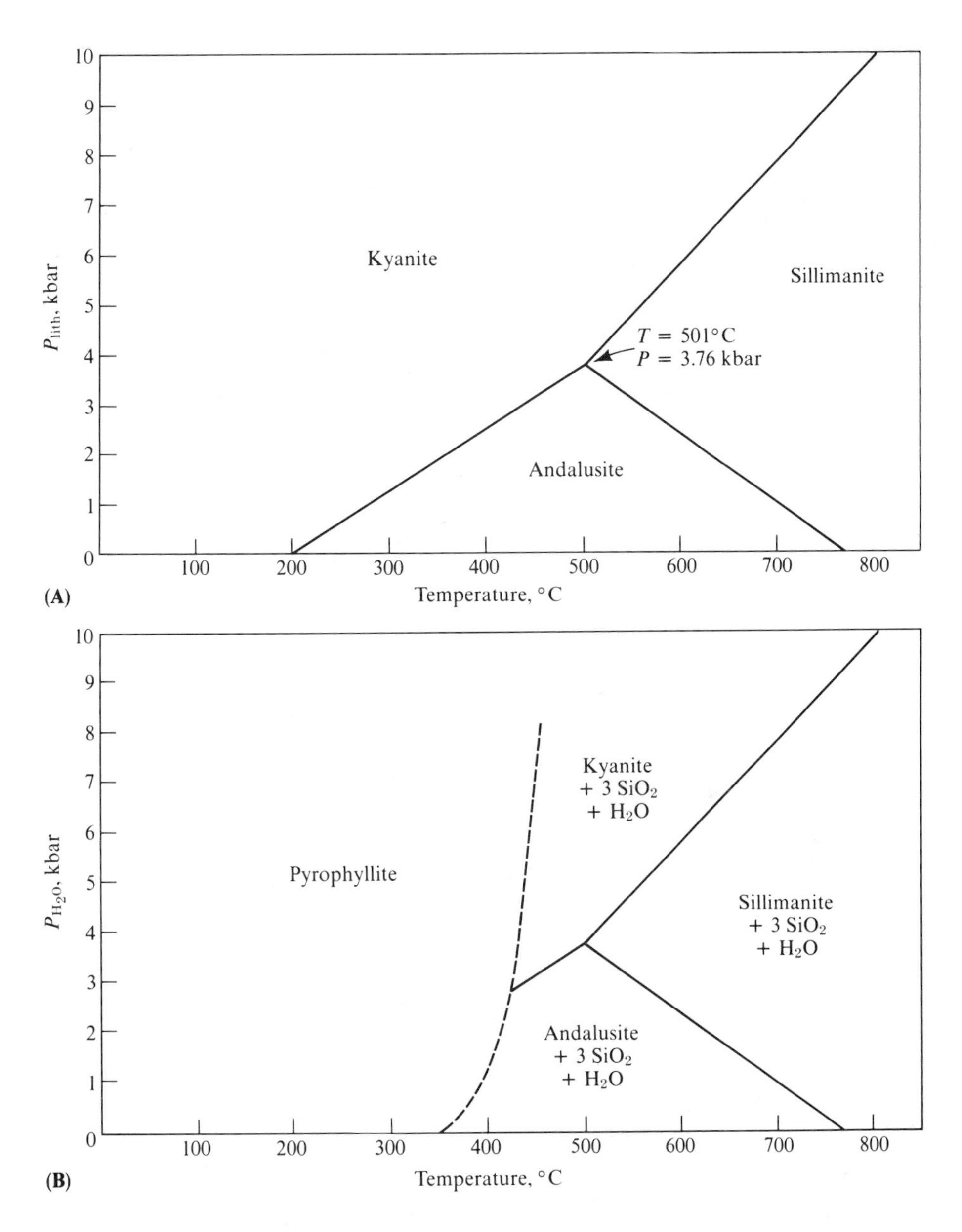

Figure 19-1
(A) The system Al_2SiO_5 as a function of P_{lith}. **(B)** The stability of $Al_2SiO_5 + 3SiO_2$ as a function of P_{H_2O}. Addition of silica to the aluminum silicate polymorphs does not change their stability relations, but the use of P_{H_2O} rather than P_{lith} results in the formation of pyrophyllite, $Al_2Si_4O_{10}(OH)_2$, at low temperatures. [From M. J. Holdaway, 1971, *Amer. Jour. Sci., 271,* Fig. 5, and D. M. Kerrick, 1968, *Amer. Jour. Sci., 266,* Fig. 4.]

If the system were determined using P_{CO_2} the results would be the same, as CO_2 does not form compounds with any of the Al_2SiO_5 polymorphs. The phase transitions would be controlled by volume differences between the solids and the heats of transition (ΔH), as they were for lithostatic pressure.

Consider now the effect of using P_{H_2O}, with the addition of some SiO_2 to the system (see Figure 19–1B). The added SiO_2 has no effect on the univariant curves, as the Al_2SiO_5 polymorphs are already silica-saturated. However, if the pressure is P_{H_2O} rather than P_{lith}, the silica will react with Al_2SiO_5 at low temperatures to form the hydrated mineral pyrophyllite, $Al_2Si_4O_{10}(OH)_2$. At temperatures higher than the pyrophyllite stability line the Al_2SiO_5 univariant curves remain the same (as the available H_2O does not react with Al_2SiO_5); below the pyrophyllite stability limit the Al_2SiO_5 phases are eliminated. The pyrophyllite stability curve is useful to indicate the pressures and temperatures of first appearance of andalusite during prograde metamorphism of quartz-bearing rocks. During retrograde metamorphism andalusite or kyanite may persist to lower temperatures than those indicated by the pyrophyllite stability curve as a result of lack of H_2O or slow reaction rates.

The composition of the vapor phase exerts a strong effect on mineral stability. Consider the following reaction (see Figure 19–2):

$$\underset{\text{Calcite}}{CaCO_3} + \underset{\text{Quartz}}{SiO_2} \rightleftharpoons \underset{\text{Wollastonite}}{CaSiO_3} + CO_2$$

At atmospheric pressure the reaction occurs at 430°C. If the system were subjected to increasing values of P_{CO_2}, the size of the stability field of the carbonate mineral would increase (Le Chatelier's Principle); that is, the reaction temperature (curve *a*) would rise with pressure. If on the other hand the pressure is caused by a substance such as H_2O that does not react with the solid phases, the reaction curve would be forced to the left (curve *b*) because the volume of wollastonite is less than that of an equivalent amount of quartz and calcite; hence wollastonite is favored (and its stability field enlarges) with increasing P_{H_2O}. (We assume that the CO_2 escapes from the system because it is "confined" only by H_2O). If the vapor phase were a mixture of both CO_2 and H_2O, the reaction curve would occupy an intermediate position between curves *a* and *b*.

Temperature, fortunately, is simpler to consider than pressure, as there is only one type. The effect of temperature, however, is varied. A rise in temperature increases the internal energy of the rock system and encourages various types of readjustments. Atomic and molecular mobility is increased and the rock may recrystallize. Premetamorphic sedimentary grains that were in contact with each other dissolve at contact points and reprecipitate in void spaces between the grains (perhaps as overgrowths on different portions of the same grains from which they dissolved). A general "healing" of the rock may occur by the elimination of strained and deformed grains, filling of cracks and fractures, and perhaps the elimination or decrease in the amount of mechanically induced twins. The general effect of recrystallization caused by temperature increase is to cause both an equigranular texture and increase of grain sizes.

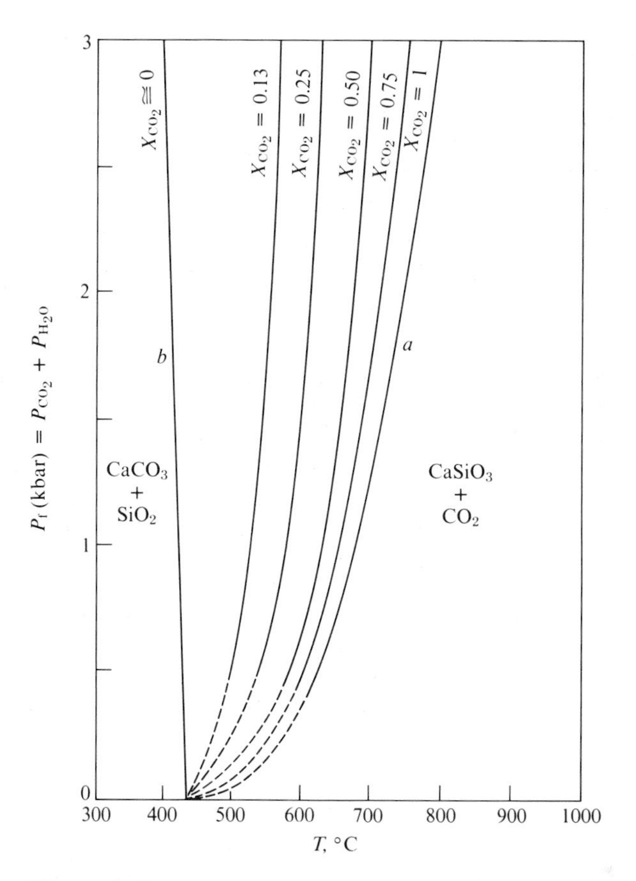

Figure 19–2
Univariant curves for the reaction $CaCO_3 + SiO_2 \rightleftarrows CaSiO_3 + CO_2$. Curve *a* shows the reaction when $P_{total} = P_{CO_2}$. Curve *b* shows the reaction when $P_{total} = P_{H_2O}$. Intermediate curves show the reaction with various ratios of P_{CO_2} and P_{H_2O}. X_{CO_2} represents that fraction of the pressure due to CO_2, with a value of 1 being a maximum. P_f = fluid pressure. [After the data of R. I. Harker and O. Tuttle, 1956, *Amer. Jour. Sci., 254,* 239–256, and H. J. Greenwood, 1967, *Amer. Miner., 52,* 1669–1680.]

A second effect of temperature rise is to eliminate some phases and create new ones. The most common occurrences are dehydration reactions. An example of this is shown in Figure 19–3A. Muscovite, the hydrated phase, decomposes with increase in temperature to yield a mixture of potassium feldspar, corundum (Al_2O_3), and water vapor. Note also that the field of stability of muscovite is enlarged with increasing P_{H_2O}; that is, the muscovite decomposes at higher temperatures. In the typical metamorphic rock, muscovite is associated with quartz. As quartz reacts with muscovite,

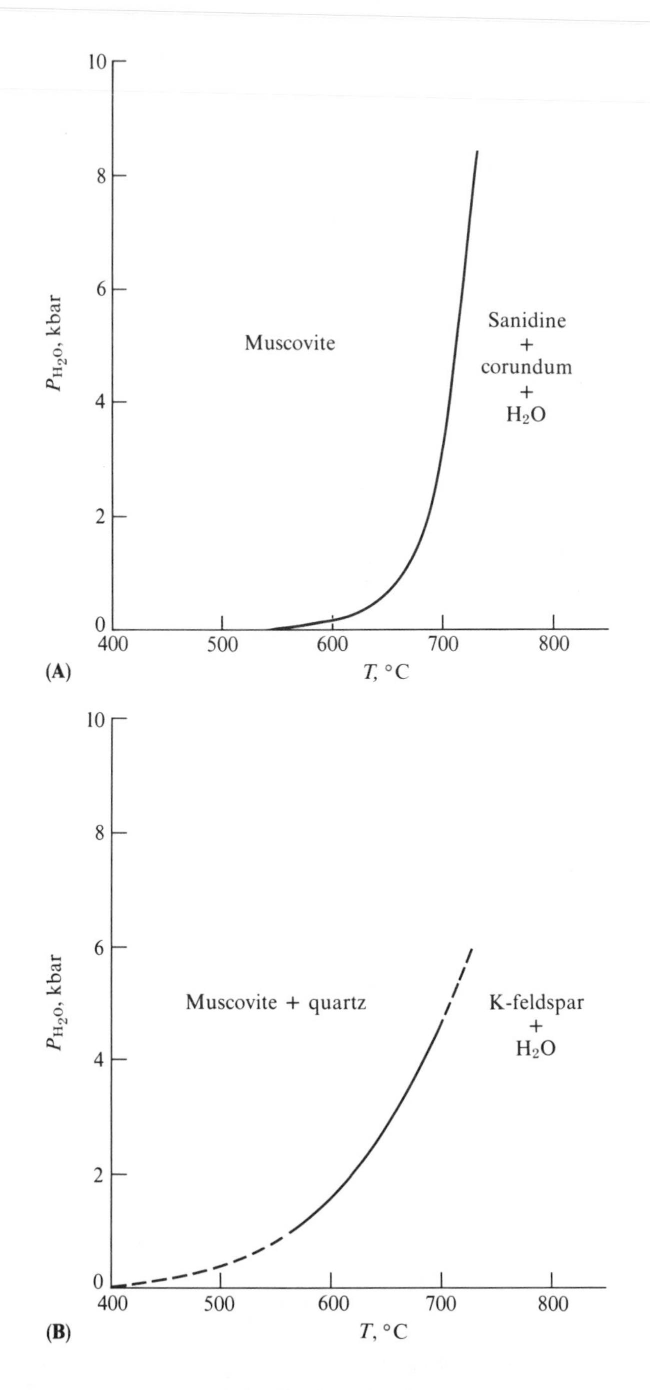

Figure 19-3
A typical dehydration reaction. The univariant curve in part A shows the decomposition temperature of muscovite as a function of P_{H_2O}. Part B shows that muscovite in the presence of quartz decomposes at a lower temperature. For both curves, however, the field of stability of the hydrate, muscovite, is enlarged as P_{H_2O} is increased. [Part A after B. Velde, 1966, *Amer. Miner., 51,* Fig. 1; part B after E. Althaus et al., 1970, *Neues Jahrb. Miner., 7,* Fig. 2.]

the decomposition curve for muscovite + quartz (see Figure 19–3B) is found at lower temperatures, but still follows approximately the general slope of the curve for pure muscovite.

A consequence of increasing temperature or pressure may be changes in the compositions of the minerals. This is illustrated for three components in Figure 19–4A, but may involve a great many more in a natural complex system. Consider the rock system *X–Y–Z*. It contains the minerals α and β, each of which can vary in composition within an area indicated by the shaded bars. At a particular *PT* condition, the lines joining the compositional bars (two-phase tie lines) indicate which compositions may coexist in equilibrium. Such lines are determined by experiment. A rock composition *h* consists of β of composition *b*, and α of composition *a*. When the system is changed to a new *PT* condition (see Figure 19–4B), the tie lines assume a different arrangement. The same rock composition, *h*, consists of the same two phases α and β; but now, due to interchange of materials by diffusion, they are of composition a' and b'. Compositional changes in metamorphic mineral pairs are now being examined by metamorphic petrologists, and compositions of coexisting minerals are being related to the various levels of metamorphism (see Appendix). Systematic changes in mineral composition are being used to decide whether metamorphic minerals have formed in equilibrium with imposed metamorphic *PT* conditions.

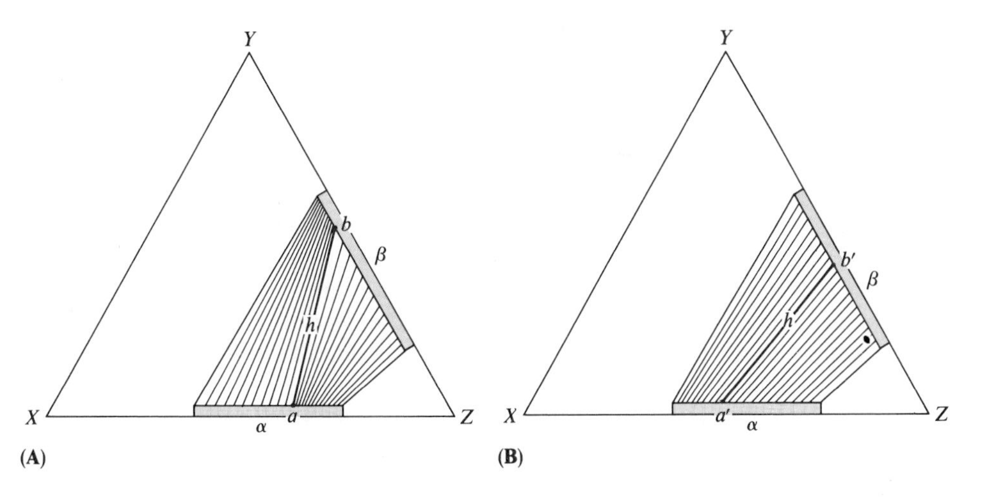

Figure 19–4
The hypothetical system *X–Y–Z*. Phases α and β exist over composition ranges indicated by the shaded bars. Coexisting compositions of α and β at a particular pressure and temperature are shown by two-phase tie lines. **(A)** A rock composition *h* consists of two phases—α of composition *a*, and β of composition *b*. **(B)** Change of *PT* conditions results in a change in position of the two-phase tie lines between phases α and β. Compositional readjustment in the rock results in composition *h* converting to α of composition a' and β of composition b'.

Another aspect of rising metamorphic level has to do with general changes in composition of rocks during metamorphism. Studies of regional metamorphism have indicated that with increasing grade of *P* and *T*, slates are converted in turn to phyllites, schists, and gneisses. The gneisses are quartz–feldspar rocks and approach the composition of granites. Earlier studies, which attempted to demonstrate that these gneissic rocks of granitic composition formed in the solid state as an ultimate metamorphic process, have indicated a relation of metamorphic grade to composition, with the higher grades becoming somewhat more granitic in composition. At present most petrologists believe that the chemical compositions of metamorphic rocks in the low to middle grades remain essentially unchanged, with the exception of losses of H_2O and CO_2 from dehydration and decarbonation reactions; compositional changes at high grade are mainly due to partial melting at high temperatures or less commonly to diffusion processes.

METAMORPHIC PROCESSES

Metamorphic processes refer to the effects of pressure, temperature, and compositional changes on the fabric and mineralogy of metamorphic rocks. In some cases all of these changes may operate simultaneously, and in others a single factor may be dominant. We shall first examine these factors somewhat separately, and then consider combined effects.

Initiation of Metamorphism

The minimum temperature at which typical regional metamorphic processes begin in sediments is about 150–200°C, with pressures on the order of 0.5–1 kbar and depth within the crust of about 4–5 km. At these pressures and temperatures diagenetic processes are complete. During burial the rock has become compacted and most of the pore spaces have been filled as a result of solution and reprecipitation. Sedimentary textures are present, and most of the H_2O in the rock is confined to hydrous minerals (with some clays containing well over 10% H_2O).

Initiation of metamorphism, as indicated by the formation of nonsedimentary phases, begins at pressures and temperatures that vary with the type of rock involved. Quartzites commonly show few changes until the higher levels of metamorphism are reached; reactive metastable materials, such as tuffs, glass, or sedimentary zeolites, respond easily to minimal metamorphic conditions. It is on the basis of such reactive materials that the idea of burial metamorphism was developed; mineral assemblages created are listed in the zeolite facies (see Figure 18–1). Glassy materials may crystallize to form laumontite ($CaAl_2Si_4O_{12} \cdot 4H_2O$) near 200°C. Zeolites of sedimentary origin may react with the loss of water to form plagioclases. One of the unique aspects of this type of low-level metamorphism is that many of the reactions occur within limited portions of the rock, thus preserving much of the original sedimentary textures.

For many years it was assumed that changes occurring in clay-rich mudrocks were not significant until the level of the greenschist facies had been reached, and the rock was obviously metamorphic. This idea probably was perpetuated because sedimentary petrologists assumed that such intermediate rocks were metamorphic and ignored them, and the metamorphic petrologists thought of them as sedimentary and did the same. Recent studies have revealed a series of gradational changes within these transitional rocks. Frey's study of metamorphism of Mesozoic clays and marls in northeastern Switzerland (1970) revealed an almost continuous series of changes from diagenesis through metamorphism to the greenschist facies. Textural and mineralogical changes are shown in Figures 19–5, 19–6, and 19–7. The nonmetamorphic rock contained illite, kaolinite, chlorite, and mixed-layer illite/montmorillonite. Mineralogical changes that were determined are:

Kaolinite and Quartz → Pyrophyllite
Mixed-layer illite/montmorillonite → Phengite and Al-rich chlorite
Irregular mixed-layer illite/montmorillonite → Regular mixed-layer mica/montmorillonite → Mixed-layer paragonite/phengite → Paragonite

(Phengite is an impure muscovite, stable at low temperatures; paragonite is a Na-rich mica.) In addition to these changes, the crystallinity of illite continuously increases during metamorphism; what this means is that the illite has lost interlayered water and absorbed potassium. This is easily observed by a slight shift and decrease in width/height ratio of the 10–20 Å X-ray diffraction peak of illite. All of these changes indicate that the transition from sediment to metamorphic rock begins within individual grains, and then gradually spreads out to encompass larger domains as the level of metamorphism increases.

Contact Metamorphism

Temperature is the dominant factor in those metamorphic rocks classified as contact type. Most contact metamorphism occurs in rocks near igneous contacts at fairly shallow depths, in the absence of differential stress. The minerals produced are characteristic of high temperatures and low pressure.

With increasing temperature and a low confining pressure, as is the case with contact metamorphic rocks, one of the most common changes is increase in grain size (see Figure 19–8). As indicated by a number of investigators, the grain size of the metamorphic rocks is greatest at the igneous contact, and decreases (approximately exponentially) with distance from the contact. It has been established also in ceramic and metallurgical studies that increase in temperature up to the melting point encourages grain-size increase by recrystallization. Temperature is only one of the fac-

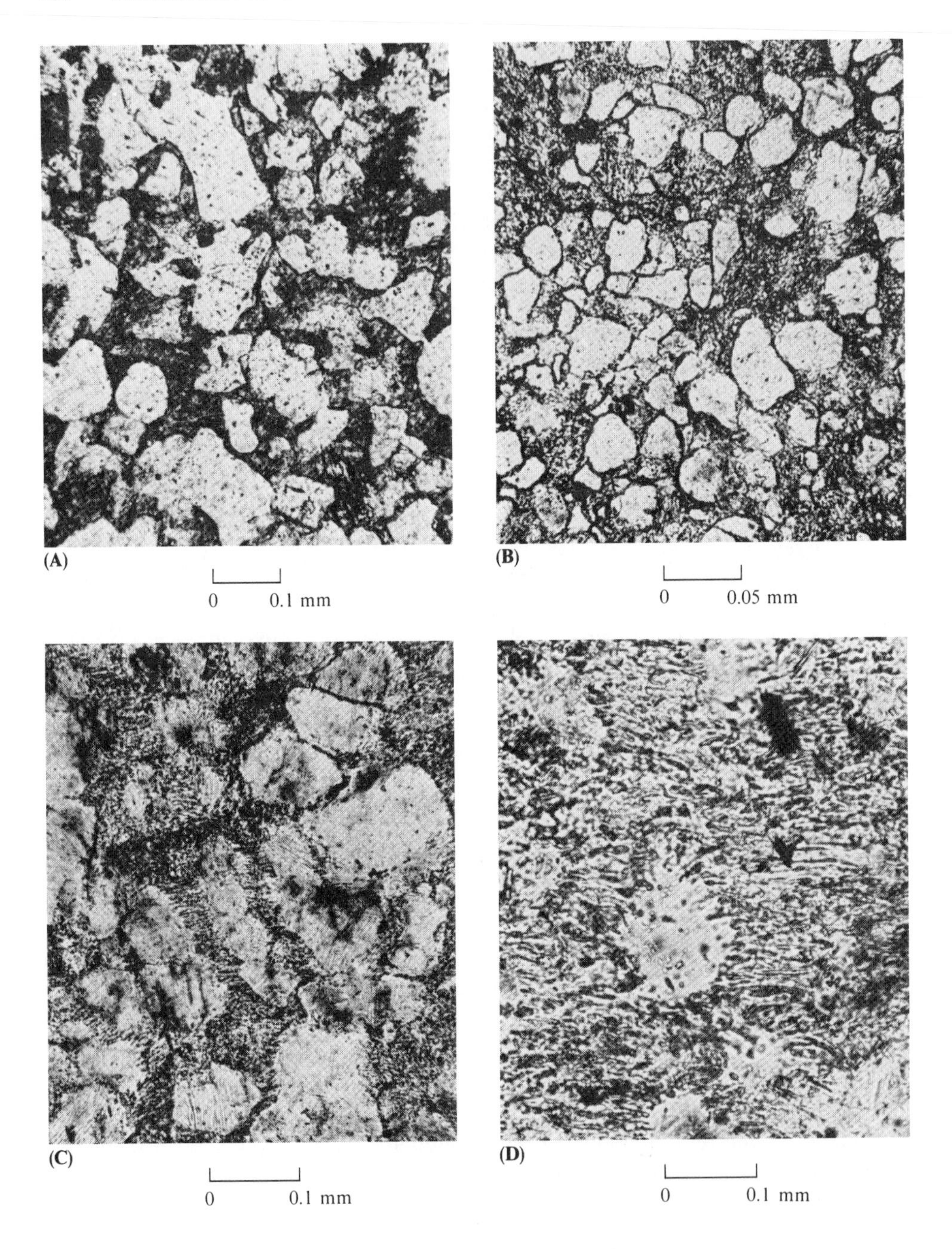
(A)
0 0.1 mm
(B)
0 0.05 mm
(C)
0 0.1 mm
(D)
0 0.1 mm

Figure 19–5 (*facing page*)
Textural changes in the transition zone from diagenesis to metamorphism. (**A**) Zone of unaltered clay cement in marl. (**B**) Zone of altered argillaceous and authigenic quartz cement in shale. (**C**) Zone of chlorite–hydromica cement and quartzite-like structures in slate. (**D**) Zone of spiny structures and chlorite–micaceous cement at the transition from slate to phyllite. [From M. Frey, 1970, *Sedimentology, 15,* Fig. 9.]

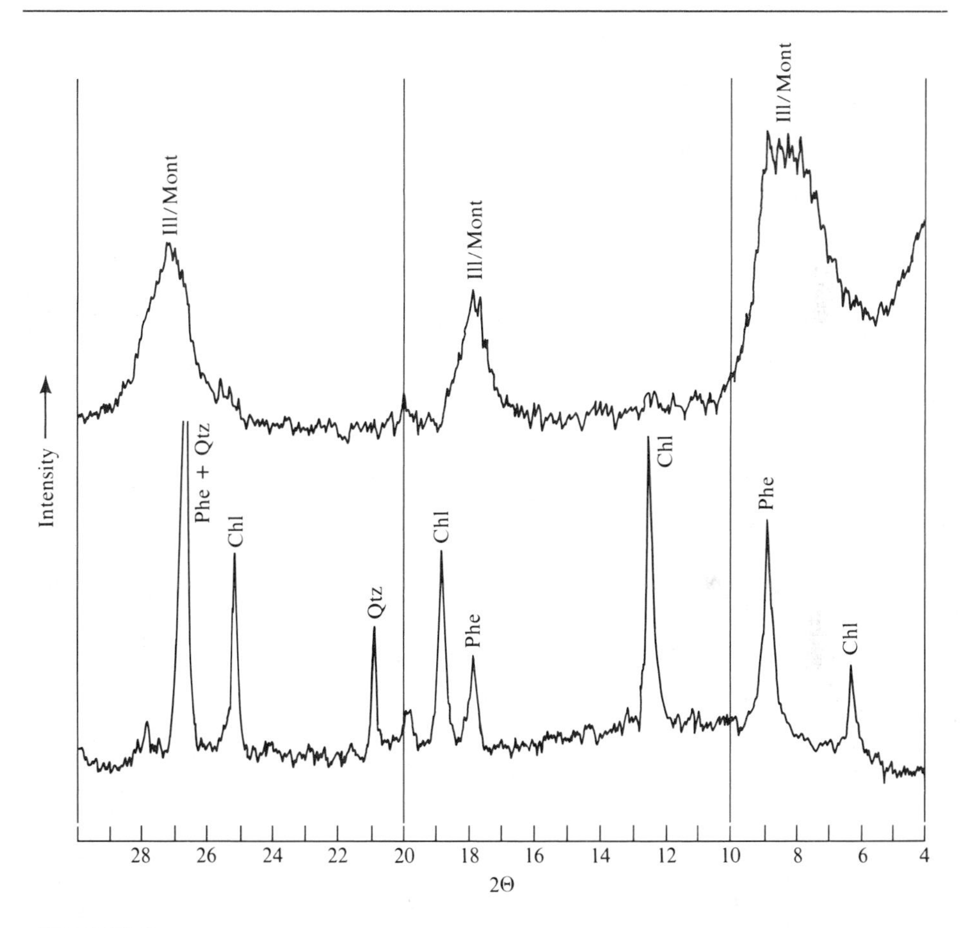

Figure 19–6
X-ray diffraction patterns of the Upper Middle Keuper sample with size fractions $< 2\mu$ and CuKα radiation. The upper pattern shows the unmetamorphosed mudrock. A broad illite/montmorillonite (Ill/Mont) peak is present between 8–10° 2θ. With initiation of metamorphism (lower pattern), the mixed-layer illite/montmorillonite peak (8–10° 2θ) becomes sharper as the illite component is converted to phengite (Phe). Chlorite (Chl) also forms at the expense of mixed-layer illite/montmorillonite. [From M. Frey, 1970, *Sedimentology, 15,* Fig. 6.]

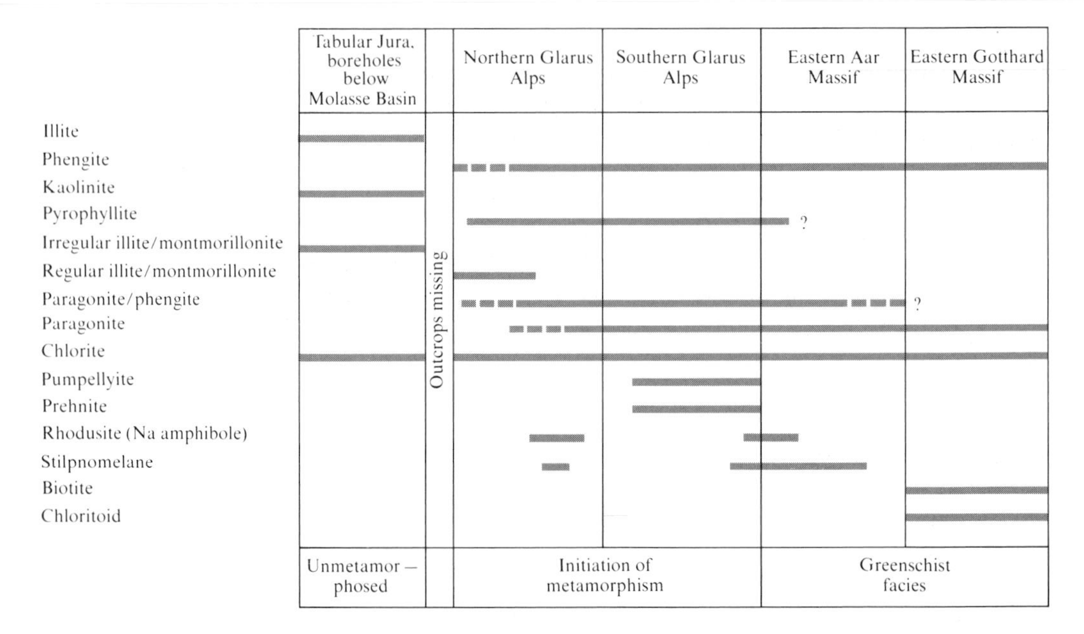

Figure 19–7
Stability ranges of some important minerals in the transition zone from diagenesis (left) to greenschist facies (right). [After M. Frey, 1970, *Sedimentology, 15,* Table 1.]

tors; the duration of heating and the amount of strain in the original material are also very important (as strained crystals tend to recrystallize into unstrained equivalents).

The textures that develop are varied. If the rock is monomineralic, as with a relatively pure sedimentary sandstone, a *granoblastic texture* is developed, with most grains equidimensional, xenoblastic, and of similar size. Often this is combined with *polygonal texture* (see Figure 19–9), where most of the grains show five or six sides in cross section and commonly intersect at *triple points* (three grain boundaries intersecting at about 120°). These types of texture arise with minerals that have little tendency to form crystal faces—that is, the surface energies of various planes within the crystal are similar. In other rocks, such as marbles, a stronger tendency toward idiomorphism exists, and recrystallization proceeds with the formation of hypidioblastic or idioblastic crystals. Such textures are called respectively *decussate* or *idiotropic* (see Figure 19–10). Triple point intersections are still very common, but the angles between edges are now commonly not 120°. As polymineralic rocks are the general rule, several minerals will commonly recrystallize simultaneously. As each has its own idiomorphic tendency (modified by the presence of other phases), this often results in a texture in which some of the minerals are idioblastic and others hypidioblastic or xenoblastic (see Figure 19–11).

The distribution and orientation of minerals often remain similar to that of the original premetamorphic rock in the early stages of contact metamorphism (see Figure 19–8). In some cases a preferred orientation of grains (as with clay and mica fragments in a pelite) may appear to be increased due to recrystallization of those phases that originally possessed a preferred primary orientation. But in general as metamorphism proceeds with recrystallization, increase in grain size, and formation of new phases, original differences in mineral distribution as well as preferred orientations are eliminated. Because of this it is difficult to judge whether a high-grade contact rock originally possessed directional or nonrandom distributional features in its premetamorphic state; often it is necessary to trace these rocks back through the lower *PT* facies and to observe the changes in texture.

Metamorphism of Igneous Rocks

Metamorphosed igneous rocks are found in many tectonic regions, as intrusive and extrusive activity can occur before, during, or after the orogenic cycle. As with sediments, the igneous rocks that have formed at one set of conditions are subjected to a new set. Igneous rocks have formed at high temperatures and are subjected to lower temperature conditions at various pressures during metamorphism. Mechanical deformation alone might result in the formation of augen and other cataclastic textures, but deformation combined with access of an aqueous or CO_2-rich fluid serves to develop a new mineralogy consistent with the imposed conditions. Igneous rocks can be brought to any facies level by metamorphism. For example, metamorphism of a granite gneiss of original igneous origin in New Hampshire resulted in the formation of sillimanite and garnet, with the garnet derived from igneous biotite. Other granites are known to contain metamorphically created zoisite and epidote. Metamorphism of mafic and ultramafic rocks is more common, as these rocks normally form part of the geosynclinal pile, and often are present in the earliest stage of orogeny. Most such metamorphism is the result not only of deformation at various temperatures, but also of the access of H_2O and CO_2 from adjacent sediments. The resultant metamorphic rocks are found at every facies level, and often are difficult to distinguish from metamorphosed sedimentary rocks.

Submarine Metamorphism

Metamorphism and low-level alteration of igneous rocks may occur under submarine conditions. Recent results obtained from deep sea drilling have confirmed this. Alteration of igneous rocks on the sea floor is facilitated by the presence of fractures. These range in size from large-scale transform faults perpendicular to rift zones, through minor faults parallel to mid-ocean ridges, fractures within median valleys spaced several meters apart (which are produced by the flow of lava over volcanic

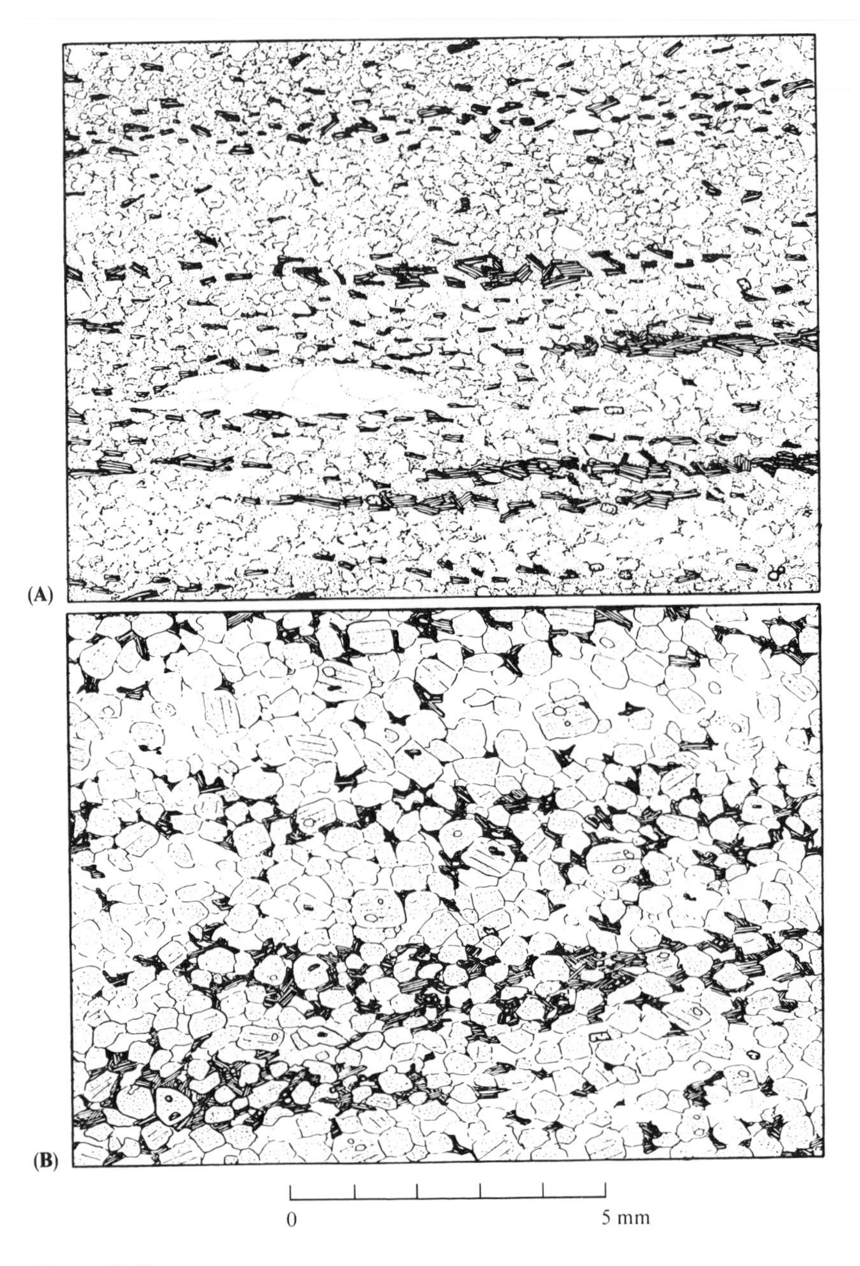

Figure 19–8
Textural changes resulting from increase in grain size during metamorphism. **(A)** A metagraywacke with a general sedimentary character. **(B, C)** Progressive grain growth of plagioclase, biotite, and quartz results in the elimination of sedimentary textures and the development of a coarsely

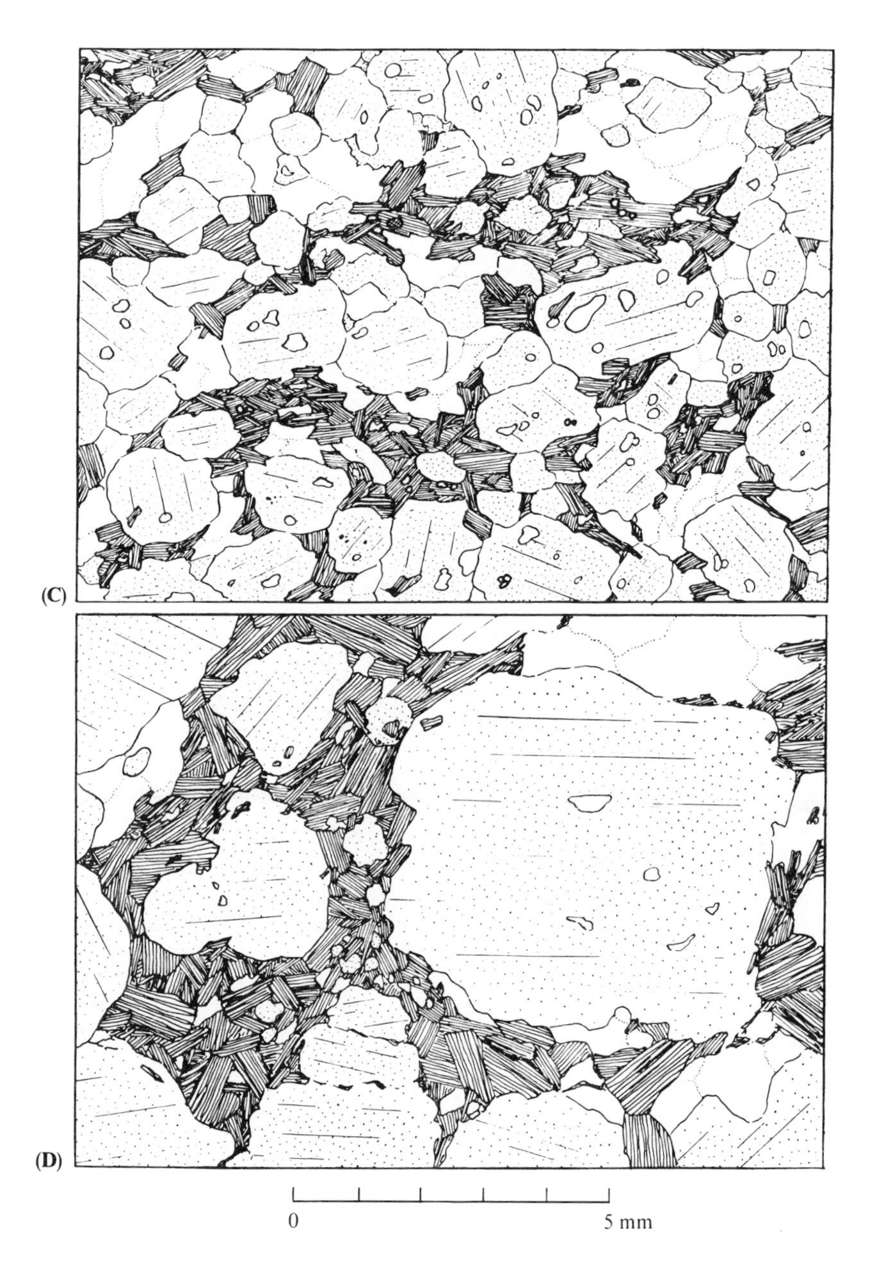

crystalline metamorphic texture. **(D)** In the final stage biotite and quartz occupy an interstitial position to the larger plagioclase grains. [From K. R. Mehnert, 1968, *Migmatites and the Origin of Granitic Rocks,* (Amsterdam: Elsevier, Fig. 23.]

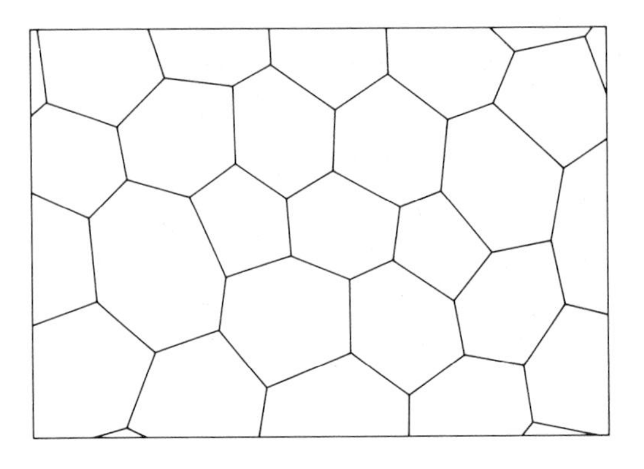

Figure 19-9
Polygonal texture. Most of the grains have five to six relatively straight sides, and intersections are usually made between three grains, with grain boundaries at about 120°. If the grains had been xenoblastic as well as equidimensional the texture would also be granoblastic.

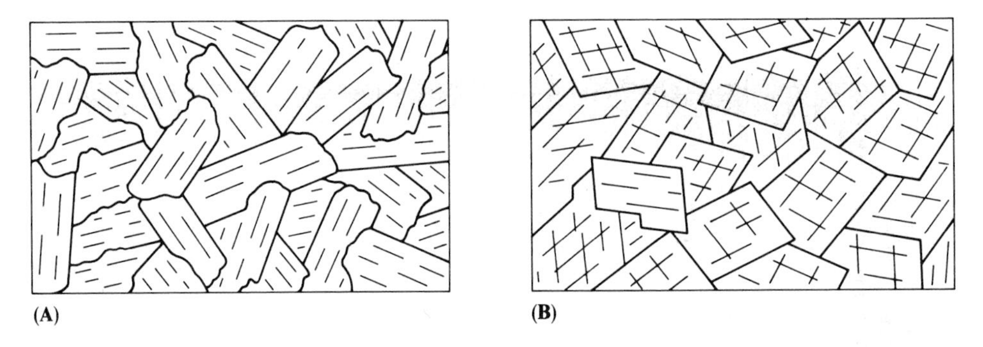

(A) **(B)**

Figure 19-10
(A) Decussate texture. Crystals are hypidioblastic, prismatic, and randomly oriented.
(B) Idiotropic texture. Crystals are generally idioblastic and randomly oriented.

rubble and unconsolidated sediment), and finally cooling cracks in pillow lavas, which are common at intervals of about 2 cm.

As a result of this extensive fracturing, the relatively shallow crust on or near a ridge crest is permeable to sea water and may have, initially, up to 25% open space. Many of the void spaces near the ridge crest become partially filled with $CaCO_3$ by alteration of basaltic rock and precipitation from flowing sea water at standard ocean bottom temperatures (a few degrees above 0°C); it has been estimated that void spaces are decreased about 5% in 110 million years. Hydration of basaltic glass results in the formation of montmorillonite clays (often called smectite). Weak but pervasive alteration is present throughout all parts of the oceanic crust examined to date.

As the igneous rocks move outward from the ridge areas, they are covered with

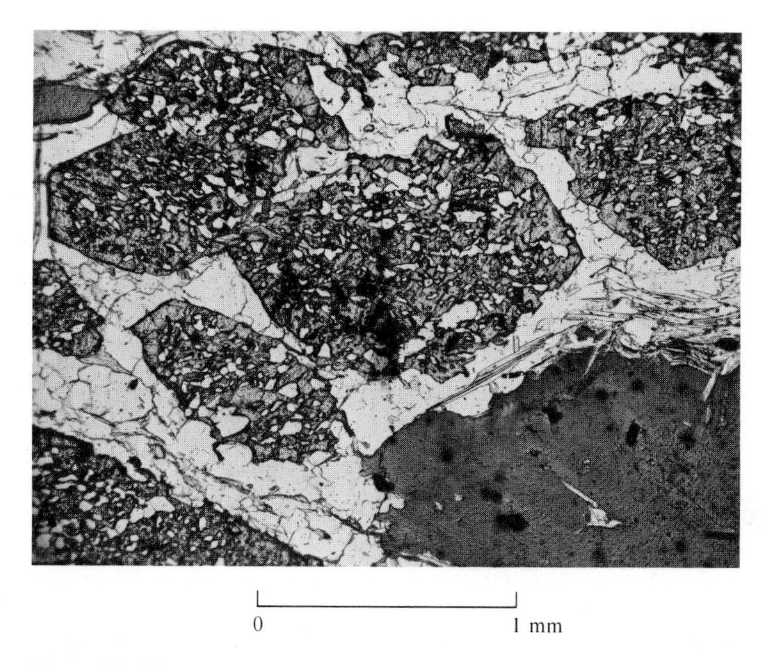

Figure 19–11
Minerals with different tendencies toward euhedralism. The large poikiloblastic staurolite is idioblastic to hypidioblastic. Inclusions are composed of quartz. The matrix is xenoblastic quartz and idioblastic muscovite. A large biotite crystal (with pleochroic halos) is at the lower right. Uncrossed nicols. [From W. R. Hansen, 1965, U.S. Geol. Surv. Prof. Paper 490, Fig. 7 (photo by W. Walker).]

increasing amounts of fine sediment and planktonic debris. Recent studies of heat flow data beneath the sedimentary blanket indicate a disturbance in the regular conductive heat flow pattern as a result of the convective action of sea water. Movement of sea water through the sedimentary cover and into the igneous crust (with consequent alteration) is indicated for periods up to 25 million years. With greater age and thicker sediment cover, additional changes (such as increase of Mg and loss of Ca) may occur by diffusion of ions between the igneous crust and bottom sea water.

The oceanic crust near spreading centers is hotter than elsewhere because of the proximity of relatively shallow magma chambers. Crustal rocks are warmed, which results in a rise of sea water (with temperatures sometimes greater than 300°C) through the typically fractured rocks above; cold sea water is drawn in to establish a convective pattern. The rising hot water is cooled and diluted by the addition of cold sea water, resulting in the precipitation of Cu, Ni, and Cd from sea water as sulfides in fractures. The Cu and Zn deposits of Cyprus may have formed in this manner. The high Fe and Mn contents of rocks found along the ridge crests (as well as increase of K with loss of Ca) may result from variants of this same process.

Examination of the ophiolite suite of rocks (see Chapter 8), thought to represent fragments of sea floor crust and mantle, indicates that metamorphism is extensive at depth; the levels of metamorphism reached are usually zeolite, prehnite–pumpellyite–graywacke, greenschist, and occasionally the epidote–amphibolite facies.

Porphyroblasts

One of the most interesting things that occurs during recrystallization is the formation of porphyroblasts. Porphyroblasts (see Figure 19–12) are crystals created during metamorphism that are conspicuously larger than other mineral grains in the rock (the metamorphic analog of igneous phenocrysts). Contrary to the tendency of metamorphism to develop an even-sized texture, the formation of porphyroblasts is fairly

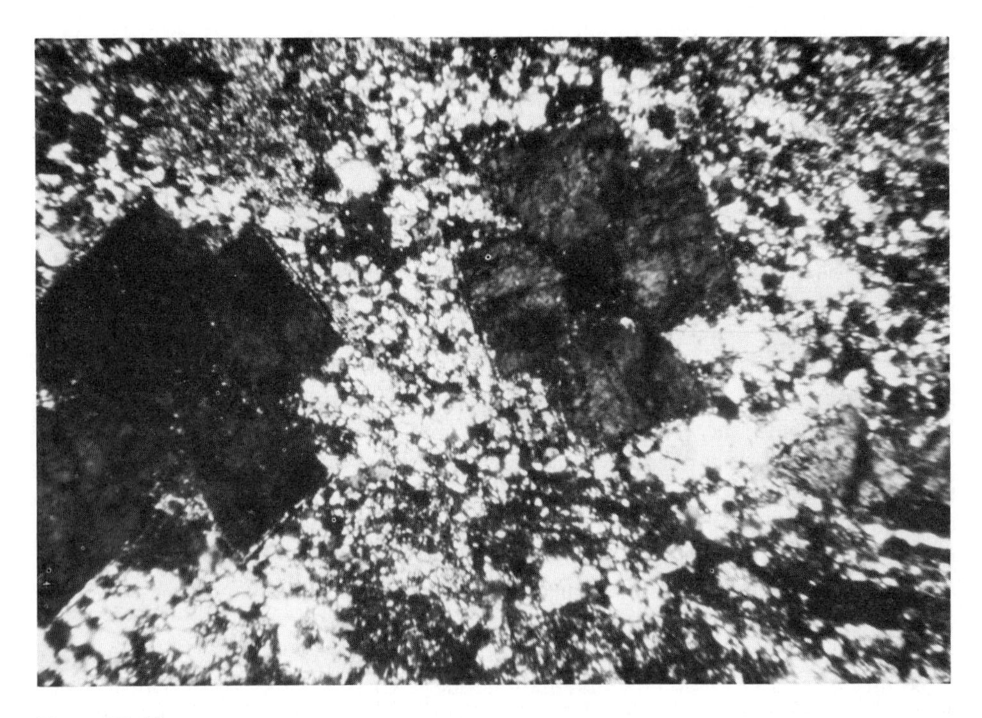

Figure 19–12
The large dark areas are porphyroblasts of andalusite (Al_2SiO_5). Both porphyroblasts are idioblastic and show a good development of penetration twins. The matrix is a very fine-grained mixture of muscovite and quartz. Crossed nicols. Width of photo is about 2.3 mm.

common, particularly in the early or late stages of a metamorphic event. Porphyroblasts are not found commonly in monomineralic rocks, as here all the crystals are of the same mineral, and each has an equal tendency to recrystallize. Some metamorphic rocks may contain the same mineral in both porphyroblasts and in the fine-grained groundmass; in others no fine-grained equivalent exists. Certain minerals, such as garnets and staurolite, occur almost exclusively as porphyroblasts, whereas others, such as quartz, have little tendency to form porphyroblasts.

The mechanism of formation is not completely understood, but seems to be related to both migration of chemical constituents and nucleation rates. If, as is often the case, the composition of the porphyroblast is quite different from the matrix materials, it is necessary for at least some of the atomic or molecular constituents to migrate (through diffusion or fluid migration) to an appropriate site. The sites for crystal growth may be determined by the location of certain chemical constituents that have limited migration rates (such as alumina in low- to medium-grade rocks) and/or mechanically disturbed portions of the rock that are relatively high-energy (and hence relatively unstable) areas. If the number of sites for nucleation is low and if migration rates are fast and of long duration, large porphyroblasts will tend to form. Alternatively, if sites for nucleation are abundant and the migration rates of constituents are slow, a large number of smaller grains are formed. The relation of time of formation of porphyroblasts to periods of deformation is discussed in Chapter 21.

Porphyroblasts vary in their degree of euhedralism; certain minerals that form porphyroblasts are characteristically idioblastic, others hypidioblastic or xenoblastic. This general sequence has been called the idioblastic series. With decreasing degrees of euhedralism the series goes from sphene, rutile, magnetite, garnet, tourmaline, staurolite, kyanite, epidote, zoisite, pyroxenes, hornblende, dolomite, albite, mica, chlorite, calcite, quartz, plagioclase, orthoclase, to microcline. A mineral higher in the series will form crystal faces in growth competition with a mineral lower in the series. This series is correct in a statistical sense only, as many exceptions have been noted as a result of different circumstances of growth. It has been used occasionally in an attempt to distinguish between rocks of igneous versus metamorphic origin.

Preferred Orientation

One of the most obvious features of metamorphic rocks is the preferred orientation shown by platy or elongate minerals or fragments. This preferred orientation, although occasionally due to an original anisotropy in the premetamorphic source rock, usually is developed as a result of deformation, recovery, and recrystallization.

Materials undergoing deformation may exhibit brittle or ductile behavior. Brittle behavior is characterized by fracturing, whereas ductility is the capacity of a material to change its shape without large-scale fracturing. The difference, however, is one of scale; detailed examination of materials that have shown ductile behavior often reveals fine-scale brittle deformation or slip within single grains.

As most materials have a tendency to retain their crystallinity, even under conditions of strong deformation, deformation is often accomplished by twinning and slip. Slip (translational gliding) is accomplished by layers within a crystalline substance gliding over other layers; the amount of gliding is equal to the unit pattern of the lattice (or a multiple of it), such that all parts of the deformed crystal are joined together in optical continuity. Movement along slip planes is greatly facilitated by imperfections within the crystal structure, such as point defects or dislocations. Twin gliding (deformation twinning) is accomplished by shearing within each layer of the crystal structure, such that the deformed crystal may possess two parts that are separated by a mirror plane of symmetry. Again the crystal structure is maintained in spite of a change in orientation of one part of the crystal with respect to the other. Silicate minerals will deform by other mechanisms as well, such as kinking (due to repeated planar concentration of shearing strain), cataclasis (with the production of microfractures), or creep by diffusion.

The deformation of grains by slip is impeded by the presence of grain boundaries, as indicated by the fact that single crystals deform more easily than aggregates of the same material. If crystals cannot accommodate the strains of their neighbors, the result will be either fracturing at crystal boundaries or (at higher temperatures) crystal boundary sliding.

During and after deformation, recovery processes are operative. This involves the elimination of strained crystals by *polygonization*—the formation of a number of smaller unstrained crystals from larger strained individuals. Recrystallization during deformation is effective at high temperatures, and results in the nucleation of stable strain-free crystals and the growth of preexisting strain-free crystals at the expense of strained crystals.

The various factors described above can result in grain rotation, which could lead to the development of a crystallographic preferred orientation within a metamorphic rock. During recrystallization (during or after deformation) the formation of crystal nuclei may result in a preferred orientation controlled by the stress or strain of deformation. Crystals in an unfavorable orientation may be eliminated. The deformed fabric may possess sufficient anisotropy to permit growth of crystals only in particular orientations.

The idea of "favorable" and "unfavorable" orientations of a crystal in a stress field refers to the fact that crystals are anisotropic in most of their properties. If stress is applied unequally to a crystal, certain crystal orientations are more unstable than others.

Consider the simple thermodynamic equation:

$$\left(\frac{dG}{dP}\right)_T = V$$

This states that the rate of change of free energy (G) as a function of pressure (P) is equal to volume (V) when the system is at constant temperature (T). Free energy is a measure of the stability of a system. A system at equilibrium (its most stable condition) has a minimum free energy consistent with the imposed *PT* conditions. Furthermore, a system will attempt spontaneously to reach a state of minimum free energy.

The concept of free energy can be applied to a system under pressure. If a system (or rock) of uniform composition (and uniformly applied constraints) contains a variety of different mineral assemblages (such as kyanite in one portion and sillimanite in another), the portion at the lower free energy level is the more stable of the two; the least stable portion will attempt to adjust itself (by polymorphic inversion in this case) to also be in the lowest free energy state. This can be seen for schematic curves for kyanite and sillimanite in Figure 19–13. The two curves show the rise in free energy for the polymorphs as a function of pressure. We have seen that because $(dG/dP)_T = V$, a phase with a higher molar volume will show a larger rise in free energy than one with a smaller molar volume. It is therefore logical that the phase with the smaller molar volume is more stable at high pressures. The sillimanite curve rises more steeply than that of kyanite. At low pressures (less than *a* on the pressure coordinate) the sillimanite curve is below that of kyanite. Therefore, sillimanite has a lower free energy than kyanite and is stable. At pressures above *a*, kyanite has a lower free energy than sillimanite and hence is the more stable. The intersection of the two curves shows the pressure of transition between the two polymorphs.

But consider a crystal under uniaxial stress (perhaps squeezed between two pistons). All crystals, because of their regular internal atomic arrangement, have some properties that are directionally dependent; one of these properties is compressibility. Calcite, for example, is more easily compressed parallel to the *c* axis, whereas quartz has its most incompressible direction parallel to the *c* axis; micas are most easily compressed when the applied stress is perpendicular to the major cleavage direction.

A quartz crystal placed between two convergent pistons will undergo the greatest compression (and therefore decrease in volume) when its *c* axis is perpendicular to the length of the pistons; when the *c* axis is parallel to the pistons the crystal will

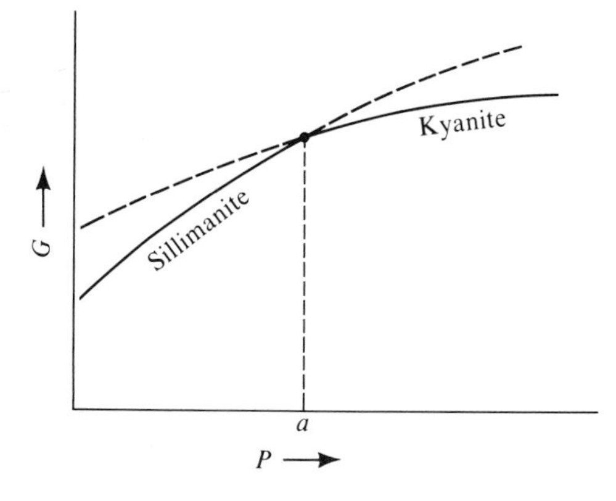

Figure 19–13
Schematic diagram showing the relative increase of free energy of sillimanite and kyanite as a function of pressure. Kyanite, having the smaller molar volume, rises at a lesser rate than sillimanite. At low pressures (below *a*) sillimanite has a lower free energy than kyanite and is the stable phase. At pressures above *a* kyanite has the lower free energy and is stable.

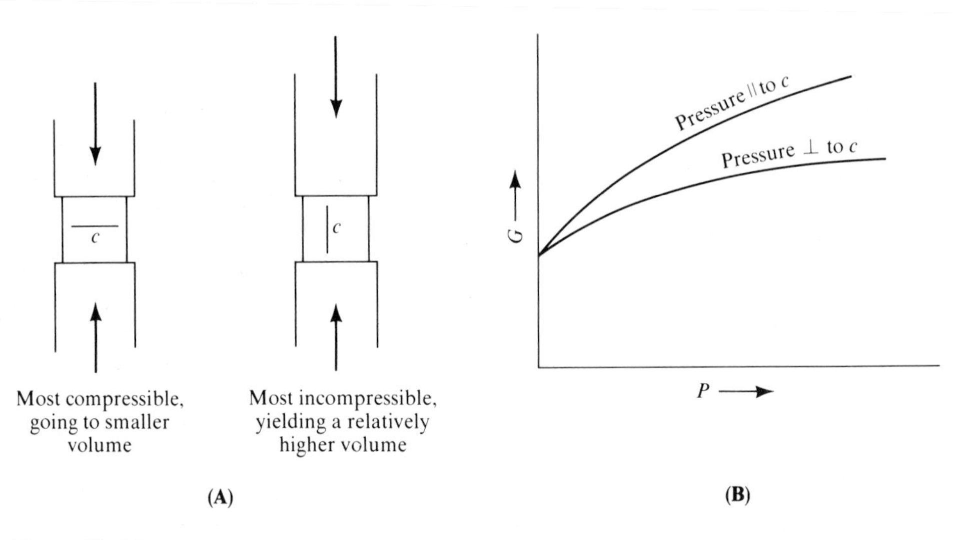

Figure 19–14
(**A**) A single crystal (whose *c* axis direction is shown) is subjected to uniaxial pressure. The mineral is most compressible perpendicular to the *c* axis and least compressible parallel to the *c* axis. (**B**) The free energy rise (as a function of pressure) will be greatest when the pressure is applied parallel to *c*, since the crystal maintains a large volume (as compared to similar pressure being applied perpendicular to *c*).

suffer a lesser degree of compression when subjected to the same pressure (see Figure 19–14A). As the volume of the quartz under compression in these two orientations is different, it follows that the rise of free energy with pressure will be different as well (see Figure 19–14B). The direction that is most compressible (and yields the smaller volume) is the most stable orientation under uniaxial stress. If the stress can be maintained for long periods of time at moderate temperatures, crystals in the relatively unstable orientations will tend to be eliminated (by diffusion or solution), and this material will be reprecipitated on crystals in the more stable orientation. Thus, without any crystal rotation, a fabric can be developed in which there exists a crystallographic preferred orientation. Note that in this case, even though the crystallographic directions of the crystals become aligned, the shape of the crystals may be completely irregular; alternatively, parallelism developed by plastic flow often results in a parallelism in the shapes of the crystals, with crystallographic parallelism being incidental. Platy minerals such as micas, which have their most compressible direction perpendicular to the plates, will assume the same parallelism as a result of either mechanical flow or recrystallization under directed stress, making it difficult to determine the mechanism of alignment.

Metamorphic Differentiation

The term metamorphic differentiation was used originally to describe the various metamorphic processes by which diverse mineral assemblages are produced from an originally homogeneous parent rock. A uniform fine-grained rock might become heterogeneous by the formation of porphyroblasts, which are often both compositionally and mineralogically different from the groundmass. Within amphibolites, segregations of biotite–hornblende or labradorite–epidote may develop. Quartz–albite veins are often developed abundantly in low-grade pelitic or quartz–feldspar schists. The conversion of high-grade schists into banded gneisses leads to considerable separation of minerals into somewhat diffuse bands rich in either quartz–feldspar or mafic constituents.

The differentiation of a metamorphic rock into mineralogical and/or chemically different portions is contrary to the generally accepted concept that metamorphism usually tends to obliterate compositional differences of premetamorphic textures. A variety of explanations for metamorphic differentiation can be suggested; these are based either on mineral compatibilities or on gradients of composition, temperature, or pressure within the rock system.

The formation of new phases during prograde metamorphism might result in making the rock more heterogeneous. This may occur due to changes in mineral compatibilities and solid solution limits. An example of this might be the formation of biotite. Suppose the composition of the original rock is microcline, chlorite, quartz, and the "impure" variety of muscovite known as phengite. With a rise in temperature the reaction

$$\text{Microcline} + \text{Chlorite} \rightarrow \text{Biotite} + \text{Phengite} + \text{Quartz} + H_2O$$

takes place. At higher temperatures,

$$\text{Phengite} + \text{Chlorite} \rightarrow \text{Idealized muscovite} + \text{Biotite} + \text{Quartz} + H_2O$$

The number of newly formed biotite crystals will depend upon the number of possible nucleation sites and the ability of constituents to migrate to these sites. If the number of sites is limited and the migration rates are high, compositional heterogeneity (a form of metamorphic differentiation) develops by the formation of a small number of large biotite crystals. If many nucleation sites are available and diffusion rates are low, many small biotite crystals will form. As the reaction continues toward a more stoichiometric muscovite composition (with the "impurities" present in phengite going into the biotite structure), chlorite is gradually eliminated, and more muscovite and biotite form. This may enlarge the earlier biotite crystals as well as create new ones.

Compositional Gradients

A compositional gradient may be created in a uniform rock due to the presence of an adjacent rock of incompatible composition. This occurs with the formation of talc–magnesite and chlorite zones adjacent to serpentine bodies. Another example was proposed relative to the formation of some amphibolites. Orville (1969) noted that although the chemical composition of amphibolites is such that they are similar to either basalts or marls, the common location of amphibolite between beds of metapelites and marbles indicated that many amphibolites may have formed because of the compositional gradient between these two incompatible rock types. Figure 19–15, an

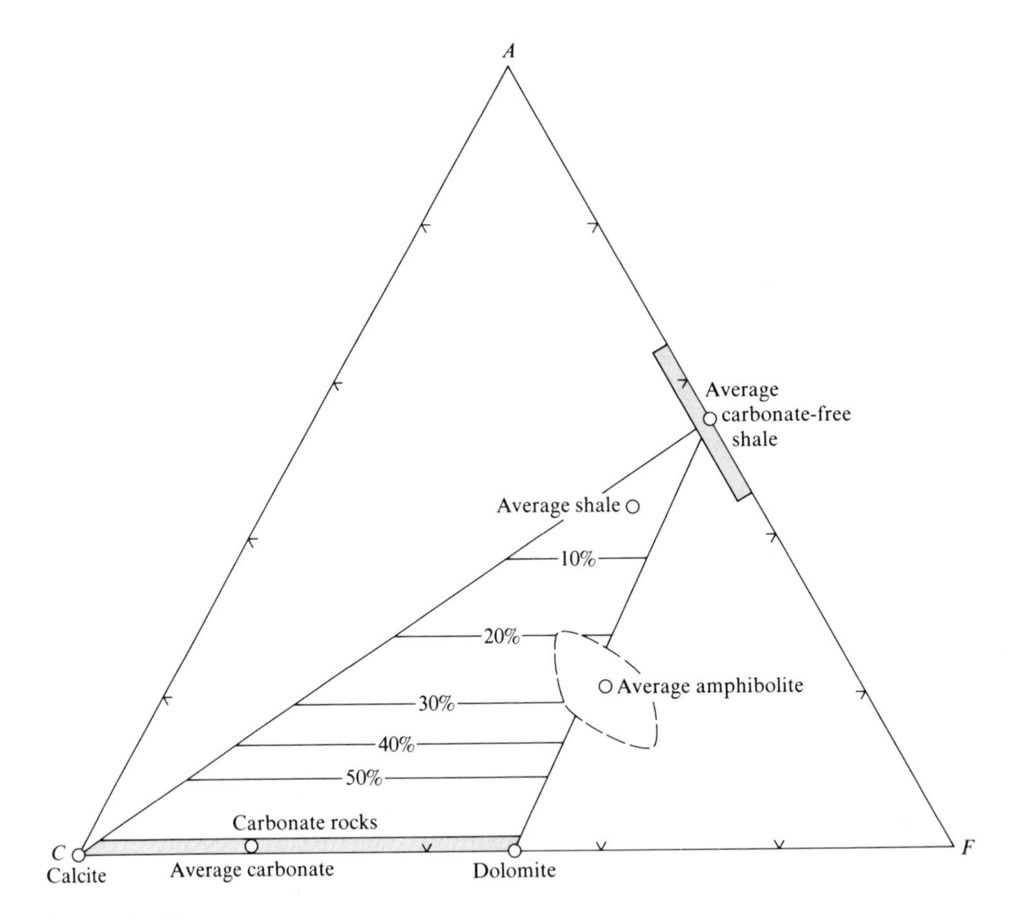

Figure 19–15
An *ACF* diagram showing the range of composition of typical pelites, carbonate rocks, and basic igneous rocks (dashed lines). The average amphibolite composition is indicated within the basic rock area. The percentage lines show locations of various weight percentages of carbonate rocks mixed with carbonate-free shales. [From P. M. Orville, 1969, *Amer. Jour. Sci., 267,* Fig. 5.]

ACF diagram, shows the composition of typical amphibolites in relation to average shales and carbonates. The average amphibolite lies within the field of mafic igneous rocks, whereas most mixtures of carbonate-free shales and carbonate rocks (except for mixtures of 20–40% carbonates with shale by weight) do not. In spite of this, a metasomatic origin for many amphibolites is indicated because of the common geographic sequence of marble → calc silicates (mainly grossularite and diopside) → amphibolite → micaceous gneisses or schists. This can be explained by consideration of the phase assemblages at the amphibolite level of metamorphism. Figure 19–16 shows the possible mineral assemblages that exist in equilibrium with an excess of

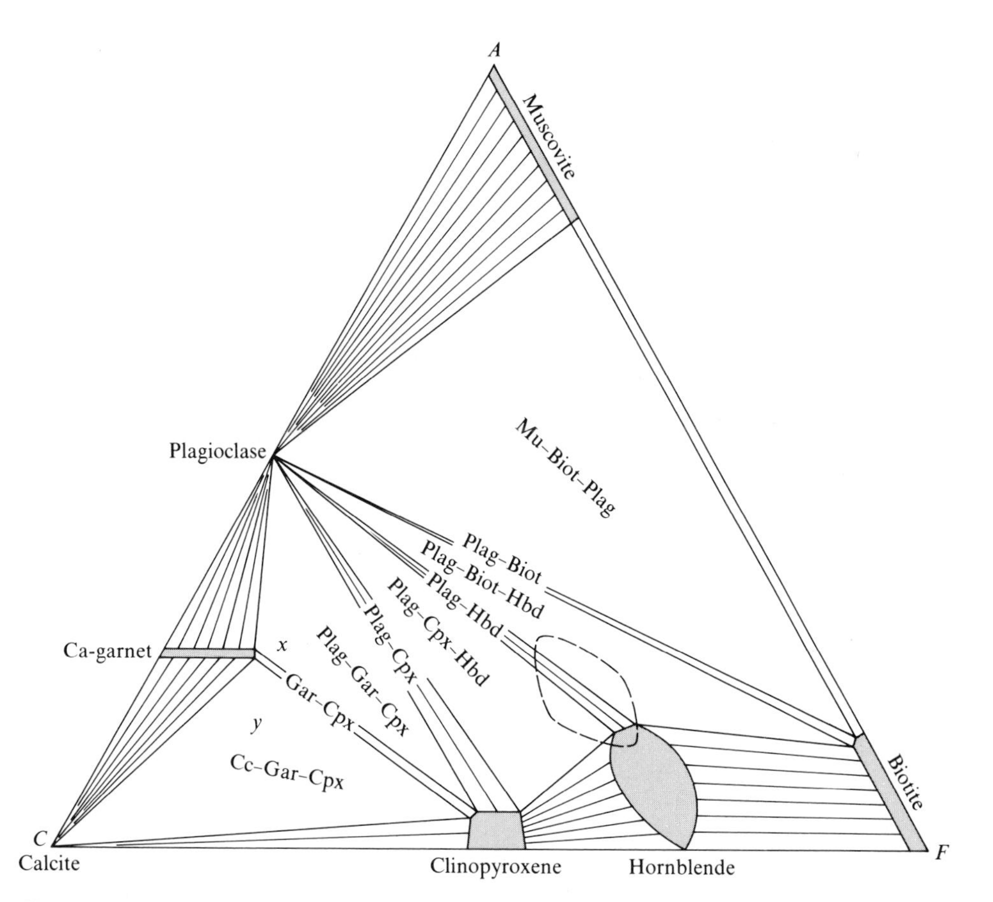

Figure 19–16
An *ACF* diagram showing minerals and compatibilities within medium-grade (amphibolite facies) rocks with excess quartz and alkali feldspar. Basic igneous rock compositions are indicated with a dashed line. Note that muscovite and biotite are plotted on the diagram, contrary to our earlier technique. [From P. M. Orville, 1969, *Amer. Jour. Sci., 267,* Fig. 9.]

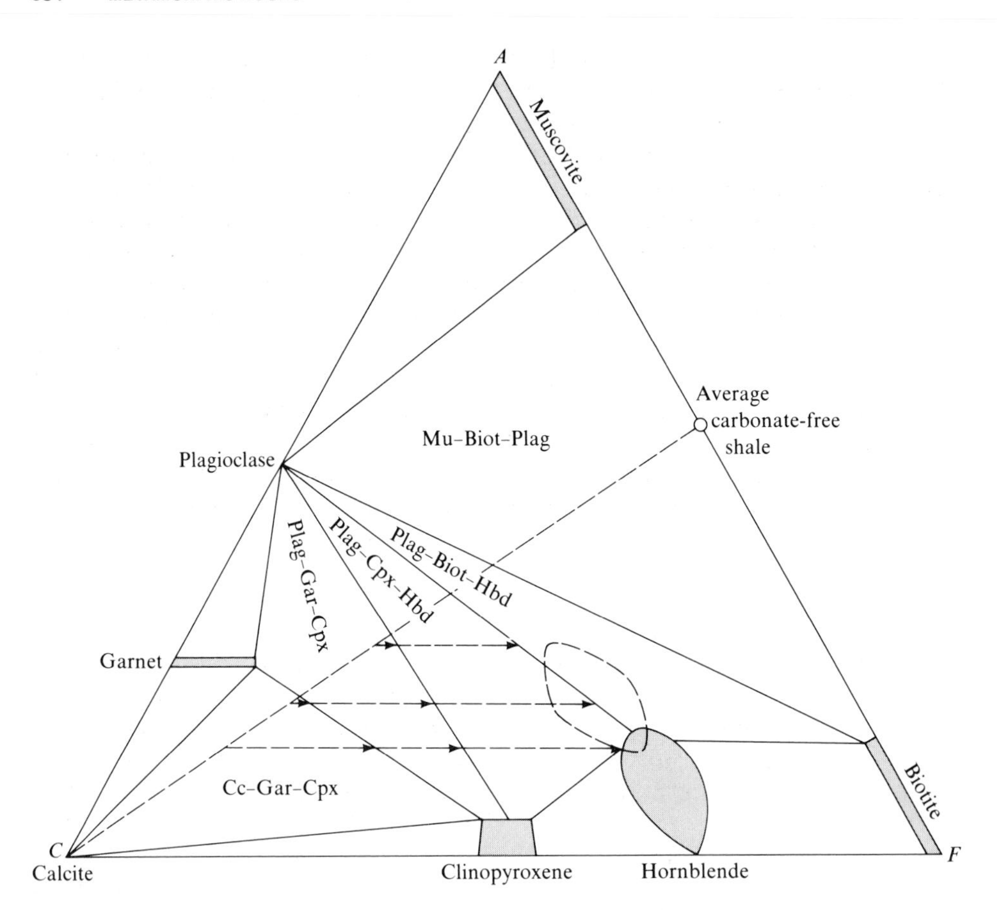

Figure 19–17
ACF diagram for medium-grade rocks with an excess of quartz and alkali feldspar. Mixtures of carbonate-free shale and $CaCO_3$ are located along the diagonal dashed line. With reciprocal exchange of (Mg, Fe) for Ca at constant Al content, shale–carbonate mixtures change (as seen by lines with arrows) to produce amphibole–plagioclase rocks. [From P. M. Orville, 1969, *Amer. Jour. Sci., 267,* Fig. 10.]

quartz and alkali feldspar. A composition *x* consists of plagioclase, garnet, and clinopyroxene (with quartz and potassium feldspar), and composition *y* is a mixture of calcite, garnet, clinopyroxene, quartz, and potassium feldspar. Each assemblage is in itself stable, but when the two are physically adjacent to each other they are incompatible; if the system is open to migration of components, diffusion will occur in the intergranular fluid with the elimination of at least one or both adjacent assemblages and the creation of the intermediate assemblage consisting of garnet and clinopyroxene (plus quartz and potassium feldspar) with or without excess plagioclase or calcite. As stated by Orville: "The presence of incompatible phases creates a potential for spontaneous transfer of material" (1969, p. 80). Vidale (1969) has demonstrated in laboratory experiments the mobility of cations at high temperature in saline solutions between carbonate and pelite compositions, with the formation of intermediate compatible compositional zones. Alumina has been demonstrated experimentally to be relatively immobile under most metamorphic or hydrothermal conditions, whereas other constituents, such as Mg, Fe, Ca, and K, are capable of migrating in a concentration gradient. Figure 19–17 shows how the composition of a calc-silicate rock might change due to the compositional gradient imposed by an adjacent pelite. Changes are accomplished by diffusion of calcium from the calc-silicate to the pelite, and migration of Mg and Fe from the pelite to the calc-silicate rock. As the Al content remains constant, the calc-silicate composition changes parallel to the base of the *ACF* triangle. This migration of composition can convert the original calc-silicate rock partially or completely to amphibolite. Extreme migration of these cations could bring the composition to the plagioclase–biotite tie line. Changes in carbonate composition could also follow a path directly away from the *C* corner of the *ACF* triangle if calcite is removed from the carbonate rock by solution.

Temperature Gradients

Differentiation of a homogeneous metamorphic rock can be established by maintaining a temperature gradient with sufficient time for diffusion within the intergranular fluid. In an experiment lasting only six days, Orville (1962) has demonstrated significant amounts of diffusion of Na^+ and K^+ over several centimeters as a result of a temperature gradient between 600°C and 630°C. Figure 19–18 shows the results of other hydrothermal experiments in which the liquid/vapor phase was composed of hydrous solutions containing KCl and HCl. The reactions studied were performed at 1,000 bar pressure and the results were the same for various solution strengths of KCl + HCl between 0.65 to 4.0 molar. The horizontal axis indicates increasing values of KCl/HCl to the right. A point X consists of potassium feldspar at 300°C in a solution having a ratio of 10^4. If the KCl/HCl ratio is decreased at constant temperature the following reaction occurs:

$$\underset{\text{Potassium feldspar}}{3KAlSi_3O_8} + 2H^+ \rightarrow \underset{\text{Muscovite}}{KAl_3Si_3O_{10}(OH)_2} + \underset{\text{Quartz}}{6SiO_2} + 2K^+$$

Hydrogen ions in the vapor phase convert potassium feldspar to muscovite, quartz, and potassium ions. The reaction curve between potassium feldspar and muscovite–quartz is inclined, and it is obvious that the reaction also will occur at a constant solution ratio if the temperature is decreased. Using this and similar relationships, Orville (1962) demonstrated how this effect could cause differentiation of a uniform rock subjected to a temperature gradient. The left side of Figure 19–19 shows the initial condition of the rock, and the right, the result of diffusion. The initial rock consists of a homogeneous mixture of quartz, muscovite, and potassium feldspar. A temperature gradient is set up along the rock with temperature increasing to the right.

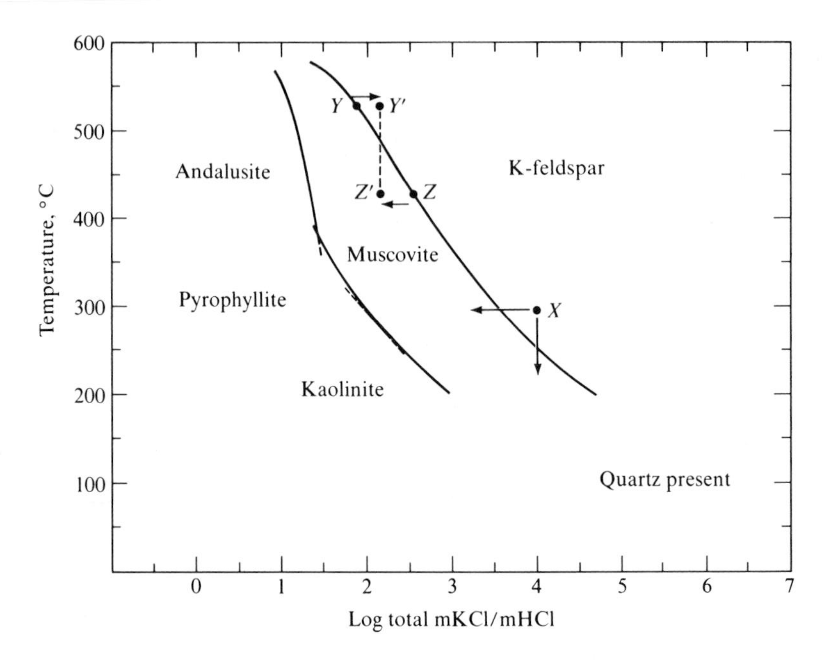

Figure 19–18
The right-hand curve shows the reaction

$$\underset{\text{K-feldspar}}{3KAlSi_3O_8} + 2H^+ \rightleftarrows \underset{\text{Muscovite}}{KAl_3Si_3O_{10}(OH)_2} + \underset{\text{Quartz}}{6SiO_2} + 2K^+$$

as a function of temperature and KC1/HC1 concentration at $P_{H_2O} = 1$ kbar. Points and arrows are described in the text. [After J. J. Hemley, 1959, *Amer. Jour. Sci., 257,* Fig. 2; 1975, *Econ. Geol., 70,* Fig. 1; and 1980, personal communication.]

As potassium feldspar is now present with both muscovite and quartz over a temperature gradient, it follows that the initial compositions of the vapor phase will not be uniform, but will lie along the univariant curve (see Figure 19–18) that joins the muscovite and potassium feldspar stability fields. A range of vapor phase compositions will develop. At the high temperature portion of the system the vapor might be Y and at the low temperature end Z, with intermediate positions having intermediate compositions. Observe that the composition of the high temperature area (at Y) has a lower K^+/H^+ ratio than Z. If substances dissolved in the fluids are free to diffuse through the rock, then in order for the fluid phase to become uniform in composition, K^+ diffuses to the right and H^+ to the left (as shown in Figure 19–19). The K^+ diffusing to the right combines with muscovite and quartz to produce potassium feldspar, and the H^+ moving left combines with potassium feldspar to produce muscovite and quartz in the low-temperature portion—the final products and conditions shown in the right side of Figure 19–19. The final equilibrium assemblages are shown

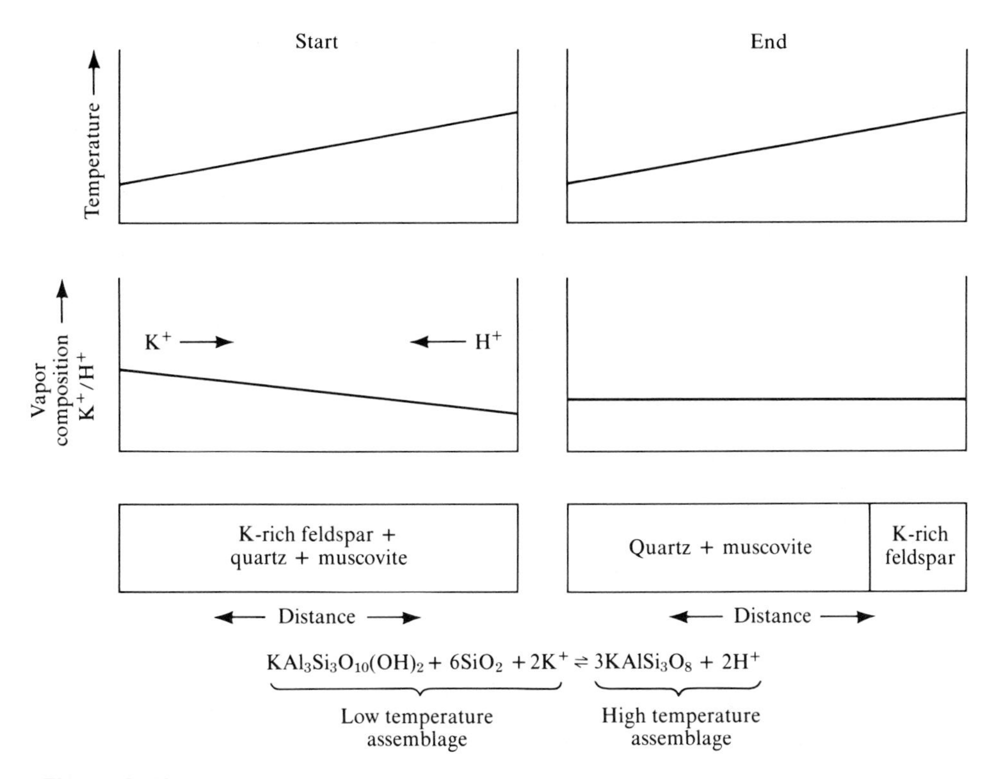

Figure 19–19
At left, a mass of rock (lower block) consists of a mixture of K-rich feldspar, quartz, and muscovite. A temperature gradient, increasing from left to right, is superimposed on the rock. Initial concentrations of K^+/H^+ are developed in the vapor phase as shown in response to the temperature gradient. At a later stage (right side), diffusion of K^+ and H^+ has resulted in a constant K^+/H^+ ratio across the rock. As a result of this migration of ions, the rock has become heterogeneous, with the hotter portion consisting of K-feldspar, and the cooler portions of a mixture of quartz and muscovite. [After P. M. Orville, 1962, *Norsk Geol. Tidsk., 42,* Fig. 7.]

along the line Y'-Z' in Figure 19–18. Reactions of this type within contact metamorphic rocks could account for the porphyroblastic growth of potassium feldspar adjacent to granitic intrusions without the benefit of alkali-rich hydrothermal solutions from the crystallizing magma.

Differentiation by Deformation

Various investigators have emphasized the importance of heterogeneous differential deformation (a result of unequal application of stress, as in shear or uniaxial pressures) in metamorphic differentiation. In many areas that have undergone regional metamorphism, compositional layering is developed that is clearly at an angle to relic bedding. This may be parallel to a metamorphically derived rock cleavage; in some cases compositional differences are developed in rocks with no evidence of heterogeneous strain. During heterogeneous stress, pressure is not equal throughout all parts of the rock (as a result of the differences in resistance of the various minerals to deformation, as well as the character of the rock fabric). This causes different portions of the rock to be temporarily subjected to higher or lower stress than the average. Constituents with the highest molar volume (and consequently the highest free energy) will be relatively unstable in the high-pressure areas and will migrate to those of low pressure. Similarly, those constituents with the smallest molar volume would have a tendency to remain in or be displaced to the higher-pressure portions. This process would operate in only a single direction (to the low-pressure areas) if open tension cracks were present. The influence of stress is to initiate recrystallization reactions, resulting in a differentiation under the existing pressure. The differentiation controls the composition of the new mineral assemblage. Low-pressure areas will favor reactions with an increase in volume and will therefore be a site of enrichment of low-density phases. In addition, a new suite of minerals of greater density will develop (through compositional changes) in the high-pressure shear zones.

These concepts provide an explanation for the observation that felsic constituents are preferentially removed from the high-stress portions of asymmetrical (small-scale) crenulation folds, and micaceous material preferentially concentrated. This situation is shown in Figure 19–20. Parts A through D show the transition from an original slaty cleavage to a crenulation cleavage as a result of rotation of the stress field with continued compression. This process could also begin by the process of kinking (as in part E), with the development of bands that are at an angle to the original rock cleavage direction. With sufficient temperature, there can be small-scale recrystallization of quartz and feldspar in the slightly crumpled areas and of dark minerals (biotite) in the steepened ones. Only in incipient cases can this process be followed. When the process has advanced far enough, only a banding of thin quartz–feldspar alternating with dark biotite remains, which could easily be mistaken for original bedding. The end result of this process is shown in Figure 19–21.

Another type of metamorphic differentiation might arise by purely mechanical means. Schmidt (1932) suggested (but could not prove) a process of tectonic unmixing, similar to that found in rolled wrought iron, which is partially unmixed during deformation into layers of relatively pure iron alternating with carbon-rich iron. Certain minerals, such as quartz, calcite, and feldspars, are known to be easily deformed during metamorphism. Deformation begins initially within these minerals as discontinuous slip surfaces, which are gradually transformed into continuous slip planes. Minerals that are easily deformed are mechanically segregated into the slip layers, whereas other minerals (mica, hornblende, pyroxene, and so on) are segregated between the slip layers, producing a somewhat rough and discontinuous mineralogical banding, which is similar to that found in schists or gneisses.

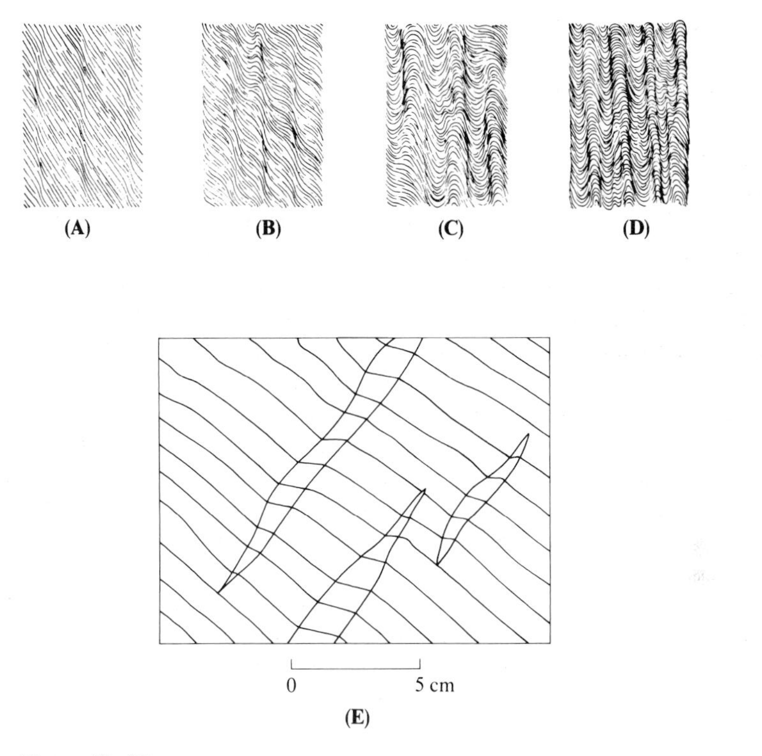

Figure 19–20
(A–D) The development of crenulation cleavage. **(E)** Kinking in phyllite. [After L. U. DeSitter, 1964, *Structural Geology* (New York: McGraw-Hill), Fig. 239.]

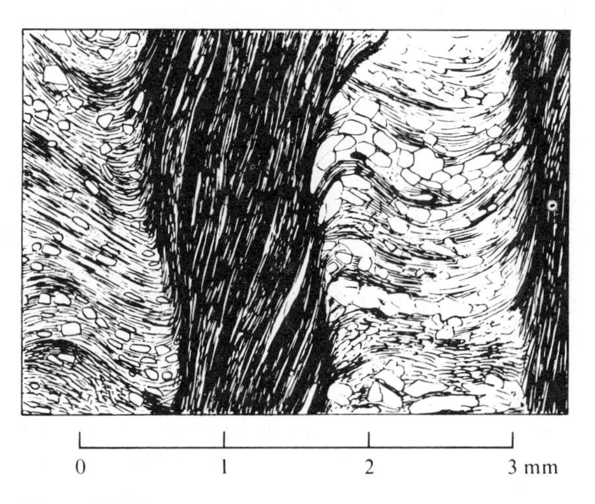

Figure 19–21
Tectonic banding developed as a result of kinking and differential concentration of minerals. [After L. U. DeSitter, 1964, *Structural Geology* (New York: McGraw-Hill), Fig. 244.]

METAMORPHIC REACTIONS

Textural evidence for mineralogical reactions is often found in metamorphic rocks. Although evidence of prograde reactions (during rising P and T) is seen much less often than evidence of retrograde (during decreasing P and T) ones, abundant indications are often present in the form of embayed crystals, reaction rims, isolated islands of a single phase in optical continuity, and so on. A good example of this is shown in Figure 19–22, where a sillimanite–quartz intergrowth has embayed a muscovite crystal, several islands of which remain in optical continuity.

As discussed earlier, most metamorphic rock assemblages appear to reflect conditions of divariant equilibrium; the persistence of various mineral associations through time and place indicate that they exist over a range of PT conditions rather than along a univariant curve or an invariant point. The various reactions that produced the mineral assemblage have generally gone to completion.

If a rock has a very simple chemical composition and was formed by the well-known low-temperature mineral A being converted to the high-temperature mineral B, it is obvious that the rock would not contain crystals of A, as A would have been consumed in the reaction. If, on the other hand, a typical pelitic rock had completed the reaction

$$\underset{\text{Reactants}}{\text{9 Staurolite} + \text{Muscovite} + \text{5 Quartz}} \xrightarrow{\Delta T} \underset{\text{Products}}{\text{17 Sillimanite} + \text{2 Garnet} + \text{Biotite} + 9H_2O}$$

it would not follow that all of the staurolite, muscovite, and quartz had been eliminated from the rock. As indicated in the equation, staurolite, muscovite, and quartz would only be completely eliminated if they were present initially in the ratio of 9:1:5. Such a reaction is typically completed by the elimination of only one of the reactants—the one in shortest supply. Thus an assemblage consisting of sillimanite, garnet, and biotite with either staurolite + muscovite, staurolite + quartz, or muscovite + quartz would suggest that the reaction had gone to completion.

If we discover a rock that contains sillimanite, garnet, and biotite with two of the above reactants it does not immediately follow that the above reaction has occurred; it is still necessary to establish that the presence of these five phases (plus various other minerals) is related to this particular reaction and not another. It is possible that the garnet may have formed from the breakdown of Fe-rich chlorite, or perhaps the sillimanite was produced from polymorphic inversion of kyanite.

Carmichael (1969), in attempting to define an aluminum silicate–garnet–biotite isograd in the Whetstone Lake area in southern Ontario, determined the mineralogical associations of a variety of metasedimentary and metavolcanic rocks (see Figure 19–23). East of the indicated line was found the lower-temperature assemblage staurolite, muscovite, and quartz, whereas to the west was found the higher-temperature

Figure 19–22
Muscovite (Mu) embayed by quartz–sillimanite (Qtz–Sill) intergrowth. [After C. F. Tozer, 1955, *Geol. Mag., 92,* 310–320; and D. M. Carmichael, 1969, *Contr. Miner. Petrology, 20,* Fig. 4b.]

assemblage sillimanite, biotite, and garnet, occasionally with the three phases of the lower-temperature assemblage, or with either muscovite + quartz or staurolite + quartz. The diversity of assemblages indicates variation in rock compositions; the somewhat wide occurrence of the six phases of the reaction indicates that the reaction probably is not strictly univariant (probably as a result of variations in composition of the phases involved); but in any case the presence of four different phase assemblages consistent with the postulated reaction strongly suggests that this reaction has occurred.

A second step in deciding whether the reaction has occurred is to examine the rocks for textural evidence. Rocks obtained above the isograd could show definite evidence of the reaction, as indicated by isolated remnants of reactant phases in optical continuity.

When determining whether a particular reaction has occurred it is also necessary to consider the relative amounts of the products that are formed. In the reaction under consideration the products are sillimanite, garnet, and biotite in the ratio of 17:2:1. This ratio, given by the formula, is in terms of the molecular proportions. In order to use this information the molecular proportions are converted into volume proportions by use of the relationship $\rho = m/V$ (or $V = m/\rho$) where ρ = density,

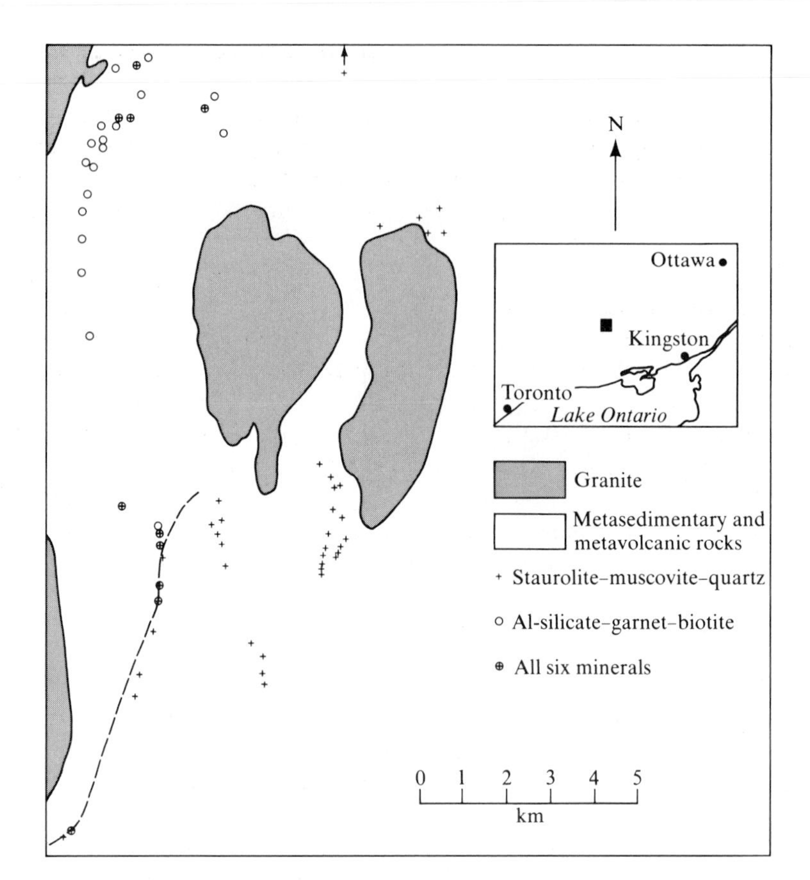

Figure 19–23
An isograd determined by the reaction

$$\text{Staurolite} + \text{Muscovite} + \text{Quartz} \rightleftarrows \text{Al-silicate} + \text{Biotite} + \text{Garnet} + H_2O$$

Whetstone Lake area, Ontario. [After D. M. Carmichael, 1969, *Contr. Miner. Petrology, 20,* Fig. 8.]

m = mass, and V = volume. In this case the mass of each of the product minerals is equal to its gram formula weight multiplied by its coefficient in the chemically balanced formula. When this term is divided by the mineral's density, its predicted volume is obtained. The volume of each of the three minerals obtained can be reset to a total of 100%. If the rock consisted of only these three phases, then the volume percent of these minerals in the rock (as determined by point-counting the grains) should correspond closely to the calculated volume percentages derived from the balanced equation.

In the real world the problem is not solved as simply as indicated above. Most rocks contain more phases than just the product minerals of a presumed reaction. This can be taken account of by determining the volume percentage of all the minerals in the rock; the volume percentages of the product minerals are extracted from the total rock percentage and reset to 100%. If none of the product minerals was preexistent in the rock (from some other reaction), these volume percentages should correspond to the calculated volume percentages. If one or more of the product phases was present before the reaction occurred, these can be "normalized" (that is, set to the correct calculated volumes), and the remaining new phases should appear in the proper ratio and amount as indicated by the reaction (see Table 19–1 for a sample calculation).

Another consideration with respect to metamorphic reactions is whether all of the required atomic constituents for the reaction are present within the minerals themselves. For example, Figure 19–24 shows isolated islands of optically continuous kyanite and quartz surrounded by muscovite and sillimanite. Neither the kyanite nor the quartz contain the potassium or H_2O necessary for muscovite formation. Consequently, a balanced equation might be written as

$$3 \text{ Kyanite} + 3 \text{ Quartz} + 2K^+ + 3H_2O \rightleftharpoons 2 \text{ Muscovite} + 2H^+$$

Where do the necessary constituents come from? An external source, such as an adjacent magma, is a possibility often called upon by many petrologists ("emanations from below"). Alternatively, as suggested by Carmichael (1969), the source for these constituents might lie in adjacent domains within the rock that are simultaneously undergoing different reactions, producing a coupling between them. Such coupling is accomplished by transfer of various constituents along concentration gradients between the reaction sites. In his analysis of a variety of such reactions Carmichael

Figure 19–24
Isolated optically continuous kyanite and quartz islands surrounded by muscovite that contains needles of sillimanite. [After G. A. Chinner, 1961, *Jour. Petrology, 2,* 312–323, and D. M. Carmichael, 1969, *Contr. Miner. Petrology, 20,* Fig. 4a.]

Table 19-1
A Method of Determining the Validity of a Proposed Metamorphic Reaction

A proposed staurolite–garnet isograd reaction

$$\underset{\text{Muscovite}}{7KAl_2AlSi_3O_{10}(OH)_2} + \underset{\text{Chlorite}}{6(Fe,Mg)_5Al_2Si_3O_{10}(OH)_8} \rightarrow$$

$$\underset{\text{Staurolite}}{6(Fe,Mg)Al_4Si_2O_{10}(OH)_2} + \underset{\text{Biotite}}{7K(Fe,Mg)_5AlSi_3O_{10}(OH)_2} + \underset{\text{Garnet}}{(Fe,Mg)_3Al_2Si_3O_{12}}$$

$$+ \underset{\text{Quartz}}{3SiO_2} + 18H_2O$$

Calculation of predicted volume percentages

$$\frac{6(\text{Molecular weight staurolite})}{\text{Density of staurolite}} : \frac{7(\text{Molecular weight biotite})}{\text{Density of biotite}} :$$

$$\frac{1(\text{Molecular weight garnet})}{\text{Density of garnet}} : \frac{3(\text{Molecular weight quartz})}{\text{Density of quartz}}$$

$$\frac{6(387)}{3.75} : \frac{7(464.6)}{3} : \frac{1(472.2)}{4.25} : \frac{3(60.05)}{2.65}$$

The relative volumes of the products

Staurolite : Biotite : Garnet : Quartz = 619.2:1084.1:111.1:68.0

The predicted relative volumes (reset to 100% volume percent)

Staurolite = 32.9%, Biotite = 57.6%, Garnet = 5.9%, Quartz = 3.6%

Modal analysis of the rock volume percent[a]

Quartz	48.3	48.3 =	55.0
Biotite	32.7	32.7 =	37.2
Muscovite	6.1		
Chlorite	1.0		
Garnet	1.4	1.4 =	1.6
Staurolite	5.4	5.4 =	6.2
Plagioclase	4.9	87.8	100.0 volume percent
Opaques	0.3		
	100.3		

	(1) Volume percent from mode after elimination of other phases	(2) Predicted volume percent from equation	(3) "Normalized" volume percent
Quartz	55.0	3.6	3.6
Biotite	37.2	57.6	57.6
Garnet	1.6	5.9	8.0
Staurolite	6.2	32.9	30.8

Table 19–1 (*continued*)

Assuming preexistent quartz and biotite, "normalize" the rock by using the predicted values for quartz and biotite (column 3), a total of 61.2%. This step eliminates the effect of preexisting quartz and biotite in the rock. The remaining 38.8% consists of garnet and staurolite. Using the relative amounts of garnet and staurolite found in the rocks, 1.6 and 6.2% (totaling 7.8% from column 1 above), set these two minerals in proper proportions to equal 38.8% (derived from column 3 above).

$$\frac{1.6}{7.8} = \frac{\text{Normalized garnet}}{38.8} \qquad \begin{array}{l}\text{Normalized}\\ \text{garnet}\end{array} = 8.0$$

$$\frac{6.2}{7.8} = \frac{\text{Normalized staurolite}}{38.8} \qquad \begin{array}{l}\text{Normalized}\\ \text{staurolite}\end{array} = 30.8$$

As the relative amounts of normalized garnet and staurolite are very similar to the volume percentages predicted by the equation, it can be assumed that the suggested reaction did occur. Exact coincidence of these values is not expected, as the compositions of the phases and their densities are probably different from the average values assumed. Attoh (1976) covers the case where only one new phase is produced; we have assumed that the other products preexist in various amounts when the reaction occurs.

[a]The product minerals are extracted and set to 100 volume percent.
Source: After K. Attoh, 1976, *Lithos, 9,* 75–85.

examined the conversion of kyanite to sillimanite. Very rarely does sillimanite ever seem to have formed directly from kyanite; it is usually formed as a result of other reactions within the rock (consider Figures 19–22 and 19–24). Considering the relative immobility of aluminum under most metamorphic conditions and the larger mobility of most other cations, Carmichael considers that the kyanite–sillimanite conversion could be accomplished by two simultaneous adjacent reactions (see Figure 19–25) such as the following:

$$3\ \text{Kyanite} + 3\ \text{Quartz} + 2K^+ + 3H_2O \rightleftarrows 2\ \text{Muscovite} + 2H^+$$

$$2\ \text{Muscovite} + 2H^+ \rightleftarrows 3\ \text{Sillimanite} + 3\ \text{Quartz} + 2K^+ + 3\ H_2O$$

Balancing the two reactions simultaneously (by crossing off equivalent constituents on opposite sides of the equations, such as $2K^+$ on the left and $2K^+$ on the right), one is left with the result 3 Kyanite $\rightleftarrows$ 3 Sillimanite. In order to accomplish this simple inversion, muscovite and quartz have entered the reaction as catalysts (being lost and reformed in different parts of the system). These subsidiary reactions can be verified by determining the relative amounts of the products as discussed above. Finally, note that although a metamorphic reaction may occur as a result of a number of subsidiary reactions, the pressure or temperature at which the reaction occurs is not changed; the subsidiary reactions merely provide an easier path than the direct reaction.

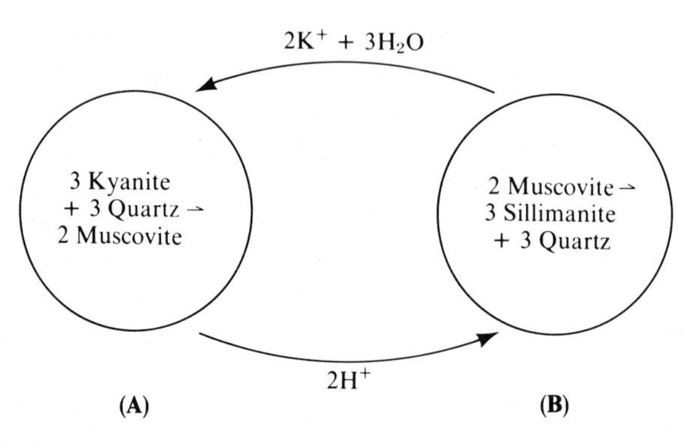

Figure 19-25
A possible mechanism for the reaction Kyanite ⇌ Sillimanite. Domains exist within the rock at *A* and *B*. In domain *A*, kyanite and quartz react to produce muscovite. In domain *B*, muscovite reacts to form sillimanite and quartz. In order for both reactions to occur, H_2O, K^+, and H^+ must migrate between the domains. The net result of both reactions is 3 Kyanite ⇌ 3 Sillimanite. [From D. M. Carmichael, 1969, *Contr. Miner. Petrology, 20,* Fig. 5.]

THE UPPER LIMIT OF METAMORPHISM

The upper limit of metamorphism is reached when melting begins. Melting is initiated after metamorphic rocks have reached the amphibolite or granulite facies. This may occur on a very limited scale in the upper levels of the crust, or on a large scale near the crust–mantle boundary.

The circumstances and degree of melting vary considerably as a function of temperature, rock type, depth of burial, and the presence of H_2O. Consider first the possibility of melting a metamorphic rock at shallow levels in the crust—about 6 km or less, with pressures not exceeding 1.5 kbar. This is a typical contact metamorphic situation. As the geothermal gradient normally is not sufficient to cause melting at this depth, we can postulate an adjacent igneous intrusion. Approaching the intrusion one passes through a series of zones indicative of higher and higher temperatures. Such zones might contain (in turn) minerals such as albite and epidote, then hornblende and plagioclase, followed by potassium feldspar, cordierite, and either sillimanite or andalusite. In the final zone, adjacent to the intrusive, the hydrous minerals amphibole and mica are eliminated, along with garnet, and melting may occur in the presence of high-temperature phases such as sanidine, tridymite, mullite, monticellite, forsterite, sillimanite, cordierite, or wollastonite. The metamorphic facies represented in metamorphism of this type is the hornfels or sanidinite facies.

The amount of melting found in such cases (if present at all) is usually limited to a thin border zone, or perhaps to xenoliths within the intrusive. There are several reasons for this general lack of melting. Although the temperature of the intrusion is high (perhaps 700°C for granitic and 900°C for basaltic magmas), the country rocks, being fairly shallow, are relatively cool. This results in considerable dissipation of magmatic heat and a steep thermal gradient, permitting only a thin border zone to be at or near igneous temperatures. Furthermore, heat from the intrusion causes vaporization of pore water in the country rock, which results in chilling of the contact zone. Dehydration and decarbonation reactions in the country rock absorb heat (endothermic) and cause a further cooling effect.

Jaeger (1957, 1968), who has calculated thermal gradients as a function of country rock heat conductivity (see Figure 19–26), shows that even neglecting the effect of endothermic reactions, the country rock often does not attain magmatic temperatures. Other factors involved in determining the amount of melting are the size of the intrusion and its water content. A larger intrusive cools over a longer period of time, allowing the formation of a wider and less steep thermal aureole. Water expelled from the intrusive could transfer considerable amounts of heat and cause great expansion of the thermal aureole. In addition, most magmas do not contain superheat—that is, heat that causes magmatic temperatures to be significantly above freezing

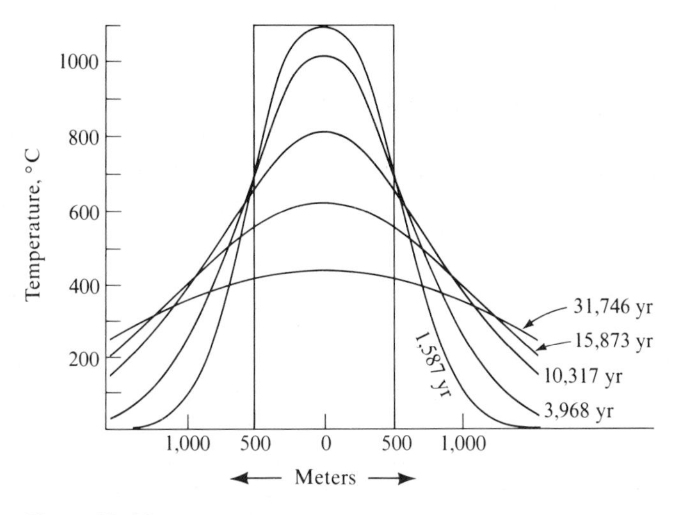

Figure 19–26
Calculated thermal gradients as a function of time. A vertical sheet 1,000 m thick is assumed. Intrusion occurs at 1100°C, crystallization range is 1100–800°C, and the country rock is initially at 0°C. Average values are assumed. [Based on data from J. C. Jaeger, 1968, in H. H. Hess and A. Poldervaart (eds.), *Basalts,* vol. 2 (New York: Interscience Publishers), Fig. 5.]

temperatures. Thus any process, such as melting of the country rock, causes crystallization and cooling of the intrusive. A final consideration is that the solidus temperature of the country rock may be above that of the temperature of the igneous intrusion.

Deep within the crust at medium to high temperatures and pressures, considerable melting can occur during orogenesis, as large masses of rock have been brought to elevated temperatures and pressures. The process of melting as a result of extreme metamorphic conditions is known as *anatexis*. If melting is complete, magma is formed that when cooled will yield a rock of typical igneous characteristics. In the case of partial melting (followed by cooling), a rock is produced with portions that retain some metamorphic characteristics and other portions that appear to be of igneous origin; such rocks are called *migmatites*. It should be pointed out that some migmatites may have formed by solid state diffusion processes (granitization), which produce an "igneous-looking" portion of the rock without the formation of melt. Hence the term migmatite should be used in a nongenetic sense. Typical migmatite structures are shown in Figure 19–27. It is obvious from these that it is very difficult in many cases to distinguish between a rock that has resulted from partial melting and one produced by plastic flow and solid state segregation.

As an example of the occurrence of migmatites we will examine some effects of the Caledonian orogeny as described by Haller (1971) in east-central Greenland. Above the silicic basement is a geosynclinal pile of lower Paleozoic and Precambrian sediments, about 16,000 m in thickness. The Caledonian orogeny caused increased temperatures and pressures in the essentially granitoid basement materials, such that this material was able to migrate by plastic and/or liquid flow as a result of orogenic pressures. The lower mobile mass, consisting of a granitoid core and metamorphic materials of originally geosynclinal-filling origin, is called the *infrastructure*. Movement of the infrastructure led to the detachment of the overlying moderately to weakly metamorphosed sedimentary materials (see Figure 19–28). This resulted in a variety of deformation patterns in the infrastructure and the overlying *superstructure* (see Figures 19–29 and 19–30). The granitoid core of the infrastructure consists of a mixture of granitic materials, migmatite, and gneissic rocks of basement origin. The attached volume of original sedimentary origin (called the *zone of detachment*) shows extreme contortion, thickening and thinning of beds, intense folds, and shear structures; rock types are mica schists and migmatite. In many areas migmatization has been so intense that the distinction between the metamorphosed sediments and the reworked basement rocks has become blurred. Materials within the infrastructure have been interpreted by Haller to be not the result of partial melting processes, but rather a migmatite front, with alkalies, alumina, and silica added from below to the basement material and converted sedimentary cover. Examination of Figure 19–31, a sketch of a typical contact, shows that the transition from granitoid to metamorphic rocks can be very gradational, making it difficult to judge whether portions of the rock have melted or whether the rock has been impregnated with incoming material and converted to granitoid composition by fluid flow or diffusion processes.

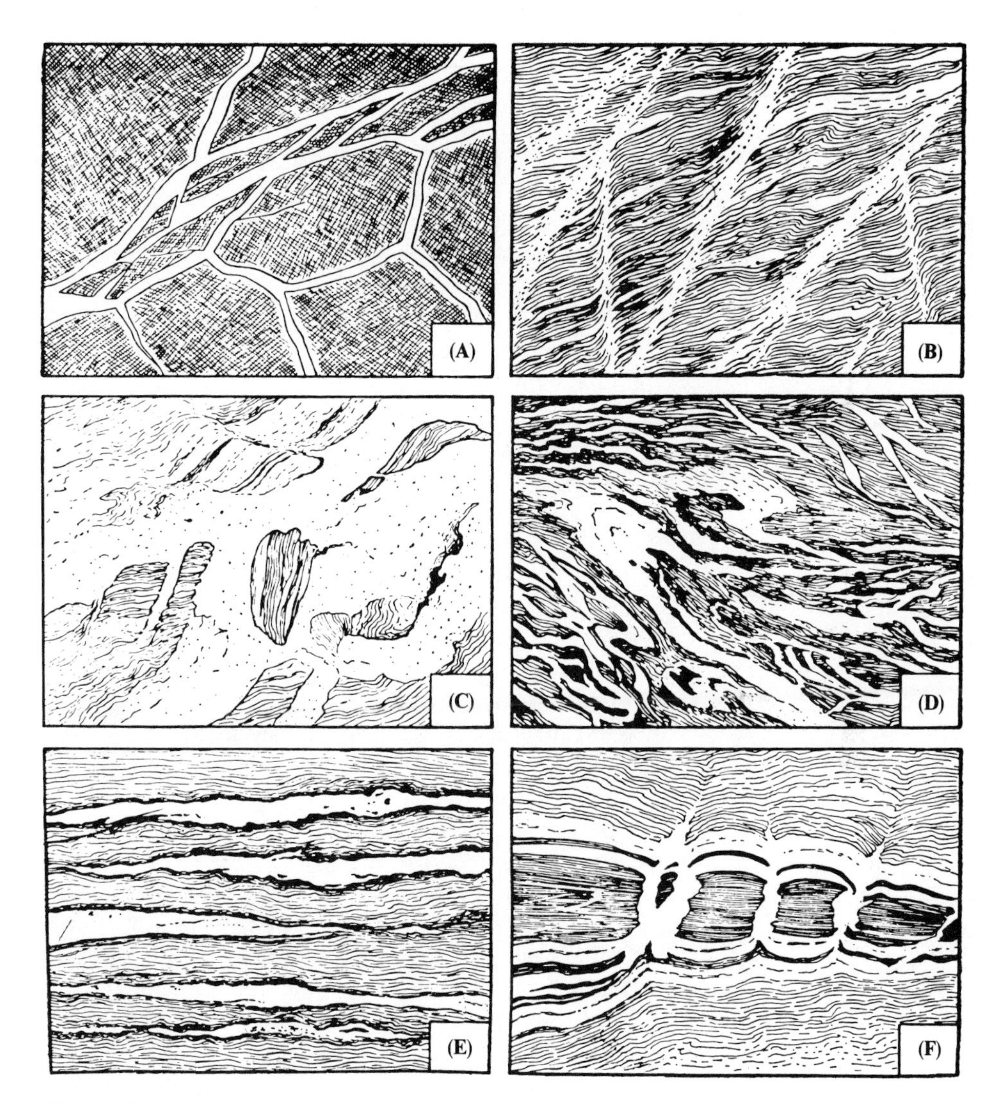

Figure 19–27
Typical structures in migmatites. The lighter areas are rich in quartz and feldspar; the darker portions are richer in mafic constituents. [From K. R. Mehnert, 1968, *Migmatites and the Origin of Granitic Rocks* (Amsterdam: Elsevier), Fig. 1a, b.]

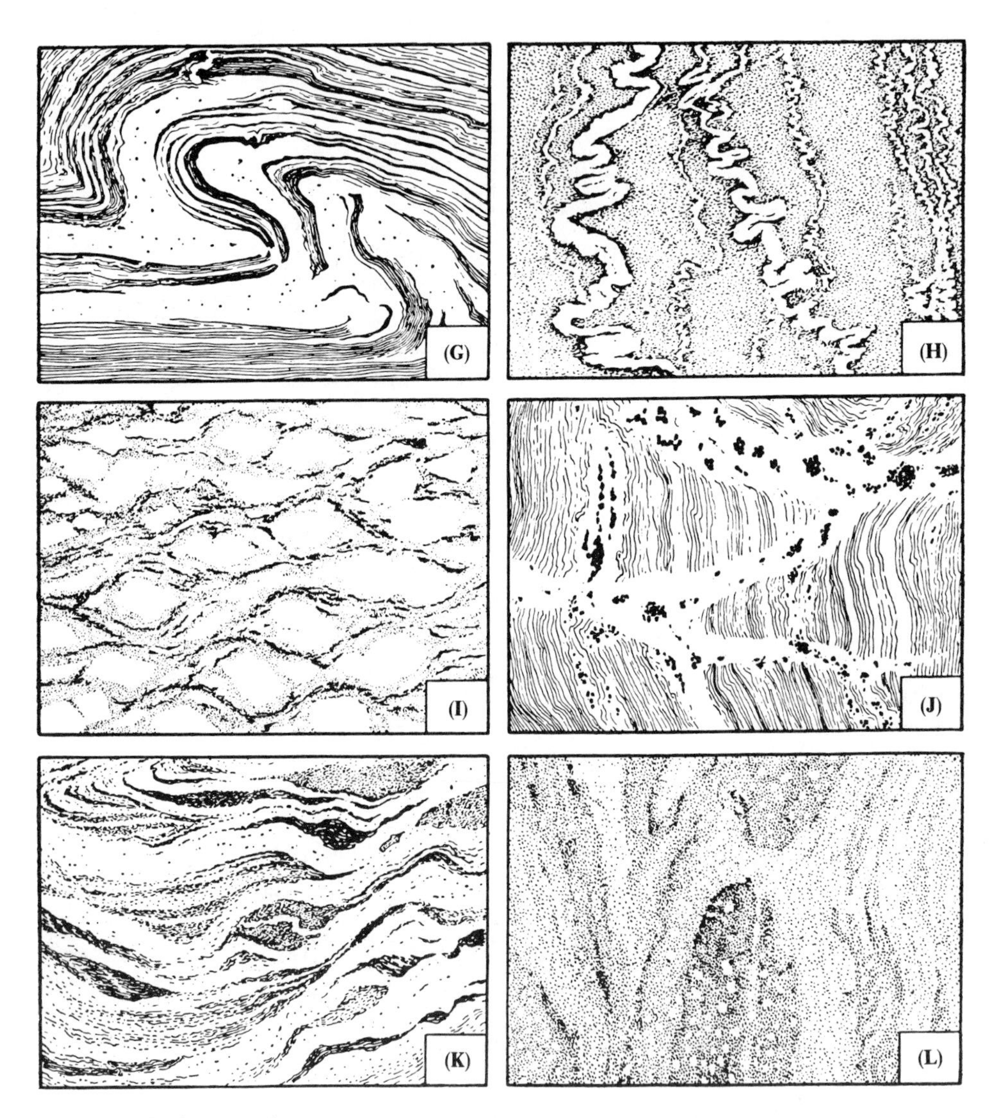

Figure 19–27 (*continued*)
Typical structures in migmatites. The lighter areas are rich in quartz and feldspar; the darker portions are richer in mafic constituents. [From K. R. Mehnert, 1968, *Migmatites and the Origin of Granitic Rocks* (Amsterdam: Elsevier), Fig. 1a, b.]

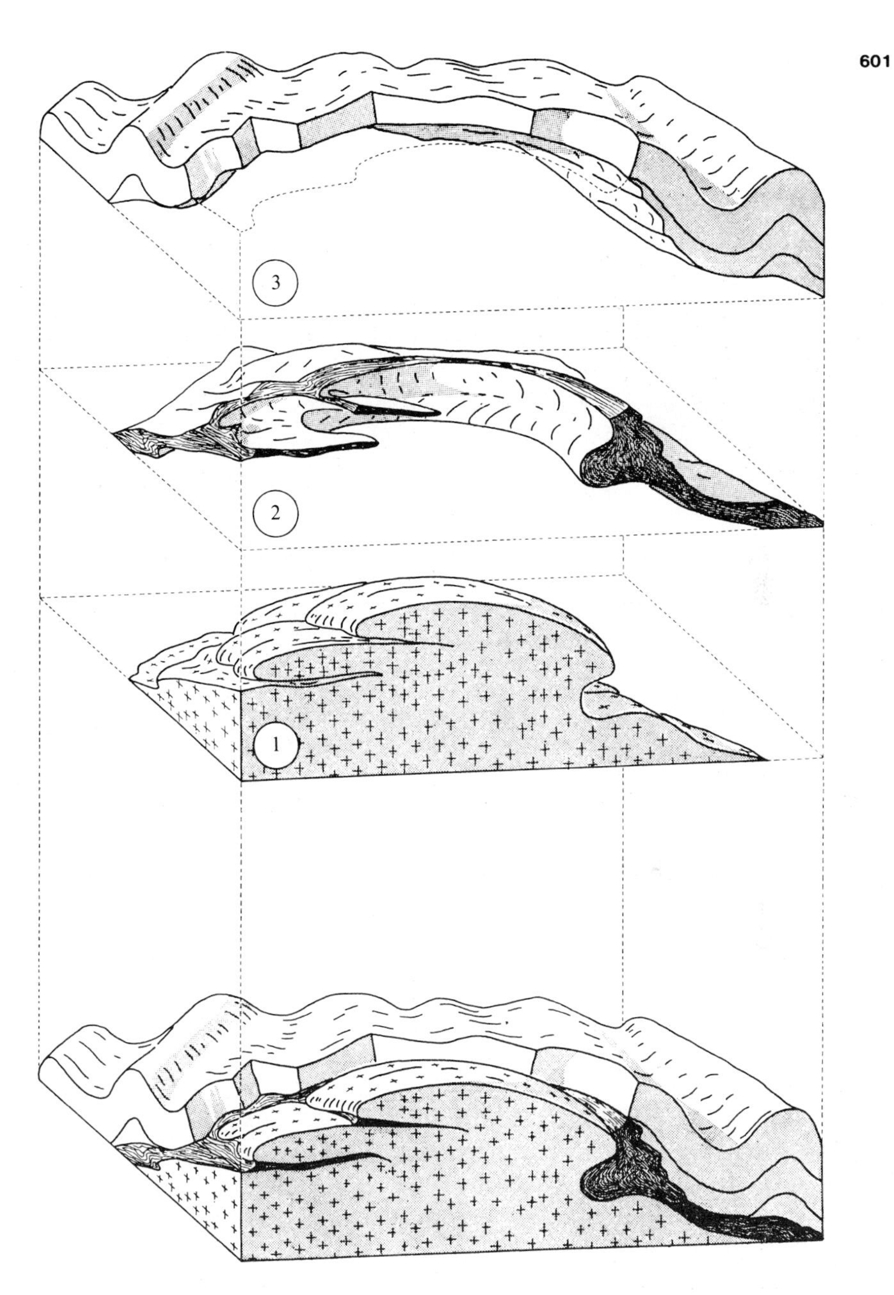

Figure 19–28
Three-dimensional representation of a migmatite complex in exploded view. Zone 1 is the granitoid basement portion of the infrastructure. Zone 2 is the zone of detachment, a portion of the infrastructure attached to the granitoid basement. Zone 3 is the weakly deformed and detached overlying superstructure. [From J. Haller, copyright © 1971, *Geology of the East Greenland Caledonides* (New York: Wiley Interscience), Fig. 51. Reprinted by permission of John Wiley and Sons, Ltd.]

Figure 19–29
A cliff at the head of Kejser Franz Joseph Fjord, Greenland, about 2,000 m high. Zones 1 and 2 show the infrastructure; zone 2 is the originally sedimentary zone of detachment; zone 3 is the superstructure. [From J. Haller, copyright © 1971, *Geology of the East Greenland Caledonides* (New York: Wiley Interscience), photo 46. Reprinted by permission of John Wiley and Sons, Ltd.]

Migmatites appear to have formed by a variety of mechanisms. If the migmatite (and adjacent granitic material) has formed by *granitization* (an influx of material in the solid state), the following characteristics may be present:

1. Veins of felsic materials show no dilation.
2. Contacts are gradational.
3. Relics of original sedimentary beds are present in an orientation compatible with surrounding rocks.
4. Feldspars present are of the subsolvus type (two alkali feldspars showing little to no perthitic exsolution).
5. The mineralogy of felsic portions is not in agreement with compositions produced by partial melting experiments.

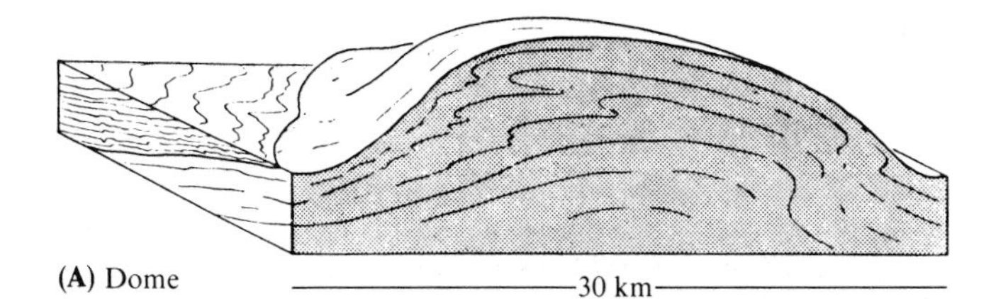

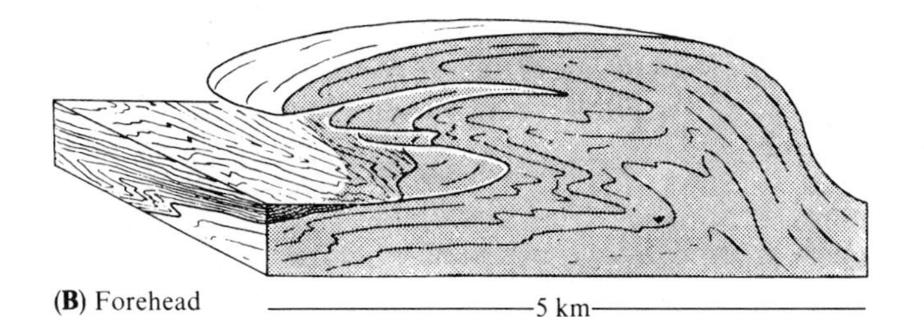

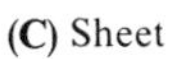

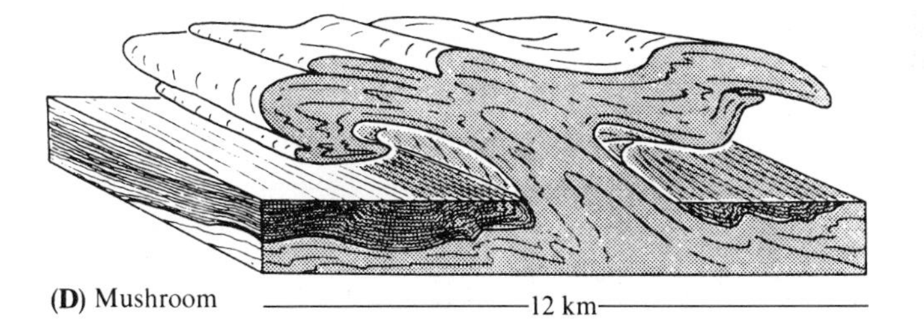

(E) Complex
35 km

Figure 19–30
Types of infrastructure upswellings. [From J. Haller, copyright © 1971, *Geology of the East Greenland Caledonides* (New York: Wiley Interscience), Fig. 64. Reprinted by permission of John Wiley and Sons, Ltd.]

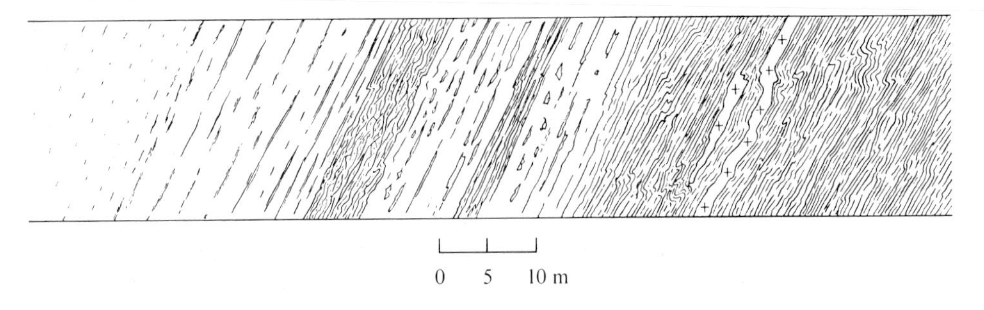

Figure 19–31
A transition zone between mica-rich gneiss and gneissose biotite granodiorite. [From J. Haller, copyright © 1971, *Geology of the East Greenland Caledonides* (New York: Wiley Interscience), Fig. 116. Reprinted by permission of John Wiley and Sons, Ltd.]

If the migmatite has formed by *anatexis,* we would expect the following:

1. The igneous and metamorphic portions should together equal the composition of adjacent metamorphic rocks.
2. Geothermometry should indicate magmatic temperatures.
3. Granitic material could be present in isolated patches.
4. The composition and phases of the igneous part of the migmatite should be compatible with partial melting experiments.

Migmatites created as a result of *metamorphic differentiation* should show the following:

1. Felsic and mafic portions together should be similar in composition to adjacent rocks.
2. The mineralogy and textures of both felsic and mafic portions should appear to be of metamorphic origin, and formed at the same pressure and temperature.

If the felsic portion of a migmatite is created by the *intrusion of igneous melt,* we would expect the following:

1. The presence of dilation veins.
2. Granitic veins often at an angle to the schistosity, and perhaps more granitic material than that expected to have formed by either anatexis or metamorphic differentiation.

Some of the more definitive information on the origin of migmatites has come from melting experiments on specific rocks or idealized compositions. Let us consider first the melting relations of a variety of igneous rocks at a variety of pressures in a water-saturated environment, where $P_{total} = P_{H_2O}$ (see Figure 19–32). The initial compositions of the various rocks examined are greatly different, but in spite of this the granitic minerals, alkali feldspar and quartz, are commonly the first phases to be eliminated during melting, with the mafic constituents hornblende and biotite (and sometimes plagioclase) being eliminated at higher temperatures. The average vapor-saturated liquidus relations of three quartz diorites over a pressure range are shown in Figure 19–32B. It is not unreasonable that the part of migmatites formed by anatexis is richer in sialic constituents than the residual part (which is more Fe–Mg rich). Other experiments show that for muscovite–plagioclase–quartz rocks, melting relations are similar to those of granite (with muscovite supplying the potassium in place of alkali feldspar). Experiments on the melting of graywackes again indicate that the first-formed liquids are granitic in composition.

All of the above experiments have been conducted with all of the pressure supplied by water vapor. Many investigators feel that such experiments cannot be applied to anatectic melting as the amount of water available near the base of the crust is insufficient to saturate the melt.

Granitic melts at high pressures are capable of dissolving large amounts of water (which results in decrease of their freezing temperatures). Figure 19–33 shows the decrease in initial melting temperature of a granitic melt (the minimum point in the system H_2O–SiO_2–$NaAlSi_3O_8$–$KAlSi_3O_8$) as a function of increasing P_{H_2O}, and the weight percent H_2O dissolved in the silicate melt; any additional water in the system would be present as a vapor phase. If a "granite" melts completely at $P_{H_2O} = 5$ kbar, 12% of the melt (by weight) will consist of H_2O. If melting is only 50% complete, then only 6% H_2O (by weight) is required to saturate the melt.

If we consider that the base of the continental crust is at a depth equivalent to a pressure of about 10 kbar, and furthermore assume that most large-scale melting occurs near this level (perhaps in the 7–10 kbar pressure range), the amount of H_2O required to saturate the melt is about 15–20% by weight. As the pore space in the rocks at this depth must be negligible, Robertson and Wyllie (1971) estimate that 1% pore fluid would be a generous estimate. This restricts additional H_2O sources to either "emanations from below" or combined water in hydrous minerals. The earth's mantle is generally considered to contain about 0.1% water (as hydrous phases are sparse), and high-grade schists and gneisses might furnish at a maximum about 3–5% water by weight. This gives us only about a third of the required water for saturation of the melt (unless some sort of alternate scheme of continuous emission of small amounts of water from the mantle can be postulated). Thus if about a third of the rock melted, the melt would be water-saturated and would fulfill the above conditions. Additional melting yields a water-undersaturated melt and a different set of melting conditions.

Melting under water-deficient (undersaturated) conditions has been studied by a number of investigators, particularly Wyllie and his coworkers. An example of the

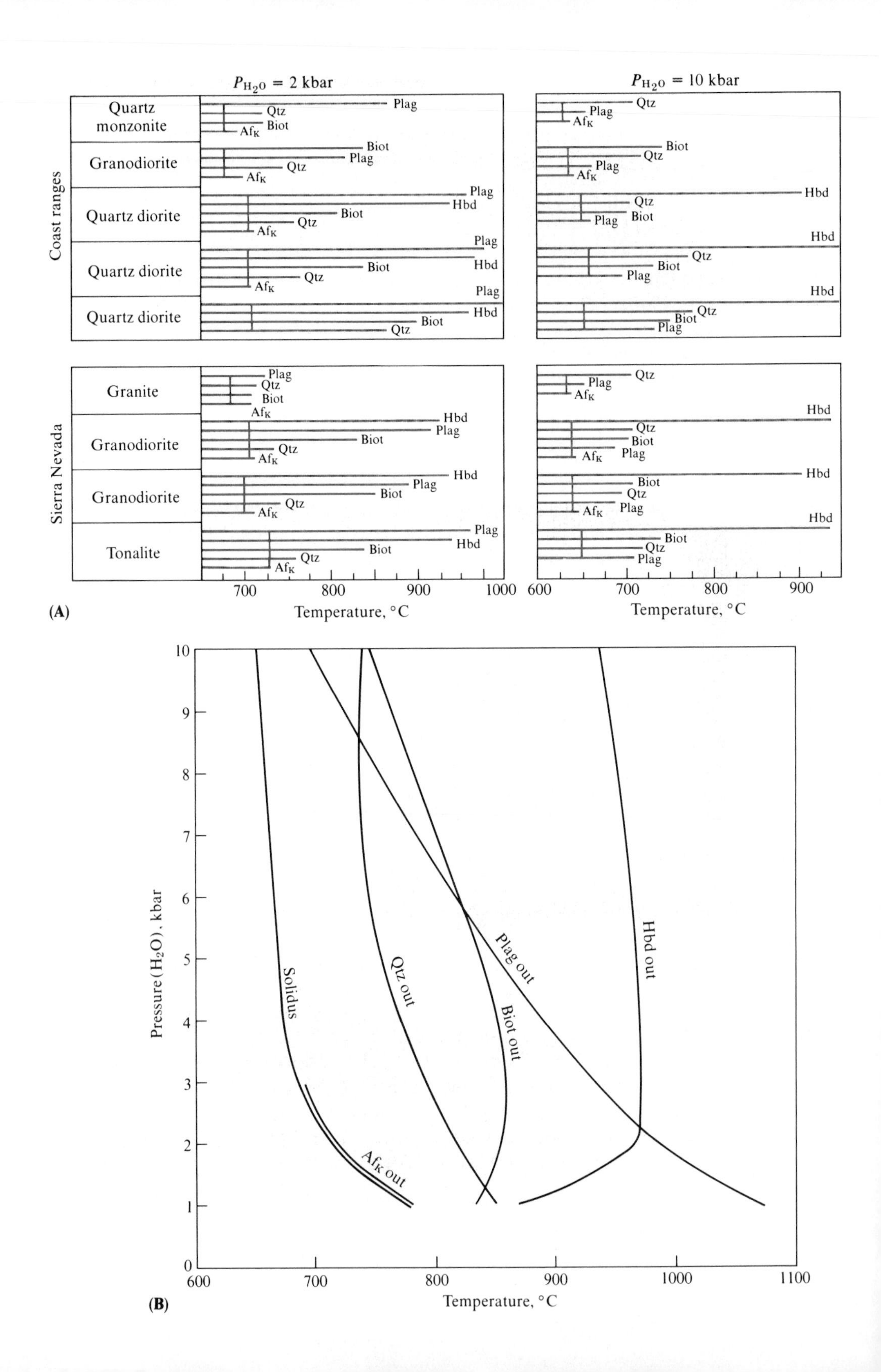

P_{H_2O} = 2 kbar
P_{H_2O} = 10 kbar
Coast ranges
Quartz monzonite
Granodiorite
Quartz diorite
Quartz diorite
Quartz diorite
Sierra Nevada
Granite
Granodiorite
Granodiorite
Tonalite
Plag
Qtz
Biot
Af_K
Hbd
Temperature, °C
(A)
Pressure (H_2O), kbar
Solidus
Af_K out
Qtz out
Biot out
Plag out
Hbd out
Temperature, °C
(B)

Figure 19–32 (*facing page*)
(**A**) Experimental determination of the crystallization/melting sequences for rocks from the California Coast Ranges and the Sierra Nevada. Magnetite and vapor were present in all experiments. Qtz = quartz, Plag = plagioclase, Af_K = potassic alkali feldspar, Biot = biotite, Hbd = hornblende. [From W. C. Luth, *The Evolution of the Crystalline Rocks,* 1976, in D. K. Bailey and R. Macdonald (eds.), (New York: Academic Press), Figs. 30, 31. Experimental data from A. J. Piwinskii, 1968, *Jour. Geol., 76,* 548–570; 1973, *Tsch. Min. Petr. Mitt., 20,* 107–130; and 1973, *Neues Jahrb. Miner., 5,* 193–215.] (**B**) Composite diagram of the vapor-saturated liquidus relations for three quartz diorites. Each curve refers to the maximum temperature limit of either a phase or the beginning of melting (solidus). [From the data of A. J. Piwinskii, 1973, *Tsch. Min. Petr. Mitt., 20,* Fig. 2.]

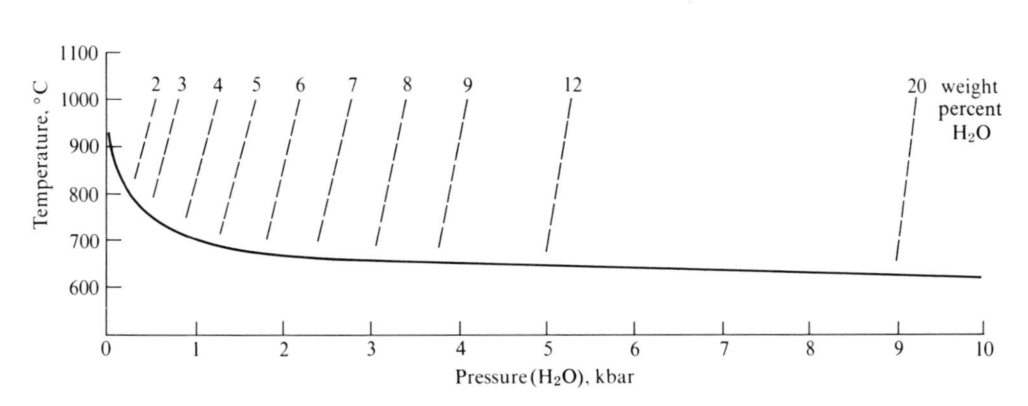

Figure 19–33
Minimum melting temperatures and H_2O saturation percentages for the simplified granite system. [After O. F. Tuttle and N. L. Bowen, 1958, Geol. Soc. Amer. Mem., 74, Fig. 28; and H. G. F. Winkler, 1976, *Petrogenesis of Metamorphic Rocks,* 4th ed. (New York: Springer-Verlag), Fig. 18–11.]

contrast in melting under saturated versus water-deficient conditions is given in Figure 19–34—the melting of a granodiorite that contains about ½% water combined in hydrous minerals. Melting under water-saturated conditions (6% H_2O over that in the granodiorite) takes place along the isopleth that contains 6½% total H_2O. Melting begins (at 2 kbar) at slightly over 700°C. The figure shows the temperatures at which various phases are eliminated. Once again, quartz and alkali feldspar are the first phases eliminated, before 750°C is reached. Melting under water-deficient conditions (1% H_2O over the ½% combined in the granodiorite) takes place along the isopleth indicated for 1½% H_2O. Melting begins at about 700°C, but temperatures near 900°C are now required to eliminate (in turn) alkali feldspar, biotite, and quartz. The melting curves of all of the solid phases are considerably raised as a result of melting with a water deficiency.

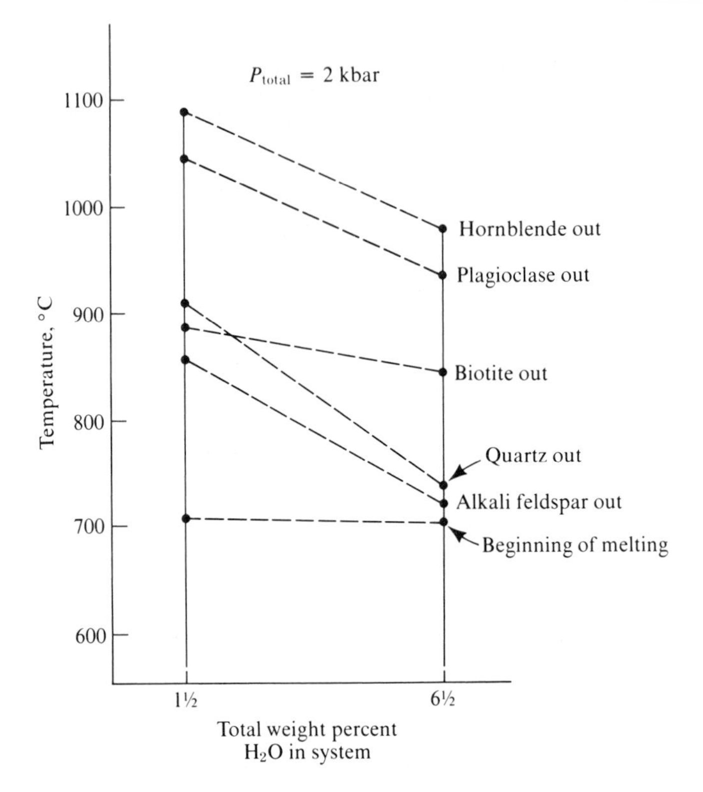

Figure 19-34
Schematic diagram at 2 kbar, illustrating the melting of a granodiorite containing ½% water (by weight) in hydrates. Water-saturated melting occurs at 6½% total weight percent H_2O; melting in a water-undersaturated condition occurs at 1½% total weight percent H_2O. The maximum stability limits of the various minerals are indicated for these percentages. Above the hornblende stability limits the system contains only liquid and vapor. With the exception of the beginning of melting, the stability limits of all phases are increased as the water content is decreased. [From the data of J. K. Robertson and P. J. Wyllie, 1971, *Amer. Jour. Sci., 271,* 252–277.]

Disagreements exist at present as to the temperatures and pressures at which melting takes place in the lower crust, with strong advocates for both water-saturated and water-deficient melting. This will probably be resolved as additional data become available from geothermometry and from further partial melting studies of melting temperatures and specific liquid compositions of migmatites.

SUMMARY

The temperature at which a metamorphic reaction occurs is dependent upon the type of pressure that is present and the permeability of the rock. The stability of hydrous minerals is increased by high P_{H_2O}, and the stability of carbonate-containing minerals is increased by high P_{CO_2}. In the event that the permeability of the rock permits escape of H_2O and CO_2, the breakdown temperatures of minerals are controlled by P_{lith}.

Metamorphism of highly reactive materials is initiated under conditions of the zeolite facies. In the absence of reactive materials, metamorphism usually is first apparent under conditions of the greenschist facies. Thorough recrystallization with increase of grain size is found within the greenschist facies. Metamorphism under submarine conditions is quite common, and has been found to be related to the presence of localized areas of thermal activity, an abundance of fractures, and the penetration of sea water as a result of convection in response to thermal differences.

In spite of a general tendency toward equalization of grain sizes during metamorphism, the formation of porphyroblasts is common. The fabric developed as a result of unequal application of stress usually shows a preferred orientation. This may occur as a result of deformation and recrystallization, or as a result of elimination of grains in unfavorable orientations during periods of anisotropic stress.

Metamorphic differentiation, the process of developing heterogeneous metamorphic rocks or fabrics from uniform source materials, may develop as a result of temperature, pressure, or compositional gradients, or as a result of deformation. Amphibolites may be formed as a result of reaction between rocks of incompatible composition, such as metapelite and marble. Temperature, pressure, and compositional gradients promote the redistribution of phases such that all parts of the rock system approach or are in equilibrium with the imposed conditions.

Phase changes in a metamorphic rock usually are not limited to a single mineral, but more commonly involve several or all of the minerals present. Coupling of reactions between several minerals is accomplished by transfer of the more mobile constituents among the reaction sites. A particular reaction can often be verified by textural evidence and consideration of the relative proportions of the products.

The upper limit of metamorphism usually is reached in either the amphibolite or the granulite facies (depending on P_{H_2O}). Melting of metamorphic rocks can be accomplished under conditions where $P_{H_2O} = P_{lith}$ or $P_{H_2O} < P_{lith}$. Transition into igneous-appearing rocks may take place as a result of solid state processes (granitization).

FURTHER READING

Attoh, K. 1976. Stoichiometric consequences of metamorphic mineral-forming reactions in pelitic rocks. *Lithos, 9,* 75–85.

Augustithis, S. S. 1978. *Atlas of Textural Patterns of Granites, Gneisses, and Associated Rock Types*. New York: Elsevier, 378 pp.

Burnham, C. W. 1959. Contact metamorphism at Crestmore, California. *Geol. Soc. Amer. Bull., 70,* 879–919.

Carmichael, D. M. 1969. On the mechanisms of prograde metamorphic reactions in quartz-bearing pelitic rocks. *Contr. Miner. Petrology, 20,* 244–267.

DeSegonzac, G. D. 1970. The transformation of clay minerals during diagenesis and low-grade metamorphism: a review. *Sedimentology, 15,* 281–346.

Frey, M. 1970. The step from diagenesis to metamorphism in pelitic rocks during Alpine orogenesis. *Sedimentology, 15,* 261–279.

Gilluly, J., chrmn. 1968. *Origin of Granite*. Geol. Soc. Amer. Mem. 28, 139 pp.

Greenwood, H. J. 1976. Metamorphism at moderate temperatures and pressures. In D. K. Bailey and R. MacDonald (eds.), *The Evolution of the Crystalline Rocks,* pp. 187–259. New York: Academic Press, 484 pp.

Haller, J. 1971. *Geology of the East Greenland Caledonides*. New York: Wiley Interscience, 413 pp.

Hemley, J. J. 1959. Some mineralogical equilibria in the system K_2O–Al_2O_3–SiO_2–H_2O, *Amer. Jour. Sci., 257,* 241–270.

Jaeger, J. C. 1957. The temperature in the neighborhood of a cooling intrusive sheet. *Amer. Jour. Sci., 255,* 306–318.

Jaeger, J. C. 1968. Cooling and solidification of igneous rocks. In H. H. Hess, and A. Poldervaart (eds.), *Basalts,* Vol. 2, pp. 503–536. New York: Wiley Interscience.

Mehnert, K. R. 1968. *Migmatites and the Origin of Granitic Rocks*. Amsterdam: Elsevier, 393 pp.

Orville, P. M. 1962. Alkali metasomatism and feldspars. *Norsk Geol. Tidsk., 42,* 283–316.

Orville, P. M. 1969. A model for metamorphic differentiation origin of thin-layered amphibolites. *Amer. Jour. Sci., 267,* 64–86.

Pitcher, W. S., and G. W. Flinn, eds. 1965. *Controls of Metamorphism*. New York: John Wiley and Sons, 368 pp.

Read, H. H. 1948. A commentary on place in plutonism. *Quart. Jour. Geol. Soc. London, 104,* 155–206.

Read, H. H. 1949. A contemplation of time in plutonism. *Quart. Jour. Geol. Soc. London, 105,* 101–156.

Robertson, J. K., and P. J. Wyllie. 1971. Rock-water systems with special reference to water-deficient systems. *Amer. Jour. Sci., 271,* 252–277.

Schmidt, W. 1932. *Tektonik und Verformungslehre*. Berlin: Borntraeger, 208 pp.

Spry, A. 1969. *Metamorphic Textures*. New York: Pergamon, 350 pp.

Vidale, R. 1969. Metamorphism in a chemical gradient and the formation of calc-silicate bands. *Amer. Jour. Sci., 267,* 857–874.

Walton, M. 1955. The emplacement of granite. *Amer. Jour. Sci., 253,* 1–18.

20

Mineral Changes During Metamorphism

Two approaches are used in the consideration of progressive metamorphism. One of these examines the characteristics of each of the metamorphic facies independent of rock type (see Figure 20–1), and the other traces the changes of a particular rock type through the various facies. We will combine both approaches, first covering the general characteristics of the major facies, followed by discussion of metamorphism of the major rock types.

The mineral assemblages of the various facies are shown in the following sections in terms of *ACF*, *A'KF*, and *AFM* diagrams (described in Chapter 18). Recall that these diagrams apply only to rocks in which a silica polymorph is present and, in the case of the *AFM* diagram, muscovite or potassium feldspar as well. The diagrams represent a shorthand summary of the mineral associations of the common rock types. Fitting a rock into one of these groupings permits not only an approximation of the pressure–temperature conditions at which the rock formed (by use of Figure 20–1), but also gives a fairly good idea of the general composition and in some cases a clue to the type of rock that existed before metamorphism. With this last possibility in mind, the average compositions of the common nonmetamorphic rocks are plotted on an *ACF* diagram in Figure 20–2.

The approach first involves determining the minerals present in a metamorphic rock and their relative amounts. This is done best with thin sections rather than hand specimens, as significant minerals may be overlooked in hand specimens due to fine grain size or limited amounts. After the minerals and their approximate amounts are determined, check Figures 18–2, 18–3 and 18–5 to see which, if any, of the minerals are represented in the *ACF*, *A'KF*, or *AFM* diagrams. Most common metamorphic rocks will contain several of the minerals on at least one of these diagrams.

As an example of the approach, consider a rock that contains plagioclase (An_{32}), quartz, almandine garnet, hornblende, epidote, and chlorite, in addition to various minor constituents. The presence of quartz indicates that the diagrams probably can be used. Examination of the *ACF*, *A'KF*, and *AFM* diagrams (see above) reveals that all of these minerals can be plotted on the *ACF* diagram, except for quartz, which is present in excess. Elimination of quartz leaves us with five minerals that can be plotted on an *ACF* diagram. Recalling the discussion in Chapter 18, in the general case of divariant equilibrium we should expect to have a maximum of only three of these minerals. The two additional minerals might be explained in several ways. Perhaps the assemblage is univariant or invariant; alternatively, relics may have persisted during prograde metamorphism, or late phases may have formed during retrograde metamorphism.

The way to check this is to examine the textural relations of the rock in thin section. In our case we will assume that this examination reveals that porphyroblasts of garnet are partially replaced by irregular rims of chlorite; additional chlorite is found in partial replacement of hornblende. Large hypidioblastic crystals of plagioclase contain masses of epidote along twin bands, adjacent to cleavage cracks, and at grain edges. Such evidence indicates that chlorite is later than the garnet and hornblende. Similarly, epidote has formed by replacement of plagioclase. Study of the diagrams in this chapter for the various facies reveals that chlorite and epidote typically are found in the greenschist or epidote–amphibolite facies, whereas amphibole, garnet, and plagioclase of intermediate composition are characteristic of the higher-level amphibolite facies. The presence of chlorite and epidote is indicative therefore of decreasing *PT* conditions, retrograde metamorphism, and hence does not reflect the maximum facies level to which the rock has been subjected.

Examination of the three remaining phases, hornblende, plagioclase, and garnet, shows that these minerals occur adjacent to each other in various portions of the thin section, with no evidence of reaction. These three minerals (with excess quartz) apparently represent an assemblage in equilibrium with the maximum conditions of pressure and temperature to which the rock was subjected. As the plagioclase composition was determined as An_{32}, this places the rock in the kyanite zone of the amphibolite facies (see Figure 20–8B). The composition point for this rock falls within the compatibility triangle plagioclase, hornblende, almandine; its specific location depends upon the relative proportions of these three phases. If the rock had contained only two of the *ACF* minerals, its composition point would lie between the composition points of these two minerals. Its position within the amphibolite facies permits us to conclude that the maximum temperature of metamorphism probably was between about 600–700°C (see Figure 20–1). Retrograde effects probably occurred within the epidote–amphibolite or greenschist facies. This approach is certainly only a rough determination of metamorphic conditions. More sophisticated *PT* determinations are discussed in the Appendix. The nature of the premetamorphic source rock can often be determined by comparison of the rock composition point (as determined above) with average nonmetamorphic rock compositions (see Figure 20–2); if for instance the rock consisted mainly of hornblende with perhaps 20% plagioclase and a few percent almandine garnet, it would correspond to a mafic rock composition.

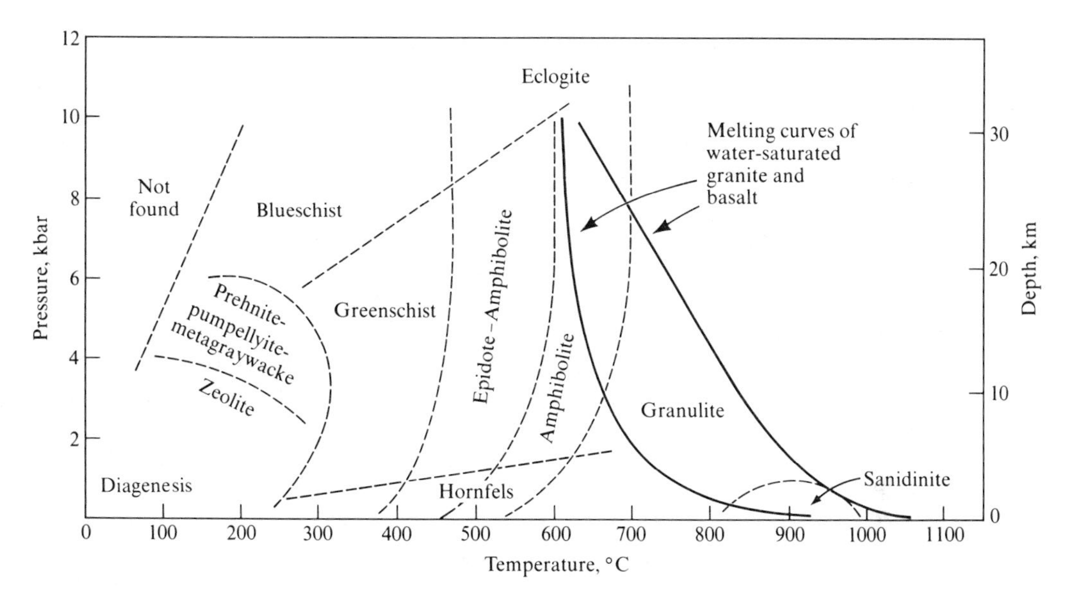

Figure 20–1
The metamorphic facies plotted as a function of pressure and temperature. The hornfels facies is occasionally subdivided into pyroxene hornfels, hornblende hornfels, and albite–epidote hornfels. Hornfels facies rocks are distinguished from higher-pressure facies mainly on the basis of texture. [The granite and basalt melting curves are from W. C. Luth, R. H. Jahns, and O. F. Tuttle, 1964, *Jour. Geophys. Res., 69,* Fig. 1; and H. S. Yoder, Jr., and C. E. Tilley, 1962, *Jour. Petrology, 3,* Fig. 33.]

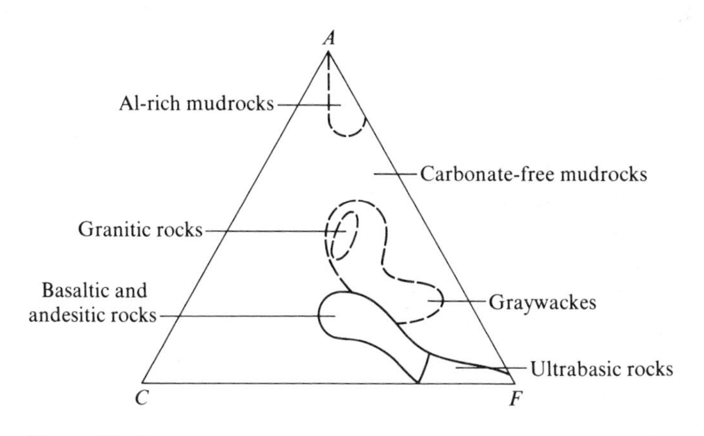

Figure 20–2
The composition of common nonmetamorphic rocks plotted on an *ACF* diagram. [Based on data from S. R. Nockolds, 1954, *Geol. Soc. Amer. Bull., 66,* 1007–1032; and A. B. Ronov and Z. V. Khlevnikov, 1957, *Geochemistry, 6,* 527–552.]

MINERAL VARIATION AND METAMORPHIC FACIES

Zeolite and Prehnite–Pumpellyite–Metagraywacke Facies

Sediments and volcanic rocks tens of thousands of meters thick are formed at convergent plate margins. Burial results in a low level of metamorphism known as burial metamorphism. This includes both the zeolite and prehnite–pumpellyite–metagraywacke facies. Evidence of these facies will be found when the rocks contain relatively reactive materials (fine-grained and glassy phases). When such reactive materials are not present, metamorphism usually is initiated at the level of the greenschist facies. Although many instances of burial metamorphism are recorded throughout the world, these facies generally are not regarded as typical of regional metamorphism.

Zeolite metamorphism begins at depths of 1–5 km, depending upon the geothermal gradient and the reactivity of the rocks involved. Transitions to the prehnite–pumpellyite–metagraywacke facies are reached at 3–13 km. Nonmetamorphic or diagenetic assemblages often contain glass, volcanic materials, analcite, and zeolites of sedimentary origin such as mordenite, clinoptilolite, stilbite, and heulandite; the zeolites typically are identified by X-ray diffraction analysis. Transition into the zeolite facies is indicated by the appearance of laumontite, albite, or perhaps wairakite. Occasionally prehnite may be found. These minerals may be identified with the microscope if sufficiently coarse-grained; otherwise X-ray diffraction analysis is necessary. Minerals of metamorphic origin are usually found in veins, cavity fillings, and in the fine-grained matrix. Typical zeolite facies assemblages are given in Figure 20–3. Note once again that rock composition points such as *a* indicate that the rock contains calcite and chlorite (plus quartz), whereas points such as *b* indicate the presence of calcite, chlorite, and prehnite (plus quartz). The proportions of these minerals (with the exception of quartz) determine the location of the composition point. The rocks of the zeolite and prehnite–pumpellyite–metagraywacke facies show considerable departure from equilibrium in both their textures and mineralogy. Relic grains and sedimentary textures usually are present.

Transition into the prehnite–pumpellyite–metagraywacke facies is marked by the elimination of laumontite and the presence of prehnite, pumpellyite, and quartz. Epidote with or without actinolite may appear, as well as calcite or aragonite. Pumpellyite may be present with actinolite if prehnite is not present. The more common associations are given in Figure 20–4.

Greenschist Facies

Rocks of the greenschist facies typically are completely recrystallized, in contrast to those of lower facies that often contain abundant relic textures and mineralogy. Foliation is well developed.

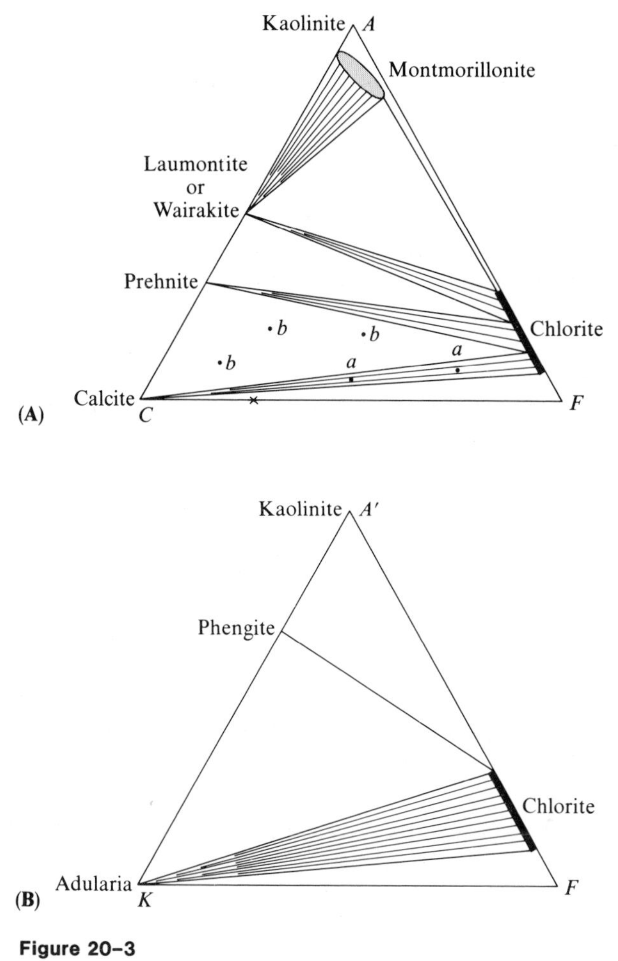

Figure 20–3
Some mineral associations of the zeolite facies. Letters are explained in the text.

The greenschist facies rocks are the most common of all the facies and are found in every regionally metamorphosed terrane. Metamorphosed pelites are present as slate, phyllite, and fine-grained schists. Basalts and andesites are converted to greenschists; ultramafic rocks are represented as serpentinites, and quartz sandstones and carbonate rocks are converted respectively to quartzite and marble. The characteristic green minerals that give this facies its name are chlorite, actinolite, and epidote. Other common minerals are quartz, muscovite, biotite, plagioclase of composition An_{0-5}, and graphite. Chloritoid or stilpnomelane, although uncommon, may be present.

The greenschist facies comprises Barrow's chlorite and biotite zones, and for this reason has been broken into subfacies, which we will call the lower and upper greenschist facies. The general mineralogy of the lower greenschist facies is given in Figure 20–5. The upper greenschist facies is marked by the appearance of biotite and elimi-

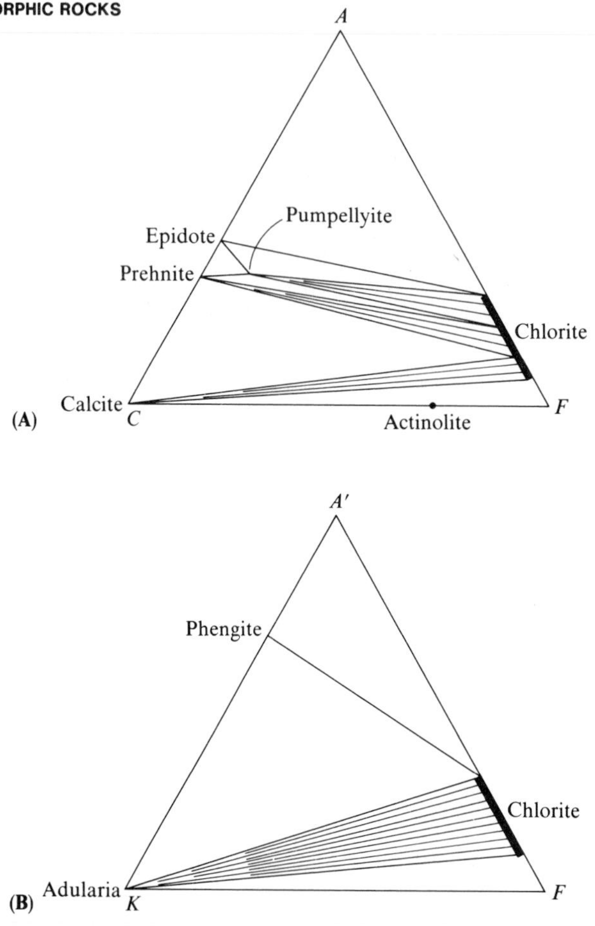

Figure 20–4
Mineral associations of the prehnite–pumpellyite–metagraywacke facies.

Figure 20–5 (*facing page*)
Mineral associations of the lower greenschist facies (chlorite zone). Tie lines in this and other diagrams are dashed to the chloritoid compositional point because chloritoid, requiring special conditions of rock composition, commonly does not form within the indicated facies diagram.

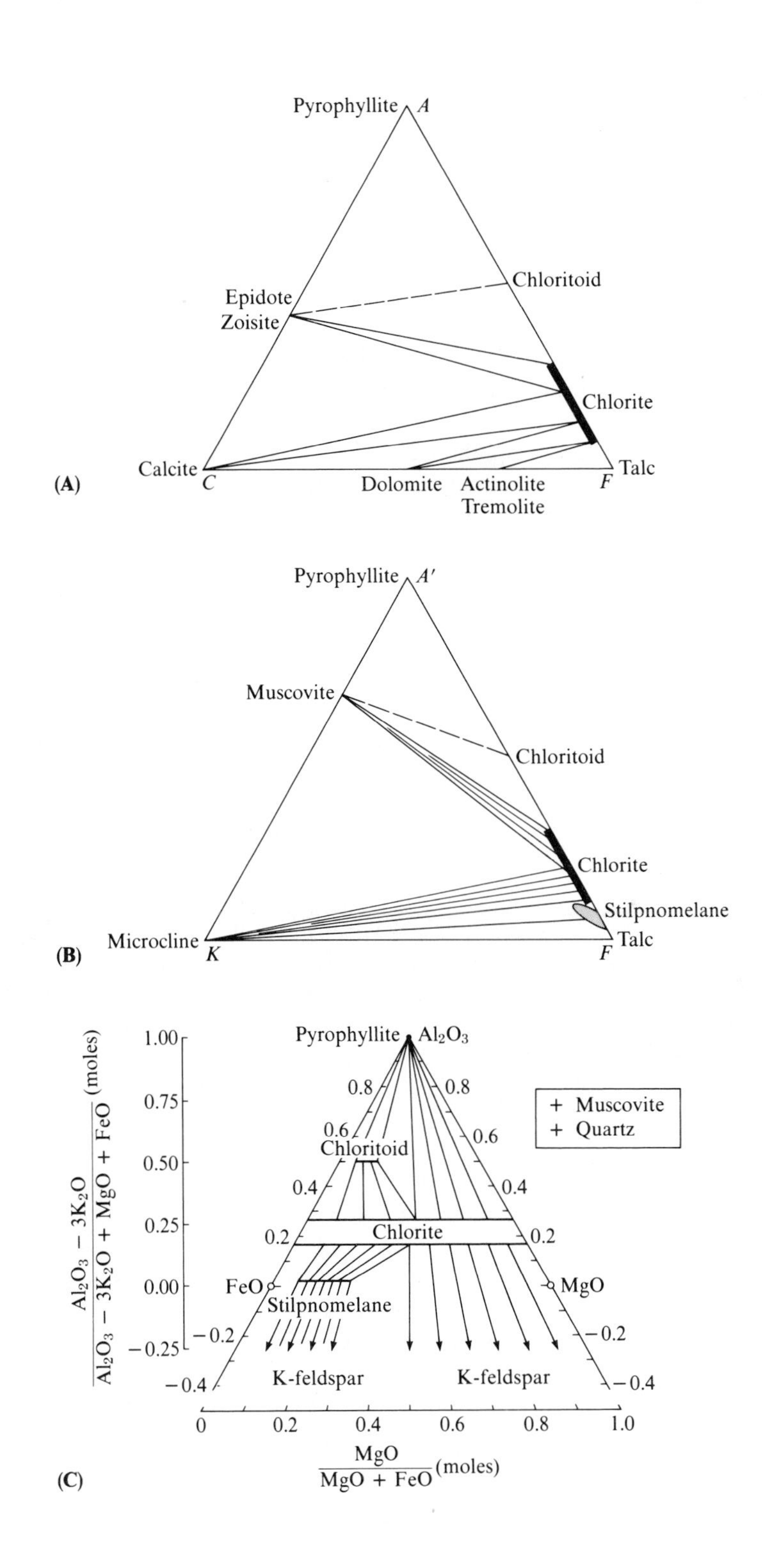
(A)
Pyrophyllite
A
Chloritoid
Epidote
Zoisite
Chlorite
Calcite
C
Dolomite
Actinolite
Tremolite
F
Talc
(B)
Pyrophyllite
A′
Muscovite
Chloritoid
Chlorite
Stilpnomelane
Microcline
K
F
Talc
(C)
Pyrophyllite
Al_2O_3
0.8
0.6
0.4
0.2
Chloritoid
Chlorite
FeO
MgO
Stilpnomelane
−0.2
−0.4
K-feldspar
K-feldspar
+ Muscovite
+ Quartz
1.00
0.75
0.50
0.25
0.00
−0.25
$\frac{Al_2O_3 - 3K_2O}{Al_2O_3 - 3K_2O + MgO + FeO}$ (moles)
0
0.2
0.4
0.6
0.8
1.0
$\frac{MgO}{MgO + FeO}$ (moles)

nation of dolomite and (usually) any stilpnomelane that might be present (see Figure 20–6).

Epidote–Amphibolite Facies

This facies has been eliminated by many geologists and listed as a third subdivision within the greenschist facies. This is the facies that characterizes Barrow's garnet zone.

Entrance to the epidote–amphibolite facies is marked by several significant changes. The Fe-rich garnet almandine makes its appearance in place of Fe-rich chlorite. Reactions suggested from petrographic observations are:

Chlorite + Muscovite + Quartz → Almandine-rich garnet + Biotite + H_2O
Chlorite + Chloritoid + Quartz → Almandine-rich garnet + H_2O

Mg-rich chlorite may coexist with this garnet. In mafic rock compositions blue-green hornblende now shows up in place of actinolite. Commonly kyanite or andalusite is present in lieu of pyrophyllite. The composition of plagioclase remains Na-rich. Typical assemblages are shown in Figure 20–7.

Amphibolite Facies

A significant number of changes mark the transition into the amphibolite facies. These changes do not all occur at precisely the same pressure and temperature, but are close enough to each other to constitute a significant break in mineralogy. Transition from the epidote–amphibolite facies results in the elimination of a number of minerals; these are pyrophyllite, chloritoid, albite, talc, blue-green hornblende, and chlorite. New minerals that might appear in their place are an aluminum silicate, plagioclase of intermediate composition ($> An_{20}$), green-brown hornblende, diopside, staurolite, cordierite, grossularite, or andradite.

Pelitic rocks in the amphibolite facies consist of micaceous schists and gneisses. Feldspathic sandstones have converted to quartzo-feldspathic schists and gneisses, and basalts and andesites to amphibolites. Calcareous sandstones are converted to calc-silicate rocks; carbonate rocks and quartz-rich sandstones are present as medium- to coarse-grained marbles and quartzites.

The amphibolite facies is complicated for several reasons. One of these is that a sequence of changes occurs that permits subdivision into zones or subfacies; a second is that a wide range of pressures is possible, with somewhat different mineral assemblages in each. The sequences that would be expected at moderate pressures are shown in Figures 20–8, 20–9, and 20–10—a standard Barrovian pattern; this is

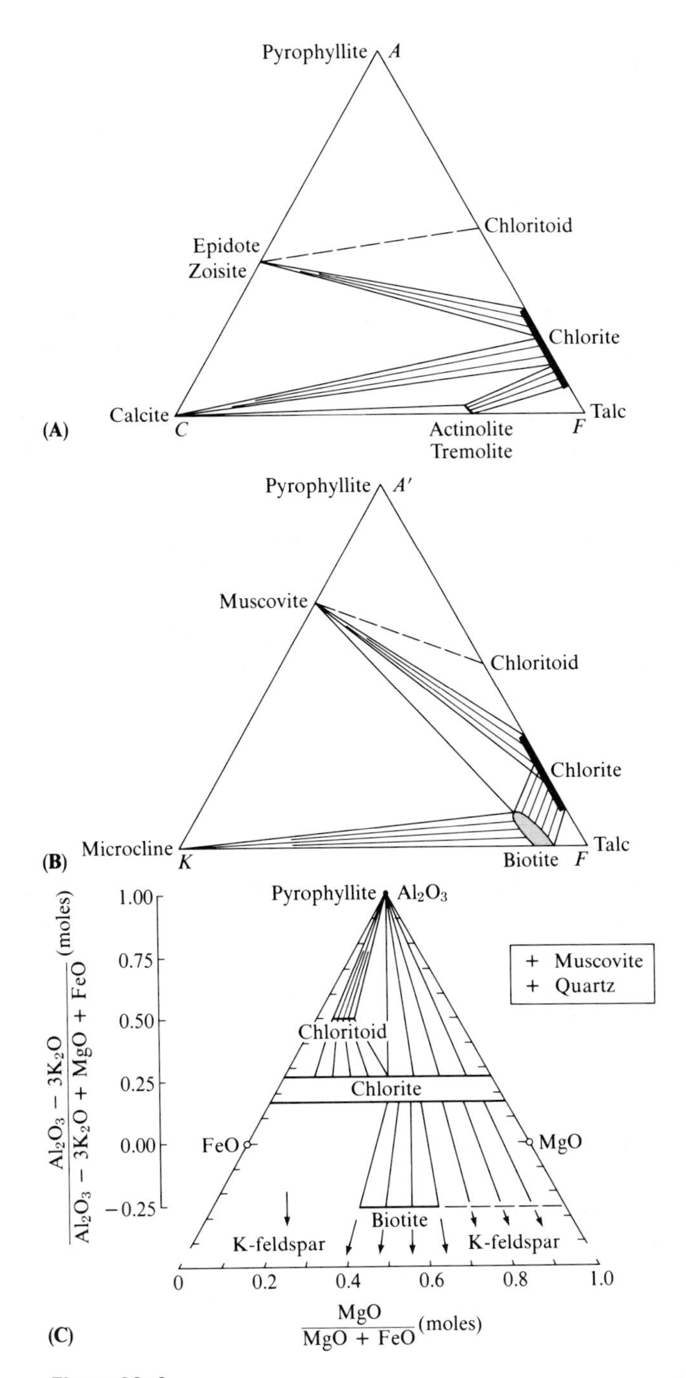

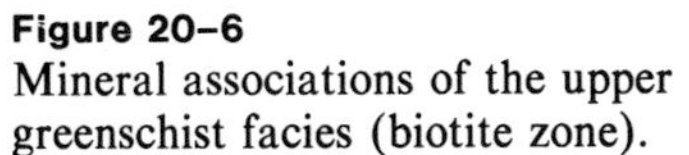
Figure 20–6
Mineral associations of the upper greenschist facies (biotite zone).

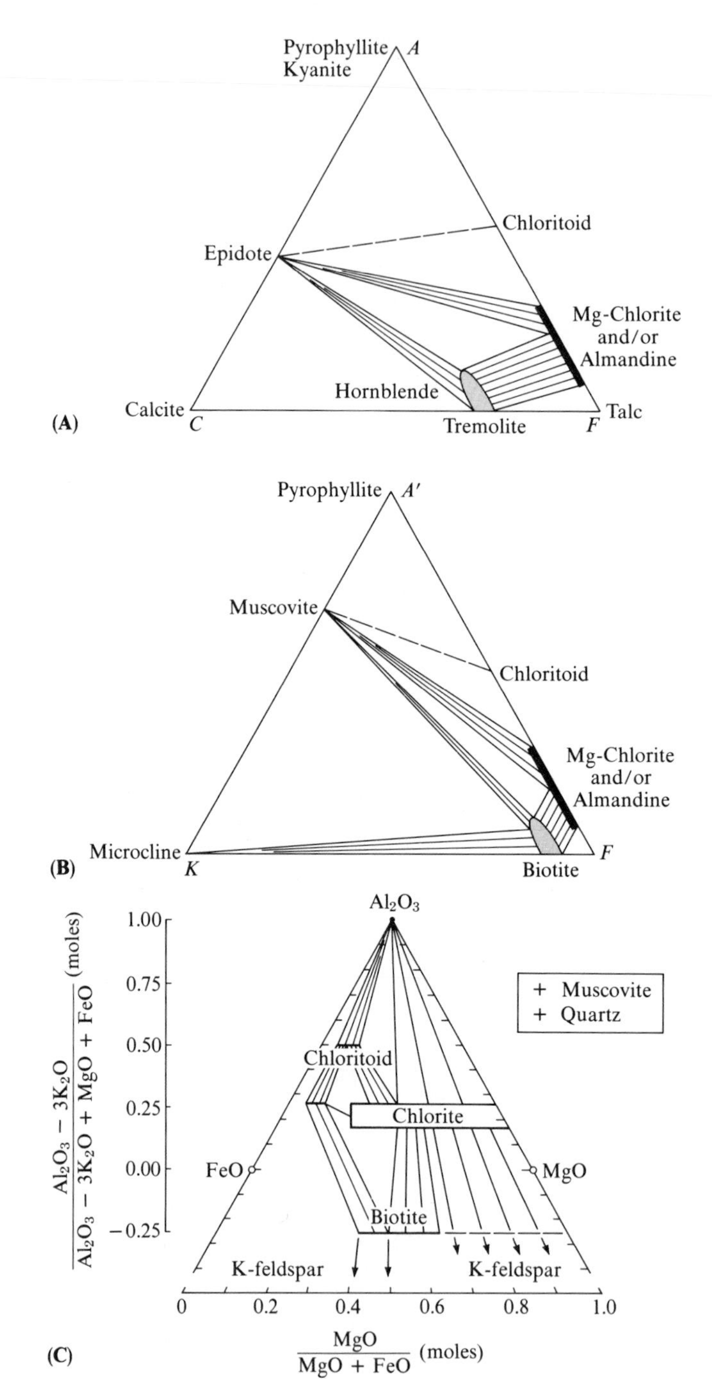

Figure 20–7
Mineral associations of the epidote–amphibolite facies. This facies is occasionally eliminated and considered as a subfacies of the greenschist facies.

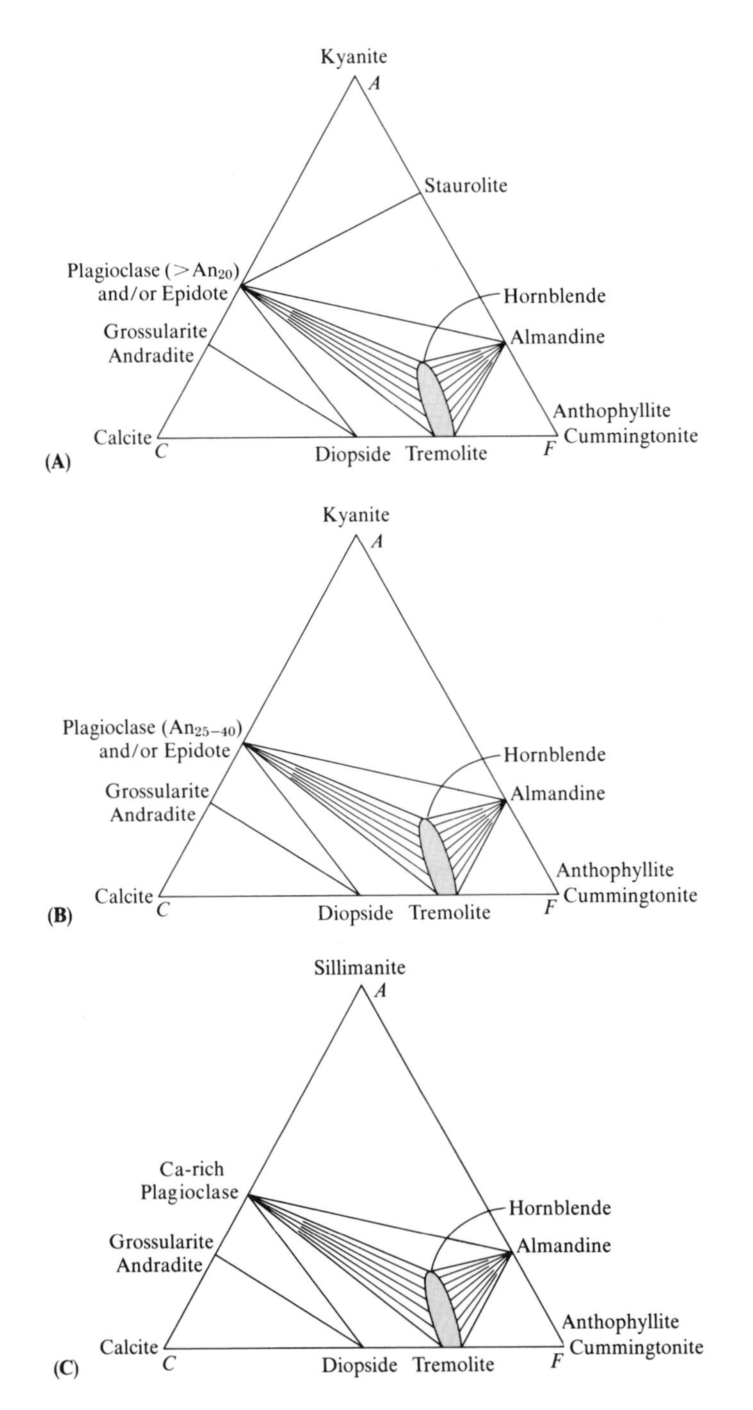

Figure 20–8
Mineral associations of the **(A)** staurolite, **(B)** kyanite, and **(C)** sillimanite zones of the amphibolite facies as seen on *ACF* diagrams. Mineralogy is consistent with medium pressure.

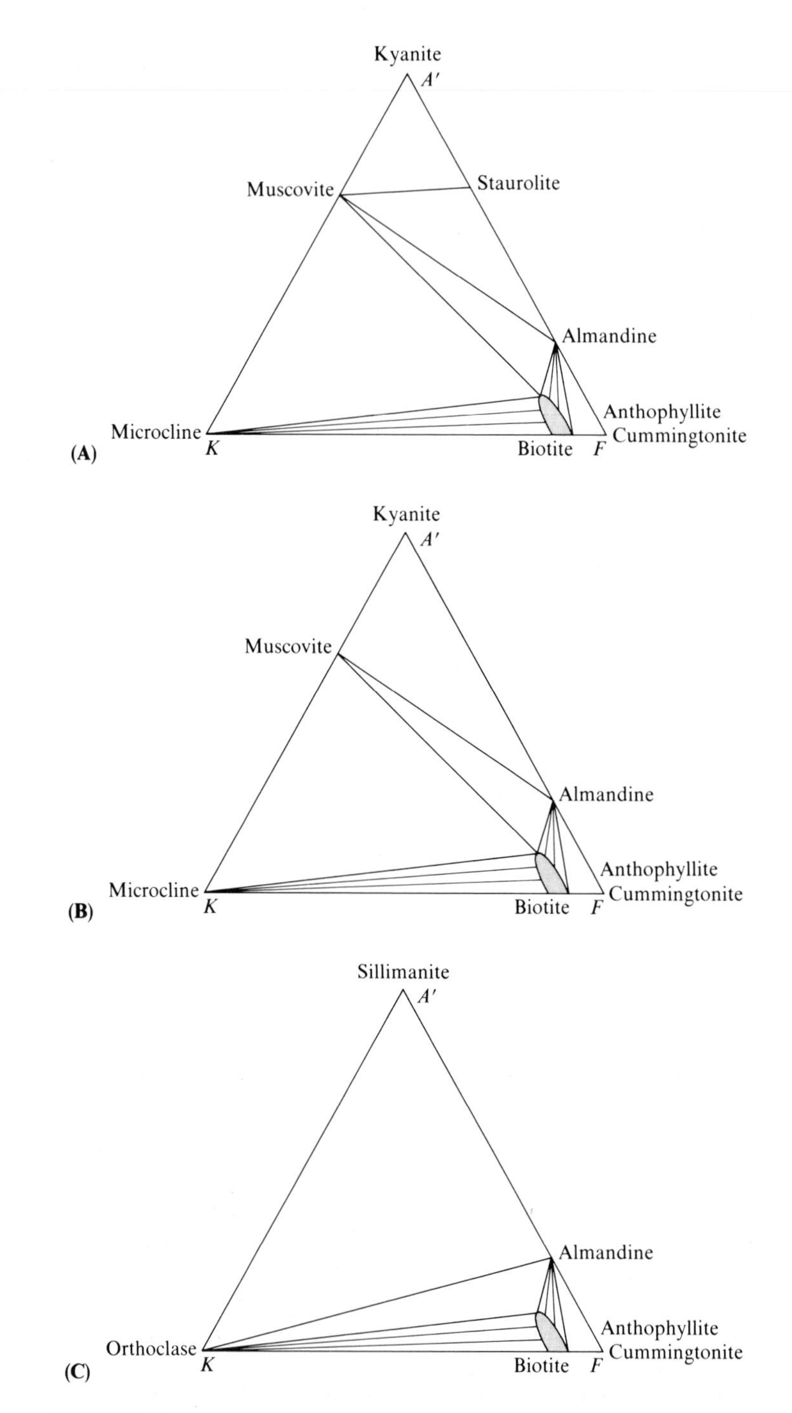

Figure 20–9
Mineral associations of the **(A)** staurolite, **(B)** kyanite, and **(C)** sillimanite zones of the amphibolite facies as seen on *A′KF* diagrams. Mineralogy is consistent with medium pressure.

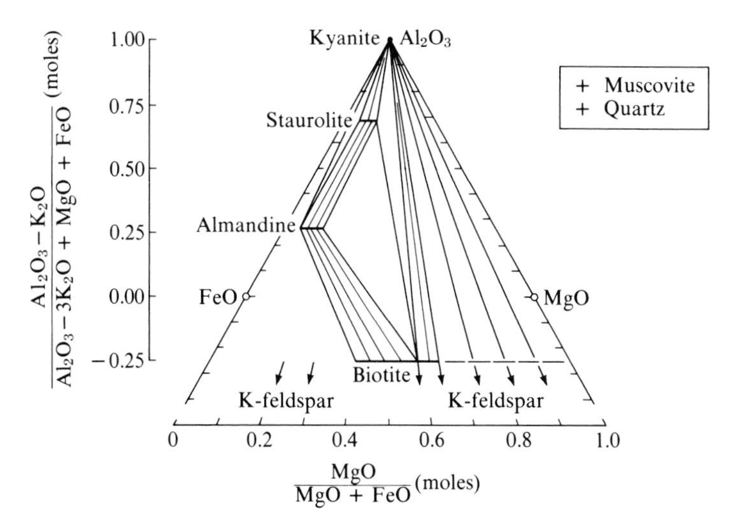

Figure 20–10
Mineral associations of the staurolite zone of the amphibolite facies as seen on an *AFM* diagram. Mineralogy is consistent with medium pressure.

marked in pelitic rocks by the development in turn of staurolite, kyanite, and sillimanite. The staurolite-forming reaction may be

$$\text{Chlorite} + \text{Muscovite} \longrightarrow \text{Staurolite} + \text{Biotite} + \text{Quartz} + H_2O$$

or

$$\text{Chlorite} + \text{Muscovite} + \text{Almandine} \longrightarrow \text{Staurolite} + \text{Biotite} + \text{Quartz} + H_2O$$

Pelitic rocks metamorphosed at relatively low pressures would show significant differences. Staurolite and almandine would not be present. Rather than the staurolite-forming reaction immediately above, chlorite, muscovite, and quartz would react to produce cordierite, biotite, Al_2SiO_5, and H_2O. The Al_2SiO_5 polymorph would be andalusite or sillimanite rather than kyanite. The minerals cordierite and andalusite are stable at moderate to high temperatures and relatively low pressures. Sillimanite is stable at high temperatures over a wide variety of pressures (see Figure 14–1A).

The amphibolite facies in many metamorphic terranes represents the maximum level of metamorphism. Transitions into migmatites may be found. As discussed in Chapter 19, this is thought to occur in those situations where $P_{H_2O} = P_{lith}$. If water pressure is low, however, transition will be made into the granulite facies.

Granulite Facies

The granulite facies is found at the extreme upper limit of both pressure and temperature. These rocks are usually developed in the deepest part of the continental crust. Amphibolite facies rocks, migmatites, or various igneous rocks may be associated with them. Most granulite facies rocks are relatively coarse-grained and are richer in quartz and feldspar than rocks of lower facies. In the lower *PT* portion of the granulite facies are found micaceous schists and gneisses that may contain hornblende, orthopyroxene and biotite. In the higher *PT* range the stability levels of biotite and hornblende are exceeded. The rocks are essentially anhydrous and often contain hypersthene (with about 10% Al_2O_3 in solid solution). Diopside, almandine garnet, and/or cordierite may be present. The rocks may or may not be gneissic in appearance. Gradual transitions are made into hypersthene granite (charnockite), hypersthene granodiorite, or hypersthene gabbro.

Within the Grenville Group of the Canadian Shield the granulite facies includes quartzofeldspathic rocks showing little schistosity or layering; associated rocks are gneisses, some of which lack schistosity but show some compositional layering. Both of these rocks contain irregular masses and small dikes of granitic material that probably originated through anatexis.

Blueschist Facies

The blueschist facies is found in environments of high pressure and low temperature—typically in materials along converging continental margins. Minerals typical of this facies are blue to blue-black glaucophane, colorless or pale green jadeite, white to pale green lawsonite, and aragonite. Common rocks vary from fine- to medium-grained schists to more massive types. Most of the rocks rich in glaucophane have a distinctive blue color, which gives the facies its name. The chemical composition of most blueschist facies rocks is similar to oceanic tholeiites.

Eclogite Facies

Eclogite facies rocks are rare, and usually of basaltic composition. The two characteristic minerals in these high-pressure rocks are omphacite and pyropic garnet. Omphacite, a soda-rich clinopyroxene close to jadeite in composition, is green to dark green and usually is medium- or fine-grained. Garnet is present as red-brown porphyroblasts. The texture generally is massive and occasionally layered or foliated. Associated low-temperature minerals may be chlorite, muscovite, epidote, or glaucophane. Eclogites have been classified into three different types based on their mode of occurrence: (1) inclusions in kimberlites or basalts, or in layers within ultramafic rocks; (2) bands or lenses within (alpine-type) peridotite–serpentinite associations in folded geosynclinal sediments of orogenic belts, often within glaucophane schists; and (3) bands or lenses within migmatites.

Hornfels and Sanidinite Facies

Rocks of these two facies are usually (but not always) developed as a result of contact metamorphism. They usually are massive in appearance due to formation without deformation; for this same reason some premetamorphic textures may be preserved.

The minerals present in the hornfels facies are similar to those at somewhat higher pressures in the corresponding greenschist, epidote–amphibolite, and amphibolite facies; the major difference in mineralogy is that the hornfels facies contains only those minerals characteristic of a low-pressure environment (see Figure 20–1). Consequently, minerals requiring high pressure, such as kyanite and staurolite, will not be present. Instead we expect minerals usually associated with lower pressures, such as cordierite, biotite, garnet, andalusite, sillimanite, or wollastonite. Some of these minerals may be present as porphyroblasts; in other cases they may form a dense fine-grained structureless rock (hornfels).

The sanidinite facies is quite rare and usually is found at igneous contacts or within xenoliths. Characteristic minerals are sanidine, mullite, pigeonite, or tridymite. Hypersthene, quartz, akermanite, monticellite, or glass may also be present. The calc-silicate minerals spurrite, tilleyite, rankinite, and larnite are very rarely found.

MINERAL VARIATION RELATED TO INITIAL ROCK COMPOSITION

Carbonate Rocks

Most carbonate rocks consist of various amounts of calcite, dolomite, and quartz, with or without significant amounts of clays. Textural and mineralogical observations of metamorphosed carbonates along a rising temperature gradient commonly reveal a progressive series of changes that demonstrate in turn the formation of talc, tremolite, forsterite, diopside, periclase, wollastonite, and monticellite. At very high temperatures, adjacent to igneous intrusives, the rare minerals akermanite, tilleyite, spurrite, rankinite, merwinite, and larnite are encountered. With the exception of spurrite, none of the above minerals contains carbonate, but many contain combined water. The common presence of hydrous silicates indicates that water vapor is often present during metamorphism of carbonate rocks in addition to the CO_2 produced by decarbonation reactions.

Experimenters, starting with Bowen (1940), have duplicated the various reactions that occur in carbonate rocks and have examined them as functions of pressure and vapor phase composition. Much of this literature is summarized in Winkler (1976) and Slaughter, Kerrick, and Wall (1975). We will examine a few of these so that the approach can be understood.

Consider rocks initially containing calcite ($CaCO_3$), dolomite [$CaMg(CO_3)_2$], and quartz (SiO_2); metamorphic reactions cause the formation of hydrous phases such as talc [$Mg_3Si_4O_{10}(OH)_2$] and tremolite [$Ca_2Mg_5Si_8O_{22}(OH)_2$]. The chemical system under consideration must therefore contain five components, CaO–MgO–SiO_2–CO_2–H_2O. All phases within the system can be represented within a compositional tetrahedron if CO_2 and H_2O are grouped together at one corner, with the other three oxides, CaO, MgO, and SiO_2, making up the other corners. Plotting the minerals common to this type of metamorphism within a tetrahedron (see Figure 20–11) is difficult to visualize, so an additional simplification is made; all mineral compositions within the tetrahedron are projected from the H_2O, CO_2 corner to the triangular base CaO–MgO–SiO_2. This permits both hydrous and carbonated minerals to be plotted together on the same two-dimensional triangular representation. Such a plot is shown in Figure 20–12, together with some of the pertinent minerals.

The phases calcite, dolomite, and quartz are stable together in the presence of a fluid phase to temperatures up to about 300–400°C (depending upon fluid composition and pressure). The fluid phase in this situation can be pure CO_2, pure H_2O, or any intermediate mixture. This equilibrium assemblage can be shown on the compositional triangle CaO–MgO–SiO_2 (with excess CO_2 and H_2O), as in Figure 20–13.

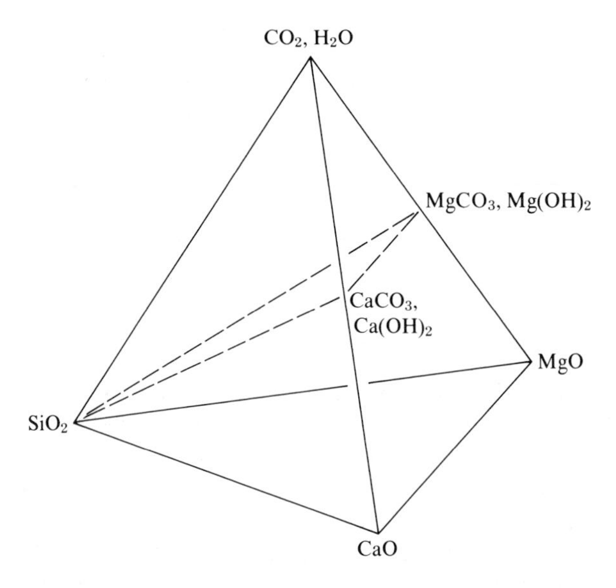

Figure 20–11
The compositional tetrahedron MgO–CaO–SiO_2–(H_2O + CO_2). The interior dashed plane includes the carbonate minerals.

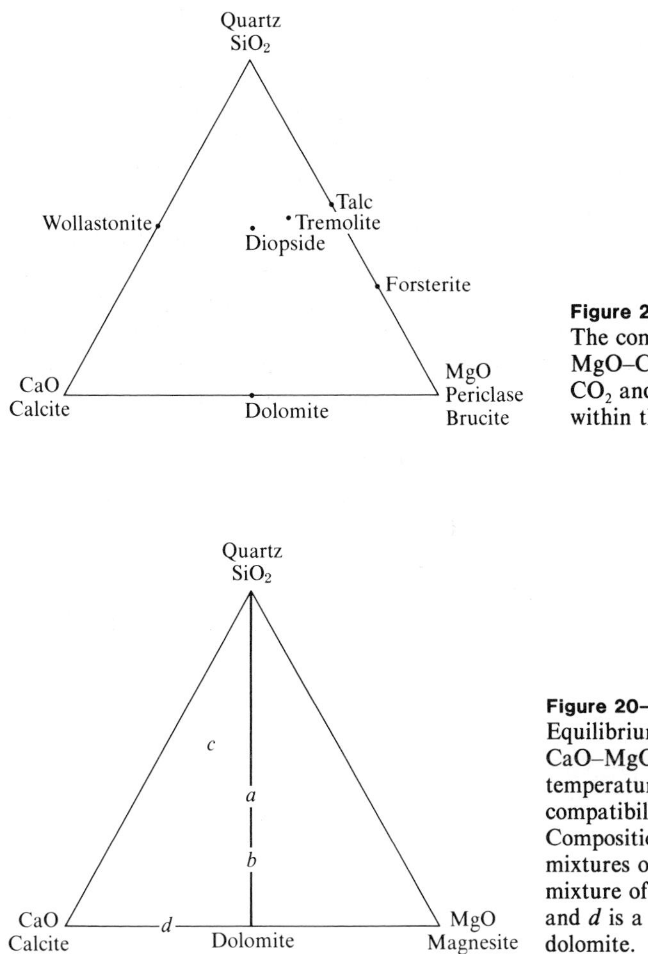

Figure 20–12
The compositional triangle MgO–CaO–SiO_2. Phases containing CO_2 and H_2O in addition are plotted within the triangle.

Figure 20–13
Equilibrium relations within the system CaO–MgO–SiO_2–(H_2O + CO_2) at low temperatures. Solid lines indicate compatibility between phases. Compositions *a* and *b* consist of mixtures of quartz and dolomite; *c* is a mixture of calcite, dolomite, and quartz, and *d* is a mixture of calcite and dolomite.

The placement of all solid lines between mineral composition points now expresses the fact that the phases joined by these lines are stable together. Thus a rock whose composition locates it at point *d* consists of calcite and dolomite, and rocks of compositions *a* and *b* are composed of quartz and dolomite. A rock composition such as *c* falls between the three mutually compatible minerals (quartz, calcite, and dolomite). Such a composition falling within a compatibility triangle indicates that the three phases whose compositions are represented by the corners of the triangles can coexist together stably. Inasmuch as magnesite-bearing rocks are relatively uncommon, we will not consider compatibilities near the magnesite corner of the compositional triangle.

Heating of a rock with compositions falling within the compositional triangle CaO–MgO–SiO_2 leads to reaction

$$\underset{\text{Dolomite}}{3CaMg(CO_3)_2} + \underset{\text{Quartz}}{4SiO_2} + H_2O \rightarrow \underset{\text{Talc}}{Mg_3Si_4O_{10}(OH)_2} + \underset{\text{Calcite}}{3CaCO_3} + 3CO_2$$

The rock composition *a* (see Figure 20–14), consisting originally of quartz, dolomite (in the ratio indicated by the formula above), and a fluid phase, reacts to form calcite, talc, and a fluid phase. All of the quartz and dolomite are eliminated. If a larger amount of dolomite were present than that required for the reaction, such as at point *b*, the same reaction would occur, but dolomite would remain in excess. This can be seen on the diagram, as composition *b* falls within the new compatibility triangle joining the phases calcite, talc, and dolomite. Similarly the composition *c*, which consisted before the reaction (see Figure 20–13) of calcite, quartz, and dolomite, now is converted to a mixture of calcite, quartz, and talc.

Additional heating (under most circumstances) leads to the formation of tremolite by reaction of talc, calcite, and quartz:

$$\underset{\text{Talc}}{5Mg_3Si_4O_{10}(OH)_2} + \underset{\text{Calcite}}{6CaCO_3} + \underset{\text{Quartz}}{4SiO_2} \rightarrow \underset{\text{Tremolite}}{3Ca_2Mg_5Si_8O_{22}(OH)_2} + 6CO_2 + 2H_2O$$

Tremolite now appears as a stable phase within the compositional triangle (see Figure 20–15). As tremolite is compatible with quartz, calcite, and talc, compatibility lines are drawn from the tremolite composition point to these minerals. This serves to subdivide the system into a greater number of compatibility triangles. Rock compositions *a* and *b* would remain unchanged, whereas composition *c* reacts to form a mixture of calcite, quartz, and tremolite.

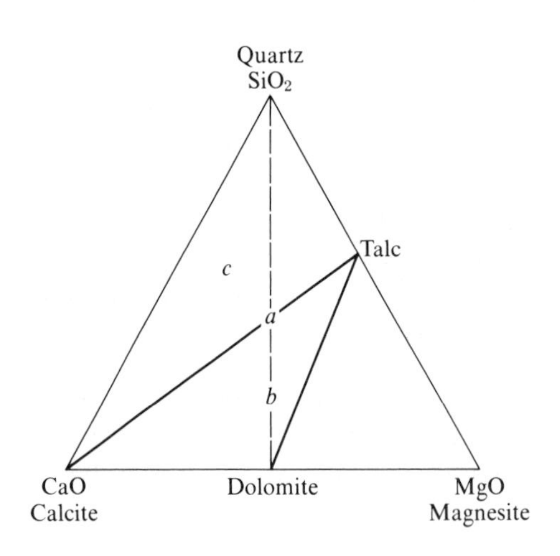

Figure 20–14
Equilibrium relations within the system CaO–MgO–SiO_2–(H_2O + CO_2) above about 350–400°C, after the formation of talc.

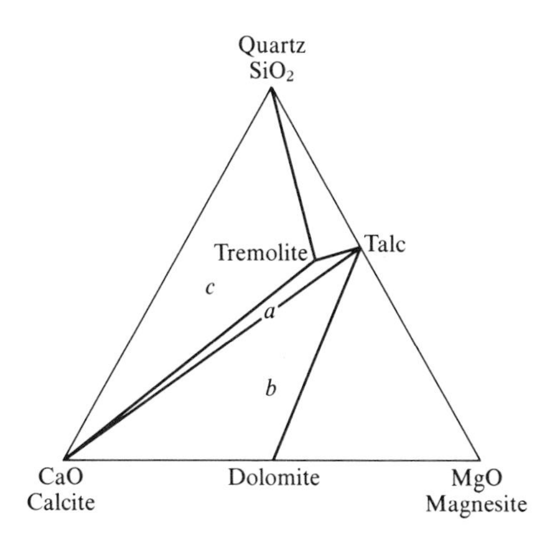

Figure 20–15
Equilibrium relations within the system CaO–MgO–SiO_2–(H_2O + CO_2) after the formation of tremolite.

A variety of such reactions occurs with increasing temperature, causing changes of the various equilibrium assemblages as phases disappear or are created. It is worthwhile to consider the particular circumstances controlling these reactions in order to assess their value in terms of estimating the geological conditions of formation of natural assemblages. The two reactions discussed above are shown in Figure 20–16. Temperature is shown on the vertical axis. The composition of the fluid phase changes continuously in composition along the horizontal coordinate, being composed of H_2O at the left and CO_2 at the right, with P_{fluid} at all times equal to 1 kbar; that is,

$$P_{\text{fluid}} = \overline{P}_{CO_2} + \overline{P}_{H_2O} = 1 \text{ kbar}$$

The reason for this unusual representation is that the temperatures of the various reactions are dependent upon the composition of the fluid phase (as well as its total pressure). The temperature at which dolomite, quartz, and H_2O react to form talc, calcite, and CO_2 increases as the mole fraction of CO_2 increases to the right side of the diagram. This is because the assemblage that contains the greatest amount of combined CO_2 (dolomite rather than calcite) is favored as the mole fraction of CO_2 is increased in the fluid phase. Looking at the reaction in the opposite manner, it is seen that the higher-temperature assemblage, which contains the hydrous phase talc, is favored as the mole fraction of H_2O is increased in the fluid phase.

The second reaction, in which tremolite is produced from the combination of talc, calcite, and quartz, yields both CO_2 and H_2O as reaction products. Inasmuch as both components of the fluid phase are released during the formation of tremolite, the temperature of the reaction varies to a considerably smaller extent than the previous reaction (which has H_2O on one side of the equation and CO_2 on the other).

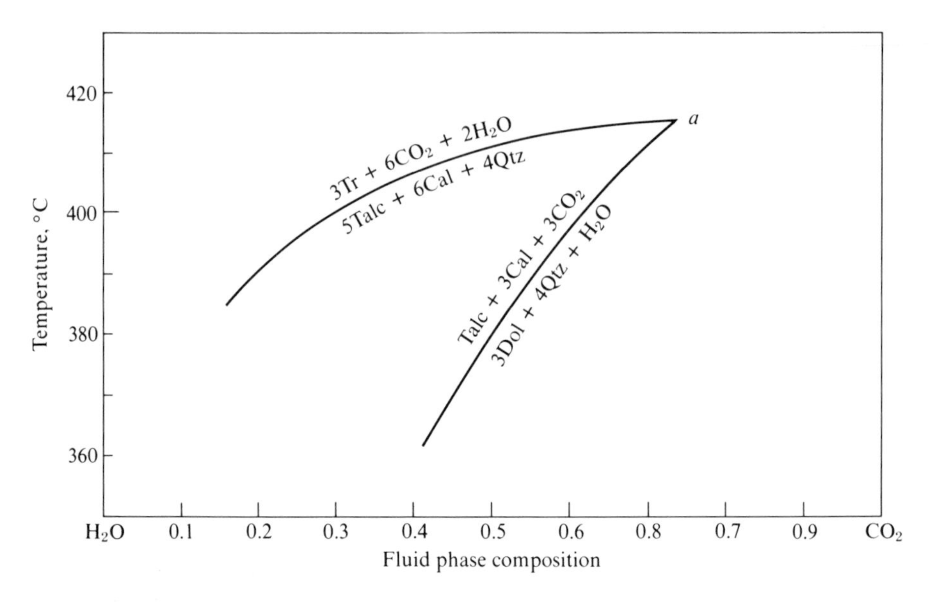

Figure 20–16
The talc- and tremolite-forming reactions at 1 kbar as a function of temperature and CO_2/H_2O. Tr = tremolite, Cal = calcite, Qtz = quartz, and Dol = dolomite. [After J. Slaughter, D. M. Kerrick, and V. J. Wall, 1975, *Amer. Jour. Sci., 275,* Fig. 69.]

It is obvious from the diagram that an attempt to assign a specific *P* and *T* to a mineral assemblage is very difficult. A rock consisting of talc and calcite, for example, merely assures one of being above the lower reaction curve in temperature. The situation is somewhat improved by a rock composed of the assemblage talc–calcite–quartz–tremolite, as this would indicate that the assemblage fell on the upper reaction curve. Still further improvement of interpretation is possible; note that the two reaction curves intersect at point ***a***. At point ***a***, an isobaric invariant point, the products and reactants of both reactions are found together—a stable assemblage containing dolomite, quartz, talc, calcite, and tremolite. If the condition of the diagram is satisfied, namely 1 kbar fluid pressure, the association of these five phases indicates a precise temperature for this assemblage as well as a definite composition of the fluid phase. Unfortunately, not all carbonate rocks are metamorphosed at 1 kbar fluid pressure. As noted earlier, the fluid pressure varies as a function of depth; similarly the temperature of the five-phase assemblage also varies (see Figure 20–17). If the pressure can be estimated by other means, such as an approximate estimate of depth of burial during metamorphism, perhaps based on the amount of overburden removed, the temperature can be determined. Or if the temperature can be estimated by other means, the pressure can be obtained from the diagram. The

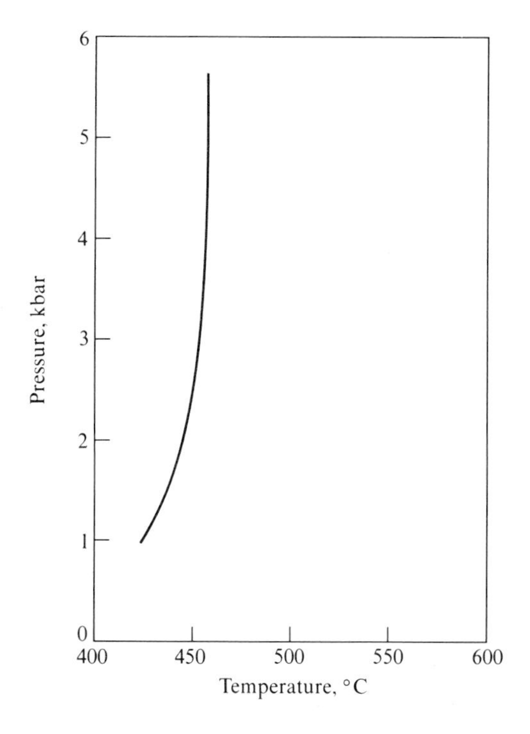

Figure 20–17
Change in the stability conditions of the assemblage tremolite–dolomite–calcite–quartz–talc as a function of fluid pressure and temperature. This assemblage is stable only along the curve. [After J. Slaughter, D. M. Kerrick, and V. J. Wall, 1975, *Amer. Jour. Sci., 275,* from data on Figs. 8, 9, 10.]

five-phase assemblage can be estimated to have formed between 400°C and 460°C, even if no pressure estimate is possible.

Many reactions occurring in carbonate rocks have been determined experimentally, and a number of isobaric invariant points have been determined. These points, which are characterized by maximum numbers of phases existing stably together, offer the greatest potential for furnishing information on *PT* conditions. This is so because such assemblages are subjected to the maximum number of physical constraints. Assemblages with a smaller number of phases occur under a much wider variety of conditions and are therefore much less useful in defining the conditions of metamorphism. When using such an approach, note that it is necessary to establish textural equilibrium among the coexisting minerals.

A type of rock related to the carbonates is *marl;* marls are composed of mixtures of carbonates and clays, with minor amounts of quartz. With initiation of metamorphism, the significant minerals present may be grossularite, zoisite/clinozoisite, or margarite (margarite is a Ca-bearing mica, virtually indistinguishable by optical methods from muscovite). Increasing temperature leads to the breakdown of grossularite and formation of Ca-rich plagioclase and calcite or wollastonite. Increase in the Ca-content of plagioclase occurs with increasing grade, but the changes occur irregularly due to the formation of other Ca-bearing phases. Vesuvianite (idocrase) may be present throughout a wide range of metamorphic conditions.

Mudrocks

Metamorphism of a pelitic rock (metapelite) results in a series of unique mineralogical changes that are a result of the particular combination of pressure and temperature to which the rock has been subjected. Miyashiro (1961) was the first to develop this concept on a regional basis for a variety of rock types. Using the fact that certain minerals are favored in high-pressure environments, and others in low, he was able to categorize various rock suites according to the level of pressure to which they had been subjected. This led in turn to classification of metamorphic belts on the basis of

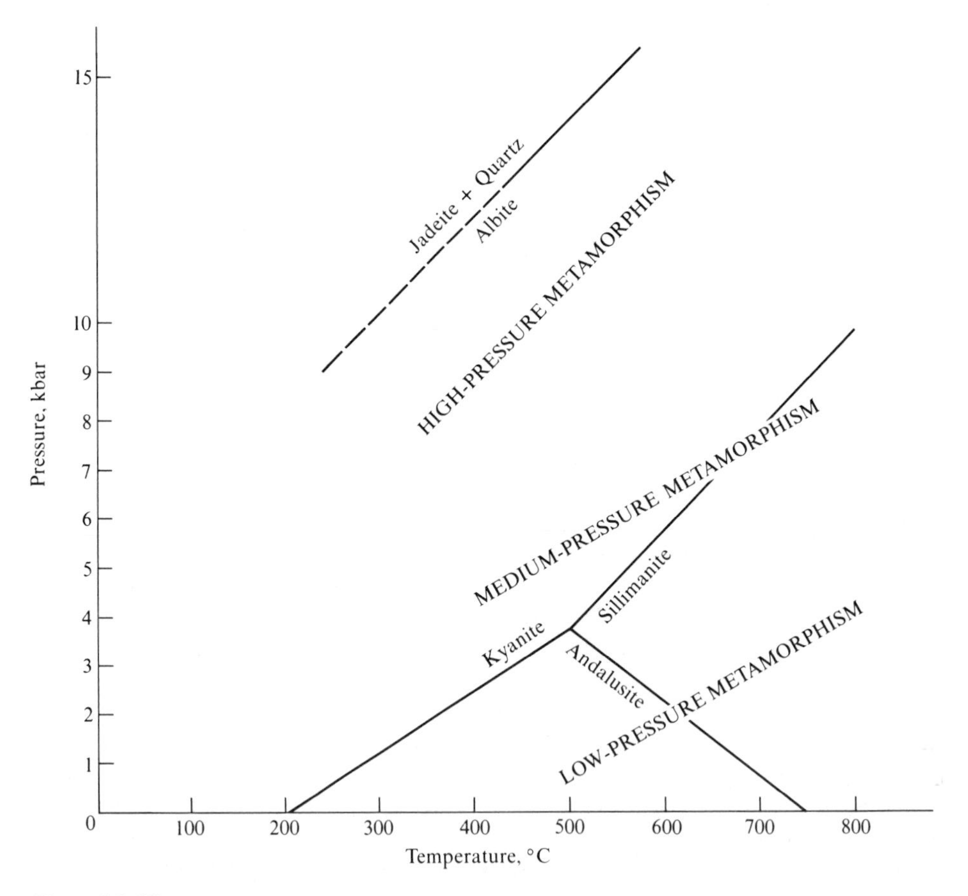

Figure 20–18
The three major baric types of metamorphism, shown in relation to the stability areas of the aluminum silicates (Al_2SiO_5) and the jadeite–quartz assemblage. [Based on data from F. Birch and P. Le Conte, 1960, *Amer. Jour. Sci., 258,* 209–217; A. L. Boettcher and P. J. Wyllie, 1968, *Geochim. Cosmochim. Acta, 32,* 999–1012; M. J. Holdaway, 1971, *Amer. Jour. Sci., 271,* 97–131; A. Miyashiro, 1961, *Jour. Petrology, 2,* 277–311; and R. C. Newton and J. V. Smith, 1967, *Jour. Geol., 75,* 268–286.]

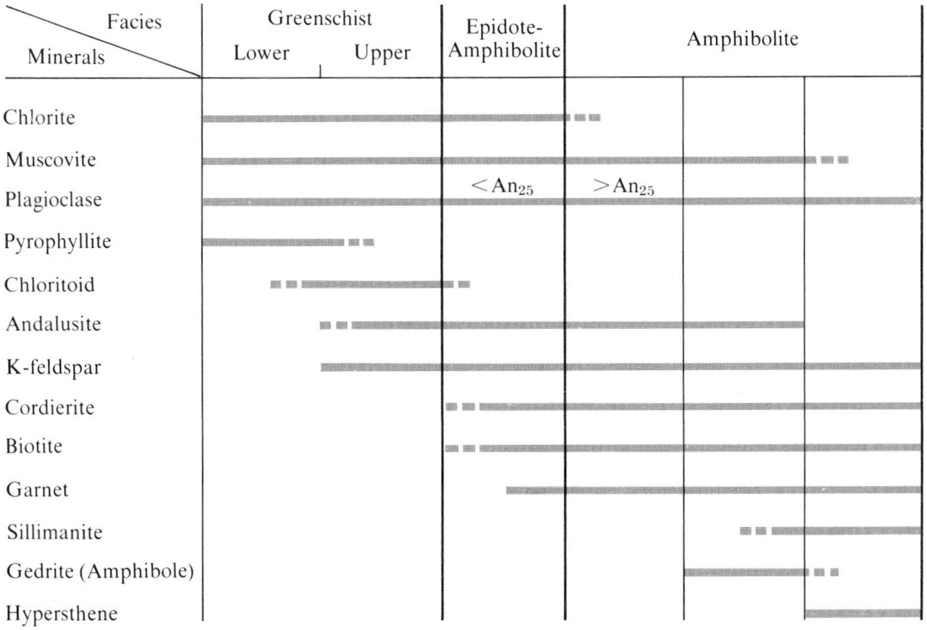

Facies / Minerals | Greenschist (Lower, Upper) | Epidote-Amphibolite | Amphibolite

Chlorite
Muscovite
Plagioclase <An_{25} >An_{25}
Pyrophyllite
Chloritoid
Andalusite
K-feldspar
Cordierite
Biotite
Garnet
Sillimanite
Gedrite (Amphibole)
Hypersthene

Figure 20–19
Mineralogical changes in low-pressure metamorphism of a metapelite, Aracena region, southwestern Spain. [From the data of J. P. Bard, 1969, *Le Metamorphisme Regional Progressif des Sierras d'Aracena en Andalousie Occidentale (Espagne)* (Univ. of Montpellier).]

baric (pressure) types. Low-pressure metamorphic belts show a transition from andalusite to sillimanite. In medium-pressure belts, kyanite converts to sillimanite. High-pressure belts contain dense phases such as jadeite or glaucophane. The general pressure–temperature regions of high-, medium-, and low-pressure metamorphism are shown in Figure 20–18. We will examine differences that arise when pelitic rocks are metamorphosed in these three *PT* environments.

Metamorphism of pelitic rocks under low-pressure conditions has been described for the Aracena region in southwestern Spain (see Figure 20–19). Mineralogic changes begin in the lower portion of the greenschist facies with chlorite, muscovite, sodic plagioclase, and pyrophyllite. Increasing levels of metamorphism result in destruction of pyrophyllite, with subsequent development of andalusite, and then sillimanite. Garnet takes the place of chlorite in the epidote–amphibolite facies; the relatively low-pressure phase cordierite shows up at this same level. In the upper levels of the amphibolite facies, the hydrous minerals muscovite and gedrite (an Al-rich amphibole) are eliminated, and the pyroxene hypersthene appears.

Metamorphism of pelitic rocks at moderate pressure is typified by the pelitic rocks of the Scottish Dalradian—the standard Barrovian sequence (see Figure 20–20).

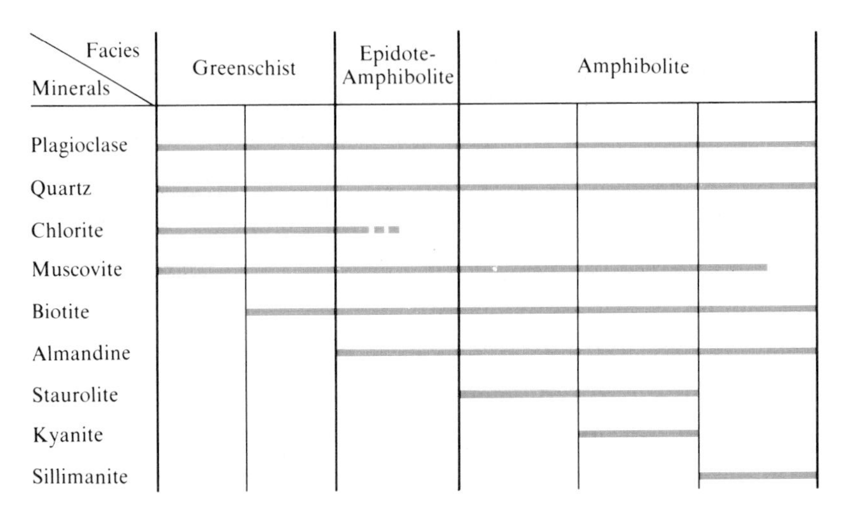

Figure 20–20
Mineralogical changes in medium-pressure metamorphism of a metapelite in the Barrovian zones of the Scottish Highlands. [From the data of A. Harker, 1932, *Metamorphism* (London: Methuen), and J. D. H. Wiseman, 1934, *Quart. Jour. Geol. Soc. London, 90,* 354–417.]

Here the rocks begin with sodic plagioclase, quartz, chlorite, and muscovite. The sequence of formation follows the pattern biotite, garnet, staurolite, kyanite, and sillimanite. Notice that staurolite and kyanite are present in these rocks, in contrast to low-pressure metamorphism ($<$ 4 kbar; see Figure 20–19), where andalusite and cordierite were found.

An example of metamorphism of pelites at high pressure is found in the Sanbagawa belt (see Figure 20–21). Here we see the formation of lawsonite and jadeite, as well as garnet and piemontite (a relatively dense mineral of the epidote group). Temperatures are not high enough to eliminate stilpnomelane or create biotite.

It is clear from these examples that the mineralogy of a metapelite and the sequence of changes observed not only show a progressive series of changes, but also reveal the pressure–temperature regime of metamorphism. Some of this can be revealed in examination of hand specimens when characteristic minerals can be identified, but more often it is necessary to determine the mineralogy with the use of the petrographic microscope.

Mafic Igneous Rocks and Tuffs

As with the metapelites, the mineralogy of metamorphosed mafic igneous rocks (metabasites) varies as a function of the pressure and temperature of metamorphism.

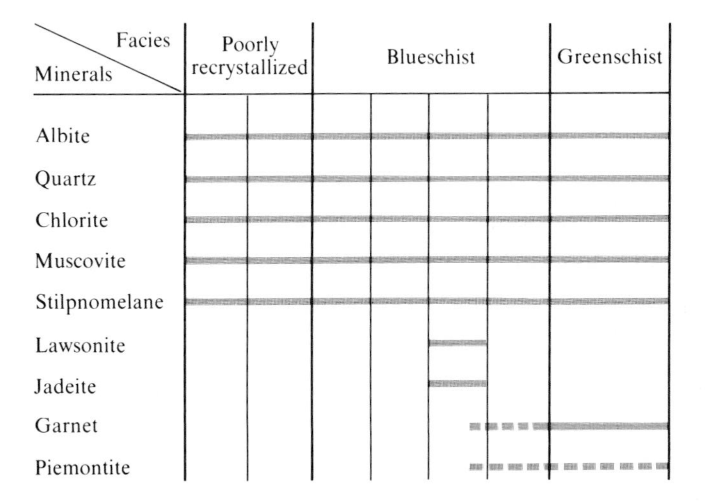

Figure 20–21
Mineralogical changes in high-pressure metamorphism of a metapelite in the Sanbagawa metamorphic belt, Kanto Mountains, Japan. [From the data of Y. Seki, 1958, *Jap. Jour. Geol. Geog., 29,* 233–259; 1960, *Amer. Jour. Sci., 258,* 705–715; 1961, *Jour. Petrology, 2,* 407–423.]

We will examine the behavior of mafic igneous rocks subjected to low-, medium-, and high-pressure metamorphism.

One thing immediately noticeable in the mineralogy of the three different environments (see Figures 20–22, 20–23, and 20–24) is the absence of the aluminosilicates and K-rich minerals that are characteristic of the metapelites. Instead we find an abundance of plagioclase, pyroxene, and amphibole. This should, of course, be expected, as mafic igneous rocks are rich in MgO, CaO, and Al_2O_3. The low- and medium-pressure mineral assemblages are rather similar, with elimination of chlorite, epidote, and Na-rich plagioclase at or near the level of the amphibolite facies. In rocks undergoing prograde metamorphism, epidote [$Ca_2Al_3Si_3O_{10}(OH)_2$] breakdown furnishes Ca for the formation of a more calcic plagioclase; retrograde reactions eliminate calcic plagioclase in favor of epidote and sodic plagioclase. Amphiboles in both low- and medium-pressure metamorphism consist of tremolite/actinolite in the greenschist facies, which are converted with increasing grade to hornblende and perhaps cummingtonite (a Ca-poor amphibole). Almandine garnet may be present in the epidote–amphibolite or amphibolite facies, depending upon rock composition. Both low- and medium-pressure metamorphism of metabasites may be brought to the granulite level of metamorphism, in which case both ortho- and clinopyroxenes may be present.

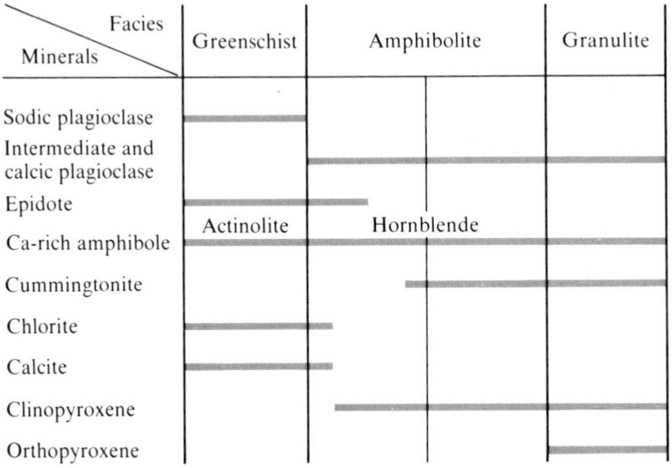

Figure 20–22
Mineralogical changes in low-pressure metamorphism of a metabasite, Abukuma Plateau, Japan. [After A. Miyashiro, 1961, *Jour. Petrology, 2,* Fig. 2. © Oxford University Press.]

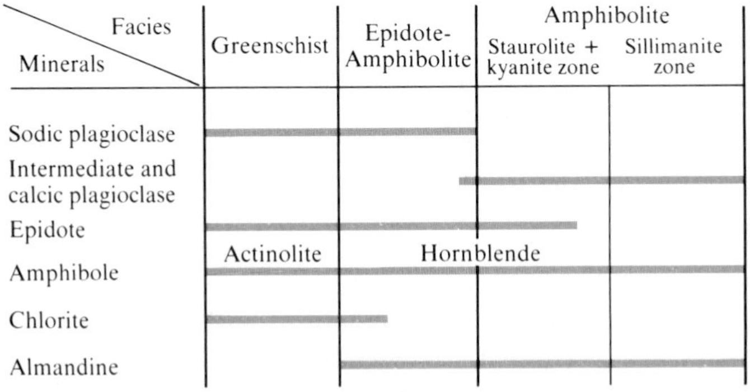

Figure 20–23
Mineralogical changes in medium-pressure metamorphism of a metabasite, Grampian Highlands, Scotland. [After A. Miyashiro, 1961, *Jour. Petrology, 2,* Fig. 1. © Oxford University Press.]

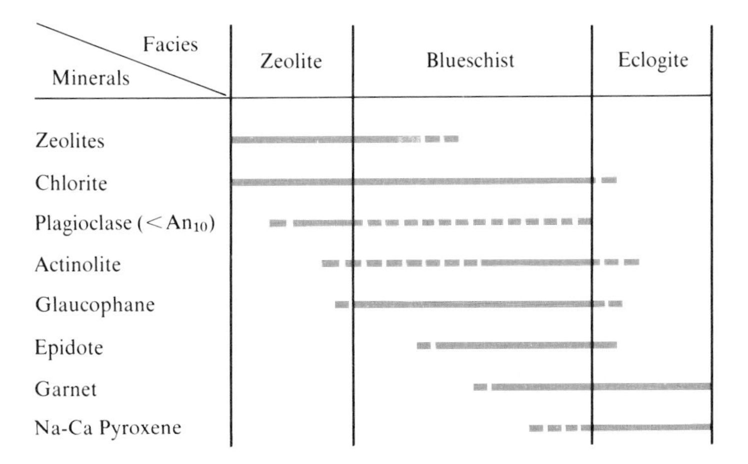

Figure 20-24
Mineralogical changes in high-pressure metamorphism of a metabasite, Kanto Mountains, Japan. [After A. Miyashiro, 1961, *Jour. Petrology, 2,* Fig. 3. © Oxford University Press.]

Metamorphism under high-pressure conditions leads to the formation of typical blueschist facies minerals such as glaucophane or lawsonite. Entrance into the eclogite facies results in the formation of the very dense phases omphacite (a pyroxene) and a pyropic garnet. The limits of stability of plagioclase have been exceeded at the *PT* limits of the eclogite facies, so it is never found here. A general plagioclase decomposition reaction has been given as

$$\underbrace{NaAlSi_3O_8 + CaAl_2Si_2O_8}_{\text{Plagioclase}} + \underbrace{CaMg(SiO_3)_2}_{\text{Diopside}} + \underbrace{Mg_2SiO_4}_{\text{Forsterite}} \rightarrow$$

$$+ \underbrace{CaMg_2Al_2Si_3O_{12}}_{\text{Garnet}} + \underbrace{NaAlSi_2O_6 + CaMgSi_2O_6}_{\text{Omphacite}} + \underbrace{SiO_2}_{\text{Quartz}}$$

Ultramafic Rocks

Most ultramafic rock complexes are made up largely of dunite (mostly olivine), harzburgite (olivine and orthopyroxene), and lherzolite (a mixture of olivine, clinopyroxene, and orthopyroxene). As a consequence, the main chemical constituents are MgO and SiO_2, with lesser amounts of FeO and CaO.

Metamorphism of these high-temperature mafic minerals usually involves reaction with H_2O or occasionally CO_2. The typical minerals that are produced can therefore be shown on a compositional triangle whose corners are MgO, SiO_2, and $H_2O + CO_2$

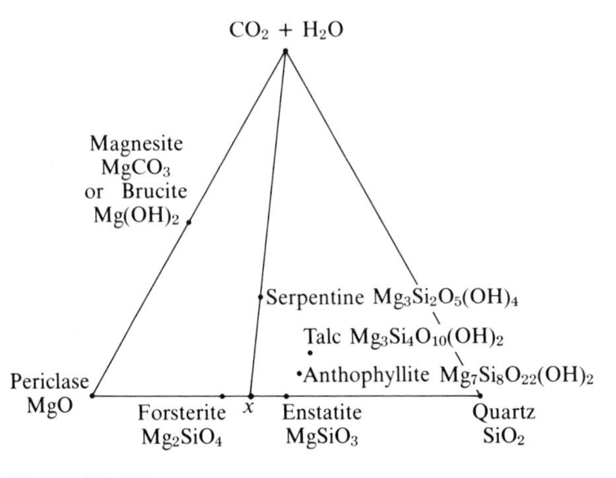

Figure 20–25
Minerals present in the system MgO–SiO_2–(CO_2, H_2O). A composition equivalent to dehydrated serpentine is located at point *x* on the base of the triangle. The composition of an average ultrabasic rock falls between Mg_2SiO_4 and point *x*.

(see Figure 20–25). The two igneous minerals forsterite and enstatite are shown on the base between the MgO and SiO_2 corners. Minerals formed as a result of hydration and carbonation are found above the base line.

One of the most common reactions that occurs during metamorphism is the formation of serpentine due to the access of H_2O. Looking again at Figure 20–25, consider the line drawn through the serpentine composition point from the H_2O apex of the triangle; the lower end of this line intersects the MgO–SiO_2 edge at *x* between the composition points of forsterite and enstatite. This intersection point indicates the ratio of MgO and SiO_2 that is contained in serpentine. If an ultramafic rock contained this same ratio of MgO and SiO_2 (in the form of forsterite and enstatite), metamorphism in the presence of water could produce a rock consisting entirely of serpentine. If, on the other hand, the ultramafic rock was a dunite that contained only forsterite, serpentinization would leave an excess of MgO. In this case, serpentinization would produce mainly serpentine and minor amounts of a more MgO-rich phase such as brucite or periclase (or perhaps magnesite if some CO_2 were also present). Serpentinization of an ultramafic rock composed of enstatite (or a mixture of enstatite and forsterite that fell between the enstatite composition point and point *x*) would produce serpentine and a more silica-rich mineral such as talc. Although both types of serpentine assemblages may be present in the same ultramafic body if it is compositionally layered, the average ultramafic rocks fall in composition between forsterite and the point *x*.

The term serpentine is the name given to a mineral group that consists of three major minerals—lizardite, chrysotile, and antigorite. Lizardite tends to be platy in habit, chrysotile is fibrous, and antigorite may be either fibrous or platy. As these minerals may look very similar to each other in hand specimen and are usually too fine-grained to distinguish microscopically, identification depends upon the use of X-ray diffraction. This is unfortunate because easy distinction would be useful. Most serpentine-rich rocks are composed mainly of lizardite; lizardite–chrysotile combinations are the next most common, and antigorite-bearing rocks are least common. Experimental and isotopic data indicate that lizardite and lizardite–chrysotile serpentinites have formed below about 350–400°C, whereas the less common antigorite types have formed between about 350–400°C and 600°C. The presence of antigorite therefore indicates either that the serpentinite has undergone progressive metamorphism or that it has been serpentinized at higher temperatures than lizardite and chrysotile.

The circumstances under which serpentinization of ultramafic rocks occurs have been debated for many years. Detailed textural examination indicates that the rate of serpentinization is greatest in olivine, less in orthopyroxene, and least in clinopyroxene. The mechanism of serpentinization presents a problem. Introduction of water into a rock and consequent hydration should result in an increase of volume and decrease in density of the solid phases. Strong evidence has been presented (in the form of perfect serpentine pseudomorphs after olivine and pyroxene) in favor of a constant-volume hydration; this requires a loss of materials (Mg and Si) during hydration. Equally strong evidence has been presented in favor of volume increase (such as internal fracturing and lack of evidence for Mg and Si metasomatism in surrounding country rock). It is now generally agreed that serpentinization either with constant volume or with volume increase probably takes place under differing conditions.

Moody (1976, p. 129) lists the following factors that may indicate a volume increase (with constant composition):

1. No evidence of Si or Mg metasomatism in surrounding country rocks.
2. Chemical composition of hydrous and anhydrous parent material is the same except for the addition of water.
3. Internal fracturing.

He also lists the following factors that may indicate a constant volume replacement:

1. Perfect pseudomorphic replacement of euhedral olivine by serpentine.
2. Density decreased and porosity increased in serpentine compared to the parent ultramafic body.
3. A portion of the Mg, Ca, and Si removed during serpentinization may be deposited far from the source.

Present data indicate that serpentinites formed within ophiolite complexes probably were serpentinized prior to emplacement. The oxygen isotope ratio of Alpine-type serpentinites suggests that the fluid phase is of meteoric–hydrothermal origin and therefore that the serpentinization probably occurred within the crust. Antigoritic serpentinites probably formed during regional metamorphism, because the oxygen isotope ratios indicate that the fluid is nonmeteoric.

After emplacement, serpentinite bodies are susceptible to alteration by CO_2. The presence of more than 5% CO_2 in the fluid phase eliminates chrysotile and lizardite; antigorite can survive a fluid with almost 20% CO_2. As a result of the presence of CO_2 in the fluid after emplacement, a talc–magnesite zone is created, which extends into the serpentinite body. This is seen texturally by pseudomorphs of talc and magnesite after serpentine. Various degrees of alteration may be found, from a thin rim of talc–magnesite around the serpentinite body to almost complete replacement. The amount of alteration is related to the degree of shear within the serpentinite intrusion, and in some cases probably to the presence of carbonate-rich country rocks.

0 1 mm

Figure 20-26
Flattened quartz pebbles in a quartzite. All of the larger grains are quartz. The finer matrix consists of muscovite, biotite, and traces of iron ore and zircon. Crossed nicols. Southern Black Hills, South Dakota. [From J. A. Redden, 1968, U.S. Geol. Surv. Prof. Paper 297-F, Fig. 123-B.]

Additional minor metasomatic alterations are associated with serpentinized ultramafic rocks. One of these is the formation of a rock called rodingite. This is a calc-silicate rock usually containing hydrated grossularite garnet, idocrase, diopside, prehnite, or epidote. This may be present as very thin rims surrounding the serpentinite body or as replacement of mafic igneous veins within the serpentinite. In addition a very thin border of chlorite-rich rock may be present within the country rock surrounding the serpentinite.

PHOTOMICROGRAPHS

Figures 20–26 through 20–42 are thin sections of metamorphic rocks and textures. Figures 20–43 through 20–50 give examples of textural evidence for metamorphic reactions.

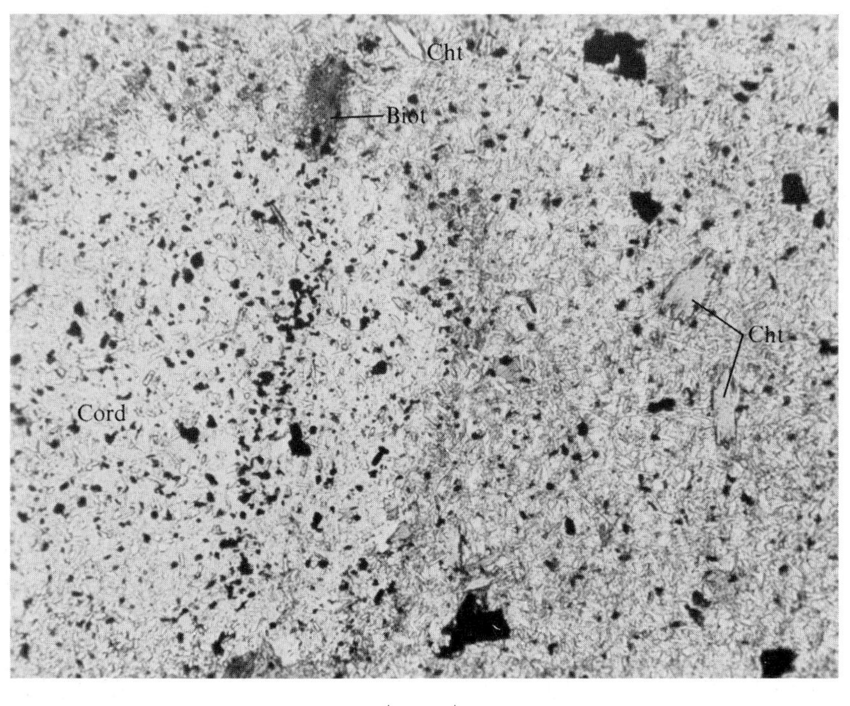

Figure 20–27
Hornfels from the biotite zone of contact aureole of the Albee Formation adjacent to the Cupsuptic pluton. The lighter area at the lower left (Cord) is a porphyroblast of cordierite. It is in a fine-grained matrix of muscovite, quartz, and albite. Cht = chlorite, Biot = biotite. Plane-polarized light. Oxford County, Maine. [From D. S. Harwood, 1973, *U.S. Geol. Surv. Bull., 1346,* Fig. 7.]

Figure 20–28
Serpentinite, composed of slip and cross-fiber asbestos. Minor oxides (usually magnetite) are present in the finer matrix. Crossed nicols. Minas Gerais, Brazil. [From N. Herz, 1970, U.S. Geol. Surv. Prof. Paper 641-B, Fig. 22.]

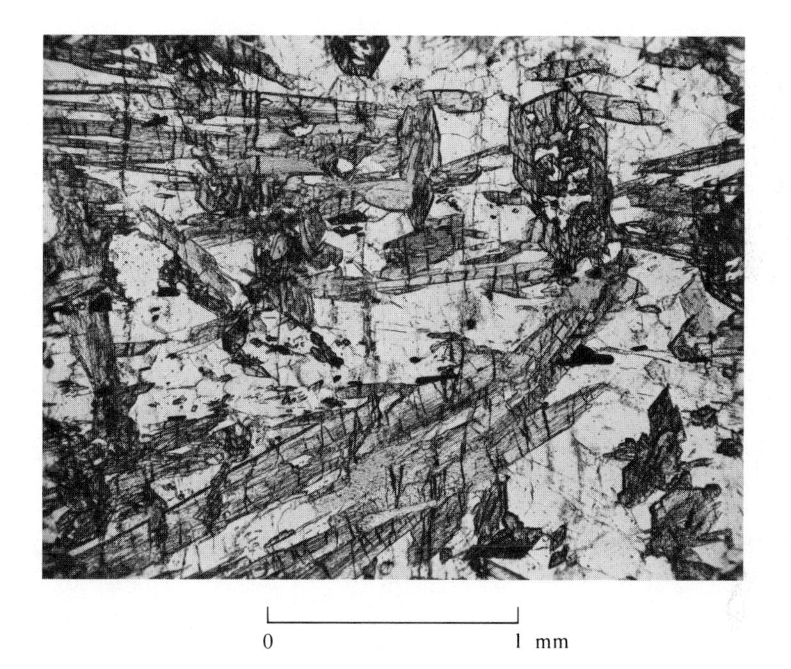

Figure 20–29
An anthophyllite–plagioclase schist. The darker elongate idioblastic grains are anthophyllite (distinguished from other amphiboles by their parallel extinction). The matrix consists of mainly untwinned plagioclase with accessory hematite and sphene. Anthophyllite is most common during regional metamorphism of ultrabasic rocks. Crossed nicols. Daggett County, Utah. [From W. R. Hansen, 1965, U.S. Geol. Surv. Prof. Paper 490, Fig. 8 (photo courtesy W. Walker).]

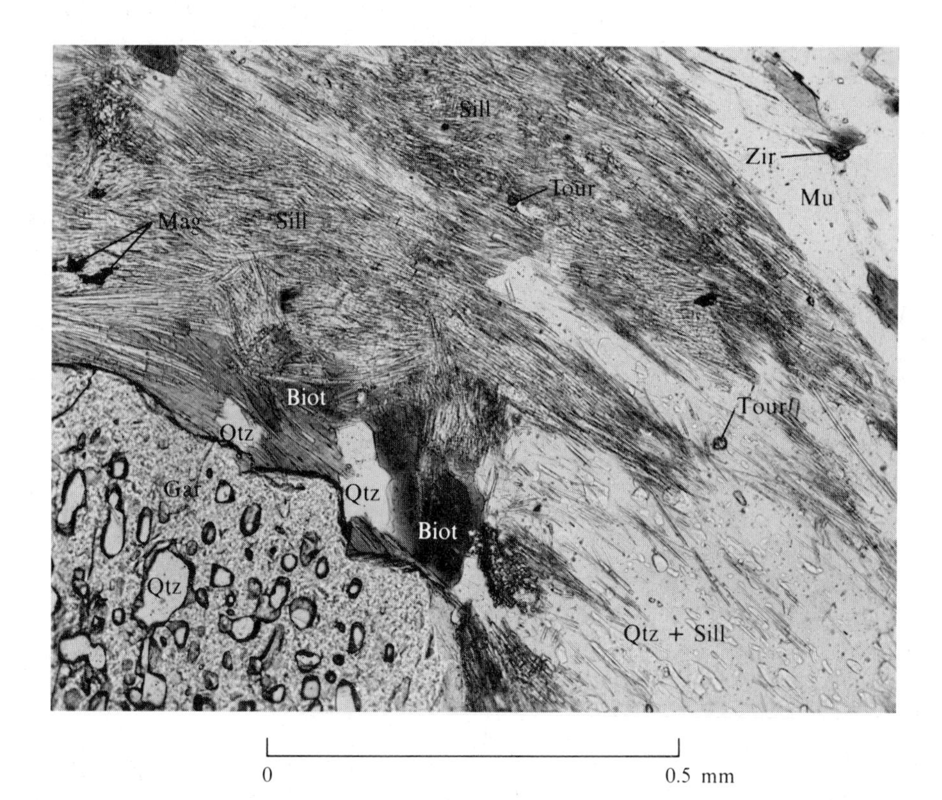

Figure 20–30
Mica schist. The large grain at the lower left is poikiloblastic garnet (Gar). The mass of fibrous crystals is mainly fibrolite [the name given to the acicular variety of sillimanite (Sill)]. Dark areas within and adjacent to the fibrolite are composed of biotite. Other minerals present in the rock are quartz (Qtz), muscovite (Mu), tourmaline (Tour), zircon (Zir), and magnetite–ilmenite (Mag). Plane-polarized light. Jefferson County, Colorado. [From D. M. Sheridan, 1967, U.S. Geol. Surv. Prof. Paper 520, Fig. 6.]

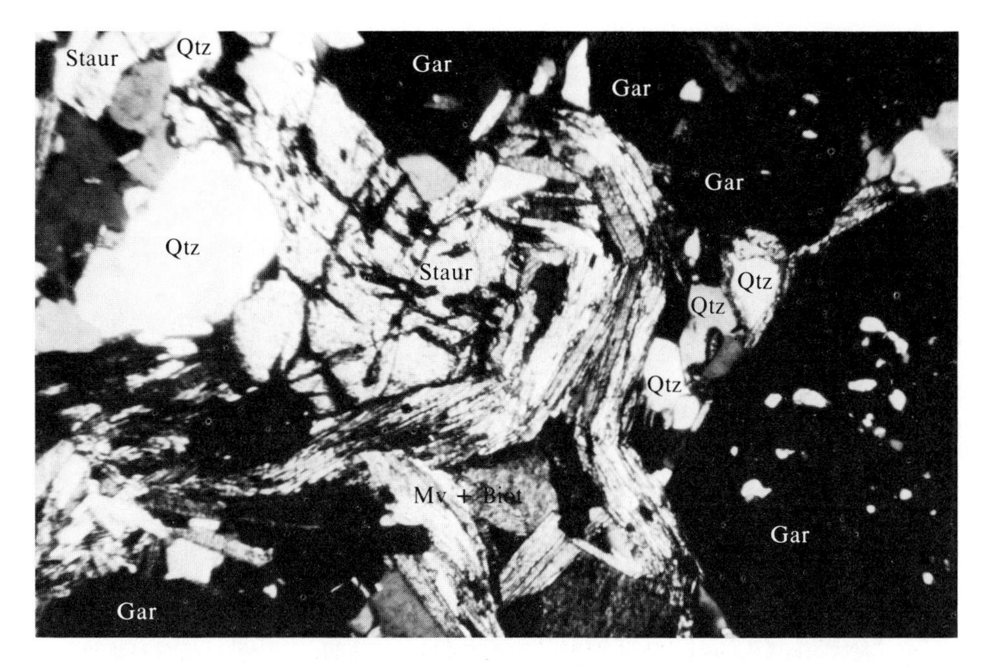

Figure 20–31
Staurolite–garnet mica schist. Large idioblastic to hypidioblastic garnet porphyroblasts make up the large black areas. Staurolite (Staur), near the center, is partly surrounded by laths of muscovite and biotite. Quartz is present as xenoblastic grains. Crossed nicols. Width of photo is about 1.9 mm.

Figure 20–32
Staurolite–biotite schist. The large central crystal is idioblastic staurolite; the smaller staurolite grain to the right has been partly sericitized. Most of the darker matrix areas are biotite, and lighter areas are quartz. Crossed nicols. Width of photo is about 2.3 mm.

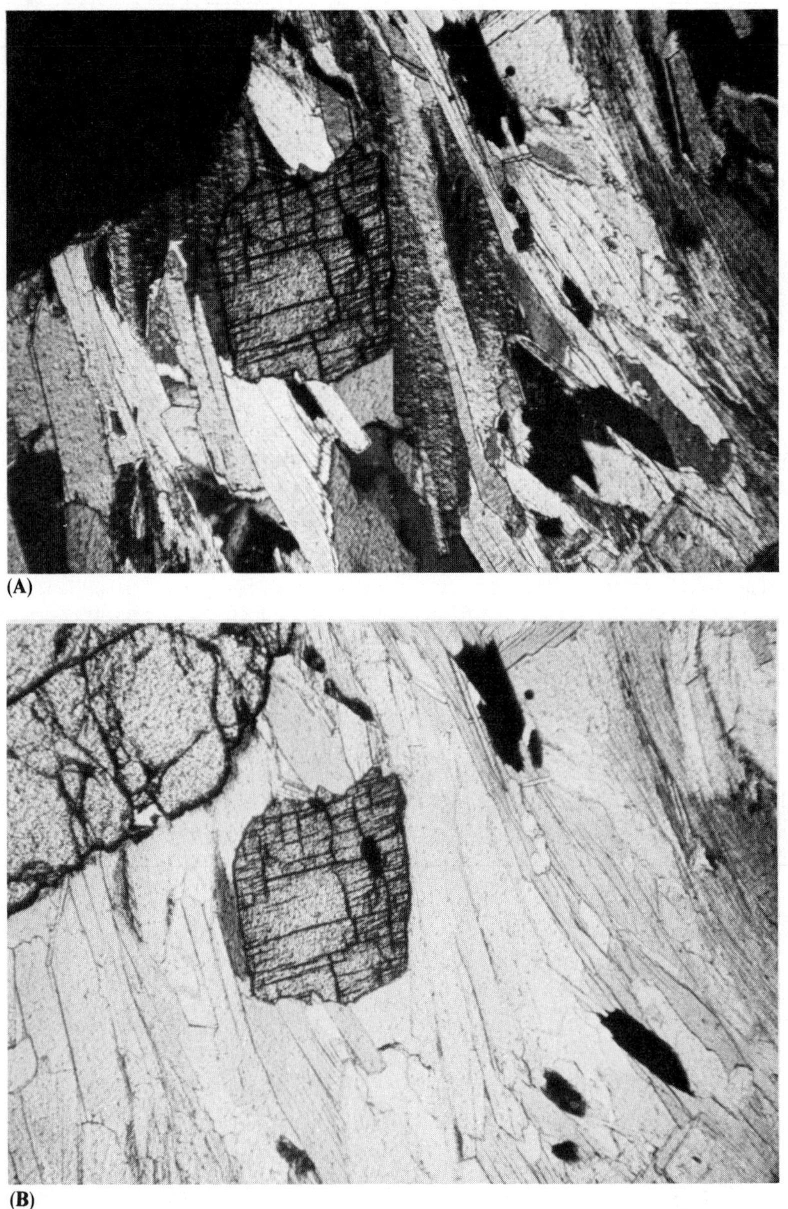

Figure 20–33
Kyanite–garnet–muscovite schist. The central blocky grain showing abundant cleavage is kyanite. Elongate vertically oriented grains are muscovite. A garnet porphyroblast is at the upper left. Width of both photos is about 2.3 mm. **(A)** Crossed nicols. Shows the "bird's-eye" appearance of mica and the isotropic nature of garnet. **(B)** Plane-polarized light. Emphasizes the high relief of both garnet and kyanite.

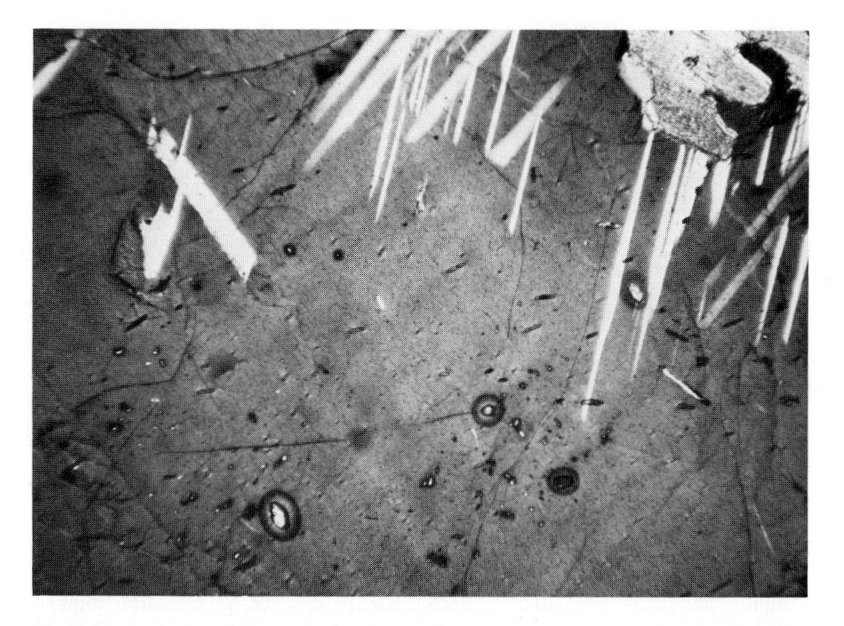

Figure 20–34
A single crystal of cordierite. Discontinuous polysynthetic twinning and lack of good cleavage distinguishes this mineral from feldspar. Inclusions of zircon showing pleochroic halos are also common in cordierite. Muscovite is present at the upper right. Crossed nicols. Width of photo is about 2.3 mm.

Figure 20–35
Glaucophane schist. The elongate high-relief crystals are idioblastic glaucophane crystals; this amphibole is easily recognized in plane-polarized light by its distinctive blue to purplish pleochroism. Smaller equant high-relief grains are garnet. The clear matrix is mainly quartz. Plane-polarized light. Width of photo is about 2.3 mm.

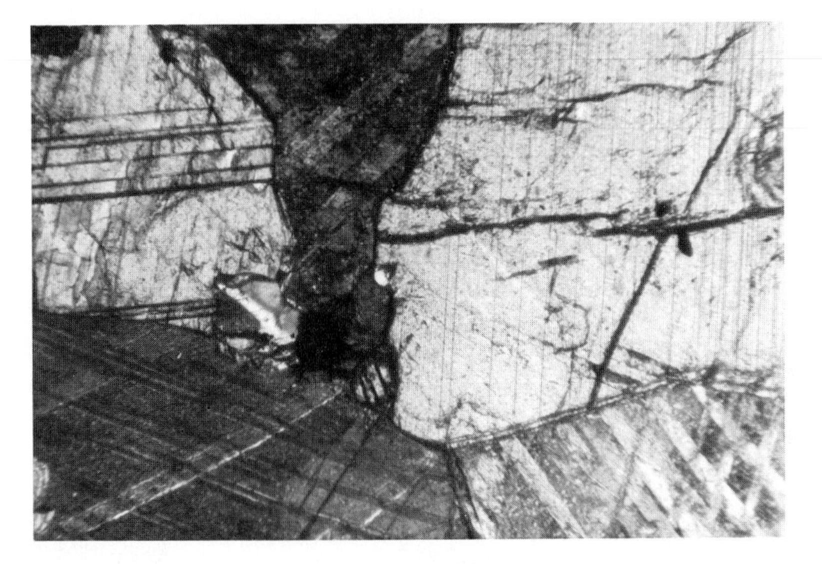

Figure 20–36
Marble. All of the crystals consist of calcite, recognized by extremely high birefringence and rhombohedral cleavage cracks. Polysynthetic twin bands are present on the three darker grains. Crossed nicols. Width of photo is about 2.3 mm.

Figure 20–37
Metamorphism of a magnesium-rich limestone has resulted in the formation of this diopsidic marble. The idioblastic crystals are diopside. Lighter areas are calcite (C), which is interstitial to diopside. Crossed nicols. Width of photo is about 2.3 mm.

Figure 20–38
Post-tectonic chlorite crystal in a fine-grained mica quartz schist. The idioblastic chlorite formed after the matrix, as it is undeformed and perpendicular to the direction of schistosity. Crossed nicols. Width of photo is about 2.3 mm.

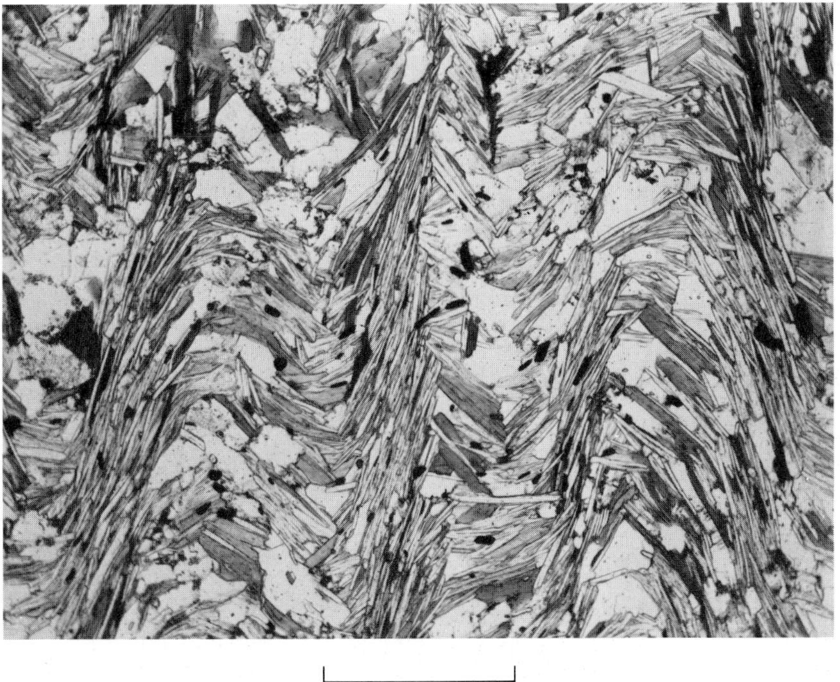

Figure 20–39
Muscovite–quartz schist. The initial foliation direction of the muscovite (laths) extends from left to right. A change in stress produced microfolding (vertical axial planes), with concentrations of mica flakes parallel to the limbs of folds. Continuation of this process could result in total elimination of any trace of the initial foliation direction. Plane-polarized light. [U.S. Geological Survey photo by D. M. Sheridan.]

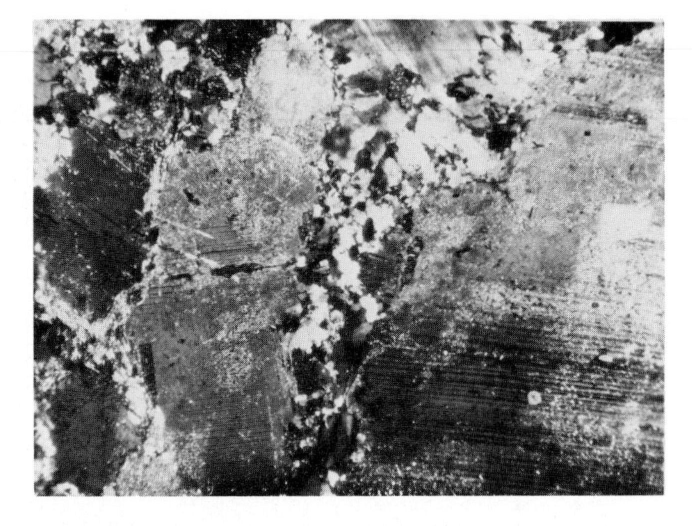

Figure 20–40
Cataclastic texture. A fine-grained mortar composed of quartz is visible between large grains of plagioclase (partially altered to muscovite and clinozoisite). Crossed nicols. Minas Gerais, Brazil. [U.S. Geological Survey photo by N. Herz.]

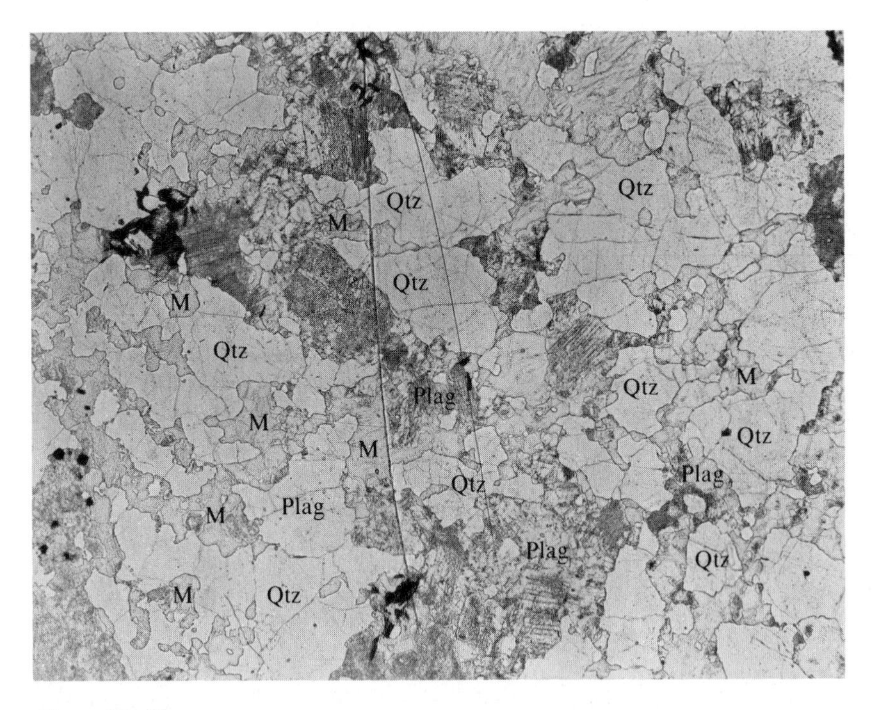

Figure 20–41
Granulite. Typical texture showing the relationship between plagioclase (Plag), microcline (M) and quartz (Qtz). The feldspar forms an irregular network enclosing clusters and individual grains of quartz. The two vertical lines are cracks in the thin section. Width of photo is about 5.5 mm. Tenmile Range, Summit County, Colorado. [From H. H. Koschmann, 1960, *Geol. Soc. Amer. Bull., 71,* Plate 2.]

Figure 20–42
Amphibolite. This rock is mainly green hornblende, which is seen here as gray high-relief grains. Hornblende has strong pleochroism, and shows (110) cleavage in most orientations. Light areas are plagioclase and quartz. Crossed nicols. Width of photo is about 2.3 mm.

Figure 20–43
An olivine phenocryst (gray) partially converted to serpentine (white). The serpentine forms rims and cross-cuts the olivine. The dark groundmass is largely fine-grained serpentine. Crossed nicols. Width of photo is about 2.3 mm.

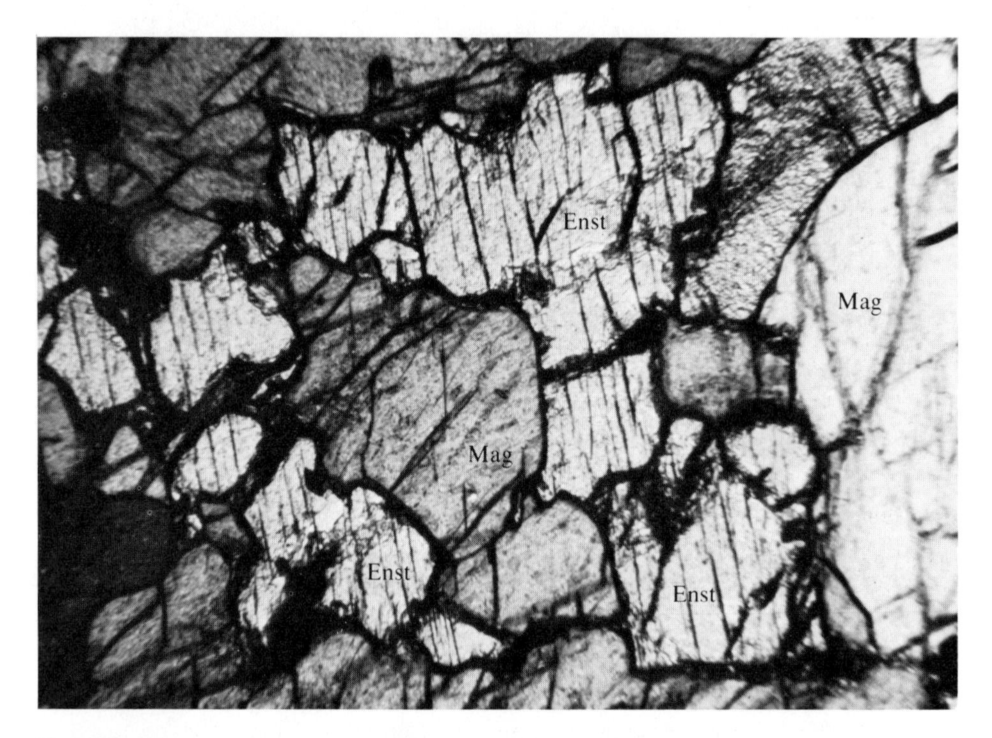

Figure 20–44
A single crystal of enstatite (Enst; light with vertical cleavage) has been converted to isolated islands (which remain in optical continuity) as a result of replacement by magnesite (Mag). Crossed nicols. Width of photo is about 2.3 mm.

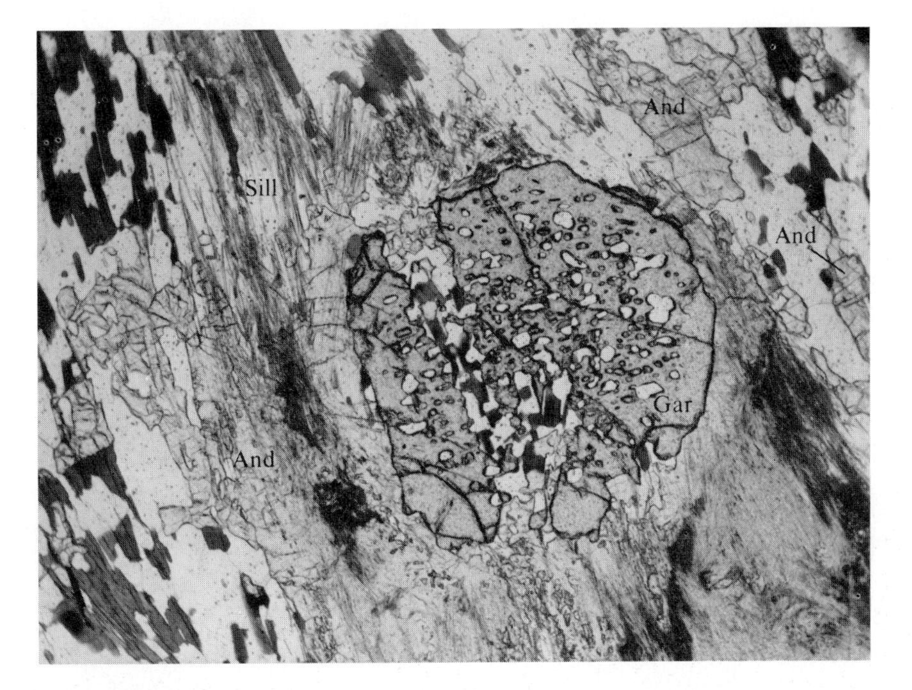

Figure 20–45
The central poikiloblast is composed of garnet. To the left of the garnet is a xenoblastic poikiloblast of andalusite (And; gray, low relief). The andalusite has been partially replaced by fibrolite [the acicular variety of sillimanite (Sill)]. Other minerals are quartz (light) and biotite (dark). Direct conversion of andalusite to sillimanite or kyanite is the exception rather than the rule. Plane-polarized light. Jefferson County, Colorado. [From D. M. Sheridan, 1967, U.S. Geol. Surv. Prof. Paper 520, Fig. 7.]

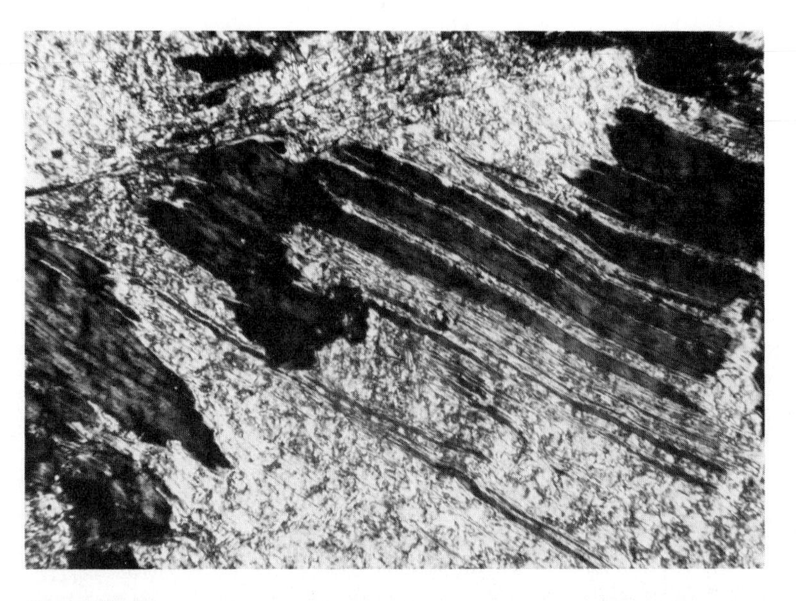

Figure 20–46
Chlorite (dark areas) has been converted to isolated optically continuous islands by replacement by fine-grained talc. The matrix is mainly talc (which is indistinguishable from muscovite in thin section). Crossed nicols. Width of photo is about 0.8 mm.

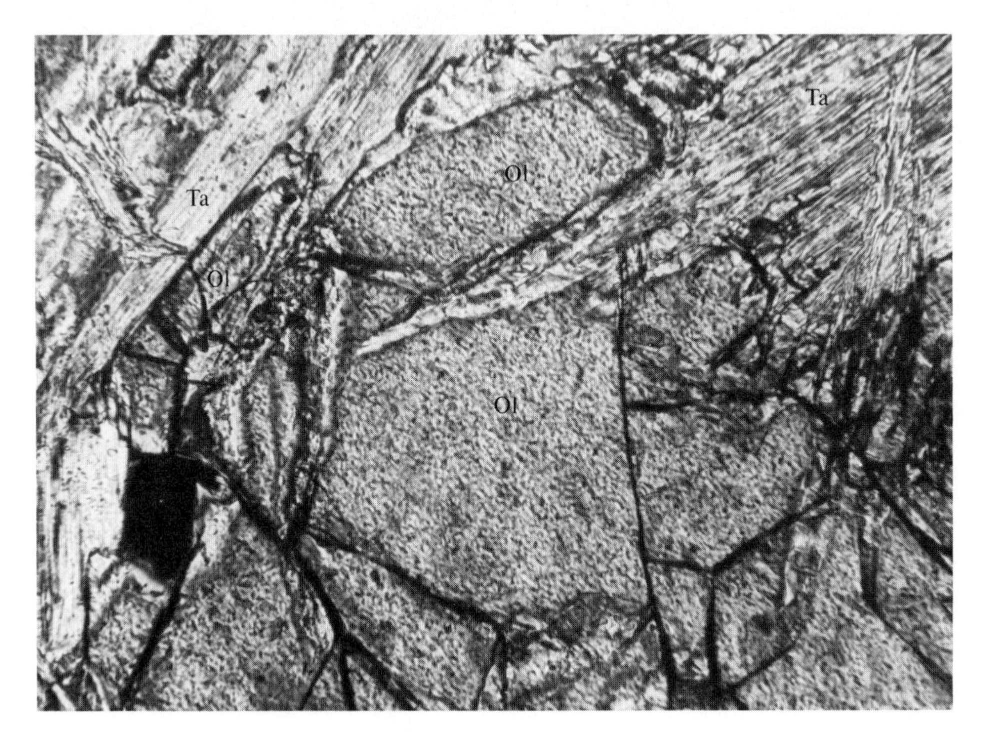

Figure 20–47
Olivine crystal (Ol) partially replaced by talc (Ta). The olivine has been deeply embayed. An optically continuous island is present at the left. Crossed nicols. Width of photo is 0.8 mm.

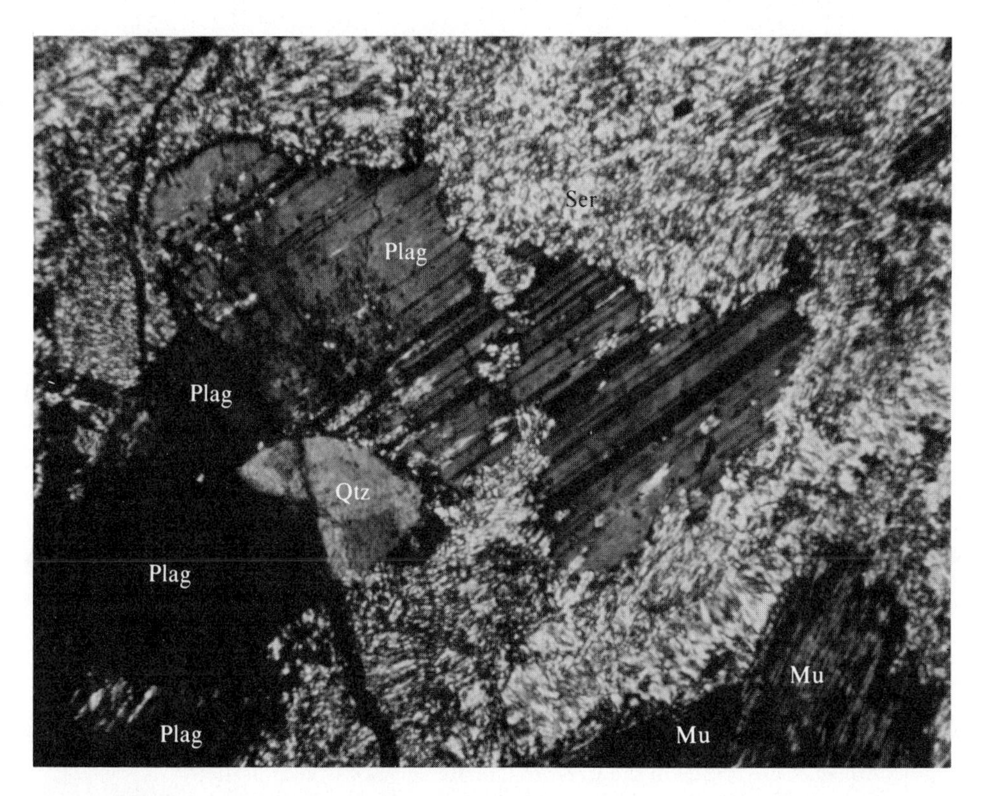

Figure 20-48
The large twinned crystal is plagioclase (Plag). Partial replacement of this crystal by sericite (Ser) has left a very irregularly shaped remnant. Qtz = quartz, Mu = muscovite. Crossed nicols. Width of photo is 0.9 mm.

Figure 20–49
A plagioclase crystal in the Itabirito Granite Gneiss has been partially replaced by fine-grained clinozoisite and muscovite. The unaltered rim of the plagioclase crystal is about An_{10}, and the core is more calcic. Calcium-rich plagioclase usually shows more susceptibility to reaction or alteration than sodium-rich varieties. Crossed nicols. Minas Gerais, Brazil. [From N. Herz, 1970, U.S. Geol. Surv. Prof. Paper 641-B, Fig. 15.]

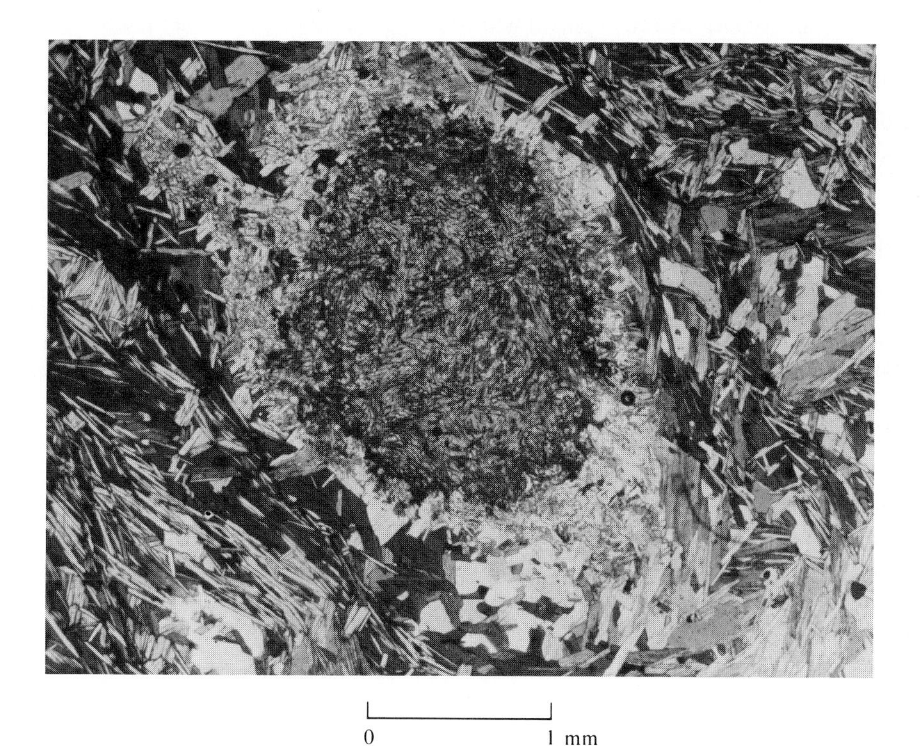

Figure 20–50
The central dark area is a porphyroblast composed of sillimanite needles (called a glomeroporphyroblast). The outer rim has been altered to fine-grained muscovite as a result of retrograde metamorphism. Coarser-grained muscovite and biotite define a foliation in the groundmass. Irregularly shaped grains of (mainly) quartz and tourmaline are present in the groundmass. Plane-polarized light. Jefferson County, Colorado. [From D. M. Sheridan, 1967, U.S. Geol. Surv. Prof. Paper 520, Fig. 4.]

SUMMARY

Prograde metamorphism of sediments may begin in the zeolite facies if reactive materials are present; in most instances, however, the first evidence for metamorphism is found in the greenschist facies. Most metamorphic reactions are characterized by dehydration and decarbonation.

Rocks undergoing metamorphism will produce characteristic mineral assemblages as a function of their composition, the relative intensity of pressure and temperature, and the composition of the associated vapor phase; the latter factor is particularly important in the case of carbonate-bearing rocks, as the vapor phase may be rich in either H_2O or CO_2.

The stable mineral assemblages produced by rocks of varied composition are summarized in *ACF*, *A'KF* and *AFM* diagrams. These diagrams have been established from worldwide observations of metamorphic rocks over a long period of time. Their use permits many metamorphic rocks to be assigned to a particular metamorphic facies. The approximate compositions, and often the metamorphic precursor, can sometimes be established by means of these diagrams. Experimental determinations of the *PT* conditions under which various mineral assemblages exist permits these assemblages to be assigned approximate temperatures and pressures of formation.

FURTHER READING

Bowen, N. L. 1940. Progressive metamorphism of siliceous limestone and dolomite. *Jour. Geol., 48,* 225–274.

Miyashiro, A. 1961. Evolution of metamorphic belts. *Jour. Petrology, 2,* 277–311.

Miyashiro, A. 1973. *Metamorphism and Metamorphic Belts.* New York: Halsted Press, John Wiley and Sons, 492 pp.

Moody, J. B. 1976. Serpentinization: a review. *Lithos, 9,* 125–138.

Slaughter, J., D. M. Kerrick, and V. J. Wall. 1975. Experimental and thermodynamic study of equilibria in the system $CaO–MgO–SiO_2–H_2O–CO_2$. *Amer. Jour. Sci., 275,* 143–162.

Turner, F. J. 1980. *Metamorphic Petrology,* 2d Ed. New York: McGraw-Hill, 512 pp.

Turner, F. J., and J. Verhoogen. 1960. *Igneous and Metamorphic Petrology.* New York: McGraw-Hill, 694 pp.

Winkler, H. G. F. 1979. *Petrogenesis of Metamorphic Rocks,* 5th Ed. New York: Springer-Verlag, 347 pp.

21

Time, Temperature, and Deformational Relationships

In those areas where detailed studies of metamorphism have been carried out, it has generally emerged that a typical metamorphic event, lasting well over a million years, is not a simple situation in which temperature and pressure have smoothly increased to a maximum, and then smoothly decreased. Commonly the temperature rise has been irregular through time (see Figure 21–1), and maximum temperatures are not necessarily related to maximum pressures or periods of deformation. In addition, many regions have been subjected to more than one metamorphic event (polymetamorphism). The evidence for the relative sequence is often revealed by the mineralogy and textures of the rock.

PORPHYROBLASTS AND TECTONISM

We have already examined some of the criteria useful in establishing relative time relationships within metamorphic rocks. These include such things as textural evidence for prograde and retrograde reactions (see Figure 21–2), the formation of melt in migmatites, and carbonate and serpentinization reactions. Another topic that has long been of great importance as a relative dating technique is the relationship of porphyroblasts to tectonism.

Textural evidence is overwhelming that porphyroblasts can grow by replacement of the groundmass. The growth–replacement process may be complete or incomplete. If complete, no trace of the former matrix material remains within the porphyroblast; if incomplete, inclusions may be found within the porphyroblast. If the number of inclusions is very large, the porphyroblast is called a poikiloblast. The inclusions may be randomly or regularly distributed. In rare cases, such as the chiastolite variety of

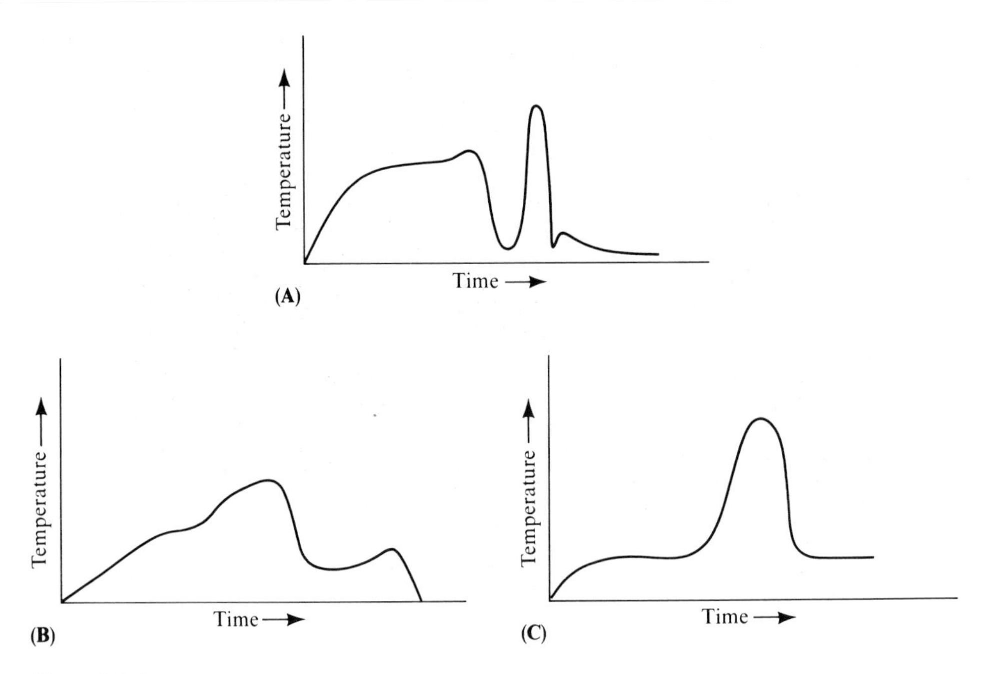

Figure 21-1
Suggested time–temperature relationships for three different regions in Scotland. [From N. Rast, 1966, in W. S. Pitcher and G. W. Flinn (eds.), *Controls of Metamorphism* (New York: John Wiley and Sons), Fig. 1.]

Figure 21-2 (*facing page*)
(**A**) A prograde reaction. Clear areas of the photomicrograph are quartz. Gray diffuse regions are calcite. Reaction of quartz and calcite to produce wollastonite ($CaSiO_3$) is shown by encroachment of prismatic wollastonite into both quartz and calcite grains. Plane-polarized light. Qtz = quartz, Plag = plagioclase, Diop = diopside, Ksp = K-feldspar, Mono County, California. [From C. D. Rinehart, 1964, U.S. Geol. Surv. Prof. Paper 385, Fig. 6-B.] (**B**) A retrograde reaction. Photomicrograph of biotite (dark) partially replaced by epidote (high-relief, prismatic grains) and muscovite (diffuse low-relief gray material). Plane-polarized light. Mu = muscovite, Biot = biotite, Epid = epidote. Minas Gerais, Brazil. [From N. Herz, 1970, U.S. Geol. Surv. Prof. Paper 641-B, Fig. 16.]

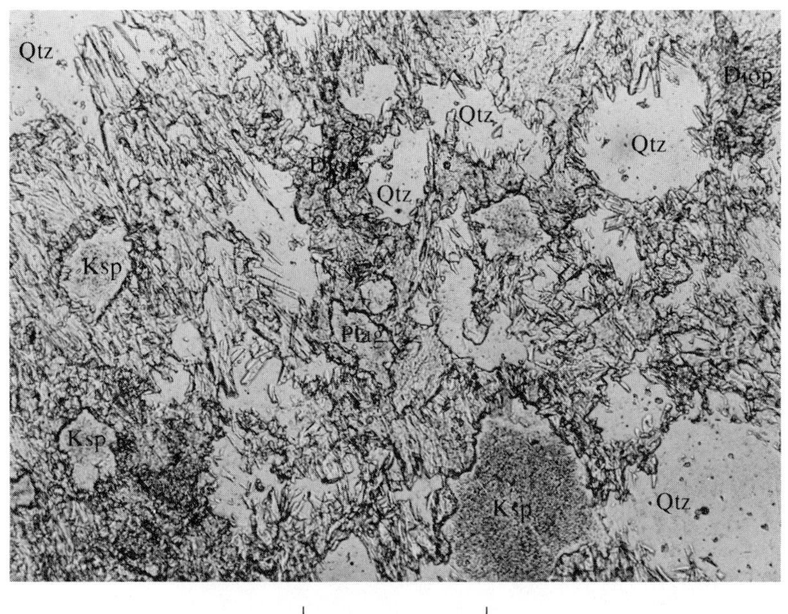

(A) 0 0.25 mm

(B) 0 0.1 mm

Figure 21–3
Oriented inclusions in andalusite (var. chiastolite). The large crystal at the upper left has been ground and polished to show the distribution of inclusions. [From J. Sinkankas, 1966, *Mineralogy: A First Course* (Princeton, N.J.: D. Van Nostrand), Fig. 309.]

andalusite, the distribution of inclusions is related to specific crystallographic directions within the porphyroblast (see Figure 21–3). In the usual situation, however, regular distribution of inclusions is a reflection of compositional layering or preferred crystal orientations that existed in the matrix before the porphyroblast was developed. The direction of lines or planes of inclusions within a porphyroblast is referred to as S_i (internal schistosity), whereas the banding or foliation that is present in the matrix (outside of the porphyroblast) is called S_e (external schistosity).

Let us consider criteria that can be used to determine whether a porphyroblast has grown before, during, or after a period of deformation in the rock—situations that are called pretectonic, syntectonic, and post-tectonic.

Pretectonic Porphyroblasts

A porphyroblast grown before a period of deformation may contain inclusions distributed in a way that indicates an earlier directional feature such as bedding plane orientation; this may show up as parallel lines of inclusions, or perhaps a preferred orientation of inclusions. During later deformation the porphyroblast often acts as a relatively rigid part of the rock in comparison to the common mica-rich matrix. The porphyroblast may be rotated during the deformation, such that the S_i is no longer in correspondence with S_e. It is equally possible that the matrix may be flattened or microfolded around the porphyroblast (see Figure 21–4). In all of these cases S_i is maintained within the porphyroblast in its original parallel arrangement, as a new direction of foliation (S_e) is developed in the matrix. At the porphyroblast edges, S_i and S_e grade into each other with no discontinuities.

The porphyroblast is not always immune to deformative processes, and it is common to find that some breakage, bending, or kinking has occurred. This is an unambiguous criterion that the porphyroblast was in existence before at least a part of the deformation. As the porphyroblast appears to act as a relatively rigid portion of the rock, it is common to find the development of pressure shadows—areas of apparently lower pressure on opposite sides of the grain (parallel to the S_e direction), where minerals such as quartz may have precipitated; another common situation is to find retrograde chlorite in the pressure shadows of porphyroblasts of garnet or biotite.

Syntectonic Porphyroblasts

The most common situation seems to be that of porphyroblast growth during deformation. If the porphyroblast has acquired inclusions during growth, the inclusions will be in different orientations in different portions of the porphyroblast; those in the older central portion of the porphyroblast show a distribution characteristic of the undeformed rock. As additional porphyroblast growth occurs during deformation, later inclusions show orientations and distributions that have been developed as a result of the continued deformation. Hence parallelism of the inclusion trails will not be found

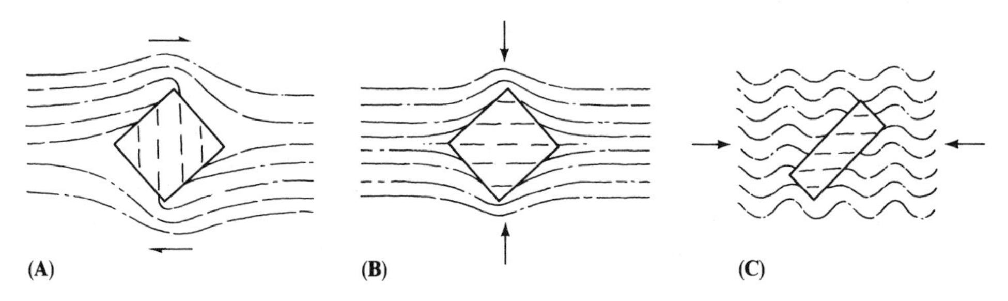

Figure 21–4
Straight lines of S_i are retained within the porphyroblast in spite of later deformation of the rock. (**A**) The porphyroblast has been rotated. (**B**) The matrix about the porphyroblast has been flattened. (**C**) The matrix about the porphyroblast has been deformed by microfolding. The porphyroblasts are classified as pretectonic. [After H. J. Zwart, 1962, *Geol. Rund., 52,* Fig. 1.]

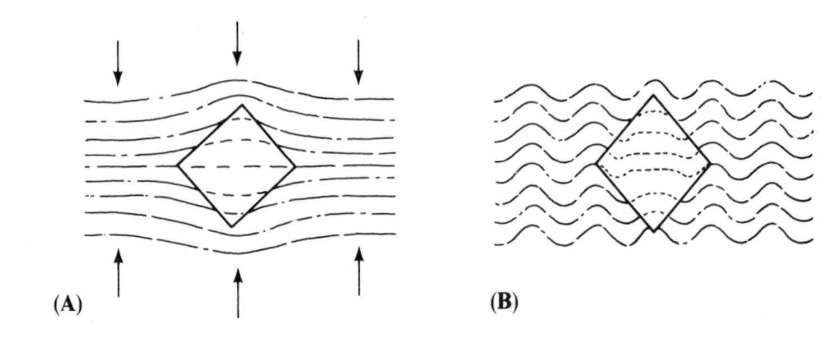

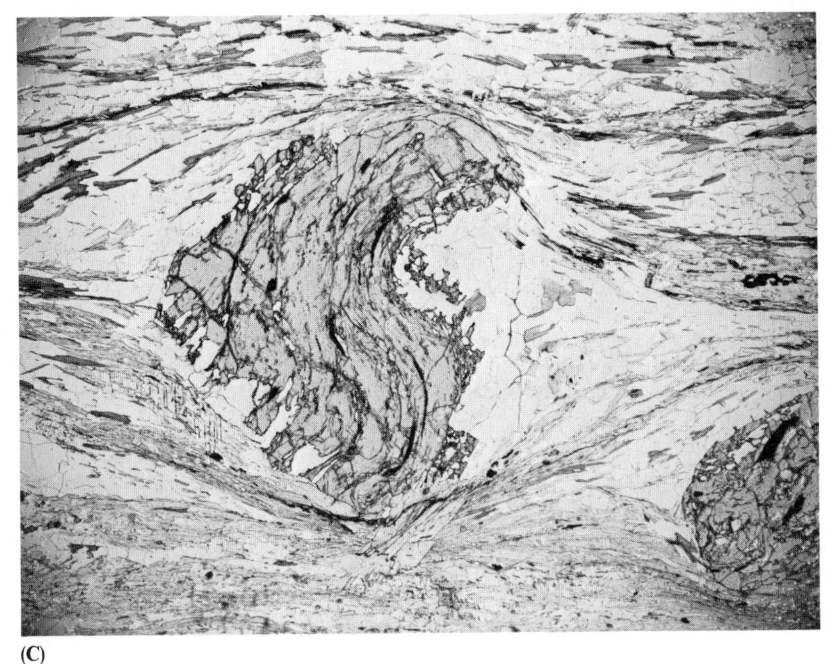

Figure 21–5
Syntectonic porphyroblasts. The porphyroblasts have grown during tectonism. (**A**) Growth during flattening of the matrix as evidenced by the change in orientation of S_i from the center to the edge of the porphyroblast. (**B**) Growth of porphyroblast during microfolding. S_i shows increasing amounts of folding from center to edge of porphyroblast. (**C**) Growth of porphyroblast during rotation. The porphyroblast has a tendency to grow elongated parallel to the schistosity. Optical continuity is retained in spite of crystal rotation. [Parts A and B after H. J. Zwart, 1962, *Geol. Rund., 52,* Fig. 1. Part C from J. L. Rosenfeld, 1970, Geol. Soc. Amer. Special Paper 129, Plate 1, Fig. 2.]

(see Figure 21–5). In part A the crystal grew during a period of matrix flattening; the growing crystal, acting as a rigid body, caused bending of the adjacent schistosity, which was incorporated into the growing porphyroblast. In part B the crystal grew during microfolding; the central earlier portion of the crystal incorporates unfolded inclusions; additional growth incorporates S_e with increasing degrees of folding. Part C shows a unique situation. Here the porphyroblast has grown with its direction of elongation parallel to S_e. During growth the crystal was rotated due to shearing stress; the principal growth direction is partially maintained along S_e, producing an S-shaped crystal, which maintains optical continuity throughout.

Pressure shadows may be developed during syntectonic porphyroblast growth. However, as the crystal is growing in a stable environment, there is little evidence of the fracturing or deformation common in pretectonic growth.

Post-Tectonic Porphyroblasts

Post-tectonic growth probably is the easiest to recognize, as the porphyroblast commonly has grown with blatant disregard of foliation directions within the matrix. If the porphyroblast has an elongate shape (as with micas and chlorites), the direction of elongation of the porphyroblast will usually be random with respect to S_e. However, perfect correspondence is found between S_i and S_e (see Figure 21–6). Furthermore, porphyroblasts will not show pressure shadows or evidence of deformation, because they have formed after the cessation of deformation.

More Complex Situations

The situations described above are agreed upon by most metamorphic petrologists; such relationships can provide a rational sequence of growth–deformation episodes.

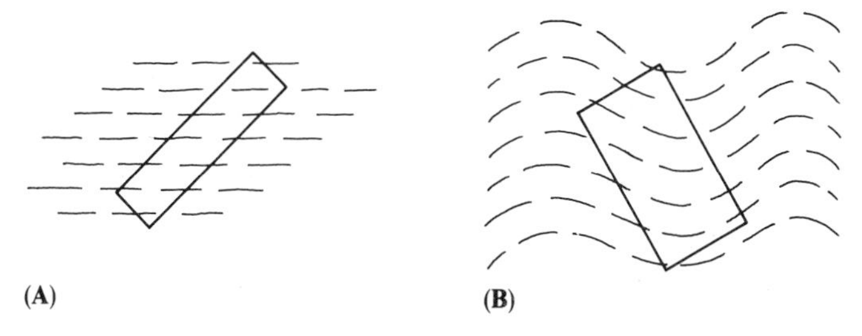

Figure 21–6
Post-tectonic porphyroblasts. The direction of elongation of the porphyroblast is independent of S_e. $S_i = S_e$.

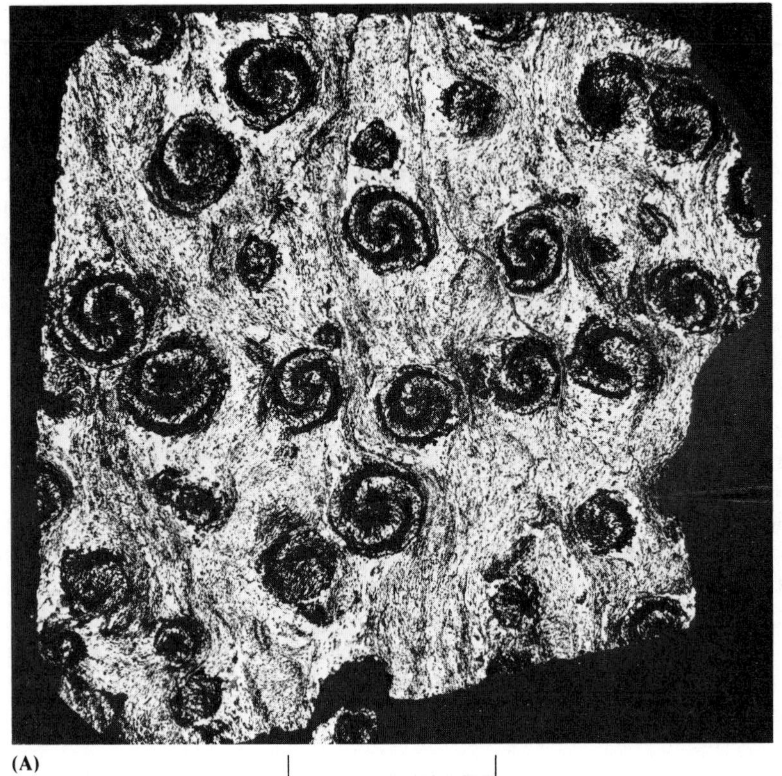

(A)

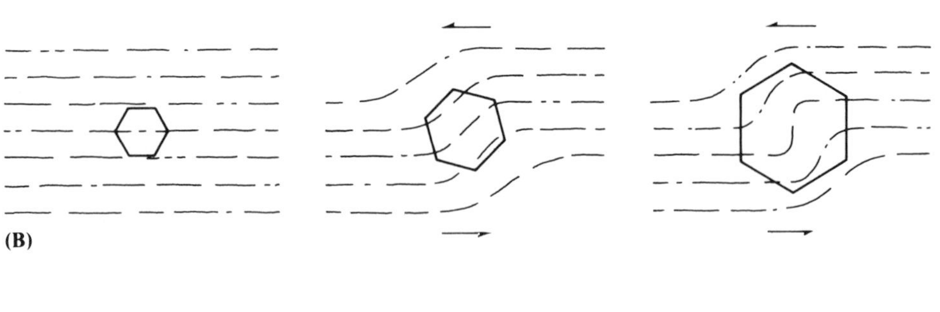

(B)

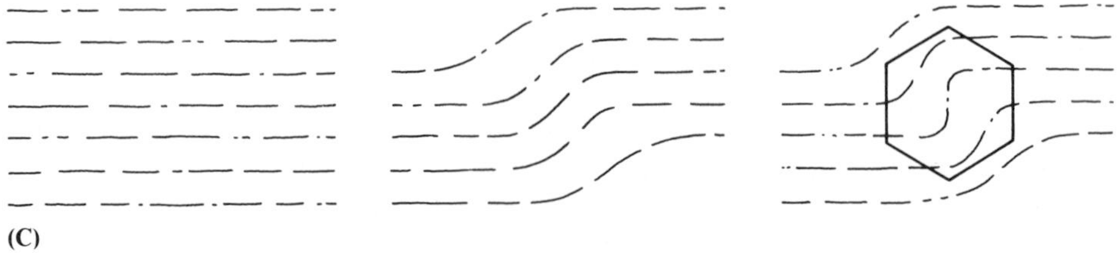

(C)

Unfortunately, many rocks contain porphyroblasts whose time relationships are open to debate or difficult to interpret.

During syntectonic growth, garnets commonly acquire inclusions of quartz (see Figure 21–7A). If, during growth, the rigid garnet is rolled within the more plastic groundmass as a result of shear, the inclusion pattern developed is curved, often S-shaped (see Figure 21–7B). Note also that S_i (although curved) joins with S_e at the edges of the crystal. The same relationship can occur as a result of post-tectonic garnet growth. In Figure 21–7C the rock is deformed and develops a schistosity that contains small folds or crenulations. After tectonism ceases, porphyroblast growth may incorporate some inclusions from the curved S_e, yielding an S-shaped inclusion pattern that mimics a rotational garnet. In the case of extreme garnet rotation (360° or more), it is usually obvious that the garnet is syntectonic rather than post-tectonic, as the inclusions show a spiral rather than S-shaped distribution; this is known as a *snowball* (pinwheel) *garnet.*

Many post-tectonic (helicitic) garnets are not distinguishable from rotational (syntectonic) types if only a single porphyroblast is examined. Examination of many porphyroblasts from the same rock or thin section is necessary. If the degree and direction of apparent rotation is more or less consistent on many porphyroblasts, they are generally considered to be syntectonic; note that the amount of apparent rotation will vary among garnets, as some porphyroblasts may be cut through their centers, and others through their edges. Alternatively, if the sense of rotation is different from one porphyroblast to the other, they are most likely post-tectonic.

A more debatable case is shown in Figure 21–8A. Here the porphyroblast is surrounded by a schistosity that bends completely around it; the schistosity is not truncated against the larger crystal. Ramberg (1947) and others have argued that this represents an example of "concretionary growth." After formation of the schistosity, material diffused to the site and the crystal grew, forcing apart the surrounding materials. The ability of a growing crystal to force aside its surroundings has been called the force of crystallization. Others have argued that a crystal growing in a metamorphic environment does not have the ability to overcome the pressure of the system and force aside the surrounding material. That this is true in many cases is indicated in Figure 21–8B. Evidence provided by inclusion trails (S_i) and by the strong develoment of schistosity adjacent to the porphyroblast indicates that with some porphyroblasts of this type, the porphyroblast has formed first, followed by the

---

Figure 21–7 *(facing page)*
(**A**) Well-developed snowball garnets. (**B**) Sequence showing growth and rotation of a garnet during tectonism to produce S-shaped S_i. (**C**) Sequence showing deformation followed by post-tectonic garnet growth leading to development of S-shaped S_i similar to part B. [Part A from J. L. Rosenfeld, 1968, in E-An Zen, W. S. White, and J. S. Hadley (eds.), *Studies of Appalachian Geology: Northern and Maritime* (New York: Wiley Interscience), Fig. 14–2, part E.]

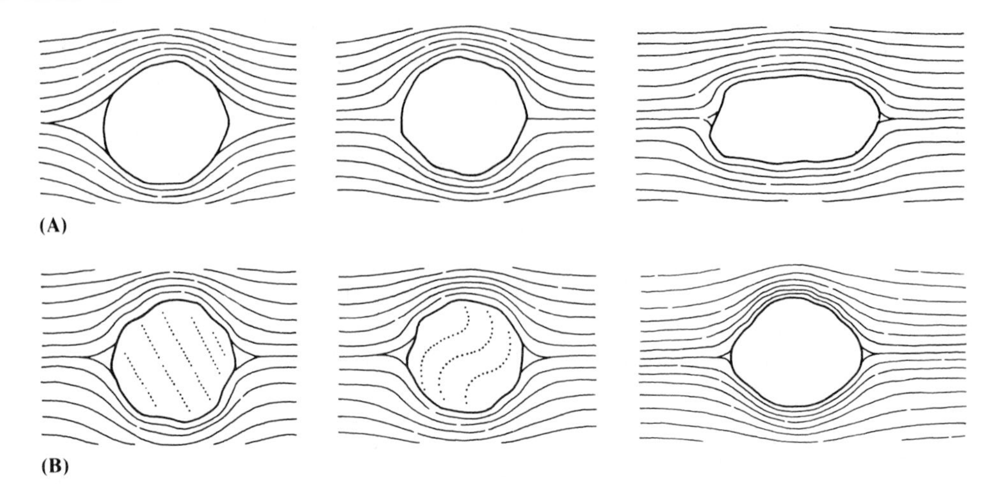

Figure 21–8
(**A**) Porphyroblasts of debatable origin. As the porphyroblasts do not truncate S_e, they may have grown either before or after development of S_e. Most investigators regard such porphyroblasts as pretectonic. (**B**) Although the S_e is not truncated, the porphyroblasts are clearly pretectonic, as indicated by the rotated S_i and increasing intensity of S_e adjacent to the porphyroblasts. [From P. Misch, 1971, *Geol. Soc. Amer. Bull., 82,* Fig. 1.]

schistosity. Strongly developed schistosity indicates movement in the matrix, suggesting that the porphyroblast acted as a rigid volume within the more plastic matrix.

In the case where no inclusion trails are present within the porphyroblast (see Figure 21–8A), debate on age relationships continues. Misch (1971) has suggested that the growing porphyroblast does not have to overcome the total pressure of the system. He suggests that porphyroblast growth results from a local chemical transfer with mechanical readjustment. Confirmatory evidence has come from a recent study by Ferguson (1980), who has demonstrated in a study of volumetric relationships within a hornfels that garnet porphyroblasts have displaced adjacent muscovite and graphite. A study in progress by Fisher and Ehlers on simulated porphyroblasts has demonstrated displacement of graphite flakes in a solid matrix. This study (see Figure 21–9) utilizes a reaction between KCl and $PbCl_2$ to form the intermediate compounds $2PbCl_2 \cdot KCl$ and $PbCl_2 \cdot 2KCl$. Pellets were made which consisted of alternate layers of $PbCl_2$ and KCl. Discontinuous layers of graphite flakes were located at the interfaces between the KCl and $PbCl_2$ layers. As the interfaces were smooth surfaces, the graphite initially formed a perfectly flat layer. The composite pellets kept in a furnace for several days at about 355°C reacted to form the intermediate compounds at the interfaces. These compounds, simulating porphyroblasts in appearance, caused a mechanical displacement of the originally flat graphite layers. Similar results were obtained at both atmospheric pressure and in a device with confining pressures up to

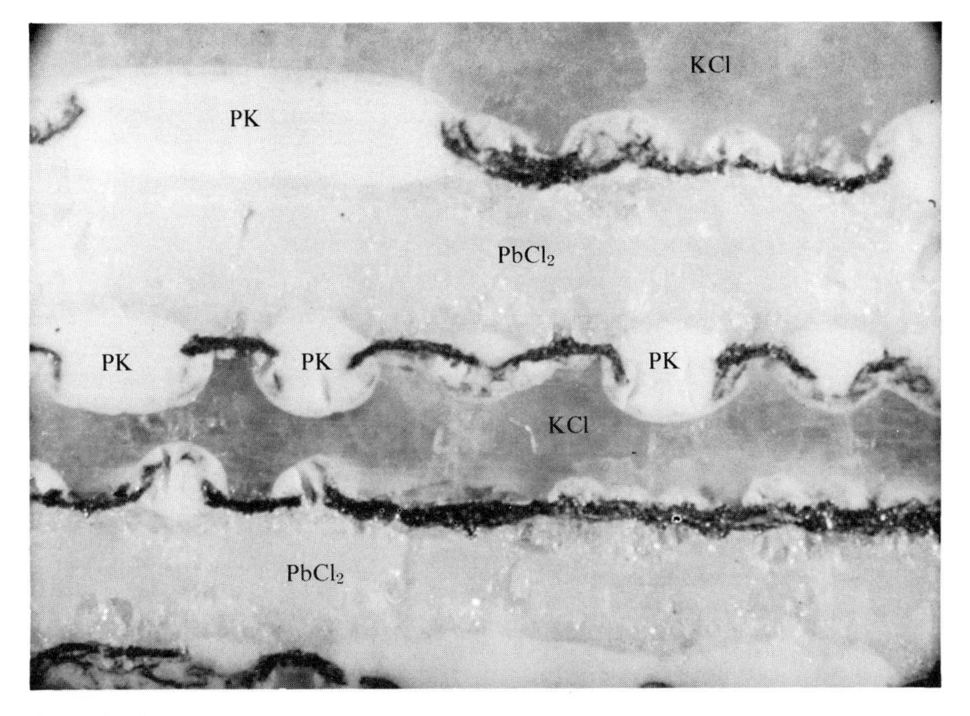

Figure 21–9
Simulated porphyroblasts created in the laboratory. A composite pellet was made, composed of alternate layers of KCl, $PbCl_2$, and graphite. All interfaces were originally flat. The chloride layers were thick and continuous across the specimen, but the graphite was thin and discontinuous. After being kept at 355°C for several days the $PbCl_2$ and KCl reacted to form intermediate compounds at the layer interfaces. Growth of these compounds between and around graphite flakes and transfer of material between the original chloride layers resulted in displacement of the graphite flakes. Compare to Figure 21–8A. PK = intermediate lead potassium chloride compounds. Black areas are graphite. Width of photo is 5.5 mm.

70 bar. Although these experiments were not conducted with metamorphic minerals at metamorphic temperatures and pressures, the analogy is clear. Crystals growing in an environment under pressure have the ability to displace matrix materials.

Another debatable example (and one of the most common porphyroblastic textures) is that shown in Figure 21–10A. Here S_e is both bowed and truncated by the porphyroblast. Misch (1971) has argued that this arrangement is due to a combination of both replacement and concretionary growth. The fact that the porphyroblast truncates S_e indicates a replacement relation, since S_e must have been preexistent; the bowing out of the matrix S_e indicates that the porphyroblast also forced aside the surrounding matrix by concretionary growth, thus indicating that the porphyroblast is

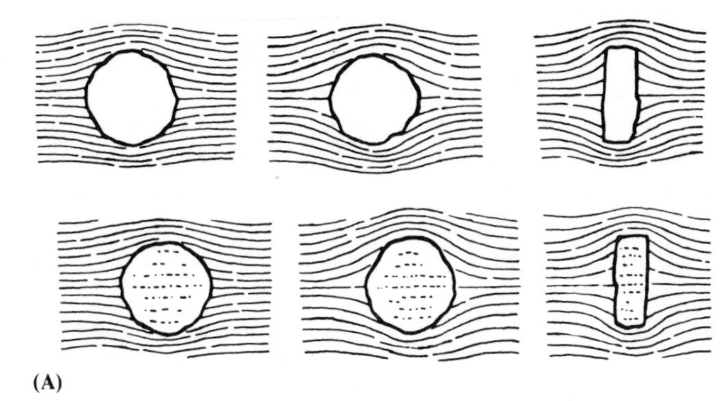

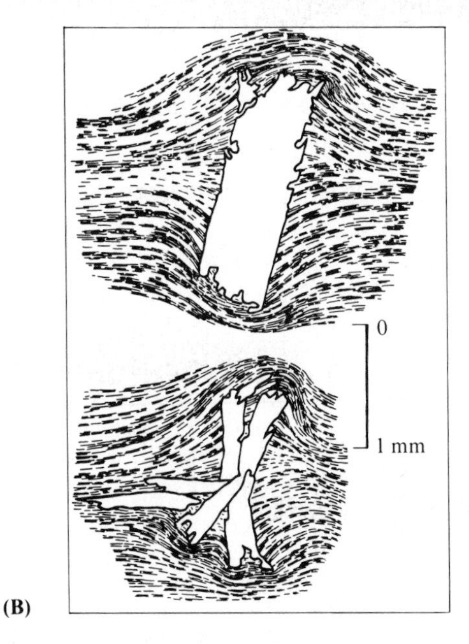

Figure 21–10
(**A**) Porphyroblasts of debatable origin (top line). Truncation of schistosity implies a post-tectonic origin, whereas bending of S_e around porphyroblasts implies a pretectonic origin. In the second line the relations of the porphyroblasts to S_e is the same as above, but the parallel S_i definitely indicates that the porphyroblasts are older than S_e. (**B**) Similar examples of chloritoid porphyroblasts within a phyllite. [From P. Misch, 1971, *Geol. Soc. Amer. Bull., 82,* Figs. 1, 5.]

post-tectonic. A mineralogical example is shown in Figure 21–10B. Spry (1972), taking what is probably a more conventional view, has argued that textures of this type are a result of a sequence involving: (1) formation of the schistosity, (2) growth of the porphyroblast by replacement, and (3) a second deformation causing distortion of the schistosity around the porphyroblast. The porphyroblast is thus post-tectonic relative to the first deformation and pretectonic relative to the second. An alternative to this is that the porphyroblast may have formed by replacement during the last stages of tectonism.

The whole issue of porphyroblast formation has been opened to serious debate by de Witt (1976), who has developed a wholly new approach (see Figure 21–11) as a result of his investigations of garnet porphyroblasts in deformed metamorphic rocks in Newfoundland. His argument, based upon textural evidence, is as follows: during deformation, fluid-filled veins are opened, particularly in folds associated with crenulation cleavages and kink bands. Garnet nucleates in these veins as they develop. Depending upon the type of deformation, the garnet may form either closely spaced

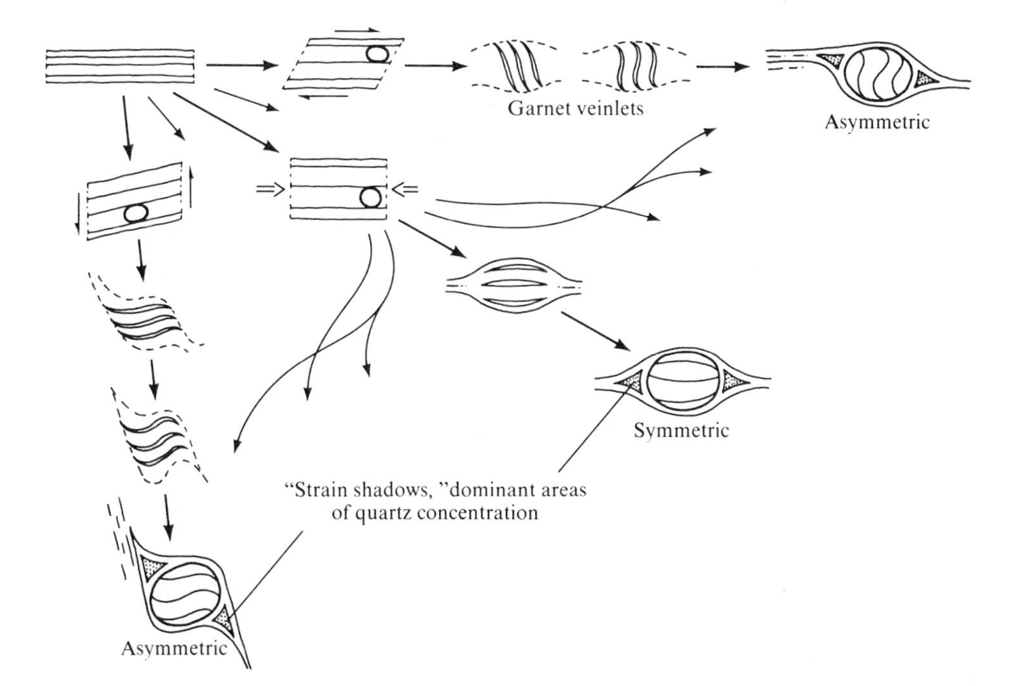

Figure 21–11
An alternative method of accounting for inclusions in porphyroblasts. During shear or uniaxial compression garnet nucleates in S-shaped or straight veins. Continued growth during deformation leads to coalescence of veins to form single porphyroblasts containing variously oriented inclusions. [Modified from M. J. de Witt, 1976, *Geol. Jour., 11,* Fig. 11.]

or widely spaced short veins. Buckling and shear may develop S-shaped veins. Continued growth of garnet leads to a coalescence of the veins to form single porphyroblasts (which may contain variously oriented inclusions). Continued deformation permits the porphyroblasts to be rotated because the units form augenlike textures. If this mechanism is found to have taken place in many rocks, serious revision of current ideas of porphyroblast formation will be necessary.

POLYMETAMORPHISM

Textural and mineralogical evidence often indicates that metamorphic terranes have been subjected to a sequence of metamorphic events involving several periods of deformation or mineral formation. The repetition of metamorphic events is called polymetamorphism.

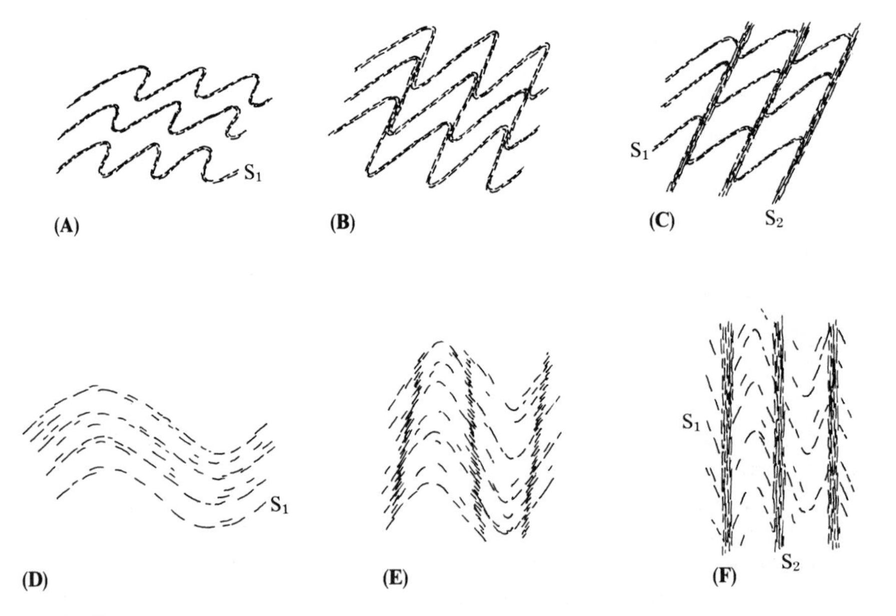

Figure 21–12
The development of a secondary planar foliation. The sequence (**A, B, C**) shows this during asymmetrical folding. The sequence (**D, E, F**) shows a similar development in symmetrical folding. S_1 = first foliation. S_2 = second foliation. [From A. Spry, 1969, *Metamorphic Textures* (Oxford: Pergamon Press), Fig. 62.]

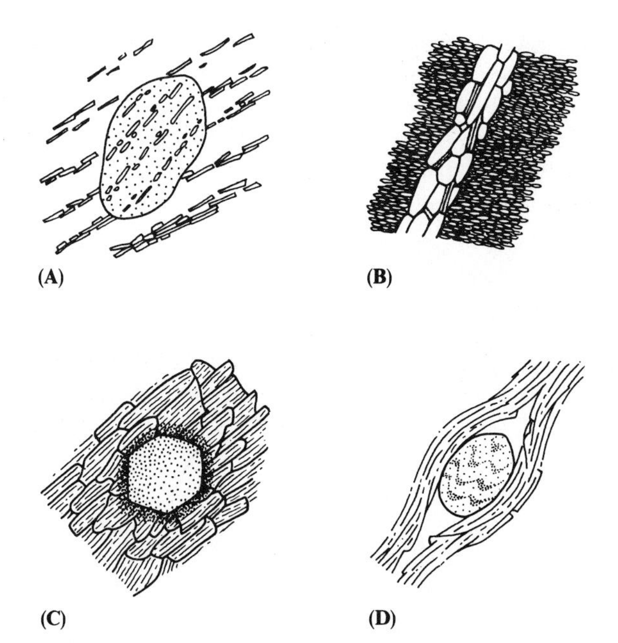

Figure 21–13
Characteristic textures in polymetamorphic rocks. (**A**) Helicitic porphyroblast. S_i and S_e are discordant. (**B**) Layer of coarse mica and quartz cutting a fine foliation. (**C**) Blue-green color in actinolite rim around garnet. The actinolite has formed after the garnet porphyroblast and derived certain constituents from it. (**D**) Helicitic crystal with S-shaped S_i, discontinuous with S_e. [From A. Spry, 1969, *Metamorphic Textures* (Oxford: Pergamon Press), Fig. 63.]

The effects of polymetamorphism may be observed in the groundmass by the development of secondary foliation (see Figure 21–12) that cross-cuts the earlier-formed foliation. With respect to porphyroblasts, if deformation follows porphyroblast formation, textures of the type shown in Figure 21–13 are developed. In these examples it should be noted that S_i does not correspond in direction with S_e, indicating either rotation of the porphyroblast or the development of a new S_e. The development of a polymetamorphic texture is well shown in Figure 21–14. It is a good exercise to examine part C of this sketch first and attempt to unravel the sequence of events before studying the development from A through C. Note that S_0 is used to indicate the original bedding plane, S_1 is the first foliation, S_2 the second foliation, and S_3 the third foliation.

(A)

(B)

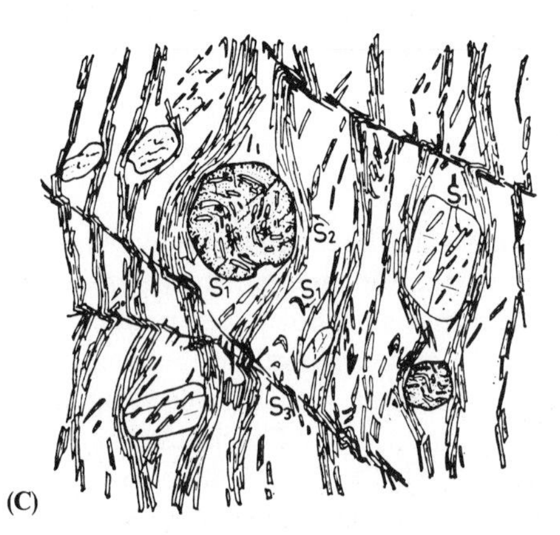

(C)

Figure 21–14
Textural evolution of a polymetamorphic schist. G = garnet, Ab = albite, S_0 = original bedding plane, S_1 = first foliation, S_2 = second foliation, S_3 = third foliation. **(A)** The bedding (S_0) is folded and a foliation (S_1) has formed. Mica, quartz, and garnet are syntectonic, whereas albite is post-tectonic to S_1. **(B)** A second deformation produces S_2. S_1 exists as folded relics and eliminates S_0. S_2 is discordant with S_1 in garnet and albite. Mica and groundmass quartz are syntectonic. **(C)** A third minor deformation folds S_2 and produces strain slip cleavage S_3 without additional crystal growth. [From A. Spry, 1969, *Metamorphic Textures* (Oxford: Pergamon Press), Fig. 64.]

THE DETERMINATION OF PRESSURE AND TEMPERATURE

Use of the Total Assemblage

We have already covered a number of aspects of pressure–temperature determination during discussions of the formation of various metamorphic rocks and minerals. A general approach to determination of *PT* relations has been demonstrated by the use of experimentally determined mineral stability diagrams.

Consider this method for a metapelite containing staurolite, almandine, kyanite, plagioclase, muscovite, and quartz, interlayered with a granite gneiss. The experimentally determined stability curves for various mineral assemblages are plotted on a *PT* diagram in Figure 21–15. Quartz is not included, as its stability range covers

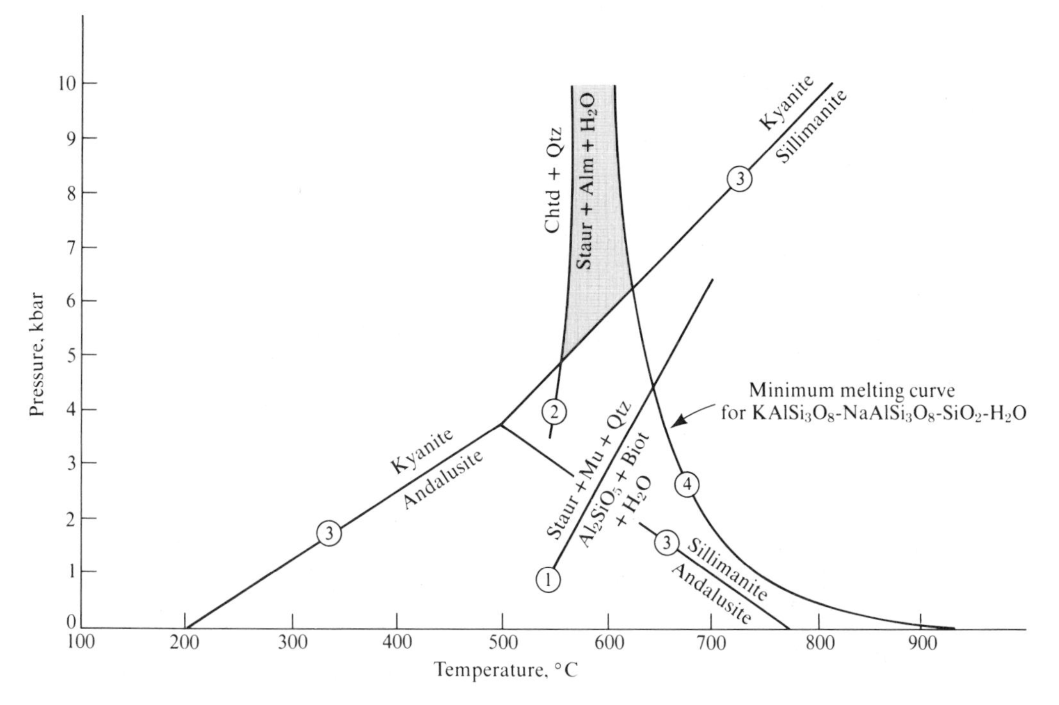

Figure 21–15
The stability limits of various rocks and minerals as a function of *P* and *T*. Chtd = chloritoid, Qtz = quartz, Biot = biotite, Alm = almandine, Mu = muscovite, Staur = staurolite. The shaded area shows the stability region of a rock (interbedded with a granite gneiss) that contains a mixture of staurolite, almandine, kyanite, plagioclase, muscovite, and quartz. [Curve 1 from G. Hoschek, 1969, *Contr. Miner. Petrology, 22,* Fig. 1. Curve 2 from S. W. Richardson, 1968, *Jour. Petrology, 9,* Fig. 5. Curves 3 from M. J. Holdaway, 1971, *Am. Jour. Sci., 271,* Fig. 5. Curve 4 from W. C. Luth, R. H. Jahns, and O. F. Tuttle, 1964, *Jour. Geophys. Res., 69,* Fig. 1.]

most of the diagram. The presence of staurolite with muscovite and quartz restricts the assemblage to temperatures below curve 1. The presence of staurolite with almandine indicates that the rock must have formed at higher temperatures than curve 2. The Al_2SiO_5 polymorph kyanite indicates that the rock formed at pressures higher than those indicated in the curves labeled 3. If water saturation is assumed and no evidence of melting is present, a general limit is set by curve 4 (an anatexis curve); this is a minimum melting curve for mixtures of quartz, albite, and potash feldspar. Under water-deficient conditions this curve is displaced to higher temperatures. Assuming that the various minerals formed together under equilibrium conditions and water saturation was maintained, the shaded area on the diagram indicates the pressure–temperature region within which the rock formed.

This approach demonstrates that moderately close approximations of the conditions of formation of mineral assemblages can be achieved. Notice that this method depends upon the coexistent minerals being in equilibrium with each other at the pressures and temperatures of metamorphism. This must be demonstrated by textures. Note also that rocks of complex mineralogy offer the greatest number of *PT* constraints, and will therefore yield the most precise estimates of pressure and temperature. As more experimental work is done on complex assemblages, this approach will continue to improve.

One of the biggest problems that occurs with this method is that most of the published stability curves of various minerals and mineral assemblages are of pure end-member minerals. However, the minerals within the rock show considerable solid solution, which changes their stability areas. For this reason the method can only yield approximate values. Better results are obtained by thermodynamic calculations.

An example of the use of mineral stability curves in the determination of *PT* conditions in the field is the studies of the metamorphism of the Greek island of Naxos (see Figure 21–16). The western side of the island consists of a granite–granodiorite intrusion; this is separated from the eastern metamorphic portion by an extensive shear zone. A north-central migmatite area can be outlined on the basis of an isograd characterized by the elimination of muscovite. The kyanite and sillimanite isograds are crossed further out from this in lower-temperature areas. The biotite isograd lies to the south of these, and diaspore occurs in the southeastern part of the island. Temperatures and pressures estimated on the basis of *PT* stability curves are shown in Figure 21–17.

Changes in Mineral Composition

In addition to determining *PT* conditions based on the presence or absence of phases, many studies have demonstrated the relationship between mineral composition and metamorphic grade. An example of this is the increase in calcium content of plagioclase with intensity of metamorphism. This is shown in the Swiss Central Alps (see

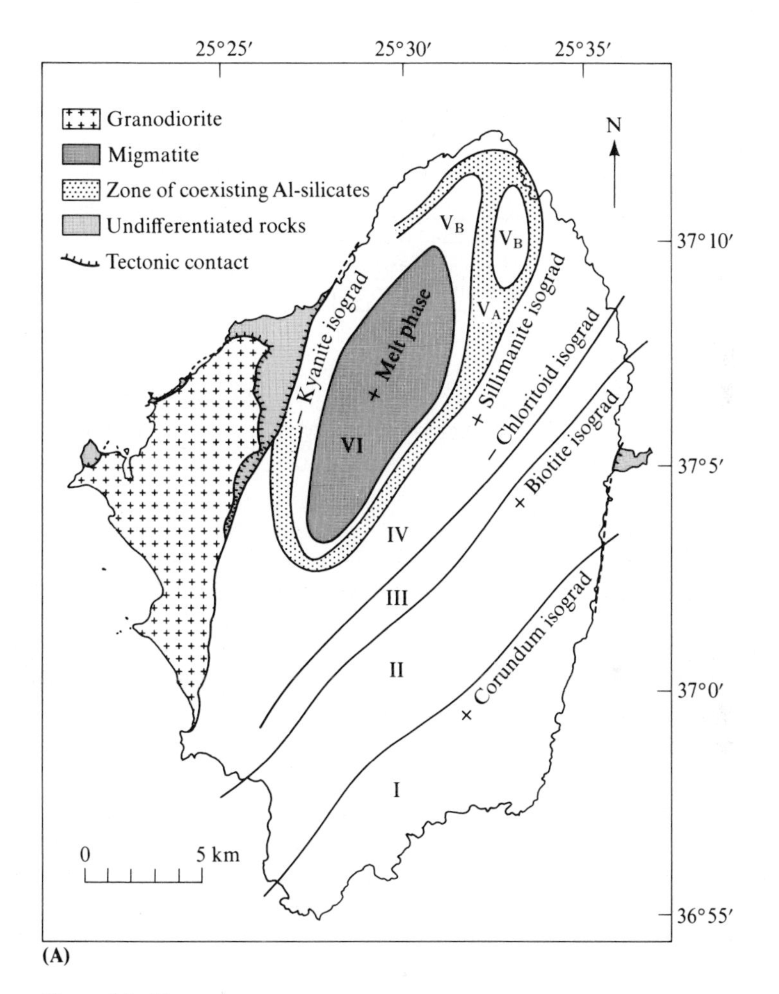

Figure 21-16
(A) Petrologic map of the island of Naxos, Greece. The various metamorphic zones are as follows: I = diaspore, II = chlorite–sericite, III = biotite–chloritoid, IV = kyanite, V_A = kyanite–sillimanite transition, V_B = sillimanite, VI = migmatite. The + or − sign at each isograd indicates the appearance or disappearance of a mineral.

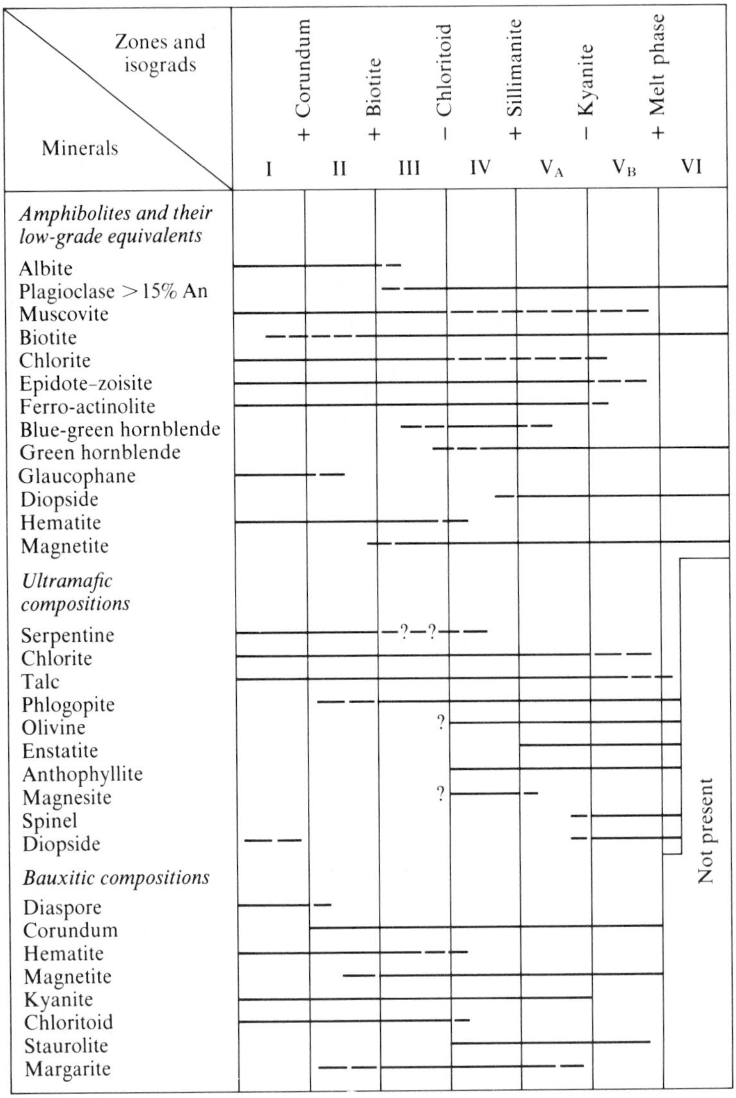

Zones and isograds

Minerals

+ Corundum
+ Biotite
− Chloritoid
+ Sillimanite
− Kyanite
+ Melt phase

I II III IV V_A V_B VI

Amphibolites and their low-grade equivalents

Albite
Plagioclase > 15% An
Muscovite
Biotite
Chlorite
Epidote–zoisite
Ferro-actinolite
Blue-green hornblende
Green hornblende
Glaucophane
Diopside
Hematite
Magnetite

Ultramafic compositions

Serpentine
Chlorite
Talc
Phlogopite
Olivine
Enstatite
Anthophyllite
Magnesite
Spinel
Diopside

Bauxitic compositions

Diaspore
Corundum
Hematite
Magnetite
Kyanite
Chloritoid
Staurolite
Margarite

?
?
?
?

Not present

(B)

Figure 21–16 (*continued*)
(B) The range of occurrence of distinctive minerals in amphibolites, ultramafic compositions, bauxitic compositions, pelites, and carbonate-rich rocks. [After J. B. H. Jansen and R. D. Schuiling, 1976, *Amer. Jour. Sci., 276,* Figs. 3, 4, 5.]

Zones and isograds: I | + Corundum | II | + Biotite | III | − Chloritoid | IV | + Sillimanite | V_A | − Kyanite | V_B | + Melt phase | VI

Minerals	I	II	III	IV	V_A	V_B	VI
Pelitic compositions							
Albite	—	—	- -				
Oligoclase–andesine			-	—	—	—	—
Paragonite	—	— - -	-				
Biotite		-	—	—	—	—	—
Muscovite	—	—	—	—	—	—	-
Chlorite	—	—	— - -	- -	- -		
Epidote–zoisite	—	—	—	—	- -	-	
Kyanite			—	—	—		
Sillimanite					—	—	—
Staurolite				- —	- -		
Chloritoid	—	—	— -				
Garnet	—	—	—	—	—	—	—
K-feldspar					- -	—	—
Carbonate-rich compositions							
Albite	—	—	- -				
Plagioclase >15% An		- -	—	—	—	—	—
Muscovite	—	—	—	- -			
Phlogopite		- -	—	—	—	—	—
Chlorite	—	—	—	- -			
Epidote–zoisite	—	—	—	—	—	- -	
Talc		—	—	— -			
Tremolite			-	- -	—	—	—
Diopside				- -	- -	- - —	—
Scapolite				- -	—	—	—
Grossularite		- -	- -	—	—	—	—
Vesuvianite					- -	—	—
Hematite	—	—	—	-			
Magnetite		-	—	—	—	—	—

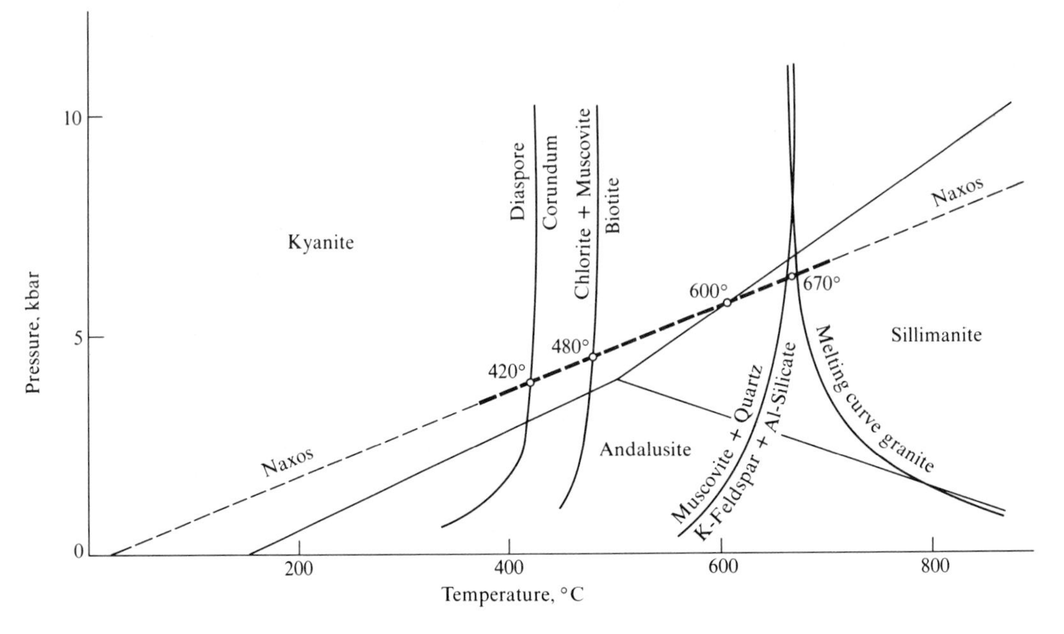

Figure 21–17
Schematic reconstruction of temperature–pressure relations on Naxos, Greece as indicated by some of the observed mineral reactions in Figure 21–16. [From R. D. Schuiling and M. G. Oosterom, 1967, *Proc. Kon. Ned. Akad. Wetensch., B70,* Fig. 2.]

Figure 21–18), where both isograds and An-percent of plagioclase show a similar geographic arrangement. The An-content increases by a series of abrupt changes, indicating a variety of reactions. More typically, plagioclase changes gradually. At the lower grades sodium-rich plagioclase coexists with calcium-rich epidote or zoisite. With increase in the degree of metamorphism, epidote minerals become unstable; the released calcium goes into the plagioclase, causing it to increase in An content.

Analyses of a particular mineral from a wide variety of metamorphic areas often reveal a relationship between composition and grade. Muscovites (see Figure 21–19) show a simultaneous increase in Al_2O_3 and decrease in $FeO + Fe_2O_3$ as grade increases. Metamorphic hornblendes (studied by Raase, 1974) show an increase in Ti with grade; the silica content has been shown to decrease as a function of pressure, while the aluminum in sixfold coordination increases.

Although studies of this type are of significance in showing general compositional changes of minerals with metamorphic conditions, they do not provide precise *PT* data, because they usually deal with a wide variety of rock compositions and in general provide only statistical correlations. More precise methods are discussed in the Appendix.

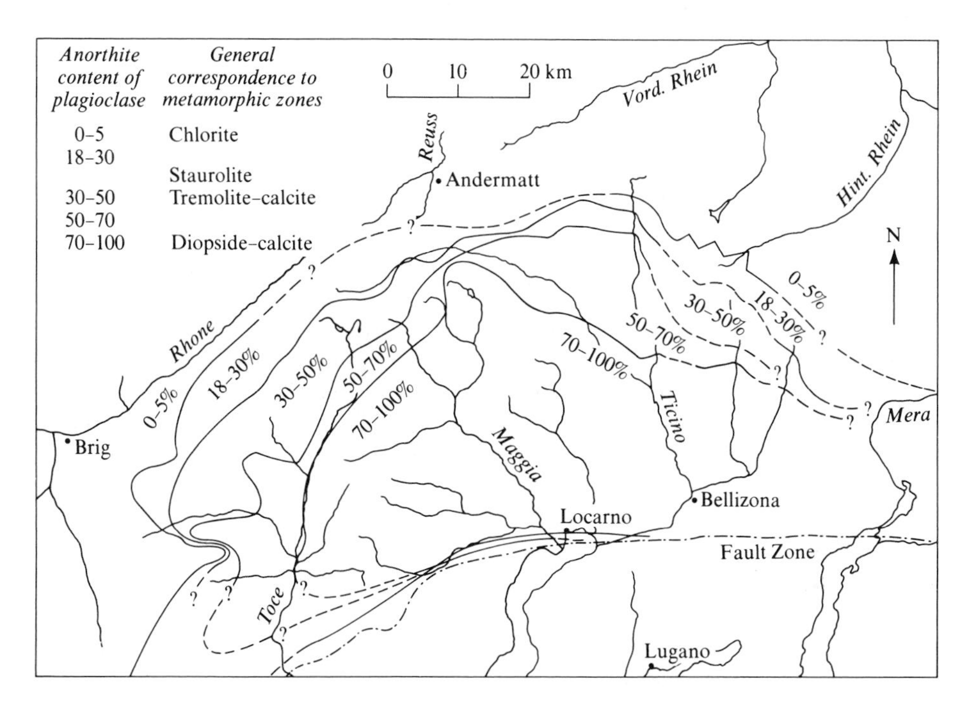

Figure 21–18
Tertiary metamorphism in the Swiss Central Alps. There is a close correspondence between isograds and anorthite percentage in plagioclase. [After E. Wenk, 1962, *Schweiz Min. Petr. Mitt., 42,* Plate 1.]

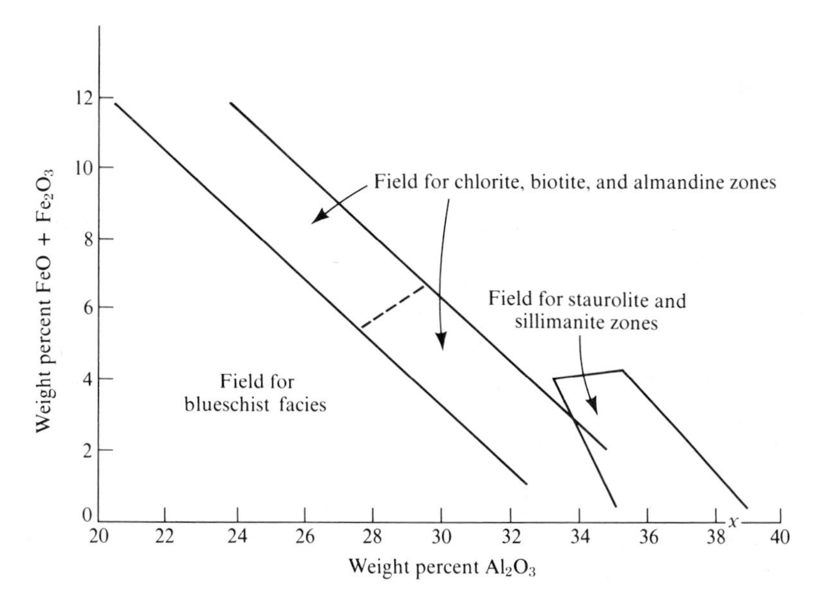

Figure 21–19
The composition of muscovites in metapelites. The x on the abscissa at 38.4% Al_2O_3 represents the composition of ideal muscovite. [From A. Miyashiro, 1973, *Metamorphism and Metamorphic Rocks* (New York: Halsted Press), Fig. 7B-2.]

SUMMARY

The relationship between a porphyroblast and the foliation of the surrounding matrix can in many cases be used to show the relative time of formation of the porphyroblast and the matrix. Inclusions within a porphyroblast may be used to indicate whether the porphyroblast formed before, during, or after a period of tectonism.

The approximate pressure and temperature of formation of a metamorphic rock may often be estimated by comparison of the observed mineral assemblage with experimentally determined limits of similar or simplified experimentally produced assemblages. This approach, although useful, is often inexact if precise mineral compositions are not determined for the rock sample, or if experimentally determined stability limits have been determined only for end-member compositions of phases in which extensive solid solution occurs.

FURTHER READING

de Witt, M. J. 1976. Metamorphic textures and deformation: a new mechanism for the development of syntectonic porphyroblasts and its implication for interpreting timing relationships in metamorphic rocks. *Geol. Jour., 11,* 71–100.

Ferguson, C. C. 1980. Displacement of inert mineral grains by growing porphyroblasts: a volume balance constraint. *Geol. Soc. Amer. Bull., 91,* 541–544.

Jansen, J. B. H., and R. D. Schuiling. 1976. Metamorphism on Naxos: petrology and geothermal gradients. *Amer. Jour. Sci., 276,* 1225–1253.

Misch, P. 1971. Porphyroblasts and crystallization force: some textural criteria. *Geol. Soc. Amer. Bull., 82,* 245–252.

Miyashiro, A. 1973. *Metamorphism and Metamorphic Belts.* New York: Halsted Press, 492 pp.

Raase, P. 1974. Al and Ti contents of hornblende, indicators of *P* and *T* of regional metamorphism. *Contr. Miner. Petrology, 45,* 231–237.

Ramberg, H. 1947. Force of crystallization as a well-definable property of crystals. *Geol. Fören. i. Stockh. Förh., 69,* 189–194.

Schuiling, R. D., and M. G. Oosterom. 1967. The metamorphic complex on Naxos (Greece) and the strontium and barium content of its carbonate rocks. *Proc. Kon. Ned. Akad. Wetensch., B70,* 165–175.

Schuiling, R. D., and H. Wensink. 1962. Porphyroblastic and poikioblastic textures. The growth of large crystals in a solid medium. *Neues Jahrb. Miner.,* 11/12, 247–254.

Spry, A. 1969. *Metamorphic Textures.* New York: Pergamon Press, 350 pp.

Spry, A. 1972. Porphyroblasts and "Crystallization Force": some textural criteria: discussion. *Geol. Soc. Amer. Bull., 83,* 1201–1202.

Zwart, H. J. 1962. On the determination of polymetamorphic mineral associations and its application of the Bosost area (central Pyrenees). *Geol. Rund., 52,* 38–65.

22

Metamorphic Rocks and Global Tectonics

The major development of metamorphic rocks occurs near the boundaries of lithospheric plates. Less abundant metamorphism occurs as a result of deep burial in sedimentary basins and thermal effects directly adjacent to igneous intrusions.

METAMORPHISM AT TRANSFORM FAULTS AND DIVERGENT JUNCTIONS

Lithospheric plates are in contact at transform faults and at divergent and convergent junctions. Movement at transform (strike slip) faults is generally not accompanied by magmatic activity. Consequently the type of metamorphic rock produced is mainly cataclastic. Studies of the San Andreas Fault in California and the Alpine fault in New Zealand indicate the presence of zones of cataclastic rocks that may be up to several kilometers in width. Cataclastic rocks have also been found in dredge samples at transform faults that cut across the Mid-Atlantic Ridge. To date most of these cataclastic rocks have not been studied extensively.

Metamorphism at divergent junctions (essentially sea-floor metamorphism) was presumed to be present before the rocks themselves were obtained and studied. This presumption was based on the results of geophysical measurements. The linear strips of magnetic anomalies that parallel the mid-ocean ridges are known to be the result of remnant magnetism of only an upper layer of basaltic rocks whose maximum thickness is about 2 km. Sheeted dikes beneath have undergone hydrothermal alteration that has left them weakly magnetized. Deeper intrusions retain an induced magnetization that doesn't contribute to the anomalies. The remnant magnetization of the intrusions, even when unaltered, is generally weak.

We have seen earlier that sea-floor metamorphism has been studied by an examination of dredge samples as well as detailed investigation of metamorphism of rocks of the ophiolite suite, which are now generally thought to represent obducted slices of oceanic crust and mantle. Rocks from these sources are of basaltic or ultramafic composition and are metamorphosed to the levels of the zeolite, greenschist, epidote–amphibolite, or amphibolite facies. Much of this metamorphism probably was initiated in the general vicinity of the spreading center. With cooling and transport from the vicinity of the ridge, reaction with sea water through fractures occurred, producing a variety of both prograde and retrograde reactions. Following this the process of obduction resulted in additional metamorphism, resulting in considerable deformation near the base of the ophiolite sequence.

METAMORPHISM AT CONVERGENT JUNCTIONS

The major source of our information about metamorphism related to plate tectonics is found at convergent junctions. Metamorphic rocks are found in wide belts that may extend for hundreds of kilometers. Major belts are found worldwide in such diverse locations as the Ural Mountains of the Soviet Union, the Appalachians of the United States, western Europe (through Scotland and Norway), and many areas surrounding the Pacific Ocean.

Miyashiro published a paper in 1961 in which he emphasized two major ideas. The first of these was that metamorphic belts could be characterized as to baric type—that is, low, medium, and high pressure (see Chapter 20). From this he developed the idea of a metamorphic *facies series*—an association of several metamorphic facies whose mineralogy reflects a particular gradient of pressure and temperature. Carrying this concept forward by examination of the petrologic literature, the second major idea evolved—that of paired belts. Miyashiro noted that in many parts of the world, and particularly around the Pacific basin, metamorphic belts often consist of two more or less parallel parts, one of which is a high-pressure type and the other a low-pressure type. Medium-pressure types are also found within the high- and low-pressure belts. Miyashiro called this combination of different baric types *paired belts*.

The location of paired belts around the Pacific Ocean is given in Figure 22–1. Notice that in almost every case the paired belts are more or less parallel to the ocean circumference, with a high-pressure facies series facing the ocean, and a low-pressure facies series on the landward side. The distribution of these metamorphic belts is considered to be due to subduction beneath island arcs and continental margins.

Subduction of a cool oceanic plate beneath an island arc or continental edge results in a local depression of isotherms, with the result that a low geothermal gradient ($< 10°C/km$) is present near the oceanic trench. This geothermal gradient, created by a rapid rate of descent, is sufficient to produce high-pressure metamorphism when combined with the high pressures involved in plate convergence. At greater depths within the subduction zone, melting in the mantle above the descending slab

Figure 22–1
Paired metamorphic belts around the Pacific Ocean. Broken lines represent high-pressure belts. Full lines represent low-pressure belts. [From A. Miyashiro, 1973, *Tectonophysics, 17,* Fig. 1.]

and possibly in the slab itself results from the release of volatiles from the slab and from the generation of frictional heat. This hot material moving upward results in a large vertical transfer of heat, producing a geothermal gradient that may be greater than 25°C/km. Vertical movement of magma to shallow depths produces high-temperature–low-pressure metamorphism in the continental crust. The result of the subduction process is often a high-pressure low-temperature metamorphic belt in the vicinity of the trench; farther inland, above deeper portions of the subduction zone, a low-pressure–high-temperature metamorphic belt is developed. When the rate of underthrusting is slow, the geothermal gradient may be too great to permit formation of the low-temperature–high-pressure belt, and the paired nature of the belt is obscured.

We discussed the origin and distribution of igneous rocks relative to subduction zones in Chapter 8. It is now obvious that if both igneous and metamorphic rocks can occur in some definite relationship to subduction zones, their occurrences should also be related to each other. This is in fact the case. Ophiolites, mafic, and ultramafic rocks often are associated with high pressure metamorphic rocks in the vicinity of oceanic trenches. Farther inland, above deeper portions of the subducting plate, it is common to find low-pressure metamorphic rocks associated with granitic plutonism and andesitic volcanism.

In modern island arc regions above subduction zones it is usual to find a nonvolcanic, relatively undisturbed region (about 100 to 250 km wide) between the high-pressure metamorphic rocks near the oceanic trench and the low-pressure metamorphic rocks associated with the volcanic belt. This has been called the arc-trench gap. This gap may be present as either uplifted mountains or a trough in which sedimentation occurs. Such a separation of paired belts is found in the western United States between the high-pressure Franciscan coastal rocks and the Sierra Nevada igneous–metamorphic complex farther inland. However in some older metamorphic belts, the high- and low-pressure metamorphic pair may be adjacent to each other. In such cases it is common to find a major fault located at a slight angle to the tectonic trend. Strike-slip movement of several hundred kilometers along such a fault could bring the two metamorphic belts into contact.

Time of Formation of Paired Belts

Examination of the occurrence of baric varieties of metamorphic rocks through time has brought out the fact that low- and medium-pressure metamorphic rocks are more or less evenly distributed throughout the geological record. High-pressure metamorphic rocks are, however, essentially limited to Phanerozoic metamorphic belts, and with increasing abundance in younger rocks (see Figure 22–2). This has naturally led to all sorts of interesting speculation.

One idea is that continents grow by seaward accretion; through time the paired belts move seaward, enlarging the areas of high-temperature–low-pressure metamorphic rocks, while maintaining the high-pressure types along a thin, easily eroded active edge. Although this hypothesis sounds fairly reasonable, two other major hypotheses are currently in vogue. One of these, alluded to above, emphasizes the rate of plate descent. Rapid plate descent favors the formation of high-pressure metamorphic belts. It has been suggested that the more rapid plate motion now observed in the Pacific region (2–6 cm/yr at the East Pacific Rise) as compared to the slower Atlantic rate (1–2 cm/yr at the Mid-Atlantic Ridge) may have existed since the late Paleozoic or early Mesozoic. Older plate motion in the Pacific was moderately rapid in the Paleozoic, with some tendency to produce belts with some high-pressure characteristics; in the Mesozoic, rates increased sufficiently to produce glaucophane-containing belts. Convergence in the Atlantic, which produced the Appalachian orogenic belt in North America and the Caledonian orogenic belt in Europe, was too

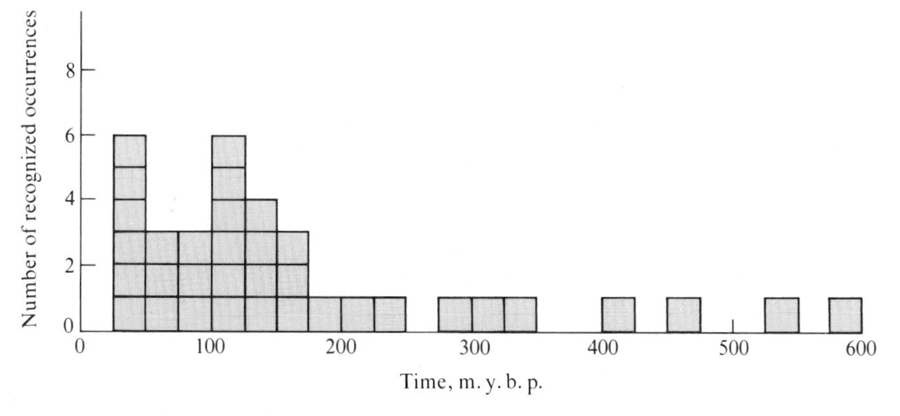

Figure 22–2
Histogram showing the incidence of the blueschist facies through time. [Modified from W. G. Ernst, 1972, *Amer. Jour. Sci., 272,* Fig. 1.]

slow to produce high-pressure metamorphic rocks except for a limited area in the southern highlands of Scotland.

Occasionally, underthrusting of an oceanic plate at a continental edge or island arc results in the subduction of a rift zone. It is thought by some investigators (Miyashiro, 1973) that prior to the descent of the rift zone the subduction rate may be particularly fast. High-pressure metamorphic belts are formed until the rift zone begins to descend; descent of the rift zone results in a high geothermal gradient, which ends high-pressure metamorphism. This situation may have occurred in both the western United States and in Japan. In the western United States this has resulted in the formation of the Franciscan high-pressure metamorphic belt, and in Japan, the Sanbagawa belt (which is located in the southeastern portion of the main island of Honshu). A comparison of these events is shown in Figure 22–3.

The second major argument relating to the abundance of high-pressure rocks in Phanerozoic time maintains that plate velocities are probably the same now as in the past; however, it is hypothesized that the earth's geothermal gradient has decreased through time as the result of the gradual depletion of elements capable of producing heat through radioactive decay. If this is correct, it means that the temperature at a particular depth in the earth was hotter in the past than it is at present.

Recall that at the base of the lithosphere is found the asthenosphere—a zone where the geothermal gradient intersects the region of incipient melting. With a lowering geothermal gradient through time, the incipient melting zone will be encountered at greater depths; that is, the lithosphere is thickening through time (see Figure 22–4). According to this argument, Precambrian lithospheric plates were thinner than later

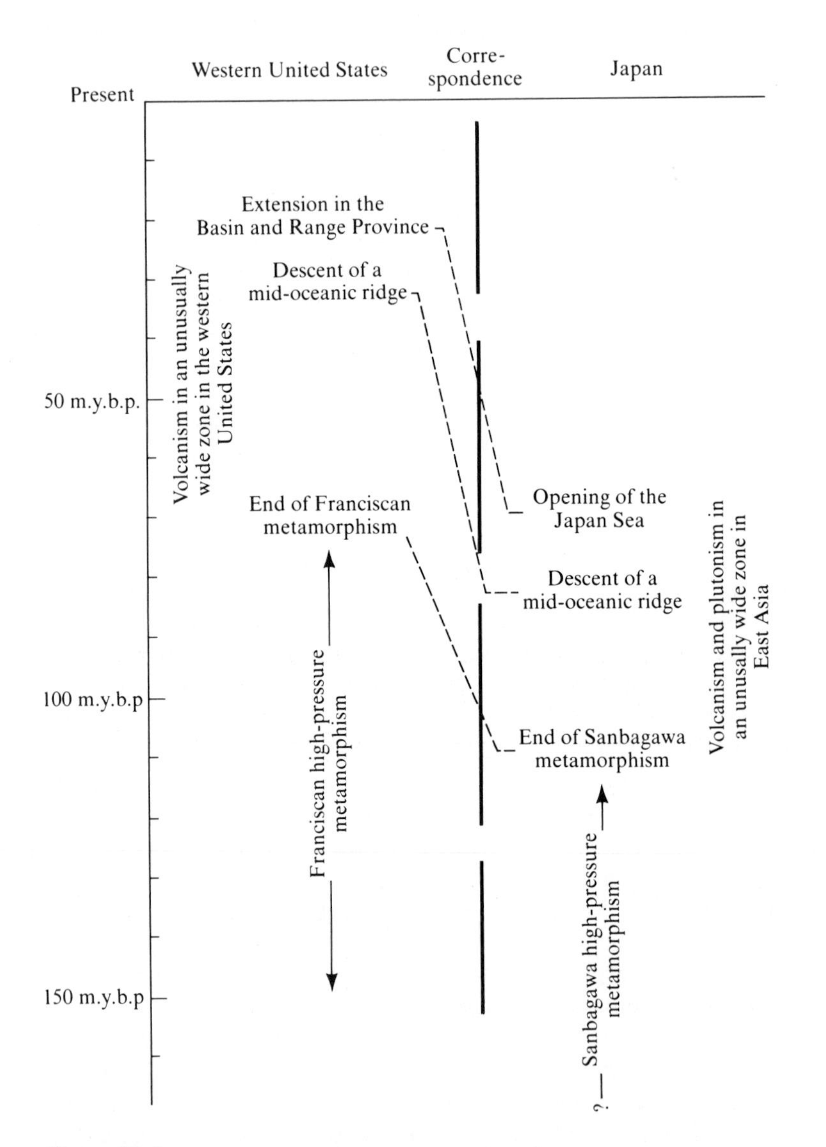

Figure 22–3
Comparison of geologic events in the western United States and Japan. In both cases mid-ocean ridges were subducted beneath continental plates, leading to termination of high-pressure metamorphism. [From A. Miyashiro, 1973, *Tectonophysics, 17,* Fig. 4.]

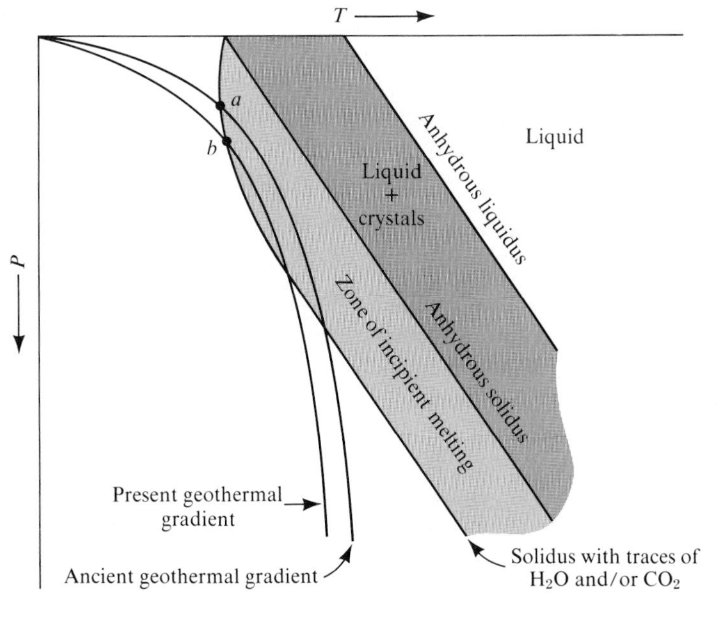

Figure 22–4
Idealized sketch showing the effects of changing geothermal gradients on the depth of the asthenosphere. The ancient (hotter) geothermal gradient encounters the zone of incipient melting at a depth (pressure) such as *a*. The present cooler geothermal gradient encounters the zone of incipient melting at *b*. As *b* is deeper than its equivalent *a*, the asthenosphere thickens through time.

ones; the higher geothermal gradients minimized the formation of the high-pressure facies. In thicker Phanerozoic plates, geothermal gradients are lower and high-pressure facies are favored. The presence or absence of high-pressure facies thus gives us evidence relating to the evolution of the earth.

SUMMARY

Metamorphic rocks are created most commonly as a result of divergence and convergence of lithospheric plates. Those metamorphic rocks formed as a result of rifting are limited in outcrop to obducted oceanic crust and mantle (metamorphosed ophiolites), and are present mainly beneath the ocean floor. Convergent plate junctions result in broad belts of metamorphic rocks. These orogenic regions (particularly around the Pacific Ocean) commonly contain parallel paired belts of metamorphic rocks, which are characterized by high-pressure–low-temperature mineral assemblages near the ocean trench, and low-pressure–high-temperature mineral assemblages above deeper portions of the subduction zone.

Low-pressure metamorphic belts are common throughout the entire geological record, whereas high-pressure metamorphic belts decrease in frequency of occurrence in older rocks. This probably is due to a change in either the geothermal gradient or the rate of plate movement through time.

FURTHER READING

Ernst, W. G. 1972. Occurrence and mineralogic evolution of blueschist belts with time. *Amer. Jour. Sci., 272,* 657–668.

Miyashiro, A. 1961. Evolution of metamorphic belts. *Jour. Petrology, 2,* 277–311.

Miyashiro, A. 1972. Metamorphism and related magmatism in plate tectonics. *Amer. Jour. Sci., 272,* 629–656.

Miyashiro, A. 1973. Paired and unpaired metamorphic belts. *Tectonophysics, 17,* 241–254.

Appendix:

Precise Determination of the Pressure, Temperature, and Age of Rocks

We have seen that the examination of mineralogical and textural features of igneous, sedimentary, and metamorphic rocks provides us with the general conditions under which these rocks have formed. This has been accomplished by use of the doctrine of uniformitarianism and the duplication of rock features in the laboratory. These approaches, while furnishing general information about the conditions of rock formation, usually do not provide specific information on temperature, pressure, and time of formation.

As an example, consider estimating the crystallization temperature of an igneous rock based upon the type of minerals present. The presence of sanidine indicates that the magma has crystallized at high temperature within the experimentally determined stability field of this phase. Consideration of the total mineral assemblage further delineates crystallization temperatures; simultaneous crystallization of a number of phases occurs in a temperature region where the stability fields of the minerals overlap. An exception to this is the relatively uncommon situation where a phase may crystallize or persist metastably.

Exact estimates of crystallization temperatures by the above approach are not successful for a variety of reasons. Recall that crystallization of a magma occurs over a range of temperatures. Crystallization usually begins with the formation of a single phase; with decreasing temperatures and change of melt composition, the stability field of a second phase is encountered. Simultaneous crystallization occurs over a temperature range as additional phases begin precipitation. The temperature at which all of the major phases crystallize is usually not a constant, as this crystallization temperature varies in response to changing composition of the magma, as a result of the changes in composition of the various precipitating phases (for example, the Na/Ca ratio in plagioclase), as well as to changes in the type and amount of pressure in the system. Additional complications result from changes in mineralogy and mineral composition as a result of deuteric (secondary) alteration or later recrystallization. As a result of such complexities, modern igneous geothermometry has concentrated mainly on a number of techniques that involve the attainment of a compositional equilibrium between pairs of coexisting minerals formed during relatively late stages of crystallization. Thus the final conditions of crystallization of the rock are determined by a knowledge of the crystallization conditions of certain specific minerals. The total crystallization history is obtained from both textural interpretations and a general knowledge of phase relationships.

THE TWO-FELDSPAR GEOTHERMOMETER

The two-feldspar geothermometer was first proposed by T. F. W. Barth (Norwegian) in 1934. This has the potential of being a very useful method, because all the user must do in order to obtain the crystallization temperatures of an igneous rock is determine the composition of the feldspars and provide from field evidence an approximate depth at which crystallization of the feldspars occurred. The approximate compositions of feldspars are determined routinely in most petrographic laboratories by optical and X-ray diffraction methods. Electron probe microanalysis is necessary for precise determination. Depth of intrusion is often estimated from contact effects and estimates of overburden removed. Problems are encountered if the feldspars have crystallized over a temperature range, have reequilibrated with decreasing temperature, or are difficult to separate and analyze.

The method relies on the phase relationships in the system $CaAl_2Si_2O_8$–$NaAlSi_3O_8$–$KAlSi_3O_8$–H_2O. In our earlier discussions of crystallization of magmas we noted that the feldspars at high temperatures showed complete solid solution for plagioclase compositions (albite to anorthite) and alkali feldspar compositions (albite to potassium feldspar). A melt whose composition places it within one of these solid solution regions will cool to crystallize a single feldspar (see Figure A–1); a melt of composition *a* will crystallize to a plagioclase of composition *a*. Melts located out of the solid solution regions will cool to crystallize two feldspars whose compositions together equal that of the melt. Thus melt *b* might cool to crystallize feldspars *c* and *d* in proportions equal to the composition of the original melt *b*. Note that *b*, *c*, and *d* must all lie on a straight line in order to make things add up correctly.

It would seem to follow from the above approach that for every feldspar composition there is another feldspar whose composition is also fixed—that is, if we have plagioclase *c* and a second feldspar, the composition of the second feldspar is always *d*. Of if we have feldspar *d* a second feldspar is always *c*. This would simplify our petrographic analysis, as determination of one feldspar would simultaneously reveal the composition of the second.

In the real world, of course, things don't work out so simply and pleasantly. It turns out that the compositions of feldspars that have crystallized together are not con-

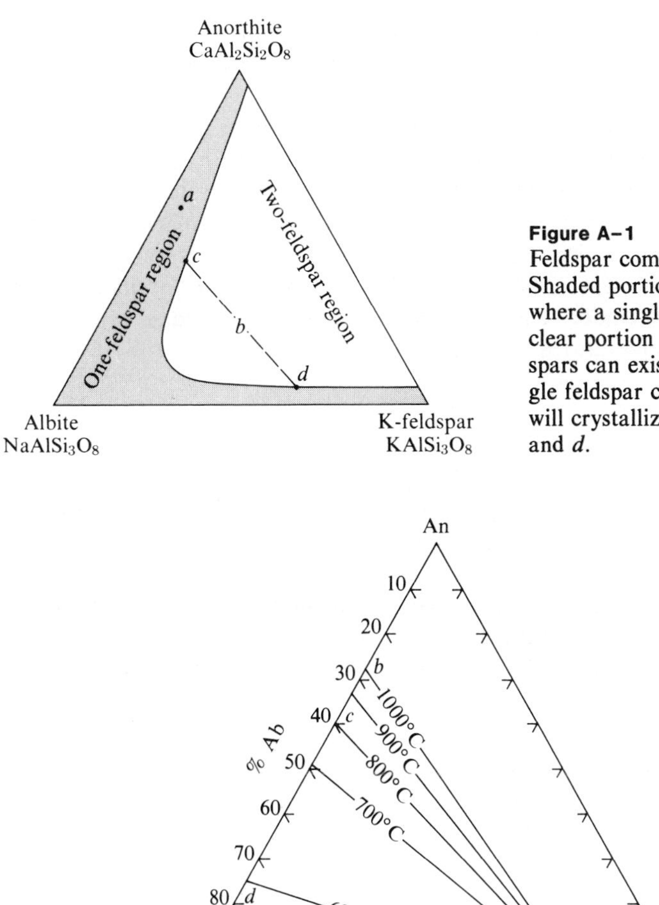

Figure A–1
Feldspar compositional diagram. Shaded portion is the general area where a single feldspar may form. The clear portion is an area where two feldspars can exist. Point *a* represents a single feldspar composition. Composition *b* will crystallize two feldspars such as *c* and *d*.

Figure A–2
An alkali feldspar *a* crystallizing at 1,000 bar pressure will coexist with various plagioclase compositions (such as *b*, *c*, or *d*) as a function of the temperature of crystallization. [Data from J. C. Stormer, 1975, *Amer. Miner., 60,* 667–674.]

stant. An alkali feldspar such as *a* (see Figure A–2) may be found in different igneous rocks containing plagioclase with a variety of different Ca/Na ratios (such as *b*, *c*, *d*, and so on). This wasn't a big shock to anyone, as the same kinds of relationships had been found in other coexistent mineral pairs. It occurs because the partitioning of elements between coexistent minerals varies as a function of the temperatures and pressures at which they formed.

Knowing that such variability exists, petrologists have studied the freezing relationships of feldspar pairs at various combinations of pressure and temperature. It is now possible to deduce feldspar crystallization temperatures by using these data. This is shown in Figure A–2. The alkali feldspar contains K to Na in an 8:2 ratio. The diagram shows the relation between the composition of the coexisting plagioclase and the temperature of formation (at 1,000 bar). Figure A–3 demonstrates another way to look at these data. Here we have assumed that a pair of feldspars has a bulk composition shown at point α. The composition of coexisting feldspar pairs is shown as a function of temperature (at 1,000 bar). Details of the method and its uses are given by Stormer (1975). Under the proper conditions, the method is capable of yielding crystallization temperatures with an accuracy of 30°C.

THE IRON–TITANIUM OXIDE GEOTHERMOMETER

The iron–titanium oxide method is based upon the very common occurrence of minor amounts of coexistent magnetite and ilmenite in most igneous and metamorphic rocks. This is visible in reflected light microscopy as fine lamellae of magnetite alternating with ilmenite, $FeTiO_3$ (see Figure A–4). It is known that at high temperatures magnetite (Fe_3O_4) is able to retain titanium in solid solution. With cooling, both oxidation and exsolution of titanium occur, with the formation of an

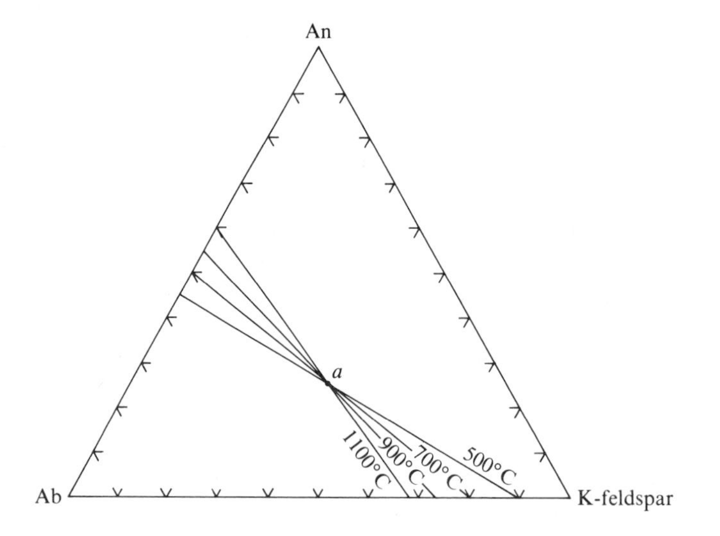

Figure A–3
An intermediate composition *a* will crystallize (at 1,000 bar) to form various plagioclase–alkali feldspar pairs as a function of the temperature of crystallization. [Data after J. C. Stormer, 1975, *Amer. Miner., 60,* 667–674.]

Figure A–4
Titanomagnetite (gray-white) showing coarse (dark gray) exsolution lamellae of ilmenite, as seen in reflected light on a polished surface. Width of photo is about 0.5 mm. [From P. Ramdohr, 1969, *The Ore Minerals and Their Intergrowths* (New York: Pergamon Press), Fig. 531.]

ilmenite–hematite ($FeTiO_3$–Fe_2O_3) phase. The compositions of the resultant magnetite and ilmenite–hematite are dependent upon both temperature and the oxygen pressure (fugacity) in the environment. Composition of these phases normally is determined using the electron microprobe; various additional factors must be taken into account based upon the complexities of the rock composition. In spite of these complexities the method is worthwhile, as it furnishes an accuracy of at least ± 30°C as well as providing a fairly accurate estimate of the oxygen pressure of the rock system.

THE PYROXENE GEOTHERMOMETER–GEOBAROMETER

The pyroxene geothermometer–geobarometer has been widely used to furnish pressures and temperatures of origin of rocks having both crustal and mantle origin. Most of the estimates of depth and temperature of origin of mantle rocks described earlier utilize this method.

The basis of the method is that pyroxenes change their limits of solid solution of Ca and Al as a function of temperature and pressure. We have seen earlier (see Chapter 6 and Figure 6–5) that pyroxenes derived from high-pressure environments are rich in Al. Knowing that such differences in chemical substitutions existed, petrologists have experimentally synthesized both orthorhombic and monoclinic pyroxenes in recent years at various pressures and temperatures in order to determine the variability of solid solution. For the best information on both pressure and temperature of origin, detailed analyses should be made of coexisting orthorhombic and monoclinic pyroxenes. In addition to the pyroxenes, it is necessary for the rock to possess other Ca- and Al-containing phases in order to be assured that the pyroxenes are at their maximum limits of solid solution.

MINERAL PAIRS IN METAMORPHIC ROCKS

Precise determination of the pressure–temperature conditions of metamorphism is being attempted by an examination of partitioning of elements between mineral pairs. The reaction

$$\text{Cordierite} \rightleftharpoons \text{Almandine} + \text{Sillimanite} + \text{Quartz}$$

is such a case. A number of rocks exist that contain these four minerals (± biotite, ± potassium feldspar) in apparent equilibrium. Both cordierite and almandine contain Fe and Mg. It has been noted that the distribution of Fe and Mg in these minerals differs in different rocks. Syntheses of these phases with different Fe/(Fe + Mg) ratios have been performed at a wide variety of pressures and temperatures, and the generalities of the Fe and Mg distribution between cordierite and garnet are known. An example of this is given in Figure A–5A. Garnet and cordierite were synthesized at various temperatures. The bulk Fe/(Fe + Mg) ratio was kept at 0.6 and the pressure at 6 kbar. It can be seen that with increasing temperature the garnet becomes more Fe-rich as the cordierite becomes more Fe-poor. The same information plotted on an AFM diagram (see Figure A–5B) shows this as a pivoting tie line. Thus experimental studies indicate that the iron/magnesium ratio in coexistent garnet and cordierite varies as a function of pressure, bulk composition, and temperature. This relationship serves as both a geothermometer and a geobarometer.

Although this approach offers considerable promise and has produced *PT* information that is consistent with other methods, the effects of other components, such as Mn and Ca, are not yet well known. In addition, the effect of changes in P_{H_2O} on element partitioning has yet to be determined.

Similar approaches consider the Mg/Fe^{2+} ratios between garnets and biotites, orthorhombic and monoclinic pyroxenes, and garnets and clinopyroxenes. Additional information can be obtained in a special issue of the *American Mineralogist* (July–August 1976) devoted to geothermometry and geobarometry.

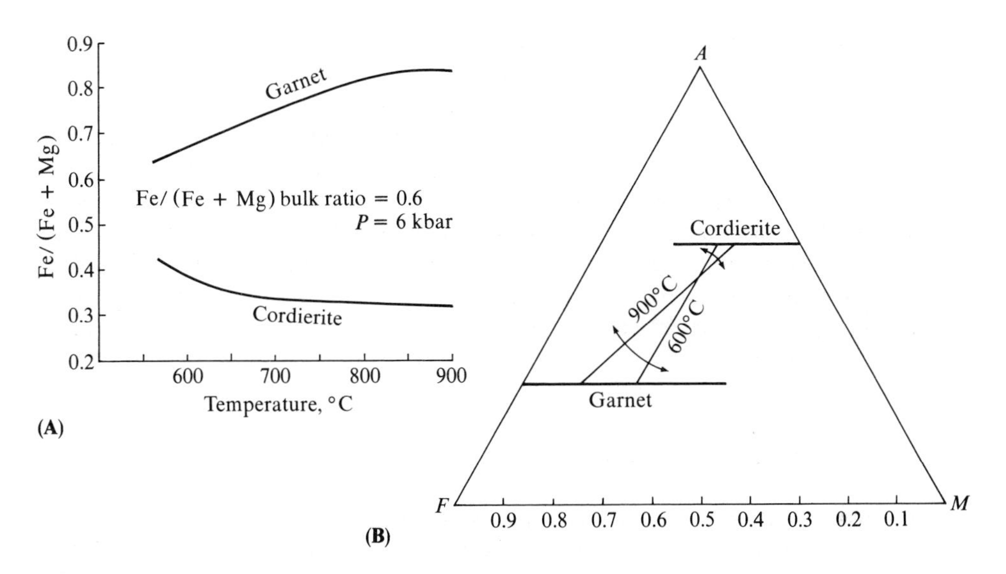

Figure A-5
(**A**) Changing compositions of coexisting garnet and cordierite at 6 kbar as a function of temperature. With increasing temperature garnet becomes more iron-rich and cordierite more iron-poor. (**B**) Coexisting compositions at 600°C and 900°C are shown plotted in terms of an *AFM* diagram. [Data taken from Currie, 1971, *Contr. Miner. Petrology, 33,* 215–226.]

THE USES OF ISOTOPES

Isotopic studies have been used to investigate a number of aspects of igneous, metamorphic, and sedimentary petrogenesis. A variety of radioactive schemes has been used to date the time of crystallization of igneous rocks, as well as the time at which metamorphism took place. Other studies have indicated the temperatures at which igneous and metamorphic rocks formed. Isotope studies have also been used to determine the site of origin of magmatic liquids. A recent text on various methods and applications is that by Faure (1977).

Age Determination Using Radioactivity

Many of the isotopic dating procedures depend upon determination of the relative amounts of daughter isotopes produced by the decay of a radioactive parent. The rubidium–strontium method will be used as an example of this technique. Rb^{87} decays (with a half-life of $5.0 \pm 0.2 \times 10^{10}$ years) to produce Sr^{87}. At the time of crystallization of a magma it can be assumed that some Sr^{87} exists within the melt that was produced by earlier decay of Rb^{87}. As a magma can be considered to be a homogeneous material, it can also be assumed that the Sr^{87} present is evenly dispersed throughout the melt. After cooling and crystallization of the magma, it therefore follows that the ratio of radiogenically produced Sr^{87} to that of any nonradiogenic Sr^{86} will be a constant value in all of the minerals present.

Rb, which has an ionic radius and charge similar to that of potassium, will tend to be concentrated in K-rich minerals during the crystallization process, whereas Sr, because of an ionic radius and charge similar to Ca, will tend to be concentrated in Ca-rich minerals. The ratio of Rb^{87} to Sr^{86} will therefore be different for each mineral present. The situation for Sr^{87}/Sr^{86} as compared to Rb^{87}/Sr^{86} at the time of crystallization is shown in Figure A–6A. The minerals (m) and the average composition for the whole rock (R) fall on a horizontal line that indicates an identical composition in terms of Sr^{87}/Sr^{86}.

After completion of crystallization (and assuming that the rock system remains closed to migration of Sr and Rb), continuation of radioactive decay of Rb^{87} to produce Sr^{87} results in a change of isotopic composition of both the minerals, and the rock as a whole. As the amount of Sr^{87} produced must equal the amount of Rb^{87} lost, this results in compositional changes shown in Figure A–6B as arrows inclined at 45° to the coordinate system. The inclined line joining the composition points is called an isochron. Determination of its slope (by means of isotopic analyses using a mass spectrometer) reveals the age of the rock (assuming that no metamorphism has occurred between the time of formation of the rock and the present).

If metamorphism has occurred between the time of formation of the rock and the present, the technique permits the age of metamorphism to be determined. Assume now that Figure A–6B shows the situation that existed when metamorphism occurred. Metamorphism permits mobilization of many of the atoms in the rock; redistribution of Sr occurs, such that the Sr^{87}/Sr^{86} ratio once again becomes equalized in all of the minerals in the rock. As metamorphism does not affect the Sr^{87}/Sr^{86} ratio in the whole rock (as a closed system is assumed), the total effect is to pivot the isochron about the whole rock composition (R) as shown in Figure A–6C. At the completion of metamorphism the isochron is once again horizontal. During the interval between the period of metamorphism and the present, radioactive decay continues, with the development of a new inclined isochron (see Figure A–6D). The slope of this isochron, as determined in the laboratory, dates the metamorphic event.

We have seen that this technique permits the dating of (1) the time of crystallization of an unmetamorphosed igneous rock, or (2) the time at which a metamorphic event occurred. It is also possible to date both the time of crystallization and a later metamorphic event. What is required is a group of igneous rocks that are comagmatic—that is, a group of igneous rocks of diverse composition that originated by fractionation of a common parent magma. If these rocks crystallized at essentially the same time and have therefore maintained an identical Sr^{87}/Sr^{86} ratio, their whole rock compositions (R_1, R_2, R_3) at the time of crystallization will plot as a horizontal line (see Figure A–7A)—similar to the various minerals within a single rock. Assume

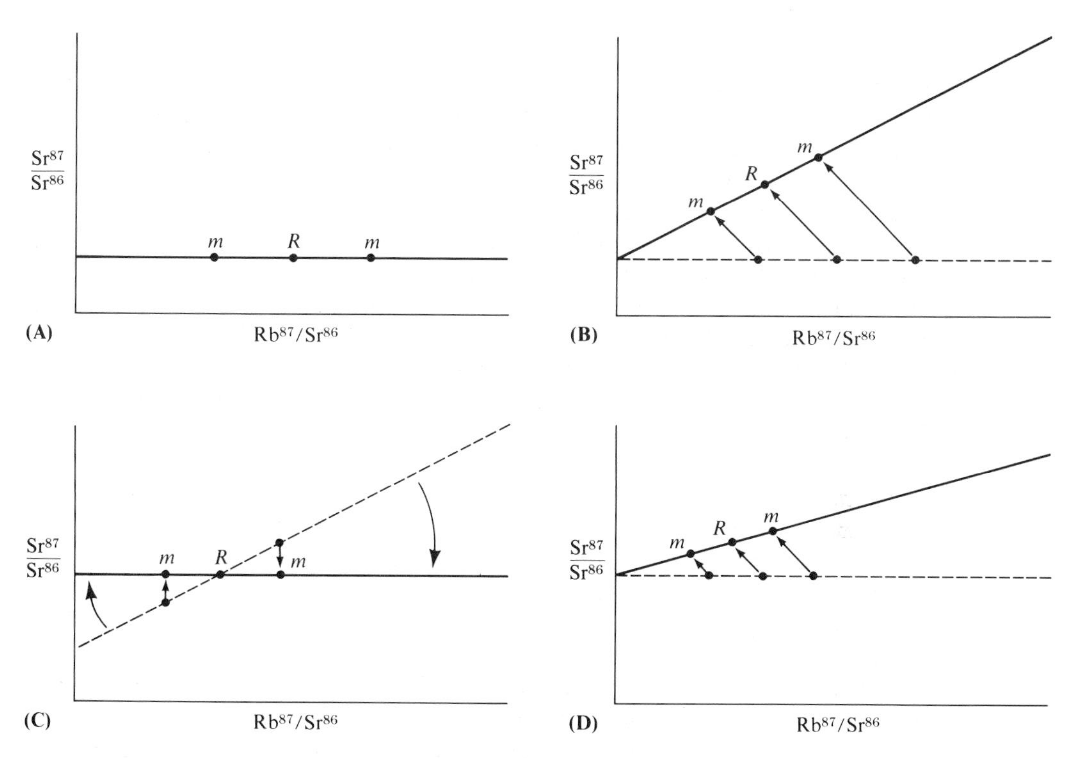

Figure A–6
(**A**) The distribution of Sr^{87}/Sr^{86} and Rb^{87}/Sr^{86} after crystallization of an igneous rock. The Sr^{87}/Sr^{86} ratio is the same for each mineral (*m*) in the rock. Point *R* represents a whole rock analysis. Each mineral has a different Rb^{87}/Sr^{86} ratio. (**B**) After time has elapsed the composition of each mineral and the whole rock composition have shifted due to conversion of Rb^{87} to Sr^{87}. The inclination of the isochron is related to elapsed time since initial crystallization. (**C**) A metamorphic event causes equalization of the Sr^{87}/Sr^{86} ratios in the minerals. The isochron is now horizontal. Note that this does not change the composition of the whole rock (*R*), as the system has remained closed. (**D**) Development of a new isochron after metamorphism. The slope of the isochron is related to the age of the metamorphic event.

now that after some time has elapsed, the isochron looks like Figure A–7B. If metamorphism occurs at this time, it does *not* affect the isochron formed by the various whole rock compositions. Recall that we saw in the previous example (Figure A–6C) that metamorphism affects only the isochron formed by the minerals within a single rock. It does not affect the whole rock compositions such as R_1, R_2, and R_3. Hence after metamorphism the isochron formed by the whole rock analyses remains unchanged. With the passage of additional time, the isochron becomes steeper (see Figure A–7C), and determination of its slope by whole rock analyses provides a date

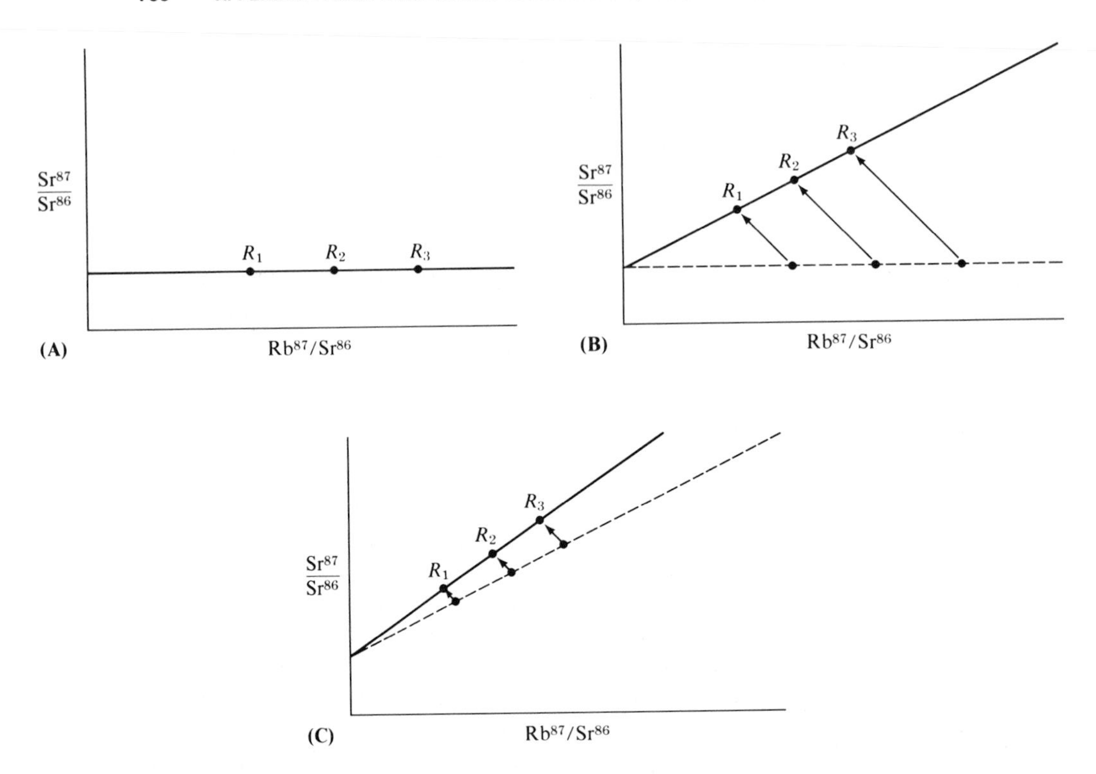

Figure A–7
(A) Isotopic composition of a comagmatic suite of rocks (R_1, R_2, R_3) that has just completed crystallization. **(B)** With elapsed time and conversion of Rb^{87} to Sr^{87}, an inclined isochron is developed. A metamorphic event at this time does not affect the whole rock compositions if they are separated from each other and the total system is closed. **(C)** The isochron is steepened in the interval from the metamorphic event to the present. The slope of the isochron is related to the time of initial crystallization, not the metamorphic event.

for the time of crystallization of the comagmatic suite; determination of an isochron by analyses of minerals from a single rock dates the metamorphic event.

Determination of Temperature of Formation Using Stable Isotopes

Determination of the temperature of formation of an igneous or metamorphic rock and many carbonate sediments and rocks can be accomplished with the use of oxygen isotopes. Oxygen is composed of three principal stable isotopes—O^{16}, O^{17}, and O^{18}—whose relative abundance in air is 99.759, 0.0372, and 0.2039. The temperature of formation of many rocks can be determined by the measurement of O^{18}/O^{16} ratios in minerals. The O^{18}/O^{16} ratio of a rock sample is compared to a fixed value, called standard mean ocean water (SMOW). The equation

$$\delta = \left[\frac{O^{18}/O^{16}\text{ Sample}}{O^{18}/O^{16}\text{ SMOW}} - 1\right]100$$

yields parts per thousand (ppt) values of δ that show the deviation from the standard value ($\delta = +10$ means the sample is 1 percent richer in O^{18} than SMOW; $\delta = +20$ is 2 percent richer; and so on). Silicate rocks generally vary in δ values from $+5$ to $+15$ ppt; sedimentary carbonate rocks are between -15 and $+2$ ppt.

Igneous and Metamorphic Rocks

It has been established that at igneous and metamorphic temperatures a partitioning of O^{18} and O^{16} occurs between the oxygen of the available water (the "water reservoir") and the crystallizing minerals. This partitioning can be described in terms of an equilibrium constant as in a standard chemical reaction. Each of the crystallizing minerals acquires a somewhat different ratio of O^{18}/O^{16} from the same water source—meaning that each mineral has a different equilibrium constant. If the water reservoir had a different O^{18}/O^{16} ratio, the crystallizing minerals would assume a different O^{18}/O^{16} ratio as well, but the equilibrium constant between each mineral and the water source would be maintained. It follows that minerals equilibrated with respect to the same water reservoir are also equilibrated with each other. The equilibrium constant for each water–mineral pair is not pressure-sensitive, as O^{18} and O^{16} occupy essentially the same volume; it does, however, vary as a function of temperature. At low temperatures the partitioning (fractionation) effect is generally greatest; differences in O^{18}/O^{16} partitioning between coexistent mineral–water pairs decrease with increasing temperature, but are measurable to at least 1300°C. The O^{18}/O^{16} partitioning between minerals and an adjacent water reservoir has been determined by hydrothermal laboratory studies. Some values are shown in Figure A–8. The vertical coordinate, $10^3 \ln \alpha$ (where α is the mineral-water fractionation constant), is similar to an equilibrium constant; the fractionation constant is the O^{18}/O^{16} ratio in one chemical species divided by that in another. A significant number of minerals have now been calibrated against water over a wide range of temperatures. For igneous and metamorphic rocks, quartz usually contains the greatest relative amount of O^{18}, followed by calcite, alkali feldspar, intermediate plagioclase, muscovite, anorthite, pyroxene, hornblende, olivine, garnet, biotite, chlorite, ilmenite, and magnetite.

In order to use the above data to determine the temperature of formation of a rock, it is not sufficient to determine the O^{18}/O^{16} ratio for a single mineral, as this ratio will vary not only with temperature, but also as a function of the O^{18}/O^{16} ratio of the water reservoir within which the mineral formed; the latter ratio is initially unknown. It can be seen in Figure A–8 that the equilibrium constants for various minerals (against water) vary in a nonparallel manner as a function of temperature. Therefore

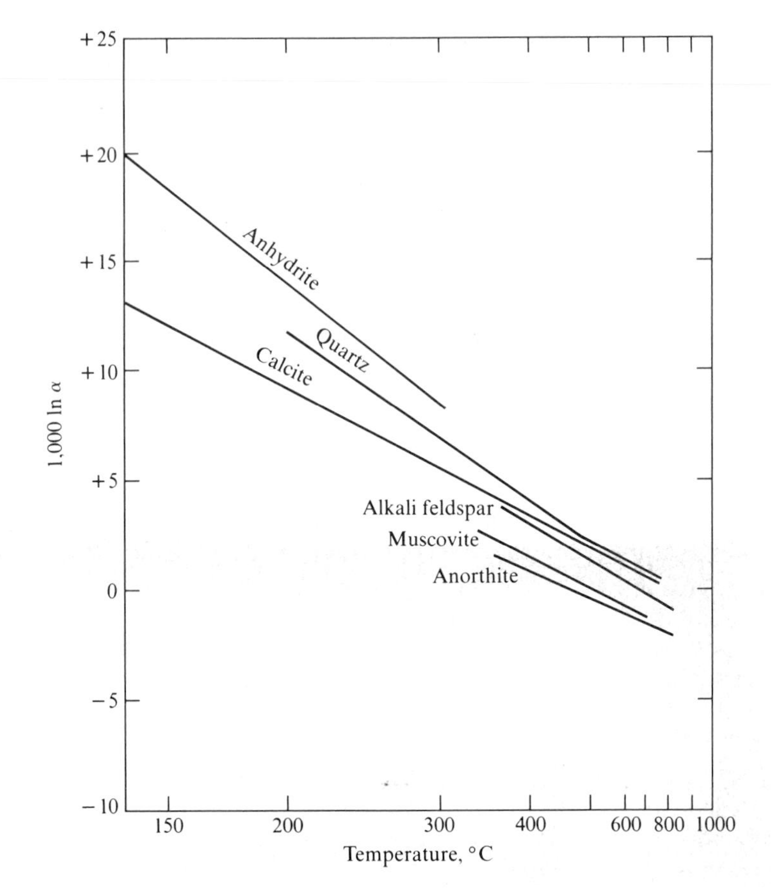

Figure A–8
Oxygen isotope calibration curves determined for mineral–water systems. Coordinates show variation in the mineral–water fractionation factor α as a function of temperature. The temperature scale is based on $10^6/T^2$, where T is the temperature in degrees Kelvin. [From G. Faure, 1977, *Isotope Geology* (New York: John Wiley and Sons), Fig. 19.1, based on earlier data.]

if the mineral–water fractionation constant of quartz were compared to that of muscovite, the *difference* would be a function of temperature, and independent of the O^{18}/O^{16} reservoir ratio. Curves showing the differences between the values of the natural logarithm of the mineral–water fractionation constants for the various coexistent minerals as a function of temperature are shown in Figure A–9. After the rock formation temperature has been determined by measuring the differences in fractionation constants between a coexistent mineral pair, it then becomes possible to return to the curves of the individual minerals, as in Figure A–8, and determine the O^{18}/O^{16} ratio of the reservoir by use of the known temperature of formation of the mineral

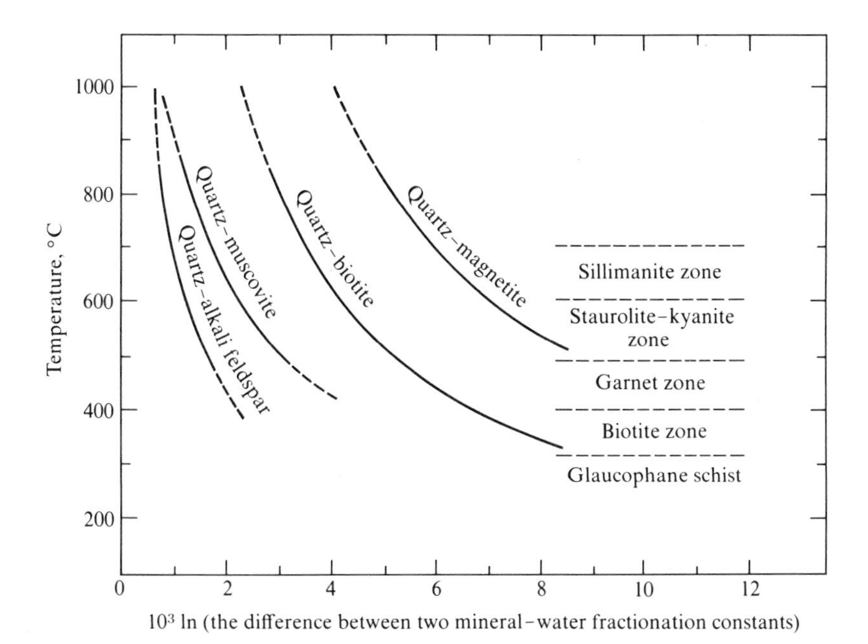

Figure A–9
Oxygen isotope calibration curves for various mineral pairs (prepared from mineral–water fractionation curves of the type shown in Figure A–8). The horizontal coordinate is a measure of the difference between various mineral–water fractionation curves. If more than one mineral pair is determined within a single rock, the temperature of formation of both pairs should be identical (concordant). Differences between pairs indicate some degree of nonequilibrium. [From G. Faure, 1977, *Isotope Geology* (New York: John Wiley and Sons), Fig. 19.8; based on data from Y. Bottinga, and M. Javoy, 1973, *Earth Planet. Sci. Letters, 20,* 250–265; S. Epstein and H. P. Taylor, 1967, in P. H. Abelson (ed.), *Researches in Geochemistry,* vol. 1 (New York: John Wiley and Sons), pp. 217–240; and Y. N. Shieh and H. P. Schwarcz, 1974, *Geochim. Cosmochim. Acta, 38,* Fig. 7.]

and the experimentally determined equilibrium constant between that mineral and water.

The following situations can cause problems with this method: (1) the minerals did not form simultaneously, (2) the minerals reequilibrated during the temperature decrease following the metamorphic or igneous event, or (3) the minerals did not completely equilibrate with the water source. All of these considerations can be taken into account if the O^{18}/O^{16} ratios are determined for more than one mineral pair. Each mineral combination would yield identical temperatures if equilibration were perfect at all stages.

Typical crystallization temperatures determined for metamorphic rocks are the following:

Sillimanite zone	600–700°C
Staurolite–Kyanite zone	510–580°C
Garnet zone	410–490°C
Biotite zone	310–410°C
Glaucophane schists	250–310°C
Marbles	Up to 660°C

Temperatures determined for igneous rocks include the following:

Granite	730–785°C
Granitic pegmatites	580–730°C
Hydrothermal veins	135–470°C

Knowledge of the O^{18}/O^{16} ratios of the water reservoir associated with metamorphic rocks gives a clue to the possible water sources. Most metamorphic rocks have relatively high values due to the fact that their sedimentary sources also have high values. At the higher grades of regional metamorphism there is a tendency for the fractionation constant to decrease, possibly due to the presence of magmatic fluids.

Sedimentary Rocks

Temperatures of formation of shallow marine carbonate sediments can be determined using the same principles as those applied to igneous and metamorphic rocks. Shallow ocean water is well stirred by waves and currents and is, therefore, a homogeneous reservoir with respect to the O^{18}/O^{16} ratio. (Exceptions exist adjacent to areas where large rivers enter the sea and in polar regions where large volumes of glacial ice dilute the saline sea water.) Thus, knowledge of oxygen isotope partitioning as a function of temperature combined with the oxygen isotopic composition of shell carbonate can be used to determine the temperature of the water in which the shells were formed.

The great sensitivity of this technique was first demonstrated in 1951 by Urey in a detailed isotopic study of a belemnite, an extinct type of cephalopod. A Jurassic belemnite shell (see Figure A–10) was sampled at 32 spots from the central axis of the cylindrical shell outward to a distance of about 1.4 cm, and indicated a life span of three summers and four winters after its youth, which contained too little carbonate for isotopic determinations to be made. The water apparently was warmer during its youth than during its old age, and it died in the spring at an age of about four years. The maximum seasonal temperature variation was about 6°C; mean temperature was 17.6°C. Although the geologic age of the belemnite is known only to within ± 3 million years, its life span is known to within about one month.

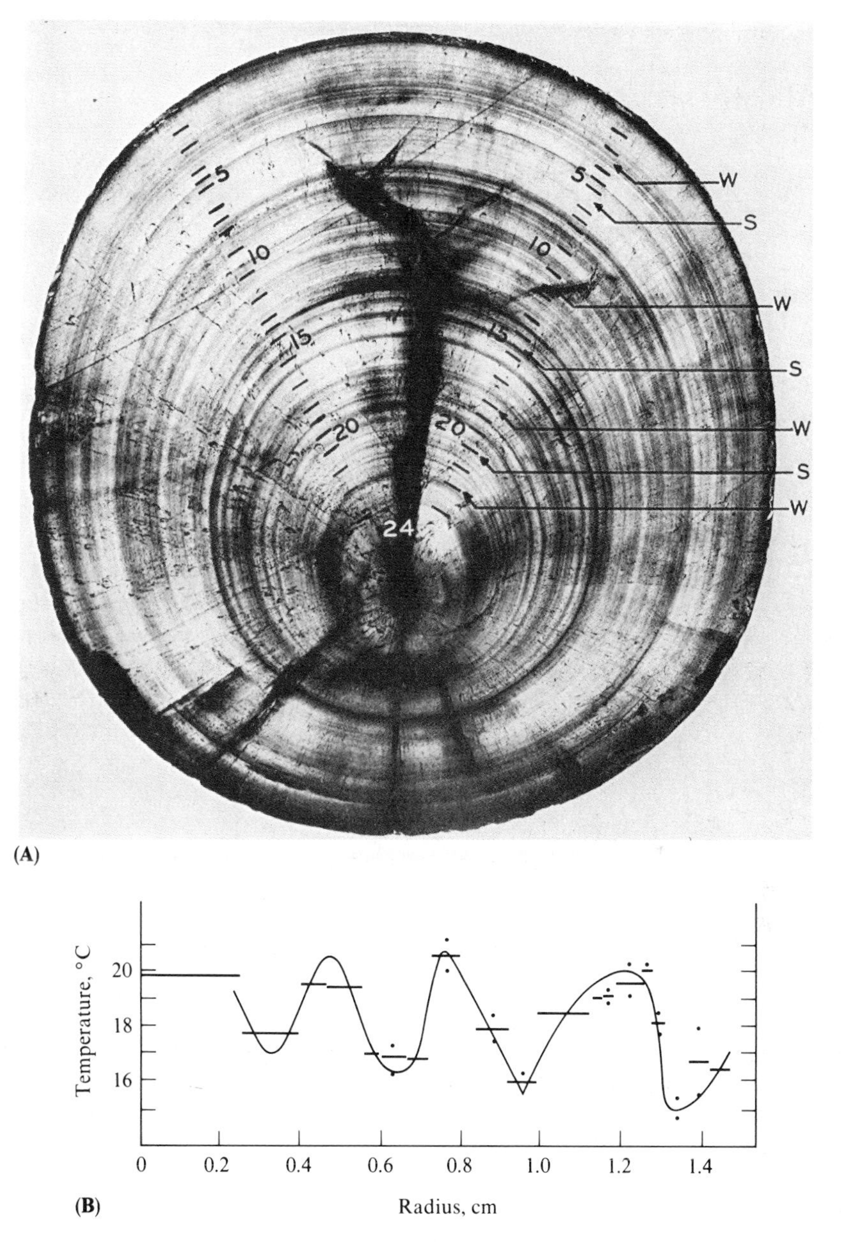

Figure A–10
(A) Cross section of Jurassic belemnite showing relation of sampling sites along two traverses to growth rings of the shell. W and S refer to winter and summer regions. **(B)** Growth temperatures of shell material as a function of distance outward from the center of the shell, as deduced from O^{18}/O^{16} values. The dots are data points. Horizontal lines located between two data points along a vertical axis represent the average position of the two data points. Horizontal lines located where there are no data points indicate that two samples at that particular radius had to be combined to obtain sufficient material for the isotopic analysis. [From H. C. Urey et al., 1951, *Geol. Soc. Amer. Bull., 62,* Plate 1.]

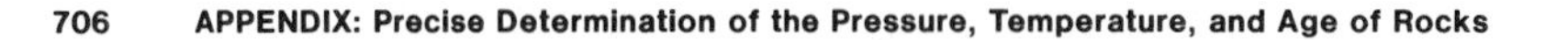

Figure A–11
Composite results from eight cores from the Caribbean Sea and Pacific Ocean showing the times and relative intensities of Pleistocene glaciations. [From C. Emiliani, 1978, *Earth Planet. Sci. Letters, 37,* Fig. 2.]

Our ability to determine paleotemperatures in ancient limestones is limited by the relatively high solubility and resultant ease of recrystallization of calcium carbonate. The bulk of original shell material is composed of either aragonite or highly magnesian calcite, both of which are unstable relative to pure calcite in most diagenetic waters. Recrystallization changes the O^{18}/O^{16} ratio and destroys the evidence of depositional paleotemperature. Some shells are resistant to recrystallization, however, and it is these that must be used to determine paleotemperatures in ancient carbonate rocks. Belemnites and brachiopods are particularly useful.

Paleotemperature determinations also have greatly increased our knowledge of Pleistocene glaciations. The O^{18}/O^{16} ratio in the shell of a planktonic organism such as *Globigerinoides sacculifera* will reflect variations in the temperature of ocean surface waters when the waters become colder and warmer as the extent of glacial ice waxes and wanes (see Figure A–11). Samples are obtained by coring the ocean floor; ages of the sediment at different depths in the cores are determined using a radioactive technique. There have been nine major glaciations during the past 730,000 years, with the spacing between glacial advances ranging between 50,000 and 100,000 years. It is clear that this method of detecting glacial and interglacial times is more sensitive than criteria used on the continents, such as the presence of ancient soil horizons or migratory patterns of land animals. These methods have detected only four major ice advances during Pleistocene time.

SUMMARY

Many of the relatively precise methods of geothermometry in igneous and metamorphic rocks currently deal with chemical equilibria between mineral pairs that formed contemporaneously during freezing or metamorphic recrystallization. Syntheses of mineral pairs under precisely controlled conditions of temperature and pressure (and perhaps oxygen fugacity) furnish information on the distribution of various cations between the pairs as a function of *P* and *T*. These data (obtained in both simple and complex systems) can be applied to chemically analyzed mineral pairs obtained from both igneous and metamorphic rocks in order to furnish the temperature (and perhaps pressure) at which these minerals equilibrated. Commonly used methods are the two-feldspar geothermometer, the iron–titanium oxide geothermometer (which also yields the oxygen fugacity), and the pyroxene geothermometer–geobarometer. Much of our information on *PT* conditions in the mantle has been obtained by use of the pyroxene method.

Analyses of igneous and metamorphic rocks to determine the amounts of radioactive isotopes and their daughter elements have proved useful in dating the time of crystallization of magmas as well as of subsequent metamorphic episodes. Determination of the partitioning of stable oxygen isotopes in laboratory experiments has permitted the estimation of temperature of formation of igneous, sedimentary, and metamorphic rocks by examination of isotope ratios of minerals or fossils from these rocks.

FURTHER READING

American Mineralogist. July–August issue, 1976. Devoted to articles on geothermometry and geobarometry.

Barth, T. F. W. 1951. The feldspar geological thermometer. *Neues Jahrb. Miner., 82,* 145–154.

Barth, T. F. W. 1962. The feldspar geologic thermometer. *Norsk Geol. Tidsk., 42,* 330–339.

Bohlen, S. R., and E. J. Essene. 1977. Feldspar and oxide thermometry of granulites in the Adirondack Highlands. *Contr. Miner. Petrology, 62,* 152–169.

Buddington, A. F., and D. H. Lindsley. 1964. Iron–titanium oxide minerals and synthetic equivalents. *Jour. Petrology, 5,* 310–357.

Carmichael, I. S. E. 1967. The iron–titanium oxides of salic volcanic rocks and their associated ferromagnesian silicates. *Contr. Miner. Petrology, 14,* 36–64.

Faure, G. 1977. *Isotope Geology.* New York: John Wiley and Sons, 464 pp.

MacGregor, I. D. 1974. The system $MgO–Al_2O_3–SiO_2$: solubility of Al_2O_3 in enstatite for spinel and garnet peridotite compositions. *Amer. Miner., 59,* 110–119.

O'Hara, M. J. 1967. Mineral paragenesis in ultrabasic rocks. In P. J. Wyllie (ed.), *Ultramafic and Related Rocks,* pp. 393–403. New York: John Wiley and Sons, 464 pp.

Stormer, Jr., J. C. 1975. A practical two-feldspar geothermometer. *Amer. Miner., 60,* 667–674.

Wells, P. R. A. 1977. Pyroxene thermometry in simple and complex systems. *Contr. Miner. Petrology, 62,* 129–139.

Whitney, J. A., and J. C. Stormer, Jr. 1977. The distribution of $NaAlSi_3O_8$ between coexisting microcline and plagioclase and its effects on geothermometric calculations. *Amer. Miner., 62,* 687–691.

Index of Names

Index of Topics

A boldface page number indicates a definition. An italic page number indicates material in an illustration.
A lowercase letter "t" after a page number indicates material in a table.